Published for
OXFORD INTERNATIONAL
AQA EXAMINATIONS

International A Level
PHYSICS

AS and A LEVEL

Jim Breithaupt

OXFORD
UNIVERSITY PRESS

OXFORD
UNIVERSITY PRESS

Great Clarendon Street, Oxford, OX2 6DP, United Kingdom

Oxford University Press is a department of the University of Oxford. It furthers the University's objective of excellence in research, scholarship, and education by publishing worldwide. Oxford is a registered trade mark of Oxford University Press in the UK and in certain other countries

First published in 2016

British Library Cataloguing in Publication Data
Data available

978-0-19-837603-3

13

Paper used in the production of this book is a natural, recyclable product made from wood grown in sustainable forests. The manufacturing process conforms to the environmental regulations of the country of origin.

Printed in United Kingdom

Acknowledgements
The publishers would like to thank the following for permissions to use their photographs:

Cover: JUAN CARLOS CASADO (starryearth.com) / SCIENCE PHOTO LIBRARY

p3: Rafe Swan/Cultura/Science Photo Library; **p4:** Shutterstock; **p10:** Gina Randall/U.S. Air Force; **p11 & p12:** Martyn F. Chillmaid/Science Photo Library; **p39 - p68:** Shutterstock; **p70:** Trl Ltd./Science Photo Library; **p79l:** Stephen Dalton/ Science Photo; **p87:** Shutterstock; **p115:** Drs A. Yazdani & D.J. Hornbaker/Science Photo Library; **p120:** Shutterstock; **p121:** National Institute on Aging/Science Photo Library; **p123:** Lawrence Berkeley National Laboratory/Science Photo Library; **p127:** Science Photo Library; **p129:** Shutterstock; **p130:** Science Photo Library; **p137:** Martyn F. Chillmaid/Science Photo Library; **p143t:** Ria Novosti/Science Photo Library; **p143b:** Fertnig/iStockphoto; **p152 - p154r:** Shutterstock; **p187t:** Library of Congress/Science Photo Library; **p187b:** Adrian Bicker/Science Photo Library; **p190:** Shutterstock; **p193:** Andrew Lambert Photography/Science Photo Library; **p195:** Berenice Abbott/Science Photo Library; **p196 & p199:** Shutterstock; **p209:** Sheila Terry/Science Photo Library; **p219:** Sciencephotos/Alamy; **p220t:** Giphotostock/Science Photo Library; **p220m:** Dept. Of Physics, Imperial College/ Science Photo Library; **p220b:** Physics Dept Imperial College/ Science Photo Library; **p225 - p240:** Shutterstock; **p248:** Hank Morgan/Science Photo Library; **p273 & p280:** Shutterstock; **p299:** Science Museum/Science & Society Picture Library; **p300 & p310:** NASA; **p311:** George Clerk/iStockphoto; **p313:** Viktar/iStockphoto; **p322:** Shutterstock; **p328:** Corbis; **p359:** Shutterstock; **p361:** Goronwy Tudor Jones, University Of Birmingham/Science Photo Library; **p362:** Andrew Lambert Photography/Science Photo Library; **p372:** Wynnter/ iStockphoto; **p375:** Shutterstock; **p391:** Stephen Harrison/ Alamy; **p394:** Pascal Goetgheluck/Science Photo Library; **p429:** EFDA-Jet/Science Photo Library; **p433:** Digital Globe/ Science Photo Library; **p453**: jbcn / Alamy Stock Photo; **p455**: 3Dsculptor / Shutterstock; **p434:** Paul Shambroom/Science Photo Library; **p467**: Igor Grochev / Shutterstock; **p461 & p469:** Shutterstock.

Although we have made every effort to trace and contact all copyright holders before publication this has not been possible in all cases. If notified, the publisher will rectify any errors or omissions at the earliest opportunity.

AS/A Level course structure

This book has been written to support students studying for the Oxford International AQA Examinations A Level Physics. The sections covered are shown in the Contents list, which also shows you the page numbers for the main topics within each section. There is also an Index at the back to help you find what you are looking for. If you are studying for AS Level Physics, you will only need to know the content in the blue box.

AS exam

Year 1 content

Skills for starting AS and A Level Physics
Section 1 Mechanics, materials, and atoms
Section 2 Electricity, waves, and particles
Skills in AS Level Physics

Year 2 content

Section 3 Further mechanics, fields and their consequences
Section 4 Energy and energy resources
Section 5 Skills in A Level Year 2 Physics

A Level exam

The A Level course is designed as a two-year course and covers all the content and skills in all the five units above. Each unit counts for 20% of the A Level marks and is tested in a two-hour examination paper. Units 1 and 2 are designed as the first year of the full two-year A Level course.

The AS Level course is designed as a one-year course and covers Sections 1 and 2 only. Each section counts for 50% of the AS Level marks and is assessed in a two-hour paper for each section. The AS Level marks may be carried forward as part of the A Level mark. The Section 1 paper includes the assessment of AS Level practical skills, which are described in Chapter 14 of this book.

Contents

Section 2

Section 3

Contents

Section 4

Section 5

Answers to the Practice Questions and Section Questions are available at www.oxfordsecondary.com/oxfordaqaexams-alevel-physics

How to use this book

Learning objectives

- → At the beginning of each topic, there is a list of learning objectives.
- → These are matched to the specification and allow you to monitor your progress.
- → A specification reference is also included.
  Specification reference: 3.1.1

This book contains many different features. Each feature is designed to foster and stimulate your interest in physics, as well as to support and develop the skills you will need for your examination.

Terms that you will need to be able to define and understand are highlighted in **bold orange text**. You can look these words up in the glossary.

Sometimes a word appears in **bold**. These are words that are useful to know but are not used on the specification. They therefore do not have to be learnt for examination purposes.

Synoptic link

These highlight how the sections relate to each other. Linking different areas of physics together is important, and you will need to be able to do this.

There are also links to the maths section to support the development of these skills.

Application features

These features contain important and interesting applications of physics in order to emphasise how scientists and engineers have used their scientific knowledge and understanding to develop new applications and technologies. There are also application features to develop your maths skills, and to develop your practical skills.

Extension features

These features contain material that is beyond the specification, designed to stretch and provide you with a broader knowledge and understanding, and to lead the way into the types of thinking and areas you might study in further education. As such, neither the detail nor the depth of questioning will be required for the examinations. But this book is about more than getting through the examinations.

1 Extension and application features have questions that link the material with concepts that are covered in the specification. Short answers are inverted at the bottom of the feature, whilst longer answers can be found in the answers section at the back of the book.

Study tips

Study tips contain prompts to help you with your revision. They can also support the development of your practical skills and your mathematical skills.

Hint

Hint features give other information or ways of thinking about a concept to support your understanding. They can also relate to practical or mathematical skills.

Summary questions

1 These are short questions that test your understanding of the topic and allow you to apply the knowledge and skills you have acquired. The questions are ramped in order of difficulty.

Skills for starting AS and A Level Physics

1 Using a calculator

Practice makes perfect when it comes to using a calculator. For AS and A Level Physics, you need no more than a scientific calculator. You need to make sure you master the technicalities of using a scientific calculator as early as possible in your Physics course. At this stage, you should be able to use a calculator to add, subtract, multiply, divide, find squares and square roots, and to calculate sines, cosines, and tangents of angles. But remember, when using a calculator, it's all too easy to make a mistake, for example to press the wrong button. So when carrying out a calculation using a calculator, check the answer by making an order-of-magnitude **estimate** of the answer in your head.

Maths link

Further important calculator functions are described in Topics 14.1 and 27.1.

2 Making measurements

You should know at this stage how to make measurements using basic equipment such as metre rules, protractors, stopwatches, thermometers, balances (for weighing an object), and ammeters and voltmeters. During the course, you will also be expected to use equipment such as micrometers, verniers, oscilloscopes, and data loggers.

Practical link

The use of these items is described for reference in Practical work in physics.

Here are some useful reminders about making measurements:

- check the zero reading when you use an instrument to make a measurement, for example, a metre ruler worn away at one end might give a zero error
- when a multi-range instrument is used, start with the highest range and switch to a lower range if the reading is too small to measure precisely
- make sure you record all your measurements in a logical order, stating the correct unit of each measured quantity
- don't pack equipment away until you are sure you have enough measurements or you have checked unexpected readings.

Practical link

See Topic 14.2 for anomalous measurements.

3 Using measurements in calculations

Whenever you make a record of a measurement, you must always note the correct unit as well as the numerical value of the measurements.

The scientific system of units is called the SI system. The fundamental or base units of the SI system you need to remember are listed below. All other units are derived from the SI base units. See Pages xii–xiv for more details about the SI system.

1. The metre (m) is the SI unit of length. Note also that 1 m = 100 cm = 1000 mm.
2. The kilogram (kg) is the SI unit of mass. Note that 1 kg = 1000 grams.
3. The second (s) is the SI unit of time.
4. The ampere (A) is the SI unit of current.
5. The Kelvin (K) is the SI unit of temperature
6. The mole (mol) is the SI unit of the amount of a substance.

Powers of ten and numerical prefixes are used to avoid unwieldy numerical values. For example:

- 1 000 000 = 10^6 which is 10 raised to the power 6 (usually stated as '10 to the 6').
- 0.000 000 1 = 10^{-7} which is 10 raised to the power −7 (usually stated as '10 to the −7').

Prefixes are used with units as abbreviations for powers of ten. For example, a distance of 1 kilometre may be written as 1000 m or 10^3 m or 1 km. The most common prefixes are shown in Table 1.

▼ **Table 1** *Prefixes*

Prefix	femto-	pico-	nano-	micro-	milli-	kilo-	mega-	giga-	tera-
Value	10^{-15} m	10^{-12}	10^{-9}	10^{-6}	10^{-3}	10^3	10^6	10^9	10^{12}
Prefix symbol	f	p	n	μ	m	k	M	G	T

Note that the centimetre (cm), the cubic centimetre (cm^3) and the gram (g) are in common use and are therefore allowed as exceptions to the prefixes shown.

Standard form is usually used for numerical values smaller than 0.001 or larger than 1000.

- The numerical value is written as a number between 1 and 10 multiplied by the appropriate power of ten. For example,

  64 000 m = 6.4×10^4 m

  0.000 005 1 s = 5.1×10^{-6} s

- A prefix may be used instead of some or all of the powers of ten. For example:

  35 000 m = 35×10^3 m = 35 km

  0.000 000 59 m = 5.9×10^{-7} m = 590 nm

To convert a number to standard form, count how many places the decimal point must be moved to make the number between 1 and 10. The number of places moved is the power of ten that must accompany the number between 1 and 10. Moving the decimal place:

- to the left gives a positive power of ten (e.g., 64 000 = 6.4×10^4)
- to the right gives a negative power of ten (e.g., 0.000 005 1 = 5.1×10^{-6}).

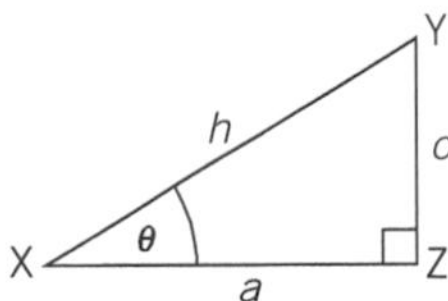

▲ **Figure 1** *The right-angled triangle*

4 Using the right-angled triangle

The sine, cosine, and tangent of an angle are defined from the right-angled triangle. Figure 1 shows a right-angled triangle XYZ in which side XY is the hypotenuse (i.e., the side opposite the right angle), side YZ is opposite angle θ, and side XZ is adjacent to angle θ.

$$\sin\theta = \frac{YZ}{XY} = \frac{o}{h}$$

$$\cos\theta = \frac{XZ}{XY} = \frac{a}{h}$$

$$\tan\theta = \frac{YZ}{XZ} = \frac{o}{a}$$

where o = YZ the side opposite angle θ

h = XY, the hypotenuse

and a = XZ the side adjacent to angle θ

To remember these formulae, recall SOHCAHTOA.

Pythagoras's theorem states that for any right-angled triangle,

the square of the hypotenuse = the sum of the squares of the other two sides

Applying Pythagoras's theorem to the right-angled triangle XYZ in Figure 1 gives:

$$(XY)^2 = (XZ)^2 + (YZ)^2$$

5 Symbols in science

Symbols are used in equations and formulae to represent physical quantities. In your previous course, you may have used equations with words instead of symbols to represent physical quantities. For example, you will have met the equation distance moved = speed × time. Perhaps you were introduced to the same equation in the symbolic form $s = vt$, where s is the symbol used to represent distance, v is the symbol used to represent speed, and t is the symbol used to represent time. It is much easier to apply the rules of algebra to an equation written in symbolic form.

If you used symbols in your GCSE course, you might have met the use of s for distance and I for current. Maybe you wondered why we don't use d for distance instead of s – or C for current instead of I. The answer is that physics discoveries have taken place in many countries. The first person to discover the key ideas about speed was Galileo, the great Italian scientist – he used the word *scale* from his own language for distance, and therefore assigned the symbol s to distance. Important discoveries about electricity were made by Ampère, the great French scientist – he wrote about the intensity of an electric current, so he used the symbol I for electric current. The symbols we now use are used in all countries in association with the **SI system** of units.

▼ **Table 2**

Physical quantity	Symbol	Unit	Unit symbol
Distance	s	metre	m
Speed or velocity	v	metre per second	$\mathrm{m\,s^{-1}}$
Acceleration	a	metre per second per second	$\mathrm{m\,s^{-2}}$
Mass	m	kilogram	kg
Force	F	newton	N
Energy or work	E	joule	J
Power	P	watt	W
Density	ρ	kilogram per cubic metre	$\mathrm{kg\,m^{-3}}$
Current	I	ampere	A
Potential difference or voltage	V	volt	V
Resistance	R	ohm	Ω
Temperature	T	kelvin	K

6 Using equations

Equations often need to be rearranged. This can be confusing if you don't learn the following basic rules at an early stage:

- **Read an equation properly**. For example, the equation $v = 3t + 2$ is not the same as the equation $v = 3(t + 2)$. If you forget the brackets when you use the second equation to calculate v when $t = 1$, then you will get $v = 5$ instead of the correct answer $v = 9$. The first equation tells you to multiply t by 3 then add 2. The second equation tells you to add t and 2, then multiply the sum by 3.
- **Rearrange an equation properly**. In simple terms, always make the same change to both sides of an equation. For example, to make t the subject of the equation $v = 3t + 2$,

  Step 1: Subtract 2 from both sides of the equation, so
  $v - 2 = 3t + 2 - 2 = 3t$

  Step 2: The equation is now $v - 2 = 3t$ and can be written $3t = v - 2$

  Step 3: Divide both sides of the equation by 3, so $\frac{3t}{3} = \frac{v - 2}{3}$

  Step 4: Cancel 3 on the top and the bottom of the left-hand side, to finish with $t = \frac{v - 2}{3}$

To use an equation as part of a calculation:

- Start by making the quantity to be calculated the **subject** of the equation.
- Write the equation out with the **numerical values** in place of the symbols.
- Carry out the calculation and make sure you give the answer with the **correct unit**.

Unless the equation is simple (e.g., $V = IR$), don't insert numerical values then rearrange the equation. Rearrange, then insert the numerical values – you are less likely to make an error if the numbers are inserted later in the process.

Estimating the result of a measurement or calculation is a useful technique to ensure big errors have not been made. For example, in a density measurement of a metal object, you would expect the result to be several times the density of water. A result of 800 kg m^{-3} compared with a value of 1000 kg m^{-3} for water means an error must have been made. Before carrying out an exact calculation of a physical quantity, a mental estimate of the approximate result of the calculation gives an *order-of-magnitude* value. Combined with an awareness of the magnitudes of physical quantities, a mental estimate can be used to spot errors before or after carrying out an accurate calculation. See Topic 14.3 for more about signs, equations, and formulae.

7 More about SI units

In the SI system of units, every derived unit can be expressed in terms of the **base units** of the SI system. Table 3, on the next page, shows how the derived units you meet at A Level are related to the SI base units and, in some cases, how they relate to each other. For example, in terms of base units, the coulomb (C) is the ampere second (A s). Such knowledge can prove useful. For example, if you can't quite recall the formula for:

- **Density**, but you know that its unit is the kilogram per cubic metre ($kg\,m^{-3}$), you should be able to see that $kg\,m^{-3}$ is the unit for mass divided by volume.
- **Centripetal acceleration**, remembering that the unit of acceleration (i.e. $m\,s^{-2}$) is the same as the unit of speed^2 divided by the unit of distance (i.e., $m^2\,s^{-2}/m$) can lead you to the formula v^2/r for centripetal acceleration. See Topic 15.2 if necessary.

▼ **Table 3** *Links between units*

Quantity	Symbol	Unit	Unit symbol	Other forms of unit	Base unit
Acceleration	a	metre per second^2	$m\,s^{-2}$		$m\,s^{-2}$
Angle	θ	radian or degree	rad or °		
Velocity, speed	v	metre per second	$m\,s^{-1}$		$m\,s^{-1}$
Force	F	newton	N		$kg\,m\,s^{-2}$
Gravitational field strength	g	newton per kilogram	$N\,kg^{-1}$		$m\,s^{-2}$
Momentum	p	kilogram metre per second	$kg\,m\,s^{-1}$	N s	$kg\,m\,s^{-1}$
Energy, work	E, W	joule	J		$kg\,m^2\,s^{-2}$
Power	P	watt	W	$J\,s^{-1}$	$kg\,m^2\,s^{-3}$
Density	ρ	kilogram per cubic metre	$kg\,m^{-3}$		$kg\,m^{-3}$
Pressure	p	pascal	Pa	$N\,m^{-2}$	$kg\,m^{-1}\,s^{-2}$
Spring (or stiffness) constant	k	newton per metre	$N\,m^{-1}$		$kg\,s^{-2}$
Stress, Young modulus	σ, E	pascal	Pa	$N\,m^{-2}$	$kg\,m^{-1}\,s^{-2}$
Charge	Q	coulomb	C		A s
Potential difference, e.m.f.	V, ε	volt	V	$J\,C^{-1}$	$kg\,m^2\,s^{-3}\,A^{-1}$
Resistance	R	ohm	Ω	$V\,A^{-1}$	$kg\,m^2\,s^{-3}\,A^{-2}$
Resistivity	ρ	ohm metre	$\Omega\,m$		$kg\,m^3\,s^{-3}\,A^{-2}$
Electric field strength	E	newton per coulomb	$N\,C^{-1}$	$V\,m^{-1}$	$kg\,m\,s^{-3}\,A^{-1}$
Frequency	f	hertz	Hz	s^{-1}	s^{-1}
Phase difference	π	radian or degree	rad or °		
Wavelength	λ	metre	m		m
Angular displacement	θ	radian or degree	rad or °		
Angular frequency	ω	radian per second	$rad\,s^{-1}$		
Angular speed or velocity	ω	radian per second	$rad\,s^{-1}$		
Gravitational potential	V	joule per kilogram	$J\,kg^{-1}$		$m^2\,s^{-2}$
Electric potential	V	volt	V	$J\,C^{-1}$	$kg\,m^2\,s^{-3}\,A^{-1}$
Permittivity of free space	ε_0	farad per metre	$F\,m^{-1}$		$kg^{-1}\,m^{-3}\,s^4\,A^2$
Capacitance	C	farad	F	$C\,V^{-1}$	$kg^{-1}\,m^{-2}\,s^4\,A^2$
Magnetic flux density	B	tesla	T	$N\,A^{-1}\,m^{-1}\,Wb\,m^{-2}$	$kg\,s^{-2}\,A^{-1}$
Magnetic flux	Φ	weber	Wb	$T\,m^2$, V s	$kg\,m^2\,s^{-2}\,A^{-1}$
Permeability of free space	μ_0	henry per metre	$H\,m^{-1}$		$kg\,m\,s^{-2}\,A^{-2}$
Activity	A	becquerel	Bq	s^{-1}	s^{-1}
Decay constant	λ	second^{-1}	s^{-1}		s^{-1}

Derived units written in terms of their base units can be used to check equations. The physical quantities on each side of an equation must match in terms of base units.

If they don't match, the equation cannot be correct. For example, consider

- the equation $v = \sqrt{2gR}$, which is used to calculate the escape speed v of an object from the surface of a planet of radius R and surface gravitational field strength g

  Left-hand side base units = m s^{-1}

  Right-hand side base units = $\sqrt{\text{m s}^{-2} \times \text{m}} = \text{m s}^{-1}$

The equation has the same combination of base units on each side, so it is correct – we say it is **homogeneous** in terms of the base units.

- the equation $W = QV$ is used to calculate the work done W to move a charge Q through a potential difference V

  Left-hand side base units (see Table 3) = $\text{kg m}^2\,\text{s}^{-2}$

  Right-hand side base units = $(\text{A s}) \times (\text{kg m}^2\,\text{s}^{-3}\,\text{A}^{-1}) = \text{kg m}^2\,\text{s}^{-2}$

The equation has the same combination of base units on each side, so it is homogeneous. Note that for simple equations such as this, homogeneity can be checked faster by recalling basic relationships between physical quantities. In this example, one volt is one joule per coulomb, so the unit of QV is the joule per coulomb × the coulomb, which is the joule.

The links between different units do not need to be made through the SI base units. For example, the volt (V) is the joule per coulomb (J C^{-1}), which is a useful link to remember as it helps you to develop your understanding of potential difference.

There are some units in the A Level specification that are not SI units but they are used in specific situations for convenience. Those listed below are in common use.

1 The **atmosphere** (atm) is a unit of pressure equal to the mean pressure of the atmosphere at sea level and is equal to 101 kPa.
2 The **electron volt** (eV) is a unit of energy defined as the work done when an electron moves through a p.d. of 1 V.
$1\,\text{eV} = 1.6 \times 10^{-19}\,\text{J}$.
Note that $1\,\text{MeV} = 10^6\,\text{eV} = 1.6 \times 10^{-13}\,\text{J}$

3 The **kilowatt hour** (kWh) is a unit of energy equal to the energy supplied to a one kilowatt appliance in one hour, which is 3.6 MJ.
4 The **light year** is the distance travelled in space by light in one year.
5 The **litre** is a unit of volume equal to $10^{-3}\,\text{m}^3$.

Practical work in physics
Moving on from GCSE

In the laboratory

The experimental skills you will develop during your course are part of the tools of the trade of every physicist. Data loggers and computers are commonplace in modern physics laboratories, but awareness on the part of the user of precision, reliability, errors, and accuracy are just as important as when measurements are made with much simpler equipment. Let's consider in more detail what you need to be aware of when you are working in the physics laboratory.

Safety and organisation

Your teacher will give you a set of safety rules and should explain them to you. You must comply with them at all times. You must also use your common sense and organise yourself so that you work safely. For example, if you set up an experiment with pulleys and weights, you need to ensure they are stable and will not topple over.

Working with others

Most scientists work in teams, each person cooperating with other team members to achieve specific objectives. This is effective because, although each team member may have a designated part to play, the exchange of ideas within the team often gives greater insight and awareness as to how to achieve the objectives.

In your AS Level practical activities, you will often work in a small group in which you need to cooperate with the others in the group so everyone understands the objectives of the practical activity and everyone participates in planning and carrying out the activity.

Planning

At AS Level, you may be asked to plan an experiment or investigation. The practical activities you carry out during your course should enable you to prepare a plan. Here are the key steps in drawing up a plan:

1 Decide in detail what you intend to investigate. Note the independent and dependent variables you intend to measure, and note the variables that need to be controlled. The other variables need to be controlled to make sure they do not change. A **control variable** that can't be kept constant would cause the **dependent variable** to alter.

2 Select the equipment necessary for the measurements. Specify the range of any electrical meters you need.

3 List the key stages in the method you intend to follow and make some preliminary measurements to check your initial plans. Consider safety issues before you do any preliminary tests. If necessary, modify your plans as a result of your preliminary tests.

4 If the aim of your investigation is to test a hypothesis or theory or to use the measurements to determine a physical quantity (e.g., resistivity), you need to know how to use the measurements you make. See Analysis and evaluation on page 7 for notes about how to process and use measurement data.

Carrying out instructions and recording your measurements

In some investigations, you will be expected to follow instructions supplied to you either verbally or on a worksheet. You should be able to follow a sequence of instructions without guidance. Part of the direct assessment of your practical work is on how well you follow instructions. However, always remember safety first and, if the instructions are not clear, ask your teacher to clarify them.

When you record your measurements, tabulate them with a column for the independent variable and one or more columns for the dependent variable to allow for repeat readings and average values, if appropriate. The table should have a clear heading for each of the measured variables, with the unit shown after the heading, as below.

Single measurements of other variables (e.g., control variables) should be recorded together, immediately before or after the table. In addition, you should record the precision (i.e., the least detectable reading) of each measurement. This information is important when you come to analyse and evaluate your measurements.

▼ **Table 1** *Tabulating the measurements from an investigation of p.d. against current for a wire*

Potential difference / V	Current / A			Average current / A
	1st set	2nd set	3rd set	

length of wire / m = ______

diameter of wire / mm = ______, ______, ______

average diameter of wire / mm = ______

Making careful measurements

Measurements and errors

Measurements play a key role in science, so they must be:

1. **valid** – the measurements are of the required data or can be used to give the required data and have been obtained by an acceptable method. For example, a voltmeter connected across a variable resistor in series with a lamp and a battery would not measure the potential difference across the lamp.
2. **repeatable** and **reproducible** – the same results are obtained if the original experimenter repeats the investigation using the same method and equipment, or if the investigation is reproduced by another person or by using different equipment or techniques.

Errors of measurement are important in finding out how accurate a measurement is. We need to consider errors in terms of differences from the mean value. Consider the example of measuring the diameter, *d*, of a uniform wire using a micrometer. Suppose the following diameter readings are taken for different positions along the wire from one end to the other

0.34 mm, 0.33 mm, 0.36 mm, 0.33 mm, 0.35 mm

- The **range** of the measurements is given by the maximum and minimum values of the measurements. Here the range is from 0.33 mm to 0.36 mm. We will see later we can use this to estimate the **uncertainty** or probable error of the measurement.
- The **mean value**, $<d>$, is 0.34 mm, calculated by adding the readings together and dividing by the number of readings. If the difference between each reading and $<d>$ changed regularly from one end of the wire to the other, it would be reasonable to conclude that the wire was non-uniform. Such differences are called **systematic errors**. If there is no obvious pattern or bias in the differences, the differences are said to be **random errors**.

What causes random errors? In the case of the wire, vibrations in the machine used to make the wire might have caused random variations in its diameter along its length. The experimenter might not use or read the micrometer correctly consistently.

The range of the diameter readings above is from 0.33 mm to 0.36 mm. The readings lie within 0.015 mm (i.e., half the range) of the mean value, which we will round up to 0.02 mm. The diameter can therefore be written as 0.34 ± 0.02 mm. The diameter is accurate to ±0.02 mm. The uncertainty or probable error in the mean value of the diameter is therefore ±0.02 mm.

Using instruments

Instruments used in the laboratory range from the very basic (e.g., a millimetre scale) to the highly sophisticated (e.g., a multichannel data

Hint

A measurement checklist

At AS Level, you should be able to:

- measure length using a ruler, vernier calipers, and a micrometer
- weigh an object and determine its mass using a spring balance, a lever balance, or a top pan balance, use a newton-meter to measure force
- use a protractor to measure an angle and use a set square
- measure time intervals using clocks, stopwatches, light gates, and the time base of an oscilloscope
- measure temperature using a thermometer
- use ammeters, voltmeters and multimeters (including resistance measurements) with appropriate scales
- use an oscilloscope.

In addition you should be able to:

- distinguish between systematic errors (including zero errors) and random errors
- understand what is meant by accuracy, sensitivity, linearity, reliability, precision, and validity
- read analogue and digital displays.

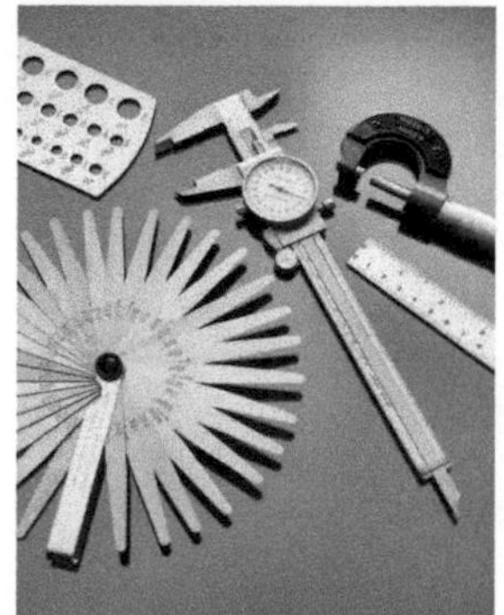

▲ **Figure 1** *Physics instruments*

recorder). Whatever type of instrument you use, you need to know what the following terms mean:

Zero error

Does the instrument read zero when it is supposed to? If not, the zero reading must be taken into account when taking measurements otherwise there will be a systematic error in the measurements.

Uncertainty

The uncertainty is the interval within which the true value can be expected to lie, expressed as a ± value (e.g., $I = 2.6 \pm 0.2$ A). If the readings are the same, the instrument precision should be used as the uncertainty. Uncertainty can be given with a level of confidence or probability (e.g., $I = 2.6 \pm 0.2$ A, at a level of confidence of 90%).

Accuracy

An **accurate** measurement is one that is close to the accepted value. **Accuracy** is a measure of confidence in a measurement and is usually expressed as the uncertainty in the measurement.

Precision

The **precision of a measurement** is the degree of exactness of the measurement. The precision is given by the extent of the random errors – a precise set of measurements will have little spread around the mean.

- If the reading of an instrument fluctuates when it is being taken, take several readings and calculate the mean and range of the measurements. The precision of the measurement is then given by the range of the readings.
- If the reading is constant, estimate the precision of a measurement directly from the instrument (or use its specified precision).

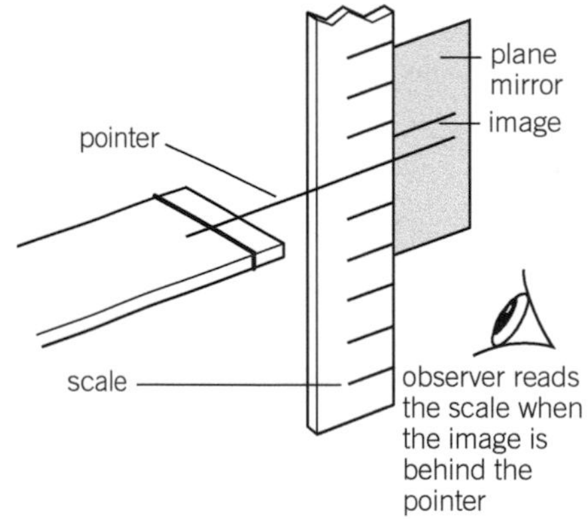

▲ **Figure 2** *Reading a scale*

Precise readings are not necessarily accurate readings because systematic errors could make precise readings higher or lower than they ought to be.

Linearity

This is a design feature of many instruments – it means the reading is directly proportional to the magnitude of the variable that causes the reading to change. For example, if the scale of a moving coil meter is **linear**, the reading of the pointer against the scale should be proportional to the current.

Reading a scale

Measurement errors are caused in analogue instruments if the pointer on the scale is not observed correctly. The observer must be directly in front of the pointer when the reading is made. Figure 2 shows how a plane mirror is used for this purpose. The image of the pointer must be directly behind the pointer to ensure the observer views the scale directly in front of the pointer.

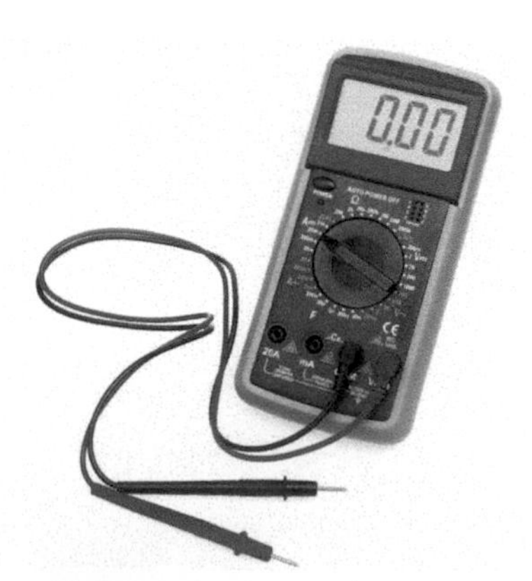

▲ **Figure 3** *A multimeter*

Instrument range

Multirange instruments such as multimeters have a range dial that needs to be set according to the maximum reading to be measured. For example, if the dial can be set at 0–0.10 A, 0–1.00 A, or 0–10.0 A, you would use the 0–1.00 A range to measure the current through a 0.25 A torch bulb as the 0–0.10 A range is too low and the 0–1.00 A range has greater precision than the 0–10.0 A range.

Everyday physics instruments

Rulers and scales

Metre rulers are often used as vertical or horizontal scales in mechanics experiments.

To set a metre ruler in a vertical position:

- use a set square perpendicular to the ruler and the bench, if the bench is known to be horizontal, or
- use a plumb line (a small weight on a string) to see if the ruler is vertical. You need to observe the ruler next to the plumb line from two perpendicular directions. If the ruler appears parallel to the plumb line from both directions, then it must be vertical.

To ensure a metre ruler is horizontal, use a set square to align the metre ruler perpendicular to a vertical metre ruler.

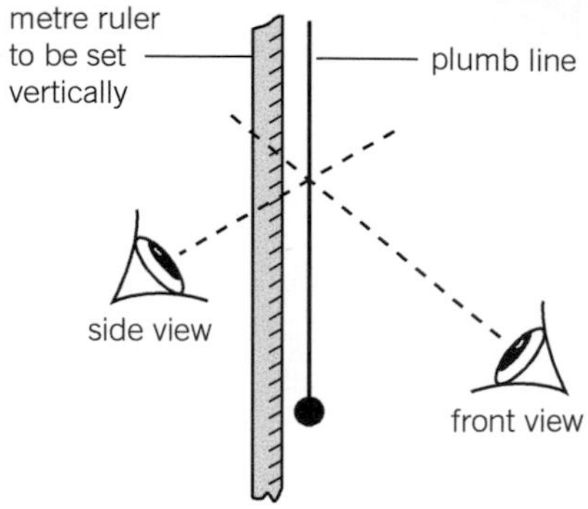

metre ruler to be set vertically
plumb line
side view
front view

▲ **Figure 1** *Finding the vertical – if the metre ruler appears parallel to the plumb line from the front and the side, the ruler must be vertical*

Micrometers and verniers

Micrometers give readings to within 0.01 mm. A **digital** micrometer gives a read-out equal to the width of the micrometer gap. An **analogue** micrometer has a barrel on a screw thread with a pitch of 0.5 mm. For such a micrometer:

- the edge of the barrel is marked in 50 equal intervals so each interval corresponds to changing the gap of the micrometer by 0.5/50 mm = 0.01 mm
- the stem of the micrometer is marked with a linear scale graduated in 0.5 mm marks
- the reading of a micrometer is where the linear scale intersects the scale on the barrel.

Figure 2 shows a reading of 4.06 mm. Note that the edge of the barrel is between the 4.0 and 4.5 mm marks on the linear scale. The linear scale intersects the sixth mark after the zero mark on the barrel scale. The reading is therefore 4.00 mm from the linear scale +0.06 mm from the barrel scale.

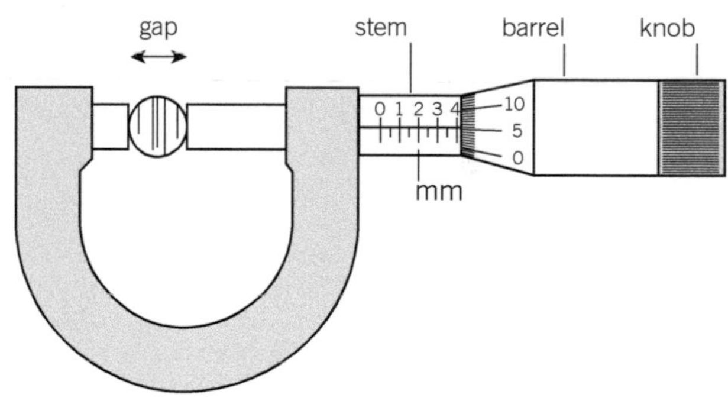

▲ **Figure 2** *Using a micrometer*

To use a micrometer correctly:

1. Check its zero reading and note the zero error if there is one.
2. Open the gap (by turning the barrel if analogue) then close the gap on the object to be measured. Turn the knob until it slips. Don't overtighten the barrel.
3. Take the reading and note the measurement after allowing if necessary for the zero error.
4. Note that the precision of the measurement is ±0.01 mm because the precision of the reading and the zero reading are both ±0.005 mm. So the difference between the two readings (i.e., the measurement) has a precision of ±0.01 mm.

Vernier calipers are used for measurements of distances up to 100 mm or more. Readings can be made to within 0.1 mm. The sliding

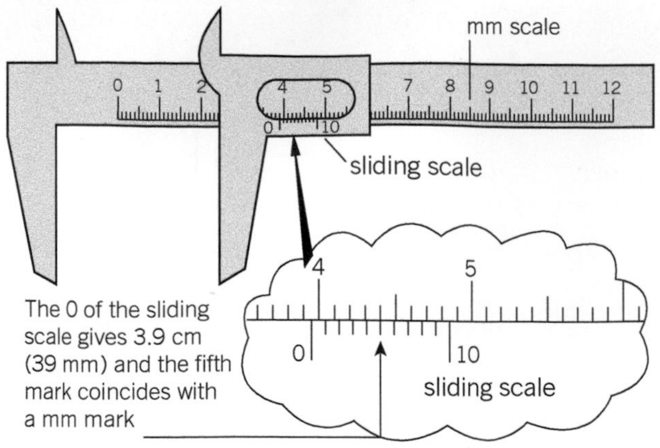

▲ **Figure 3** *Using a vernier*

scale of an analogue vernier has ten equal intervals covering a distance of exactly 9 mm so each interval of this scale is 0.1 mm less than a 1 mm interval. To make a reading:

1. The zero mark on the sliding scale is used to read the main scale to the nearest millimetre. This reading is rounded down to the nearest millimetre.
2. The mark on the sliding scale closest to a mark on the millimetre scale is located and its number noted. Multiplying this number by 0.1 mm gives the distance to be added on to the rounded-down reading.

Figure 3 shows the idea. The zero mark on the sliding scale is between 39 and 40 mm on the mm scale. So the rounded-down reading is 39 mm. The fifth mark after the zero on the sliding scale is nearest to a mark on the millimetre scale. So the extra distance to be added on to 39 mm is 0.5 mm (= 5 × 0.1 mm). Therefore, the reading is 39.5 mm.

Timers

Stopwatches used for interval timings are subject to human error because reaction time, about 0.2 s, is variable for any individual. With practice, the delays when starting and stopping a stopwatch can be reduced. Even so, the precision of a single timing is unlikely to be better than 0.1 s. Digital stopwatches usually have read-out displays with a resolution of 0.01 s, but reaction time makes such precision unrealistic, and the precision of a single timing is the same as for an analogue stopwatch.

Timing oscillations requires timing for as many cycles as possible. The timing should be repeated several times to give an average (mean) value. Any timing that is significantly different to the other values is probably due to miscounting the number of oscillations so that timing should be rejected. For accurate timings, a fiducial mark is essential. The mark should be lined up with the centre of the oscillations so it provides a reference position to count the number of cycles as the object swings past it each cycle in a certain direction.

Synoptic link

You will meet the use of light gates with data loggers in more detail in Topic 4.3, Conservation of momentum.

Electronic timers use automatic switches, or gates, to start and stop the timer. However, just as with a digital stopwatch, a timing should be repeated, if possible several times, to give an average value. Light gates may be connected via an interface unit to a microcomputer. Interrupt signals from the light gates are timed by the microcomputer's internal clock. A software program is used to provide a set of instructions to the microcomputer.

Balances

A balance is used to measure the weight of an object. Spring balances are usually less precise than lever balances. Both types of balance are usually much less precise than an electronic top pan balance. The scale or read-out of a balance may be calibrated for convenience in kilograms or grams. The accuracy of an electronic top pan balance can easily be tested using accurately known masses.

Analysis and evaluation

Data processing

For a single measurement, the precision of the measuring instrument determines the precision of the measurement. A micrometer with a precision of 0.01 mm gives readings that each have a precision of 0.01 mm.

For several readings, the number of significant figures of the mean value should be the same as the precision of each reading. For example, consider the following measurements of the diameter of a wire: 0.34 mm, 0.33 mm, 0.36 mm, 0.33 mm, 0.35 mm. The mean value of the diameter readings works out at 0.342 mm, but the third significant figure cannot be justified as the precision of each reading is 0.01 mm. Therefore the mean value is rounded down to 0.34 mm.

Note:

The uncertainty in the mean value is ±0.02 mm (i.e., half the range) as explained in Topic 14.2.

Using error estimates

How confident can you be in your measurements and any results or conclusions you draw from your measurements? If you work out what each uncertainty is, as a percentage of the measurement (the percentage uncertainty), you can then see which measurement is least accurate. You can then think about how that measurement could be made more accurately.

Worked example: Percentage uncertainty

The mass and diameter of a ball bearing were measured and the uncertainty of each measurement was estimated.

The mass, m, of a ball bearing = $4.85 \times 10^{-3} \pm 0.02 \times 10^{-3}$ kg

The diameter, d, of the ball bearing = $1.05 \times 10^{-2} \pm 0.01 \times 10^{-2}$ m

Calculate and compare the percentage uncertainty of these two measurements.

Solution

The percentage uncertainty of the mass $m = \frac{0.02}{4.85} \times 100\% = 0.4\%$

The percentage uncertainty of the diameter $d = \frac{0.01}{1.05} \times 100\% = 1.0\%$

The percentage uncertainty in the diameter measurement is therefore more than twice as large as that of the mass measurement.

Study tip

At the end of a calculation, don't give the answer to as many significant figures as shown on your calculator display. Give your answer to the same number of significant figures as the data with the least number of significant figures.

More about errors

1 When two measurements are added or subtracted, the uncertainty of the result is the sum of the uncertainties of the measurements. For example, the mass of a beaker is measured when it is empty and then when it contains water:
   - the mass of an empty beaker = 65.1 ± 0.1 g
   - the mass of the beaker and water = 125.6 ± 0.1 g.

Then the mass of the water could be as much as

$(125.6 + 0.1) - (65.1 - 0.1)\,g = 60.7\,g$, or as little as
$(125.6 - 0.1) - (65.1 + 0.1)\,g = 60.3\,g$.

The mass of water is therefore $60.5 \pm 0.2\,g$.

2 When a measurement in a calculation is raised to a power n, the percentage uncertainty is increased n times. For example, suppose you need to calculate the area A of cross section of a wire that has a diameter of $0.34 + 0.01$ mm. You will need to use the equation $A = \pi d\frac{2}{4}$. The calculation should give an answer of $9.08 \times 10^{-8}\,m^2$. The percentage uncertainty of d is $\frac{0.01}{0.34} \times 100\% = 2.9\%$. So the percentage uncertainty of A is 5.8% (= 2 × 2.9%). The consequence of this rule is that in any calculation where a quantity is raised to a higher power, the uncertainty of that quantity becomes much more significant.

Hint

Percentage uncertainty

To work out the percentage uncertainty of A, you could:

- Calculate the area of cross section for $d = 0.34 - 0.01$ mm = 0.33 mm.

  This should give an answer of $8.55 \times 10^{-8}\,m^2$.

- Calculate the area of cross section for $d = 0.34 + 0.01$ mm = 0.35 mm.

  This should give an answer of $9.62 \times 10^{-8}\,m^2$.

  Therefore, the area lies between $8.55 \times 10^{-8}\,m^2$ and $9.62 \times 10^{-8}\,m^2$.

  In other words, the area is $(9.08 \pm 0.53) \times 10^{-8}\,m^2$ (as $9.08 - 0.53 = 8.55$ and $9.08 + 0.53 = 9.62$).

  The percentage uncertainty of A is $\frac{0.53}{9.08} \times 100\% = 5.8\%$.

  This is twice the percentage uncertainty of d.

  It can be shown as a general rule that for a measurement x, the percentage uncertainty in x^n is n times the percentage uncertainty in x.

Graphs and errors

Straight-line graphs are important because they are used to establish the relationship between two physical quantities. Consider a set of measurements of the distance fallen by an object released from rest and the time it takes. A graph of distance fallen against $(\text{time})^2$ should be a straight line through the origin. If the line is straight, the theoretical equation $s = \frac{1}{2}gt^2$ (where s is the distance fallen and t is the time taken) is confirmed. The value of g can be calculated, as the gradient of the graph is equal to $\frac{1}{2}g$. If the straight line does not pass through the origin, there is a systematic error in the distance measurement. Even so, the gradient is still $\frac{1}{2}g$.

Synoptic link

You will meet line graphs in Topic 2.4, Free fall.

A best-fit test

Suppose you have obtained your own measurements for an experiment and you use them to plot a graph that is predicted to be a straight line. The plotted points are unlikely to be exactly straight in line with each other. The next stage is to draw a straight line of best fit so that the points are on average as close as possible to the line. Some problems may occur at this stage:

1 There might be a point much further than any other point from the line of best fit. The point is referred to as an **anomaly**. Methods for dealing with an anomalous point are as follows:

- If possible, the measurements for that point should be repeated and used to replace the anomalous point, if the repeated measurement is much nearer the line.
- If the repeated measurement confirms the anomaly, there could be a fault in the equipment or the way it is used. For example, in an electrical experiment, it could be caused by a change of the range of a meter to make that measurement. If no fault is found, make more measurements near the anomaly to see if these measurements lead towards the anomaly. If they do, it is no longer an anomaly and the measurements are valid.
- If a repeat measurement is not possible, the anomalous point should be ignored when drawing the best-fit line and a comment made in your report (or on the graph) about it.

2 The points might seem to curve away from the straight line of best fit. The uncertainty of each measurement can be used to give a small range or **error bar** for each measurement. Figure 1 shows the idea. The straight line of best fit should pass through all the error bars. If it doesn't, the following notes might be helpful. You could use the error bars to draw a straight line of maximum gradient and a straight line of minimum gradient.

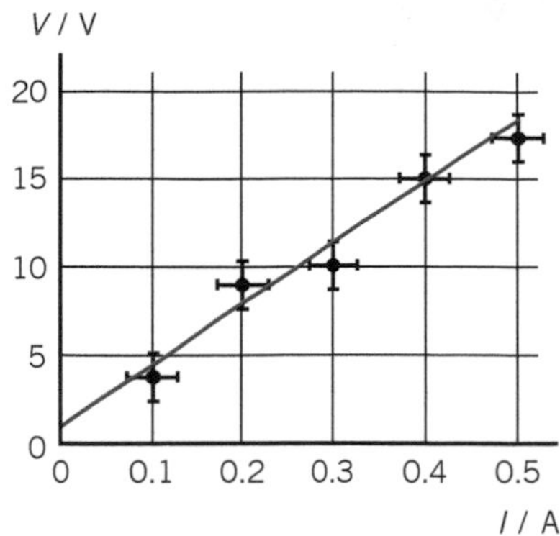

▲ **Figure 1** *Error bars*

- Suppose the points lie along a straight line over most of the range but curve away further along the line. This would indicate that a straight-line relationship between the plotted quantities is valid only over the range of measurements which produced the straight part of the line.
- Only two or three points in Figure 2 seem to lie on a straight line. In this case, it cannot be concluded that there is a linear relationship between the plotted quantities. You might need to plot further graphs to find out if a different type of graph would give a straight-line relationship. A data analysis software package on a computer could be used to test different possible relationships (or for the second year of your course, a log graph could be plotted).

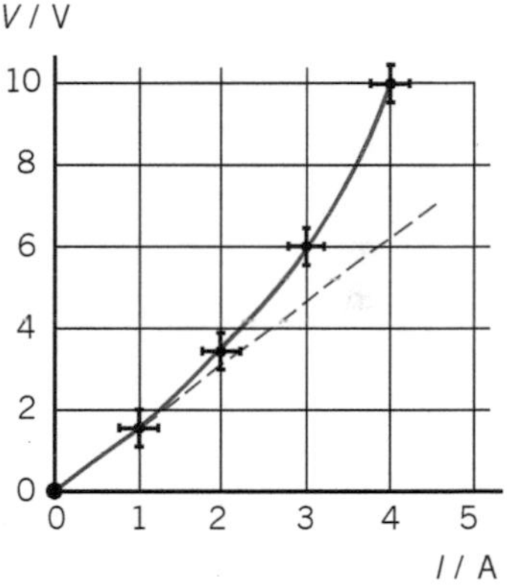

▲ **Figure 2** *Curves*

Evaluating your results

You should be able to form a conclusion from the results of an investigation. This might be a final calculation of a physical quantity or property (e.g., resistivity) or a statement of the relationship established between two variables. As explained earlier, the degree of accuracy of the measurements could be used as a guide to the number of significant figures in a 'final result' conclusion. Mathematical links established or verified between quantities should be stated in a 'relationship' conclusion.

You always need to evaluate the conclusion(s) of an experiment or investigation to establish its validity. This evaluation could start with a discussion of the strength of the experimental evidence used to draw the conclusions:

- Discuss the reliability of the data and suggest improvements, where appropriate, that would improve the reliability. You may need to consider the effect of the control variables, if the experimental evidence is not as reliable as it should be.
- Discuss the methods taken (or proposed) to eliminate or reduce any random or systematic errors. Describe the steps taken to deal with anomalous results.
- Evaluate the accuracy of the results by considering the percentage uncertainties in the measurements. These can be compared to identify the most significant sources of error in the measurements, which can then lead to a discussion of how to reduce the most significant sources of error.
- Propose improvements to the strategy or experimental procedures, referring to the above discussion on validity as justification for the proposals.
- Suggest further experimental work, based on the strength of the conclusions. Strong conclusions could lead to a prediction and how to test it.

Study tip

A final result, including its uncertainty, can be compared with the accepted (accurate) value, if known. Such a comparison can be used to evaluate the method used.

For example, for an obtained resistivity value of $4.8 \times 10^{-7} \pm 0.3 \times 10^{-7}\,\Omega\,m$ for a certain metal wire, with an accepted value of $5.2 \times 10^{-7}\,\Omega\,m$, you could review your method to see where improvements can be made (e.g., by making more measurements and taking a mean value).

More about measurements

At A level, you will learn how to:

- use vernier callipers and micrometers to measure small distances
- use appropriate analogue apparatus to measure angles, length and distance, volume, force, temperature, and pressure
- use appropriate digital instruments to measure mass, time, current, resistance, and voltage
- use a stopwatch or light gates for timing
- use methods to increase accuracy of measurements, such as repeating and averaging timings or using a fiduciary marker, set square, or plumb line
- correctly construct circuits from own or given circuit diagrams using dc power supplies, cells, and a range of circuit components, including those where polarity is important
- generate and measure waves, using a microphone and loudspeaker, or ripple tank, or vibration transducer, or microwave / radio wave source
- use a suitable light source to investigate characteristics of light, including interference and diffraction
- use a data logger with a variety of sensors to collect data, or use software to process data.

In addition to using the instruments and techniques listed above, you are also expected to be able to use instruments as described below that are more complex. Such instruments include the oscilloscope, the Geiger–Müller tube with a scaler counter or ratemeter, and data loggers and/or light gates in some investigations (e.g., oscillations, capacitor discharge). You should also know how to time multiple oscillations and how to avoid parallax errors when reading a scale. Two further instruments, the travelling microscope (see page 12) and the spectrometer (see page 219), are also described in this book – although you do *not* need to know how to use these in your A Level Physics course, they give more experience of using very accurate instruments.

Synoptic link

If you are unsure about how to use any of the listed instruments, see Everyday physics instruments, and consult your teacher.

Synoptic link

The spectrometer is covered in Topic 12.4, The diffraction grating.

An oscilloscope

An oscilloscope is used to display waveforms and to measure p.ds and time intervals. You will probably use an oscilloscope in the Year 1 part of your A Level course when you sound waves. In Year 2 topics such as capacitor discharge, you will use an oscilloscope to measure the p.d. across a discharging capacitor. When you use an oscilloscope, you should assume that the control dials for its time base and voltage gain are calibrated accurately. However,

- always check if the oscilloscope has a variable control for either the time base or the voltage gain in addition to the fixed settings of each control dial. If so, you need to ensure that the variable control is at the correct setting (e.g., fully clockwise) for the calibration figure for each of the fixed settings to apply. See Topic 11.10, Using an oscilloscope, if you need to revise the use of an oscilloscope.

▲ **Figure 1** *The oscilloscope*

- if you are using the oscilloscope as a dc voltmeter, make sure the input is set for dc measurements rather than for ac measurements. Likewise, if you are measuring an ac waveform, you should check the input is set for ac measurements rather than dc measurements.

In addition, when measuring

- an ac waveform, ensure the Y-gain is adjusted so the vertical height of the waveform is as large as possible with the full waveform from top to bottom on the screen. When measuring a time period, ensure several cycles are displayed across the screen and that you measure across as many cycles as possible to reduce experimental uncertainty.
- a dc voltage, for example in a capacitor discharge experiment, ensure the zero reading is correct for zero input p.d. and check it has not drifted during the investigation.

A data logger

A data logger enables routine or remote measurements to be made as well as measurements over very long or short time scales. Electronic sensors connected to a data logger are necessary to record the variation of a physical property such as temperature. Ammeter and voltage sensors are necessary to measure currents and potential differences.

Data loggers vary considerably in complexity and ease of use. Assuming the data logger and sensors are set up, before using a data logger, you may need to choose:

- the most appropriate time scale for the recording
- the time interval between successive recordings (or the number of recordings per second/minute/hour)
- the most appropriate range of each sensor.

If a recording is too fast or too long or the sensors are out of range, the recording should be repeated if possible.

Most data loggers will be linked to computers which are loaded with appropriate software for recording, processing, and/or plotting graphs of the results. You may need to print a graph out if you intend to use it to measure, for example, the gradient if it is a straight-line graph. However, the computer software may do such measurements for you.

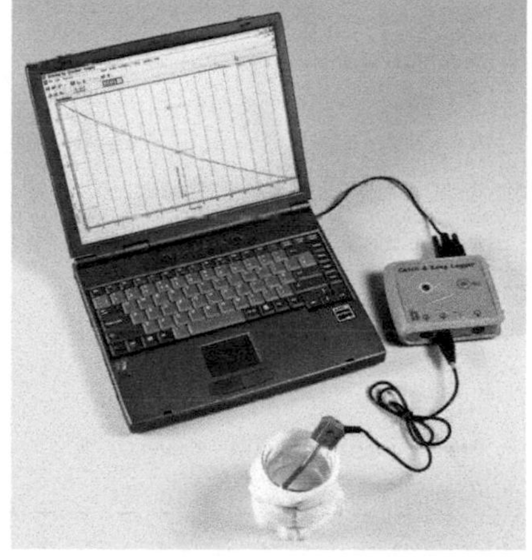

▲ **Figure 2** *Using a data logger*

Light gates

Light gates are used with a computer or a data logger or timer to remove some of the random errors associated with personal judgements when a moving object passes a certain position, for example, if you have to time an object to move from rest through a certain distance.

The effect of using light gates should be to reduce the range of the readings for a given measurement. However, light gates may not be suitable for every experiment in which a moving object has to be timed. For example, the time period of an oscillating object that repeatedly moves backwards and forwards through a light gate could only be timed for one half-cycle of the object's motion, corresponding to the object moving through the light gate in one direction to start the timing and then in the opposite direction to stop the light gate.

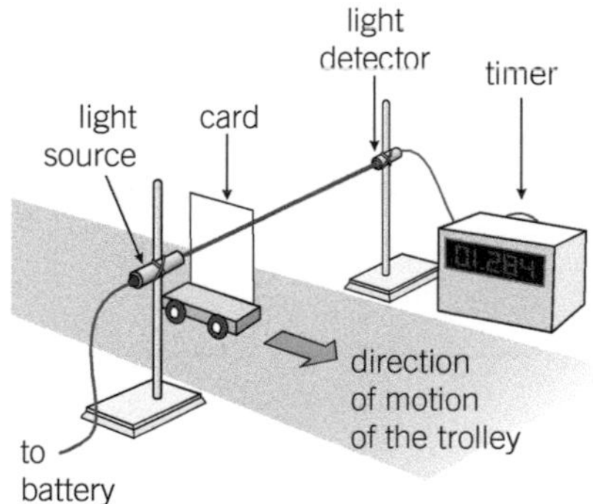

▲ **Figure 3** *Using a light gate*

Study tip

After starting the counter and the timer simultaneously, observe the timer and stop the counter exactly when the time interval has elapsed. Don't try to stop the timer at the same time as stopping the counter and watching the timer.

Extension

A travelling microscope

This is a microscope on an adjustable platform that can be moved vertically or horizontally by turning an adjustment screw. The platform is fitted with a horizontal and a vertical vernier scale, so its horizontal or vertical position can be measured to within ±0.1 mm. A travelling microscope would be used, for example, to measure the internal diameter of a glass tube (e.g., 1 mm bore), because a micrometer or a conventional vernier could not access the internal surface of the tube. Before use, a travelling microscope should be levelled using a spirit level in two perpendicular directions so that its platform is horizontal. Otherwise, there could be a systematic error in the measurements.

The light gate would need to be exactly at the centre of the oscillations, otherwise the timing would not be exactly one half-cycle. Repeated measurements of one half-cycle could be made to give a more reliable mean value, and this might give better results than using a stopwatch if the oscillations are too fast to time manually.

An ionising radiation detector

The ionising radiation detector that you use will probably be a Geiger–Müller tube. This may be used with a scaler counter, which counts the number of ionising particles that enter the tube or it may be used with a ratemeter which gives a read-out of the count rate (i.e., number of counts per unit time) of the particles entering the tube. The tube p.d. must be set at its operating p.d., which is normally in excess of 300 V. The number of counts in a certain time interval is measured by setting the counter to zero, then starting the counter and stopping it after a certain time.

Figure 4b shows how the count rate varies with the tube p.d. The operating p.d. corresponds to the plateau of the graph sufficiently far from the minimum p.d. necessary for the tube to operate (i.e., the threshold p.d.) as to be unaffected by random fluctuations in the tube p.d.

When using the tube with a scaler counter, the number of counts in a given time (e.g., 100 s) should be measured several times to give a mean value of the count rate (i.e., counts per second). The bigger the total number of counts, the smaller the uncertainty in the measurement. If the time interval is too short, random errors that may occur in starting and stopping the counter could be more significant than if a longer time interval were used. If the time interval is too long, it would be difficult to tell if the activity of the source is decreasing or if an error in starting or stopping the timer has occurred.

When using the tube with a ratemeter, ensure the ratemeter is set on the range which gives the largest reading. For example, if the range dial has three positions, 1, 10, and 100 counts per second, the range may need to be set at 1 count per second if the reading is very small on the '10' and '100' positions.

When using either a scaler counter or a ratemeter, remember to measure the background count rate and subtract it from the measurements made when the source is present.

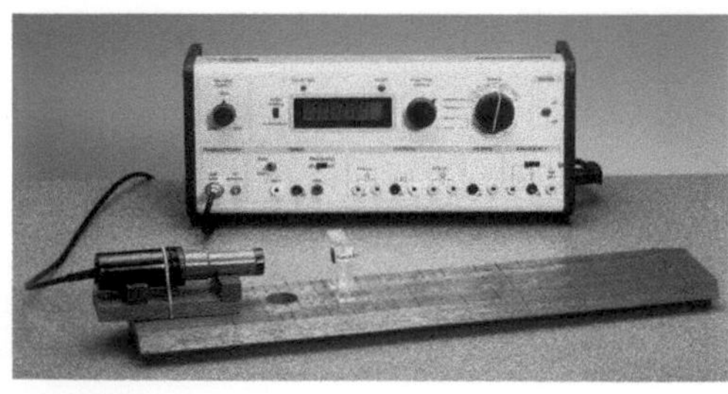

a *A Geiger–Müller tube connected to a scaler counter*

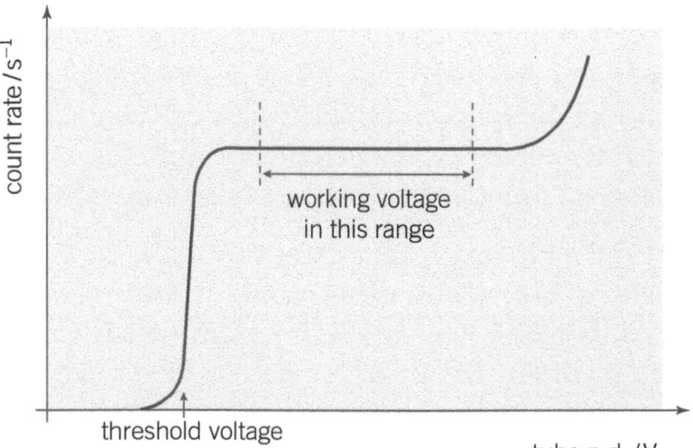

b *A graph of count rate against tube p.d.*

▲ **Figure 4** *Using a Geiger–Müller tube*

About practical assessment

Assessment outline

Practical work is a vital part of any physics course. It helps you to understand new ideas and difficult concepts and it helps you to appreciate the importance of experiments in testing and developing scientific theories. Practical work also develops the skills that scientists use in their everyday work. Such skills involve planning, researching, making and processing measurements, and analysing and evaluating experimental results. The notes in this section provide information about how your proficiency in practical skills is assessed at A Level.

The Oxford AQA International A Level two-year course is designed so that the first year course and the AS Physics course cover the same course content.

Papers 1 and 2 test content from Sections 1 and 2 of the AS / first year of the A Level course.

Papers 3 and 4 test content from Sections 3 and 4 of the second year of the A Level course.

Paper 5 tests Section 5, which covers practical and analytical skills, and content from the full two-year A Level course.

Practical skills acquired during each year of the course are assessed as part of a written examination in the AS and A Level examination papers. Practical skills questions in the examination papers will expect students to have completed a set of required practical experiments successfully in their course. The list of required practicals is shown in the Table below.

In carrying out these required practical activities, you should become proficient in all the practical skills assessed in your A level examination papers.

▼ **Table 1** *Required practical activities*

Practical number	Practical title	Section number
AS Level / Year 1 A Level		
1	Measuring the acceleration of a freely falling object	2.4
2	Measuring the Young modulus of elasticity of a wire	6.3
3	Measuring the e.m.f. and the internal resistance of a cell or a battery	10.3
4	Investigation of the oscillations of a mass–spring system and a simple pendulum	11.1
5	Investigating interference using Young's double slit experiment	12.1
	Using a diffraction grating to measure the wavelength of different colours of light	12.4
Year 2 A Level		
6	Investigation of the charge and discharge of capacitors	19.4
7	Investigation of the efficiency of a transformer	21.5
8	Determination of the specific heat capacity using an electrical method	22.2
9	Investigation of Boyle's law and Charles's law	23.1
10	Investigation of the inverse square law for light using an LDR and a point source	26.2

Extension

Safety first

Before you carry out a practical task, you should eliminate (if possible) or minimise any health and safety hazards. A risk assessment requires you to think about the possible hazards in an activity and plan to eliminate or minimise them. Your teacher ought to have made a risk assessment of every practical activity in advance to ensure the practical activities you undertake are safe. For example, if you are about to stretch springs or wires, you should be provided with eye protection which you should wear. However, you should also carry out your own risk assessment to ensure you use the apparatus you are given safely.

To be able to answer examination questions on practical and analytical skills, you need to carry out the required practical activities. Questions may be set on these practicals directly, or on the skills contained within the practicals.

These skills could include, but are not limited to:

- planning experiments, including identifying and understanding how to control variables
- choosing equipment, or evaluating the use of specified pieces of equipment
- skills required for carrying out experiments such as taking readings or recording data
- choosing, constructing, and interpreting appropriate graphical displays for data
- analysing and interpreting data, including carrying out calculations on data
- evaluating experimental procedures.

In carrying out your practical work, you should know how to:

- use callipers and micrometers to measure small distances, using digital and vernier scales
- use appropriate analogue apparatus to measure angles, length and distance, volume, force, temperature, and pressure,
- use appropriate digital instruments including electrical multimeters to measure mass, resistance, time, current, and voltage
- use a stopwatch or light gates for timing
- use methods to increase accuracy of measurements, such as timing over multiple oscillations, use of a fiduciary marker, or a set square, and plumb line
- design, check, and correctly construct circuits from circuit diagrams using dc power supplies, cells, and a range of circuit components, including those where polarity is important
- use a signal generator and oscilloscope to measure voltages and time intervals or periods
- generate and measure waves, using a microphone and loudspeaker, or ripple tank, or vibration transducer, or microwave / radio wave source
- use suitable light sources including a laser to investigate characteristics of light, including interference and diffraction
- use ICT for computer modelling and to process data
- use a data logger with a variety of sensors to collect data.

The practical experiments and investigations you carry out during your course may include additional practical activities as well as the required practical activities. These additional practical activities will enable you to develop and reinforce your competence in the practical skill areas you will be assessed on. For example, in the experiment to measure the resistivity of the material of a wire, you need to appreciate at the outset that the resistance of a wire depends on its length and its diameter so you may decide to measure the resistance of wires of different diameters which are all the same length and material. Then you need to decide:

- how to measure the resistance and diameter of each wire and also their lengths (as this is a control variable and you need to ensure it is the same for each wire)
- what measuring instruments to use, how to use the instruments, and how to ensure your measurements are as accurate as possible
- how to avoid systematic errors (e.g., check zero errors) and random errors (e.g., measure the diameter at several places along each wire and obtain an average value), and how to assess the uncertainty in each of your measurements.

Once you have a set of results, you need to know how to process and use the results. This particular step may require you to do some research to find the theoretical relationship between the two variables you have measured in order to plot a graph. In this example, you could plot a graph of the resistance of the wire against the reciprocal of the area of cross section. This graph should be a straight line through the origin, as in Figure 1. The theory should tell you that the gradient is equal to the resistivity × the length of the wire. Therefore the resistivity of the wire can be determined. Finally, you could use your uncertainty estimates to determine the overall uncertainty in your value of resistivity. See page 7–8, Analysis and evaluation, if necessary.

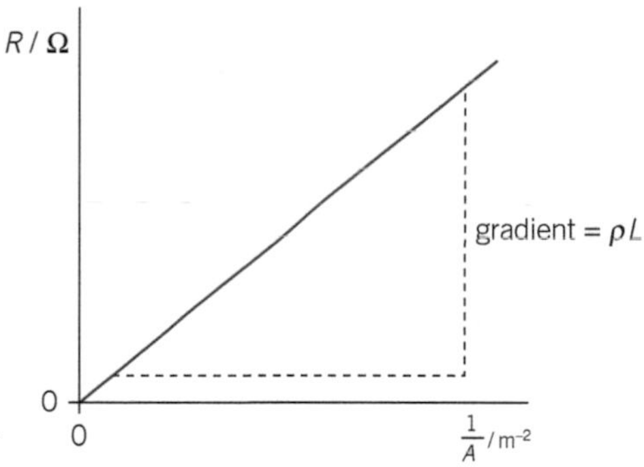

▲ **Figure 1** *Graph of resistance R against* $\frac{1}{\text{area of cross-section } A}$ *for wires of the same material of resistivity ρ and length L*

The practical experiments and investigations you will carry out during the course will enable you to develop and practice your practical skills. Each of the four content sections in the book includes Practice questions designed to test your practical skills. So at the end of your course, you should be well prepared for any questions on practical skills in your written examination papers. In addition, you should also be well prepared for physics courses beyond A Level.

Extension

The laboratory apparatus you could use to make measurements might include

- basic apparatus (metre rule, set square, protractors)
- electrical meters and multimeters (analogue or digital) for measuring current, voltage, and resistance
- a micrometer, vernier callipers
- a top-pan electronic balance, newtonmeters
- measuring cylinders
- a digital stopwatch or light gates for timing
- thermometers
- a data logger with sensors
- a pressure gauge, a signal generator, an oscilloscope, a Geiger–Müller counter.

1 Forces in equilibrium

1.1 Vectors and scalars

Learning objectives:

- → Define a vector quantity.
- → Describe how we represent vectors.
- → Explain how we add and resolve vectors.

Specification reference: 3.2.1

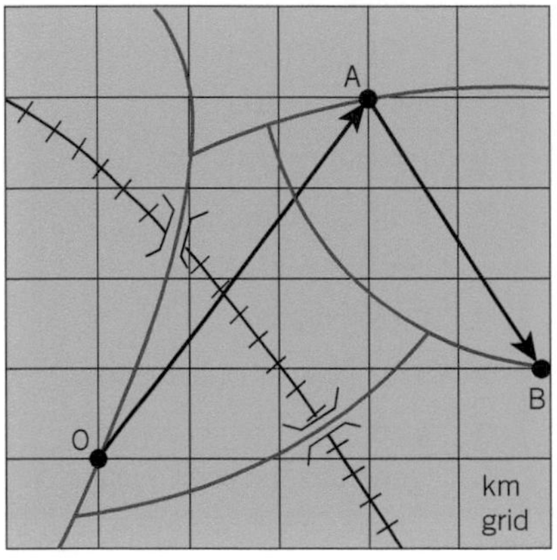

▲ **Figure 1** *Map of locality*

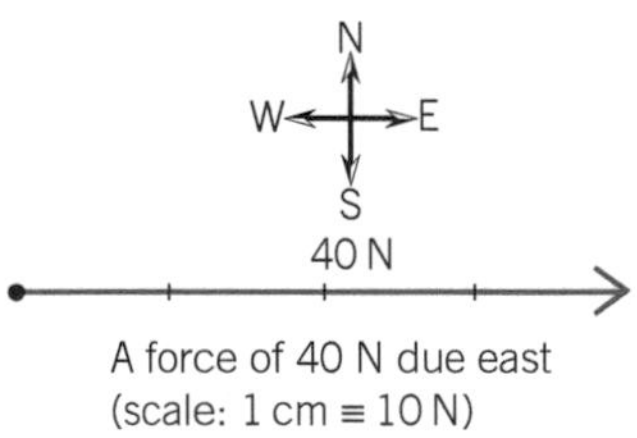

▲ **Figure 2** *Representing vectors*

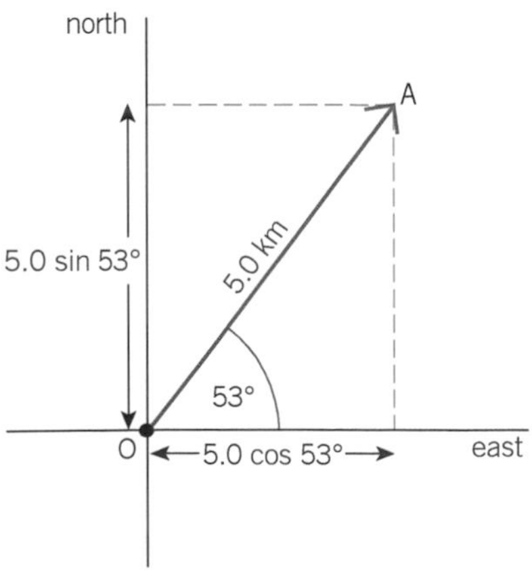

▲ **Figure 3** *Resolving a vector*

Representing a vector

Imagine you are planning to cycle to a friend's home several kilometres away from your home. The **distance** you travel depends on your route. However, the direct distance from your home to your friend's home is the same, whichever route you choose. Distance in a given direction, or **displacement**, is an example of a **vector** quantity because it has magnitude and direction. In Figure 1, suppose your home is at point O on the map and your friend's home is at A. The road distance you would cycle from O to A is greater than the direct distance or displacement from O to A. This is represented by the arrow from O to A.

A vector is any physical quantity that has a direction as well as a magnitude.

Further examples of vectors include velocity, acceleration, weight and force.

A scalar is any physical quantity that is not directional.

For example, distance is a scalar because it takes no account of direction. Further examples of scalars include mass, density, volume, and energy.

Any vector can be represented as an arrow. The length of the arrow represents the magnitude of the vector quantity. The direction of the arrow gives the direction of the vector.

- **Displacement** is distance in a given direction. As shown in Figure 1, the displacement from one point to another can be represented on a map or a scale diagram as an arrow from the first point to the second point. The length of the arrow must be in proportion to the least distance between the two points.
- **Velocity** is speed in a given direction. The velocity of an object can be represented by an arrow of length in proportion to the speed pointing in the direction of motion of the object.
- **Force and acceleration** are both vector quantities and therefore can each be represented by an arrow in the appropriate direction and of length in proportion to the magnitude of the quantity.

Resolving a vector into two perpendicular components

This is the process of working out the components of a vector in two perpendicular directions from the magnitude and direction of the vector. Figure 3 shows the displacement vector OA represented on a scale diagram that also shows lines due north and due east. The components of this vector along these two lines are $5.0\cos 53°$ km (= 3.0 km) along the line due east and $5.0\sin 53°$ km (= 4.0 km) along the line due north.

In general, to resolve any vector into two perpendicular components, draw a diagram showing the two perpendicular directions and an arrow to represent the vector. Figure 4 shows this diagram for a vector OP. The components are represented by the projection of the vector onto each line. If the angle θ between the vector OP and one of the lines is known,

- the component along that line = OP $\cos\theta$ and
- the component perpendicular to that line (i.e., along the other line) = OP $\sin\theta$.

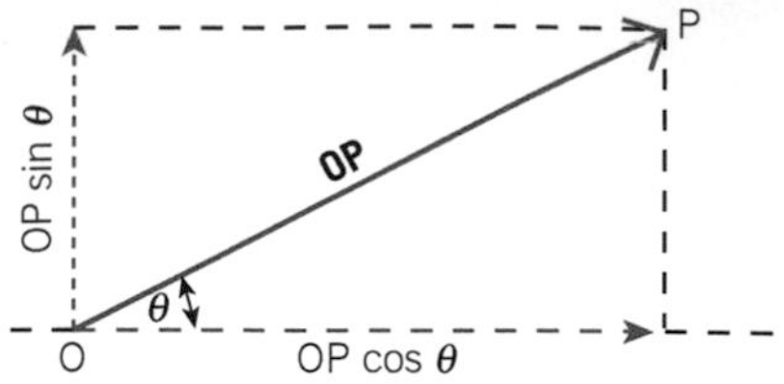

▲ **Figure 4** *The general rule for resolving a vector*

Thus a force F can be resolved into two perpendicular components:

- **$F\cos\theta$ parallel to a line at angle θ to the line of action of the force and**
- **$F\sin\theta$ perpendicular to the line.**

Worked example

A paraglider is pulled along at constant height at steady speed by a cable attached to a speedboat as shown in Figure 5. The cable pulls on the paraglider with a force of 500 N at an angle of 35° to the horizontal. Calculate the horizontal and vertical components of this force.

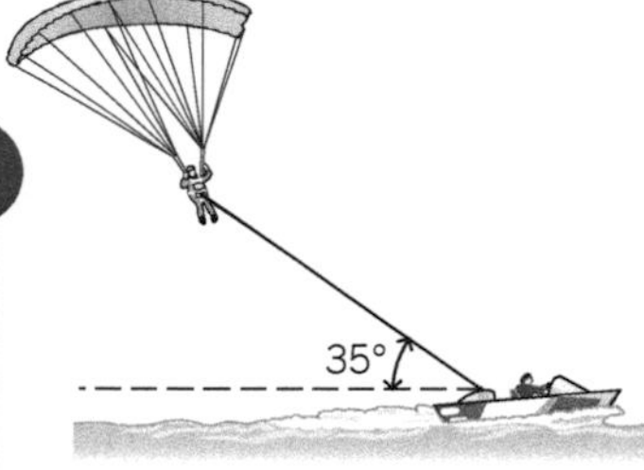

▲ **Figure 5**

Solution

Because the force on the paraglider is at an angle of 35° below the horizontal, the horizontal and vertical components of this force are:

- $500\cos 35° = 410$ N horizontally to the right
- $500\sin 35° = 287$ N vertically downwards.

Addition of vectors using a scale diagram

Let's go back to the cycle journey in Figure 6. Suppose when you reach your friend's home at A, you then go on to another friend's home at B. Your journey is now a two-stage journey:

- **Stage 1, from O to A,** is represented by the displacement vector $\overrightarrow{\mathbf{OA}}$.
- **Stage 2, from A to B,** is represented by the displacement vector $\overrightarrow{\mathbf{AB}}$.

Figure 6 shows how the overall displacement from O to B, represented by vector $\overrightarrow{\mathbf{OB}}$, is the result of adding vector $\overrightarrow{\mathbf{AB}}$ to vector $\overrightarrow{\mathbf{OA}}$. The **resultant** is the third side of a triangle, where OA and AB are the other two sides.

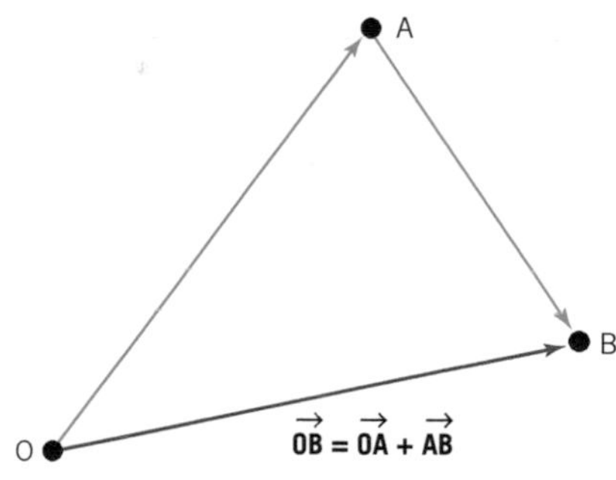

▲ **Figure 6** *Displacement from O to B*

$$\overrightarrow{\mathbf{OB}} = \overrightarrow{\mathbf{OA}} + \overrightarrow{\mathbf{AB}}$$

Use Figure 6 to show that the resultant displacement $\overrightarrow{\mathbf{OB}}$ is 5.1 km in a direction 11° North of due East.

Hint

Since $\overrightarrow{\mathbf{OA}}$ and $\overrightarrow{\mathbf{AB}}$ are vectors, the lengths do not add directly.

Any two vectors of the same type can be added together using a scale diagram. For example, Figure 7a shows a ship pulled via cables by two tugboats. The two pull forces F_1 and F_2 acting on the ship are at 40° to each other. Suppose the forces are both 8.0 kN. Figure 7b shows how

Study tip

Drawing vector diagrams requires a ruler and careful drawing.

you can find the **resultant *R*** (combined effect) of the two forces using a scale diagram.

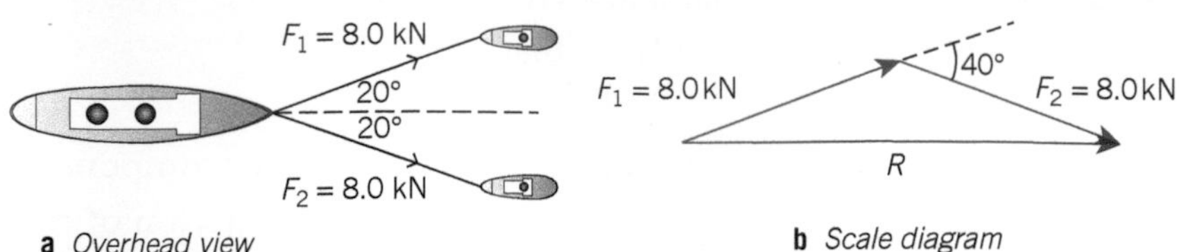

a *Overhead view* **b** *Scale diagram*

▲ **Figure 7** *Adding two forces using a scale diagram*

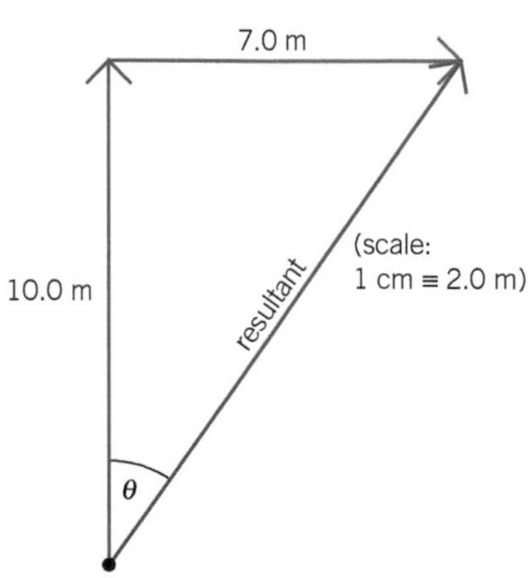

▲ **Figure 8** *Adding two displacements at right angles to each other*

Addition of two perpendicular vectors using a calculator

1 Adding two displacement vectors that are at right angles to each other

Suppose you walk 10.0 m forward then turn through exactly 90° and walk 7.0 m. At the end, how far will you be from your starting point? The vector diagram to add the two displacements is shown in Figure 8, drawn to a scale of 1 cm to 2.0 m. The two displacements form the two shorter sides of a right-angled triangle with the overall displacement, the resultant, as the hypotenuse. Using a ruler and a protractor, the resultant displacement can be shown to be a distance of 12.2 m at an angle of 35° to the initial direction. You can check this using

- Pythagoras's theorem for the distance $(= (10.0^2 + 7.0^2)^{\frac{1}{2}})$
- the trigonometry equation $\tan\theta = \dfrac{7.0}{10.0}$ for the angle θ between the resultant and the initial direction.

2 Two forces acting at right angles to each other

Figure 9 shows an object O acted on by two forces F_1 and F_2 at right angles to each other. The vector diagram for this situation is also shown. The two forces in the vector diagram form two of the sides of a right-angled triangle in which the resultant force is represented by the hypotenuse.

As explained above, the magnitude and direction of the resultant force can be worked out to give a magnitude of 12.2 N and a direction of 35° to the 10 N force.

In general, if the two perpendicular forces are F_1 and F_2, then

- **the magnitude of the resultant $F = (F_1^{\,2} + F_2^{\,2})^{\frac{1}{2}}$ and**
- **the angle θ between the resultant and F_1 is given by $\tan\theta = \dfrac{F_2}{F_1}$**

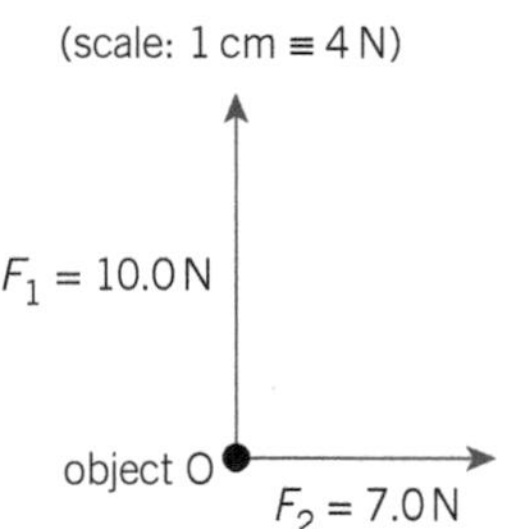

a *An object acted on by two perpendicular forces*

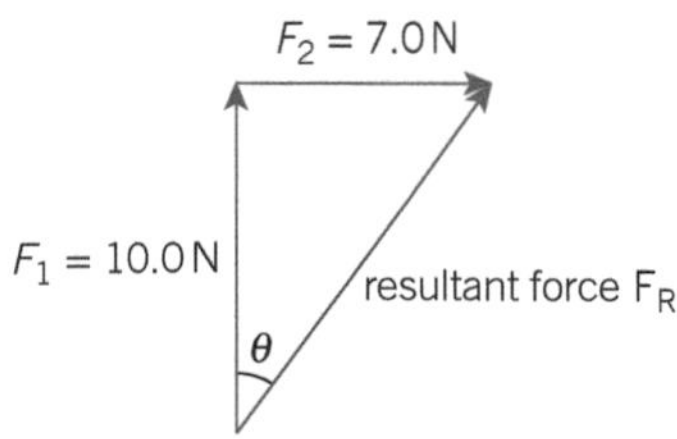

b *Vector diagram for* **a**

▲ **Figure 9**

Adding two vectors that are at angle θ to each other

Consider an object, O, acted on by forces F_1 and F_2 at angle θ to each other, as shown in Figure 10a. The magnitude and direction of the resultant force F_R can be found by resolving one of the forces into components that are parallel and perpendicular to the other force, as shown in Figure 10b.

- Resolving F_1 parallel and perpendicular to F_2 gives $F_1 \cos\theta$ for the parallel component and $F_1 \sin\theta$ for the perpendicular component.
- Adding the components in each direction therefore gives the parallel component of F_R as $F_1 \cos\theta + F_2$ and the perpendicular component as $F_1 \sin\theta$.

Using Pythagoras's theorem to find the magnitude of the resultant force gives

$$F_R = [(F_1 \cos\theta + F_2)^2 + (F_1 \sin\theta)^2]^{\frac{1}{2}}$$

Because $\sin^2\theta + \cos^2\theta = 1$ for all angles of θ, it can be shown that

$$F_R{}^2 = F_1{}^2 + F_2{}^2 + 2F_1F_2\cos\theta$$

Using the trigonometry rule for tan θ to find θ_R, the angle between the resultant force and F_2, gives

$$\tan\theta_R = \frac{F_1 \sin\theta}{(F_1 \cos\theta + F_2)}$$

Note:

The resultant of two vectors that act along the same line has a magnitude that is:

- the **sum,** if the two vectors are in the *same* direction. For example, if an object is acted on by a force of 6.0 N and a force of 4.0 N, both acting in the same direction, the resultant force is 10.0 N
- the **difference,** if the two vectors are in *opposite* directions. For example, if an object is acted on by a 6.0 N force and a 4.0 N force in opposite directions, the resultant force is 2.0 N in the direction of the 6.0 N force.

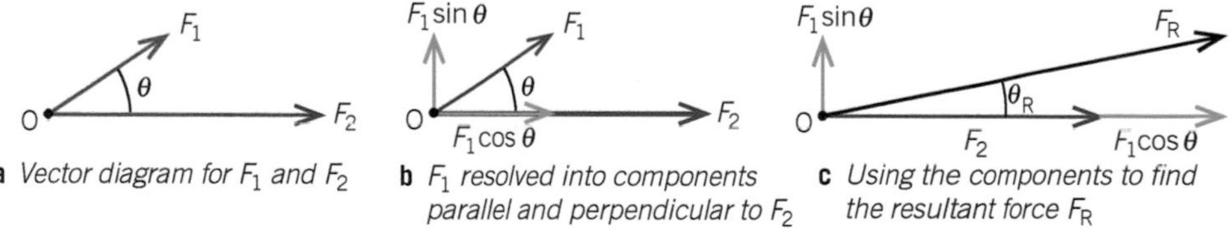

a *Vector diagram for F_1 and F_2* **b** *F_1 resolved into components parallel and perpendicular to F_2* **c** *Using the components to find the resultant force F_R*

▲ **Figure 10** *Using a calculator to find a resultant force*

Summary questions

1 Figure 11 shows three situations **a–c** where two or more known forces act on an object. For each situation, calculate the magnitude and direction of the resultant force.

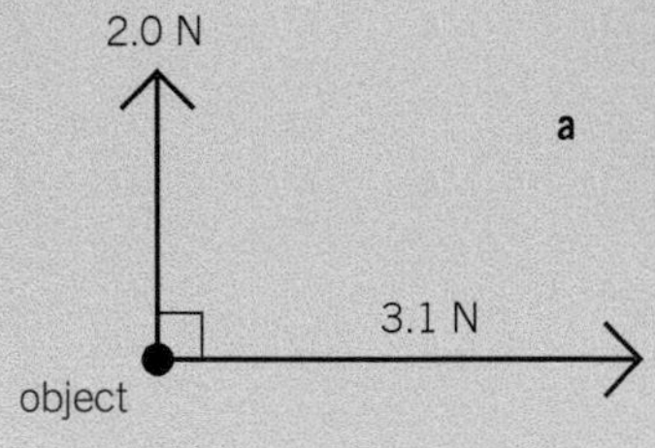

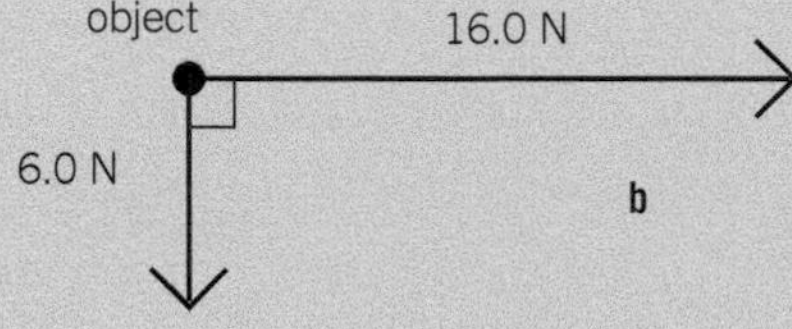

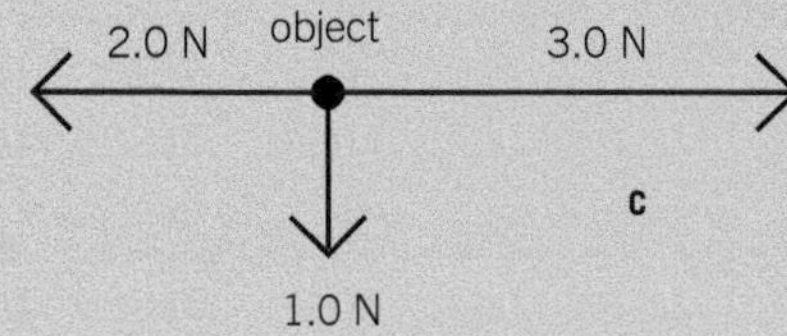

▲ **Figure 11**

2 Calculate the magnitude and direction of the resultant force on an object which is acted on by a force of 4.0 N and a force of 10 N that are

a in the same direction

b in opposite directions

c at right angles to each other

d at 35° to each other.

3 A crane is used to raise one end of a steel girder off the ground, as shown in Figure 12. When the cable attached to the end of the girder is at 20° to the vertical, the force of the cable on the girder is 6.5 kN. Calculate the horizontal and vertical components of this force.

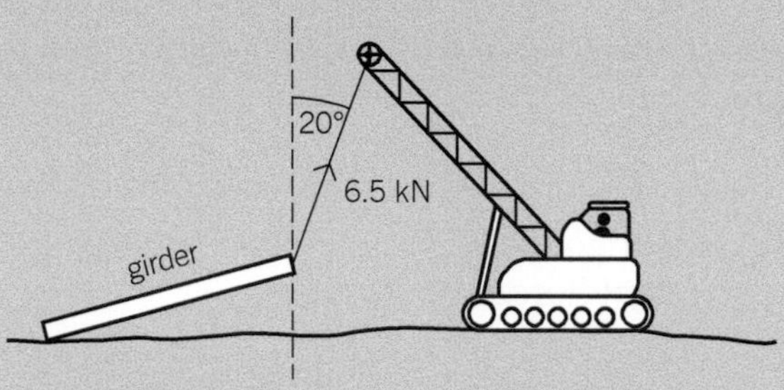

▲ **Figure 12**

4 A yacht is moving due north as a result of a force, due to the wind, of 350 N in a horizontal direction of 40° east of due north, as shown in Figure 13.

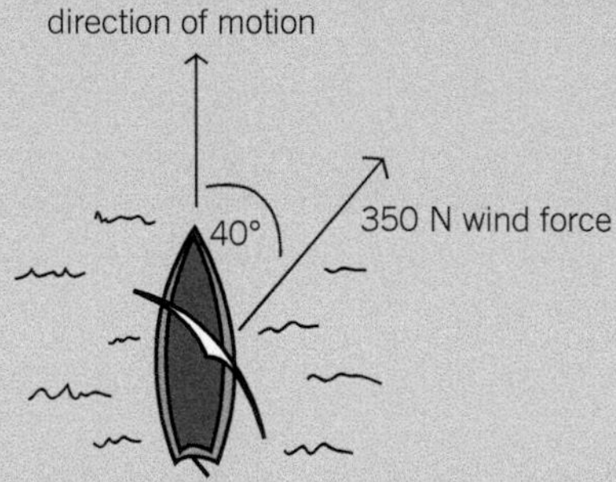

▲ **Figure 13**

Calculate the component of the force of the wind:

a in the direction the yacht is moving

b perpendicular to the direction in which the yacht is moving.

1.2 Balanced forces

Equilibrium of a point object

When two forces act on a point object, the object is in **equilibrium** (at rest or moving at constant velocity) only if the two forces are equal and opposite to each other. The resultant of the two forces is therefore zero. The two forces are said to be **balanced**.

For example, an object resting on a horizontal surface is acted on by its weight W (i.e., the force of gravity on it) acting downwards, and a support force S from the surface, acting upwards. The S is equal and opposite to W, provided the object is at rest (or moving at constant velocity).

$$S = W$$

When three forces act on a point object, their combined effect (the resultant) is zero only if the resultant of any two of the forces is equal and opposite to the third force. To check the combined effect of the three forces is zero:

- resolve each force along the same parallel and perpendicular lines
- balance the components along each line.

Worked example 1

A child of weight W on a swing is at rest due to the swing seat being pulled to the side by a horizontal force F_1. The rope supporting the seat is then at an angle θ to the vertical, as shown in Figure 2.

Solution

Assuming the swing seat is of negligible weight, the swing seat is acted on by three forces, which are: the weight of the child, the horizontal force F_1, and the tension T in the rope. Resolving the tension T vertically and horizontally gives $T\cos\theta$ for the vertical component of T (which is upwards) and $T\sin\theta$ for the horizontal component of T. Therefore, the balance of forces gives:

1 horizontally: $F_1 = T\sin\theta$

2 vertically: $W = T\cos\theta$

Because $\sin^2\theta + \cos^2\theta = 1$, then $F_1^2 + W^2 = T^2\sin^2\theta + T^2\cos^2\theta = T^2$

$\therefore T^2 = F_1^2 + W^2$

Also, because $\tan\theta = \dfrac{\sin\theta}{\cos\theta}$, then $\dfrac{F_1}{W} = \dfrac{T\sin\theta}{T\cos\theta} = \tan\theta$

$\therefore \tan\theta = \dfrac{F_1}{W}$

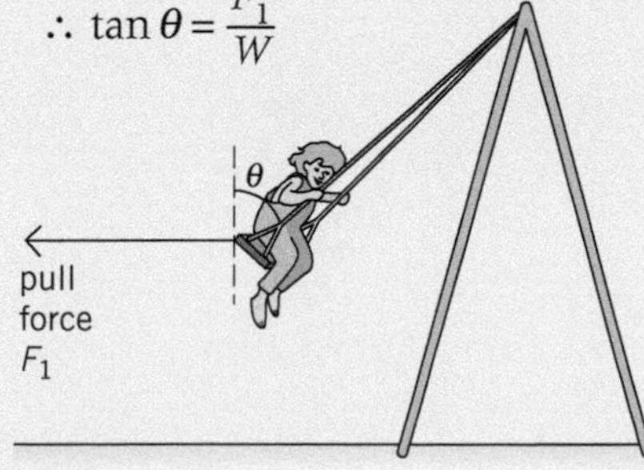

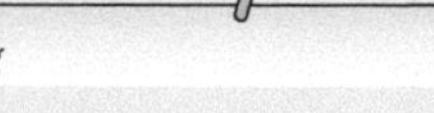

a *On a swing*

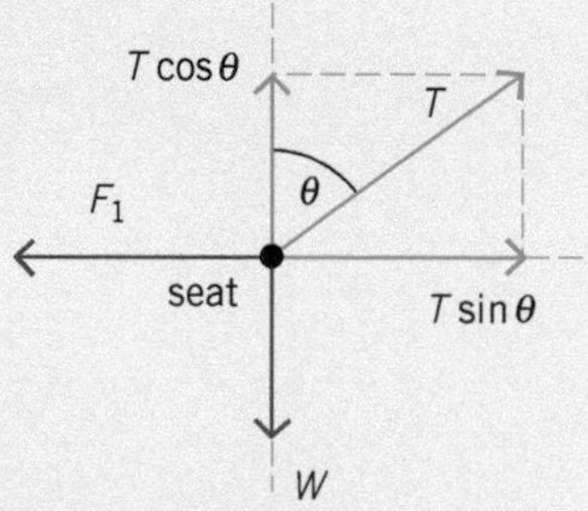

b *Force diagram*

▲ **Figure 2**

Learning objectives:

- → Explain why we have to consider the direction in which a force acts.
- → Demonstrate when two (or more) forces have no overall effect on a point object.
- → Explain the parallelogram of forces.

Specification reference: 3.2.1

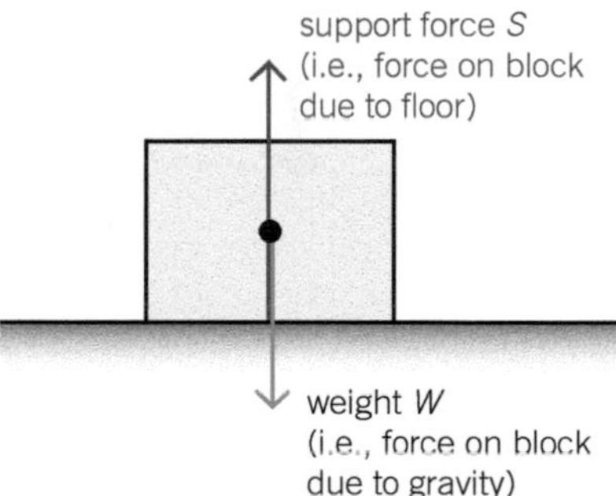

▲ **Figure 1** *Balanced forces*

Study tip

Forces can often be calculated using several different methods, for example, using trigonometric functions or Pythagoras's theorem.

You will meet another method of solving these problems, by using the triangle of forces, later on in this chapter.

Synoptic link

Further guidance on trigonometric relationships can be found in Topic 14.2 Trigonometry.

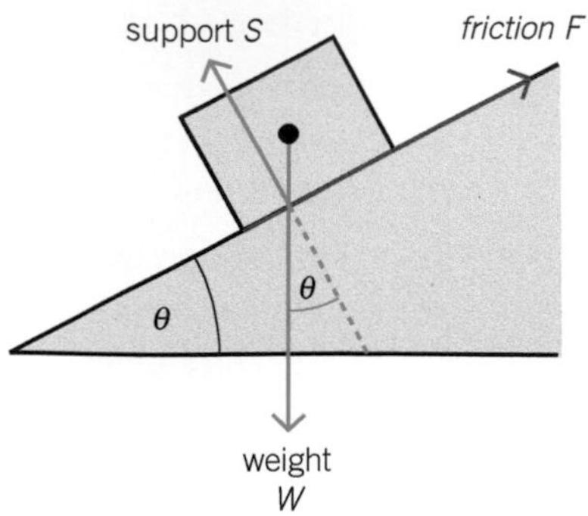

▲ **Figure 3**

Worked example 2

An object of weight W is at rest on a rough slope (i.e., a slope with a roughened surface as in Figure 3). The object is acted on by a frictional force F, which prevents it sliding down the slope, and a support force S from the slope, perpendicular to the slope.

Resolving the three forces parallel and perpendicular to the slope gives:

1 horizontally: $F = W\sin\theta$

2 vertically: $S = W\cos\theta$

Because $\sin^2\theta + \cos^2\theta = 1$
then $F^2 + S^2 = W^2\sin^2\theta + W^2\cos^2\theta = W^2$

$\therefore\ W^2 = F^2 + S^2$

Also, because $\tan\theta = \dfrac{\sin\theta}{\cos\theta}$, then $\dfrac{F}{S} = \dfrac{W\sin\theta}{W\cos\theta} = \tan\theta$

$\therefore\ \tan\theta = \dfrac{F}{S}$

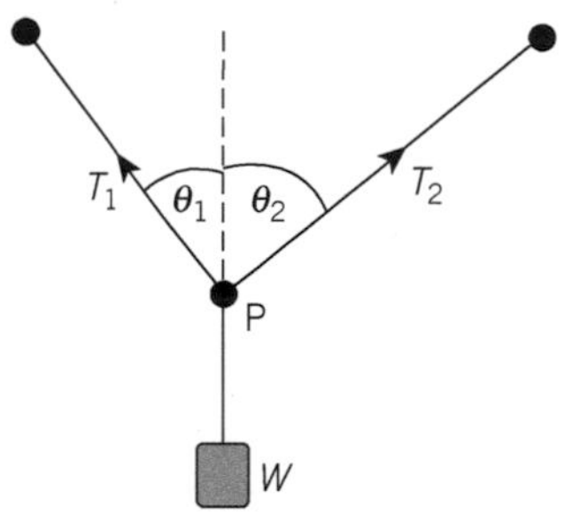

▲ **Figure 4** *A suspended weight*

Worked example 3

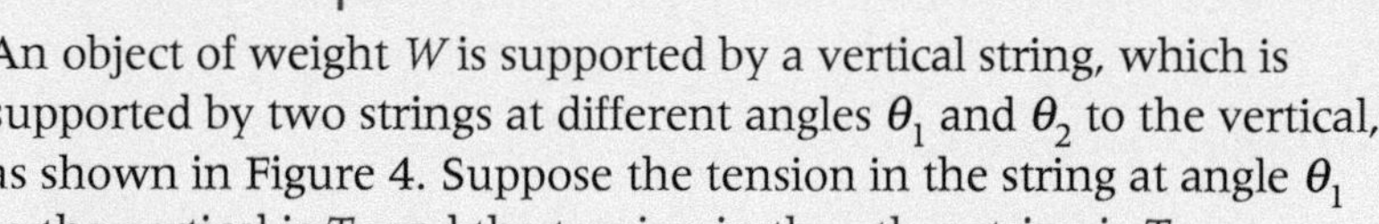

An object of weight W is supported by a vertical string, which is supported by two strings at different angles θ_1 and θ_2 to the vertical, as shown in Figure 4. Suppose the tension in the string at angle θ_1 to the vertical is T_1 and the tension in the other string is T_2.

At the point P where the strings meet, the forces T_1, T_2, and W are in equilibrium.

Resolving T_1 and T_2 vertically and horizontally gives:

1 horizontally: $T_1\sin\theta_1 = T_2\sin\theta_2$

2 vertically: $T_1\cos\theta_1 + T_2\cos\theta_2 = W$.

Study tip

Always draw force diagrams clearly, representing the magnitude and direction of each force as accurately as possible.

Testing three forces in equilibrium

Figure 5a shows a practical arrangement to test three forces acting on a point object P. The tension in each string pulls on P and is due to the weight it supports, either directly or indirectly over a pulley. Provided the pulleys are frictionless, each tension force acting on P is equal to the weight supported by its string.

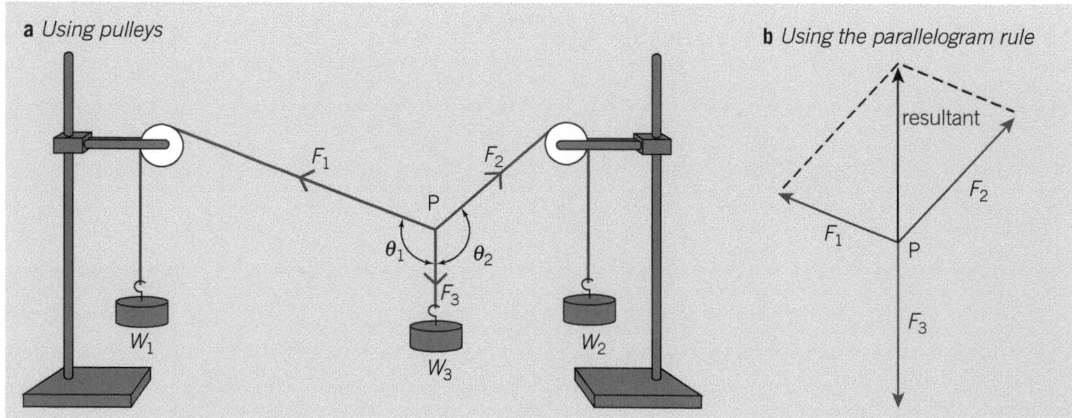

▲ **Figure 5** *Testing three forces*

The three forces F_1, F_2, and F_3 acting on P are in equilibrium, so any two should give a resultant equal and opposite to the third force. For example, the resultant of F_1 and F_2 is equal and opposite to F_3. You can test this by measuring the angle between each of the upper strings and the lower string which is vertical. A scale diagram as shown in Figure 5b, can then be constructed using the fact that the magnitudes of F_1, F_2, and F_3 are equal to W_1, W_2, and W_3, respectively. Your diagram should show that the resultant of F_1 and F_2 is equal and opposite to F_3.

Note:

Greater accuracy can be obtained by drawing a parallelogram, using the two force vectors F_1 and F_2 as adjacent sides. The resultant is the diagonal of the parallelogram *between* the two force vectors. This should be equal and opposite to F_3.

Summary questions

1 A point object of weight 6.2 N is acted on by a horizontal force of 3.8 N.

a Calculate the resultant of these two forces.

b Determine the magnitude and direction of a third force acting on the object for it to be in equilibrium.

2 A small object of weight 5.4 N is at rest on a rough slope, which is at an angle of 30° to the horizontal.

a Sketch a diagram and show the three forces acting on the object.

b Calculate

i the frictional force on the object

ii the support force from the slope on the object.

Hint

Remember that *rough* means friction is involved.

3 An archer pulls a bowstring back until the two halves of the string are at 140° to each other (Figure 6). The force needed to hold the string in this position is 95 N.

Calculate:

a the tension in each part of the bowstring in this position

b the resultant force on an arrow at the instant the bowstring is released from this position.

4 An elastic string is stretched horizontally between two fixed points 0.80 m apart. An object of weight 4.0 N is suspended from the midpoint of the string, causing the midpoint to drop a distance of 0.12 m. Calculate:

a the angle of each part of the string to the vertical

b the tension in each part of the string.

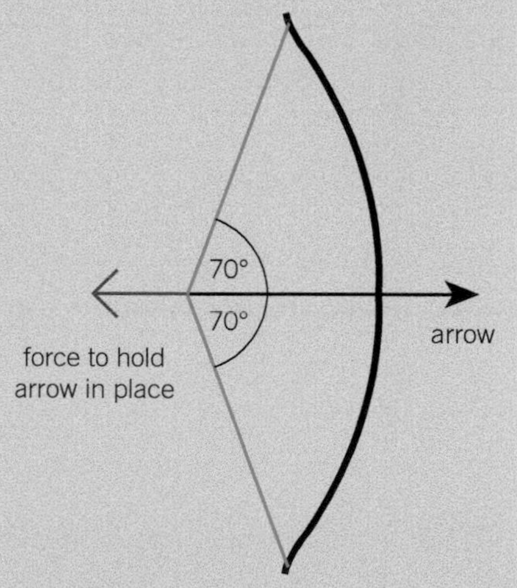

▲ **Figure 6**

1.3 The principle of moments

Learning objectives:

- → Describe the conditions under which a force produces a turning effect.
- → Explain how the turning effect of a given force can be increased.
- → Explain what is required to balance a force that produces a turning effect.
- → Explain why the centre of mass is an important idea.

Specification reference: 3.2.2

Turning effects

Whenever you use a lever or a spanner, you are using a force to turn an object about a pivot. For example, if you use a spanner to loosen a wheel nut on a bicycle, you need to apply a force to the spanner to make it turn about the wheel axle. The effect of the force depends on how far it is applied from the wheel axle. The longer the spanner, the less force needed to loosen the nut. However, if the spanner is too long and the nut is too tight, the spanner could snap if too much force is applied to it.

The moment of a force about any point is defined as the force × the perpendicular distance from the line of action of the force to the point.

The unit of the moment of a force is the newton metre (N m).

For a force F acting along a line of action at perpendicular distance d from a certain point,

the moment of the force = $F \times d$

Notes:

1 The greater the distance d, the greater the moment.

2 The distance d is the **perpendicular** distance from the line of action of the force to the point.

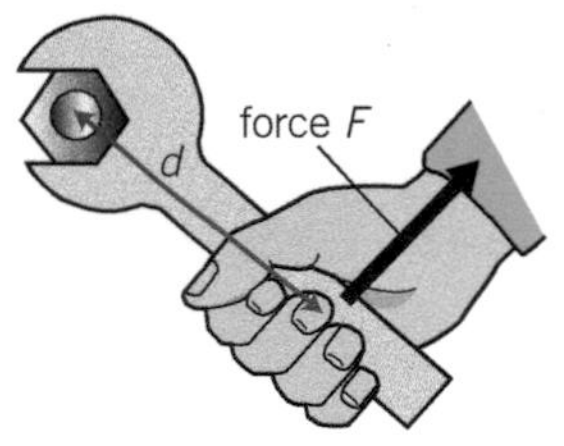

▲ **Figure 1** *A turning force. The force F produces an anticlockwise moment*

The principle of moments

An object that is not a point object is referred to as a body. Any such object turns if a force is applied to it anywhere other than through its centre of mass. If a body is acted on by more than one force and it is in equilibrium, the turning effects of the forces must balance out. In more formal terms, considering the **moments** of the forces about any point, for equilibrium

the sum of the clockwise moments = the sum of the anticlockwise moments

This statement is known as the **principle of moments**.

For example, consider a uniform metre rule balanced on a pivot at its centre, supporting weights W_1 and W_2 suspended from the rule on either side of the pivot.

- Weight W_1 provides an anticlockwise moment about the pivot = $W_1 d_1$, where d_1 is the distance from the line of action of the weight to the pivot.
- Weight W_2 provides a clockwise moment about the pivot = $W_2 d_2$, where d_2 is the distance from the line of action of the weight to the pivot.

For equilibrium, applying the principle of moments:

$$W_1 d_1 = W_2 d_2$$

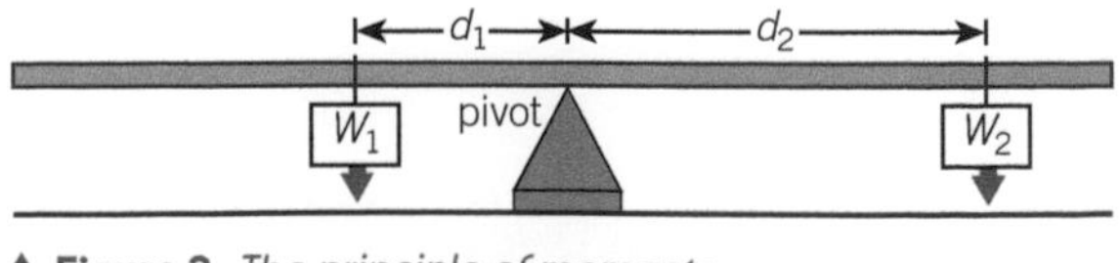

▲ **Figure 2** *The principle of moments*

Study tip

When there are several unknown forces, take moments about a point through which one of the unknown forces acts. This point will give the force which acts through it a moment of zero, thereby simplifying your calculations.

Note:

If a third weight W_3 is suspended from the rule on the same side of the pivot as W_2 at distance d_3 from the pivot, then the rule can be rebalanced by increasing distance d_1.

At this new distance d_1' for W_1

$$W_1 d_1' = W_2 d_2 + W_3 d_3$$

Centre of mass

A tightrope walker knows just how important the centre of mass of an object can be. One slight off-balance movement can be catastrophic. The tightrope walker uses a horizontal pole to ensure his or her overall centre of mass is always directly above the rope. The support force from the rope then acts upwards through the centre of mass of the walker.

The centre of mass of a body is the point through which a single force on the body has no turning effect. In effect, it is the point where we consider the weight of the body to act when studying the effect of forces on the body. For a uniform regular solid, for example, a wooden block, the centre of mass is at its centre.

Centre of mass tests

1 Balance a ruler at its centre on the end of your finger. The centre of mass of the ruler is directly above the point of support. Tip the ruler too much and it falls off because the centre of mass is no longer above the point of support.

2 To find the centre of mass of a triangular card, suspend the piece of card on a clamp stand as shown in Figure 3. Draw pencil lines along the plumb line. The centre of mass is where the lines drawn on the card cross.

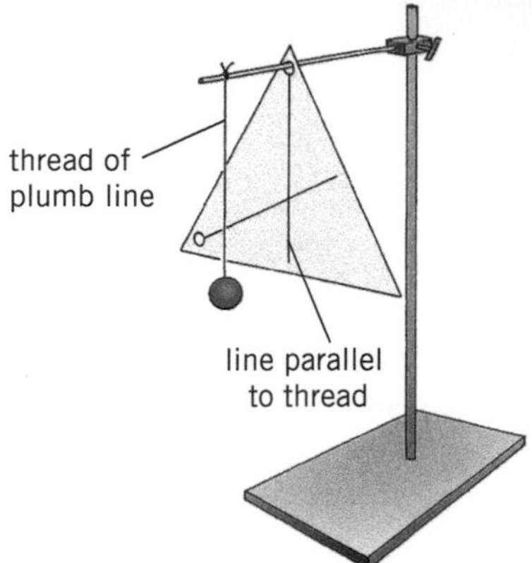

▲ **Figure 3** *A centre of mass test*

Calculating the weight of a metre rule

1 Locate the centre of mass of a metre rule by balancing it horizontally on a horizontal knife-edge. Note the position of the centre of mass. The rule is **uniform** if its centre of mass is exactly at the middle of the rule.

2 Balance the metre rule off-centre on a knife-edge, using a known weight W_1 as shown in Figure 4. The position of the known weight needs to be adjusted gradually until the rule is exactly horizontal.

At this position

- the known weight W_1 provides an anticlockwise moment about the pivot = $W_1 d_1$, where d_1 is the perpendicular distance from the line of action of W_1 to the pivot
- the weight of the rule W_0 provides a clockwise moment = $W_0 d_0$, where d_0 is the distance from the centre of mass of the rule to the pivot.

Applying the principle of moments,

$$W_0 d_0 = W_1 d_1$$

By measuring distance d_0 and d_1, the weight W_0 of the rule can therefore be calculated.

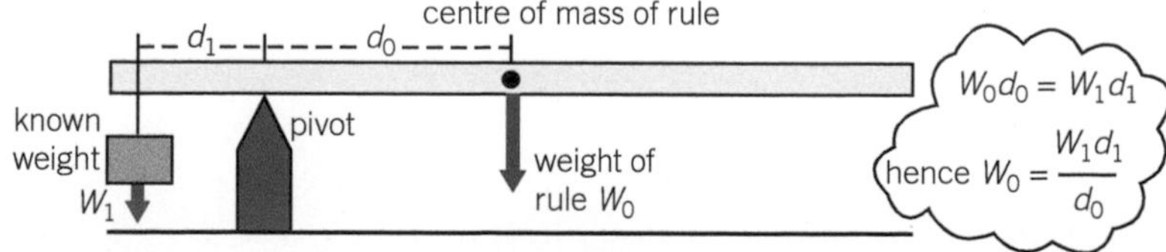

▲ **Figure 4** *Finding the weight of a metre rule*

Summary questions

1 A child of weight 200 N sits on a seesaw at a distance of 1.2 m from the pivot at the centre. The seesaw is balanced by a second child sitting on it at a distance of 0.8 m from the centre. Calculate the weight of the second child.

2 A metre rule, pivoted at its centre of mass, supports a 3.0 N weight at its 5.0 cm mark, a 2.0 N weight at its 25 cm mark, and a weight *W* at its 80 cm mark.

   a Sketch a diagram to represent this situation.

   b Calculate the weight *W*.

3 In **2**, the 3.0 N weight and the 2.0 N weight are swapped with each other. Sketch the new arrangement and calculate the new distance of weight *W* from the pivot to balance the metre rule.

4 A uniform metre rule supports a 4.5 N weight at its 100 mm mark. The rule is balanced horizontally on a horizontal knife-edge at its 340 mm mark. Sketch the arrangement and calculate the weight of the rule.

1.4 More on moments

Learning objectives:

- → Describe the support force on a pivoted body.
- → When a body in equilibrium is supported at two places, state how much force is exerted on each support.
- → Explain what is meant by a couple.

Specification reference: 3.2.2

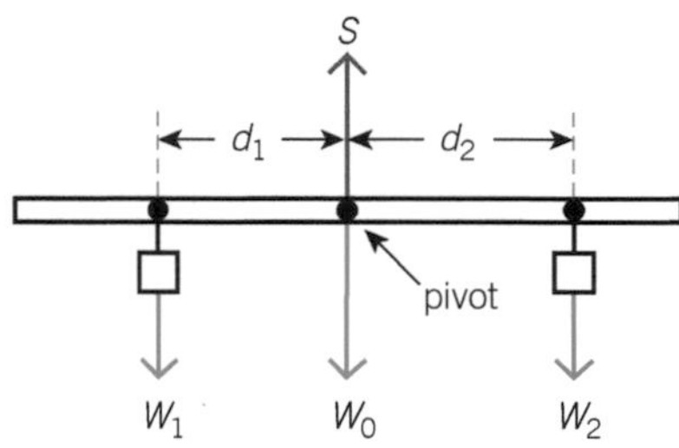

▲ **Figure 1** *Support forces*

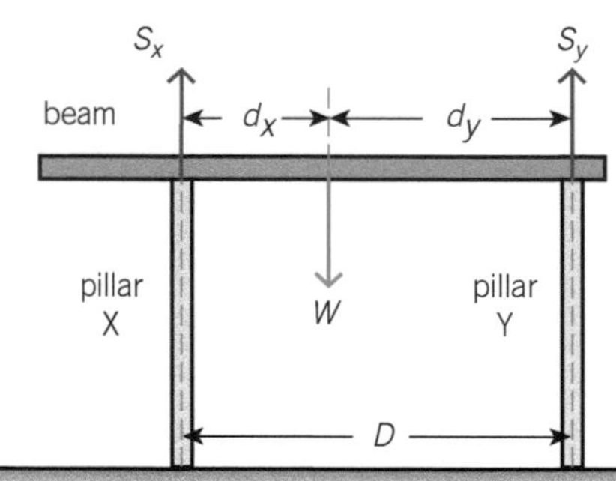

▲ **Figure 2** *A two-support problem*

Study tip

In calculations, you can always eliminate the turning effect of a force by taking moments about a point through which it acts.

Support forces

Single-support problems

When an object in equilibrium is supported at one point only, the support force on the object is equal and opposite to the total downward force acting on the object. For example, in Figure 1, a uniform rule is balanced on a knife-edge at its centre of mass, with two additional weights W_1 and W_2 attached to the rule. The support force S on the rule from the knife-edge must be equal to the total downward weight. Therefore

$$S = W_1 + W_2 + W_0, \text{ where } W_0 \text{ is the weight of the rule}$$

As explained in Topic 1.3, taking moments about the knife-edge gives

$$W_1 d_1 = W_2 d_2$$

Note:

What difference is made by taking moments about a different point? Consider moments about the point where W_1 is attached to the rule.

The sum of the clockwise moments = $W_0 d_1 + W_2 (d_1 + d_2)$

The sum of the anticlockwise moments = $S d_1 = (W_1 + W_2 + W_0) d_1$

$\therefore (W_1 + W_2 + W_0) d_1 = W_0 d_1 + W_2 (d_1 + d_2)$

Multiplying out the brackets gives:

$W_1 d_1 + W_2 d_1 + W_0 d_1 = W_0 d_1 + W_2 d_1 + W_2 d_2$

which simplifies to become $W_1 d_1 = W_2 d_2$

This is the same as the equation obtained by taking moments about the original pivot point. So moments can be taken about any point. It makes sense therefore to choose a point through which one or more unknown forces act, as such forces have zero moment about this point.

Two-support problems

Consider a uniform beam supported on two pillars X and Y, which are a distance D apart. The weight of the beam is shared between the two pillars according to how far the beam's centre of mass is from each pillar. For example:

- If the centre of mass of the beam is midway between the pillars, the weight of the beam is shared equally between the two pillars. In other words, the support force on the beam from each pillar is equal to half the weight of the beam.
- If the centre of mass of the beam is at distance d_x from pillar X and distance d_y from pillar Y, as shown in Figure 2, then taking moments about

  1 where X is in contact with the beam,
  $S_y D = W d_x$, where S_y is the support force from pillar Y
  gives $S_y = \frac{W d_x}{D}$

  2 where Y is in contact with the beam,
  $S_x D = W d_y$, where S_x is the support force from pillar X
  gives $S_x = \frac{W d_y}{D}$.

Therefore, if the centre of mass is closer to X than to Y, $d_x < d_y$ so $S_y < S_x$.

Couples

A **couple** is a pair of equal and opposite forces acting on a body, but not along the same line. Figure 3 shows a couple acting on a beam. The couple turns, or tries to turn, the beam.

The moment of a couple = force × perpendicular distance between the lines of action of the forces.

To prove this, consider the couple in Figure 3. Taking moments about an arbitrary point P between the ends at distance x along the beam from one end,

1 the moment due to the force F at that end = Fx clockwise
2 the moment due to the force F at the other end = $F(d-x)$ clockwise, where d is the perpendicular distance between the lines of action of the forces.

Therefore the total moment = $Fx + F(d-x) = Fx + Fd - Fx = Fd$.

The total moment is the same, regardless of the point about which the moments are taken.

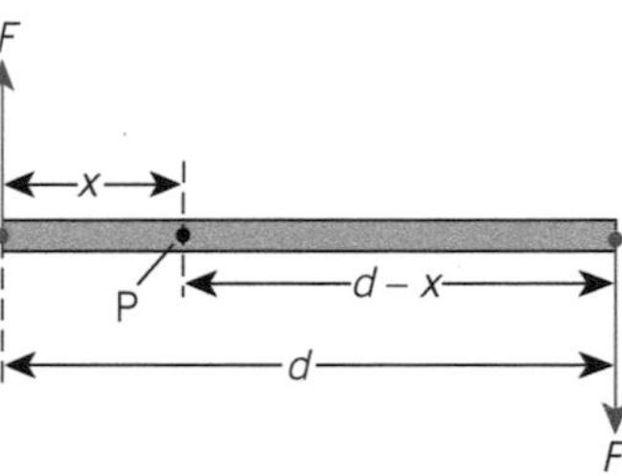

▲ **Figure 3** *A couple*

Summary questions

1 A uniform metre rule of weight 1.2 N rests horizontally on two knife-edges at the 100 mm mark and the 800 mm mark. Sketch the arrangement and calculate the support force on the rule due to each knife-edge.

2 A uniform beam of weight 230 N and of length 10 m rests horizontally on the tops of two brick walls, 8.5 m apart, such that a length of 1.0 m projects beyond one wall and 0.5 m projects beyond the other wall. Figure 4 shows the arrangement.

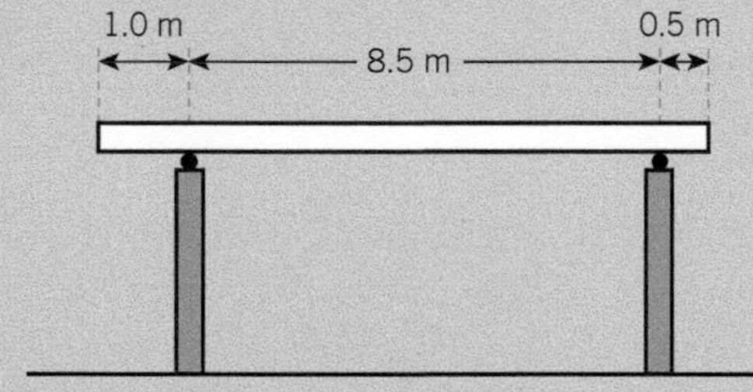

▲ **Figure 4**

Calculate:

a the support force of each wall on the beam
b the force of the beam on each wall.

3 A uniform bridge span of weight 1200 kN and of length 17.0 m rests on a support of width 1.0 m at either end. A stationary lorry of weight 60 kN is the only object on the bridge. Its centre of mass is 3.0 m from the centre of the bridge.

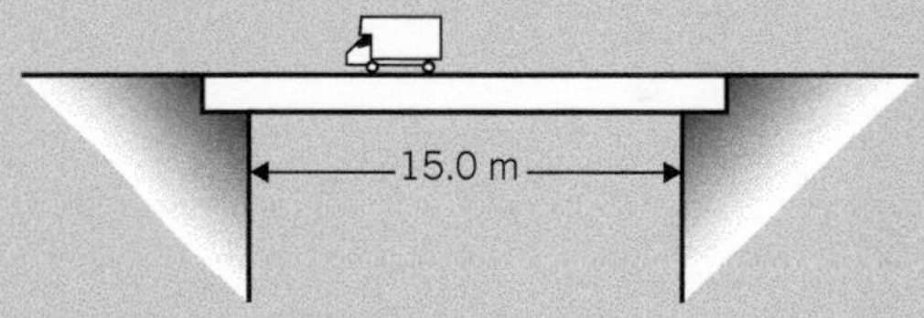

▲ **Figure 5**

Calculate the support force on the bridge at each end. Assume the support forces act where the bridge meets its support.

4 A uniform plank of weight 150 N and of length 4.0 m rests horizontally on two bricks. One of the bricks is at the end of the plank. The other brick is 1.0 m from the other end of the plank.

a Sketch the arrangement and calculate the support force on the plank from each brick.
b A child stands on the free end of the plank and just causes the other end to lift off its support. Sketch this arrangement and calculate the weight of the child.

1.5 Stability

Learning objectives:

- → Explain the difference between stable and unstable equilibrium.
- → Assess when a tilted object will topple over.
- → Explain why a vehicle is more stable when its centre of mass is lower.

Specification reference: 3.2.2

Stable and unstable equilibrium

If a body in **stable equilibrium** is displaced then released, it returns to its equilibrium position. For example, if an object such as a coat hanger hanging from a support is displaced slightly, it swings back to its equilibrium position.

Why does an object in stable equilibrium return to equilibrium when it is displaced and then released? The reason is that the centre of mass of the object is directly below the point of support when the object is at rest. The weight of the object is considered to act at its centre of mass. Thus the support force and the weight are directly equal and opposite to each other when the object is in equilibrium. However, when it is displaced, at the instant of release, the line of action of the weight no longer passes through the point of support, so the weight returns the object to equilibrium.

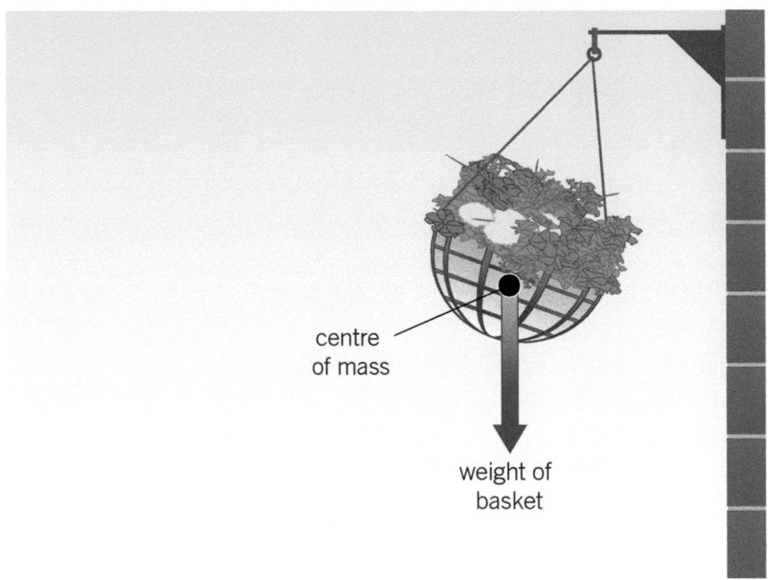

▲ **Figure 1** *Returning to equilibrium*

A plank balanced on a drum is in **unstable equilibrium** (Figure 2). If it is displaced slightly from equilibrium then released, the plank will roll off the drum. The reason is that the centre of mass of the plank is directly above the point of support when it is in equilibrium. The support force is exactly equal and opposite to the weight. If the plank is displaced slightly, the centre of mass is no longer above the point of support. The weight therefore acts to turn the plank further from the equilibrium position.

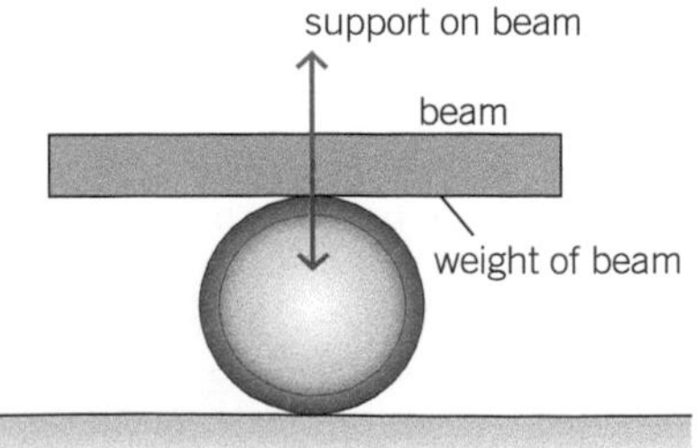

▲ **Figure 2** *Unstable equilibrium*

Tilting and toppling

Skittles at a bowling alley are easy to knock over because they are tall, so their centre of mass is high and the base is narrow. A slight nudge from a ball causes a skittle to tilt then tip over.

Tilting

This is where an object at rest on a surface is acted on by a force that raises it up on one side. For example, if a horizontal force F is applied to the top of a tall free-standing bookcase, the force can make the bookcase tilt about its base along one edge.

In Figure 3, to make the bookcase tilt, the force must turn it clockwise about point P. The entire support from the floor acts at point P. The weight of the bookcase provides an anticlockwise moment about P.

1 The clockwise moment of F about P = Fd, where d is the perpendicular distance from the line of action of F to the pivot.
2 The anticlockwise moment of W about P = $\frac{Wb}{2}$ where b is the width of the base.

Therefore, for tilting to occur $Fd > \frac{Wb}{2}$.

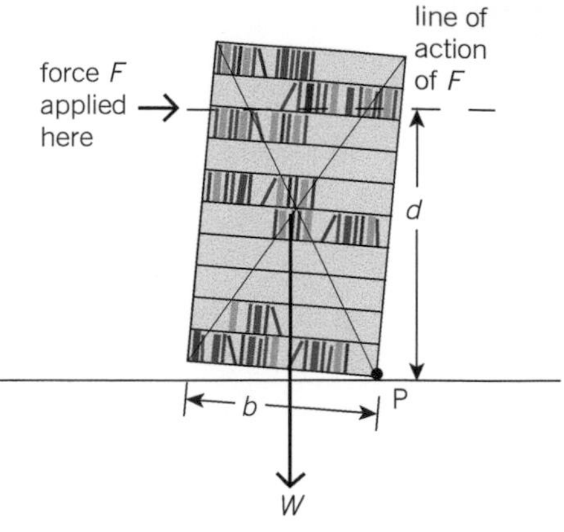

▲ **Figure 3** *Tilting*

Toppling

A tilted object will topple over if it is tilted too far. If an object on a flat surface is tilted more and more, the line of action of its weight (which is through its centre of mass) passes closer and closer to the pivot. If the object is tilted so much that the line of action of its weight passes beyond the pivot, the object will topple over if allowed to. The position where the line of action of the weight passes through the pivot is the furthest it can be tilted without toppling. Beyond this position, it topples over if it is released.

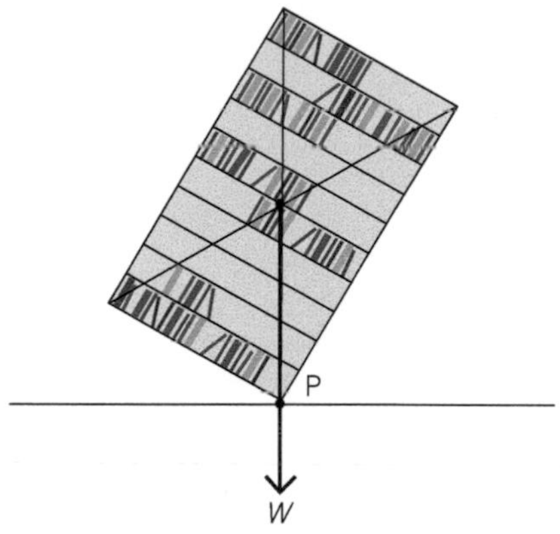

▲ **Figure 4** *Toppling*

On a slope

A tall object on a slope, for example, a high-sided vehicle on a road with a sideways slope, will topple over if the slope is too great. This will happen if the line of action of the weight (passing through the centre of mass of the object) lies outside the wheelbase of the vehicle. In Figure 5, the vehicle will not topple over because the line of action of the weight lies within the wheelbase.

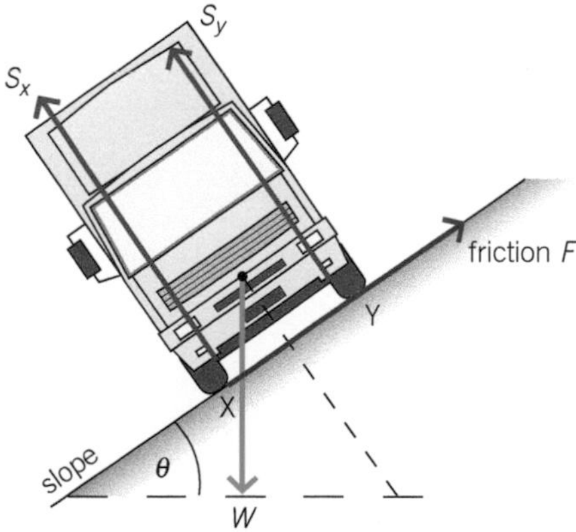

▲ **Figure 5** *A high-sided vehicle on a slope*

Consider the forces acting on the vehicle on a sideways slope when it is at rest. The sideways friction F, the support forces S_x and S_y, and the force of gravity on the vehicle (i.e., its weight) act as shown in Figure 5.

Hint

The lower the centre of mass of an object, the more stable it is.

For equilibrium, resolving the forces parallel and perpendicular to the slope gives:

1 parallel to the slope
$F = W \sin\theta$

2 perpendicular to the slope
$S_x + S_y = W\cos\theta.$

Note that S_x is greater than S_y because X is lower than Y.

Summary questions

1 Explain why a bookcase with books on its top shelf only is less stable than if the books were on the bottom shelf.

2 An empty wardrobe of weight 400 N has a square base 0.8 m × 0.8 m and a height of 1.8 m. A horizontal force is applied to the top edge of the wardrobe to make it tilt. Calculate the force needed to lift the wardrobe base off the floor along one side.

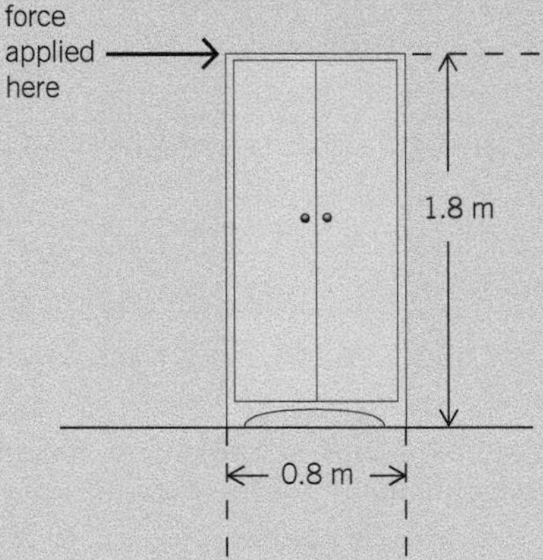

▲ **Figure 6**

3 A vehicle has a wheelbase of 1.8 m and a centre of mass, when unloaded, which is 0.8 m from the ground.

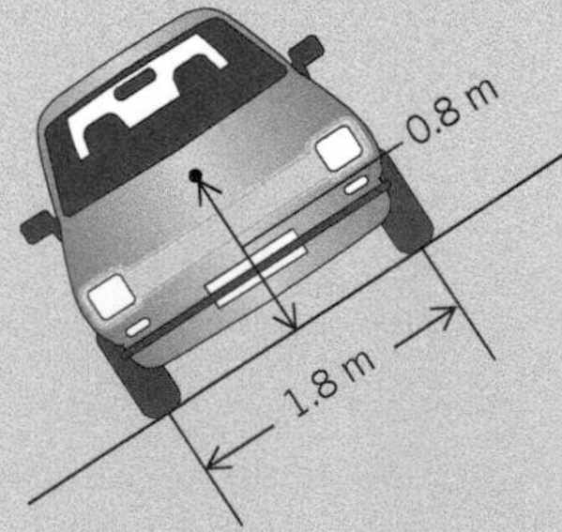

▲ **Figure 7**

a The vehicle is tested for stability on an adjustable slope. Calculate the maximum angle of the slope to the horizontal if the vehicle is not to topple over.

b If the vehicle carries a full load of people, will it be more or less likely to topple over on a slope? Explain your answer.

1.6 Equilibrium rules

Learning objectives:

→ Explain what condition must apply to the forces on an object in equilibrium.

→ Explain what condition must apply to the turning effects of the forces.

→ Describe how we can apply these conditions to predict the forces acting on a body in equilibrium.

Specification reference: 3.2.1 and 3.2.2

Free body force diagrams

When two objects interact, they always exert equal and opposite forces on one another. A diagram showing the forces acting on an object can become very complicated, if it shows the forces the object exerts on other objects as well. A **free body force diagram** shows only the forces acting on the object.

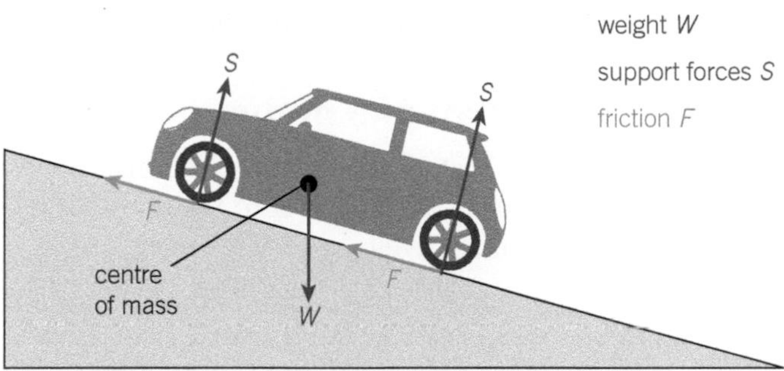

▲ **Figure 1** *A free body diagram of a car parked on a slope*

The triangle of forces

For a point object acted on by three forces to be in equilibrium, the three forces must give an overall resultant of zero. The three forces as vectors should form a triangle. In other words, for three forces F_1, F_2, and F_3 to give zero resultant,

$$\textbf{their vector sum } \vec{F_1} + \vec{F_2} + \vec{F_3} = 0$$

As explained in Topic 1.2, any two of the forces give a resultant that is equal and opposite to the third force. For example, the resultant of $\vec{F_1} + \vec{F_2}$ is equal and opposite to $\vec{F_3}$ (i.e., $\vec{F_1} + \vec{F_2} = -\vec{F_3}$).

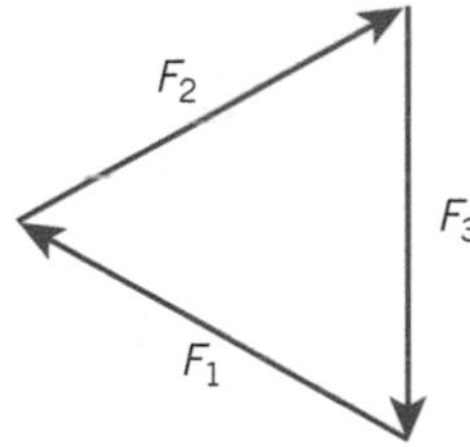

▲ **Figure 2** *The triangle of forces for a point object*

The same rule applies to a body in equilibrium acted on by three forces. In addition, their lines of action must intersect at the *same* point, otherwise the body cannot be in equilibrium, as the forces will have a net turning effect. Consider the example of a rectangular block on a rough slope, as shown in Figure 3.

- The weight W of the block acts vertically down through the centre of mass of the block.
- The frictional force F on the block due to the slope prevents the block from sliding down the slope, so it acts up the slope.
- The support force S on the block due to the slope acts normal to the slope through the point where the lines of action of W and F act. Figure 3 also shows the triangle of forces for the three forces W, F, and S acting on the block.

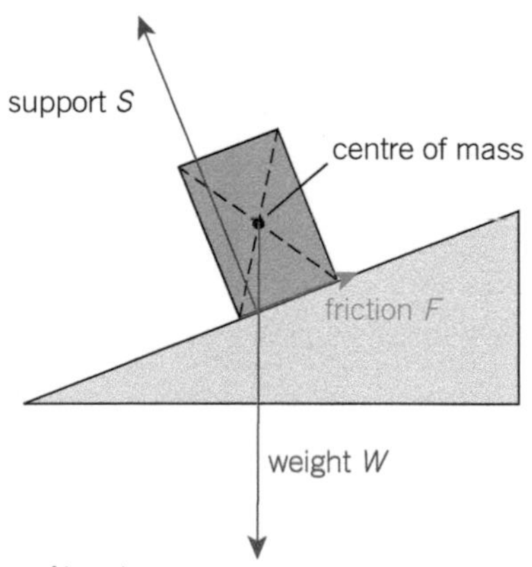

a *At rest*

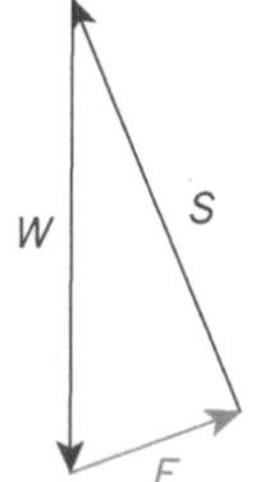

b *The triangle of forces*

▲ **Figure 3** *A block on a slope*

We can draw a scale diagram of the triangle of forces to find an unknown force or angle, given the other forces and angles in the triangle. For example, to find the unknown force, F_3, in the triangle of forces in Figure 4:

1 Draw one of the known force vectors, F_1, to scale, as one side of the force triangle.

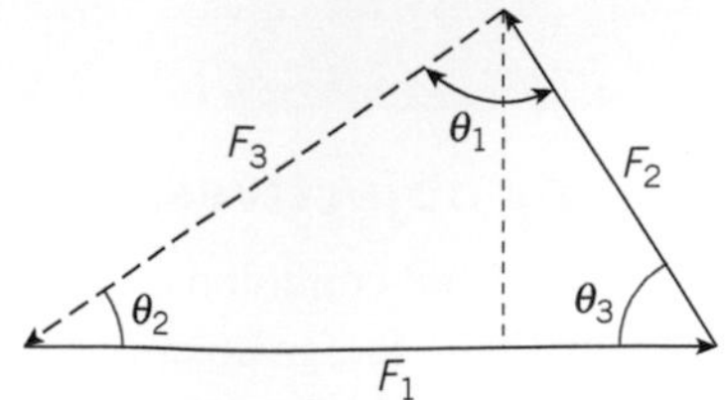

▲ **Figure 4** *Constructing a scale diagram*

Study tip

Rearranging the equation $F_2 \sin\theta_3 = F_3 \sin\theta_2$ gives

$$\frac{F_2}{\sin\theta_2} = \frac{F_3}{\sin\theta_3}$$

By applying the same theory to F_1 and either F_2 or F_3, it can be shown that

$$\frac{F_1}{\sin\theta_1} = \frac{F_2}{\sin\theta_2} = \frac{F_3}{\sin\theta_3}$$

This rule, known as the *sine rule*, will be useful if you study A Level Maths.

2 Use a protractor and ruler to draw the other known force (e.g., F_2) at the correct angle to F_1 as the second side of the triangle. The third side of the triangle can then be drawn in to give the unknown force (F_3).

You could calculate the unknown force by resolving it and F_2 parallel and perpendicular to the base force F_1. Labelling θ_2 as the angle opposite F_2 and θ_3 as the angle opposite F_3, the perpendicular components of the resolved forces, $F_2 \sin\theta_3$ and $F_3 \sin\theta_2$, correspond to the height of the triangle and are equal and opposite to each other:

$$\mathbf{F_2 \sin\theta_3 = F_3 \sin\theta_2}$$

Therefore the unknown force can be calculated if F_2, θ_2, and θ_3 are known.

The conditions for equilibrium of a body

An object in equilibrium is either at rest or moving with a constant velocity. The forces acting on the object must give zero resultant, and their turning effects must balance out as well.

Therefore, for a body in equilibrium:

1 The resultant force must be zero. If there are only three forces, they must form a closed triangle.
2 The principle of moments must apply (i.e., the moments of the forces about the same point must balance out).

See Topics 1.3 and 1.4 again if necessary.

Worked example

A uniform shelf of width 0.6 m and of weight 12 N is attached to a wall by hinges and is supported horizontally by two parallel cords attached at two corners of the shelf, as shown in Figure 5. The other end of each cord is fixed to the wall 0.4 m above the hinge. Calculate:

a the angle between each cord and the shelf

b the tension in each cord.

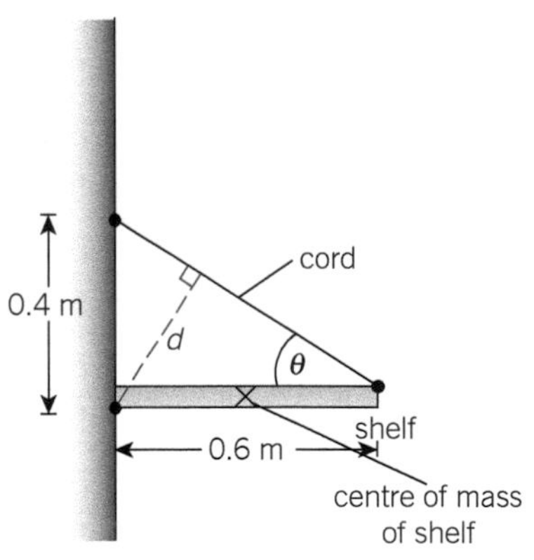

▲ **Figure 5**

Solution

a Let the angle between each cord and the shelf = θ.

From Figure 5, $\tan\theta = \dfrac{0.4}{0.6}$, so $\theta = 34°$.

b Taking moments about the hinge eliminates the force at the hinge (as its moment is zero) to give:

1 The sum of the clockwise moments = weight of shelf × distance from hinge to the centre of mass of the shelf = 12 × 0.3 = 3.6 N m.

2 The sum of the anticlockwise moments = $2Td$, where T is the tension in each cord and d is the perpendicular distance from the hinge to either cord.

From Figure 5 it can be seen that $d = 0.6 \sin\theta = 0.6 \sin 34° = 0.34$ m.

Applying the principle of moments gives:

$$2 \times 0.34 \times T = 3.6$$

$$T = 5.3\,\text{N}$$

Extension

The physics of lifting

Lifting is a process that can damage your spine if it is not done correctly. Figure 6 shows how *not* to lift a heavy suitcase. The spine pivots about the hip joints and is acted upon by the tension T in the back muscles, the weight W_0 of the upper part of the body, the weight W of the suitcase, and the reaction force R from the hip.

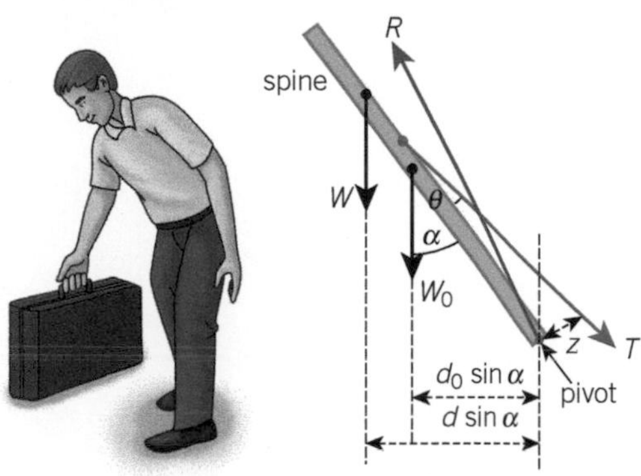

▲ **Figure 6** *Lifting forces*

In Figure 6, the spine is at angle α to the vertical direction, and weight W acts on the spine at distance d from the pivot. Assume that weight W_0 acts on the spine at distance d_0 from the pivot.

Before the suitcase is lifted off the ground:

- Weight W_0 creates an anticlockwise moment about the hip joints that is equal to $W_0 d_0 \sin\alpha$. This moment is because W_0 has a component $W_0 \sin\alpha$ perpendicular to the spine acting on the spine at distance d_0 from the pivot.
- Tension T_0 is necessary to provide an equal and opposite clockwise moment $T_0 \times z$ about the hip joints, where z is the perpendicular distance from the pivot to the line of action of the tension.

Applying the principle of moments to the spine in this position gives $T_0 z = W_0 d_0 \sin\alpha$.

Typically, $d_0 = 10\,z$, and therefore, $\boldsymbol{T_0 = 10\,W_0 \sin\alpha}$.

When the suitcase is lifted off the ground:

- Its weight W creates an extra anticlockwise moment about the hip joints that is equal to $W d \sin\alpha$. This moment is because W has a component $W \sin\alpha$ perpendicular to the spine acting on the spine at distance d from the pivot.
- Extra tension ΔT is needed in order to provide an equal and opposite clockwise moment $\Delta T \times z$ to lift the suitcase off the ground, where z is the perpendicular distance from the pivot to the line of action of the tension.

Applying the principle of moments gives $\Delta T z = W d \sin\alpha$.

Typically, $d = 15\,z$, and therefore, $\boldsymbol{\Delta T = 15 W \sin\alpha}$.

The components of T, W_0, and W parallel to the spine act down the spine and are opposed by the parallel component of the reaction R acting up the spine. The effect of these forces is to compress the spine. Therefore:

- *Before* the suitcase is lifted off the ground, the compressive force in the spine $= T_0 \cos\theta + W_0 \cos\alpha$, where θ is the angle between the spine and the line of action of tension T. Because $\theta < 10°$, $\cos\theta \approx 1$, the compressive force $\approx T_0 + W_0 \cos\alpha$
  $= \boldsymbol{10\,W_0 \sin\alpha + W_0 \cos\alpha}$.
- *When* the suitcase is lifted off the ground, the extra compressive force in the spine $= \Delta T \cos\theta + W \cos\alpha$. Because $\theta < 10°$, $\cos\theta \approx 1$, the extra compressive force $\approx \Delta T + W \cos\alpha$
  $= \boldsymbol{15\,W \sin\alpha + W \cos\alpha}$.

Questions

1. Estimate the compressive force in the spine of a person leaning forward at angle α equal to **a** 30°, **b** 60° *just before* lifting a 20 kg suitcase off the ground. Assume $W_0 = 400$ N.
2. Estimate the total compressive force in the spine in **Q1a** and **b** when the suitcase is lifted off the ground.
3. Figure 7 shows a suitcase being lifted by a person who is not leaning over. Explain why the compressive force in the spine is considerably less than if the person had leaned forward to lift the suitcase.

▲ **Figure 7**

Summary questions

1 A uniform plank of length 5.0 m rests horizontally on two bricks that are 0.5 m from either end. A child of weight 200 N stands on one end of the plank and causes the other end to lift, so it is no longer supported at that end. Calculate:

a the weight of the plank

b the support force acting on the plank from the supporting brick.

▲ Figure 8

2 A security camera is supported by a frame that is fixed to a wall and ceiling as shown in Figure 9. The support structure must be strong enough to withstand the effect of a downward force of 1500 N acting on the camera (in case the camera is gripped by someone below it). Calculate:

a the moment of a force of 1500 N on the camera about the point where the support structure is attached to the wall

b the extra force on the vertical strut supporting the frame, when the camera is pulled with a downward force of 1500 N.

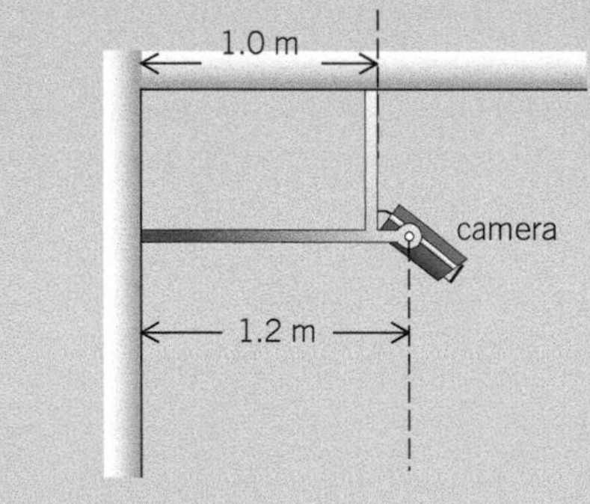

▲ Figure 9

3 A crane is used to raise one end of a 15 kN girder of length 10.0 m off the ground. When the end of the girder is at rest 6.0 m off the ground, the crane cable is perpendicular to the girder, as shown in Figure 10.

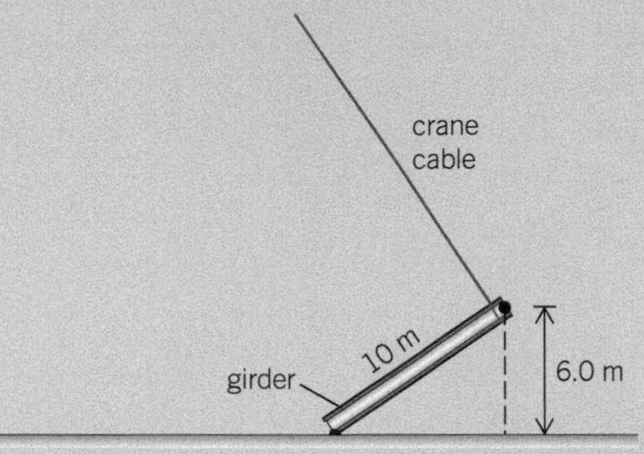

▲ Figure 10

a Calculate the tension in the cable.

b Show that the support force on the girder from the ground has a horizontal component of 3.6 N and a vertical component of 10.2 kN. Calculate the magnitude of the support force.

4 A small toy of weight 2.8 N is suspended from a horizontal beam by means of two cords that are attached to the same point on the toy (Figure 11). One cord makes an angle of 60° to the beam and the other makes an angle of 40° to the vertical. Calculate the tension in each cord.

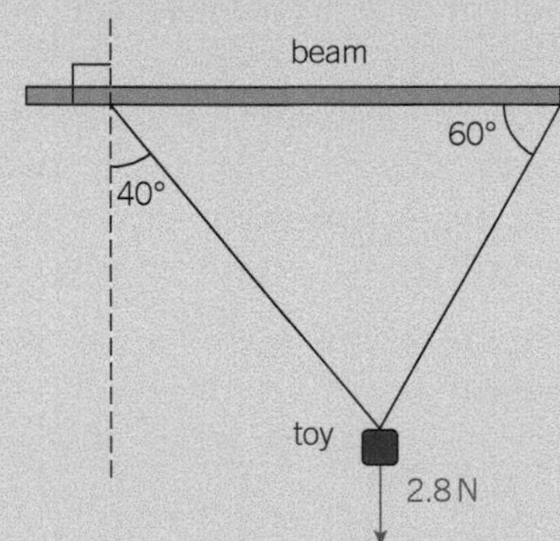

▲ Figure 11

1.7 Statics calculations

Learning objective:

→ State the important principles that always apply to a body in equilibrium.

Specification reference: 3.2.1 and 3.2.2

Study tip

Get as far as you can using force diagrams, resolving, and common sense.

1 Calculate the magnitude of the resultant of a 6.0 N force and a 9.0 N force acting on a point object when the two forces act:
  a in the same direction
  b in opposite directions
  c at 90° to each other.

2 A point object in equilibrium is acted on by a 3 N force, a 6 N force, and a 7 N force. What is the resultant force on the object if the 7 N force is removed?

3 A point object of weight 5.4 N in equilibrium is acted on by a horizontal force of 4.2 N and a second force *F*.
  a Draw a free body force diagram for the object and determine the magnitude of *F*.
  b Calculate the angle between the direction of *F* and the horizontal.

4 An object of weight 7.5 N hangs on the end of a cord, which is attached to the midpoint of a wire stretched between two points on the same horizontal level, as shown in Figure 1. Each half of the wire is at 12° to the horizontal. Calculate the tension in each half of the wire.

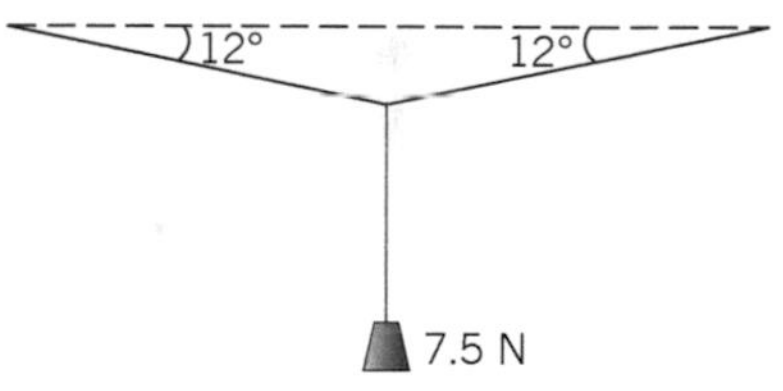

▲ **Figure 1**

5 A ship is towed at constant speed by two tugboats, each pulling the ship with a force of 9.0 kN. The angle between the tugboat cables is 40°, as shown in Figure 2.

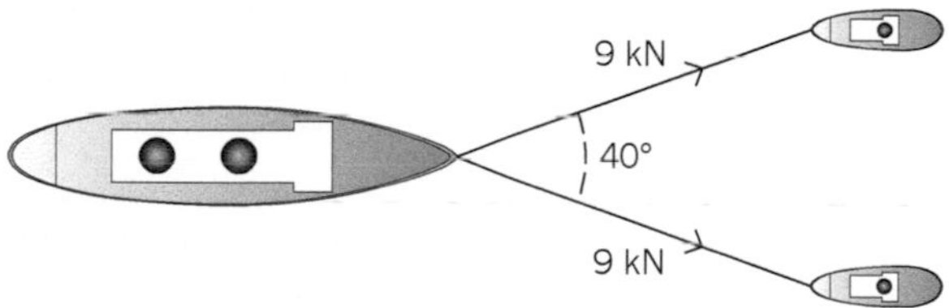

▲ **Figure 2**

  a Calculate the resultant force on the ship due to the two cables.
  b Calculate the drag force on the ship.

6 A metre rule of weight 1.0 N is pivoted on a knife-edge at its centre of mass, supporting a weight of 5.0 N and an unknown weight *W* as shown in Figure 3. To balance the rule horizontally with the unknown weight on the 250 mm mark of the rule, the position of the 5.0 N weight needs to be at the 810 mm mark.
  a Calculate the unknown weight.
  b Calculate the support force on the rule from the knife-edge.

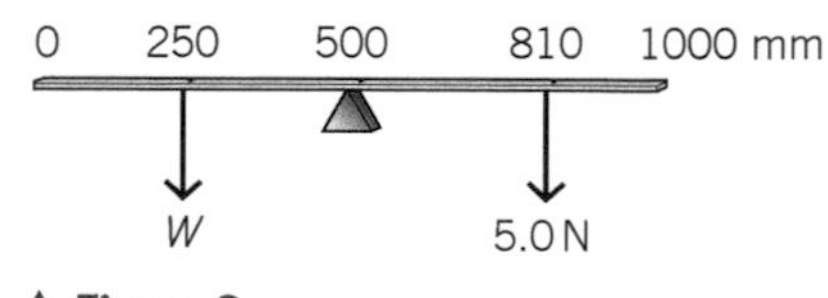

▲ **Figure 3**

7 In Figure 3, a 2.5 N weight is also suspended from the rule at its 400 mm mark. What adjustment needs to be made to the position of the 5.0 N weight to rebalance the rule?

8 A uniform metre rule is balanced horizontally on a knife-edge at its 350 mm mark, by placing a 3.0 N weight on the rule at its 10 mm mark.
  a Sketch the arrangement and calculate the weight of the rule.
  b Calculate the support force on the rule from the knife-edge.

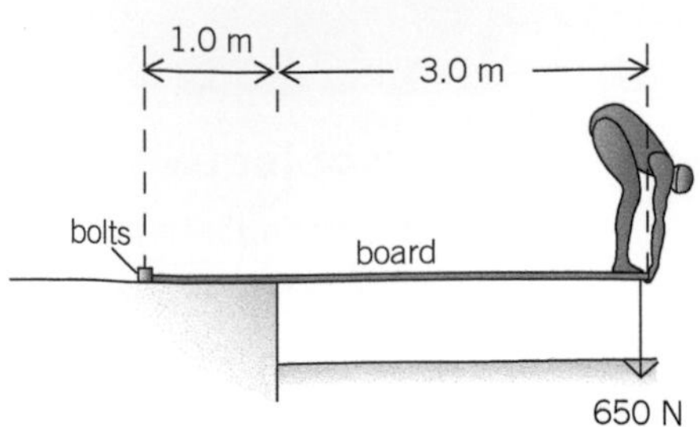

▲ Figure 4

9 A uniform diving board has a length 4.0 m and a weight of 250 N, as shown in Figure 4. It is bolted to the ground at one end and projects by a length of 3.0 m beyond the edge of the swimming pool. A person of weight 650 N stands on the free end of the diving board. Calculate:

   a the force on the bolts

   b the force on the edge of the swimming pool.

10 A uniform beam XY of weight 1200 N and of length 5.0 m is supported horizontally on a concrete pillar at each end. A person of weight 500 N sits on the beam at a distance of 1.5 m from end X.

   a Sketch a free body force diagram of the beam.

   b Calculate the support force on the beam from each pillar.

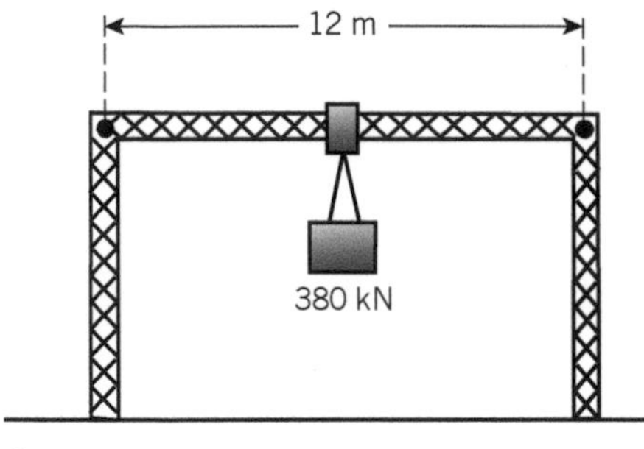

▲ Figure 5

11 A bridge crane used at a freight depot consists of a horizontal span of length 12 m fixed at each end to a vertical pillar, as shown in Figure 5.

   a When the bridge crane supports a load of 380 kN at its centre, a force of 1600 kN is exerted on each pillar. Calculate the weight of the horizontal span.

   b The same load is moved by the bridge crane to the left across a horizontal distance of 2.0 m. Sketch a free body force diagram of the horizontal span and calculate the force exerted on each pillar.

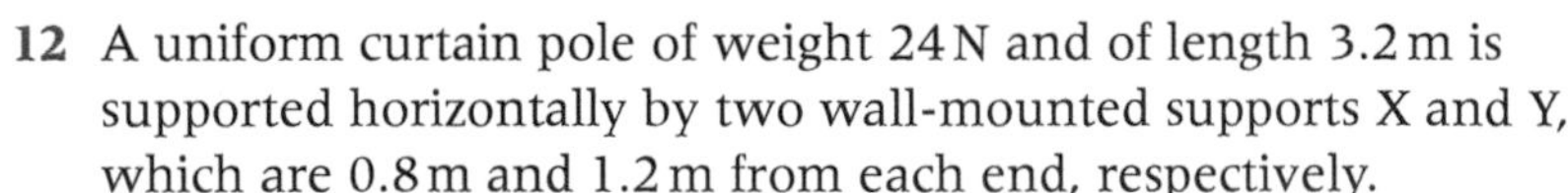

12 A uniform curtain pole of weight 24 N and of length 3.2 m is supported horizontally by two wall-mounted supports X and Y, which are 0.8 m and 1.2 m from each end, respectively.

   a Sketch the free body force diagram for this arrangement and calculate the force on each support when there are no curtains on the pole.

   b When the pole supports a pair of curtains of total weight 90 N drawn along the full length of the pole, what will be the force on each support?

13 A uniform steel girder of weight 22 kN and of length 14 m is lifted off the ground at one end by means of a crane. When the raised end is 2.0 m above the ground, the cable is vertical.

   a Sketch a free body force diagram of the girder in this position.

   b Calculate the tension in the cable at this position and the force of the girder on the ground.

14 A rectangular picture 0.80 m high and 1.0 m wide, of weight 24 N, hangs on a wall, supported by a cord attached to the frame at each of the top corners, as shown in Figure 6. Each section of the cord makes an angle of 25° with the picture, which is horizontal along its width.

   a Copy the diagram and mark the forces acting on the picture on your diagram.

   b Calculate the tension in each section of the cord.

You will find the answers to these questions at the back of the book.

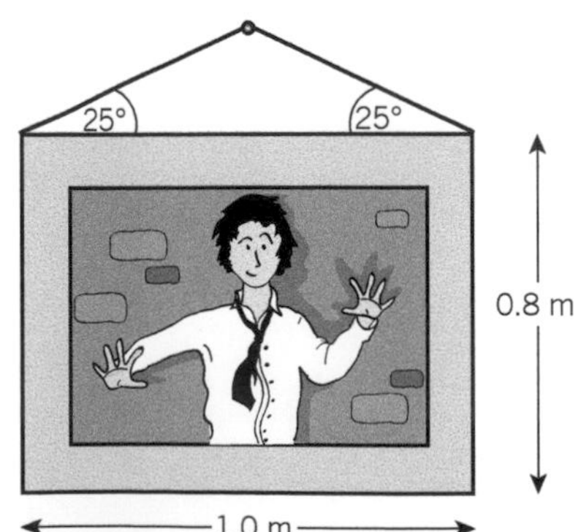

▲ Figure 6

Practice questions: Chapter 1

1 A student set up a model bridge crane to find out how the support forces change with the position of a load of weight W on the horizontal beam of the crane. With the load at different measured distances d from the fixed support at X, she used a newtonmeter to measure the support force S at Y near the other end, when the beam was horizontal. She repeated the measurements and also measured the distance D from X to Y.

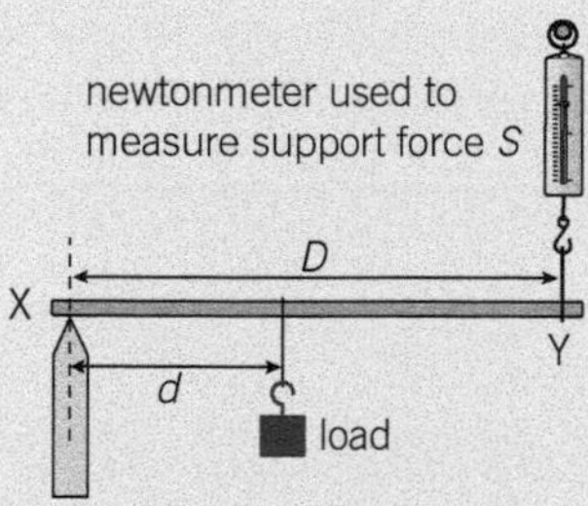

▲ **Figure 1**

The measurements she obtained are in the table below.
distance XY = 480 mm

Distance d / mm		40	120	200	280	360	440
Support force S / N	1st set	1.4	2.6	4.4	5.6	7.3	8.6
	2nd set	1.2	2.9	4.3	5.9	7.0	8.8
	mean						

(a) (i) Copy and complete the table.
(ii) Plot a graph of S on the y-axis against d on the x-axis. *(5 marks)*

(b) (i) By taking moments about X, show that $S = \frac{Wd}{D} + 0.5\,W_0$, where W_0 is the weight of the beam.
(ii) Use your graph to determine the weight of the load, W. *(6 marks)*

(c) The distances were measured to an accuracy of ±1 mm and the newtonmeter readings to an accuracy of ±0.1 N.
(i) Show that the percentage uncertainty in S is significantly greater than the percentage uncertainty in d.
(ii) Without using additional apparatus, discuss what further measurements you could make to improve the accuracy of your measurements. *(3 marks)*

2 **(a)** An object is acted upon by forces of 9.6 N and 4.8 N, with an angle of 40° between them. Draw a vector diagram of these forces, using a scale of 1 cm representing 1 N. Complete the vector diagram to determine the magnitude of the resultant force acting on the object. Measure the angle between the resultant force and the 9.6 N force. *(3 marks)*

(b) Calculate the magnitude of the resultant force when the same two forces act at right angles to each other.
You must not use a scale diagram for this part. *(2 marks)*

AQA, 2007

3 Figure 2 shows a uniform steel girder being held horizontally by a crane. Two cables are attached to the ends of the girder and the tension in each of these cables is T.

(a) If the tension, T, in each cable is 850 N, calculate:
(i) the horizontal component of the tension in each cable
(ii) the vertical component of the tension in each cable
(iii) the weight of the girder. *(4 marks)*

(b) Describe the line of action of the weight of the girder. *(1 mark)*

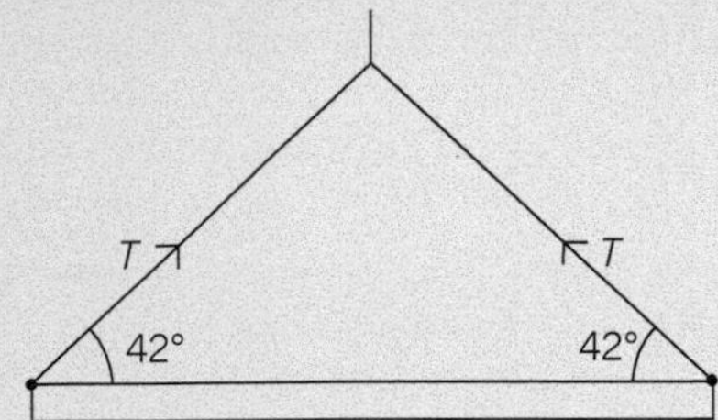

▲ **Figure 2**

AQA, 2005

4 **(a)** Define the moment of a force. *(2 marks)*

(b) Figure 3 shows the force, F, acting on a bicycle pedal.

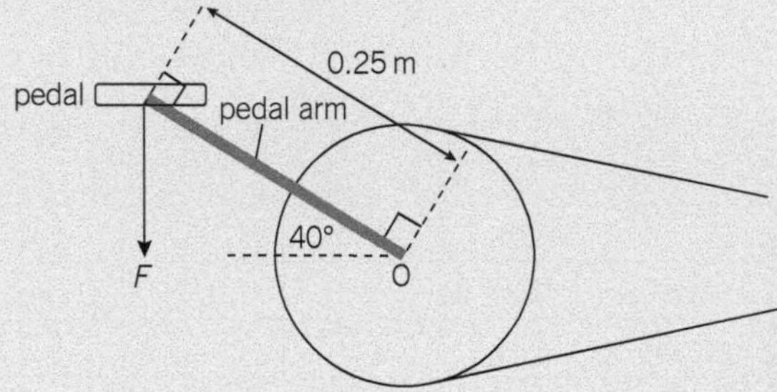

▲ **Figure 3**

(i) The moment of the force about O is 46 N m in the position shown. Calculate the value of the force, F.

(ii) Force F is constant in magnitude and direction while the pedal is moving downwards. State and explain how the moment of F changes as the pedal arm moves down through 80°, from the position shown. *(4 marks)*

AQA, 2007

5 Figure 4 shows a student standing on a plank that pivots on a log. The student intends to cross the stream.

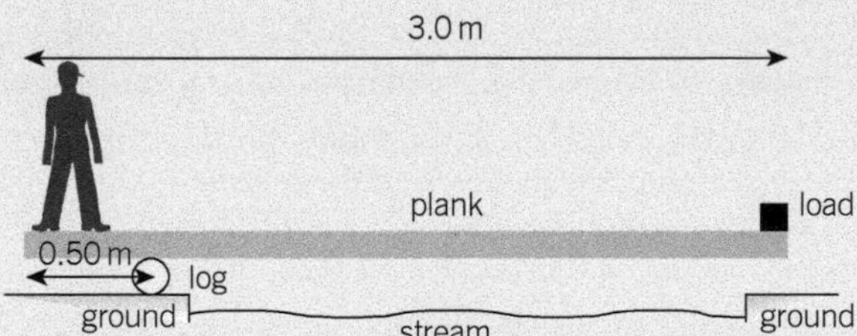

▲ **Figure 4**

(a) The plank has a mass of 25 kg and is 3.0 m long with a uniform cross section. The log pivot is 0.50 m from the end of the plank. The student has a mass of 65 kg and stands at the end of the plank. A load is placed on the far end in order to balance the plank horizontally.
Draw on a copy of Figure 4 the forces that act on the plank. *(3 marks)*

(b) By taking moments about the log pivot, calculate the load, in N, needed on the right-hand end of the plank in order to balance the plank horizontally. *(3 marks)*

(c) Explain why the load will eventually touch the ground as the student walks toward the log. *(2 marks)*

AQA, 2003

2 On the move

2.1 Speed and velocity

Speed

Displacement is distance in a given direction.

Speed is defined as change of distance per unit time.

Velocity is defined as change of displacement per unit time. In other words, velocity is speed in a given direction.

Speed and distance are scalar quantities. Velocity and displacement are vector quantities.

The unit of speed and of velocity is the metre per second ($\mathrm{m\,s^{-1}}$).

Motion at constant speed

An object moving at a constant speed travels equal distances in equal times. For example, a car travelling at a speed of $30\,\mathrm{m\,s^{-1}}$ on a motorway travels a distance of 30 m every second or 1800 m every minute. In 1 hour, the car would therefore travel a distance of 108 000 m or 108 km. So $30\,\mathrm{m\,s^{-1}} = 108\,\mathrm{km\,h^{-1}}$.

For an object which travels distance s in time t at constant speed,

$$\textbf{speed } v = \frac{s}{t}$$

$$\textbf{distance travelled } s = vt$$

For an object moving at constant speed on a circle of radius r, its speed

$$v = \frac{2\pi r}{T}$$

where T is the time to move round once and $2\pi r$ is the circumference of the circle.

Motion at changing speed

There are two types of speed cameras. One type measures the speed of a vehicle as it passes the camera. The other type is linked to a second speed camera and a computer, which works out the average speed of the vehicle between the two cameras. This will catch drivers who slow down for a speed camera then speed up again!

For an object moving at changing speed that travels a distance s in time t,

$$\textbf{average speed} = \frac{s}{t}$$

In a short time interval Δt, the distance Δs it travels is given by $\Delta s = v\Delta t$, where v is the speed at that time (i.e., its instantaneous speed). Rearranging this equation gives:

$$v = \frac{\Delta s}{\Delta t}$$

In physics and maths, the delta Δ notation is often used to mean *a change of* something.

Distance–time graphs

For an object moving at **constant speed**, its graph of distance against time is a straight line with a constant gradient.

Learning objectives:

- → Explain how displacement differs from distance.
- → Explain the difference between instantaneous speed and average speed.
- → Describe when it is necessary to consider velocity rather than speed.

Specification reference: 3.2.3

Hint

1 km = 1000 m, 1 hour = 3600 s, $108\,\mathrm{km\,h^{-1}} = 30\,\mathrm{m\,s^{-1}}$

▲ **Figure 1**

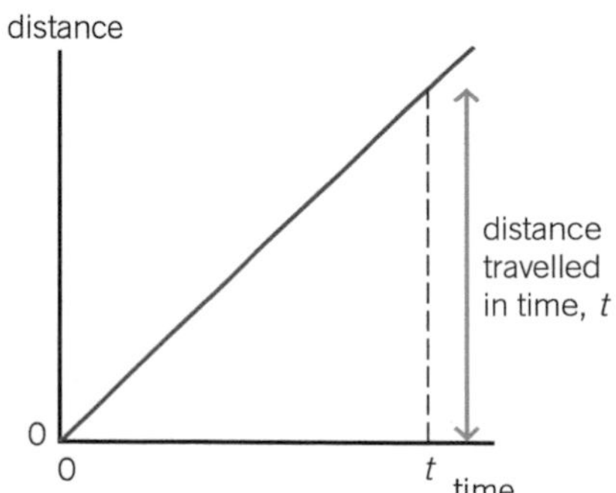

▲ **Figure 2** *Constant speed*

Summary questions

1 kilometre (km) = 1000 m

1. A car travels a distance of 60 km in 45 min at constant speed. Calculate its speed in:
   a $\mathrm{km\,h^{-1}}$ **b** $\mathrm{m\,s^{-1}}$.
2. Use Figure 3 to calculate
   a the instantaneous speed of the object at X
   b the average speed of the object between O and Y.
3. A satellite moves round the Earth at constant speed on a circular orbit of radius 8000 km with a time period of 120 min. Calculate its orbital speed in **a** $\mathrm{km\,h^{-1}}$, **b** $\mathrm{m\,s^{-1}}$.
4. A vehicle joins a motorway and travels at a steady speed of $25\,\mathrm{m\,s^{-1}}$ for 30 min then it travels a further distance of 40 km in 20 min before leaving the motorway. Calculate **a** the distance travelled in the first 30 min, **b** its average speed on the motorway.
5. **a** Explain the difference between speed and velocity.
   b A police car joins a straight motorway at Junction 4 and travels for 12 km at a constant speed for 400 s. It then leaves at Junction 5 and rejoins on the opposite side and travels for 8 km at a constant speed for 320 s to reach the scene of an accident. Calculate **i** the displacement from Junction 4 to the accident, **ii** the velocity of the car on each side of the motorway.
   c Sketch a displacement–time graph for the journey.

The speed of the object = $\dfrac{\textbf{distance travelled}}{\textbf{time taken}}$ **= gradient of the line**

For an object moving at **changing speed**, the gradient of the line changes. The gradient of the line at any point can be found by drawing a tangent to the line at that point and then measuring the gradient of the tangent. This is shown in Figure 3 where PR is the tangent at point Y on the line. Show for yourself that the speed at point X on the line is $2.5\,\mathrm{m\,s^{-1}}$.

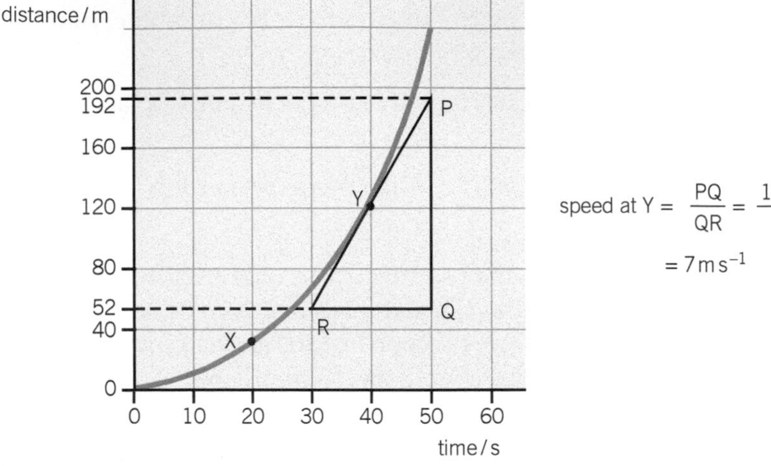

▲ **Figure 3** *Changing speed*

Velocity

An object moving at **constant velocity** moves at the same speed without changing its direction of motion.

If an object changes its direction of motion or its speed or both, its velocity changes. For example, the velocity of an object moving on a circular path at constant speed changes continuously because its direction of motion changes continuously.

Displacement–time graphs

An object travelling along a straight line has two possible directions of motion. To distinguish between the two directions, we need a direction code where positive values are in one direction and negative values are in the opposite direction.

For example, consider an object thrown vertically into the air. The direction code is + for upwards and – for downwards. Figure 4 shows how its displacement changes with time. The object has an initial positive velocity and a negative return velocity. The displacement and velocity are both positive when the object is ascending. However, when it is descending, its velocity is negative and its displacement is positive until it returns to its initial position. We will consider displacement–time graphs in more detail in Topic 2.5.

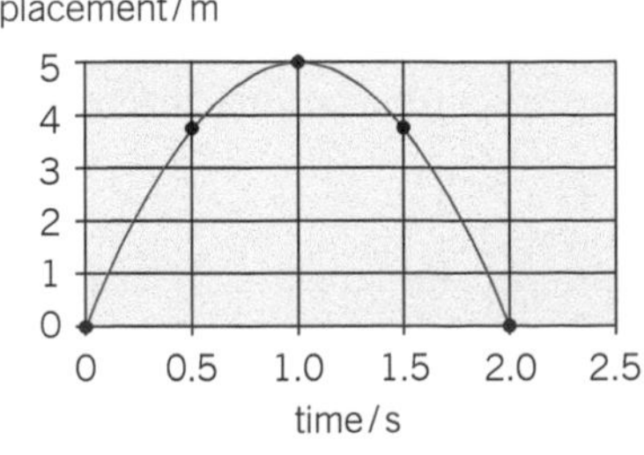

▲ **Figure 4** *A displacement–time graph*

2.2 Acceleration

Performance tests

A car maker wants to compare the performance of a new model with the original model. To do this, the velocity of each car is measured on a test track. Each vehicle accelerates as fast as possible to top velocity from a standstill. The results are listed in Table 1.

▼ **Table 1**

Time from a standing start / s	0	2	4	6	8	10
Velocity of old model / m s^{-1}	0	8	16	24	32	32
Velocity of new model / m s^{-1}	0	10	20	30	30	30

Which car accelerates fastest?

The old model took 8 s to reach its top velocity of 32 m s^{-1}. Its velocity must have increased by 4.0 m s^{-1} every second when it was accelerating.

The new model took 6 s to reach its top velocity of 30 m s^{-1}. Its velocity must have increased by 5.0 m s^{-1} every second when it was accelerating.

Clearly, the velocity of the new model increases at a faster rate than that of the old model. In other words, its acceleration is greater.

Acceleration is defined as change of velocity per unit time.

- The unit of acceleration is the metre per second per second (m s^{-2}).
- Acceleration is a vector quantity.
- Deceleration values are negative and signify that the velocity decreases with respect to time.

For a moving object that does not change direction, its acceleration at any point can be worked out from its rate of change of velocity, (ie. its change of velocity per second) because there is no change of direction. For example:

The acceleration of the old model above is 4.0 m s^{-2} because its velocity increased by 4.0 m s^{-1} every second.

The acceleration of the new model above is 5.0 m s^{-2} because its velocity increased by 5.0 m s^{-1} every second.

Uniform acceleration

Uniform acceleration is where the velocity of an object moving along a straight line changes at a constant rate. In other words, the acceleration is constant. Consider an object that accelerates uniformly from velocity u to velocity v in time t along a straight line. Figure 2 shows how its velocity changes with time.

The acceleration, a, of the object $= \dfrac{\text{change of velocity}}{\text{time taken}} = \dfrac{v - u}{t}$

$$\boldsymbol{a = \frac{v - u}{t}}$$

To calculate the velocity v at time t, rearranging this equation gives

$$at = v - u$$

$$\boldsymbol{v = u + at}$$

Learning objectives:

- → Describe what is meant by acceleration and deceleration.
- → Explain why uniform acceleration is a special case.
- → Explain why acceleration is considered to be a vector.

Specification reference: 3.2.3

▲ **Figure 1** *A racing car on a test track*

Study tip

For a change of velocity Δv in a time interval Δt, the acceleration $a = \dfrac{\Delta v}{\Delta t}$

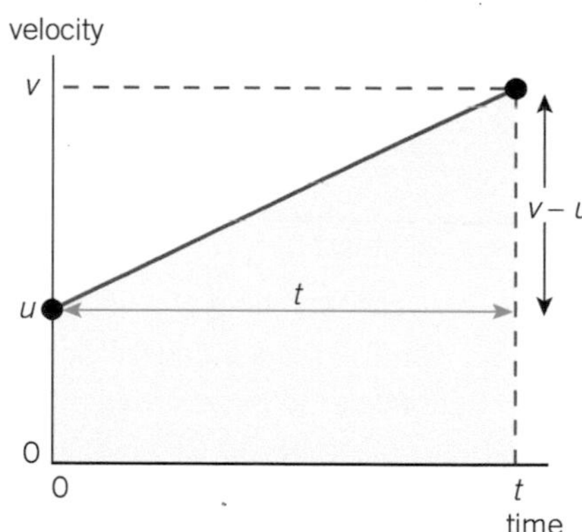

▲ **Figure 2** *Uniform acceleration on a velocity–time graph*

Worked example

The driver of a vehicle travelling at $8\,\mathrm{m\,s^{-1}}$ applies the brakes for 30 s and reduces the velocity of the vehicle to $2\,\mathrm{m\,s^{-1}}$. Calculate the deceleration of the vehicle during this time.

Solution

$u = 8\,\mathrm{m\,s^{-1}}$, $v = 2\,\mathrm{m\,s^{-1}}$, $t = 30$ s

$$a = \frac{v-u}{t} = \frac{2-8}{30} = \frac{-6}{30} = -0.2\,\mathrm{m\,s^{-2}}$$

Hint

Remember that acceleration is a vector quantity. It is dependent on both speed and direction.

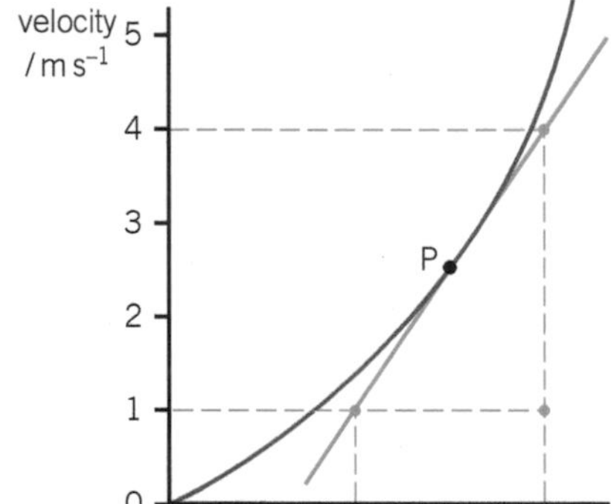

▲ Figure 3

Non-uniform acceleration

Non-uniform acceleration is where the direction of motion of an object changes, or its speed changes, at a varying rate. Figure 3 shows how the velocity of an object increases for an object moving along a straight line with an increasing acceleration. This can be seen directly from the graph because the gradient increases with time (the graph becomes steeper and steeper) and the gradient represents the acceleration.

The acceleration at any point is the gradient of the tangent to the curve at that point. In Figure 3,

$$\text{the gradient of the tangent at point P} = \frac{\text{height of gradient triangle}}{\text{base of gradient triangle}}$$

$$= \frac{4.0\,\mathrm{m\,s^{-1}} - 1.0\,\mathrm{m\,s^{-1}}}{2.0\,\mathrm{s}} = 1.5\,\mathrm{m\,s^{-2}}$$

Therefore the acceleration at P is $1.5\,\mathrm{m\,s^{-2}}$.

Acceleration = gradient of the line on the velocity–time graph.

Summary questions

1 **a** An aeroplane taking off accelerates uniformly on a runway from a velocity of $4\,\mathrm{m\,s^{-1}}$ to a velocity of $64\,\mathrm{m\,s^{-1}}$ in 40 s. Calculate its acceleration.

b A car travelling at a velocity of $20\,\mathrm{m\,s^{-1}}$ brakes sharply to a standstill in 8.0 s. Calculate its deceleration, assuming its velocity decreases uniformly.

2 A cyclist accelerates uniformly from a velocity of $2.5\,\mathrm{m\,s^{-1}}$ to a velocity of $7.0\,\mathrm{m\,s^{-1}}$ in a time of 10 s. Calculate **a** its acceleration, **b** its velocity 2.0 s later if it continues to accelerate at the same rate.

3 A train on a straight journey between two stations accelerates uniformly from rest for 20 s to a velocity of $12\,\mathrm{m\,s^{-1}}$. It then travels at constant velocity for a further 40 s before decelerating uniformly to rest in 30 s.

a Sketch a velocity–time graph to represent its journey.

b Calculate its acceleration in each part of the journey.

4 The velocity of an object released in water increases as shown in Figure 4.

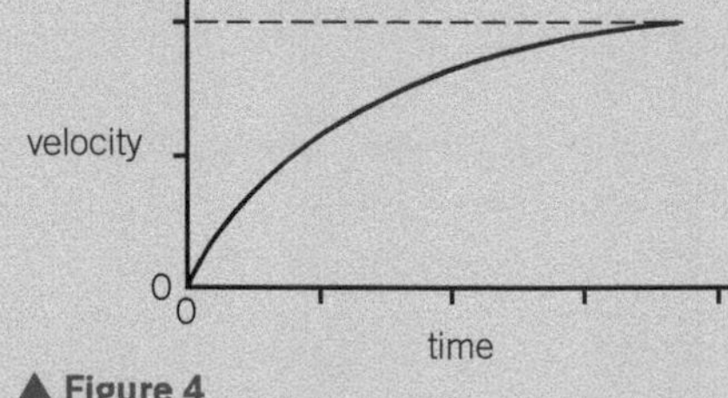

▲ Figure 4

Describe how **a** the velocity of the object changed with time, **b** the acceleration of the object changed with time.

2.3 Motion along a straight line at constant acceleration

The dynamics equations for constant acceleration

Consider an object that accelerates uniformly from initial velocity u to final velocity v in time t without change of direction. Figure 2 in Topic 2.2 shows how its velocity changes with time.

1 The acceleration $a = \frac{(v - u)}{t}$, as explained in Topic 2.2, can be rearranged to give

$$v = u + at \qquad \textbf{(Equation 1)}$$

2 The displacement s = average velocity × time taken
Because the acceleration is uniform, average velocity $= \frac{(u + v)}{2}$

Combining these two equations gives

$$s = \frac{(u + v)t}{2} \qquad \textbf{(Equation 2)}$$

3 By combining the two equations above to eliminate v, a further useful equation is produced.
To do this, substitute $u + at$ in place of v in Equation 2. This gives

$$s = \frac{(u + (u + at))}{2}t = \frac{(u + u + at)t}{2} = \frac{(2ut + at^2)}{2}$$

$$s = ut + \frac{1}{2}at^2 \qquad \textbf{(Equation 3)}$$

4 A fourth useful equation is obtained by combining Equations 1 and 2 to eliminate t. This can be done by multiplying

$a = \frac{(v - u)}{t}$ and $s = \frac{(u + v)t}{2}$ together to give

$$as = \frac{(v - u)}{t} \times \frac{(u + v)t}{2}$$

This simplifies to become

$$as = \frac{(v - u)\,(v + u)}{2} = \frac{(v^2 - uv + uv - u^2)}{2} = \frac{(v^2 - u^2)}{2}$$

Therefore $2as = v^2 - u^2$ or

$$v^2 = u^2 + 2as \qquad \textbf{(Equation 4)}$$

The four equations, sometimes referred to as the *suvat* equations, are invaluable in any situation where the acceleration is constant.

Worked example

A driver of a vehicle travelling at a speed of $30\,\text{m s}^{-1}$ on a motorway brakes sharply to a standstill in a distance of 100 m. Calculate the deceleration of the vehicle.

Solution

$u = 30\,\text{m s}^{-1}$, $v = 0$, $s = 100\,\text{m}$, $a = ?$

To find a, use $v^2 = u^2 + 2as$

Learning objectives:

- → Distinguish between u and v.
- → Calculate the displacement of an object moving with uniform acceleration.
- → Explain what else we need to know to calculate the acceleration of an object if we know its displacement in a given time.

Specification reference: 3.2.3

Maths link

Q: If a space rocket accelerated from rest at $10\,\text{m s}^{-2}$, estimate how many years it would take to reach half the speed of light. The speed of light is $3.0 \times 10^8\,\text{m s}^{-1}$.

Answer: 0.5 years.

Hint

If $a = 0$ (i.e., constant velocity), the equations all reduce to $s = vt$ or $v = u$.

Therefore $0 = u^2 + 2as$ because $v = 0$

Rearranging this equation gives

$2as = -u^2$

$$a = -\frac{u^2}{2s} = -\frac{30^2}{2 \times 100} = -4.5\,\text{m s}^{-2}$$

Note:

The acceleration is negative which means it is a deceleration as it is in the opposite direction to the velocity.

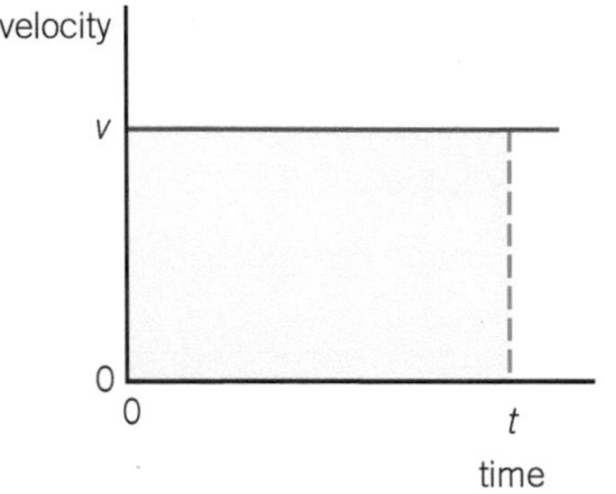

▲ **Figure 1**

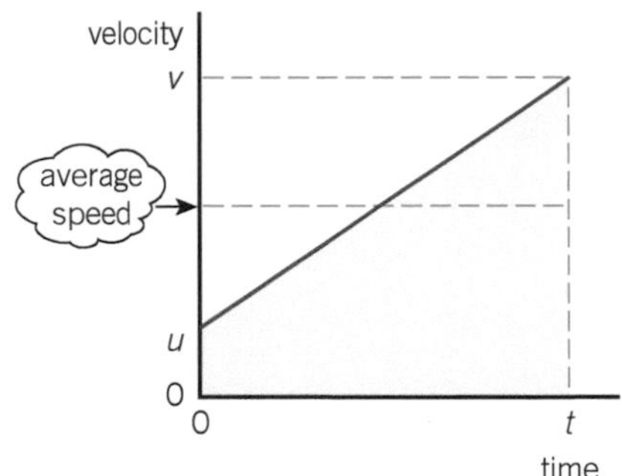

▲ **Figure 2**

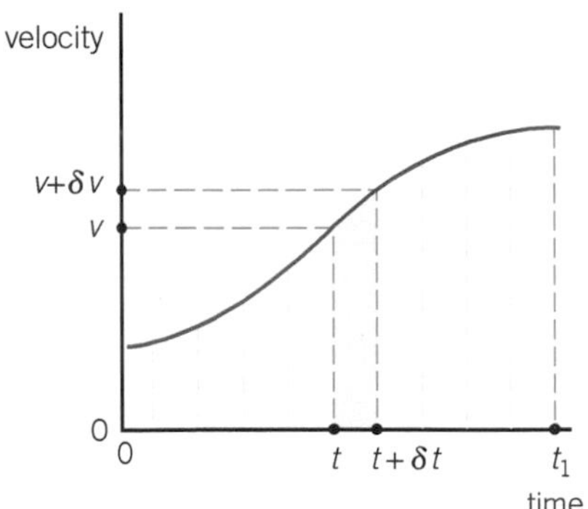

▲ **Figure 3**

Using a velocity–time graph to find the displacement

1 **Consider an object moving at constant velocity, *v*,** as shown in Figure 1. The displacement in time *t*, *s* = velocity × time taken = *vt*. This displacement is represented on the graph by the area under the line between the start and time *t*. This is a rectangle of height corresponding to velocity *v* and of base corresponding to the time *t*.

2 **Consider an object moving at constant acceleration, *a*,** from initial velocity *u* to velocity *v* at time *t*, as shown in Figure 2. From Equation 2, the displacement *s* moved in this time is given by

$$s = \frac{(u + v)t}{2}$$

This displacement is represented on the graph by the area under the line between the start and time *t*. This is a trapezium which has a base corresponding to time *t* and an average height corresponding to the average velocity $\frac{1}{2}(u + v)$.

Therefore the area of the trapezium (= average height × base) corresponds to $\frac{1}{2}(u + v)t$, and gives the displacement.

3 **Consider an object moving with a changing acceleration,** as shown in Figure 3. Let *v* represent the velocity at time *t* and let $v + \delta v$ represent the velocity a short time later at $t + \delta t$. (δ is pronounced 'delta'.)

Because the velocity change δv is small compared with the velocity *v*, the displacement δs in the short time interval δt is $v\delta t$.

This is represented on the graph by the area of the shaded strip under the line, which has a base corresponding to δt and a height corresponding to *v*. In other words, $\delta s = v\delta t$ is represented by the area of this strip.

By considering the whole area under the line in strips of similar width, the total displacement from the start to time *t* can be estimated from the sum of the area of every strip. Note that the smaller the value of δt, the narrower the strips are so the closer their area is to the total area under the curve, giving a better estimate.

Whatever the shape of the line of a velocity–time graph,

displacement = area under the line of a velocity–time graph.

The equations for constant acceleration are:

$$v = u + at$$

$$s = \frac{1}{2}(u + v)t$$

$$s = ut + \frac{1}{2}at^2$$

$$v^2 = u^2 + 2as$$

Study tip

Always begin a *suvat* calculation by identifying which three variables are known. Remember the acceleration must be constant for these equations to apply.

Summary questions

1 A vehicle accelerates uniformly along a straight road, increasing its velocity from 4.0 $m\,s^{-1}$ to 30.0 $m\,s^{-1}$ in 13 s. Calculate:

a its acceleration

b the displacement.

2 An aircraft lands on a runway at a velocity of 40 $m\,s^{-1}$ and brakes to a halt in a distance of 860 m. Calculate:

a the braking time

b the deceleration of the aircraft.

3 A cyclist accelerates uniformly from rest to a speed of 6.0 $m\,s^{-1}$ in 30 s then brakes at uniform deceleration to a halt in a distance of 24 m.

a For the first part of the journey, calculate **i** the acceleration, **ii** the displacement.

b For the second part of the journey, calculate **i** the deceleration, **ii** the time taken.

c Sketch a velocity–time graph for this journey.

d Use the graph to determine the average velocity for the journey.

4 The velocity of an athlete for the first 5 s of a sprint race is shown in Figure 4. Use the graph to determine:

a the initial acceleration of the athlete

b the displacement in the first 2 s

c the displacement in the next 2 s

d the average speed over the first 4 s.

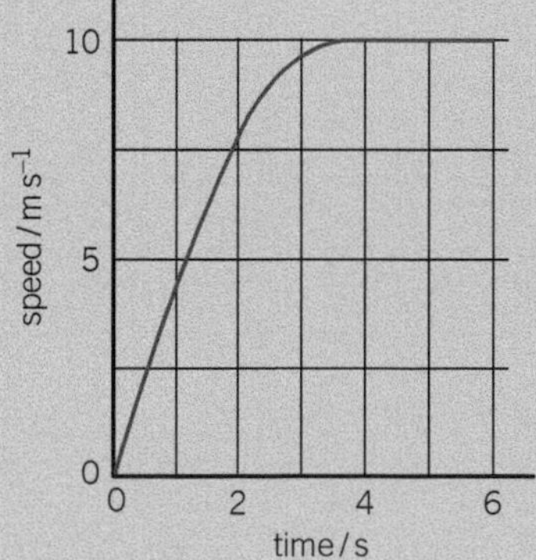

▲ **Figure 4**

2.4 Free fall

Learning objectives:

- → Define free fall.
- → Explain how the velocity of a freely falling object changes as it falls.
- → Discuss if objects of different masses or sizes all fall with the same acceleration.

Specification reference: 3.2.3

▲ **Figure 1** *Galileo Galilei 1564–1642*

Experimental tests

Does a heavy object fall faster than a lighter object?

Release a stone and a small coin at the same time. Which one hits the ground first? The answer to this question was first discovered by Galileo Galileli about four centuries ago. He reasoned that because any number of identical objects must fall at the same rate, then any one such object must fall at the same rate as the rest put together. So he concluded that any two objects must fall at the same rate, regardless of their relative weights. He was reported to have demonstrated the correctness of his theory by releasing two different weights from the top of the Leaning Tower of Pisa.

The inclined plane test

Galileo wanted to know if a falling object speeds up as it falls, but clocks and stopwatches were devices of the future. The simplest test he could think of was to time a ball as it rolled down a plank. He devised a dripping water clock, counting the volume of the drips as a measure of time. He measured how long the ball took to travel equal distances down the slope from rest. His measurements showed that the ball gained speed as it travelled down the slope. In other words, he showed that the ball accelerated as it rolled down the slope. He reasoned that the acceleration would be greater the steeper the slope. So he concluded that an object falling vertically accelerates.

Acceleration due to gravity

One way to investigate the free fall of a ball is to make a multiflash photo or video clip of the ball's flight as it falls after being released from rest. A vertical metre rule can be used to provide a scale. To obtain a multiflash photo, an ordinary camera with a slow speed shutter may be used to record the ball's descent in a dark room illuminated by a stroboscope (a flashing light). The flashing light needs to flash at a known constant rate of about 30 flashes per second. Figure 2 shows a possible arrangement using a steel ball and a multiflash photo taken with this arrangement. If you make a video clip, you need to be able to rerun the clip at slow speed with the time displayed.

Practical link

A digital camera or a similar device can also be used to obtain a video clip that can be analysed to obtain g.

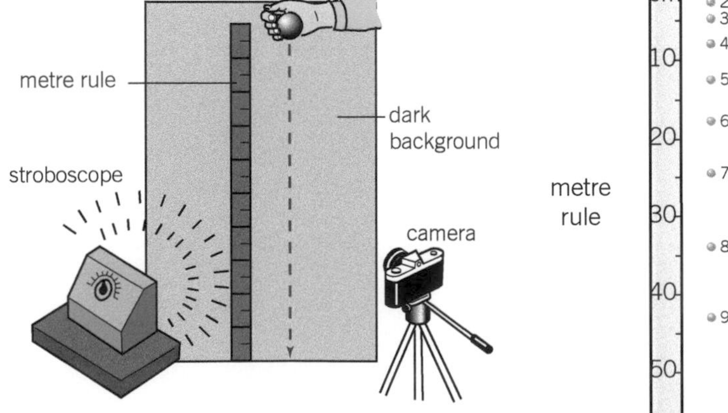

▲ **Figure 2** *Investigating free fall*

For each image of the ball on the photograph, the time of descent of the ball and the distance fallen by the ball from rest can be measured directly. The photograph shows that the ball speeds up as it falls, because it travels further between successive images. Measurements from this photograph are given in Table 1.

▼ **Table 1** *Free fall measurements*

Number of flashes after start	0	2	3	4	5	6	7	8	9
Time of descent t / s	0	0.06	0.10	0.13	0.16	0.19	0.23	0.26	0.29
Distance fallen s / m	0	0.02	0.04	0.07	0.12	0.17	0.24	0.33	0.42

How can you tell if the acceleration is constant from these results? One way is to consider how the distance fallen, s, would vary with time, t, for constant acceleration. From Topic 2.3, we know that

$s = ut + \frac{1}{2}at^2$, where u = the initial speed and a = acceleration.

In this experiment, $u = 0$.

Therefore $s = \frac{1}{2}at^2$ for constant acceleration, a.

Compare this equation with the general equation for a straight-line graph, $y = mx + c$, where m is the gradient and c is the y-intercept. If we let y represent s and let x represent t^2, then $m = \frac{1}{2}a$ and $c = 0$.

So a graph of s against t^2 should therefore give a straight line through the origin. In addition, the gradient of the line $(= \frac{1}{2}a)$ can be measured, enabling the acceleration (= 2 × gradient) to be calculated.

Figure 3 shows this graph. As you can see, it is a straight line through the origin. We can therefore conclude that the equation $s = \frac{1}{2}at^2$ applies here so the acceleration of a falling object is constant. Show for yourself that the gradient of the line is $5.0\,\mathrm{m\,s^{-2}}$ ($\pm 0.2\,\mathrm{m\,s^{-2}}$), giving an acceleration of $10\,\mathrm{m\,s^{-2}}$.

Because there are no external forces acting on the object apart from the force of gravity, this value of acceleration is known as the **acceleration of free fall** and is represented by the symbol g. Accurate measurements give a value of $9.8\,\mathrm{m\,s^{-2}}$ near the Earth's surface.

The *suvat* equations from Topic 2.3 may be applied to any free fall situation where air resistance is negligible.

The equations can also be applied to situations where objects are thrown vertically upwards. As a general rule, apply the direction code + for *upwards* and – for *downwards* when values are inserted into the *suvat* equations.

Extension

What Galileo did for science

In addition to his discoveries on motion, Galileo made other important scientific discoveries such as the four innermost moons of Jupiter. He became a big supporter of the Copernican, or Sun-centred, model of the Solar System. This upset the Catholic Church, which believed that the Earth was at the centre of the Universe, and Galileo was banned from teaching the Copernican model. He wrote a book about it in protest, which became a bestseller and upset the Church even more. Galileo was condemned to spend the rest of his life under house arrest. However, his trial attracted a lot of publicity all over Europe, and his work was taken up enthusiastically by other scientists as a result. Galileo showed the importance of valid observations and measurements in developing our understanding of the natural world. He also showed that science is powerful enough to change established beliefs.

Q: Describe another situation when scientific observations caused a scientific theory to be altered or replaced.

Synoptic link

You will meet different graphs in more detail in Topic 14.4, Straight-line graphs, and Topic 14.5, More on graphs.

Summary questions

$g = 9.8\,\text{m s}^{-2}$

1 A pebble, released at rest from a canal bridge, took 0.9 s to hit the water. Calculate:

  a the distance it fell before hitting the water

  b its speed just before hitting the water.

2 A spanner was dropped from a hot air balloon when the balloon was at rest 50 m above the ground. Calculate:

  a the time taken for the spanner to hit the ground

  b the speed of impact of the spanner on hitting the ground.

3 A bungee jumper jumped off a platform 75 m above a lake, releasing a small object at the instant she jumped off the platform.

  a Calculate **i** the time taken by the object to fall to the lake, **ii** the speed of impact of the object on hitting the water, assuming air resistance is negligible.

  b Explain why the bungee jumper would take longer to descend than the time taken in part **a**.

4 An astronaut on the Moon threw an object 4.0 m vertically upwards and caught it again 4.5 s later. Calculate:

  a the acceleration due to gravity on the Moon

  b the speed of projection of the object

  c how high the object would have risen on the Earth, for the same speed of projection.

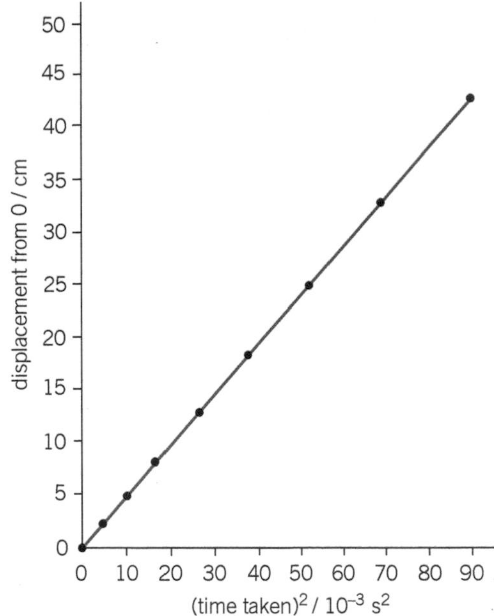

▲ **Figure 3** *A graph of s against t^2*

Measuring *g*

Use an electronic timer or motion sensor to make your own measurements of *s* and *t*. Repeat your measurements to obtain an average timing for each measured distance and plot a suitable graph to find *g*.

Worked example

$g = 9.8\,\text{m s}^{-2}$

A coin was released at rest at the top of a well. It took 1.6 s to hit the bottom of the well. Calculate the distance fallen by the coin and its speed just before impact.

Solution

$u = 0$, $t = 1.6$ s, $a = -9.8\,\text{m s}^{-2}$ (– as *g* acts downwards)

To find distance, *s*, use $s = \frac{1}{2}at^2$ as $u = 0$

Therefore $s = \frac{1}{2} \times -9.8 \times 1.6^2$

$= -12.5\,\text{m}$ (– indicates 12.5 m downwards)

To find its speed before impact, *v*, use $v = u + at$

$= 0 + (-9.8 \times 1.6) = -15.7\,\text{m s}^{-1}$ (– indicates downward velocity)

2.5 Motion graphs

Learning objectives:

- → Distinguish between a distance–time graph and a displacement–time graph.
- → Describe what the gradient of a velocity–time graph represents.
- → Describe what the area under a velocity–time graph represents.

Specification reference: 3.2.3

The difference between a distance–time graph and a displacement–time graph

Displacement is distance in a given direction from a certain point. Consider a ball thrown directly upwards and caught when it returns. If the ball rises to a maximum height of 2.0 m, on returning to the thrower its displacement from its initial position is zero. However, the distance it has travelled is 4.0 m.

The displacement of the ball changes with time as shown by Figure 1. The line in this graph fits the equation $s = ut - \frac{1}{2}gt^2$, where s is the displacement, t is the time taken and u is the initial velocity of the object.

The gradient of the line in Figure 1 represents the velocity of the object.

- Immediately after leaving the thrower's hand, the velocity is positive and large so the gradient is positive and large.
- As the ball rises, its velocity decreases so the gradient decreases.
- At maximum height, its velocity is zero so the gradient is zero.
- As the ball descends, its velocity becomes increasingly negative, corresponding to increasing speed in a downward direction. So the gradient becomes increasingly negative.

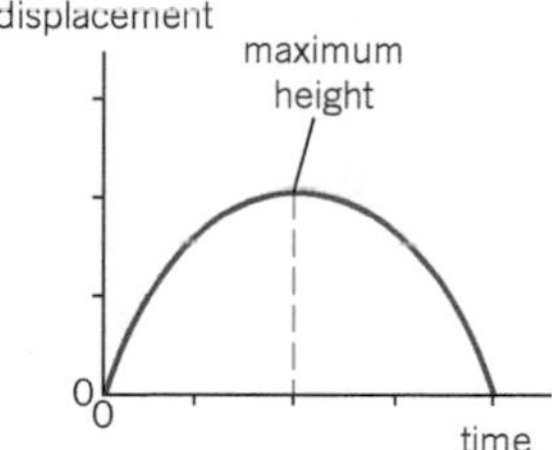

▲ **Figure 1** *Displacement–time graph for an object projected upwards*

The distance travelled by the object changes with time as shown by Figure 2. The gradient of this line represents the speed. From projection to maximum height, the shape is exactly the same as in Figure 1. After maximum height, the distance continues to increase so the line curves up, not down like in Figure 1.

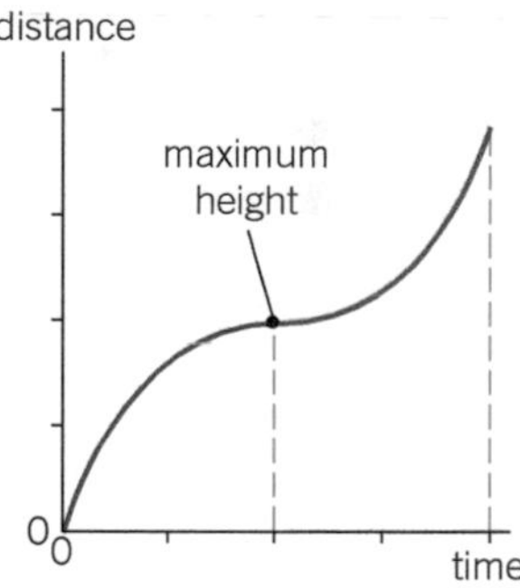

▲ **Figure 2** *Distance–time graph for an object projected upwards*

The difference between a speed–time graph and a velocity–time graph

Velocity is speed in a given direction. Consider how the velocity of an object thrown into the air changes with time. The object's velocity decreases from its initial positive (upwards) value to zero at maximum height then increases in the negative (downwards) direction as it falls.

Figure 3 shows how the velocity of the object changes with time.

1. **The gradient of the line represents the object's acceleration.** This is constant and negative, equal to the acceleration of free fall, g. The acceleration of the object is the same when it descends as when it ascends so the gradient of the line is always equal to $-9.8\,\mathrm{m\,s^{-2}}$.
2. **The area under the line represents the displacement of the object from its starting position.**
   - The area between the positive section of the line and the time axis represents the displacement during the ascent.
   - The area under the negative section of the line and the time axis represents the displacement during the descent.

Taking the area for **a** as positive and the area for **b** as negative, the total area is therefore zero. This corresponds to zero for the total displacement.

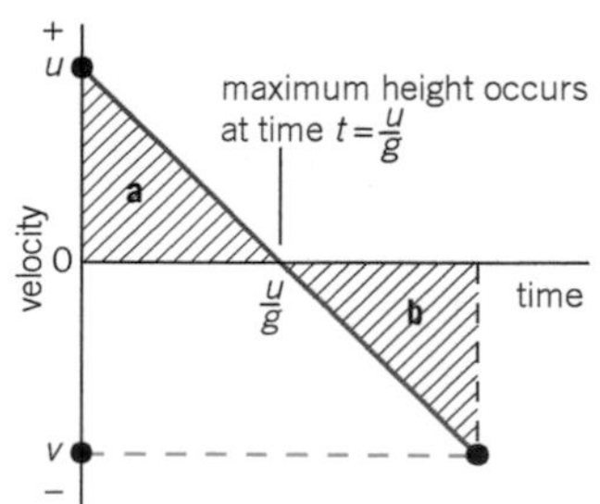

▲ **Figure 3** *Velocity–time graph*

Worked example

$g = 9.8\,\text{m}\,\text{s}^{-2}$

A ball released from a height of 1.20 m above a concrete floor rebounds to a height of 0.82 m.

a Calculate **i** the time of descent of the ball, **ii** its speed immediately before it hits the floor.

b Calculate **i** the speed of the ball immediately after it leaves the floor, **ii** its time of ascent.

c Sketch a velocity–time graph for the ball. Assume the contact time is negligible compared with the time of descent or ascent and that positive velocity is upwards.

Solution

a $u = 0$, $a = -9.8\,\text{m}\,\text{s}^{-2}$, $s = -1.2\,\text{m}$

i To find t, use $s = ut + \frac{1}{2}at^2$

$-1.2 = 0 + 0.5 \times -9.8 \times t^2$ so $-1.2 = -4.9t^2$

$t^2 = \frac{-1.2}{-4.9} = 0.245$ so $t = 0.49\,\text{s}$

ii To find v, use $v = u + at$

$v = 0 + -9.8 \times 0.49 = -4.8\,\text{m}\,\text{s}^{-1}$ (– because downwards)

b $v = 0$, $a = -9.8\,\text{m}\,\text{s}^{-2}$, $s = +0.82\,\text{m}$

i Using $v^2 = u^2 + 2as$ to find u gives $0 = u^2 + 2 \times -9.8 \times 0.82$

$u^2 = 16.1\,\text{m}^2\,\text{s}^{-2}$ so $u = +4.0\,\text{m}\,\text{s}^{-1}$ (+ because upwards)

ii Using $v = u + at$ to find t gives $0 = 4.0 + -9.8 \times t$

$9.8t = 4.0$ so $t = 0.41\,\text{s}$

c The sketch graph is given in Figure 4.

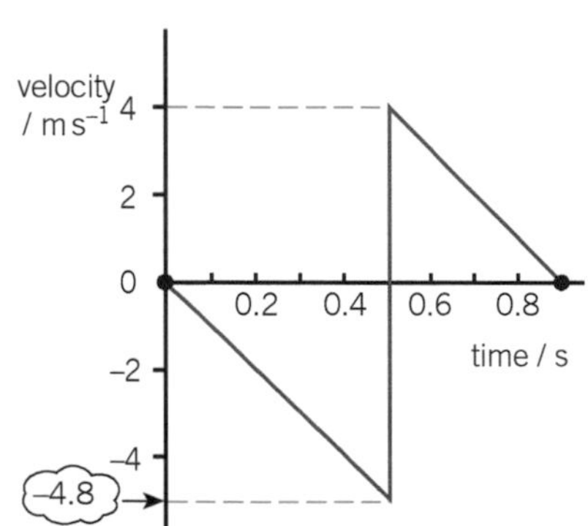

▲ **Figure 4** *Velocity–time graph for a bouncing ball, where gradient = g*

Summary questions

$g = 9.8\,\text{m}\,\text{s}^{-2}$

1 A swimmer swims 100 m from one end of a swimming pool to the other end at a constant speed of $1.2\,\text{m}\,\text{s}^{-1}$, then swims back at constant speed, returning to the starting point 210 s after starting.

  a Calculate how long the swimmer takes to swim from **i** the starting end to the other end, **ii** back to the start from the other end.

  b For the swim from start to finish, sketch **i** a displacement–time graph, **ii** a distance–time graph, **iii** a velocity–time graph.

2 A motorcyclist travelling along a straight road at a constant speed of $8.8\,\text{m}\,\text{s}^{-1}$ passes a cyclist travelling in the same direction at a speed of $2.2\,\text{m}\,\text{s}^{-1}$. After 200 s, the motorcyclist stops.

  a Calculate how long the motorcyclist has to wait before the cyclist catches up.

  b On the same axes, sketch velocity–time graphs for **i** the motorcyclist, **ii** the cyclist.

3 A hot air balloon is ascending vertically at a constant velocity of $3.5\,\text{m}\,\text{s}^{-1}$ when a small metal object falls from its base and hits the ground 3.0 s later.

  a Sketch a graph to show how the velocity of the object changed with time during its descent.

  b Show that the balloon base was 33.6 m above the ground when the object fell off the base.

4 A ball is released from a height of 1.8 m above a level surface and rebounds to a height of 0.90 m.

  a Given $g = 9.8\,\text{m}\,\text{s}^{-2}$, calculate **i** the duration of its descent, **ii** its velocity just before impact, **iii** the duration of its ascent, **iv** its velocity just after impact.

  b Sketch a graph to show how its velocity changes with time from release to rebound at maximum height.

  c Sketch a further graph to show how the displacement of the ball changes with time.

2.6 More calculations on motion along a straight line

Learning objectives:

- → Calculate the motion of an object with constant acceleration if its velocity reverses.
- → Deduce whether the overall motion should be broken down into stages.
- → Explain how we use signs to work out if an object is moving forwards or backwards.

Specification reference: 3.2.3

Motion along a straight line at constant acceleration

- The equations for motion at constant acceleration, a, are

$$v = u + at \qquad \textbf{(Equation 1)}$$

$$s = \frac{1}{2}(u + v)t \qquad \textbf{(Equation 2)}$$

$$s = ut + \frac{1}{2}at^2 \qquad \textbf{(Equation 3)}$$

$$v^2 = u^2 + 2as \qquad \textbf{(Equation 4)}$$

  where s is the displacement in time t, u is the initial velocity, and v is the final velocity.
- For motion along a straight line at constant acceleration, one direction along the line is positive and the opposite direction is negative.

Worked example

A space vehicle moving towards a docking station at a velocity of $2.5\,\text{m}\,\text{s}^{-1}$ is 26 m from the docking station when its reverse thrust motors are switched on to slow it down and stop it when it reaches the station. The vehicle decelerates uniformly until it comes to rest at the docking station when its motors are switched off.

Calculate **a** its deceleration, **b** how long it takes to stop.

Solution

Let the + direction represent motion towards the docking station and – represent motion away from the station.

Initial velocity $u = +2.5\,\text{m}\,\text{s}^{-1}$, final velocity $v = 0$, displacement $s = +26\,\text{m}$.

a To find its deceleration, a, use $v^2 = u^2 + 2as$

$$0 = 2.5^2 + 2a \times 26 \qquad \text{so } -52a = 2.5^2$$

$$a = \frac{2.5^2}{-52} = -0.12\,\text{m}\,\text{s}^{-2}$$

b To find the time taken, use $v = u + at$

$$0 = 2.5 - 0.12t \qquad \text{so } 0.12t = 2.5$$

$$t = \frac{2.5}{0.12} = 21\,\text{s}$$

▲ **Figure 1** *A space vehicle docking*

Two-stage problems

Consider an object released from rest, falling, then hitting a bed of sand. The motion is in two stages:

1 falling motion due to gravity: acceleration = g (downwards)
2 deceleration in the sand: initial velocity = velocity of object just before impact.

The acceleration in each stage is *not* the same. The link between the two stages is that the velocity at the end of the first stage is the same as the velocity at the start of the second stage.

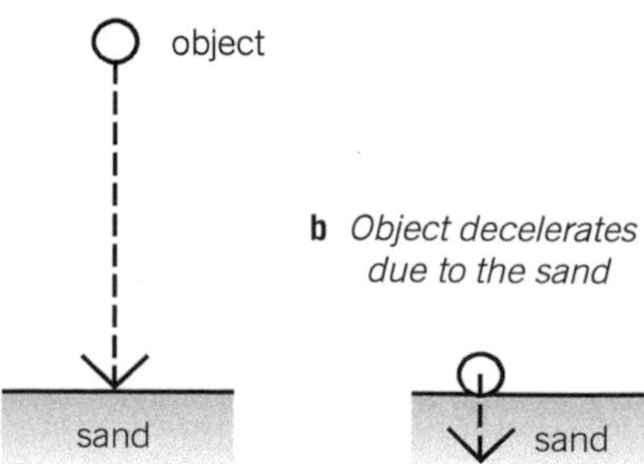

▲ **Figure 2** *A two-stage problem*

Study tip

Applying the sign convention + for forwards (or up), – for backwards (or down), for an object

- moving forwards and accelerating, $v > 0$ and $a > 0$
- moving forwards and decelerating, $v > 0$ and $a < 0$
- moving backwards and accelerating, $v < 0$ and $a < 0$
- moving backwards and decelerating, $v < 0$ and $a > 0$.

For example, a ball is released from a height of 0.85 m above a bed of sand, and creates an impression in the sand of depth 0.025 m. For directions, let + represent upwards and – represent downwards.

Stage 1: $u = 0$, $s = -0.85\,\text{m}$, $a = -9.8\,\text{m}\,\text{s}^{-2}$.

To calculate the speed of impact v, use $v^2 = u^2 + 2as$

$$v^2 = 0^2 + 2 \times -9.8 \times -0.85 = 16.7\,\text{m}^2\text{s}^{-2} \qquad v = -4.1\,\text{m}\,\text{s}^{-1}$$

Note:
$v^2 = 16.7\,\text{m}^2\text{s}^{-2}$ so $v = -4.1$ or $+4.1\,\text{m}\,\text{s}^{-1}$. The negative answer is chosen as the ball is moving downwards.

Stage 2: $u = -4.1\,\text{m}\,\text{s}^{-1}$, $v = 0$ (as the ball comes to rest in the sand), $s = -0.025\,\text{m}$

To calculate the deceleration, a, use $v^2 = u^2 + 2as$

$$0^2 = (-4.1)^2 + 2a \times -0.025$$

$$2a \times 0.025 = 16.8$$

$$a = \frac{16.8}{2 \times 0.025} = 336\,\text{m}\,\text{s}^{-2} = -340\,\text{m}\,\text{s}^{-2} \text{ to 2 significant figures}$$

Note:
$a > 0$ and therefore in the opposite direction to the direction of motion, which is downwards. Thus the ball slows down in the sand with a deceleration of $336\,\text{m}\,\text{s}^{-2}$.

Summary questions

$g = 9.8\,\text{m}\,\text{s}^{-2}$

1 A vehicle on a straight downhill road accelerates uniformly from a speed of $4.0\,\text{m}\,\text{s}^{-1}$ to a speed of $29\,\text{m}\,\text{s}^{-1}$ over a distance of 850 m. The driver then brakes and stops the vehicle in 28 s.

- **a** Calculate **i** the time taken to reach $29\,\text{m}\,\text{s}^{-1}$ from $4\,\text{m}\,\text{s}^{-1}$, **ii** the acceleration during this time.
- **b** Calculate **i** the distance travelled during deceleration, **ii** the deceleration for the last 28 s.

2 A rail wagon moving at a speed of $2.0\,\text{m}\,\text{s}^{-1}$ on a level track reaches a steady incline which slows it down to rest in 15.0 s and causes it to reverse. Calculate:

- **a** the distance the wagon moves up the incline
- **b** its acceleration on the incline
- **c** its velocity and position on the incline after 20.0 s.

3 A cyclist accelerates from rest at a constant acceleration of $0.4\,\text{m}\,\text{s}^{-2}$ for 20 s, then stops pedalling and slows to a standstill at constant deceleration over a distance of 260 m.

- **a** Calculate **i** the distance travelled by the cyclist in the first 20 s, **ii** the speed of the cyclist at the end of this time.
- **b** Calculate **i** the time taken to cover the distance of 260 m after she stops pedalling, **ii** her deceleration during this time.

4 A rocket is launched directly upwards from rest. Its motors operate for 30 s after leaving the launch pad, providing it with a constant vertical acceleration of $6.0\,\text{m}\,\text{s}^{-2}$ during this time. Its motors then switch off.

- **a** Calculate **i** its velocity, **ii** its height above the launch pad when its motors switch off.
- **b** Calculate the maximum height reached after its motors switch off.
- **c** Calculate the velocity with which it would hit the ground if it fell from maximum height without the support of a parachute.

2.7 Projectile motion 1

A **projectile** is any object projected into the air and acted upon only by the forces of gravity and air resistance. Initially we shall assume air resistance is negligible. Three key principles apply if air resistance is negligible:

1 The acceleration of the object is always equal to g and is always downwards because the force of gravity acts downwards. The acceleration therefore only affects the vertical motion of the object.
2 The horizontal velocity of the object is constant because the acceleration of the object does not have a horizontal component.
3 The motions in the horizontal and vertical directions are independent of each other.

Learning objectives:

→ Explain why the acceleration of a projectile is always vertically downwards.

→ Identify the horizontal component of a vertical vector.

→ Describe the effect of gravity on horizontal speed.

Specification reference: 3.2.4

Vertical projection

Such an object moves vertically as it has no horizontal motion. Its acceleration is $9.8\,\text{m}\,\text{s}^{-2}$ downwards. Using the direction code + is upwards, – is downwards, its displacement, y, and velocity, v, after time t are given by

$$v = u - gt$$

$$y = ut - \frac{1}{2}gt^2$$

where u is its initial velocity in the upward direction.

See Topic 2.4 for more about vertical motion with no horizontal projection.

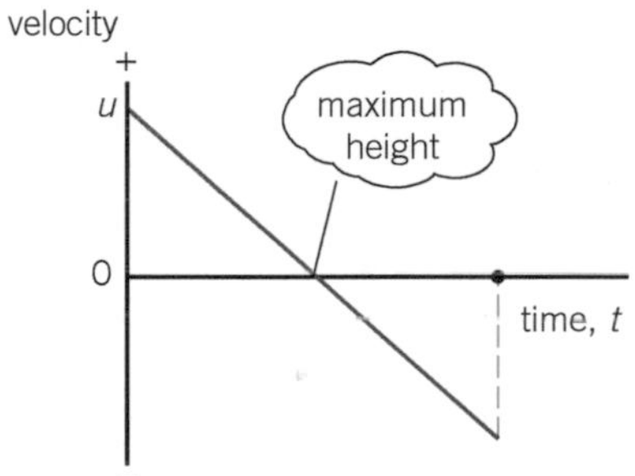

▲ **Figure 1** *Upward projection*

Horizontal projection

A stone thrown from a cliff top follows a curved path downwards before it hits the water. If its initial projection is horizontal:

- its path through the air becomes steeper and steeper as it drops
- the faster it is projected, the further away it will fall into the sea
- the time taken for it to fall into the sea does not depend on how fast it is projected.

Suppose two balls are released at the same time above a level floor such that one ball drops vertically and the other is projected horizontally. Which one hits the floor first? In fact, they both hit the floor simultaneously. Try it!

Why should the two balls in Figure 2 hit the ground at the same time? They are both pulled to the ground by the force of gravity, which gives each ball a downward acceleration g. The ball that is projected horizontally experiences the same downward acceleration as the other ball. This downward acceleration does not affect the horizontal motion of the ball projected horizontally – only the vertical motion is affected.

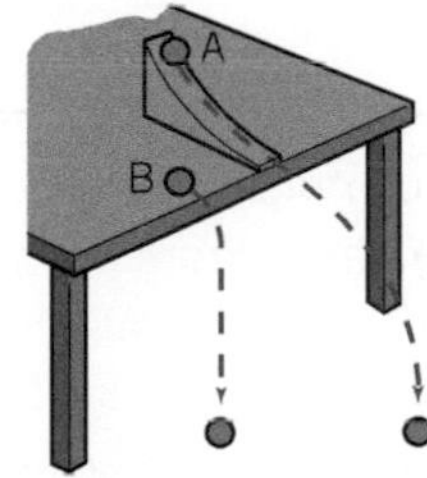

▲ **Figure 2** *Testing horizontal projection*

Investigating horizontal projection

A stroboscope and a camera with a slow shutter (or a video camera) may be used to record the motion of a projectile (see Topic 2.4). Figure 3 shows a multiflash photograph of two balls A and B released at the same time. A was released from rest and dropped vertically. B was given an initial horizontal projection so it followed a curved path. The stroboscope flashed at a constant rate so images of both balls were recorded at the same time.

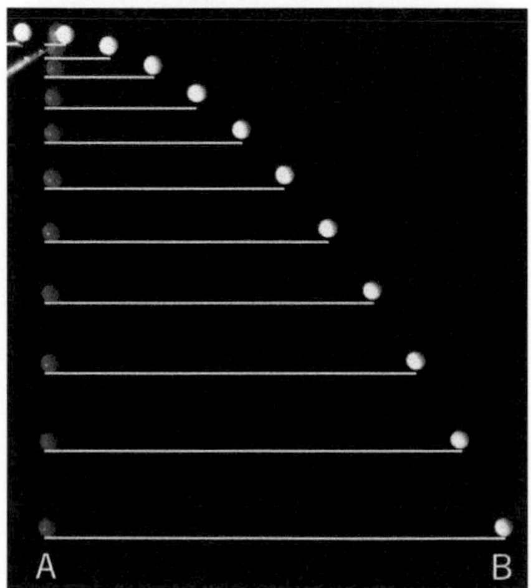

▲ **Figure 3** *Multiflash photo of two falling balls*

Summary questions

$g = 9.8\,\text{m s}^{-2}$

1 An object is released, 50 m above the ground, from a hot air balloon, which is descending vertically at a speed of $4.0\,\text{m s}^{-1}$. Calculate:
   a the velocity of the object at the ground
   b the duration of descent
   c the height of the balloon above the ground when the object hits the ground.

2 An object is projected horizontally at a speed of $16\,\text{m s}^{-1}$ into the sea from a cliff top of height 45.0 m. Calculate:
   a how long it takes to reach the sea
   b how far it travels horizontally
   c its impact velocity.

3 A dart is thrown horizontally along a line passing through the centre of a dartboard, 2.3 m away from the point at which the dart was released. The dart hits the dartboard at a point 0.19 m below the centre. Calculate:
   a the time of flight of the dart
   b its horizontal speed of projection.

4 A parcel is released from an aircraft travelling horizontally at a speed of $120\,\text{m s}^{-1}$ above level ground. The parcel hits the ground 8.5 s later. Calculate:
   a the height of the aircraft above the ground
   b the horizontal distance travelled in this time by **i** the parcel, **ii** the aircraft
   c the speed of impact of the parcel at the ground.

- The **horizontal position** of B changes by equal distances between successive flashes. This shows that the horizontal component of B's velocity is constant.
- The **vertical position** of A and B changes at the same rate. At any instant, A is at the same level as B. This shows that A and B have the same vertical component of velocity at any instant.

The projectile path of a ball projected horizontally

An object projected horizontally falls in an arc towards the ground as shown in Figure 4. If its initial velocity is U, then at time t after projection:

- The horizontal component of its displacement,

$$x = Ut$$

(**because** it moves horizontally at a constant speed).

- The vertical component of its displacement,

$$y = \frac{1}{2}gt^2$$

(**because** it has no vertical component of its initial velocity).

- Its velocity has a horizontal component $v_x = U$ and a vertical component $v_y = -gt$.

Note that, at time t, the speed $= \left(v_x^2 + v_y^2\right)^{\frac{1}{2}}$

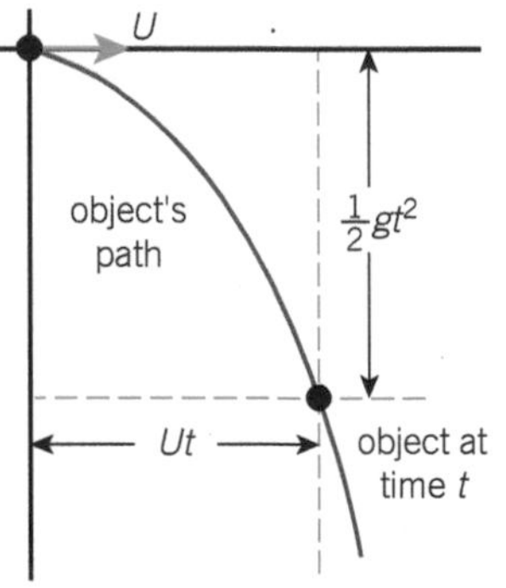

▲ **Figure 4** *Horizontal projection*

Worked example

$g = 9.8\,\text{m s}^{-2}$

An object is projected horizontally at a speed of $15\,\text{m s}^{-1}$ from the top of a tower of height 35.0 m. Calculate:

a how long it takes to fall to the ground
b how far it travels horizontally
c its speed just before it hits the ground.

Solution

a $y = -35\,\text{m}$, $a = -9.8\,\text{m s}^{-2}$ (– for downwards)

$$y = \frac{1}{2}gt^2$$

$$t^2 = \frac{2y}{g} = 2 \times \frac{-35}{-9.8} = 7.14\,\text{s}^2$$

$t = 2.67\,\text{s} = 2.7\,\text{s}$ to 2 significant figures

b $U = 15\,\text{m s}^{-1}$, $t = 2.67\,\text{s}$

$x = Ut = 15 \times 2.67 = 40\,\text{m}$

c Just before impact, $v_x = U = 15\,\text{m s}^{-1}$ and

$v_y = -gt = -9.8 \times 2.67 = -26.2\,\text{m s}^{-1}$

Therefore speed just before impact

$v = \left(v_x^2 + v_y^2\right)^{\frac{1}{2}} = (15^2 + 26.2^2)^{\frac{1}{2}} = 30.2\,\text{m s}^{-1} = 30\,\text{m s}^{-1}$ to 2 significant figures

Remember that the horizontal and vertical motions of a projectile take the same amount of time.

2.8 Projectile motion 2

Learning objectives:

- → Describe where else we meet projectile motion.
- → Describe what would happen if we could switch gravity off.
- → Describe how projectile motion is affected by the air it passes through.

Specification reference: 3.2.4

Projectile-like motion

Any form of motion where an object experiences a constant acceleration in a different direction to its velocity will be like projectile motion. For example:

- The path of a ball rolling across an inclined board will be a projectile path (Figure 1). The object is projected across the top of the board from the side. Its path curves down the board and is **parabolic**. This is because the object is subjected to constant acceleration acting down the board, and its initial velocity is across the board.
- The path of a beam of electrons directed between two oppositely charged parallel plates is also parabolic (Figure 2). Each electron in the beam is acted on by a constant force towards the positive plate, because the charge of an electron is negative. Therefore, each electron experiences a constant acceleration towards the positive plate. Its path is parabolic because its motion parallel to the plates is at **zero** acceleration whereas its motion perpendicular to the plates is at constant (non-zero) acceleration.

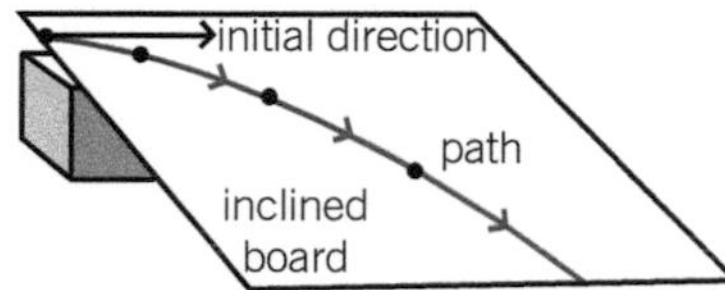

▲ **Figure 1** *Using an inclined board*

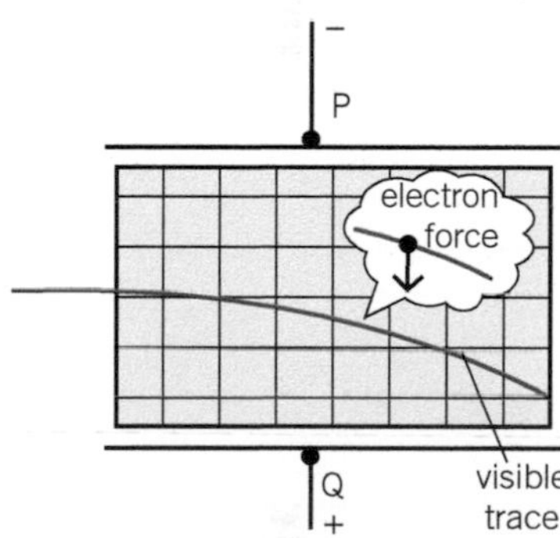

▲ **Figure 2** *An electron beam on a parabolic path*

The effects of air resistance

A projectile moving through air experiences a force that drags on it because of the resistance of the air it passes through. This **drag force** is partly caused by friction between the layers of air near the projectile's surface where the air flows over the surface. The drag force:

- acts in the opposite direction to the direction of motion of the projectile, and it increases as the projectile's speed increases
- has a horizontal component that reduces both the horizontal speed of the projectile and its range
- reduces the maximum height of the projectile if its initial direction is above the horizontal and makes its descent steeper than its ascent.

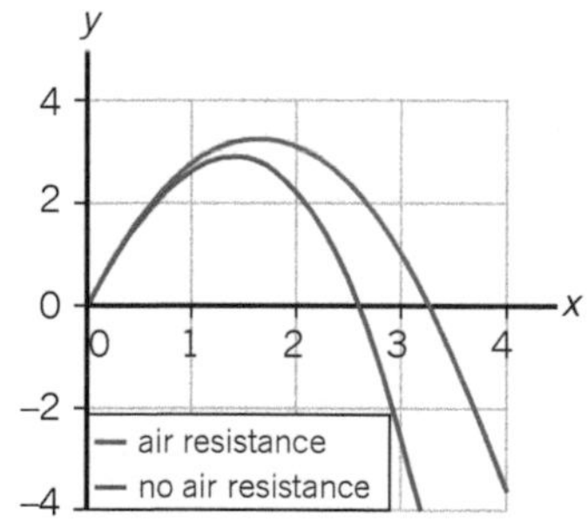

▲ **Figure 3** *Projectile paths*

Note:

The shape of the projectile is important because it affects the drag force and may also cause a **lift force** in the same way as the cross-sectional shape of an aircraft wing creates a lift force. This happens if the shape of the projectile causes the air to flow faster over the top of the object than underneath it. As a result, the pressure of the air on the top surface is less than that on the bottom surface. This pressure difference causes a lift force on the object.

A spinning ball also experiences a force due to the same effect. However, this force can be downwards, upwards, or sideways, depending on how the ball is made to spin.

Imagination at work

Imagine if gravity could be switched off, with air resistance negligible. An object projected into the air would follow a straight line with no change in its velocity (Figure 4). Suppose it had been launched at speed U in a direction at angle θ above the horizontal. Its initial velocity would therefore have a horizontal component $U\cos\theta$ and a vertical component $U\sin\theta$.

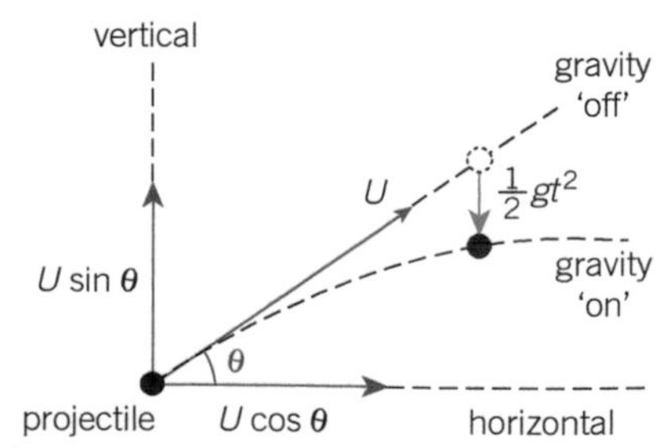

▲ **Figure 4** *Imagination at work*

Without gravity, its velocity would be unchanged, so its displacement at time t after projection would have

- a horizontal component $x = Ut \cos\theta$
- a vertical component $y = Ut \sin\theta$.

Let's now consider the difference made to its displacement and velocity if we bring back gravity. At time t after its launch,

- its vertical position with gravity on would be $\frac{1}{2}gt^2$ lower
- its vertical component of velocity would be changed by $-gt$ from its initial value $U \sin\theta$.

Hence
$$y = Ut \sin\theta - \frac{1}{2}gt^2$$
$$v_y = U \sin\theta - gt$$

Study tip

Projections below the horizontal can be analysed by assigning a negative value of θ. The speed v at time t is given by $v = (v_x^2 + v_y^2)^{\frac{1}{2}}$.

Worked example

An arrow is fired at a speed of $48\,\text{m}\,\text{s}^{-1}$ at an angle of 30° above the horizontal from a height of 1.5 m above a level field. The arrow hits a target 0.25 s later.

Calculate the height of the target above the field gained by the arrow. Assume air resistance is negligible.

Solution

At time $t = 0.25$ s, vertical displacement $y = Ut \sin\theta - \frac{1}{2}gt^2$.

Therefore, $y = (48 \times 0.25 \times \sin 30°) - (0.5 \times 9.8 \times 0.25^2) = 5.7$ m.

The maximum height = 5.7 + 1.5 = 7.2 m.

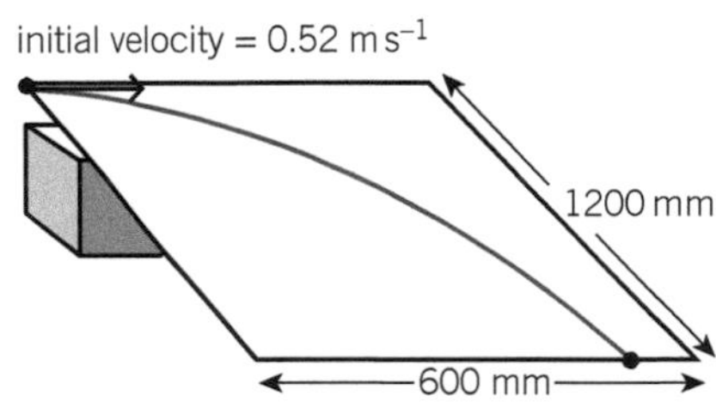

▲ Figure 5

Summary questions

$g = 9.8\,\text{m}\,\text{s}^{-2}$

1 A ball is projected horizontally at a speed of $0.52\,\text{m}\,\text{s}^{-1}$ across the top of an inclined board of width 600 mm and length 1200 mm. It reaches the bottom of the board 0.9 s later (Figure 5). Calculate:

  a the distance travelled by the ball across the board

  b its acceleration on the board

  c its speed at the bottom of the board.

2 An arrow hits the ground after being fired horizontally at a speed of $25\,\text{m}\,\text{s}^{-1}$ from the top of a tower 20 m above the ground. Calculate:

  a how long the arrow takes to fall to the ground

  b the distance travelled horizontally by the arrow.

3 An object is released from a cable car travelling at a speed of $4.6\,\text{m}\,\text{s}^{-1}$ in a direction 40° above the horizontal (Figure 6).

  a Calculate the horizontal and vertical components of velocity of the object at the instant it is released.

  b The object takes 5.8 s to fall to the ground. Calculate i the distance fallen by the object from the point of release, ii the horizontal distance travelled by the object from the point of release to where it hits the ground.

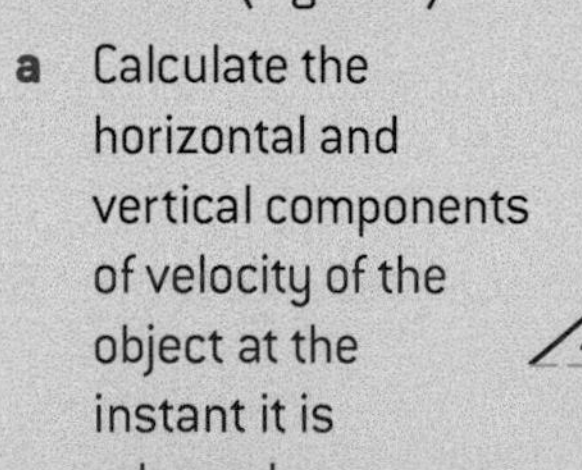

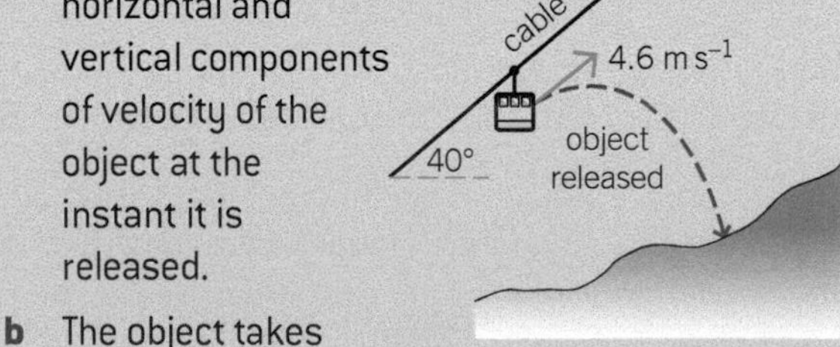

▲ Figure 6

Practice questions: Chapter 2

1 A car accelerates uniformly from rest to a speed of $100\,\mathrm{km\,h^{-1}}$ in 5.8 s.

(a) Calculate the magnitude of the acceleration of the car in $\mathrm{m\,s^{-2}}$. *(3 marks)*

(b) Calculate the distance travelled by the car while accelerating. *(2 marks)*

AQA, 2005

2 **Figure 1** shows how the velocity of a toy train moving in a straight line varies over a period of time.

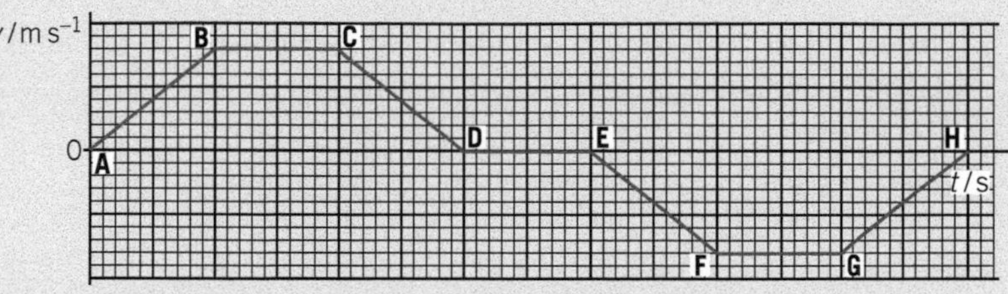

▲ **Figure 1**

(a) Describe the motion of the train in the following regions of the graph:
AB BC CD DE EF *(5 marks)*

(b) What feature of the graph represents the displacement of the train? *(1 mark)*

(c) Explain, with reference to the graph, why the distance travelled by the train is different from its displacement. *(2 marks)*

AQA, 2002

3 A vehicle accelerates uniformly from a speed of $4.0\,\mathrm{m\,s^{-1}}$ to a speed of $12\,\mathrm{m\,s^{-1}}$ in 6.0 s.

(a) Calculate the vehicle's acceleration. *(2 marks)*

(b) Sketch a graph of speed against time for the vehicle covering the 6.0 s period in which it accelerates. *(2 marks)*

(c) Calculate the distance travelled by the vehicle during its 6.0 s period of acceleration. *(2 marks)*

AQA, 2002

4 A supertanker, cruising at an initial speed of $4.5\,\mathrm{m\,s^{-1}}$, takes one hour to come to rest.

(a) Assuming that the force slowing the tanker down is constant, calculate
(i) the deceleration of the tanker
(ii) the distance travelled by the tanker while slowing to a stop. *(4 marks)*

(b) Sketch a distance–time graph representing the motion of the tanker until it stops. *(2 marks)*

(c) Explain the shape of the graph you have sketched in part (b). *(2 marks)*

AQA, 2006

5 **(a)** A cheetah accelerating uniformly from rest reaches a speed of $29\,\mathrm{m\,s^{-1}}$ in 2.0 s and then maintains this speed for 15 s. Calculate:
(i) its acceleration
(ii) the distance it travels while accelerating
(iii) the distance it travels while it is moving at constant speed. *(4 marks)*

(b) The cheetah and an antelope are both at rest and 100 m apart. The cheetah starts to chase the antelope. The antelope takes 0.50 s to react. It then accelerates uniformly for 2.0 s to a speed of $25\,\mathrm{m\,s^{-1}}$ and then maintains this speed. **Figure 2** shows the speed–time graph for the cheetah.

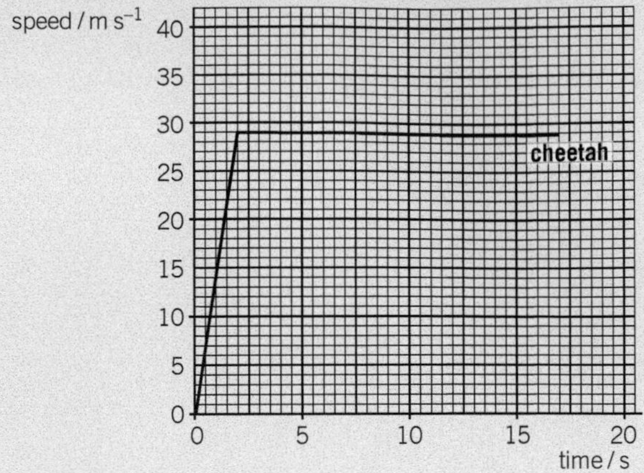

▲ **Figure 2**

(i) Using the same axes plot the speed–time graph for the antelope during the chase.
(ii) Calculate the distance covered by the antelope in the 17 s after the cheetah started to run.
(iii) How far apart are the cheetah and the antelope after 17 s? *(6 marks)*

AQA, 2007

6 The following data were obtained when two students performed an experiment to determine the acceleration of free fall. One student released a lump of lead the size of a tennis ball from a window in a tall building and the other measured the time for it to reach the ground.

distance fallen by lump of lead = 35 m
time to reach the ground = 2.7 s

(a) Calculate a value for the acceleration of free fall, g, from these observations. *(2 marks)*
(b) State and explain the effect on the value of g obtained by the students if a tennis ball were used instead of the lump of lead. *(3 marks)*
(c) The graph in **Figure 3** shows how the velocity changes with time for the lump of lead from the time of release until it hits the ground. Sketch on the same axes a graph to show how the velocity would change with time if a tennis ball were used by the students instead of a lump of lead.

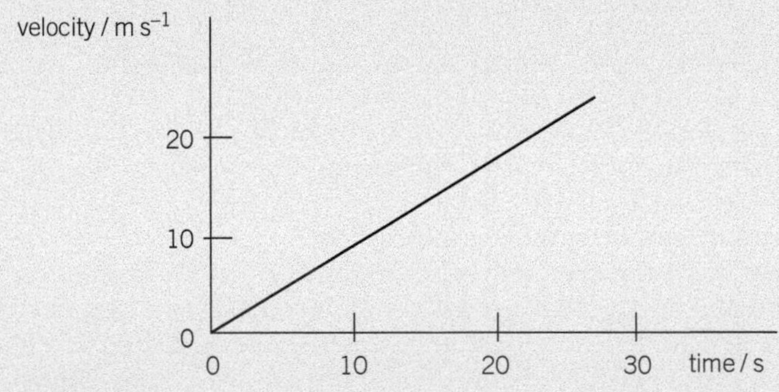

▲ **Figure 3**

(2 marks)

AQA, 2004

7 A man jumps from a plane that is travelling horizontally at a speed of $70\,\mathrm{m\,s^{-1}}$. If air resistance can be ignored, and before he opens his parachute, determine

(a) his horizontal velocity 2.0 s after jumping
(b) his vertical velocity 2.0 s after jumping
(c) the magnitude and direction of his resultant velocity 2.0 s after jumping. *(5 marks)*

AQA, 2003

Answers to the Practice Questions and Section Questions are available at www.oxfordsecondary.com/oxfordaqaexams-alevel-physics

3 Newton's laws of motion

3.1 Force and acceleration

Learning objectives:

- → Describe what effect a resultant force produces.
- → Describe what would happen to a body that was already in motion if there was no resultant force acting on it.
- → Explain how weight is different from mass.

Specification reference: 3.2.5

Motion without force

Motorists on icy roads in winter need to be very careful, because the tyres of a car have little or no grip on the ice. Moving from a standstill on ice is very difficult. Stopping on ice is almost impossible, as a car moving on ice will slide when the brakes are applied. **Friction** is a hidden force that we don't usually think about until it is absent!

If you have ever pushed a heavy crate across a rough concrete floor, you will know all about friction. The push force is opposed by friction, and as soon as you stop pushing, friction stops the crate moving. If the crate had been pushed onto a patch of ice, it would have moved across the ice without any further push needed.

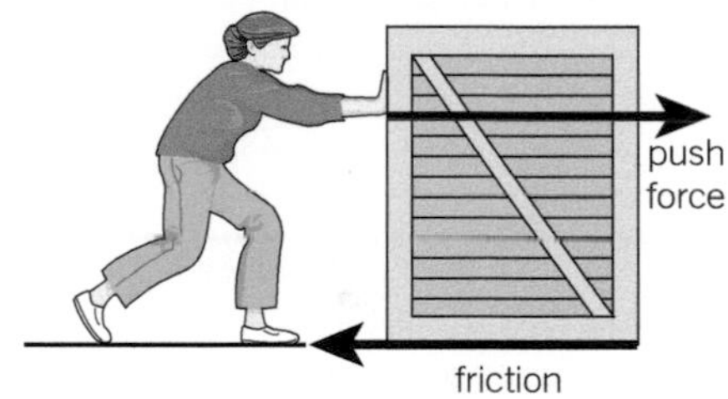

▲ **Figure 1** *Overcoming friction*

Figure 2 shows an air track which allows motion to be observed in the absence of friction. The glider on the air track floats on a cushion of air. Provided the track is level, the glider moves at constant velocity along the track because friction is absent.

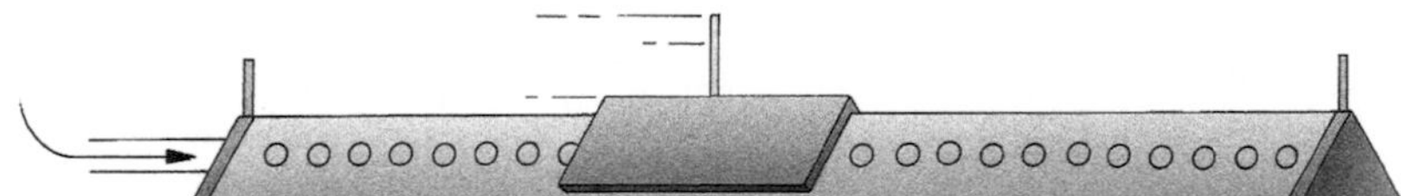
▲ **Figure 2** *The linear air track*

Newton's first law of motion

Objects either stay at rest or move with constant velocity unless acted on by a force.

Sir Isaac Newton was the first person to realise that a moving object remains in uniform motion unless acted on by a force. He recognised that when an object is acted on by a resultant force, the result is to change the object's velocity. In other words, an object moving at constant velocity is either

- acted on by no forces, or
- the forces acting on it are balanced (so the resultant force is zero).

Investigating force and motion

How does the velocity of an object change if it is acted on by a constant force? Figure 3 shows how this can be investigated, using a dynamics trolley and a motion sensor connected to a computer. The computer is used to process a signal from the motion sensor and to display a graph showing how the velocity of the trolley changes with time.

The trolley is pulled along a sloping runway using one or more elastic bands stretched to the same length. The runway is sloped just enough to compensate for friction. To test for the correct slope, the trolley should move down the runway at constant speed after being given a brief push.

Hint

The acceleration is always in the same direction as the resultant force. For example, a freely moving projectile in motion experiences a force vertically downwards due to gravity. Its acceleration is therefore vertically downwards, no matter what its direction of motion is.

Study tip

The mass m must be in kg and a in m s^{-2} when calculating a force in N.

Maths link

Why does a heavy object fall at the same rate as a lighter object?

As outlined in Topic 2.4, Free fall, Galileo proved that objects fall at the same rate, regardless of their weight. He also reasoned that if two objects joined by a string were released and they initially fell at different rates, the faster one would be slowed down by the slower one, which would be speeded up by the faster one until they fell at the same rate. Many years later, Newton explained their identical falling motion because, for an object of mass m in free fall,

$$\text{acceleration} = \frac{\text{force of gravity } mg}{\text{mass } m} = g$$

which is independent of m

Synoptic link

You learnt about Galileo in Topic 2.4, Free fall.

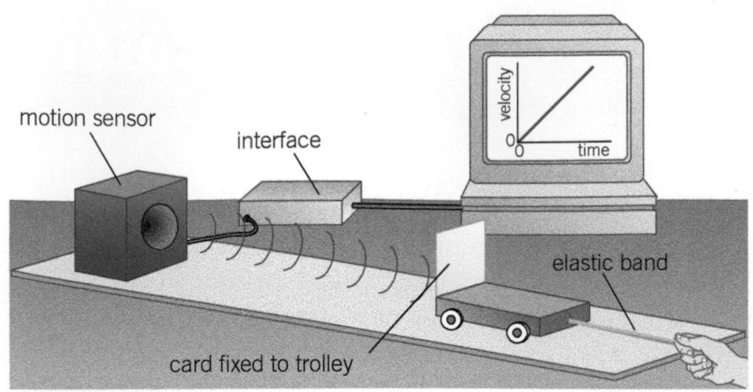

▲ **Figure 3** *Investigating force and motion*

As a result of pulling the trolley with a constant force, the velocity–time graph should show that the velocity increases at a constant rate. The acceleration of the trolley is therefore constant and can be measured from the velocity–time graph. Table 1 below shows typical measurements using different amounts of force (one, two, or three elastic bands in parallel stretched to the same length each time) and different amounts of mass (i.e., a single or double trolley).

▼ **Table 1**

Force (no. of elastic bands)	1	2	3	1	2	3
Mass (no. of trolleys)	1	1	1	2	2	2
Acceleration m s^{-2}	12	24	36	6	12	18
Mass × acceleration	12	24	36	12	24	36

The results in the table show that the force is proportional to the mass × the acceleration. In other words, if a resultant force F acts on an object of mass m, the object undergoes acceleration a such that

$$F \text{ is proportional to } ma$$

By defining the unit of force, the *newton*, as the amount of force that will give an object of mass 1 kg an acceleration of 1 m s^{-2}, the above proportionality statement can be expressed as an equation

$$F = ma$$

where F = resultant force (in N), m = mass (in kg), a = acceleration (in m s^{-2}).

This equation is known as **Newton's second law** for constant mass.

Worked example

A vehicle of mass 600 kg accelerates uniformly from rest to a velocity of 8.0 m s^{-1} in 20 s. Calculate the force needed to produce this acceleration.

Solution

Acceleration $a = \frac{v - u}{t} = \frac{8.0 - 0}{20} = 0.4\,\text{m s}^{-2}$

Force $F = ma = 600 \times 0.4 = 240\,\text{N}$

Weight

- The acceleration of a falling object acted on by gravity only is g. Because the force of gravity on the object is the only force acting on it, its **weight** (in newtons) $W = mg$, where m = the mass of the object (in kg).
- When an object is in equilibrium, the support force on it is equal and opposite to its weight. Therefore, an object placed on a weighing balance (e.g., a newtonmeter or a top pan balance) exerts a force on the balance equal to the weight of the object. Thus the balance measures the weight of the object. See Figure 4.
- g is also referred to as the **gravitational field strength** at a given position, as it is the force of gravity per unit mass on a small object at that position. So the gravitational field strength at the Earth's surface is $9.81\,\mathrm{N\,kg^{-1}}$. Note that the weight of a fixed mass depends on its location. For example, the weight of a 1 kg object is 9.81 N on the Earth's surface and 1.62 N on the Moon's surface.
- The mass of an object is a measure of its **inertia**, which is its resistance to change of motion. More force is needed to give an object a certain acceleration than to give an object with less mass the same acceleration. Figure 5 shows an entertaining demonstration of inertia. When the card is flicked, the coin drops into the glass because the force of friction on it due to the moving card is too small to shift it sideways.
- The scale of a top pan balance is usually calibrated for convenience in grams or kilograms.

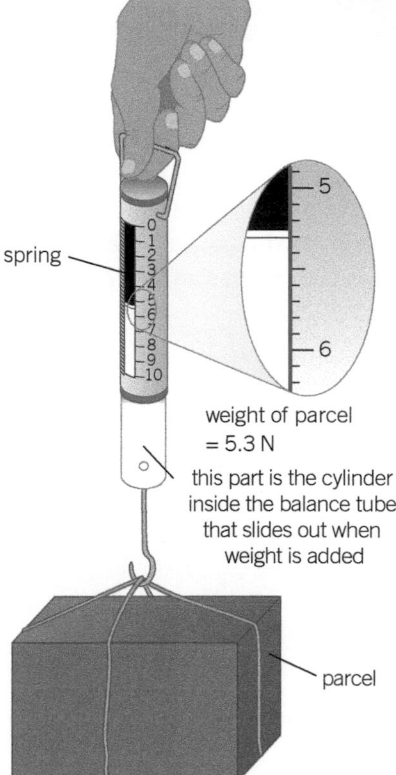

▲ **Figure 4** *Using a newtonmeter to weigh an object*

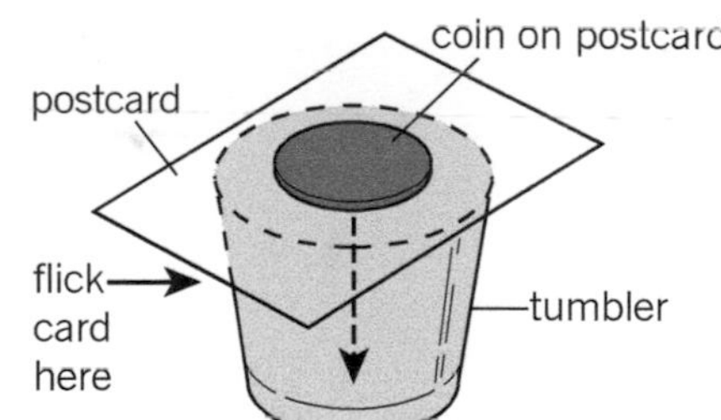

▲ **Figure 5** *An inertia trick*

Summary questions

$g = 9.81\,\mathrm{N\,kg^{-1}}$

1 A car of mass 800 kg accelerates uniformly along a straight line from rest to a speed of $12\,\mathrm{m\,s^{-1}}$ in 50 s. Calculate:

a the acceleration of the car

b the force on the car that produced this acceleration

c the ratio of the accelerating force to the weight of the car.

2 An aeroplane of mass 5000 kg lands on a runway at a speed of $60\,\mathrm{m\,s^{-1}}$ and stops 25 s later. Calculate:

a the deceleration of the aeroplane

b the braking force on the aircraft.

3 **a** A vehicle of mass 1200 kg on a level road accelerates from rest to a speed of $6.0\,\mathrm{m\,s^{-1}}$ in 20 s, without change of direction. Calculate the force that accelerated the car.

b The vehicle in **a** is fitted with a trailer of mass 200 kg. Calculate the time taken to reach a speed of $6.0\,\mathrm{m\,s^{-1}}$ from rest for the same force as in **a**.

4 A bullet of mass 0.002 kg travelling at a speed of $120\,\mathrm{m\,s^{-1}}$ hits a tree and penetrates a distance of 55 mm into the tree. Calculate:

a the deceleration of the bullet

b the impact force on the bullet.

3.2 Using $F = ma$

Learning objectives:

- → Apply $F = ma$ when the forces on an object are in opposite directions.
- → Explain why you experience less support as an ascending lift stops.
- → Describe any situations in which $F = ma$ cannot be applied.

Specification reference: 3.2.5

Two forces in opposite directions

When an object is acted on by two unequal forces acting in opposite directions, the object accelerates in the direction of the larger force. If the forces are F_1 and F_2, where $F_1 > F_2$

$$\textbf{resultant force, } F_1 - F_2 = ma$$

where m is the mass of the object and a is its acceleration, which is in the same direction as F_1.

If the object is on a horizontal surface and F_1 and F_2 are horizontal and in opposite directions, the above equation still applies. The support force on the object is equal and opposite to its weight.

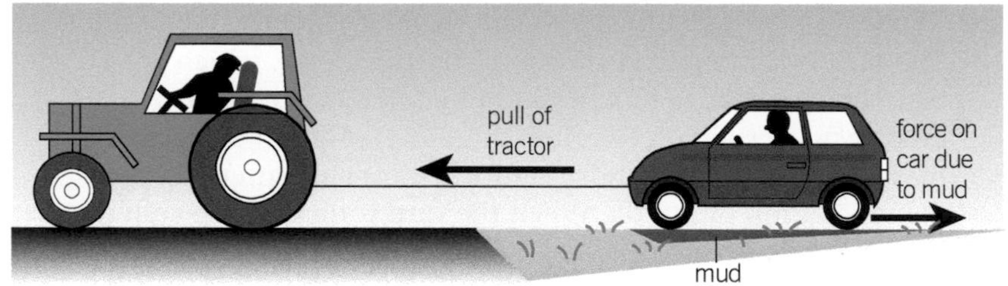

▲ **Figure 1** *Unbalanced forces*

Some examples are given below where two forces act in different directions on an object.

Towing a trailer

Consider the example of a car of mass M fitted with a trailer of mass m on a level road. When the car and the trailer accelerate, the car pulls the trailer forward and the trailer holds the car back. Assume air resistance is negligible.

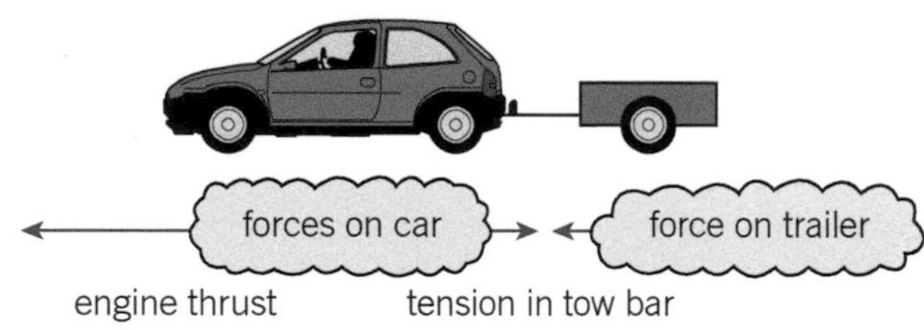

▲ **Figure 2** *Car and trailer*

- The car is subjected to a driving force F pushing it forwards (from its engine thrust) and the tension T in the tow bar holding it back.

  Therefore the resultant force on the car $= F - T = Ma$
- The force on the trailer is due to the tension T in the tow bar pulling it forwards.

  Therefore $T = ma$.

Combining the two equations gives $\boldsymbol{F = Ma + ma = (M + m)a}$

Study tip

Remember that it is the force of friction between the road and the tyres of the drive wheels of the car that enables the car to move forward. If friction was negligible, the car would not be able to move. See Topic 3.4 for more about friction.

Rocket problems

If T is the thrust of the rocket engine when its mass is m and the rocket is moving upwards, its acceleration a is given by $\boldsymbol{T - mg = ma}$.

Rocket thrust $\boldsymbol{T = mg + ma}$

The rocket thrust must therefore overcome the weight of the rocket for the rocket to take off (Figure 3).

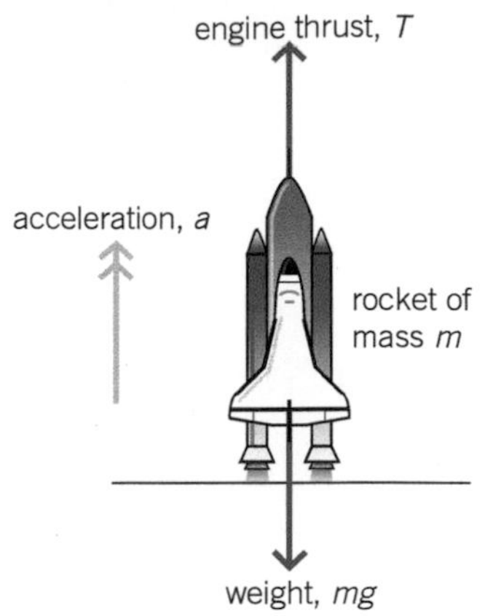

▲ **Figure 3** *Rocket launch*

Lift problems

Using 'upwards is positive' gives the resultant force on the lift as $T - mg$, where T is the tension in the lift cable and m is the total mass of the lift and its occupants (Figure 4).

$\boldsymbol{T - mg = ma}$, where a = acceleration.

a If the lift is moving at constant velocity, then $a = 0$ so $T = mg$ (tension = weight).

b If the lift is moving up and accelerating, then $a > 0$ so $T = mg + ma > mg$.

c If the lift is moving up and decelerating, then $a < 0$ so $T = mg + ma < mg$.

d If the lift is moving down and accelerating, then $a < 0$ (velocity and acceleration are both downwards, therefore negative) so $T = mg + ma < mg$.

e If the lift is moving down and decelerating, then $a > 0$ (velocity downwards and acceleration upwards, therefore positive) so $T = mg + ma > mg$.

The tension in the cable is less than the weight if

- the lift is moving up and decelerating (velocity > 0 and acceleration < 0)
- the lift is moving down and accelerating (velocity < 0 and acceleration < 0).

The tension in the cable is greater than the weight if

- the lift is moving up and accelerating (velocity > 0 and acceleration > 0)
- the lift is moving down and decelerating (velocity < 0 and acceleration > 0).

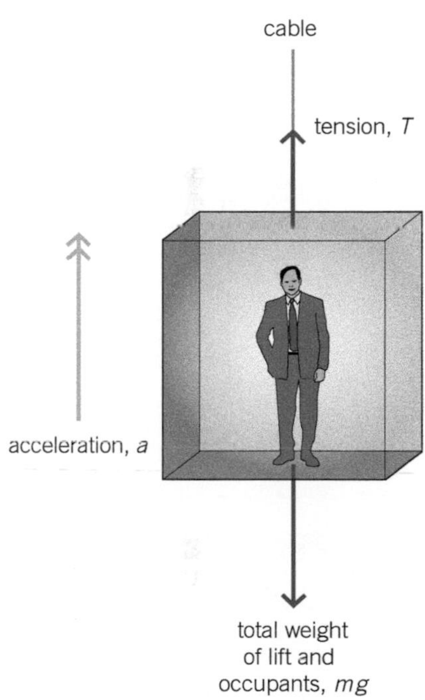

▲ **Figure 4** *In a lift*

Worked example

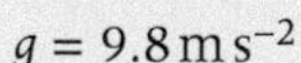

$g = 9.8\,\mathrm{m\,s^{-2}}$

A lift of total mass 650 kg moving downwards decelerates at $1.5\,\mathrm{m\,s^{-2}}$ and stops. Calculate the tension in the lift cable during the deceleration.

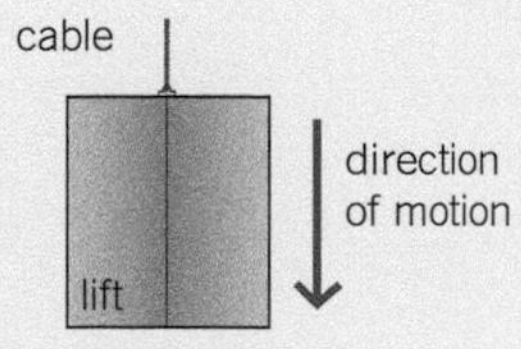

▲ **Figure 5**

Solution

The lift is moving down so its velocity $v < 0$. Since it decelerates, its acceleration a is in the opposite direction to its velocity, so $a > 0$.

Therefore, inserting $a = +1.5\,\mathrm{m\,s^{-2}}$ in the equation $T - mg = ma$ gives

$T = mg + ma = (650 \times 9.8) + (650 \times 1.5) = 7300\,\mathrm{N}$

Hint

Downward acceleration = upward deceleration.

Study tip

Identify the separate forces acting, then work out the resultant force – show these steps by clear working (usually starting in algebra, such as $T - mg = ma$).

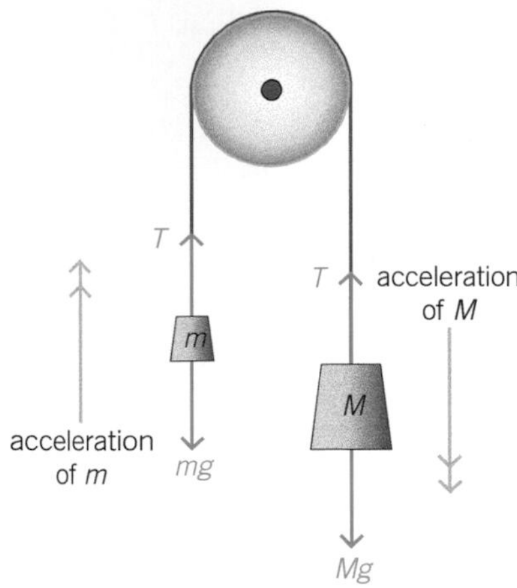

▲ **Figure 6** *Pulley problems*

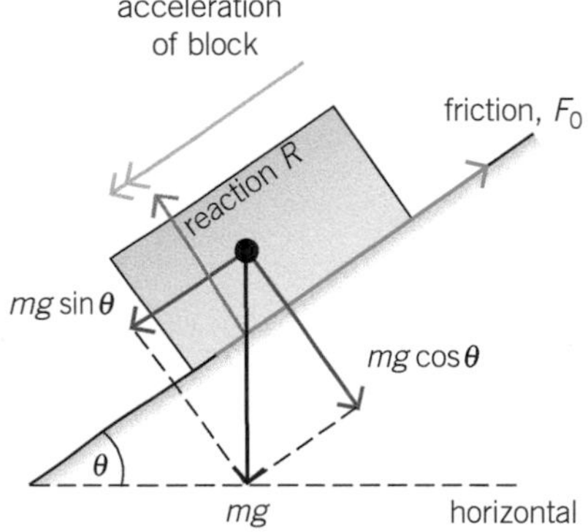

▲ **Figure 7** *Sliding down a slope*

Further $F = ma$ problems

Pulley problems

Consider two masses M and m (where $M > m$) attached to a thread hung over a frictionless pulley, as in Figure 6. When released, mass M accelerates downwards and mass m accelerates upwards. If a is the acceleration and T is the tension in the thread, then

- the resultant force on mass M, $Mg - T = Ma$
- the resultant force on mass m, $T - mg = ma$.

Therefore, adding the two equations gives

$$Mg - mg = (M + m)a$$

Sliding down a slope

Consider a block of mass m sliding down a slope, as in Figure 7.

The component of the block's weight down the slope $= mg\sin\theta$.

If the force of friction on the block $= F_0$, then the resultant force on the block $= mg\sin\theta - F_0$.

Therefore $\boldsymbol{mg\sin\theta - F_0 = ma}$, where a is the acceleration of the block.

Note:

With the addition of an engine force F_E, the above equation can be applied to a vehicle on a downhill slope of constant gradient. Thus the acceleration is given by $F_E + mg\sin\theta - F_0 = ma$, where F_0 is the combined sum of the air resistance and the braking force.

Summary questions

$g = 9.8\,\mathrm{m\,s^{-2}}$

1 A rocket of mass 550 kg blasts vertically from the launch pad at an acceleration of $4.2\,\mathrm{m\,s^{-2}}$. Calculate:

a the weight of the rocket

b the thrust of the rocket engines.

2 A car of mass 1400 kg, pulling a trailer of mass 400 kg, accelerates from rest to a speed of $9.0\,\mathrm{m\,s^{-1}}$ in a time of 60 s on a level road. Assuming air resistance is negligible, calculate:

a the tension in the tow bar

b the engine force.

3 A lift and its occupants have a total mass of 1200 kg. Calculate the tension in the lift cable when the lift is:

a stationary

b ascending at constant speed

c ascending at a constant acceleration of $0.4\,\mathrm{m\,s^{-2}}$

d descending at a constant deceleration of $0.4\,\mathrm{m\,s^{-2}}$.

4 A brick of mass 3.2 kg on a sloping flat roof, at 30° to the horizontal, slides at constant acceleration 2.0 m down the roof in 2.0 s from rest. Calculate:

a the acceleration of the brick

b the frictional force on the brick due to the roof.

3.3 Terminal speed

Any object moving through a fluid experiences a force that drags on it due to the fluid. The **drag force** depends on

- the shape of the object
- its speed
- the viscosity of the fluid, which is a measure of how easily the fluid flows past a surface.

The faster an object travels in a fluid, the greater the drag force on it.

Learning objectives:

→ Explain why the speed reaches a maximum when a driving force is still acting.

→ Explain what we mean by a drag force.

→ Explain what determines the terminal speed of a falling object or a vehicle.

Specification reference: 3.2.5

Motion of an object falling in a fluid

The speed of an object released from rest in a fluid increases as it falls, so the drag force on it due to the fluid increases. The resultant force on the object is the difference between the force of gravity on it (its weight) and the drag force. As the drag force increases, the resultant force decreases, so the acceleration becomes less as it falls. If it continues falling, it attains **terminal speed**, when the drag force on it is equal and opposite to its weight. Its acceleration is then zero and its speed remains constant as it falls (Figure 1).

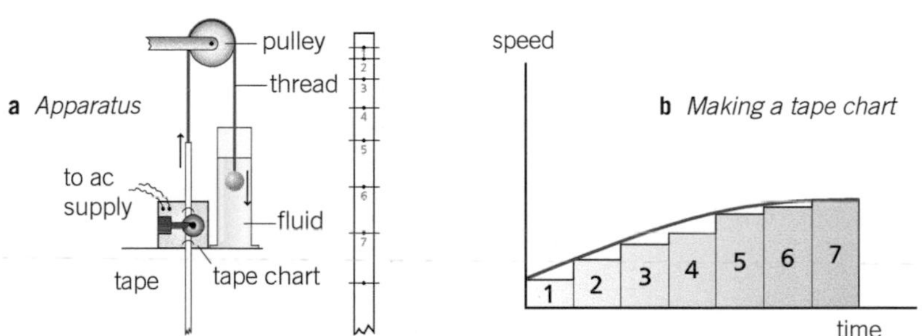

▲ **Figure 2** *Investigating the motion of an object falling in a fluid*

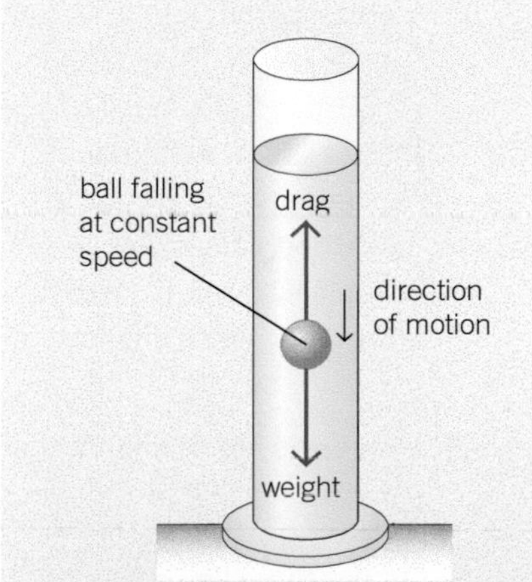

a *Falling in a fluid*

b *Skydiving*

▲ **Figure 1** *At terminal speed*

Figure 2a shows how to investigate the motion of an object falling in a fluid. When the object is released, the thread attached to the object pulls a tape through a tickertimer, which prints dots at a constant rate on the tape. The spacing between successive dots is a measure of the speed of the object as the dots are printed on the tape at a constant rate. A tape chart can be made from the tape to show how the speed changes with time. This is done by cutting the tape at each dot and pasting successive lengths side by side on graph paper, as shown in Figure 2b. The results show that:

- the speed increases and reaches a constant value, which is the terminal speed
- the acceleration at any instant is the gradient of the speed–time curve.

At any instant, the resultant force $F = mg - D$, where m is the mass of the object and D is the drag force.

Therefore, the acceleration of the object $= \dfrac{mg - D}{m} = g - \dfrac{D}{m}$

- The initial acceleration $= g$ because the speed is zero, and therefore the drag force is zero, at the instant the object is released.
- At the terminal speed, the potential energy of the object is transferred, as it falls, into internal energy of the fluid by the drag force.

Study tip

The magnitude of the acceleration at any instant is the gradient of the speed–time curve.

Motion of a powered vehicle

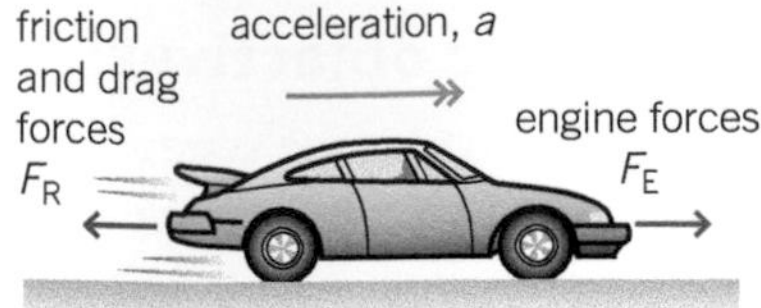

▲ **Figure 3** *Vehicle power*

The top speed of a road vehicle or an aircraft depends on its engine power and its shape. A vehicle with a streamlined shape can reach a higher top speed than a vehicle with the same engine power that is not streamlined.

For a powered vehicle of mass m moving on a level surface, if F_E represents the **motive force** (driving force or engine force) provided by the engine, the resultant force on it = $F_E - F_R$, where F_R is the **resistive force** opposing the motion of the vehicle. (F_R = the sum of the drag forces acting on the vehicle.)

Therefore its acceleration $\boldsymbol{a} = \dfrac{\boldsymbol{F_E - F_R}}{\boldsymbol{m}}$

Because the drag force increases with speed, the maximum speed (the terminal speed) of the vehicle v_{max} is reached when the resistive force becomes equal and opposite to the engine force, and $a = 0$.

Worked example

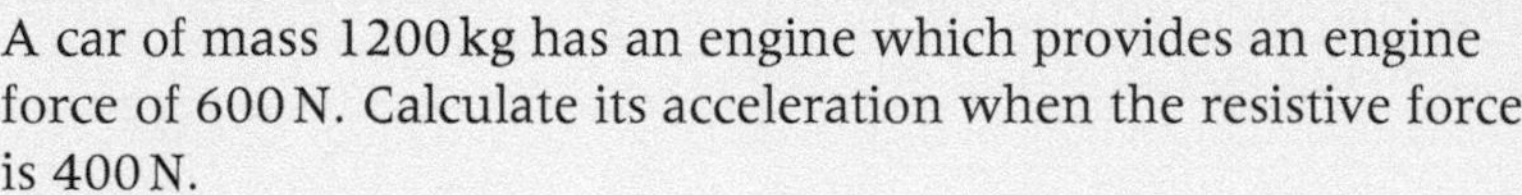

A car of mass 1200 kg has an engine which provides an engine force of 600 N. Calculate its acceleration when the resistive force is 400 N.

Solution

When the resistive force = 400 N, the resultant force = engine force − drag force = 600 − 400 = 200 N.

$$\text{Acceleration} = \frac{\text{force}}{\text{mass}} = \frac{200}{1200} = 0.17\,\text{m s}^{-2}$$

Extension

Hydrofoil physics

Some hydrofoil boats travel much faster than an ordinary boat because they have powerful jet engines that enable them to ski on the hydrofoils when the jet engines are switched on.

When the jet engine is switched on and takes over from the less powerful propeller engine, the boat speeds up and the hydrofoils are extended. The boat rides on the hydrofoils, so the drag force is reduced, as its hull is no longer in the water. At top speed, the motive force of the jet engine is equal to the drag force on the hydrofoils.

When the jet engine is switched off, the drag force on the hydrofoils reduces the speed of the boat and the hydrofoils are retracted. The speed drops further until the drag force is equal to the motive force of the propeller engine. The boat then moves at a constant speed, which is much less than its top speed with the jet engine on.

Q: Describe how the force of the jet engine compares to the drag force when the boat speeds up.

Answer: The driving force is greater than the drag force.

Summary questions

$g = 9.8\,\text{m s}^{-2}$

1. **a** A steel ball of mass 0.15 kg, released from rest in a liquid, falls a distance of 0.20 m in 5.0 s. Assuming the ball reaches terminal speed within a fraction of a second, calculate **i** its terminal speed, **ii** the drag force on it when it falls at terminal speed.
   b State and explain whether or not a smaller steel ball would fall at the same rate in the same liquid.
2. Explain why a cyclist can reach a higher top speed by crouching over the handlebars instead of sitting upright while pedalling.
3. A vehicle of mass 32 000 kg has an engine which has a maximum engine force of 4.4 kN and a top speed of $36\,\text{m s}^{-1}$ on a level road.
   Calculate **a** its maximum acceleration from rest, **b** the distance it would travel at maximum acceleration to reach a speed of $12\,\text{m s}^{-1}$ from rest.
4. Explain why a vehicle has a higher top speed on a downhill stretch of road than on a level road.

3.4 On the road

Stopping distances

Traffic accidents often happen because vehicles are being driven too fast and too close. A driver needs to maintain a safe distance between his or her own vehicle and the vehicle in front. If a vehicle suddenly brakes, the driver of the following vehicle needs to brake as well to avoid a crash.

Thinking distance is the distance travelled by a vehicle in the time it takes the driver to react. For a vehicle moving at constant speed v, the thinking distance s_1 = speed × reaction time = vt_0, where t_0 is the reaction time of the driver. The reaction time of a driver is affected by distractions, drugs, and alcohol, which is why drivers in the UK are banned from using mobile phones while driving and breathalyser tests are carried out on drivers.

Braking distance is the distance travelled by a car in the time it takes to stop safely, from when the brakes are first applied. Assuming constant deceleration, a, to zero speed from speed u, the braking distance $s_2 = \frac{u^2}{2a}$ since $u^2 = 2as_2$.

Stopping distance = thinking distance + braking distance = $ut_0 + \frac{u^2}{2a}$ where u is the speed before the brakes were applied.

Figure 1 shows how thinking distance, braking distance, and stopping distance vary with speed for a reaction time of 0.67 s and a deceleration of 6.75 m s^{-2}. Using these values for reaction time and deceleration in the above equations, Figure 1 gives the shortest stopping distances on a dry road as recommended in the Highway Code.

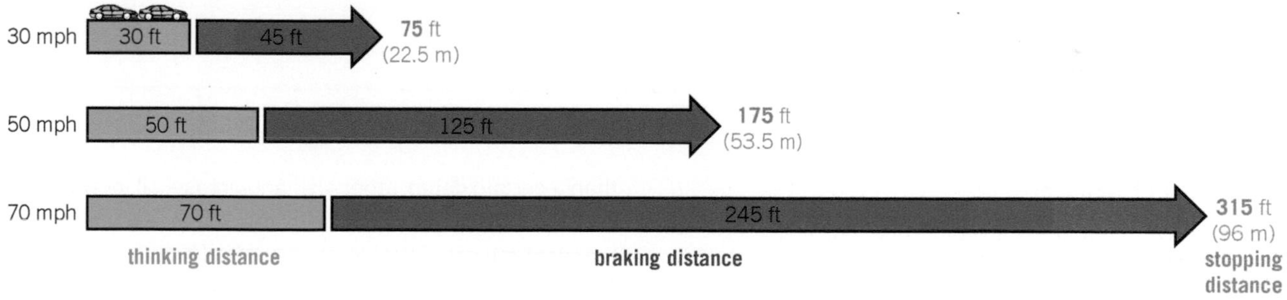

Learning objectives:

- → Describe what is happening to a vehicle's speed while a driver is reacting to a hazard ahead.
- → Distinguish between braking distance and stopping distance.
- → Describe how road conditions affect these distances.

Specification reference: 3.2.5

Speed	thinking distance	braking distance	stopping distance
30 mph	30 ft	45 ft	75 ft (22.5 m)
50 mph	50 ft	125 ft	175 ft (53.5 m)
70 mph	70 ft	245 ft	315 ft (96 m)

▲ **Figure 1** *Stopping distances*

Study tip

A car can only move forwards by pushing backwards on the road. The same applies when you take a forward step. When you push backwards on the ground, the ground exerts an equal and opposite force on you that enables you to move forward.

Extension

Skidding

On a front-wheel drive vehicle, the engine turns the front wheels via the transmission system. Friction between the tyres and the road prevents wheel spin (slipping) so the driving wheels therefore roll along the road. If the driver tries to accelerate too fast, the wheels skid. This is because there is an upper limit to the amount of friction between the tyres and the road (Figure 2).

When the brakes are applied, the wheels are slowed down by the brakes. The vehicle therefore slows down, provided the wheels do not skid. If the

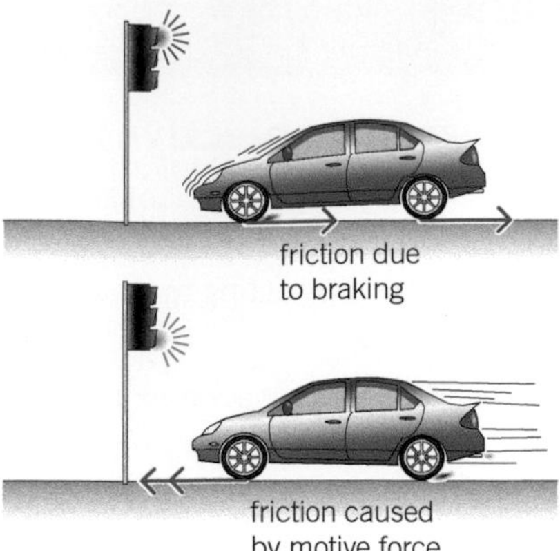

▲ **Figure 2** *Stopping and starting*

braking force is increased, the friction force between the tyres and the road increases. However, if the upper limit of friction (usually referred to as the limiting frictional force) between the tyres and the road is reached, the wheels skid. When this happens, the brakes lock and the vehicle slides uncontrollably forward. Most vehicles are now fitted with an anti-lock brake system (ABS), which consists of a speed sensor on each wheel linked to a central electronic control unit that controls hydraulic valves in the brake system. When the brakes are applied, if a wheel starts to lock, the control unit senses that the wheel is rotating much more slowly than the other wheels – as a result, the valves are activated, so the brake pressure on the wheel is reduced to stop it locking.

Q: State the direction of the frictional force of the road on the driving wheels when a car accelerates.

Answer: In the direction in which the car is moving.

Testing friction

Measure the limiting friction between the underside of a block and the surface it is on. Pull the block with an increasing force until it slides. The limiting frictional force on the block is equal to the pull force on the block just before sliding occurs. Find out how this force depends on the weight of the block.

▲ **Figure 4** *Tyre treads*

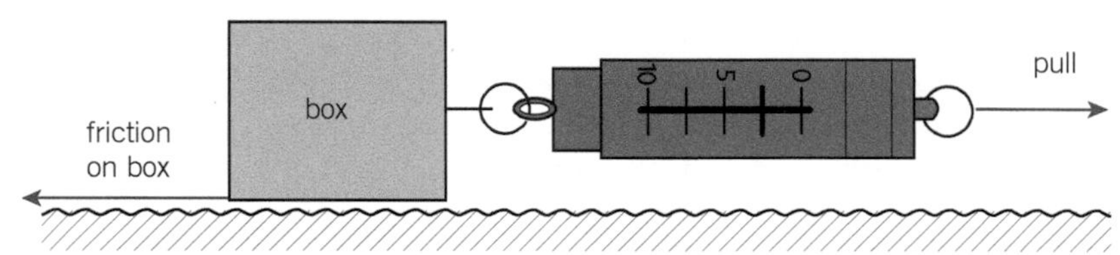

▲ **Figure 3** *Testing friction*

Application

More about braking distance

The braking distance for a vehicle depends on the speed of the vehicle at the instant the brakes are applied, on the road conditions, and on the condition of the vehicle tyres.

On a greasy or icy road, skidding is more likely because the limiting frictional force between the road and the tyres is reduced from its dry value. To stop a vehicle safely on a greasy or icy road, the brakes must be applied with less force than on a dry road, otherwise skidding will occur. Therefore the braking distance is longer than on a dry road. In fast-moving traffic, a driver must ensure there is a bigger gap to the car in front so as to be able to slow down safely if the vehicle in front slows down.

The condition of the tyres of a vehicle also affects braking distance. The tread of a tyre must not be less than a certain depth, otherwise any grease, oil, or water on the road reduces the friction between the road and the tyre very considerably. In the United Kingdom, the absolute minimum legal requirement is that every tyre of a vehicle must have a tread depth of at least 1.6 mm across three-quarters of the width of the tyre all the way round. If the pressure on a tyre is too low or too high, or the wheel is unbalanced, then the tyre will wear unevenly and will quickly become unsafe. In addition, the braking force may be reduced if the tyre pressure is too high as the tyre area in contact with the road will be reduced. Clearly, a driver must check the condition of the tyres regularly. If you are a driver or hope to become one, remember that the vehicle tyres are the only contacts between the vehicle and the road and must therefore be looked after.

Worked example

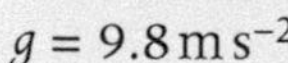

$g = 9.8\,\text{m}\,\text{s}^{-2}$

A vehicle of mass 900 kg, travelling on a level road at a speed of $15\,\text{m}\,\text{s}^{-1}$, can be brought to a standstill without skidding by a braking force equal to 0.5 × its weight. Calculate **a** the deceleration of the vehicle, **b** the braking distance.

Solution

a Weight = 900 × 9.8 = 8800 N

Braking force = 0.5 × 8800 = 4400 N

$$\text{Deceleration} = \frac{\text{braking force}}{\text{mass}} = \frac{4400}{900} = 4.9\,\text{m}\,\text{s}^{-2}$$

b $u = 15\,\text{m}\,\text{s}^{-1}$, $v = 0$, $a = -4.9\,\text{m}\,\text{s}^{-2}$

To calculate s, use $v^2 = u^2 + 2as$

$$s = \frac{-u^2}{2a} = \frac{-15^2}{2 \times -4.9} = 23\,\text{m}$$

Summary questions

$g = 9.8\,\text{m}\,\text{s}^{-2}$

1 A vehicle is travelling at a speed of $18\,\text{m}\,\text{s}^{-1}$ on a level road, when the driver sees a pedestrian stepping off the pavement into the road 45 m ahead. The driver reacts within 0.4 s and applies the brakes, causing the car to decelerate at $4.8\,\text{m}\,\text{s}^{-2}$.

- **a** Calculate **i** the thinking distance, **ii** the braking distance.
- **b** How far does the driver stop from where the pedestrian stepped into the road?

2 The braking distance of a vehicle for a speed of $18\,\text{m}\,\text{s}^{-1}$ on a dry level road is 24 m. Calculate:

- **a** the deceleration of the vehicle from this speed to a standstill over this distance
- **b** the frictional force on a vehicle of mass 1000 kg on this road as it stops.

3 **a** What is meant by the braking distance of a vehicle?

- **b** Explain, in terms of the forces acting on the wheels of a car, why a vehicle slows down when the brakes are applied.

4 The frictional force on a vehicle travelling on a certain type of level road surface is 0.6 × the vehicle's weight. For a vehicle of mass 1200 kg,

- **a** show that the maximum deceleration on this road is $5.9\,\text{m}\,\text{s}^{-2}$
- **b** calculate the braking distance on this road for a speed of $30\,\text{m}\,\text{s}^{-1}$.

3.5 Vehicle safety

Learning objectives:

- → Describe the force on a moving body if it is suddenly stopped, for example, in a road accident.
- → Explain what should be increased to give a smaller deceleration from a given speed.
- → State which design features attempt to achieve this in a modern vehicle.

Specification reference: 3.2.5

Impact forces

▲ **Figure 1** *Head-on collision*

Measuring impacts

The effect of a collision on a vehicle can be measured in terms of the acceleration or deceleration of the vehicle. By expressing an acceleration or deceleration in terms of g, the acceleration due to gravity, the force of the impact can then easily be related to the weight of the vehicle. For example, suppose a vehicle hits a wall and its deceleration is $30\,\text{m}\,\text{s}^{-2}$. In terms of g, the deceleration $\approx 3g$. So the **impact force** of the wall on the vehicle must have been three times its weight ($\approx 3mg$, where m is the mass of the vehicle). Such an impact is sometimes described as being equal to $3g$. This statement, although technically wrong because the deceleration not the impact force is equal to $3g$, is a convenient way of expressing the effect of an impact on a vehicle or a person.

Study tip

Remember that a negative acceleration is equal to a positive deceleration.

Application

How much acceleration or deceleration can a person withstand?

The duration of the impact as well as the magnitude of the acceleration affect the person. A person who is sitting or upright can survive a deceleration of $20\,g$ for a time of a few milliseconds, although not without severe injury. A deceleration of over $5\,g$ lasting for a few seconds can cause injuries. Car designers carry out tests using dummies in remote-control vehicles to measure the change of motion of different parts of a vehicle or a dummy. Sensors linked to data recorders and computers are used, as well as video cameras that record the motion to allow video clips to be analysed.

▲ **Figure 2** *A side-on impact*

Contact time and impact time

When objects collide and bounce off each other, they are in contact with each other for a certain time, which is the same for both objects. The shorter the contact time, the greater the impact force for the same initial velocities of the two objects. When two vehicles collide, they may or may not separate from each other after the collision. If they remain tangled together, they exert forces on each other until they are moving at the same velocity. The duration of the impact force is not the same as the contact time in this situation.

The **impact time** t (the duration of the impact force) can be worked out by applying the equation $s = \frac{1}{2}(u + v)t$ to one of the vehicles, where s is the distance moved by that vehicle during the impact, u is its initial velocity, and v is its final velocity. If the vehicle mass is known, the impact force can also be calculated.

For a vehicle of mass m in time t, assuming the acceleration is constant

$$\textbf{the impact time } t = \frac{2s}{u + v}$$

$$\textbf{the acceleration } a = \frac{v - u}{t}$$

$$\textbf{the impact force } F = ma$$

Worked example

A 1000 kg vehicle moving at $20\,\text{m}\,\text{s}^{-1}$ slows down in a distance of 4.0 m to a velocity of $12\,\text{m}\,\text{s}^{-1}$, as a result of hitting a stationary vehicle. Calculate

a the vehicle's acceleration in terms of g

b the impact force for this collision.

Solution

a $t = \frac{2s}{(u + v)} = \frac{2 \times 4.0}{(20 + 12)} = 0.25\,\text{s}$

The acceleration, $a = \frac{(v - u)}{t}$

$= \frac{12 - 20}{0.25}$

$= -32\,\text{m}\,\text{s}^{-2}$

$= -3.3g$ (where $g = 9.8\,\text{m}\,\text{s}^{-2}$).

b The impact force $F = ma = 1000 \times -32 = -32\,000\,\text{N}$.

Study tip

When writing answers and when you are driving (especially just after passing your driving test), don't rush – it's safer to take longer over what you are doing!

Study tip

The work done by the impact force F over an impact distance s $(= Fs)$ is equal to the change of kinetic energy of the vehicle. The impact force can also be worked out using the equation

$$F = \frac{\textbf{change of kinetic energy}}{\textbf{impact distance}}$$

Synoptic link

You will learn in Topic 5.2, Kinetic energy and potential energy, that the kinetic energy of a vehicle of mass m moving at speed $v = \frac{1}{2}mv^2$. Prove for yourself that the vehicle in the example above has 200 kJ of kinetic energy at $20\,\text{m}\,\text{s}^{-1}$ and 72 kJ at $12\,\text{m}\,\text{s}^{-1}$. You can then show the change of kinetic energy ÷ impact distance gives −32 000 N for the impact force.

Summary questions

$g = 9.8\,m\,s^{-2}$

1 A car of mass 1200 kg travelling at a speed of $15\,m\,s^{-1}$ is struck from behind by another vehicle, causing its speed to increase to $19\,m\,s^{-1}$ in a time of 0.20 s. Calculate:

a the acceleration of the car, in terms of g

b the impact force on the car.

2 The front end of a certain type of car of mass 1500 kg travelling at a speed of $20\,m\,s^{-1}$ is designed to crumple in a distance of 0.8 m, if the car hits a wall. Calculate:

a the impact time if it hits a wall at $20\,m\,s^{-1}$

b the impact force.

3 The front bumper of a car of mass 900 kg is capable of withstanding an impact with a stationary object, provided the car is not moving faster than $3.0\,m\,s^{-1}$, when the impact occurs. The impact time at this speed is 0.40 s. Calculate:

a the deceleration of the car from $3.0\,m\,s^{-1}$ to rest in 0.40 s

b the impact force on the car.

4 In a crash, a vehicle travelling at a speed of $25\,m\,s^{-1}$ stops after 4.5 m. A passenger of mass 68 kg is wearing a seat belt, which restrains her forward movement relative to the car to a distance of 0.5 m. Calculate **a** the deceleration of the passenger in terms of g, **b** the resultant force on the passenger.

Extension

Car safety features

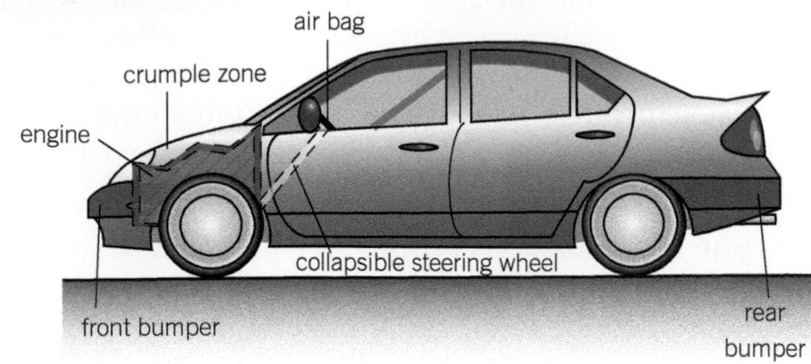

▲ **Figure 3** *Vehicle safety features*

In the example on the previous page, the vehicle's deceleration and hence the impact force would be lessened if the impact time was greater. So design features that would increase the impact time reduce the impact force. With a reduced impact force, the vehicle's occupants are less affected. The following vehicle safety features are designed to increase the impact time and so reduce the impact force.

- **Vehicle bumpers** give way a little in a low-speed impact and so increase the impact time. The impact force is therefore reduced as a result. If the initial speed of impact is too high, the bumper and/or the vehicle chassis are likely to be damaged.
- **Crumple zones** The engine compartment of a car is designed to give way in a front-end impact. If the engine compartment were rigid, the impact time would be very short, so the impact force would be very large. By designing the engine compartment so it crumples in a front-end impact, the impact time is increased and the impact force is therefore reduced.
- **Seat belts** In a front-end impact, a correctly fitted seat belt restrains the wearer from crashing into the vehicle frame after the vehicle suddenly stops. The restraining force on the wearer is therefore much less than the impact force would be if the wearer hit the vehicle frame. With the seat belt on, the wearer is stopped more gradually than without it.
- **Collapsible steering wheel** In a front-end impact, the seat belt restrains the driver without holding the driver rigidly. If the driver makes contact with the steering wheel, the impact force is lessened as a result of the steering wheel collapsing in the impact.
- **Airbags** An airbag reduces the force on a person, because the airbag acts as a cushion and increases the impact time on the person. More significantly, the force of the impact is spread over the contact area, which is greater than the contact area with a seat belt. So the pressure on the body is less.

Q: Explain why an air bag cushions an impact.

Practice questions: Chapter 3

1 A car moving at a velocity of $20\,\text{m}\,\text{s}^{-1}$ brakes to a standstill in a distance of 40 m. A child of mass 15 kg is sitting in a forward-facing child car seat fitted to the back seat of the car.

(a) Calculate (i) the deceleration of the car, (ii) the force on the child. *(3 marks)*

(b) In 2006, the UK government passed a law requiring that children in cars must travel in child car seats. If the child had not been in the child car seat, explain why the force on her would have been much greater than the value calculated in (a)(ii). *(2 marks)*

(c) Road accidents involving children are being reduced by reducing the speed limit near schools to 20 mph ($9\,\text{m}\,\text{s}^{-1}$). Discuss the effect this has on road safety where the speed limit is currently 30 mph. *(3 marks)*

2 A packing case is being lifted vertically at a constant speed by a cable attached to a crane.

(a) With reference to one of Newton's laws of motion, explain why the tension, T, in the cable must be equal to the weight of the packing case. *(3 marks)*

(b) A 12.0 N force and a 8.0 N force act on a body of mass 6.5 kg at the same time. For this body, calculate

(i) the maximum resultant acceleration that it could experience

(ii) the minimum resultant acceleration that it could experience. *(4 marks)*

AQA, 2005

3 **Figure 1** shows how the velocity of a motor car increases with time as it accelerates from rest along a straight horizontal road.

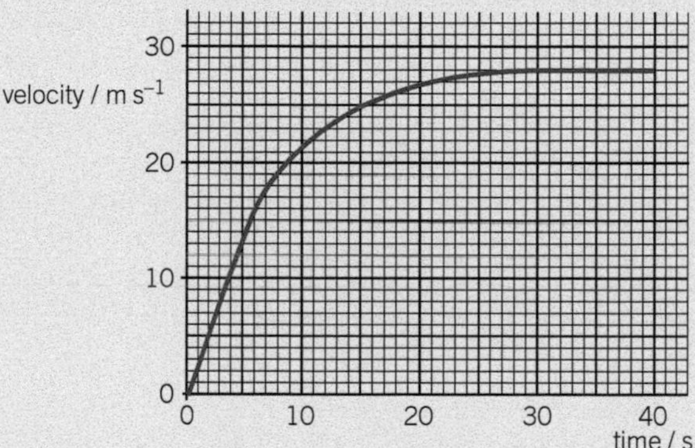

▲ **Figure 1**

(a) The acceleration is approximately constant for the first five seconds of the motion. Show that, over the first five seconds of the motion, the acceleration is approximately $2.7\,\text{m}\,\text{s}^{-2}$. *(3 marks)*

(b) Throughout the motion shown in **Figure 1** there is a constant driving force of 2.0 kN acting on the car.

(i) Calculate the mass of the car and its contents.

(ii) What is the magnitude of the resistive force acting on the car after 40 s? *(3 marks)*

(c) Find the distance travelled by the car during the first 40 s of the motion. *(3 marks)*

AQA, 2006

4 A constant resultant horizontal force of 1.8×10^3 N acts on a car of mass 900 kg, initially at rest on a level road.

(a) Calculate:

(i) the acceleration of the car

(ii) the speed of the car after 8.0 s

(iii) the distance travelled by the car in the first 8.0 s of its motion. *(5 marks)*

(b) In practice the resultant force on the car changes with time. Air resistance is one factor that affects the resultant force acting on the vehicle.

(i) Suggest, with a reason, how the resultant force on the car changes as its speed increases.

(ii) Explain, using Newton's laws of motion, why the vehicle has a maximum speed. *(5 marks)*

AQA, 2004

5 A passenger aircraft has a mass of 3.2×10^5 kg when fully laden. It is powered by four jet engines each producing a maximum thrust of 270 kN.

(a) (i) Calculate the total force of the engines acting on the aircraft.

(ii) Show that the initial acceleration of the aircraft with the engines set to full thrust is about $3.4\,m\,s^{-2}$. Ignore any frictional forces. *(3 marks)*

(b) The aircraft starts from rest at the beginning of its take-off and has a take-off speed of $90\,m\,s^{-1}$.

(i) Calculate the time taken for the aircraft to reach its take-off speed if frictional forces are ignored.

(ii) Frictional forces reduce the actual acceleration of the aircraft to $2.0\,m\,s^{-2}$. Calculate the mean total frictional force acting against the aircraft during the time taken to reach its take-off speed. *(4 marks)*

(c) Calculate the minimum runway length required by this aircraft for take-off when the acceleration is $2.0\,m\,s^{-2}$. *(2 marks)*

(d) The cruising speed of the aircraft in level flight with the engines at maximum thrust is a constant $260\,m\,s^{-1}$. The pilot adjusts the thrust so that the *horizontal* acceleration is always $2.0\,m\,s^{-2}$.
Calculate the time taken from take-off to reach its cruising speed. *(2 marks)*

(e) When it is at cruising speed, the aircraft travels at a constant velocity and at a constant height. Explain, in terms of the horizontal and vertical forces acting on the aircraft, how this is achieved. *(2 marks)*

AQA, 2002

6 A fairground ride ends with a car moving up a ramp at a slope of 30° to the horizontal.

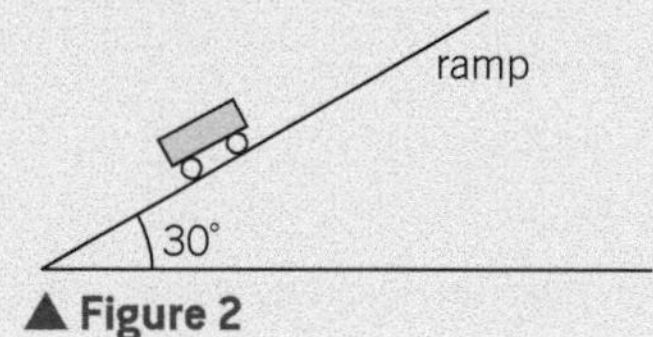

▲ **Figure 2**

(a) The car and its passengers have a total weight of 7.2×10^3 N. Show that the component of the weight parallel to the ramp is 3.6×10^3 N. *(1 mark)*

(b) Calculate the deceleration of the car assuming the only force causing the car to decelerate is that calculated in part (a). *(2 marks)*

(c) The car enters at the bottom of the ramp at $18\,m\,s^{-1}$. Calculate the minimum length of the ramp for the car to stop before it reaches the end. The length of the car should be neglected. *(2 marks)*

(d) Explain why the stopping distance is, in practice, shorter than the value calculated in part (c). *(2 marks)*

AQA, 2005

4 Force and momentum

4.1 Momentum and impulse

Learning objectives:

- → Calculate momentum.
- → Describe the connection between Newton's first and second laws of motion and momentum.
- → Define the impulse of a force, and calculate it from a force versus time graph.

Specification reference: 3.2.6

Momentum

If you have ever run into someone on the sports field, you will know something about momentum. If the person you ran into was more massive than you, then you probably came off worse than the other person. When two bodies collide, the effect they have on each other depends not only on their initial velocities, but also on the mass of each object. You can easily test the idea using coins, as shown in Figure 1. You might already have developed your skill in this area! It is not too difficult to show that when a large coin and a small coin collide, the motion of the small coin is affected more.

Sir Isaac Newton was the first person to realise that a **force** was needed to change the velocity of an object. He realised that the effect of a force on an object depended on its mass as well as on the amount of force. He defined the **momentum** of a moving object as its mass × its velocity and showed how the momentum of an object changes when a force acts on it. In this chapter, you will consider the ideas that Newton established in full.

▲ **Figure 1** *Momentum games*

Although Newton put forward his ideas over 300 years ago, his laws continue to provide the essential mathematical rules for predicting the motion of objects in any situation except inside the atom (where the rules of quantum physics apply) and at speeds approaching the speed of light or in very strong gravitational fields (where Einstein's theories of relativity apply). For example, the launch of a rocket is carefully planned using **Newton's laws of motion** and his law of gravitation. However, the laws do not, for example, predict the existence of black holes, which were a confirmed prediction of Einstein's theory of general relativity. In fact, Einstein showed that his theories of relativity simplify into Newton's laws where gravity is weak and the speed of objects is much less than the speed of light.

Synoptic link

You have met Newton's laws of motion in Chapter 3, and will study Newton's laws of gravitation later on if you're studying A Level Physics.

The momentum of an object is defined as its mass × its velocity.

- The unit of momentum is $\mathrm{kg\,m\,s^{-1}}$. The symbol for momentum is p.
- Momentum is a vector quantity. Its direction is the same as the direction of the object's velocity.
- For an object of mass m moving at velocity v, its momentum $p = mv$.

For example, a ball of mass 2.0 kg moving at a velocity of $10\,\mathrm{m\,s^{-1}}$ has the same amount of momentum as a person of mass 50 kg moving at a velocity of $0.4\,\mathrm{m\,s^{-1}}$.

Momentum and Newton's laws of motion

Newton's first law of motion: An object remains at rest or in uniform motion unless acted on by a force.

In effect, Newton's first law tells us that a force is needed to change the momentum of an object. If the momentum of an object is constant, there is no resultant force acting on it. Clearly, if the mass of an object is constant and the object has constant momentum, it follows that the velocity of the object is also constant. If a moving object with constant momentum gains or loses mass, however, its velocity would change to keep its momentum constant. For example, a cyclist in a race who collects a water bottle as he or she speeds past a service point gains mass (i.e., the water bottle) and therefore loses velocity.

Synoptic link

You have met the force equation in Topic 3.2, Using $F = ma$.

Newton's second law of motion: The rate of change of momentum of an object is proportional to the resultant force on it. In other words, the resultant force is proportional to the change of momentum per second.

You can write Newton's second law in the form force = mass × acceleration. Now, you will look at how this equation is derived from Newton's second law in its general form as stated above.

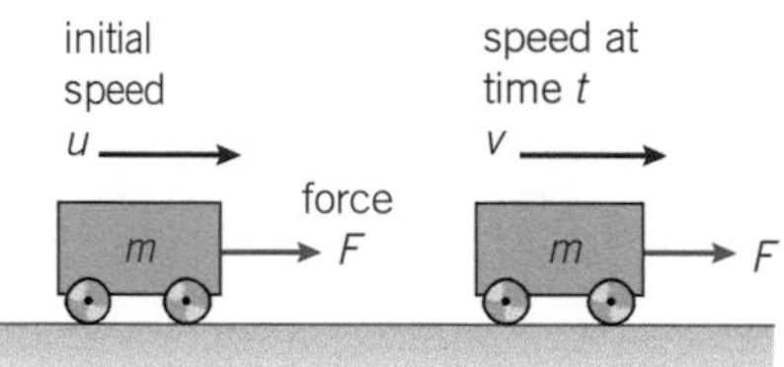

▲ **Figure 2** *Force and momentum*

Consider an object of constant mass *m* acted on by a constant force *F*. Its acceleration causes a change of its speed from initial speed u to speed v in time t without change of direction:

- its initial momentum = mu, and its final momentum = mv
- its change of momentum = its final momentum (mv) – its initial momentum (mu).

According to Newton's second law, the force is proportional to the change of momentum per second.

Therefore, force $F \propto \dfrac{\text{change of momentum}}{\text{time taken}} = \dfrac{mv - mu}{t}$

$$= \frac{m(v-u)}{t} = ma$$

where $a = \dfrac{v-u}{t}$ = the acceleration of the object.

This proportionality relationship (i.e., $F \propto ma$) can be written as $F = kma$, where k is a constant of proportionality.

Study tip

The unit of momentum may be either kg m s^{-1} or (more neatly) N s. The equation $F = \dfrac{\Delta(mv)}{\Delta t}$ always applies but $F = ma$ applies only to objects of constant mass.

The value of k is made equal to 1 by defining the unit of force, **the newton**, as the amount of force that gives an object of mass 1 kg an acceleration of 1 m s^{-2} (i.e., force F = 1 N, mass m = 1 kg, acceleration a = 1 m s^{-2} so k = 1).

Therefore, with $k = 1$, the equation $\boldsymbol{F = ma}$ follows from Newton's second law provided the mass of the object is constant.

In general, the change of momentum of an object may be written as $\Delta(mv)$, where the symbol Δ means change of. Therefore, if the momentum of an object changes by $\Delta(mv)$ in time Δt, the force F on the object is given by the equation

$$F = \frac{\Delta(mv)}{\Delta t}$$

1 **If *m* is constant,** then $\Delta(mv) = m\Delta v$, where Δv is the change of velocity of the object.

$\therefore F = \dfrac{m\Delta v}{\Delta t} = ma$ where acceleration $a = \dfrac{\Delta v}{\Delta t}$

2 **If m changes at a constant rate** as a result of mass being transferred at constant velocity, then $\Delta(mv) = v\Delta m$, where Δm is the change of mass of the object.

$$\therefore F = \frac{v\Delta m}{\Delta t} \text{ where } \frac{\Delta m}{\Delta t} = \text{change of mass per second.}$$

This form of Newton's second law is used in any situation where an object gains or loses mass continuously.

For example, if a rocket ejects burnt fuel as hot gas from its engine at speed v, the force F exerted by the engine to eject the hot gas is given by

$$F = \frac{v\Delta m}{\Delta t} \text{ where } \frac{\Delta m}{\Delta t} = \text{mass of hot gas lost per second.}$$

An equal and opposite reaction force acts on the jet engine due to the hot gas, propelling the rocket forwards.

The **impulse** of a force is defined as the force × the time for which the force acts. Therefore, for a force F which acts for time Δt,

the impulse = $F\Delta t = \Delta(mv)$

Hence the impulse of a force acting on an object is equal to the change of momentum of the object.

Force–time graphs

Suppose an object of constant mass m is acted on by a constant force F which changes its velocity from initial velocity u to velocity v in time t. As explained earlier in this topic, Newton's second law gives

$$F = \frac{mv - mu}{t}$$

Rearranging this equation gives $Ft = mv - mu$.

Figure 3 is a graph of force versus time for this situation. Because force F is constant for time t, the area under the line represents the impulse of the force Ft, which is equal to $mv - mu$. In other words,

the area under the line of a force–time graph represents the change of momentum or the impulse of the force

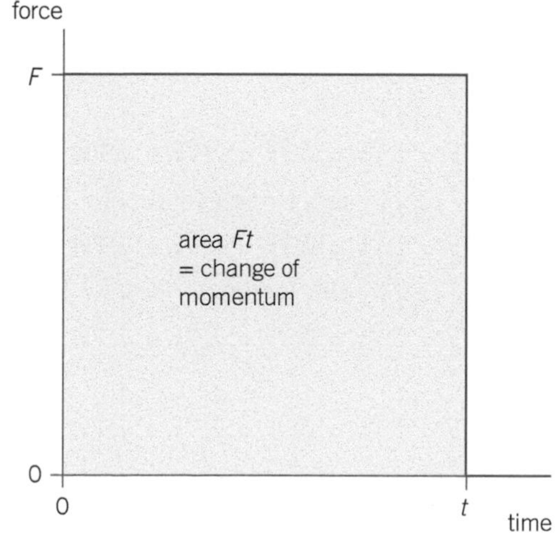

▲ **Figure 3** *Force against time for constant force*

Synoptic link

The area rule is especially useful when the force changes with time. This rule can also be used for a velocity–time graph where the velocity changes. You have met this in Topic 2.3, Motion along a straight line at constant acceleration.

Note:
The unit of momentum can be given as the newton second (N s) or the kilogram metre per second ($kg\,m\,s^{-1}$). The unit of impulse is usually given as the newton second.

Worked example

A force of 10 N acts for 20 s on an object of mass 50 kg which is initially at rest.

Calculate:

a the change of momentum of the object
b the velocity of the object at 20 s.

Solution

a Change of momentum = impulse of the force = $Ft = 10 \times 20 = 200$ N s

b Momentum at 20 s = 200 N s as the object was initially at rest.

$$\text{Velocity} = \frac{\text{momentum}}{\text{mass}} = \frac{200}{50} = 4.0\,\text{m s}^{-1}$$

Summary questions

1 a Calculate the momentum of:

i an atom of mass 4.0×10^{-25} kg moving at a velocity of 3.0×10^{6} m s^{-1}

ii a pellet of mass 4.2×10^{-4} kg moving at a velocity of 120 m s^{-1}

iii a bird of mass 0.56 kg moving at a velocity of 25 m s^{-1}.

b Calculate:

i the mass of an object moving at a velocity of 16 m s^{-1} with momentum of 96 kg m s^{-1}

ii the velocity of an object of mass 6.4 kg that has a momentum of 128 kg m s^{-1}.

2 A train of mass 24 000 kg moving at a velocity of 15.0 m s^{-1} is brought to rest by a braking force of 6000 N. Calculate

a the initial momentum of the train

b the time taken for the brakes to stop the train.

3 An aircraft of total mass 45 000 kg accelerates on a runway from rest to a velocity of 120 m s^{-1} when it takes off. During this time, its engines provide a constant driving force of 120 kN. Calculate:

a the gain of momentum of the aircraft

b the take-off time.

4 The velocity of a vehicle of mass 600 kg was reduced from 15 m s^{-1} by a constant force of 400 N which acted for 20 s then by a constant force of 20 N for a further 20 s.

a Sketch the force versus time graph for this situation.

b i Calculate the initial momentum of the vehicle.

ii Use the force versus time graph to determine the total change of momentum.

iii Show that the final velocity of the vehicle is 1 m s^{-1}.

4.2 Impact forces

▲ **Figure 1** *A golf ball impact*

Learning objectives:

- → Describe what happens to the impact force (and why) if the duration of impact is reduced.
- → Calculate $\Delta(mv)$ for a moving object that stops or reverses.
- → Describe what happens to the momentum of a ball when it bounces off a wall.

Specification reference: 3.2.6

A sports person knows that the harder a ball is hit, the further it travels. The impact changes the momentum of the ball in a very short time when the object exerting the impact force is in contact with the ball.

- If the ball is initially stationary and the impact causes it to accelerate to speed v in time t, the gain of momentum of the ball due to the impact = mv, where m is the mass of the ball.

  Therefore, the force of the impact $F = \dfrac{\text{change of momentum}}{\text{contact time}} = \dfrac{mv}{t}$

- If the ball is moving with an initial velocity u, and the impact changes its velocity to v in time t, the change of momentum of the ball = $mv - mu$.

  Therefore, the force of impact $F = \dfrac{\text{change of momentum}}{\text{contact time}}$

$$F = \frac{mv - mu}{t}$$

Synoptic link

Use of $F = ma$ and $a = \dfrac{v - u}{t}$ is another way to calculate an impact force.

You have met impact force in road accidents in Topic 3.5, Vehicle safety.

Worked example

A ball of mass 0.63 kg initially at rest was struck by a bat which gave it a velocity of $35\,\text{m}\,\text{s}^{-1}$. The contact time between the bat and ball was 25 ms. Calculate:

a the momentum gained by the ball

b the average force of impact on the ball.

(*Note:* The 'm' in ms stands for milli.)

Solution

a Momentum gained = $0.63 \times 35 = 22\,\text{kg}\,\text{m}\,\text{s}^{-1}$

b Impact force = $\dfrac{\text{gain of momentum}}{\text{contact time}} = \dfrac{22}{0.025} = 880\,\text{N}$

Application

Vehicle safety reminders

In Topic 3.5, you looked at the physics of vehicle safety features such as crumple zones. These and other features such as side-impact bars are all designed to lessen the effect of an impact on passengers in the vehicle. The essential idea is to increase the time taken by an impact so the acceleration or deceleration is less and therefore reducing the impact force. The result is the same using the idea of momentum – for a given change of momentum, the force is reduced if the impact time is increased. You will meet how these ideas can be developed further by using the concept of momentum in Topic 4.3, Conservation of momentum.

Force–time graphs for impacts

The variation of an impact force with time on a ball can be recorded using a force sensor connected using suitably long wires or a radio link to a computer. The force sensor is attached to the object (e.g., a bat) that causes the impact. Because the force on the bat is equal and opposite to the force on the ball during impact, the force on the ball due to the bat varies in exactly the same way as the force on the bat due to the ball. The variation of force with time is displayed on the computer screen.

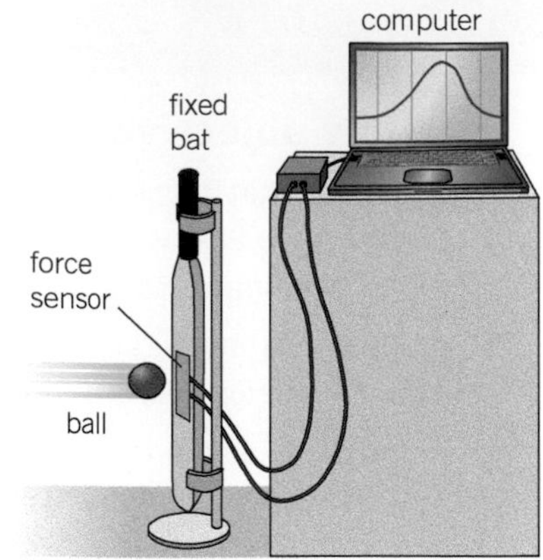

▲ **Figure 2** *Investigating an impact force on a ball*

Figure 3 shows a typical force–time graph for an impact. The graph shows that the impact force increases then decreases during the impact. As explained in Topic 4.1, the area under the graph is equal to the change of momentum. The average force of impact can be worked out from the change of momentum divided by the contact time.

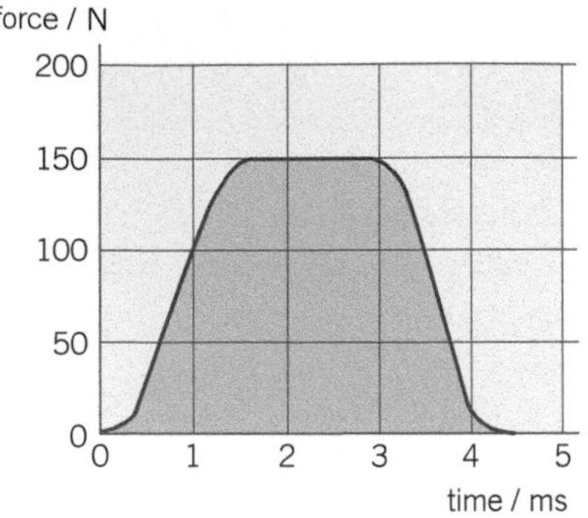

▲ **Figure 3** *Force against time for an impact*

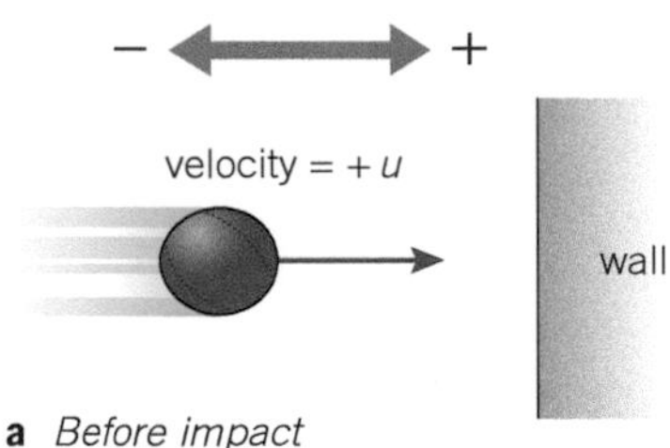

a *Before impact*

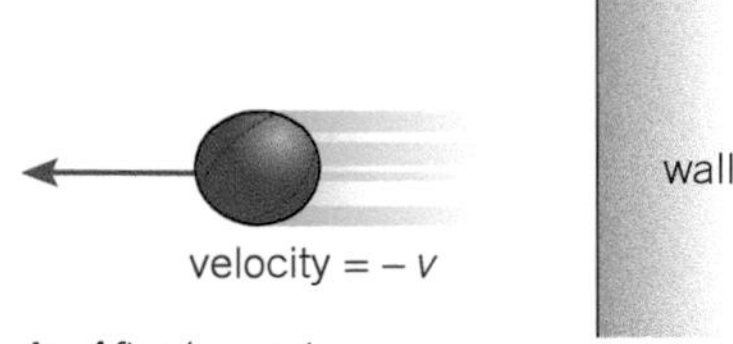

b *After impact*

▲ **Figure 4** *A rebound*

Rebound impacts

When a ball hits a wall and rebounds, its momentum changes direction due to the impact. If the ball hits the wall normally (i.e., its direction of motion is perpendicular to the wall), it rebounds normally so the direction of its momentum is reversed. The velocity and therefore the momentum after the impact is in the opposite direction to the velocity before the impact and therefore has the opposite sign. Figure 4 shows the idea.

Suppose the ball hits the wall normally with an initial speed u and it rebounds at speed v in the opposite direction. Since its direction of motion reverses on impact, a sign convention is necessary to represent the two directions. Using + for towards the wall and – for away from the wall, its initial momentum = $+mu$, and its final momentum = $-mv$.

Therefore,

its change of momentum = final momentum – initial momentum

$$= (-mv) - (mu)$$

$$\text{The impact force } F = \frac{\text{change of momentum}}{\text{contact time}} = \frac{(-mv) - (mu)}{t}$$

Notes:

1 If there is no loss of speed on impact, then $v = u$ so the impact force

$$F = \frac{(-mv) - (mu)}{t} = \frac{-2mu}{t}$$

2 If the impact is oblique (i.e., the initial direction of the ball is not perpendicular to the wall, as in Figure 5), the normal components of the velocity must be used. For an impact in which the initial and final direction of the ball are at the same angle θ to the normal and there is no loss of speed (i.e., $u = v$), the normal component of the initial velocity is $+u \cos\theta$ and the normal component of the final velocity is $-u \cos\theta$. The change of momentum of the ball is therefore $-2mu \cos\theta$.

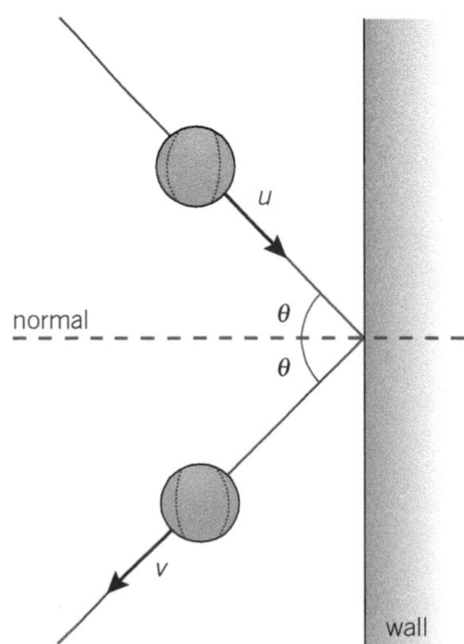

▲ **Figure 5** *An oblique impact*

Worked example

A tennis ball of mass 0.20 kg moving at a speed of $18\,\mathrm{m\,s^{-1}}$ is hit by a bat, causing the ball to go back in the direction it came from at a speed of $15\,\mathrm{m\,s^{-1}}$. The contact time is 0.12 s. Calculate:

a the change of momentum of the ball
b the impact force on the ball.

Solution

a Mass of ball $m = 0.20\,\mathrm{kg}$, initial velocity $u = +18\,\mathrm{m\,s^{-1}}$, final velocity $= -15\,\mathrm{m\,s^{-1}}$.

Change of momentum $= mv - mu = (0.20 \times -15) - (0.20 \times 18)$
$= -3.0 - 3.6 = -6.6\,\mathrm{kg\,m\,s^{-1}}$

b Impact force $= \dfrac{\text{change of momentum}}{\text{time taken}} = \dfrac{-6.6}{0.12} = -55\,\mathrm{N}$

Note: The minus sign indicates that the force on the ball is in the same direction as the velocity after the impact.

Summary questions

1 A 2000 kg lorry reversing at a speed of $0.80\,\mathrm{m\,s^{-1}}$ backs accidentally into a steel fence. The fence stops the lorry 0.5 s after the lorry first makes contact with the fence. Calculate:

a the initial momentum of the lorry

b the force of the impact.

2 A car of mass 600 kg travelling at a speed of $3.0\,\mathrm{m\,s^{-1}}$ is struck from behind by another vehicle. The impact lasts for 0.40 s and causes the speed of the car to increase to $8.0\,\mathrm{m\,s^{-1}}$. Calculate:

a the change of momentum of the car due to the impact

b the impact force.

3 A molecule of mass $5.0 \times 10^{-26}\,\mathrm{kg}$ moving at a speed of $420\,\mathrm{m\,s^{-1}}$ hits a surface at right angles to the surface and rebounds at the same speed in the opposite direction in an impact lasting 0.22 ns. Calculate:

a the change of momentum

b the force on the molecule.

4 Repeat the calculation in **Q3** for a molecule of the same mass at the same speed which hits the surface at 60° to the normal and rebounds without loss of speed at 60° to the normal, as shown in Figure 6. You will need to work out the components of the molecule's velocity parallel to the normal before and after the impact. Assume the contact time is the same.

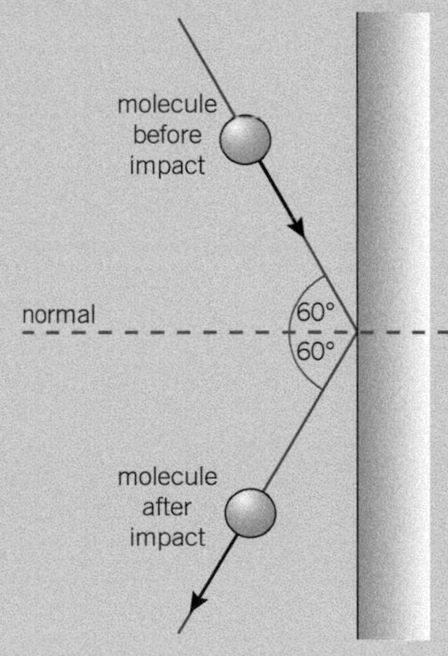

▲ **Figure 6**

4.3 Conservation of momentum

Learning objectives:

- → Consider whether momentum is ever lost in a collision.
- → Define conservation of momentum.
- → State the condition that must be satisfied if the momentum of a system is conserved.

Specification reference: 3.2.6

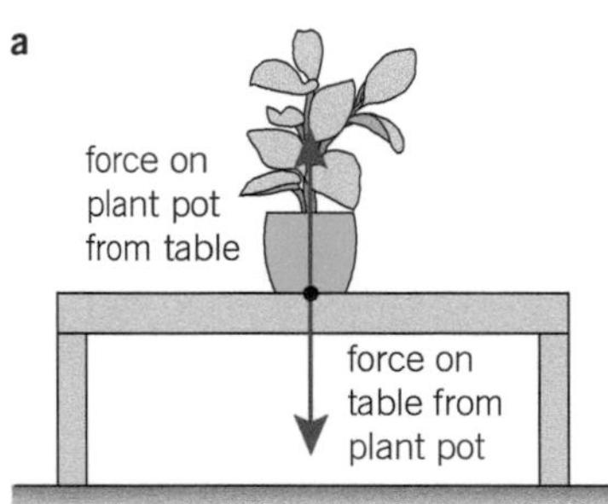

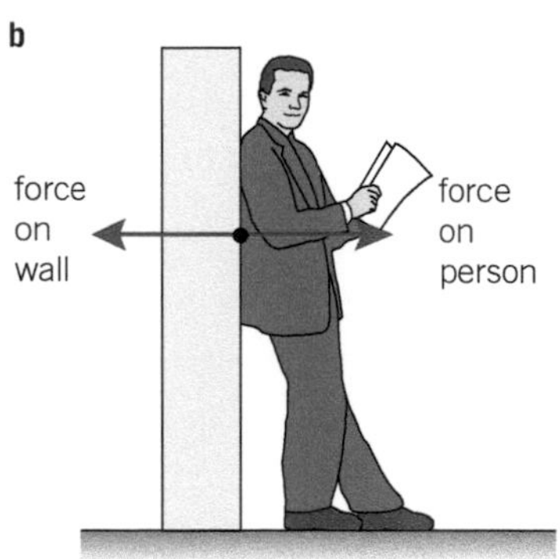

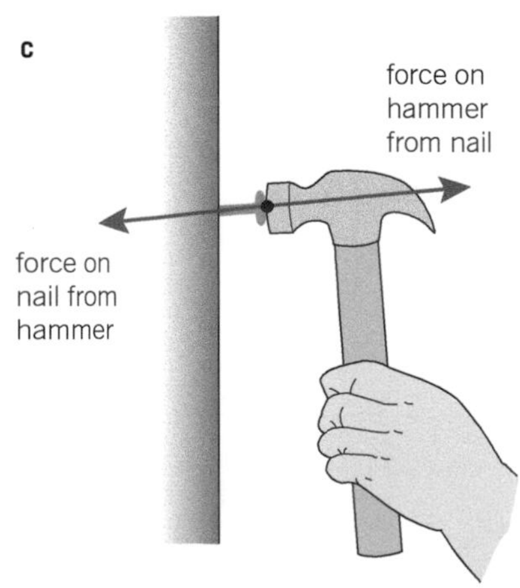

▲ **Figure 1** *Examples of Newton's third law*

Newton's third law of motion

When two objects interact, they exert equal and opposite forces on each other.

In other words, if object A exerts a force on object B, there must be an equal and opposite force acting on object A due to object B. For example,

- the Earth exerts a force due to gravity on an object which exerts an equal and opposite force on the Earth
- a jet engine exerts a force on hot gas in the engine to expel the gas – the gas being expelled exerts an equal and opposite force on the engine.

However, it is worth remembering that two forces must be of the same type, and acting on different objects, for the forces to be considered a force pair. For example, weight and normal reaction would not constitute a force pair in this case. (See Figure 1 for further examples.)

The principle of conservation of momentum

When an object is acted on by a resultant force, its momentum changes. If there is no change of its momentum, there can be no resultant force on the object. Now consider several objects which interact with each other. If no external resultant force acts on the objects, the total momentum does not change. However, interactions between the objects can transfer momentum between them. But the total momentum does not change.

The principle of conservation of momentum states that for a system of interacting objects, the total momentum remains constant, provided no external resultant force acts on the system.

Consider two objects that collide with each other then separate. As a result, the momentum of each object changes. They exert equal and opposite forces on each other when they are in contact. So the change of momentum of one object is equal and opposite to the change of momentum of the other object. In other words, if one object gains momentum, the other object loses an equal amount of momentum. So the total amount of momentum is unchanged.

Let's look in detail at the example of two snooker balls A and B in collision, as shown in Figure 2 on the next page.

The impact force F_1 on ball A due to ball B changes the velocity of A from u_A to v_A in time t.

Therefore, $F_1 = \frac{m_A v_A - m_A u_A}{t}$, where t = the time of contact between A and B, and m_A = the mass of ball A.

The impact force F_2 on ball B due to ball A changes the velocity of B from u_B to v_B in time t.

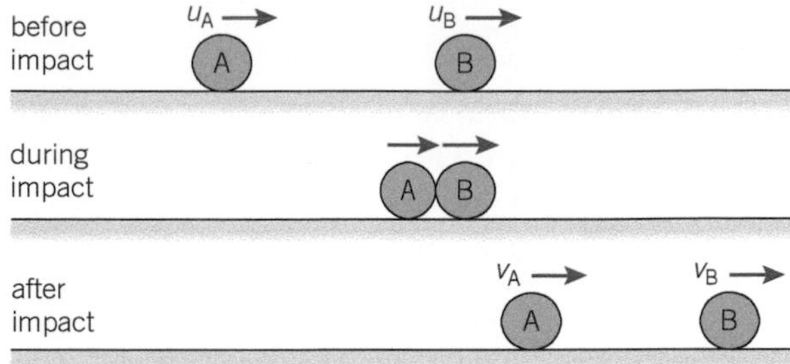

▲ **Figure 2** *Conservation of momentum*

Therefore, $F_2 = \frac{m_B v_B - m_B u_B}{t}$, where t = the time of contact between A and B, and m_B = the mass of ball B.

Because the two forces are equal and opposite to each other, $F_2 = -F_1$.

Therefore, $\frac{m_B v_B - m_B u_B}{t} = -\frac{(m_A v_A - m_A u_A)}{t}$.

Cancelling t on both sides gives

$$m_B v_B - m_B u_B = -(m_A v_A - m_A u_A)$$

Rearranging this equation gives

$$m_B v_B + m_A v_A = m_A u_A + m_B u_B$$

Therefore,

the total final momentum = the total initial momentum

Hence the total momentum is unchanged by this collision, that is, it is conserved.

Note:

If the colliding objects stick together as a result of the collision, they have the same final velocity. The above equation with V as the final velocity may therefore be written

$$(m_B + m_A)V = m_A u_A + m_B u_B$$

Testing conservation of momentum

Figure 3a shows an arrangement that can be used to test **conservation of momentum** using a motion sensor linked to a computer. The mass of each trolley is measured before the test. With trolley B at rest, trolley A is given a push so it moves towards trolley B at constant velocity. The two trolleys stick together on impact. The computer records and displays the velocity of trolley A throughout this time.

The computer display shows that the velocity of trolley A dropped suddenly when the impact took place. The velocity of trolley A immediately before the collision, u_A, and after the collision, V, can be measured. The measurements should show that the total momentum of both trolleys after the collision is equal to the momentum of trolley A before the collision. In other words,

$$(m_B + m_A)V = m_A u_A$$

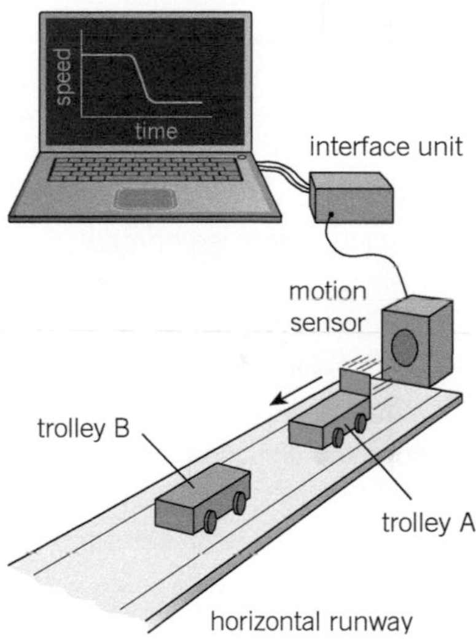

▲ **Figure 3a** *Using a motion sensor*

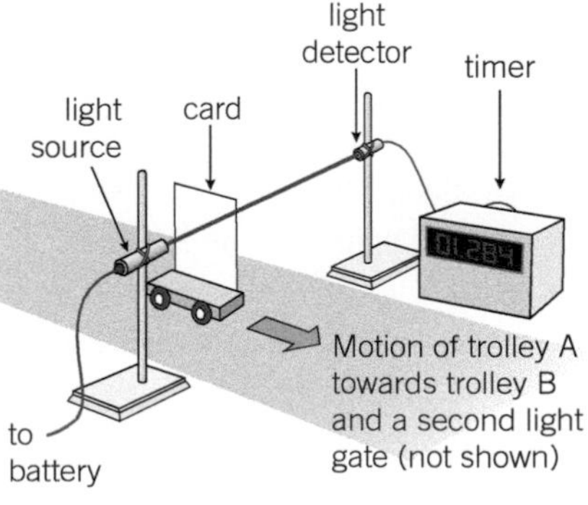

▲ **Figure 3b** *Using light gates (as described on p84)*

Worked example

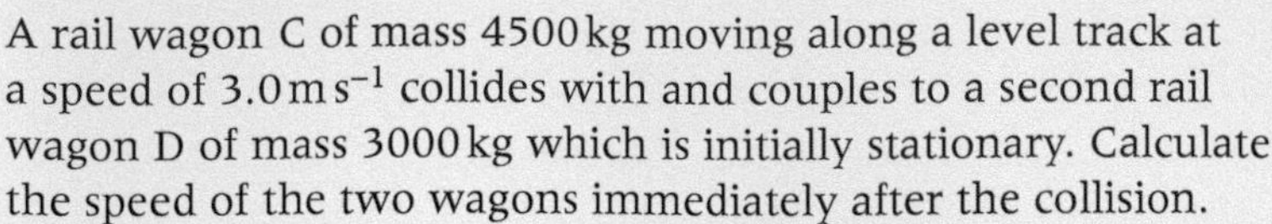

A rail wagon C of mass 4500 kg moving along a level track at a speed of 3.0 m s^{-1} collides with and couples to a second rail wagon D of mass 3000 kg which is initially stationary. Calculate the speed of the two wagons immediately after the collision.

Solution

Total initial momentum = initial momentum of C + initial momentum of D

$= (4500 \times 3.0) + (3000 \times 0) = 13\,500\ \text{kg m s}^{-1}$

Total final momentum = total mass of C and D × velocity v after the collision

$= (4500 + 3000)\,v = 7500\,v$

Using the principle of conservation of momentum,

$$7500\,v = 13\,500$$

$$v = \frac{13\,500}{7500} = 1.8\ \text{m s}^{-1}$$

Instead of a motion sensor, we may choose to use light gates linked to a computer or a data logger as shown in Figure 3b. A card attached to trolley A passes through the first of two light gates. Trolley B is positioned so that A collides with it just after passing through the first light gate. The velocity of A before the collision, u_A, is calculated by dividing the card length by the time taken to pass through the first gate. The velocity of A after the collision, V, is calculated by dividing the card length by the time taken to pass through the second gate.

Head-on collisions

Consider two objects moving in opposite directions that collide with each other. Depending on the masses and initial velocities of the two objects, the collision could cause them both to stop. The momentum of the two objects after the collision would then be zero. This could only happen if the initial momentum of one object was exactly equal and opposite to that of the other object. In general, if two objects move in opposite directions before a collision, then the vector nature of momentum needs to be taken into account by assigning numerical values of velocity + or − according to the direction.

For example, if a car of mass 600 kg travelling at a velocity of 25 m s^{-1} collides head-on with a lorry of mass 2400 kg travelling at a velocity of 10 m s^{-1} in the opposite direction, the total momentum before the collision is 9000 kg m s^{-1} in the direction the lorry was moving. As momentum is conserved in a collision, the total momentum after the collision is the same as the total momentum before the collision. Prove for yourself that if the two vehicles were to stick together after the collision, their velocity would be 3.0 m s^{-1} in the direction the lorry was moving before the impact.

Summary questions

1. A rail wagon of mass 3000 kg moving at a velocity of 1.2 m s^{-1} collides with a stationary wagon of mass 2000 kg. After the collision, the two wagons couple together. Calculate their speed immediately after the collision.
2. A rail wagon of mass 5000 kg moving at a velocity of 1.6 m s^{-1} collides with a stationary wagon of mass 3000 kg (see Figure 4). After the collision, the 3000 kg wagon moves away at a velocity of 1.5 m s^{-1}. Calculate the speed and direction of the 5000 kg wagon after the collision.

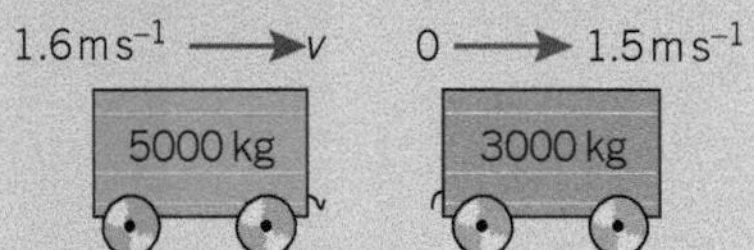

▲ **Figure 4**

3. In a laboratory experiment, a trolley of mass 0.50 kg moving at a speed of 0.25 m s^{-1} collides with a trolley of mass 1.0 kg moving in the opposite direction at a speed of 0.20 m s^{-1}. The two trolleys couple together on collision. Calculate their speed and direction immediately after the collision.
4. A ball of mass 0.80 kg moving at a speed of 2.5 m s^{-1} along a straight line collides with a ball of mass 2.5 kg which was initially stationary. As a result of the collision, the 2.5 kg ball has a velocity of 1.0 m s^{-1} along the same line. Calculate the speed and direction of the 0.80 kg ball immediately after the collision.

4.4 Elastic and inelastic collisions

Learning objectives:

- → Distinguish between an elastic collision and an inelastic collision.
- → Describe what is conserved in a perfectly elastic collision.
- → Discuss whether any real collisions are ever perfectly elastic.

Specification reference: 3.2.6

Drop a bouncy rubber ball from a measured height onto a hard floor. The ball should bounce back almost to the same height. Try the same with a cricket ball and there will be very little bounce! An elastic ball would be one that bounces back to exactly the same height. Its kinetic energy just after impact must equal its **kinetic energy** just before impact. Otherwise, it cannot regain its initial height. There is no loss of kinetic energy in an **elastic collision**.

An elastic collision is one where there is no loss of kinetic energy.

A very low speed impact between two cars is almost perfectly elastic, provided no damage is done. However, if the collision causes damage to the vehicles, some of the initial kinetic energy is transferred to the surroundings. This collision may be described as **inelastic**.

An inelastic collision occurs where the colliding objects have less kinetic energy after the collision than before the collision.

Objects that collide and couple together undergo inelastic collisions, as some of the initial kinetic energy is transferred to the surroundings.

To work out whether a collision is elastic or inelastic, the kinetic energy of each object before and after the collision must be calculated.

Examples

1. For a ball of mass m falling in air from a measured height H above the floor and rebounding to a height h,
   - i the kinetic energy immediately before impact = loss of potential energy through height $H = mgH$ (see Topic 5.2)
   - ii the kinetic energy immediately after impact = gain of potential energy through height $h = mgh$.

So the height ratio h/H gives the fraction of the initial kinetic energy that is recovered as kinetic energy after the collision.

2. For a collision between two objects, the kinetic energy of each object can be worked out using the kinetic energy equation $E_K = \frac{1}{2}mv^2$, where m is the mass of the object and v is its speed (see Topic 5.2). Using this equation, the total initial kinetic energy and the total final kinetic energy can be worked out if the mass, initial speed, and speed after collision of each object is known.

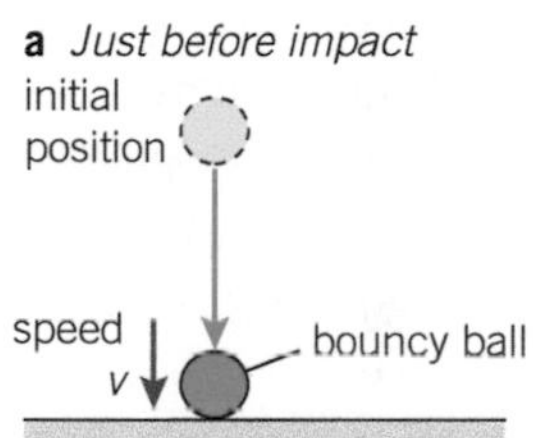

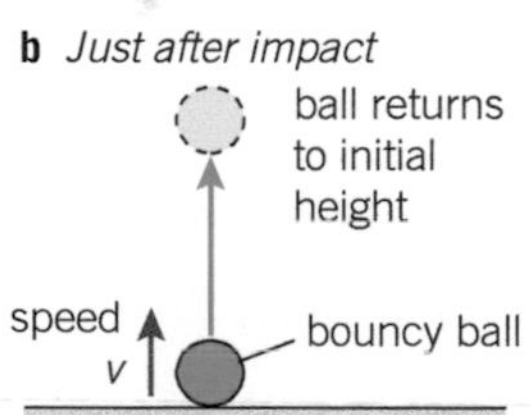

▲ **Figure 1** *An elastic impact*

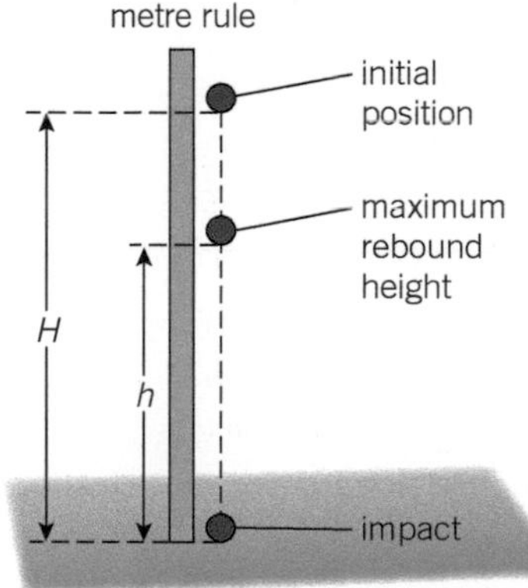

▲ **Figure 2** *Testing an impact*

Worked example

A railway wagon of mass 8000 kg moving at $3.0\,\mathrm{m\,s^{-1}}$ collides with an initially stationary wagon of mass 5000 kg. The two wagons separate after the collision. The 8000 kg wagon moves at a speed of $1.0\,\mathrm{m\,s^{-1}}$ without change of direction after the collision. Calculate:

a the speed and direction of the 5000 kg wagon after the collision

b the loss of kinetic energy due to the collision.

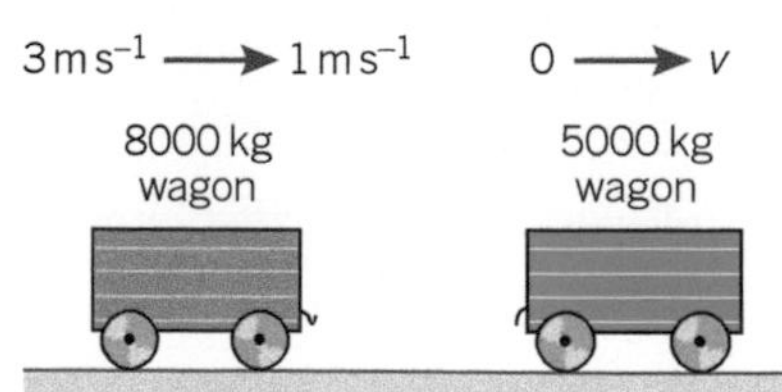

▲ **Figure 3**

Study tip

Momentum is always conserved in collisions provided no external forces act. Total energy is always conserved, but kinetic energy may be transferred into other energy stores.

Synoptic link

Excitation by collision in a gas occurs when the gas molecules undergo collisions and are excited to higher energy states. You will meet particle collisions in Topic 13.3, Collisions of electrons with atoms. You will also learn more about kinetic energy in Topic 5.2, Kinetic energy and potential energy.

Solution

a The total initial momentum = $8000 \times 3 = 24\,000\,\text{kg m s}^{-1}$.

The total final momentum = $(8000 \times 1.0) + 5000V$, where V is the speed of the 5000 kg wagon after the collision.

Using the principle of conservation of momentum

$$8000 + 5000V = 24\,000$$

$$5000V = 24\,000 - 8000 = 16\,000$$

$$V = \frac{16\,000}{5000} = 3.2\,\text{m s}^{-1}$$

b Kinetic energy of the 8000 kg wagon before the collision

$$= \frac{1}{2} \times 8000 \times 3.0^2 = 36\,000\,\text{J}$$

Kinetic energy of the 8000 kg wagon after the collision

$$= \frac{1}{2} \times 8000 \times 1.0^2 = 4000\,\text{J}$$

Kinetic energy of the 5000 kg wagon after the collision

$$= \frac{1}{2} \times 5000 \times 3.2^2 = 25\,600\,\text{J}$$

∴ loss of kinetic energy due to the collision

$$= 36\,000 - (4000 + 25\,600)$$
$$= 6400\,\text{J}$$

Summary questions

1 a A squash ball is released from rest above a flat surface. Describe how its energy changes if:

- **i** it rebounds to the same height
- **ii** it rebounds to a lesser height.

b In **a ii**, the ball is released from a height of 1.20 m above the surface and it rebounds to a height of 0.90 m above the surface. Show that 25% of its kinetic energy is lost in the impact.

2 A vehicle of mass 800 kg moving at a speed of $15.0\,\text{m s}^{-1}$ collides with a vehicle of mass 1200 kg moving in the same direction at a speed of $5.0\,\text{m s}^{-1}$. The two vehicles lock together on impact. Calculate:

- **a** the velocity of the two vehicles immediately after impact
- **b** the loss of kinetic energy due to the impact.

3 An ice puck of mass 1.5 kg moving at a speed of $4.2\,\text{m s}^{-1}$ collides head-on with a second ice puck of mass 1.0 kg moving in the opposite direction at a speed of $4.0\,\text{m s}^{-1}$. After the impact, the 1.5 kg ice puck continues in the same direction at a speed of $0.80\,\text{m s}^{-1}$. Calculate:

- **a** the speed and direction of the 1.0 kg ice puck after the collision
- **b** the loss of kinetic energy due to the collision.

4 The bumper cars at a fairground are designed to withstand low-speed impacts without damage. A bumper car of mass 250 kg moving at a velocity of $0.90\,\text{m s}^{-1}$ collides elastically with a stationary car of mass 200 kg. Immediately after the impact, the 250 kg car has a velocity of $0.10\,\text{m s}^{-1}$ in the same direction as it was initially moving in.

a **i** Calculate the velocity of the 200 kg car immediately after the impact.

ii Show that the collision was an elastic collision.

b Without further calculations, discuss the effect of the impact on the driver of each car.

4.5 Explosions

▲ **Figure 1** *The gun barrel recoils when the shell is fired. Large springs fitted to the barrel take away and store the kinetic energy of the barrel as it recoils.*

Learning objectives:

- → Describe the energy changes that take place in an explosion.
- → State what can always be said about the total momentum of a system that has exploded.
- → Describe the consequences when, after the explosion, only two bodies move apart.

Specification reference: 3.2.6

When two objects fly apart after being initially at rest, they recoil from each other with equal and opposite amounts of momentum. So they move away from each other in opposite directions. Consider Figure 2, where two spring-loaded trolleys of mass m_A and m_B respectively are initially positioned at rest and in contact. These trolleys move apart at speeds v_A and v_B respectively when the trigger is tapped to release the spring in trolley A.

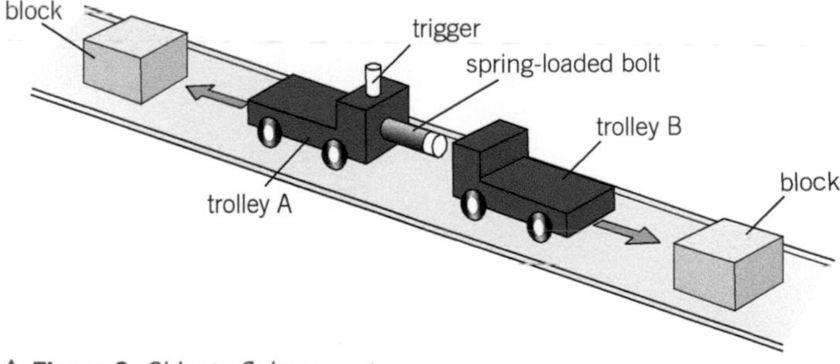

▲ **Figure 2** *Objects flying apart*

The total initial momentum = 0

The total momentum immediately after the explosion

$$= \text{momentum of A} + \text{momentum of B}$$

$$= m_A v_A + m_B v_B$$

Using the principle of the conservation of momentum, $m_A v_A + m_B v_B = 0$

$$\therefore m_B v_B = -m_A v_A$$

The minus sign means that the two masses move apart from each other in opposite directions.

For example, if $m_A = 1.0\,\text{kg}$, $v_A = 2.0\,\text{m}\,\text{s}^{-1}$, and $m_B = 0.5\,\text{kg}$,

$$\text{then } v_B = -\frac{m_A v_A}{m_B} = -4.0\,\text{m}\,\text{s}^{-1}$$

So A and B move apart at speeds of $2.0\,\text{m}\,\text{s}^{-1}$ and $4.0\,\text{m}\,\text{s}^{-1}$ in opposite directions.

Synoptic link

The α particles from a given isotope are always emitted with the same kinetic energy because they are emitted with the same speed, provided no external forces act. This is because the total energy released is always the same and each α particle and the nucleus that emits it move apart with equal and opposite amounts of momentum. You will meet these ideas in more detail in Topic 8.2 The properties of α, β and γ radiation.

Testing a model explosion

In Figure 2, when the spring is released from one of the trolleys, the two trolleys, A and B, push each other apart. The blocks are positioned so that the trolleys hit the blocks at the same moment. The distance travelled by each trolley to the point of impact with the block is equal to its speed × the time taken to travel that distance. As the time taken is the same for the two trolleys, the distance ratio is the same as the speed ratio. Because the trolleys have equal (and opposite) amounts of momentum, the ratio of their speeds is the inverse of the mass ratio. The distance ratio should therefore be equal to the inverse of the mass ratio. In other words, if trolley A travels twice as far as trolley B, then the mass of A must be half the mass of B (so they carry away equal amounts of momentum).

Note:

In this experiment, the kinetic energy of the two trolleys immediately after they separate from each other is equal to the energy stored in the spring when it was originally compressed. For two or more objects that fly apart due to an explosion, their total kinetic energy immediately after the explosion is less than the total chemical energy released in the explosion because heat, light, and sound all carry away energy.

Summary questions

1 A shell of mass 2.0 kg is fired at a speed of 140 m s^{-1} from an artillery gun of mass 800 kg. Calculate the recoil velocity of the gun.

2 In a laboratory experiment to measure the mass of an object X, two identical trolleys A and B, each of mass 0.50 kg, were initially stationary on a track. Object X was fixed to trolley A. When a trigger was pressed, the two trolleys moved apart in opposite directions at speeds of 0.30 m s^{-1} and 0.25 m s^{-1}.

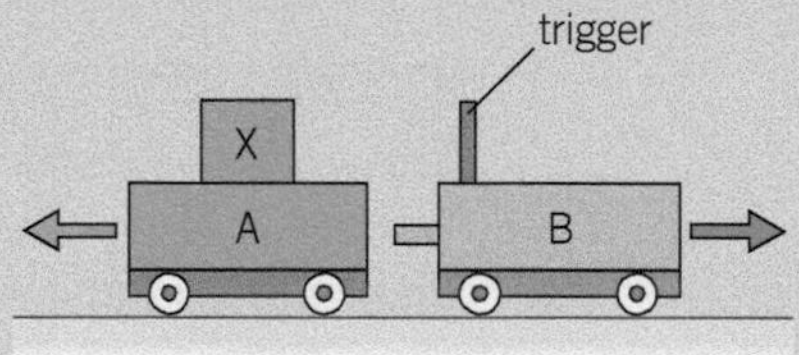

▲ **Figure 3**

a Which of the two speeds given above was the speed of trolley A? Give a reason for your answer.

b Show that the mass of X must have been 0.10 kg.

3 Two trolleys, X of mass 1.20 kg and Y of mass 0.80 kg, are initially stationary on a level track.

a When a trigger is pressed on one of the trolleys, a spring pushes the two trolleys apart. Trolley Y moves away at a velocity of 0.15 m s^{-1}.

i Calculate the velocity of trolley X.

ii Calculate the total kinetic energy of the two trolleys immediately after the explosion.

b In part **a**, if the test had been carried out with trolley X held firmly, calculate the speed at which Y would have recoiled, assuming the energy stored in the spring before release is equal to the total kinetic energy calculated in part **a ii**.

4 A person stands in a stationary boat (total mass of boat + person = 150 kg) and throws a rock of mass 2.0 kg out of the boat. As a result, the boat recoils at a speed of 0.12 m s^{-1}. Calculate:

a the speed at which the rock was thrown from the boat

b the kinetic energy gained by

i the boat

ii the rock.

Practice questions: Chapter 4

1 **(a)** Collisions can be described as *elastic* or *inelastic*.
State what is meant by an inelastic collision. *(1 mark)*

(b) A ball of mass 0.12 kg strikes a stationary cricket bat with a speed of $18\,\mathrm{m\,s^{-1}}$. The ball is in contact with the bat for 0.14 s and returns along its original path with a speed of $15\,\mathrm{m\,s^{-1}}$. Calculate:

(i) the momentum of the ball before the collision
(ii) the momentum of the ball after the collision
(iii) the total change of momentum of the ball
(iv) the average force acting on the ball during contact with the bat
(v) the kinetic energy lost by the ball as a result of the collision. *(6 marks)*

AQA, 2001

2 Two carts A and B, with a compressed spring between them, are pushed together and held at rest, as shown in **Figure 1**. The spring is not attached to either cart. The carts are then released.

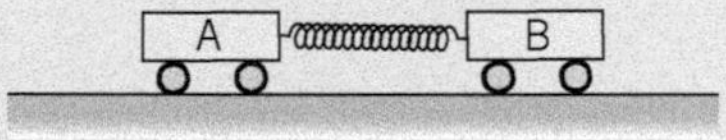

▲ Figure 1

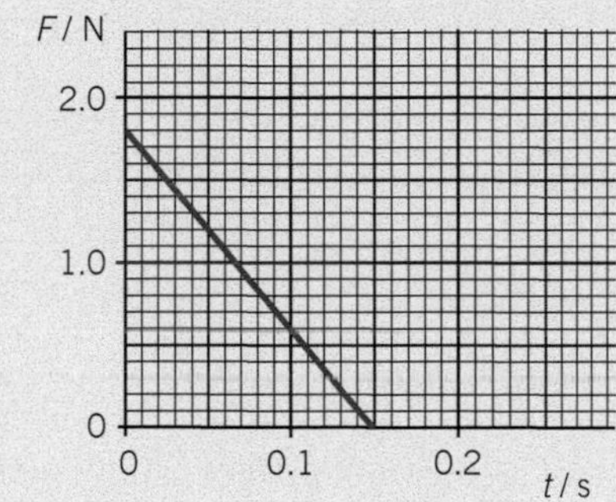

▲ Figure 2

Figure 2 shows how the force, *F*, exerted by the spring on the carts varies with time, *t*, after release.
When the spring returns to its unstretched length and drops away, cart A is moving at $0.60\,\mathrm{m\,s^{-1}}$.

(a) Calculate the impulse given to each cart by the spring as it expands.
(b) Calculate the mass of cart A.
(c) State the final total momentum of the system at the instant the spring drops away. *(5 marks)*

AQA, 2004

3 A railway engine is about to couple with a stationary carriage of mass $4.0 \times 10^4\,\mathrm{kg}$. When they have joined up, the engine and the carriage move at a constant speed. The engine has a mass of $6.2 \times 10^4\,\mathrm{kg}$ and is moving at $0.35\,\mathrm{m\,s^{-1}}$ just before coupling.

(a) Calculate the momentum of the engine.
(b) Calculate the speed of the engine and carriage after coupling. *(5 marks)*

AQA, 2007

4 **(a)** State two quantities that are conserved in an elastic collision. *(2 marks)*

(b) A gas molecule makes an elastic collision with the walls of a gas cylinder. The molecule is travelling at $450\,\mathrm{m\,s^{-1}}$ at right angles towards the wall before the collision.

(i) What is the magnitude and direction of its velocity after the collision?
(ii) Calculate the change in momentum of the molecule during the collision if it has a mass of $8.0 \times 10^{-26}\,\mathrm{kg}$. *(4 marks)*

(c) Use Newton's laws of motion to explain how the molecules of a gas exert a force on the wall of a container. *(4 marks)*

AQA, 2006

5 When an α particle is emitted from a nucleus of the polonium isotope $^{210}_{84}Po$, a nucleus of lead (Pb) is formed.

The α particle is emitted at a speed of $1.6 \times 10^7\,m\,s^{-1}$ and has a mass of 4.0 u (where 1 u = 1 atomic mass unit). The lead nucleus has a mass of 206 u.

Calculate the speed of recoil of the lead nucleus immediately after the α particle has been emitted. Assume the polonium nucleus is initially at rest. *(3 marks)*

6 **Figure 3** shows how the force, F, on a steel ball varies with time, t, when the ball is dropped onto a thick steel plate and rebounds. The kinetic energy of the ball after the collision is the same as it was before the collision.

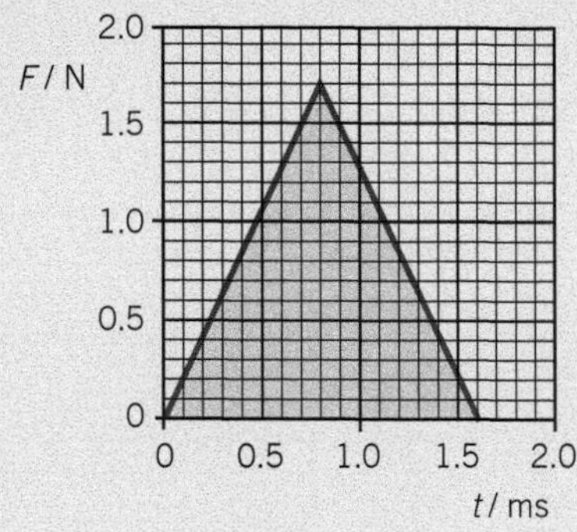

▲ **Figure 3**

(a) State the name of the quantity that is obtained by determining the shaded area.

(b) Use **Figure 3** to determine the initial momentum of the ball.

(c) Sketch a graph to show how the momentum of the ball varies between times $t = 0$ and $t = 2.0$ ms. *(6 marks)*

AQA, 2006

7 (a) Explain what is meant by the principle of conservation of momentum. *(2 marks)*

(b) A hose pipe is used to water a garden. The supply delivers water at a rate of $0.31\,kg\,s^{-1}$ to the nozzle which has a cross-sectional area of $7.3 \times 10^{-5}\,m^2$.

(i) Show that water leaves the nozzle at a speed of about $4\,m\,s^{-1}$. density of water = $1000\,kg\,m^{-3}$

(ii) Before it leaves the hose, the water has a speed of $0.68\,m\,s^{-1}$. Calculate the force on the hose.

(iii) The water from the hose is sprayed onto a brick wall, the base of which is firmly embedded in the ground. Explain why there is no overall effect on the rotation of the Earth. *(7 marks)*

AQA, 2005

Answers to the Practice Questions and Section Questions are available at www.oxfordsecondary.com/oxfordaqaexams-alevel-physics

5 Work, energy, and power

5.1 Work and energy

Learning objectives:

- → Define energy and describe how we measure it.
- → Discuss whether energy ever disappears.
- → Define work (in the scientific sense).

Specification reference: 3.2.7 and 3.2.8

Energy rules

Energy is needed to make stationary objects move, to change their shape, or to warm them up. When you lift an object, you transfer energy from your muscles to the object.

Objects can possess energy in different types of stores, including:

- gravitational potential stores (the position of objects in a gravitational field)
- kinetic stores (moving objects)
- thermal stores (hot objects)
- elastic stores (objects compressed or stretched).

Energy can be transferred between objects in different ways, including:

- by radiation (e.g., light)
- electrically
- mechanically (e.g., by sound).

Energy is measured in **joules (J)**. One joule is equal to the energy needed to raise a 1 N weight through a vertical height of 1 m.

Whenever energy is transferred, the total amount of energy after the transfer is always equal to the total amount of energy before the transfer. The total amount of energy is unchanged.

Energy cannot be created or destroyed.

This statement is known as the **principle of conservation of energy**.

Forces at work

Work is done on an object when a force acting on it makes it move. As a result, energy is transferred to the object. The amount of work done depends on the force and the distance the object moves. The greater the force or the further the distance, the greater the work done.

Work done = force × distance moved in the direction of the force whilst the force is acting on the object.

The unit of work is the joule (J), equal to the work done when a force of 1 N moves its point of application by a distance of 1 m in the direction of the force. For example, as shown in Figure 1:

- A force of 1 N is required to raise an object of weight 1 N steadily. If it is raised by 1 m, the work done by the force is 1 J (= 1 N × 1 m). The gain in potential energy of the raised object is 1 J.
- For a 2 N object raised to a height of 1 m, the work done, and therefore potential energy of the raised object, is 2 J (= 2 N × 1 m).

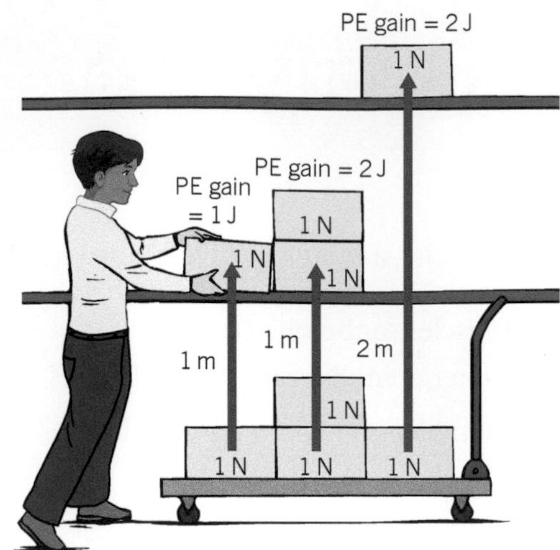

▲ **Figure 1** *Using joules*

Hint

No work is done when F and s are at right angles to each other.

Force and displacement

Imagine a yacht acted on by a wind force F at an angle θ to the direction in which the yacht moves. The wind force has a component $F\cos\theta$ in the direction of motion of the yacht and a component $F\sin\theta$ at right angles to the direction of motion. If the yacht is moved a distance s by the wind, the work done on it, W, is equal to the component of force in the direction of motion × the distance moved.

$$W = Fs\cos\theta$$

▲ **Figure 2** *Force and displacement*

Note that if $\theta = 90°$ (which means that the force is perpendicular to the direction of motion) then, because $\cos 90° = 0$, the work done is zero.

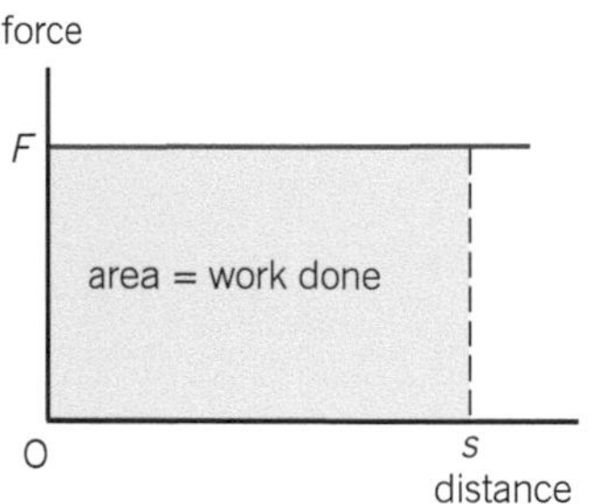

▲ **Figure 3** *Force–distance graph for a constant force*

Force–distance graphs

- If a constant force F acts on an object and makes it move a distance s in the direction of the force, the work done on the object $W = Fs$. Figure 3 shows a graph of force against distance in this situation. The area under the line is a rectangle of height representing the force and of base length representing the distance moved. **Therefore the area under the line represents the work done.**
- If a variable force F acts on an object and causes it to move in the direction of the force, the work done for a small amount of distance Δs, $\Delta W = F\Delta s$. This is represented on a graph of the force F against distance s by the area of a strip under the line of width Δs and height F (Figure 4). The total work done is therefore the sum of the areas of all the strips (i.e., the total area under the line).

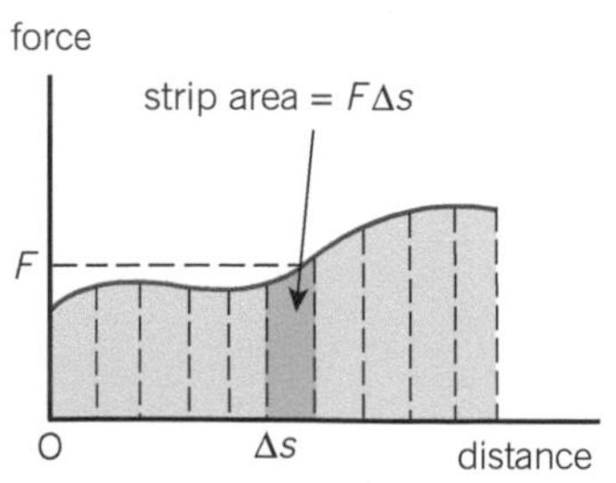

▲ **Figure 4** *Force–distance graph for a variable force*

The area under the line of a force–distance graph represents the total work done.

For example, consider the force needed to stretch a spring. The greater the force, the more the spring is extended from its unstretched length. Figure 5 shows how the force needed to stretch a spring changes with the extension of the spring. The graph is a straight line through the origin. Therefore, the force needed is proportional to the extension of the spring. This is known as **Hooke's law**. See Topic 6.2 for more about springs.

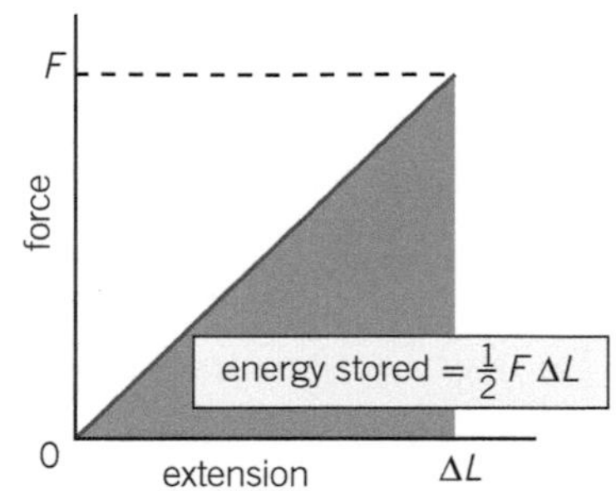

▲ **Figure 5** *Force against extension for a spring*

Figure 5 is a graph of force against distance, in this case the distance the spring is extended. Therefore, the area under the line represents the work done to stretch the spring. If F is the force needed to extend the spring to extension ΔL, the area under the line from the origin to extension ΔL represents the work done to stretch the spring to extension ΔL. As this area is a triangle, the work done $= \frac{1}{2} \times$ height $\times$ base $= \frac{1}{2} F \Delta L$.

To stretch the spring to extension ΔL, work done $= \frac{1}{2} F \Delta L$

Summary questions

$g = 9.8\,\text{m}\,\text{s}^{-2}$

1 Calculate the work done when:

 a a weight of 40 N is raised by a height of 5.0 m

 b a spring is stretched to an extension of 0.45 m by a force that increases to 20 N.

2 Calculate the energy transferred by a force of 12 N when it moves an object by a distance of 4.0 m:

 a in the direction of the force

 b in a direction at 60° to the direction of the force

 c in a direction at right angles to the direction of the force.

3 A luggage trolley of total weight 400 N is pushed at steady speed 20 m up a slope, by a force of 50 N acting in the same direction as the object moves in. At the end of this distance, the trolley is 1.5 m higher than at the start. Calculate:

 a the work done pushing the trolley up the slope

 b the gain of potential energy of the trolley

 c the energy wasted due to friction.

4 A spring that obeys Hooke's law requires a force of 1.2 N to extend it to an extension of 50 mm. Calculate:

 a the force needed to extend it to an extension of 100 mm

 b the work done when the spring is stretched to an extension of 100 mm from zero extension.

5.2 Kinetic energy and potential energy

Learning objectives:

- → Describe what happens to the work done on an object when it is lifted.
- → Describe what energy change takes place when an object is allowed to fall.
- → Describe the effect on the kinetic energy of a car if its speed is doubled.

Specification reference: 3.2.7 and 3.2.8

Kinetic energy

Kinetic energy is the energy store of an object due to its motion. The faster an object moves, the more kinetic energy it has. To see the exact link between kinetic energy and speed, consider an object of mass m, initially at rest, acted on by a constant force F for a time t.

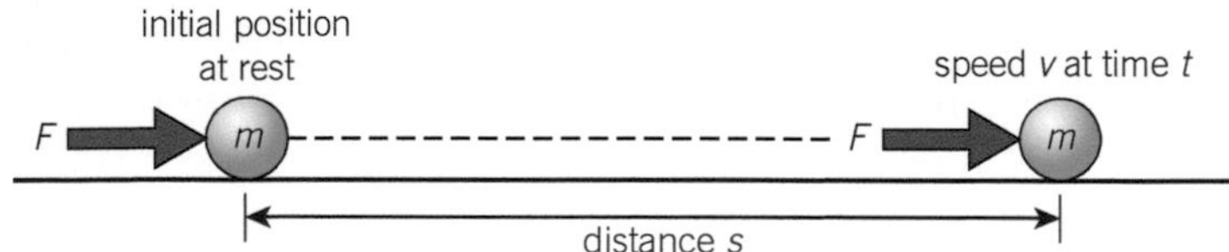

▲ **Figure 1** *Gaining kinetic energy*

Let the speed of the object at time t be v.

Therefore, distance travelled, $s = \frac{1}{2}(u + v)t = \frac{1}{2}vt$ because $u = 0$

acceleration $a = \frac{(v - u)}{t} = \frac{v}{t}$

Using Newton's second law, $F = ma = \frac{mv}{t}$

Therefore, the work done W by force F to move the object through distance s,

$$W = Fs = \frac{mv}{t} \times \frac{vt}{2} = \frac{1}{2}mv^2$$

Because the gain of kinetic energy is due to the work done, then

$$\textbf{kinetic energy, } E_K = \frac{1}{2}mv^2$$

Hint

The equation for kinetic energy does not hold at speeds approaching the speed of light. Einstein's theory of special relativity tells us that the mass of an object increases with speed and that the energy E of an object can be worked out from the equation $E = mc^2$, where c is the speed of light in free space and m is the mass of the object.

Potential energy

Potential energy is the energy store of an object due to its position.

If an object of mass m is raised through a vertical height Δh at steady speed, the force needed to raise it is equal and opposite to its weight mg. Therefore,

the work done to raise the object = force × distance moved
$= mg\Delta h$

The work done on the object increases its gravitational potential energy.

Change of gravitational potential energy $\Delta E_P = mg\Delta h$

At the Earth's surface, $g = 9.8\,\text{m}\,\text{s}^{-2}$.

Hint

The equation for gravitational potential energy only holds if the change of height h is much smaller than the Earth's radius. If height h is not insignificant compared with the Earth's radius, the value of g is not the same over height h. The force of gravity on an object decreases with increased distance from the Earth.

Energy changes involving kinetic and potential energy

Consider an object of mass m released above the ground. If air resistance is negligible, the object gains speed as it falls. Its potential energy therefore decreases and its kinetic energy increases.

After falling through a vertical height Δh, its kinetic energy is equal to its loss of potential energy:

$$\frac{1}{2}mv^2 = mg\Delta h$$

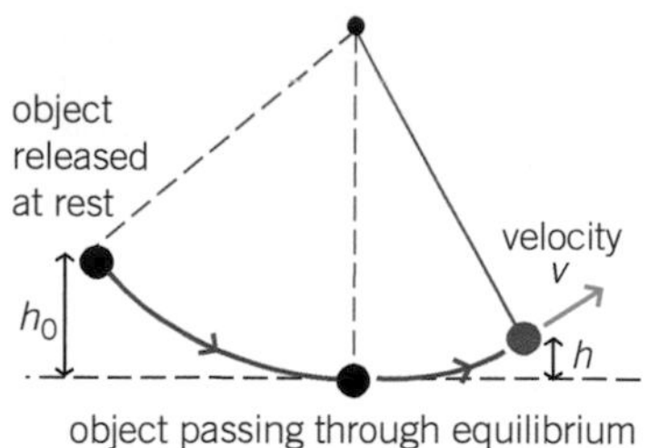

▲ **Figure 2** *A pendulum in motion*

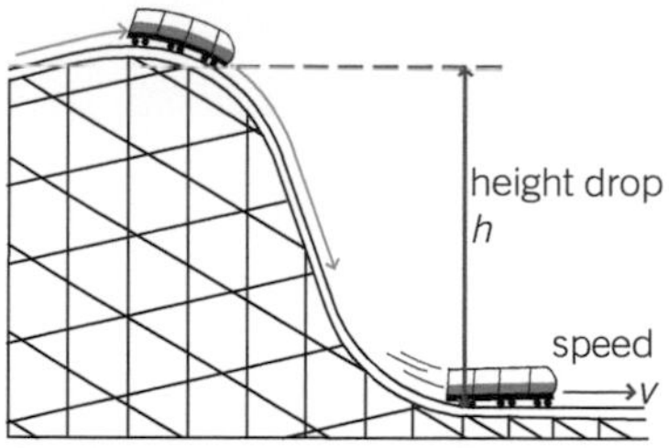

▲ **Figure 3** *At the fairground*

A pendulum bob

A pendulum bob is displaced from its equilibrium position and then released with the thread taut. The bob passes through the equilibrium position at maximum speed then slows down to reach maximum height on the other side of the equilibrium position. If its initial height above equilibrium position = h_0, then whenever its height above the equilibrium position = h, its speed v at this height is such that

kinetic energy = loss of potential energy from maximum height

$$\frac{1}{2}mv^2 = mg(h_0 - h)$$

A fairground vehicle of mass *m* on a downward track

If a fairground vehicle was initially at rest at the top of the track and its speed is v at the bottom of the track, then at the bottom of the track

- its kinetic energy = $\frac{1}{2}mv^2$
- its loss of potential energy = mgh, where h is the vertical distance between the top and the bottom of the track
- the work done to overcome friction and air resistance = $mgh - \frac{1}{2}mv^2$

Worked example

$g = 9.8\,\text{m}\,\text{s}^{-2}$

On a fairground ride, the track descends by a vertical drop of 55 m over a distance of 120 m along the track. A train of mass 2500 kg on the track reaches a speed of $30\,\text{m}\,\text{s}^{-1}$ at the bottom of the descent after being at rest at the top. Calculate **a** the loss of potential energy of the train, **b** its gain of kinetic energy, **c** the average frictional force on the train during the descent.

Solution

a Loss of potential energy = $mg\Delta h = 2500 \times 9.8 \times 55 = 1.35 \times 10^6\,\text{J}$

b Its gain of kinetic energy = $\frac{1}{2}mv^2 = 0.5 \times 2500 \times 30^2$
$= 1.13 \times 10^6\,\text{J}$

c Work done to overcome friction = $mg\Delta h - \frac{1}{2}mv^2$
$= 1.35 \times 10^6 - 1.13 \times 10^6$
$= 2.2 \times 10^5\,\text{J}$

Because the work done to overcome friction = frictional force × distance moved along track,

the frictional force = $\dfrac{\text{work done to overcome friction}}{\text{distance moved}}$

$= \dfrac{2.2 \times 10^5}{120} = 1830\,\text{N}$

Summary questions

$g = 9.8\,\text{m}\,\text{s}^{-2}$

1 A ball of mass 0.50 kg is thrown directly up at a speed of $6.0\,\text{m}\,\text{s}^{-1}$. Calculate:

a its kinetic energy at $6\,\text{m}\,\text{s}^{-1}$

b its maximum gain of potential energy

c its maximum height gain.

2 A cyclist of mass 80 kg (including the bicycle) freewheels from rest 500 m down a hill. The foot of the hill is 20 m lower than the cyclist's starting point and the cyclist reaches a speed of $12\,\text{m}\,\text{s}^{-1}$ at the foot of the hill. Calculate:

a the loss of potential energy

b the gain of kinetic energy of the cyclist and cycle

c the work done against friction and air resistance during the descent

d the average resistive force during the descent.

3 A fairground vehicle of total mass 1200 kg moving at a speed of $2\,\text{m}\,\text{s}^{-1}$ descends through a height of 50 m to reach a speed of $28\,\text{m}\,\text{s}^{-1}$ after travelling a distance of 75 m along the track. Calculate:

a its loss of potential energy

b its initial kinetic energy

c its kinetic energy after the descent

d the work done against friction

e the average frictional force on it during the descent.

5.3 Power

Learning objectives:

- → State which physical quantities are involved in power.
- → Explain how you could develop more power as you go up a flight of stairs.
- → Explain why a 100 W light bulb is more powerful than a 40 W light bulb when each works at the same mains voltage.

Specification reference: 3.2.7

Power and energy

Energy can be transferred from one object to another by means of:

- **work done** by a force due to one object making the other object move
- **heat transfer** from a hot object to a cold object. Heat transfer can be due to conduction or convection or radiation.

In addition, electricity, sound waves, and **electromagnetic radiation** such as light or radio waves, transfer energy.

In any energy transfer process, the more energy transferred per second, the greater the power of the transfer process. For example, in a tall building where there are two elevators of the same total weight, the more powerful elevator is the one that can reach the top floor fastest. In other words, its motor transfers energy from electricity at a faster rate than the motor of the other elevator. The energy transferred per second is the **power** of the motor.

Power is defined as the rate of transfer of energy.

The unit of power is the watt (W), equal to an energy transfer rate of 1 joule per second. Note that 1 kilowatt (kW) = 1000 W, and 1 megawatt (MW) = 10^6 W.

If energy ΔE is transferred steadily in time Δt,

$$\textbf{power } P = \frac{\Delta E}{\Delta t}$$

Where energy is transferred by a force doing work, the energy transferred is equal to the work done by the force. Therefore, the rate of transfer of energy is equal to the work done per second. In other words, if the force does work ΔW in time Δt,

$$\textbf{power } P = \frac{\Delta W}{\Delta t}$$

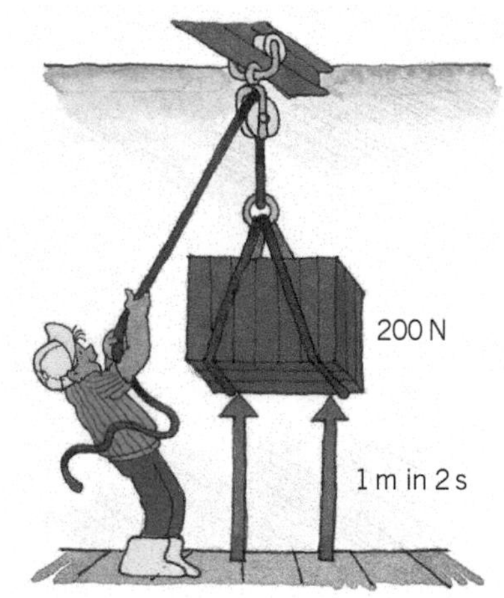

▲ **Figure 1** *A 100 watt worker*

Power measurements

Muscle power

Test your own muscle power by timing how long it takes you to walk up a flight of steps. To calculate your muscle power, you will also need to know your weight and the total height gain of the flight of steps.

- Your gain of potential energy = your weight × total height gain.
- Your muscle power = $\dfrac{\text{energy transferred}}{\text{time taken}} = \dfrac{\text{weight} \times \text{height gain}}{\text{time taken}}$.

For example, a person of weight 480 N who climbs a flight of stairs of height 10 m in 12 s has leg muscles of power 400 W (= 480 N × 10 m / 12 s). Each leg would therefore have muscles of power 200 W.

Electrical power

The power of a 12 V light bulb can be measured using a joulemeter, as shown in Figure 2. The joulemeter is read before and after the light bulb is switched on. The difference between the readings is the energy supplied to the light bulb. If the light bulb is switched on for a measured time, the power of the light bulb can be calculated from the energy supplied to it ÷ the time taken.

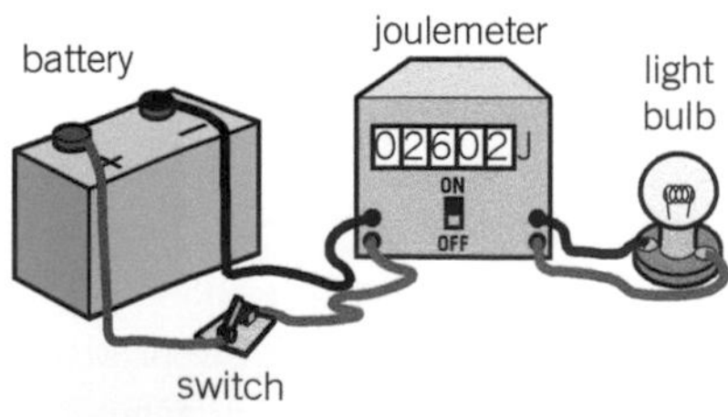

▲ **Figure 2** *Using a joulemeter*

Engine power

Vehicle engines, marine engines, and aircraft engines are all designed to make objects move. The output power of an engine is sometimes called its motive power.

When a powered object moves at constant velocity at constant height, the resistive forces (e.g., friction, air resistance, drag) are equal and opposite to the motive force.

The work done by the engine is transferred into the internal energy of the surroundings by the resistive forces.

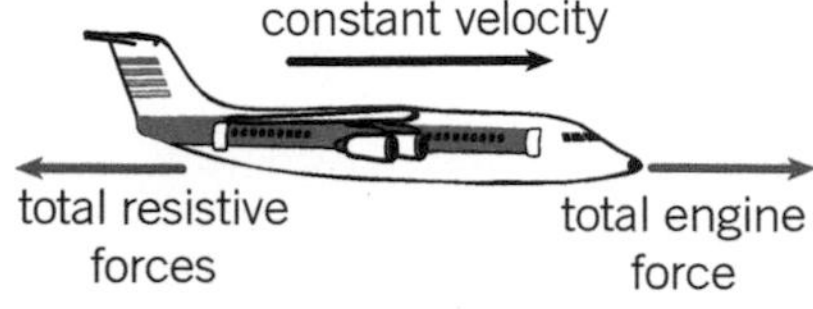

▲ **Figure 3** *Engine power*

For a powered vehicle driven by a constant force F moving at speed v,

the work done per second = force × distance moved per second

Therefore, the output power of the engine $\boldsymbol{P = Fv}$.

Worked example

An aircraft powered by engines that exert a force of 40 kN is in level flight at a constant velocity of $80\,\text{m}\,\text{s}^{-1}$. Calculate the output power of the engine at this speed.

Solution

$F = 40\,\text{kN} = 40\,000\,\text{N}$

$\text{Power} = \text{force} \times \text{velocity} = 40\,000 \times 80 = 3.2 \times 10^{6}\,\text{W}$

When a powered object gains speed, the output force exceeds the resistive forces on it.

Consider a vehicle that speeds up on a level road. The output power of its engine is the work done by the engine per second. The work done by the engine increases the kinetic energy of the vehicle and enables the vehicle to overcome the resistive forces acting on it. Because the resistive forces increase the internal energy of the surroundings,

$$\textbf{the motive power} = \frac{}{}\textbf{energy per second wasted due to the resistive force} + \textbf{the gain of kinetic energy per second}$$

Juggernaut physics

The maximum weight of a truck with six or more axles on UK roads must not exceed 44 tonnes, which corresponds to a total mass of 44 000 kg. This limit is set so as to prevent damage to roads and bridges. European Union regulations limit the output power of a large truck to a maximum of 6 kW per tonne. Therefore, the maximum output power of a 44 tonne truck is 264 kW. Prove for yourself that a truck with an output power of 264 kW moving at a constant speed of $31\,\text{m}\,\text{s}^{-1}$ (= 70 miles per hour) along a level road experiences a drag force of 8.5 kN.

Summary questions

$g = 9.8\,\text{m}\,\text{s}^{-2}$

1 A student of weight 450 N climbs 2.5 m up a rope in 18 s. Calculate:

 a the gain of potential energy of the student

 b the useful energy transferred per second.

2 Calculate the power of the engines of an aircraft at a speed of $250\,\text{m}\,\text{s}^{-1}$ if the total engine thrust to maintain this speed is 2.0 MN.

3 A rocket of mass 5800 kg accelerates vertically from rest to a speed of $220\,\text{m}\,\text{s}^{-1}$ in 25 s. Calculate:

 a its gain of potential energy

 b its gain of kinetic energy

 c the power output of its engine, assuming no energy is wasted due to air resistance.

4 Calculate the height through which a 5 kg mass would need to drop to lose potential energy equal to the energy supplied to a 100 W light bulb in 1 min.

5.4 Energy and efficiency

Learning objectives:

- → State the force that is mainly responsible for energy dissipation when mechanical energy is transferred from one store to another.
- → State the energy store that wasted energy is almost always transferred into.
- → Discuss whether any device can ever achieve 100% efficiency.

Specification reference: 3.2.7

Machines at work

A machine that lifts or moves an object applies a force to the object to move it. If the machine exerts a force F on an object to make it move through a distance s in the direction of the force, the work done, W, on the object by the machine can be calculated using the equation

work done, $W = Fs$

If the object moves at a constant velocity v due to this force being opposed by an equal and opposite force caused by friction, the object moves a distance $s = vt$ in time t.

Therefore, the output power of the machine

$$P_{OUT} = \frac{\text{work done by the machine}}{\text{time taken}} = \frac{Fvt}{t} = Fv$$

output power, $P_{OUT} = Fv$

where F = output force of the machine and v = speed of the object.

Examples

1 An electric motor operating a sliding door exerts a force of 125 N on the door, causing it to open at a constant speed of $0.40\,m\,s^{-1}$. The output power of the motor is $125\,N \times 0.40\,m\,s^{-1} = 50\,W$. The motor must therefore transfer 50 J every second to the sliding door while the door is being opened.

Friction in the motor bearings and also electrical resistance of the motor wires means that some of the electrical energy supplied to the motor is wasted. For example, if the motor is supplied with electrical energy at a rate of $150\,J\,s^{-1}$ and it transfers $50\,J\,s^{-1}$ to the door, the difference of $100\,J\,s^{-1}$ is wasted as a result of friction and electrical resistance in the motor.

2 A pulley system is used to raise a **load** of 80 N at a speed of $0.15\,m\,s^{-1}$ by means of a constant **effort** of 30 N applied to the system. Figure 1 shows the arrangement. Note that for every metre the load rises, the effort needs to act over a distance of 3 m because the load is supported by three sections of rope. The effort must therefore act at a speed of $0.45\,m\,s^{-1}$ ($= 3 \times 0.15\,m\,s^{-1}$).

- The work done on the load each second = load × distance raised per second = $80\,N \times 0.15\,m\,s^{-1} = 12\,J\,s^{-1}$
- The work done by the effort each second = effort × distance moved by the effort each second = $30\,N \times 0.45\,m\,s^{-1} = 13.5\,J\,s^{-1}$.

The difference of 1.5 W ($= 1.5\,J\,s^{-1}$) is the energy wasted each second in the pulley system. This is due to friction in the bearings and also because energy must be supplied to raise the lower pulley.

▲ **Figure 1** *Using pulleys*

Efficiency measures

Useful energy is energy transferred for a purpose. In any machine where friction is present, some of the energy transferred by the machine is wasted. In other words, not all the energy supplied to the machine is transferred for the intended purpose. For example, suppose a 500 W electric winch raises a weight of 150 N by 6.0 m in 10 s.

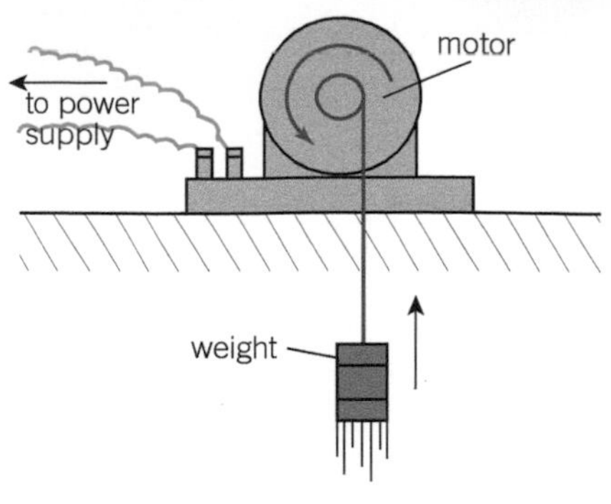

▲ **Figure 2** *Efficiency*

- The electrical energy supplied to the winch = 500 W × 10 s = 5000 J.
- The useful energy transferred by the machine = potential energy gain of the load = 150 N × 6 m = 900 J.

Therefore, in this example, 4100 J of energy is wasted.

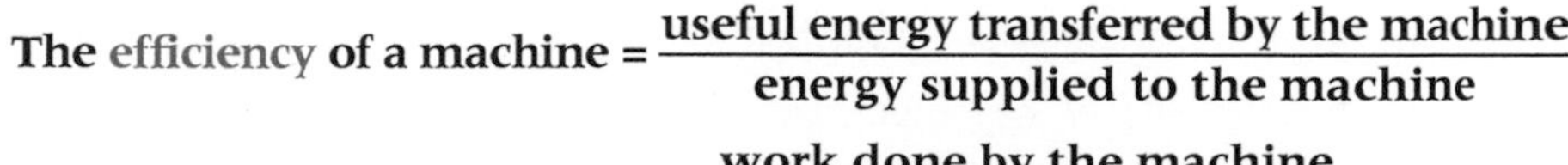

$$\textbf{The efficiency of a machine} = \frac{\textbf{useful energy transferred by the machine}}{\textbf{energy supplied to the machine}}$$

$$= \frac{\textbf{work done by the machine}}{\textbf{energy supplied to the machine}}$$

Notes:

- **Efficiency** can be expressed as $\frac{\text{the output power of a machine}}{\text{the input power to the machine}}$.
- **Percentage efficiency** = efficiency × 100%. In the above example, the efficiency of the machine is therefore 0.18 or 18%.

Improving efficiency

In any process or device where energy is transferred for a purpose, the efficiency of the transfer process or the device is the fraction of the energy supplied that is used for the intended purpose. The rest of the energy supplied is wasted, usually as heat and/or sound. The devices we use could be made more efficient. For example:

- A 100 W filament light bulb that is 12% efficient emits 12 J of energy as light for every 100 J of energy supplied to it by electricity. It therefore wastes 88 J of energy per second as heat.
- An energy efficient bulb with the same light output that is 80% efficient wastes just 3 J per second as heat. It gives the same light output for only 15 J of electrical energy supplied each second.

Is it possible to stop energy being wasted as heat? If the petrol engine of a car were insulated to stop heat loss, the engine would overheat. Less energy is wasted in an electric motor because it doesn't burn fuel so an electric car would be more efficient than a petrol car. However, the power stations where our electricity is generated are typically less than 40% efficient. This is partly because we need to burn fuel to produce the steam or hot gases used to drive turbines which turn the electricity generators. If the turbines were not kept cool, they would stop working because the pressure inside would build up and prevent the steam or hot gas from entering. Stopping the heat transfer to the cooling system would stop the generators working.

Summary questions

1 In a test of muscle efficiency, an athlete on an exercise bicycle pedals against a brake force of 30 N at a speed of 15 m s^{-1}.

  a Calculate the useful energy supplied per second by the athlete's muscles.

  b If the efficiency of the muscles is 25%, calculate the energy per second supplied to the athlete's muscles.

2 A 60 W electric motor raises a weight of 20 N through a height of 2.5 m in 8.0 s. Calculate:

  a the electrical energy supplied to the motor

  b the useful energy transferred by the motor

  c the efficiency of the motor.

3 A 200 W electric motor operating a sliding door exerts a force of 180 N on the door causing it to open at a speed of 0.35 m s^{-1}.
Calculate:

  a the output power of the motor

  b the percentage efficiency of the motor.

4 A vehicle engine has a power output of 6.2 kW and uses fuel which releases 45 MJ per kilogram when burned. At a speed of 30 m s^{-1} on a level road, the fuel usage of the vehicle is 18 km per kilogram. Calculate:

  a the time taken by the vehicle to travel 18 km at 30 m s^{-1}

  b the useful energy supplied by the engine in this time

  c the overall efficiency of the engine.

Practice questions: Chapter 5

1 **(a)** An electric car is fitted with a battery, which is used to drive an electric motor that drives the car. The battery has a maximum power output of 12 kW, which gives a maximum driving force of 600 N.
Calculate (i) the top speed of the car, (ii) the car's maximum range, if the battery lasts for 90 minutes at maximum power output without being recharged. *(4 marks)*

(b) A hybrid vehicle has a battery-driven electric motor and a petrol engine, which takes over from the electric motor when the vehicle has reached a certain speed. Above this speed, the petrol engine also recharges the battery. The vehicle has an overall fuel efficiency of 18 km per litre, compared with 10 km per litre for an equivalent petrol-only car, which has a carbon emission of 180 grams per km. Discuss the benefits of the use of such hybrid cars instead of petrol-only cars, in terms of carbon emissions, given the average annual distance travelled by a driver in the UK is about 20 000 km. The average annual carbon emission per UK household, including driving, is about 10 000 kg. *(5 marks)*

2 **Figure 1** shows apparatus that can be used to investigate energy changes.

▲ **Figure 1**

The trolley and the mass are joined by an inextensible string. In an experiment to investigate energy changes, the trolley is initially held at rest, and is then released so that the mass falls vertically to the ground.

(a) (i) State the energy changes of the falling mass.
(ii) Describe the energy changes that take place in this system. *(4 marks)*

(b) State what measurements would need to be made to investigate the conservation of energy. *(2 marks)*

(c) Describe how the measurements in part (b) would be used to investigate the conservation of energy. *(4 marks)*

AQA, 2006

3 A small hydroelectric power station uses water which falls through a height of 4.8 m.

(a) Calculate the change in potential energy of a 1.0 kg mass of water falling through a vertical height of 4.8 m. *(2 marks)*

(b) Calculate the maximum power available from the water passing through this power station when water flows through it at a rate of 6.7×10^7 kg per hour. *(3 marks)*

(c) State *two* factors that affect the usefulness of hydroelectric power stations for electricity production. *(2 marks)*

AQA, 2002

4 **Figure 2** represents the motion of a car of mass 1.4×10^3 kg, travelling in a straight line.

(a) Describe, without calculation, how the *resultant* force acting on the car varies over this ten second interval. *(2 marks)*

(b) Calculate the maximum kinetic energy of the car. *(2 marks)*

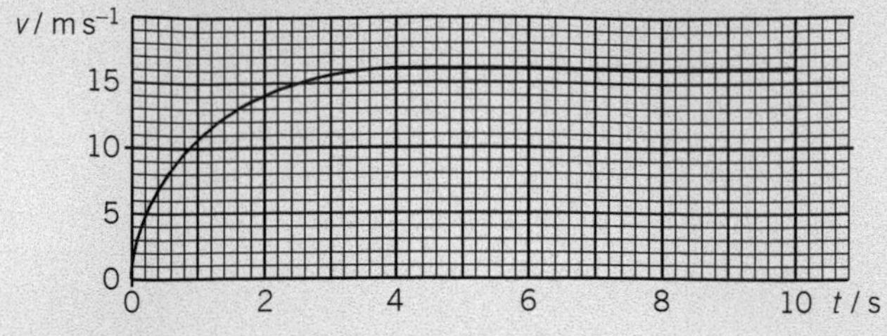

▲ **Figure 2**

(c) At some time later, when the car is travelling at a steady speed of $30\,\mathrm{m\,s^{-1}}$, the useful power developed by the engine is 20 kW. Calculate the driving force required to maintain this speed. *(2 marks)*

AQA, 2002

5 A skydiver of mass 70 kg jumps from a stationary balloon and reaches a speed of $45\,\mathrm{m\,s^{-1}}$ after falling a distance of 150 m.

(a) Calculate the skydiver's:

(i) loss of gravitational potential energy

(ii) gain in kinetic energy. *(4 marks)*

(b) The difference between the loss of gravitational potential energy and the gain in kinetic energy is equal to the work done against air resistance. Use this fact to calculate:

(i) the work done against air resistance

(ii) the average force due to air resistance acting on the skydiver. *(3 marks)*

AQA, 2004

6 A car travels at constant velocity along a horizontal road.

(a) The car has an effective power output of 18 kW and is travelling at a constant velocity of $10\,\mathrm{m\,s^{-1}}$. Show that the total resistive force acting is 1800 N. *(1 mark)*

(b) The total resistive force consists of two components. One of these is a constant frictional force of 250 N and the other is the force of air resistance, which is proportional to the square of the car's speed. Calculate:

(i) the force of air resistance when the car is travelling at $10\,\mathrm{m\,s^{-1}}$

(ii) the force of air resistance when the car is travelling at $20\,\mathrm{m\,s^{-1}}$

(iii) the effective output power of the car required to maintain a constant speed of $20\,\mathrm{m\,s^{-1}}$ on a horizontal road. *(4 marks)*

AQA, 2001

7 **(a)** A ball bearing is released when near the top of a tall cylinder containing oil. Discuss the energy changes which take place when the ball bearing (i) accelerates from rest, (ii) travels at constant velocity. *(6 marks)*

(b) A pump-operated hydraulic jack is used to raise a large object of mass 470 kg. The jack is used by pushing down on the handle of a lever connected to the pump. During each stroke, a force of 150 N is applied downward on the handle, which is moved through a distance of 0.42 m. Fifty-two strokes of the handle are needed to raise the object through a vertical height of 0.58 m.
Calculate the efficiency of this process. *(5 marks)*

6 Materials

6.1 Density

Learning objectives:

- → Define density.
- → State the unit of density.
- → Measure the density of an object.

Specification reference: 3.2.9

Density and its measurement

Lead is much more dense than aluminium. Sea water is more dense than tap water. To find how much more dense one substance is compared with another, we can measure the mass of equal volumes of the two substances. The substance with the greater mass in the same volume is more dense. For example, a lead sphere of volume 1 cm^3 has a mass of 11.3 g whereas an aluminium sphere of the same volume has a mass of 2.7 g.

The density of a substance is defined as its mass per unit volume.

For a certain amount of a substance of mass m and volume V, its density ρ (pronounced 'rho') may be calculated using the equation

$$\text{density, } \rho = \frac{m}{V}$$

The unit of density is the kilogram per cubic metre ($kg\,m^{-3}$).

Rearranging the above equation gives $m = \rho V$ or $V = \frac{m}{\rho}$.

Table 1 on the next page shows the density of some common substances in $kg\,m^{-3}$.

You can see that gases are much less dense than solids or liquids. This is because the average separation between the molecules in a gas is much greater than in a liquid or solid.

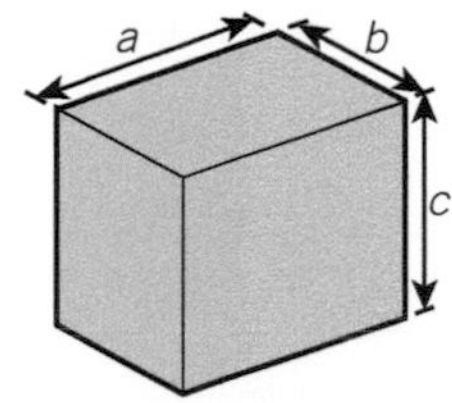

a *Volume of cuboid = a × b × c*

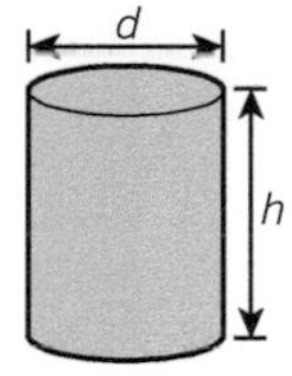

b *Volume of cylinder* $= \frac{\pi d^2}{4} \times h$

▲ **Figure 1** *Volume equations*

Density measurements

An unknown substance can often be identified if its density is measured and compared with the density of known substances. The following procedures could be used to measure the density of a substance.

1 A regular solid

- Measure its mass using a top pan balance.
- Measure its dimensions using vernier calipers or a micrometer and calculate its volume using the appropriate equation (e.g., for a sphere of radius r, volume $= \frac{4}{3}\pi r^3$ – see Figure 1 for other volume equations). Calculate the density from mass/volume.

2 A liquid

- Measure the mass of an empty measuring cylinder. Pour some of the liquid into the measuring cylinder and measure the volume of the liquid directly. Use as much liquid as possible to reduce the percentage error in your measurement.
- Measure the mass of the cylinder and liquid to enable the mass of the liquid to be calculated. Calculate the density from mass/volume.

3 An irregular solid

- Measure the mass of the object.
- Immerse the object on a thread in liquid in a measuring cylinder and observe the increase in the liquid level. This is the volume of the object.
- Calculate the density of the object from its mass/volume.

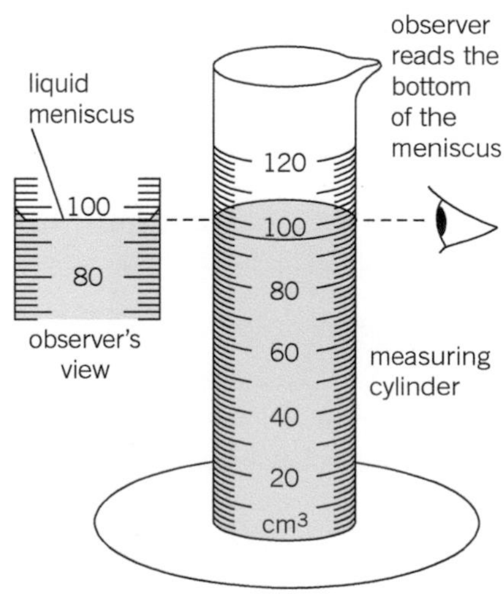

▲ **Figure 2** *Using a measuring cylinder*

▼ **Table 1** *Densities of common substances*

Substance	Density / kg m^{-3}
Air	1.2
Aluminium	2700
Copper	8900
Gold	19 300
Hydrogen	0.083
Iron	7900
Lead	11 300
Oxygen	1.3
Silver	10 500
Water	1000

Density of alloys

An alloy is a solid mixture of two or more metals. For example, brass is an alloy of copper and zinc that has good resistance to corrosion and wear.

For an alloy, of volume V, that consists of two metals A and B,

- if the volume of metal A = V_A, the mass of metal A = $\rho_A V_A$, where ρ_A is the density of metal A
- if the volume of metal B = V_B, the mass of metal B = $\rho_B V_B$, where ρ_B is the density of metal B.

Therefore, the mass of the alloy, $m = \rho_A V_A + \rho_B V_B$.

Hence the density of the alloy $\rho = \frac{m}{V} = \frac{\rho_A V_A + \rho_B V_B}{V} = \frac{\rho_A V_A}{V} + \frac{\rho_B V_B}{V}$.

Worked example

A brass object consists of $3.3 \times 10^{-5}\,m^3$ of copper and $1.7 \times 10^{-5}\,m^3$ of zinc. Calculate the mass and the density of this object. The density of copper = $8900\,kg\,m^{-3}$. The density of zinc = $7100\,kg\,m^{-3}$.

Solution

Mass of copper = density of copper × volume of copper

$= 8900 \times 3.3 \times 10^{-5}\,m^3 = 0.294\,kg$

Mass of zinc = density of zinc × volume of zinc

$= 7100 \times 1.7 \times 10^{-5}\,m^3 = 0.121\,kg$

Total mass, $m = 0.294 + 0.121 = 0.415\,kg$

Total volume, $V = 5.0 \times 10^{-5}\,m^3$

Density of alloy $\rho = \frac{m}{V} = \frac{0.415\,kg}{5.0 \times 10^{-5}\,m^3} = 8300\,kg\,m^{-3}$

Study tip

More about units

mass: $1\,kg = 1000\,g$

length: $1\,m = 100\,cm$
$= 1000\,mm$

volume: $1\,m^3 = 10^6\,cm^3$

density: $1000\,kg\,m^{-3} = \frac{10^6\,g}{10^6\,cm^3}$
$= 1\,g\,cm^{-3}$

Study tip

Unit errors are commonplace in density calculations. Avoid such errors by writing the unit and the numerical value of each quantity in your working.

Summary questions

1. A rectangular brick of dimensions 5.0 cm × 8.0 cm × 20.0 cm has a mass of 2.5 kg. Calculate **a** its volume, **b** its density.
2. An empty paint tin of diameter 0.150 m and of height 0.120 m has a mass of 0.22 kg. It is filled with paint to within 7 mm of the top. Its total mass is then 6.50 kg. Calculate **a** the mass, **b** the volume, **c** the density of the paint in the tin.
3. A solid steel cylinder has a diameter of 12 mm and a length of 85 mm. Calculate **a** its volume in m^3, **b** its mass in kg. The density of steel = $7800\,kg\,m^{-3}$.
4. An alloy tube of volume $1.8 \times 10^{-4}\,m^3$ consists of 60% aluminium and 40% magnesium by volume. Calculate **a** the mass of **i** aluminium, **ii** magnesium in the tube, **b** the density of the alloy. The density of aluminium = $2700\,kg\,m^{-3}$. The density of magnesium = $1700\,kg\,m^{-3}$.

6.2 Springs

Learning objectives:

- → Discuss whether there is any limit to the linear graph of force against extension for a spring.
- → Define the spring constant, and state its unit of measurement.
- → If the extension of a spring is doubled, calculate how much more energy it stores.

Specification reference: 3.2.9

Hooke's law

A stretched spring exerts a pull on the object holding each end of the spring. This pull, referred to as the *tension* in the spring, is equal and opposite to the force needed to stretch the spring. The more a spring is stretched, the greater the tension in it. Figure 1 shows a stretched spring supporting a weight at rest. This arrangement may be used to investigate how the tension in a spring depends on its extension from its unstretched length. The measurements may be plotted on a graph of tension against extension, as shown in Figure 2. The graph shows that the force needed to stretch a spring is proportional to the extension of the spring. This is known as **Hooke's law**, after its discoverer, Robert Hooke, a seventeenth century scientist.

Hooke's law states that the force needed to stretch a spring is directly proportional to the extension of the spring from its natural length.

Hooke's law may be written as

$$\textbf{force, } F = k\Delta L$$

where k is the spring constant (sometimes referred to as the stiffness constant) and ΔL is the extension from its natural length L.

- The greater the value of k, the stiffer the spring is. The unit of k is $\mathrm{N\,m^{-1}}$.
- The graph of F against ΔL is a straight line of gradient k through the origin.
- If a spring is stretched beyond its **elastic limit**, it does not regain its initial length when the force applied to it is removed.
- In AS/A Level Maths students may meet Hooke's law in the form $F = \lambda\frac{\Delta L}{L}$, where L is the unstretched length of the spring and λ (= kL) is the spring modulus. Note that λ is not needed on this course.

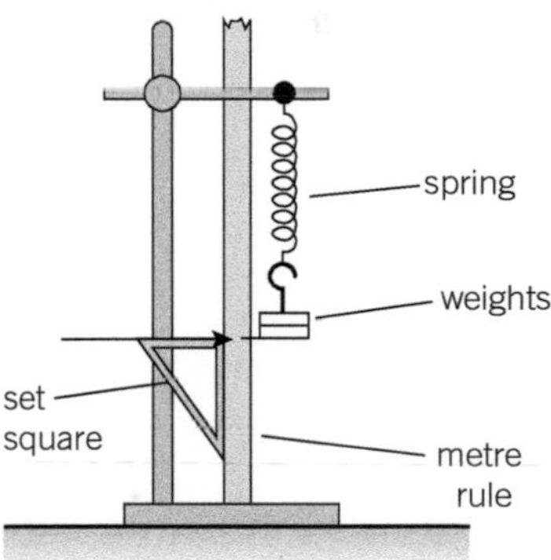

▲ **Figure 1** *Testing the extension of a spring*

Worked example

A vertical steel spring fixed at its upper end has an unstretched length of 300 mm. Its length is increased to 385 mm when a 5.0 N weight attached to the lower end is at rest. Calculate:

a the spring constant

b the length of the spring when it supports an 8.0 N weight at rest.

Solution

a Use $F = k\Delta L$ with $F = 5.0\,\mathrm{N}$ and $\Delta L = 385 - 300\,\mathrm{mm} = 85\,\mathrm{mm} = 0.085\,\mathrm{m}$.

Therefore $k = \frac{F}{\Delta L} = \frac{5.0\,\mathrm{N}}{0.085\,\mathrm{m}} = 59\,\mathrm{N\,m^{-1}}$.

b Use $F = k\Delta L$ with $F = 8.0\,\mathrm{N}$ and $k = 59\,\mathrm{N\,m^{-1}}$ to calculate ΔL.

$\Delta L = \frac{F}{k} = \frac{8.0\,\mathrm{N}}{59\,\mathrm{N\,m^{-1}}} = 0.136\,\mathrm{m}$

Therefore the length of the spring = 0.300 m + 0.136 m = 0.436 m.

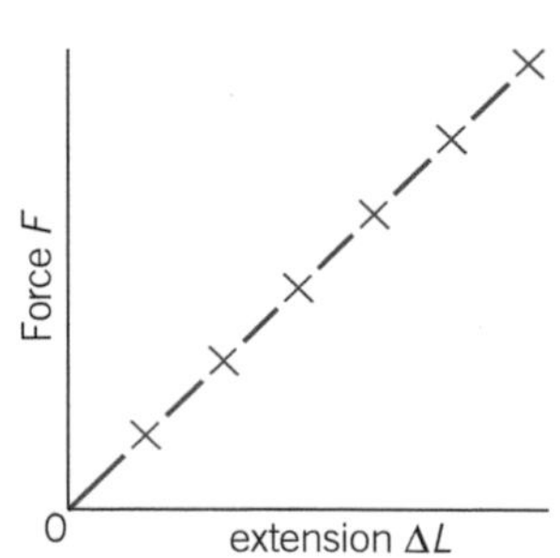

▲ **Figure 2** *Hooke's law*

Spring combinations

Springs in parallel

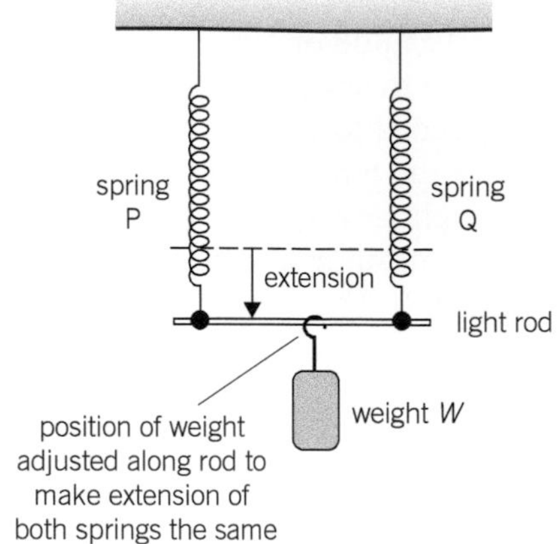

▲ **Figure 3** *Two springs in parallel*

Figure 3 shows a weight supported by means of two springs P and Q in parallel with each other. The extension, ΔL, of each spring is the same. Therefore

- the force needed to stretch P, $F_P = k_P \, \Delta L$
- the force needed to stretch Q, $F_Q = k_Q \, \Delta L$

where k_P and k_Q are the spring constants of P and Q, respectively.

Since the weight W is supported by both springs,

$W = F_P + F_Q = k_P \, \Delta L + k_Q \, \Delta L = k \, \Delta L$

where the effective spring constant, $k = k_P + k_Q$.

Springs in series

Figure 4 shows a weight supported by means of two springs joined end-on in series with each other. The tension in each spring is the same and is equal to the weight W.

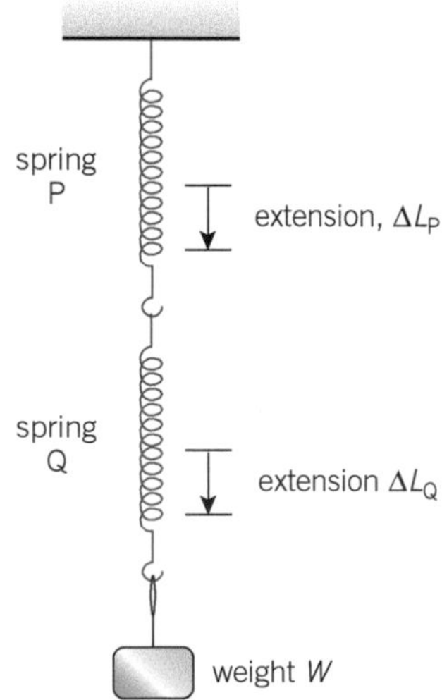

▲ **Figure 4** *Two springs in series*

Therefore:

- the extension of spring P, $\Delta L_P = \dfrac{W}{k_P}$
- the extension of spring Q, $\Delta L_Q = \dfrac{W}{k_Q}$

where k_P and k_Q are the spring constants of P and Q, respectively.

Therefore the total extension, $\Delta L = \Delta L_P + \Delta L_Q = \dfrac{W}{k_P} + \dfrac{W}{k_Q} = \dfrac{W}{k}$

where k, the effective spring constant, is given by the equation

$$\frac{1}{k} = \frac{1}{k_P} + \frac{1}{k_Q}$$

The energy stored in a stretched spring

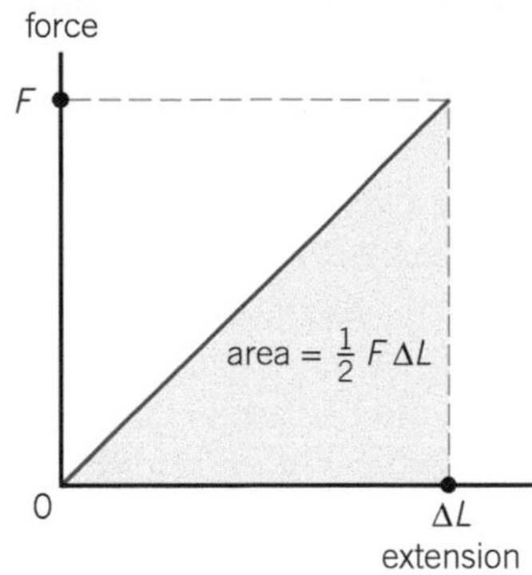

▲ **Figure 5** *Energy stored in a stretched spring*

Elastic potential energy is stored in a stretched spring. If the spring is suddenly released, the elastic energy stored in it is suddenly transferred into kinetic energy of the spring (and of any object attached to it or catapulted by it). The work done to stretch a spring by extension ΔL from its unstretched length $= \frac{1}{2} F \Delta L$, where F is the force needed to stretch the spring to extension ΔL. The work done on the spring is stored as elastic potential energy. Therefore, the elastic potential energy E_p in the spring $= \frac{1}{2} F \Delta L$. Also, since $F = k \Delta L$, where k is the spring constant, then $E_p = \frac{1}{2} k \Delta L^2$.

Elastic potential energy stored in a stretched spring,

$$E_p = \frac{1}{2} F \Delta L = \frac{1}{2} k \Delta L^2$$

Study tip

Energy stored by a spring is $\frac{1}{2} F \Delta L$, not $F \Delta L$.

Synoptic link

You have met work done in Topic 5.1, Work and energy.

Summary questions

$g = 9.8\,\text{m}\,\text{s}^{-2}$

1 A steel spring has a spring constant of $25\,\text{N}\,\text{m}^{-1}$. Calculate:

- **a** the extension of the spring when the tension in it is equal to 10 N
- **b** the tension in the spring when it is extended by 0.50 m from its unstretched length.

2 Two identical steel springs of length 250 mm are suspended vertically side by side from a fixed point. A 40 N weight is attached to the ends of the two springs. The length of each spring is then 350 mm. Calculate:

- **a** the tension in each spring
- **b** the extension of each spring
- **c** the spring constant of each spring.

3 Repeat **Q 2a** and **b** for the two springs in series and vertical.

4 An object of mass 0.150 kg is attached to the lower end of a vertical spring of unstretched length 300 mm, which is fixed at its upper end. With the object at rest, the length of the spring becomes 420 mm as a result. Calculate:

- **a** the spring constant
- **b** the energy stored in the spring
- **c** the weight that needs to be added to extend the spring to 600 mm.

6.3 Deformation of solids

Learning objectives:

- → Relate stress to force, and strain to extension.
- → Describe what is meant by the Young modulus.
- → Define tensile.
- → Explain why we bother with stress and strain, when force and extension are more easily measured.

Specification reference 3.2.9

Force and solid materials

Look around at different materials and think about the effect of force on each material. To stretch or twist or compress the material, a pair of forces is needed. For example, stretching a rubber band requires the rubber band to be pulled by a force at either end. Some materials, such as rubber, bend or stretch easily. The **elasticity** of a solid material is its ability to regain its shape after it has been deformed or distorted and the forces that deformed it have been released. Deformation that stretches an object is **tensile**, whereas deformation that compresses an object is **compressive**.

Figure 1 in Topic 6.2 shows how to test different materials to see how easily they stretch. In each case, the material is held at its upper end and loaded by hanging weights at its lower end. A set square or pointer attached to the bottom of the weights may be used to measure the extension of the material, as the weight of the load is increased in steps then decreased to zero. The extension of the strip of material at each step is its increase of length from its unloaded length. The tension in the material is equal to the weight. The measurements may be plotted as a tension–extension graph, as shown in Figure 1.

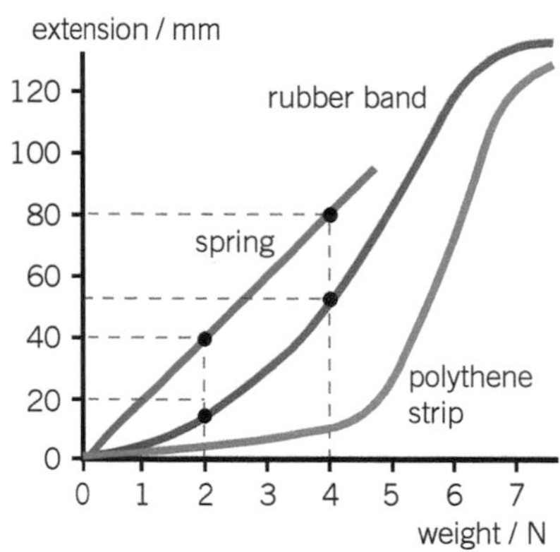

▲ **Figure 1** *Typical extension curves*

- A steel spring gives a straight line (blue line in Figure 1), in accordance with Hooke's law (see Topic 6.2).
- A rubber band at first extends easily when it is stretched. However, it becomes fully stretched and very difficult to stretch further when it has been lengthened considerably (red line in Figure 1).
- A polythene strip 'gives' and stretches easily after its initial stiffness is overcome. However, after 'giving' easily, it extends little and becomes difficult to stretch (green line in Figure 1).

Tensile stress and tensile strain

The extension of a wire under tension may be measured using Searle's apparatus, as shown in Figure 2 on the next page (or similar apparatus with a vernier scale). A micrometer attached to the control wire is adjusted so the spirit level between the control and test wire is horizontal. When the test wire is loaded, it extends slightly, causing the spirit level to drop on one side. The micrometer is then readjusted to make the spirit level horizontal again. The change of the micrometer reading is therefore equal to the extension. The extension may be measured for different values of tension by increasing the test weight in steps.

For a wire of length L and area of cross section A under tension:

- The **tensile stress** in the wire, $\sigma = \frac{T}{A}$, where T is the tension. The unit of stress is the **pascal** (Pa) equal to $1\,\mathrm{N\,m^{-2}}$.
- The **tensile strain** in the wire, $\varepsilon = \frac{\Delta L}{L}$, where ΔL is the extension (increase in length) of the wire. Strain is a ratio and therefore has no unit.

Study tip

Remember that the extension is always measured from the original (unstretched) length of the object.

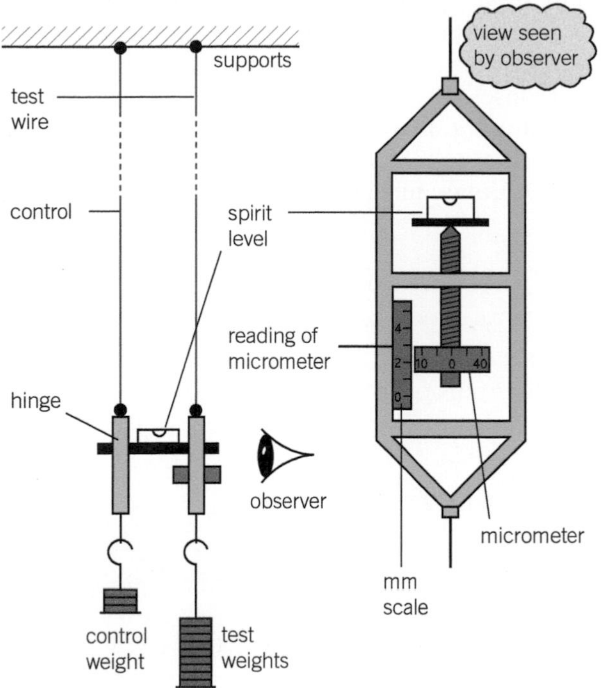

▲ **Figure 2** *Searle's apparatus*

Figure 3 shows how the tensile stress in a wire varies with tensile strain.

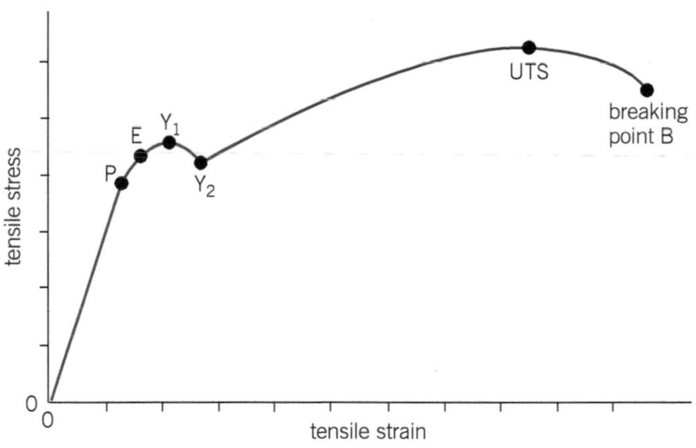

▲ **Figure 3** *Tensile stress versus tensile strain for a metal wire*

- From 0 to the limit of proportionality P, the tensile stress is proportional to the tensile strain.
  The value of stress/strain is a constant, known as the **Young modulus** of the material. This is equal to the gradient of the straight section OP of the stress strain graph.

$$\textbf{Young modulus, } E = \frac{\textbf{tensile stress, } \sigma}{\textbf{tensile strain, } \varepsilon} = \frac{T}{A} \div \frac{\Delta L}{L} = \frac{TL}{A \Delta L}$$

For a wire of uniform diameter d, the area of cross section

$$A = \frac{\pi d^2}{4}$$

- Beyond P, the line curves and continues beyond the elastic limit E to the **yield point** Y_1, which is where the wire weakens temporarily. The **elastic limit** is the point beyond which the wire is permanently stretched and suffers **plastic deformation**.

Hint

Some materials can be strong and brittle, whilst others can be plastic and tough. Toughness is a measure of the energy needed to break a material.

Study tip

Don't forget the italicised words in the definitions below:

Stress is tension per *unit cross-sectional area*.

Strain is extension per *unit length*.

- Beyond Y_2, a small increase in the tensile stress causes a large increase in tensile strain as the material of the wire undergoes plastic flow. Beyond maximum tensile stress, the **ultimate tensile stress** (UTS), the wire loses its strength, extends, and becomes narrower at its weakest point. Increase of tensile stress occurs due to the reduced area of cross section at this point until the wire breaks at point B. The ultimate tensile stress is sometimes called the **breaking stress**.

Worked example

A crane fitted with a steel cable of uniform diameter 2.3 mm and length 28 m is used to lift an iron girder of weight 3200 N off the ground. Calculate the extension of the cable when it supports the girder at rest.

The Young modulus for steel = 2.1×10^{11} Pa

Solution

Tension $T = 3200\,\text{N}$, $L = 28\,\text{m}$

Area of cross section of wire $= \dfrac{\pi(2.3 \times 10^{-3})^2}{4} = 4.15 \times 10^{-6}\,\text{m}^2$

To find the extension, rearranging the Young modulus equation

$E = \dfrac{TL}{A\Delta L}$ gives

$$\Delta L = \frac{TL}{AE} = \frac{3200 \times 28}{4.15 \times 10^{-6} \times 2.1 \times 10^{11}} = 0.103\,\text{m}$$

Stress–strain curves for different materials

The **stiffness** of different materials can be compared using the **gradient** of the stress–strain line, which is equal to the Young modulus of the material. Thus steel is stiffer than copper (Figure 4).

The **strength** of a material is its ultimate tensile stress (UTS), which is its maximum tensile stress. Steel is stronger than copper because its maximum tensile stress is greater.

A **brittle** material snaps without any noticeable yield. For example, glass breaks without any give.

A **ductile** material can be drawn into a wire. Copper is more ductile than steel.

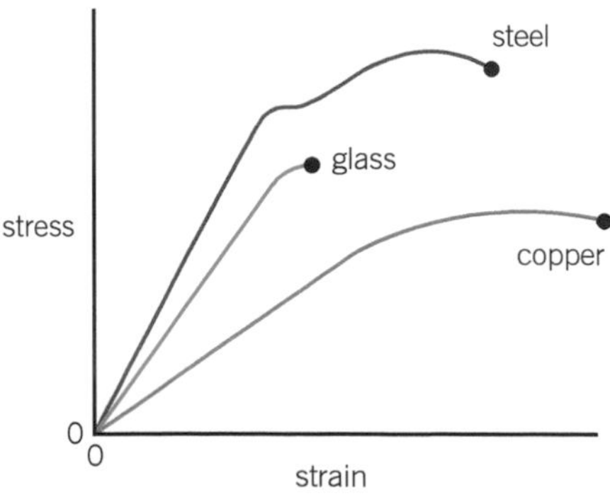

▲ **Figure 4** *Stress–strain curves*

Summary questions

1. Calculate the tensile stress in a wire of diameter 0.25 mm when the tension in the wire is 50 N.
2. A metal wire of diameter 0.23 mm and of unstretched length 1.405 m is suspended vertically from a fixed point. When a 40 N weight is suspended from the lower end of the wire, the wire stretches by an extension of 10.5 mm. Calculate the Young modulus of the wire material.
3. A vertical steel wire of length 2.5 m and diameter 0.35 mm supports a weight of 90 N. Calculate:
   - **a** the tensile stress in the wire
   - **b** the extension of the wire. The Young modulus of steel = 2.1×10^{11} Pa.
4. Compare the two stress–strain curves in Figure 5. Use the curves to identify
   - **a** the material, X or Y, that is **i** stiffest, **ii** strongest
   - **b** the material, X or Y, that is **i** brittle, **ii** ductile.

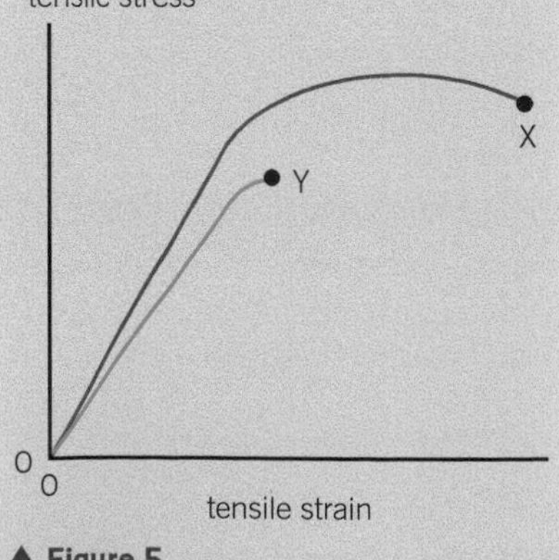

▲ **Figure 5**

6.4 More about stress and strain

Loading and unloading of different materials

How does the strength of a material change as a result of being stretched? Figure 1 in Topic 6.2 may be used to investigate this question. The tension in a strip of material is increased by increasing the weight it supports in steps. At each step, the extension of the material is measured. Typical results for different materials are shown in Figure 1. For each material, the loading curve and the subsequent unloading curve are shown.

- For a metal wire, its loading and unloading curves are the same, provided its elastic limit is not exceeded. This means the wire returns to its original length when unloaded. However, beyond its elastic limit, the unloading line is parallel to the loading line. In this case, the wire is slightly longer when unloaded – it has a **permanent extension**.
- For a rubber band, it returns to the same unstretched length when it is unloaded. However, the unloading curve is below the loading curve except at zero and maximum extensions.
- For a polythene strip, the extension during unloading is also greater than during loading. However, the strip does not return to the same initial length when it is completely unloaded. The polythene strip has a low limit of proportionality *and* suffers **plastic deformation**.

Learning objectives:

→ Predict whether a metal wire stretched below its elastic limit will return to its original length.

→ Describe what happens when a metal wire is stretched beyond its elastic limit and then unloaded.

→ Compare the deformation of other materials such as rubber and polythene with a metal wire.

Specification reference: 3.2.9

Strain energy

As explained in Topic 6.2, the area under the line of a force–extension graph is equal to the work done to stretch the wire. The work done to deform an object increases its elastic store of energy and is referred to as **strain energy**. Consider the energy transfers for each of the three materials in Figure 1 when each material is loaded then unloaded.

Application

Polythene and rubber

Polythene is an example of a polymer, which means that its molecules are long chains of atoms. Before a strip of polythene is stretched, the molecules are tangled together. Weak bonds, or cross-links, form between the molecules. When polythene is under tension, it easily stretches as the weak cross-links break. In this stretched state, new weak cross-links form, and, when the tension is removed, the polythene strip stays stretched.

Rubber is also a polymer but its molecules are curled and tangled together when it is in an unstretched state. When placed under tension, its molecules are straightened out but these curl up again when the tension is removed – the rubber regains its initial length.

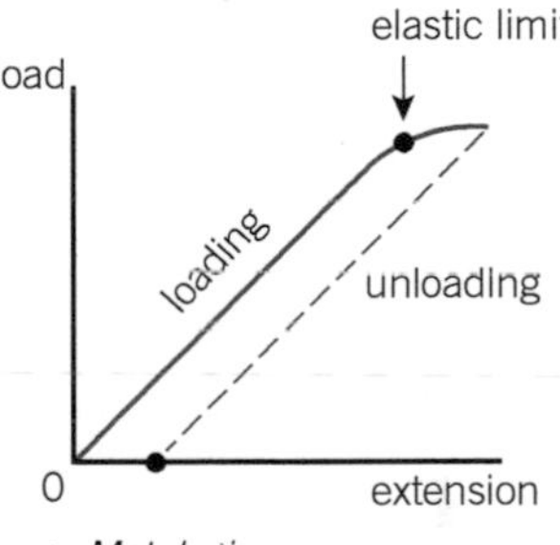

a *Metal wire*

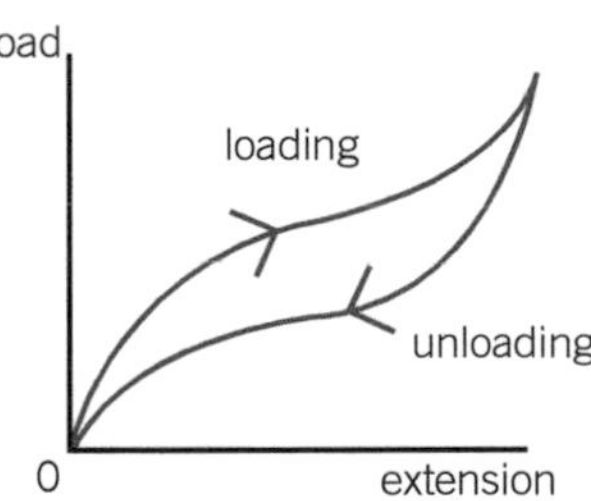

b *Rubber band*

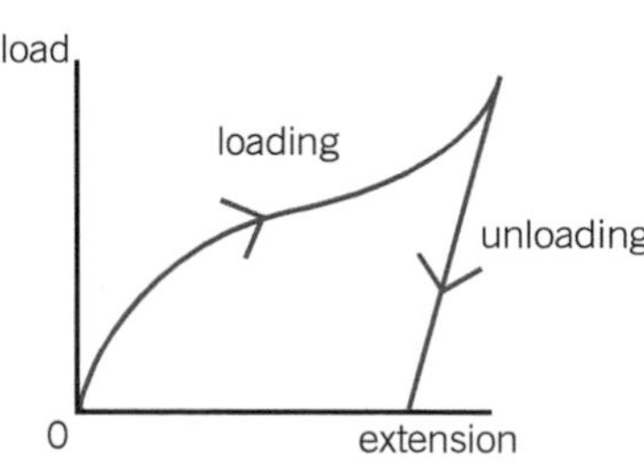

c *Polythene strip*

▲ **Figure 1** *Loading and unloading curves*

1 Metal wire (or spring)

Provided the limit of proportionality is not exceeded, to stretch a wire to an extension ΔL, the work done $= \frac{1}{2} T\Delta L$, where T is the tension in the wire at this extension. Because the elastic limit is not reached, the work done is stored as elastic energy in the wire.

the elastic energy stored in a stretched wire $= \frac{1}{2} T\Delta L$

Because the graph of tension against extension is the same for unloading as for loading, all the energy stored in the wire can be recovered when the wire is unloaded.

Worked example

A steel wire of uniform diameter 0.35 mm and of length 810 mm is stretched to an extension of 2.5 mm. Calculate **a** the tension in the wire, **b** the elastic energy stored in the wire.

The Young modulus for steel $= 2.1 \times 10^{11}$ Pa

Solution

a Extension, $\Delta L = 2.5\,\text{mm} = 2.5 \times 10^{-3}\,\text{m}$

Area of cross section of wire $= \dfrac{\pi(0.35 \times 10^{-3})^2}{4} = 9.6 \times 10^{-8}\,\text{m}^2$

To find the tension, rearranging the Young modulus equation

$E = \dfrac{TL}{A\,\Delta L}$ gives

$$T = \frac{EA\,\Delta L}{L} = \frac{2.1 \times 10^{11} \times 9.6 \times 10^{-8} \times 2.5 \times 10^{-3}}{0.810} = 62\ \text{N}$$

b Elastic energy stored in the wire $= \frac{1}{2} T\Delta L = 0.5 \times 62 \times 2.5 \times 10^{-3}$ $= 7.8 \times 10^{-2}$ J.

2 Rubber band

The work done to stretch the rubber band is represented by the area under the loading curve. The work done by the rubber band, when it is unloaded, is represented by the area under the unloading curve. The area between the loading curve and the unloading curve therefore represents the difference between energy stored in the rubber band when it is stretched and the useful energy recovered from it when it is unstretched. The difference occurs because some of the energy stored in the rubber band becomes the internal energy of the molecules when the rubber band unstretches.

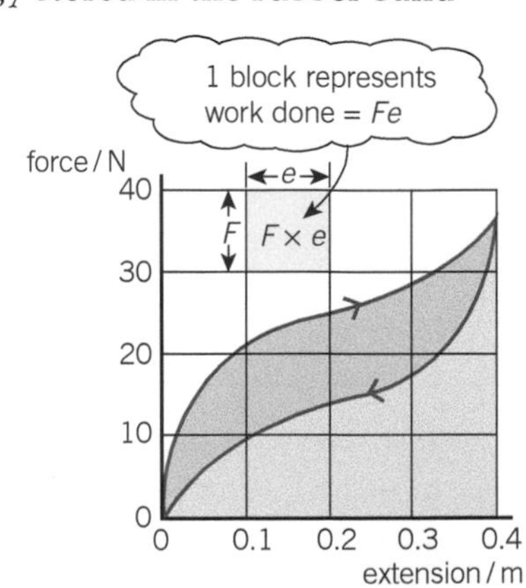

▲ **Figure 2** *Energy changes when loading and unloading rubber*

3 Polythene

As it does not regain its initial length, the area between the loading and unloading curves represents work done to deform the material permanently, as well as internal energy retained by the polythene when it unstretches.

Summary questions

The Young modulus for steel $= 2.1 \times 10^{11}$ Pa

1. A vertical steel cable of diameter 24 mm and of length 18 m supports a weight of 1500 N attached to its lower end. Calculate **a** the tensile stress in the cable, **b** the extension of the cable, **c** the elastic energy stored in the cable, assuming its elastic limit has not been reached.
2. A vertical steel wire of diameter 0.28 mm and of length 2.0 m is fixed at its upper end, and has a weight of 15 N suspended from its lower end. Calculate **a** the extension of the wire, **b** the elastic energy stored in the wire.
3. A steel bar of length 40 mm and cross-sectional area $4.5 \times 10^{-4}\,\text{m}^2$ is placed in a vice and compressed by 0.20 mm when the vice is tightened. Calculate **a** the compressive force exerted on the bar, **b** the work done to compress it.
4. Figure 2 shows a force against extension curve for rubber band. Use the graph to determine **a** the work done to stretch the rubber band to an extension of 0.40 m, **b** the internal energy retained by the rubber band when it unstretches.

Practice questions: Chapter 6

1 **(a)** Define the *density* of a material. (*1 mark*)

(b) Brass, an alloy of copper and zinc, consists of 70% *by volume* of copper and 30% *by volume* of zinc.

density of copper = $8.9 \times 10^3\,\text{kg}\,\text{m}^{-3}$
density of zinc = $7.1 \times 10^3\,\text{kg}\,\text{m}^{-3}$

(i) Determine the mass of copper and the mass of zinc required to make a rod of brass of volume $0.80 \times 10^{-3}\,\text{m}^3$.

(ii) Calculate the density of brass. (*5 marks*)

AQA, 2004

2 **Figure 1** shows a lorry of mass $1.2 \times 10^3\,\text{kg}$ parked on a platform used to weigh vehicles. The lorry compresses the spring that supports the platform by 0.030 m.

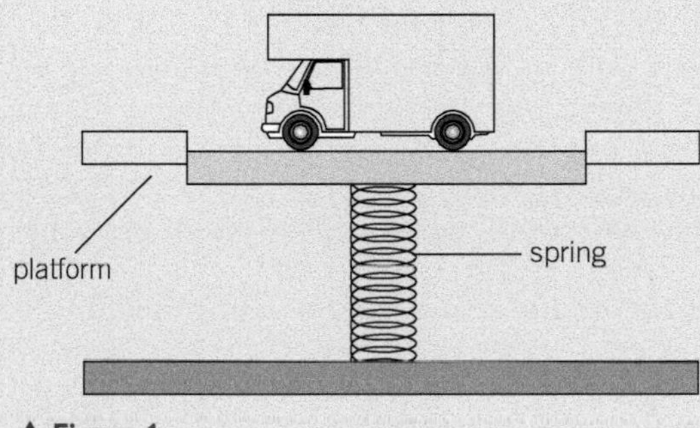

▲ **Figure 1**

Calculate the energy stored in the spring. (*3 marks*)

AQA, 2002

3 **(a)** **Figure 2** shows the variation of tensile stress with tensile strain for two wires X and Y, having the same dimensions, but made of different materials. The materials fracture at the points F_X and F_Y respectively.

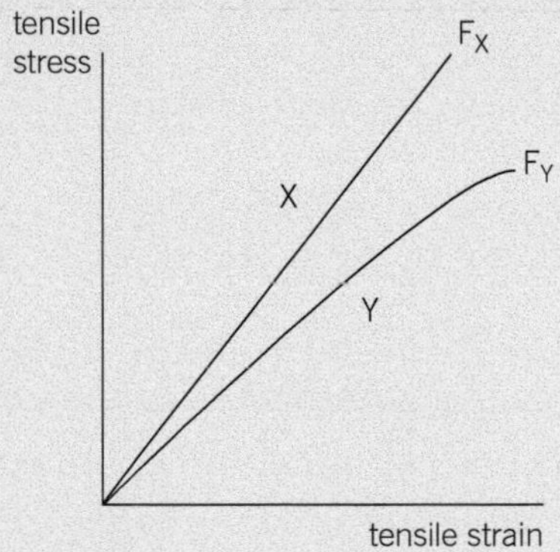

▲ **Figure 2**

State, with a reason for each, which material, X or Y,

(i) obeys Hooke's law up to the point of fracture

(ii) is the weaker material

(iii) is ductile

(iv) has the greater elastic strain energy for a given tensile stress. (*8 marks*)

(b) An elastic cord of unstretched length 160 mm has a cross-sectional area of 0.64 mm^2. The cord is stretched to a length of 190 mm. Assume that Hooke's law is obeyed for this range and that the cross-sectional area remains constant.

the Young modulus for the material of the cord = 2.0×10^7 Pa

(i) Calculate the tension in the cord at this extension.

(ii) Calculate the energy stored in the cord at this extension. *(5 marks)*

AQA, 2003

4 A material in the form of a wire, 3.0 m long and with cross-sectional area of $2.8 \times 10^{-7}\,m^2$, is suspended from a support so that it hangs vertically. Different masses may be suspended from its lower end. The table shows the extension of the wire when it is subjected to an increasing load and then a decreasing load.

Load / N	0	24	52	70	82	88	94	101	71	50	16	0
Extension / mm	0	2.2	4.6	6.4	7.4	8.2	9.6	13.0	10.2	8.0	4.8	3.2

(a) Plot a graph of load (on the *y*-axis) against extension (on the *x*-axis) for both increasing and decreasing loads. *(4 marks)*

(b) Explain what the shape of the graph tells us about the behaviour of the material in the wire. *(4 marks)*

(c) Using the graph, determine a value of the Young modulus for the material of the wire. *(3 marks)*

AQA, 2003

5 **Figure 3** shows two wires, one made of steel and the other of brass, firmly clamped together at their ends. The wires have the same unstretched length and the same cross-sectional area. One of the clamped ends is fixed to a horizontal support and a mass M is suspended from the other end, so that the wires hang vertically.

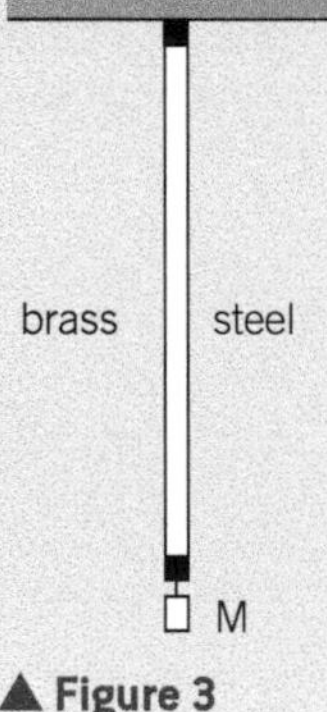

▲ **Figure 3**

(i) Since the wires are clamped together the extension of each wire will be the same. If E_S is the Young modulus for steel and E_B the Young modulus for brass, show that

$$\frac{E_S}{E_B} = \frac{F_S}{F_B}$$

where F_S and F_B are the respective forces in the steel and brass wires.

(ii) The mass M produces a total force of 15 N. Show that the magnitude of the force F_S = 10 N.

the Young modulus for steel = 2.0×10^{11} Pa

the Young modulus for brass = 1.0×10^{11} Pa

(iii) The cross-sectional area of each wire is $1.4 \times 10^{-6}\,m^2$ and the unstretched length is 1.5 m. Determine the extension produced in either wire. *(3 marks)*

AQA, 2005

Answers to the Practice Questions and Section Questions are available at www.oxfordsecondary.com/oxfordaqaexams-alevel-physics

7 Matter and radiation

7.1 Inside the atom

Learning objectives:

- → Describe what is inside an atom.
- → Explain the term isotope.
- → Represent different atoms.

Specification reference: 3.3.1

The structure of the atom

Atoms are so small (less than a millionth of a millimetre in diameter) that we need to use an electron microscope to see images of them. Although we cannot see inside them, we know, from Rutherford's alpha-scattering investigations, that every atom contains

- a positively charged nucleus composed of protons and neutrons
- electrons that surround the nucleus.

We use the word **nucleon** for a proton or a neutron in the nucleus. Note that most hydrogen atoms have a single proton as the nucleus.

Each electron has a negative charge. Because the nucleus is positively charged, the electrons are held in the atom by the electrostatic force of attraction between them and the nucleus. Rutherford's investigations showed that the nucleus contains most of the mass of the atom and its diameter is of the order of 0.000 01 times the diameter of a typical atom.

Table 1 shows the charge and the mass of the proton, the neutron, and the electron in SI units (coulomb for charge and kilogram for mass) and relative to the charge and mass of the proton. Notice that:

1. The electron has a much smaller mass than the proton or the neutron.
2. The proton and the neutron have almost equal mass.
3. The electron has equal and opposite charge to the proton. The neutron is uncharged.

▼ **Table 1** *Inside the atom*

	Charge / C	Charge relative to proton	Mass / kg	Mass relative to proton
Proton	$+1.60 \times 10^{-19}$	1	1.67×10^{-27}	1
Neutron	0	0	1.67×10^{-27}	1
Electron	-1.60×10^{-19}	−1	9.11×10^{-31}	0.0005

This means that an uncharged atom has equal numbers of protons and electrons. An uncharged atom becomes an ion if it gains or loses electrons.

Isotopes

Every atom of a given element has the same number of protons as any other atom of the same element. The **proton number** is also called the **atomic number** (symbol Z) of the element. For example:

- $Z = 6$ for carbon because every carbon atom has six protons in its nucleus
- $Z = 92$ for uranium because every uranium atom has 92 protons in its nucleus.

The atoms of an element can have different numbers of neutrons. Atoms of the same element with different numbers of neutrons are called **isotopes**.

Study tip

Don't mix up 'n' words – nucleus, neutron, nucleon, nuclide!

▲ **Figure 1** *Atoms seen using a scanning tunnelling microscope (STM)*

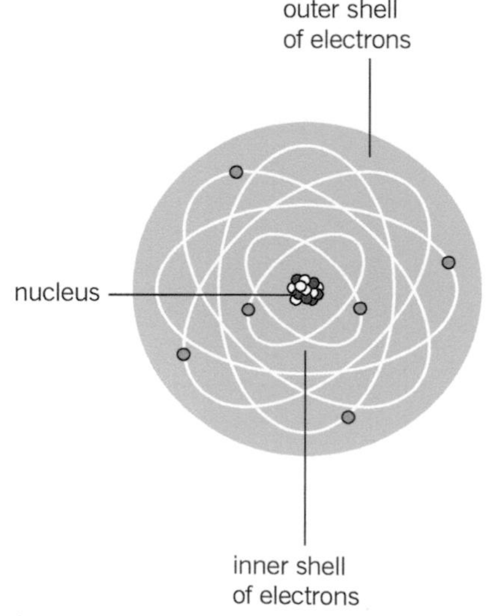

▲ **Figure 2** *Inside the atom (not to scale)*

Summary questions

You will need to use data from the Useful data for A Level Physics section on Page 497 to answer some of the questions below.

1 a State the number of protons and the number of neutrons in a nucleus of

i $^{12}_{6}C$ ii $^{16}_{8}O$

iii $^{235}_{92}U$ iv $^{24}_{11}Na$

v $^{63}_{29}Cu$.

b Which of the above nuclei has

i the smallest specific charge?

ii the largest specific charge?

2 Name the part of an atom which

a has zero charge

b has the largest specific charge

c when removed, leaves a different isotope of the element.

3 A $^{63}_{29}Cu$ atom loses two electrons. For the ion formed,

a calculate its charge in C

b state the number of nucleons it contains

c calculate its specific charge in C kg^{-1}.

4 a Calculate the mass of an ion with a specific charge of 1.20×10^7 C kg^{-1} and a negative charge of 3.2×10^{-19} C.

b The ion has eight protons in its nucleus. Calculate its number of neutrons and electrons.

For example, the most abundant isotope of natural uranium contains 146 neutrons and the next most abundant contains 143 neutrons.

Isotopes are atoms with the same number of protons and different numbers of neutrons.

The total number of protons and neutrons in an atom is called the **nucleon number** (symbol A) or sometimes the **mass number** of the atom. This is because it is almost numerically equal to the mass of the atom in relative units (where the mass of a proton or neutron is approximately 1). A nucleon is a neutron or a proton in the nucleus.

We label the isotopes of an element according to their atomic number Z, their mass number A, and the chemical symbol of the element. Figure 3 shows how we do this. Notice that:

- Z is at the bottom left of the element symbol and gives the number of protons in the nucleus
- A is at the top left of the element symbol and gives the number of protons and neutrons in the nucleus
- the number of neutrons in the nucleus = $A - Z$.

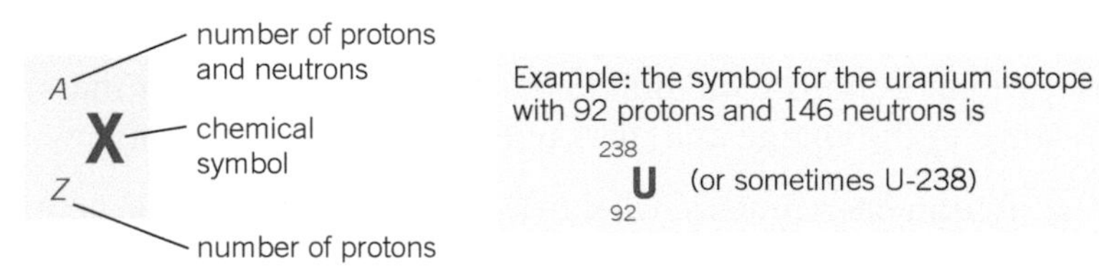

▲ **Figure 3** *Isotope notation*

Each type of nucleus is called a **nuclide** and is labelled using the isotope notation. For example, a nuclide of the carbon isotope $^{12}_{6}C$ has two fewer neutrons and two fewer protons than a nuclide of the oxygen isotope $^{16}_{8}O$.

Specific charge

The **specific charge** of a charged particle is defined as its charge divided by its mass. We can calculate the specific charge of a charged particle if we know the charge and the mass of the particle. For example:

A nucleus of $^{1}_{1}H$ has a charge of 1.60×10^{-19} C and a mass of 1.67×10^{-27} kg. Its specific charge is therefore 9.58×10^{7} C kg^{-1}.

The electron has a charge of -1.60×10^{-19} C and a mass of 9.11×10^{-31} kg. Its specific charge is therefore 1.76×10^{11} C kg^{-1}. Note that the electron has the largest specific charge of any particle.

An ion of the magnesium isotope $^{24}_{12}Mg$ has a charge of $+3.2 \times 10^{-19}$ C and a mass of 3.98×10^{-26} kg (ignoring the mass of the electrons). Its specific charge is therefore 8.04×10^{6} C kg^{-1}.

7.2 Stable and unstable nuclei

The strong nuclear force

A stable isotope has nuclei that do not disintegrate, so there must be a force holding them together. We call this force the **strong nuclear force** because it overcomes the electrostatic force of repulsion between the protons in the nucleus and keeps the protons and neutrons together.

Some further important points about the strong nuclear force are:

- Its range is no more than about 3–4 femtometres (fm), where $1\,\text{fm} = 10^{-15}\,\text{m} = 0.000\,000\,000\,000\,001\,\text{m}$. This range is about the same as the diameter of a small nucleus. In comparison, the electrostatic force between two charged particles has an infinite range (although it decreases as the range increases).
- It has the same effect between two protons as it does between two neutrons or a proton and a neutron.
- It is an attractive force from 3–4 fm down to about 0.5 fm. At separations smaller than this, it is a repulsive force that acts to prevent neutrons and protons being pushed into each other.

Figure 1 shows how the strong nuclear force varies with separation between two protons or neutrons. Notice that the equilibrium separation is where the force curve crosses the x-axis.

Learning objectives:

- → State what keeps the protons and neutrons in a nucleus together.
- → Explain why some nuclei are stable and others unstable.
- → Describe what happens when an unstable nucleus emits an alpha particle or a beta minus particle.

Specification reference: 3.3.1 and 3.3.3

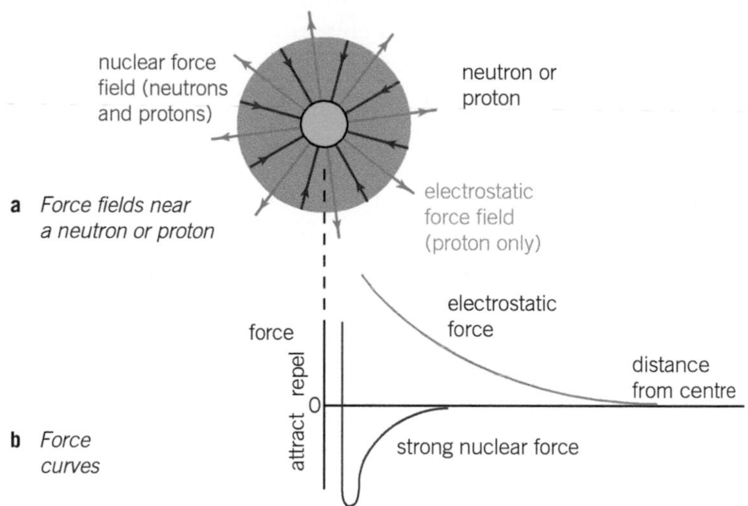

▲ **Figure 1** *The strong nuclear force*

Radioactive decay

Naturally occurring radioactive isotopes release three types of radiation.

1. **Alpha radiation** consists of alpha particles which each comprise two protons and two neutrons. The symbol for an alpha particle is ${}^{4}_{2}\alpha$ because its proton number is 2 and its mass number is 4. Figure 2 shows what happens to an unstable nucleus of an element X when it emits an alpha particle. Its nucleon number A decreases by 4 and its atomic number Z decreases by 2. As a result of the change, the product nucleus belongs to a different element Y.

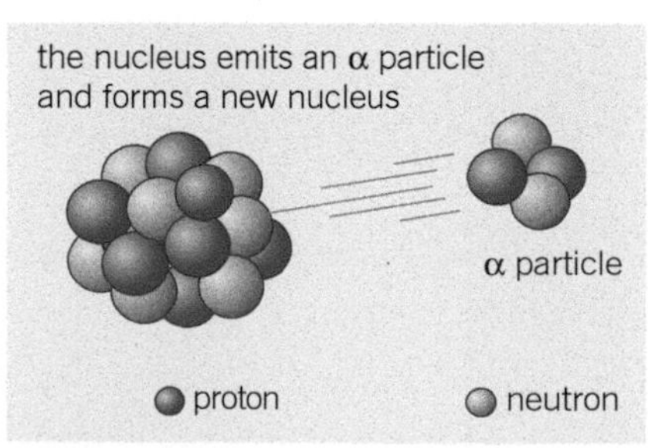

▲ **Figure 2** *Alpha particle emission (not to scale)*

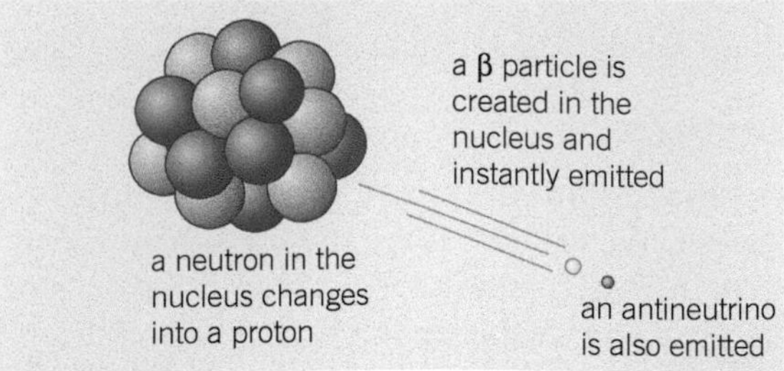

▲ **Figure 3** *Beta particle emission (not to scale)*

We can represent this change by means of the equation below:

$$^{A}_{Z}\mathrm{X} \rightarrow {}^{A-4}_{Z-2}\mathrm{Y} + {}^{4}_{2}\alpha$$

2 **Beta-minus radiation** consists of fast-moving electrons. The symbol for an electron as a beta particle is $^{0}_{-1}\beta$ (or β^-) because its charge is equal and opposite to that of the proton and its mass is much smaller than the proton's mass.

Figure 3 shows what happens to an unstable nucleus of an element X when it emits a β^- particle. This happens as a result of a neutron in the nucleus changing into a proton. The beta particle is created when the change happens and is emitted instantly. In addition, an antiparticle with no charge, called an antineutrino (symbol $\bar{\nu}$), is emitted. You will learn more about antiparticles and neutrinos in Topic 7.4. Because a neutron changes into a proton in the nucleus, the atomic number increases by 1 but the nucleon number stays the same. As a result of the change, the product nucleus belongs to a different element Y. This type of change happens to nuclei that have too many neutrons.

We can represent this change by means of the equation below:

$$^{A}_{Z}\mathrm{X} \rightarrow {}^{A}_{Z+1}\mathrm{Y} + {}^{0}_{-1}\beta + \bar{\nu}$$

3 **Gamma radiation** (symbol γ) is electromagnetic radiation emitted by an unstable nucleus. It can pass through thick metal plates. It has no mass and no charge. It is emitted by a nucleus with too much energy, following an alpha or beta emission.

Journey into the atom (part 1): A very elusive particle!

When the energy spectrum of beta particles was first measured, it was found that beta particles were released with kinetic energies up to a maximum that depended on the isotope. The scientists at the time were puzzled why the energy of the beta particles varied up to a maximum, when each unstable nucleus lost a certain amount of energy in the process. Either energy and momentum were not conserved in the change or some of each were carried away by mystery particles, which they called **neutrinos** and **antineutrinos**. This hypothesis was unproven for over 20 years until antineutrinos were detected. Antineutrinos were detected as a result of their interaction with cadmium nuclei in a large tank of water. This was installed next to a nuclear reactor as a controllable source of these very elusive particles. Now we know that billions of these elusive particles from the Sun sweep through our bodies every second without interacting! (See Topic 4.5)

Summary questions

1 Which force, the strong nuclear force or the electrostatic force,
   a does not affect a neutron
   b has a limited range
   c holds the nucleons in a nucleus
   d tends to make a nucleus unstable?

2 Complete the following radioactive decay equations:
   a $^{229}_{90}\mathrm{Th} \rightarrow \mathrm{Ra} + \alpha$
   b $^{65}_{28}\mathrm{Ni} \rightarrow \mathrm{Cu} + \beta + \bar{\nu}$.

3 A bismuth $^{213}_{83}\mathrm{Bi}$ nucleus emits a beta particle then an alpha particle then another beta particle before it becomes a nucleus X.
   a Show that X is a bismuth isotope.
   b i Determine the nucleon number of X.
   ii How many protons and how many neutrons are in the nucleus just after it emits the alpha particle?

4 The neutrino hypothesis was put forward to explain beta decay.
   a Explain the term *hypothesis*.
   b i State one property of the neutrino.
   ii Name two objects that produce neutrinos.

7.3 Photons

Electromagnetic waves

Light is just a small part of the spectrum of **electromagnetic waves**. Our eyes cannot detect the other parts. The world would appear very different to us if they could. For example, all objects emit infrared radiation. Infrared cameras enable objects to be observed in darkness.

In a vacuum, all electromagnetic waves travel at the speed of light, c, which is $3.00 \times 10^8\,\mathrm{m\,s^{-1}}$. As you know from GCSE, the wavelength λ of **electromagnetic radiation** of frequency f in a vacuum is given by the equation

$$\lambda = \frac{c}{f}$$

Note that we often express light wavelengths in nanometres (nm), where $1\,\mathrm{nm} = 0.000\,000\,001\,\mathrm{m} = 10^{-9}\,\mathrm{m}$.

The main parts of the electromagnetic spectrum are listed in Table 1.

Table 1 *The main parts of the electromagnetic spectrum*

Type	Radio	Microwave	Infrared	Visible	Ultraviolet	X-rays	Gamma rays
Wavelength range	>0.1 m	0.1 m to 10 mm	10 mm to 700 nm	700 nm to 400 nm	400 nm to 10 nm	10 nm to 0.001 nm	<1 nm

As shown in Figure 1, an electromagnetic wave consists of an electric wave and a magnetic wave which travel together and vibrate

- at right angles to each other and to the direction in which they are travelling
- in phase with each other. As you can see the two waves reach a peak together so they are in step. When waves do this we say they are in phase.

Photons

Electromagnetic waves are emitted by a charged particle when it loses energy. This can happen when

- a fast-moving electron is stopped (for example, in an X-ray tube) or slows down or changes direction
- an electron in a shell of an atom moves to a different shell of lower energy.

Electromagnetic waves are emitted as short bursts of waves, each burst leaving the source in a different direction. Each burst is a packet of electromagnetic waves and is referred to as a **photon**. The photon theory was established by Einstein in 1905, when he used his ideas to explain the **photoelectric effect**. This is the emission of electrons from a metal surface when light is directed at the surface.

Learning objectives:

- → Recall what is meant by a photon.
- → Calculate the energy of a photon.
- → Estimate how many photons a light source emits every second.

Specification reference: 3.3.3 and 3.5.10

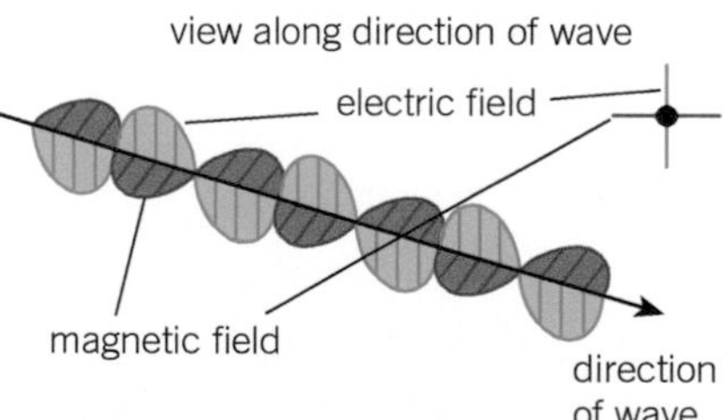

Figure 1 *Electromagnetic waves*

Synoptic link

You will meet the photoelectric effect and the photon theory in more detail in Topic 13.1, Photoelectricity.

Einstein imagined photons to be like *flying needles*, and he assumed that the energy E of a photon depends on its frequency f in accordance with the equation

$$\textbf{photon energy } E = hf$$

where h is a constant referred to as the Planck constant. The value of h is 6.63×10^{-34} J s.

Worked example

Calculate the frequency and the energy of a photon of wavelength 590 nm.
$h = 6.63 \times 10^{-34}$ J s, $c = 3.00 \times 10^{8}$ m s^{-1}

Solution

To calculate the frequency, use $f = \frac{c}{\lambda} = \frac{3.00 \times 10^{8}}{590 \times 10^{-9}} = 5.08 \times 10^{14}$ Hz.

To calculate the energy of a photon of this wavelength, we use $E = hf$.

$E = hf = 6.63 \times 10^{-34} \times 5.08 \times 10^{14} = 3.37 \times 10^{-19}$ J.

Laser power

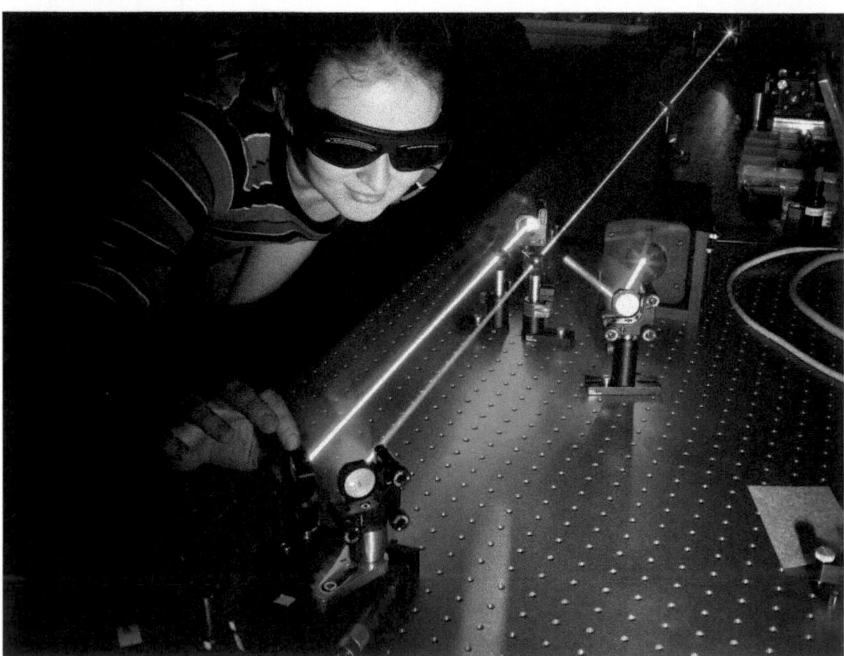

▲ **Figure 2** *A laser at work*

A laser beam consists of photons of the same frequency. The power of a laser beam is the energy per second transferred by the photons. For a beam consisting of photons of frequency f,

$$\textbf{the power of the beam} = nhf$$

where n is the number of photons in the beam passing a fixed point each second. This is because each photon has energy hf. Therefore, if n photons pass a fixed point each second, the energy per second (or power) is nhf.

Summary questions

$c = 3.00 \times 10^{8}$ m s^{-1}
$h = 6.63 \times 10^{-34}$ J s

1 a List the main parts of the electromagnetic spectrum in order of increasing wavelength.
  b Calculate the frequency of
    i light of wavelength 590 nm
    ii radio waves of wavelength 200 m.

2 With the aid of a suitable diagram, explain what is meant by an electromagnetic wave.

3 Light from a certain light source has a wavelength of 430 nm. Calculate
  a the frequency of light of this wavelength
  b the energy of a photon of this wavelength.

4 a Calculate the frequency and energy of a photon of wavelength 635 nm.
  b A laser emits light of wavelength 635 nm in a beam of power 1.5 mW. Calculate the number of photons emitted by the laser each second.

7.4 Particles and antiparticles

Antimatter

When **antimatter** and matter particles (i.e., the particles that make up everything in our universe) meet, they destroy each other and radiation is released. We make use of this effect in a positron emitting tomography (PET) hospital scanner. The P in PET stands for the **positron**, which is the antiparticle of the electron. When a PET scanner is used for a brain scan, a positron-emitting isotope is administered to the patient and some of it reaches the brain via the blood system. Each positron emitted travels no further than a few millimetres before it meets an electron and they annihilate each other. Two gamma photons, produced as a result, are sensed by detectors linked to computers. Gradually, an image is built up from the detector signals of where the positron-emitting nuclei are inside the brain.

Positron emission takes place when a proton changes into a neutron in an unstable nucleus with too many protons. The positron (symbol ${}^{0}_{+1}\beta$ or β^{+}) is the antiparticle of the electron, so it carries a positive charge. In addition, a neutrino (symbol ν), which is uncharged, is emitted.

$$ {}^{A}_{Z}\mathrm{X} \rightarrow {}^{A}_{Z-1}\mathrm{Y} + {}^{0}_{+1}\beta + \nu $$

Positron-emitting isotopes do not occur naturally. They are manufactured by placing a stable isotope, in liquid or solid form, in the path of a beam of protons. Some of the nuclei in the substance absorb extra protons and become unstable positron emitters.

Antimatter was predicted in 1928 by the English physicist Paul Dirac, before the first antiparticle, the positron, was discovered. More than 20 years earlier, Einstein had shown that the mass of a particle increases the faster it travels, and that his famous equation $E = mc^2$ related the energy supplied to the particle to its increase in mass. More significantly, Einstein said that the mass of a particle when it is stationary, its rest mass (m_0), corresponds to **rest energy** m_0c^2 locked up as mass. He showed that rest energy must be included in the conservation of energy. Dirac predicted the existence of antimatter particles (or **antiparticles**) that would unlock rest energy, whenever a particle and a corresponding antiparticle meet and annihilate each other.

Dirac's theory of antiparticles predicted that for every type of particle there is a corresponding antiparticle that:

- annihilates the particle and itself if they meet, converting their total mass into photons
- has exactly the same rest mass as the particle
- has exactly opposite charge to the particle if the particle has a charge.

In addition to the **annihilation** process described above, Dirac predicted the opposite process of **pair production**. In this process, Dirac predicted that a photon with sufficient energy passing near a nucleus or an electron can suddenly change into a particle–antiparticle pair, which would then separate from each other. Figure 1 shows both of these processes.

Learning objectives:

- → Define antimatter.
- → Describe what happens when a particle and its antiparticle meet.
- → Discuss whether anti-atoms are possible.

Specification reference: 3.3.2 and 3.3.3

Application

Brain imaging

PET scanners are used in hospitals for brain imaging of, for example, stroke patients. The T stands for tomography, which is the name for the electronic and mechanical system used to perform the scan.

Q: What type of radiation is detected in a PET scanner?

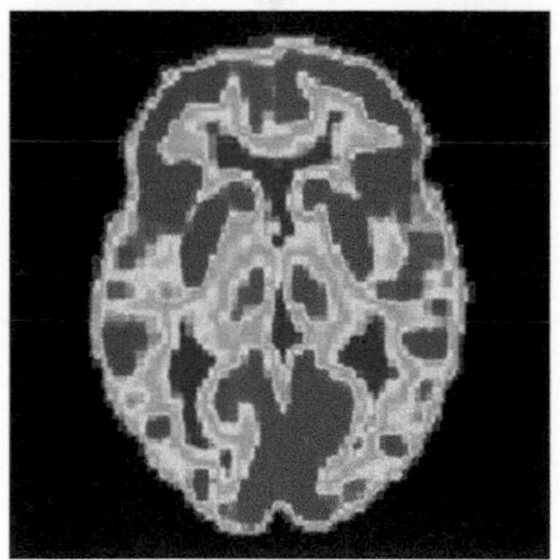

Answer: Gamma radiation.

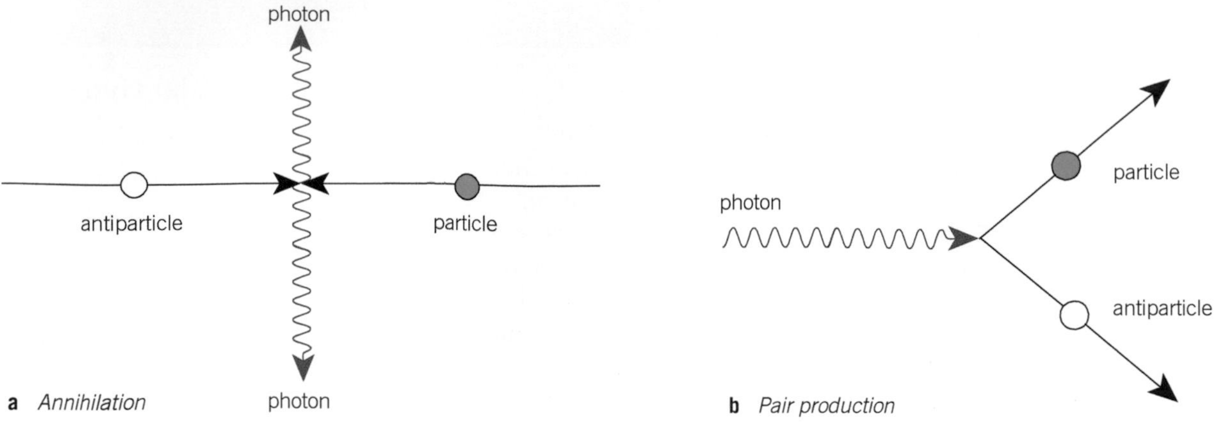

▲ **Figure 1** *Particles and antiparticles*

Particles, antiparticles, and $E = mc^2$

The energy of a particle or antiparticle is often expressed in millions of electron volts (MeV), where **1 MeV = 1.60 × 10⁻¹³ J**. One electron volt is defined as the energy transferred when an electron is moved through a potential difference of one volt. Given the rest mass of a particle or antiparticle, its rest energy in MeV can be calculated using $E = mc^2$. However, you won't need to do this type of calculation in this topic, as the rest energies of different particles are listed on page 513 of this book.

Annihilation occurs when a particle and a corresponding antiparticle meet and their mass is converted into radiation energy. Two photons are produced in this process (as a single photon cannot ensure a total momentum of zero after the collision). Therefore, the minimum energy of each photon, hf_{min}, is given by equating the energy of the two photons, $2hf_{min}$, to the rest energy of the particle and of the antiparticle (i.e., $2hf_{min} = 2E_0$, where E_0 is the rest energy of the particle).

Minimum energy of each photon produced, $hf_{min} = E_0$

In **pair production**, a photon creates a particle and a corresponding antiparticle, and vanishes in the process. For a particle and antiparticle, each of rest energy E_0, we can calculate the minimum energy and minimum frequency f_{min} that the photon must have to produce this particle–antiparticle pair. Remember, c is the speed of light in a vacuum ($3.00 \times 10^8\,m\,s^{-1}$).

Minimum energy of photon needed = $hf_{min} = 2E_0$

For example, the electron has a rest energy of 0.511 MeV. Therefore, for pair production of an electron and a positron from a photon:

Minimum energy of a photon = 2 × 0.511 MeV = 1.022 MeV

$$= 1.64 \times 10^{-13}\,J$$

A photon with less energy could not therefore create a positron and an electron.

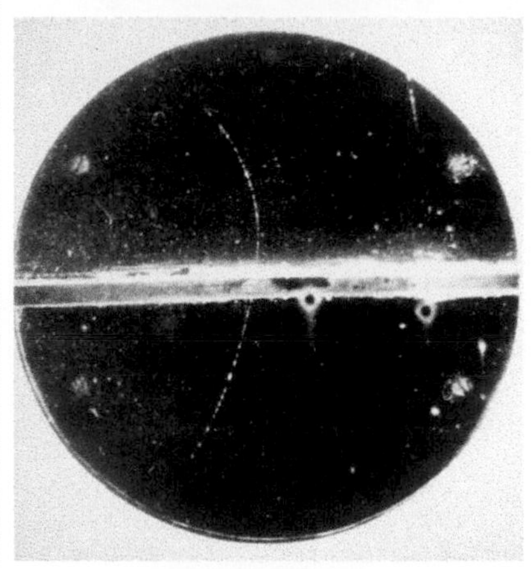

▲ **Figure 2** *The discovery of the positron*

Journey into the atom (part 2): The discovery of the positron

We can see the paths of alpha and beta particles using a cloud chamber. This is a small transparent container containing air saturated with vapour and made very cold. The same conditions exist high in the atmosphere. Ionising particles leave a visible trail of liquid droplets when they pass through the air – just like a jet plane does when it passes high overhead on a clear day. In 1932, the American physicist Carl Anderson was using a cloud chamber and a camera to photograph trails produced by cosmic rays. He decided to see if the particles could pass through a lead plate in the chamber. With a magnetic field applied to the chamber, he knew the trail of a charged particle would bend in the field.

- A positive particle would be deflected by the magnetic field in the opposite direction to a negative particle travelling in the same direction.
- The slower it went, the more it would bend.

If a particle went through the plate, he thought it would be slowed down so its trail would bend more afterwards. Imagine his surprise when he discovered a beta particle that slowed down but bent in the opposite direction to all the other beta trails he had photographed. He had made a momentous discovery – a positron, the first antiparticle to be detected (Figure 2).

Summary questions

1 MeV = 1.60×10^{-13} J

1 **a** The rest energy of a proton is 1.501×10^{-10} J. Calculate its rest energy in MeV.

b Show that a photon must have a minimum energy of 1876 MeV to create a proton–antiproton pair.

2 Explain why a photon of energy 2 MeV could produce an electron–positron pair but not a proton–antiproton pair.

3 The rest energy of an electron is 0.511 MeV.

a State the minimum energy in J of each photon created when a positron and an electron annihilate each other.

b A positron created in a cloud chamber in an experiment has 0.158 MeV of kinetic energy. It collides with an electron at rest, creating two photons of equal energies as a result of annihilation.

i Calculate the total energy of the positron and the electron.

ii Show that the energy of each photon is 0.590 MeV.

4 A positron can be produced by pair production or by positron emission from a proton-rich nucleus.

a Describe the changes that take place in a proton-rich nucleus when it emits a positron.

b State two ways in which pair production of a positron and an electron differs from positron emission.

Practice questions: Chapter 7

1 **(a)** Copy and complete the table below: *(3 marks)*

	Particle or antiparticle	Charge / proton charge
Antiproton	antiparticle	
Neutrino		
Neutron		0
Positron		

(b) The tracks of a positron and an electron created by pair production in a magnetic field curve in opposite directions as shown above.

(i) Why do they curve in opposite directions?

(ii) Both particles spiral inwards. What can you deduce from this observation about their kinetic energy?

(iii) In **Figure 2** of Topic 7.4, the track is a typical beta track. Explain how Carl Anderson deduced from the photograph that the track was created by a positron rather than an electron travelling in the opposite direction. *(5 marks)*

2 **(a)** ${}^{229}_{90}\text{Th}$ is a neutral atom of thorium. How many protons, neutrons, and electrons does it contain? *(2 marks)*

(b) ${}^{Y}_{X}\text{Th}$ is a neutral atom of a different isotope of thorium which contains Z electrons. Give possible values for X, Y, and Z. *(3 marks)*

AQA, 2001

3 An atom of argon ${}^{37}_{18}\text{Ar}$ is ionised by the removal of two orbiting electrons.

(a) How many protons and neutrons are there in this ion? *(2 marks)*

(b) What is the charge, in C, of this ion? *(2 marks)*

(c) Which constituent particle of this ion has

(i) a zero charge per unit mass ratio?

(ii) the largest charge per unit mass ratio? *(2 marks)*

(d) Calculate the percentage of the total mass of this ion that is accounted for by the mass of its electrons. *(3 marks)*

AQA, 2002

4 **(a)** A neutral atom contains 28 nucleons.
Write down a possible number of protons, neutrons, and electrons contained in the atom. *(2 marks)*

(b) A certain isotope of uranium may split into a caesium nucleus, a rubidium nucleus, and four neutrons in the following process:

$${}^{236}_{92}\text{U} \rightarrow {}^{137}_{55}\text{Cs} + {}^{X}_{37}\text{Rb} + 4\,{}^{1}_{0}\text{n}$$

(i) Explain what is meant by isotopes.

(ii) How many neutrons are there in the ${}^{137}_{55}\text{Cs}$ nucleus?

(iii) Calculate the ratio $\frac{\text{charge}}{\text{mass}}$, in C kg^{-1}, for the ${}^{236}_{92}\text{U}$ nucleus.

(iv) Determine the value of X for the rubidium nucleus. *(4 marks)*

AQA, 2003

5 An α particle is the same as a nucleus of helium, ${}^{4}_{2}\text{He}$.

The equation ${}^{229}_{90}\text{Th} \rightarrow {}^{X}_{Y}\text{Ra} + \alpha$

represents the decay of a thorium isotope by the emission of an α particle.
Determine:

(a) the values of X and Y, shown in the equation *(2 marks)*

(b) the ratio $\frac{\text{mass of } {}^{X}_{Y}\text{Ra nucleus}}{\text{mass of } \alpha \text{ particle}}$ *(1 mark)*

AQA, 2005

6 **(a)** (i) Describe an α particle and state its properties.

(ii) $^{218}_{85}At$ is an isotope of the element astatine (At) which decays into an isotope of bismuth (Bi) by emitting an α particle. Write down the equation to represent this reaction. *(5 marks)*

(b) (i) State what happens when an unstable nucleus decays by emitting a β^- particle.

(ii) Write down and complete the following equation, showing how an isotope of molybdenum decays into an isotope of technetium:

$$^{...}_{42}Mo \rightarrow {}^{99}_{...}Tc + \beta^- + ...$$ *(5 marks)*

7 In a radioactive decay of a nucleus, a β^+ particle is emitted followed by a γ photon of wavelength 8.30×10^{-13} m.

(a) (i) State the rest mass, in kg, of the β^+ particle.

(ii) Calculate the energy of the γ photon.

(iii) Determine the energy of the γ photon in MeV. *(6 marks)*

(b) Name the fundamental interaction or force responsible for β^+ decay. *(1 mark)*

AQA, 2004

8 **(a)** (i) State the name of the antiparticle of a positron.

(ii) Describe what happens when a positron and its antiparticle meet. *(3 marks)*

(b) Calculate the minimum amount of energy, in J, released as radiation energy when a particle of rest energy 0.51 MeV meets its corresponding antiparticle. *(2 marks)*

AQA, 2005

9 In a particle accelerator a proton and an antiproton, travelling at the same speed, undergo a head-on collision and produce subatomic particles.

(a) The total kinetic energy of the two particles just before the collision is 3.2×10^{-10} J.

(i) What happens to the proton and antiproton during the collision?

(ii) State why the total energy after the collision is more than 3.2×10^{-10} J. *(2 marks)*

(b) In a second experiment the total kinetic energy of the colliding proton and antiproton is greater than 3.2×10^{-10} J.
State *two* possible differences this could make to the subatomic particles produced. *(2 marks)*

AQA, 2001

10 An electron may interact with a proton in the following way:

$$e^- + p \rightarrow n + \nu_e.$$

Name the fundamental force responsible for this interaction. *(1 mark)*

AQA, 2003

11 **(a)** Give an example of an antiparticle other than a positron, and state one difference between it and a positron. *(2 marks)*

(b) State the type of event that takes place when a particle meets its corresponding antiparticle, and state what can be produced in this event. *(2 marks)*

AQA, 2006

12 Describe what happens in pair production and give *one* example of this process. *(3 marks)*

AQA, 2005

8 Radioactivity

8.1 The discovery of the nucleus

Learning objectives:

- → State how big the nucleus is.
- → Describe how the nucleus was discovered.
- → Explain why it was not discovered earlier.

Specification reference: 3.3.1

The nuclear model of the atom was proposed by Ernest Rutherford in 1911. He knew from the work of J.J. Thomson that every atom contains one or more electrons. Thomson had shown that the electron is a negatively charged particle inside every atom but no one knew until Rutherford's theory was confirmed experimentally in 1913 how the positive charge in the atom was distributed. Thomson thought the atom could be like a currant bun – with electrons dotted in the atom like currants in a bun. In this model, the positive charge was supposedly spread throughout the atom like the dough of the bun.

Rutherford knew that the atoms of certain elements were unstable and emitted radiation. It had been shown that there were three types of such radiation, referred to as **alpha radiation** (symbol α), **beta radiation** (symbol β), and **gamma radiation** (symbol γ). Rutherford knew that α radiation consisted of fast-moving positively charged particles. He and his co-workers used this type of radiation to probe the atom. He reckoned that a beam of α particles directed at a thin metal foil might be scattered slightly by the atoms of the foil if the positive charge was spread throughout each atom. He was astonished when he discovered that some of the particles bounced back from the foil – in his own words 'as incredible as if you fired a 15 inch naval shell at tissue paper and it came back'.

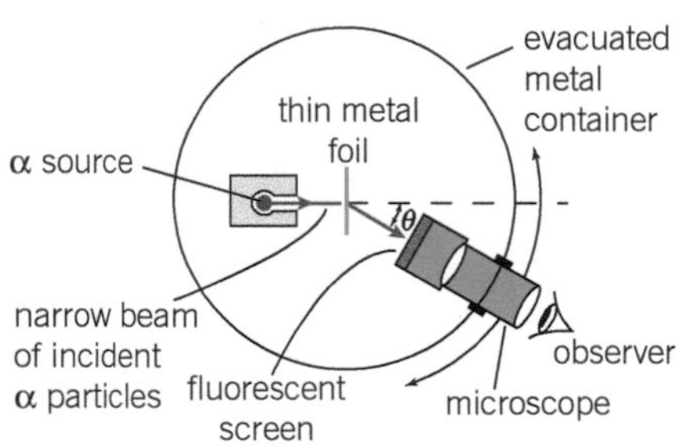

▲ **Figure 1** *Rutherford's α-scattering apparatus*

Let's consider Rutherford's experiment in more detail. Rutherford used a narrow beam of α particles, all of the same kinetic energy, in an evacuated container to probe the structure of the atom. Figure 1 shows an outline of the arrangement he used. A thin metal foil was placed in the path of the beam. Alpha particles scattered by the metal foil were detected by a detector which could be moved round at a constant distance from the point of impact of the beam on the metal foil.

Ernest Rutherford 1871–1937

Ernest Rutherford arrived in Britain from New Zealand in 1895. By the age of 28, he was a professor. He made important discoveries about radioactivity and was awarded the Nobel prize for chemistry in 1908 for his investigations into the disintegration of radioactive substances. He worked in the universities of Montreal, Manchester, and Cambridge. When at Manchester, he put forward the nuclear model of the atom and proved it experimentally using α-scattering experiments. He was knighted in 1914 and made Lord Rutherford of Nelson in 1931. His co-worker Otto Hahn described him as a 'very jolly man'. In 1915, he expressed the hope that 'no one discovers how to release the intrinsic energy of radium until man has learned to live at peace with his neighbour'. After his death in 1937, his ashes were placed close to Newton's tomb in Westminster Abbey.

Extension

What made Rutherford originate the nuclear model of the atom?

Rutherford had previously investigated the scattering of alpha particles by gas atoms and noted on rare occasions that an α particle was scattered through a very large angle when most particles were hardly scattered, if at all. He realised there could be a 'nucleus' of positively charged matter inside the atom which could explain these rare events.

1. The α particles must have the same speed otherwise slow α particles would be deflected more than faster α particles on the same initial path.
2. The container must be evacuated or else the α particles would be stopped by air molecules.
3. The source of the α particles must have a long half-life otherwise later readings would be lower than earlier readings due to radioactive decay of the source nuclei.

Q: Describe the type of force that Rutherford believed would scatter an α particle.

Answer: The electrostatic force of repulsion.

Rutherford used a microscope to observe the pinpoints of light emitted when α particles hit the atoms of a fluorescent screen. He measured the number of α particles reaching the detector per minute for different angles of deflection from zero to almost 180°. His measurements showed that:

1 most α particles passed straight through the foil with little or no deflection – about 1 in 2000 were deflected
2 a small percentage of α particles (about 1 in 10 000) were deflected through angles of more than 90°.

Imagine throwing tennis balls at a row of vertical posts separated by wide gaps. Most of the balls would pass between the posts and therefore would not be deflected much. However, some would rebound as a result of hitting a post. Rutherford realised the α-scattering measurements could be explained in a similar way by assuming every atom has a 'hard centre' much smaller than the atom. His interpretation of each result was that

- most of the atom's mass is concentrated in a small region, the **nucleus**, at the centre of the atom
- the nucleus is positively charged because it repels α particles (which carry positive charge) that approach it too closely.

▲ **Figure 2** *Ernest Rutherford*

Figure 3 shows the paths of some α particles which pass near a fixed nucleus.

- Alpha particle C collides head-on with the nucleus and rebounds back so its angle of deflection is 180°.
- Alpha particles A, B, and D are deflected through different angles. The closer the initial direction of an α particle is to the head-on direction,
  - the greater its deflection is because the electrostatic force of repulsion between the α particle and the nucleus increases with decreased separation between them
  - the smaller the least distance of approach of the α particle to the nucleus.
- α particle E does not approach the nucleus closely enough to be significantly deflected.

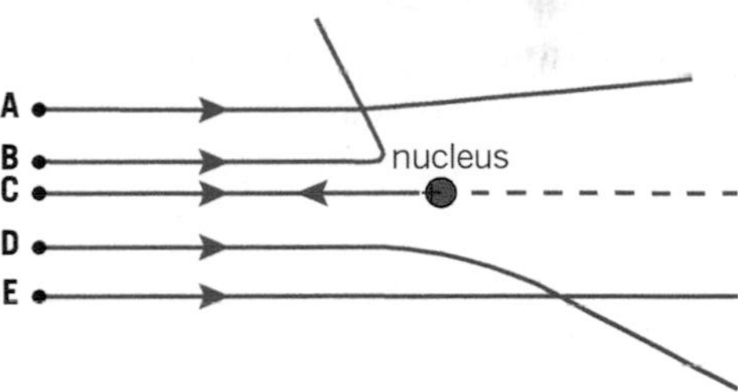

▲ **Figure 3** *α-scattering paths*

Study tip

Nuclear diameter is of the order of 10^{-15} m.

Using Coulomb's law of force (i.e., the law of force between charged objects – see Topic 18.4, Coulomb's law) and Newton's laws of motion, Rutherford used his nuclear model to explain the exact pattern of the results. By testing foils of different metal elements, he also showed that the magnitude of the charge of a nucleus is $+Ze$, where e is the charge of the electron and Z is the **atomic number** of the element.

Estimate of the size of the nucleus

About 1 in 10 000 α particles are deflected by more than 90° when they pass through a metal foil. The foil must be very thin otherwise α particles are scattered more than once.

For such single scattering by a foil that has n layers of atoms, the probability of an α particle being deflected by a given atom is therefore about 1 in 10 000n. This probability depends on the effective area

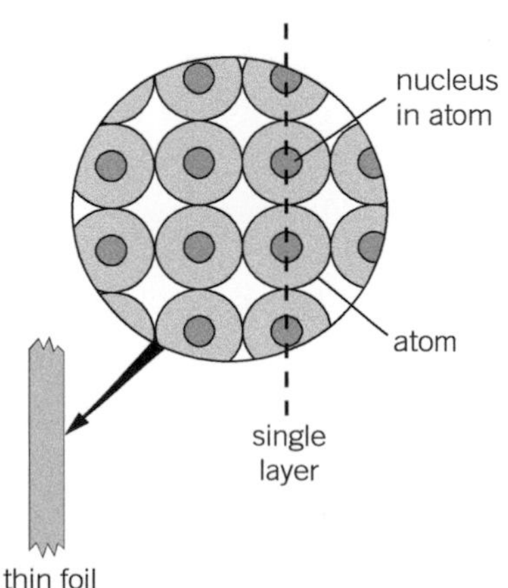

▲ **Figure 4** *Estimating nuclear size*

Notes

In a head-on impact, the α particle stops momentarily at the least distance of approach, *d*. At this point, the potential energy of the α particle in the electric field of the nucleus is equal to the initial kinetic energy, E_K, of the α particle. So, using the formula in Topic 18.5 for potential energy gives

$E_K = \frac{Q_\alpha Q_N}{4\pi\varepsilon_0 d}$, where Q_α is the charge of the α particle (+2*e*), Q_N is the charge of the nucleus (= +*Ze*), and *d* is the least distance of approach. The constant ε_0 is called the permittivity of free space. See Topic 18.4, Coulomb's law. Prove for yourself that the least distance of approach of an α particle with kinetic energy of 8×10^{-13} J to an aluminium nucleus ($Z = 13$) is about 7×10^{-15} m which is about 10^5 times smaller than the diameter of a typical atom. A more accurate method of determining the radius of a nucleus is explained in Topic 24.1.

of cross section of the nucleus to that of the atom. Therefore, for a nucleus of diameter *d* in an atom of diameter *D*, the area ratio is equal to $\frac{\frac{\pi d^2}{4}}{\frac{\pi D^2}{4}}$. So $d^2 = \frac{D^2}{10\,000n}$.

A typical value of $n = 10^4$ gives $d = \frac{D}{10\,000}$.

In other words, the size of a nucleus relative to an atom is about the same as a football to a football stadium!

Summary questions

1 a In the Rutherford α particle scattering experiment, most of the α particles passed straight through the metal foil. What did Rutherford deduce about the atom from this discovery?

b A small fraction of the α particles were deflected through large angles. What did Rutherford deduce about the atom from this discovery?

2 In Rutherford's α particle scattering experiment, why was it essential that:

a the apparatus was in an evacuated chamber?

b the foil was very thin?

c the α particles in the beam all had the same speed?

d the beam was narrow?

3 An α particle collides with a nucleus and is deflected by it, as shown in Figure 5.

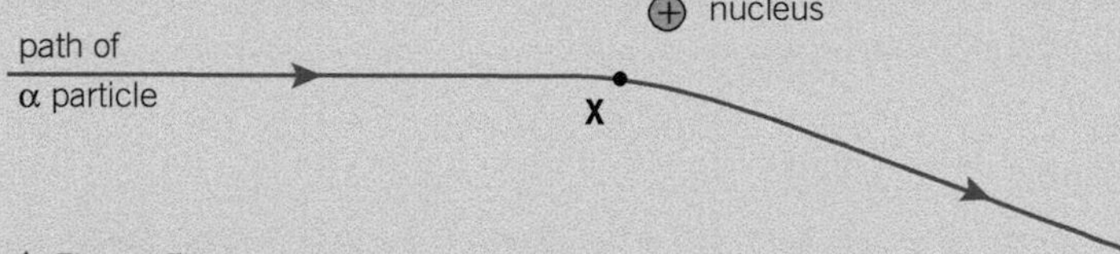

▲ **Figure 5**

a Copy the diagram and show on it

i the direction of motion of the α particle

ii the direction of the force on the α particle when it was at the position marked X.

b Describe how

i the kinetic energy of the α particle

ii the potential energy of the α particle changed during this interaction.

4 In the α particle scattering experiment, about 1 in 10 000 α particles are deflected by more than 90°.

a For a metal foil which has *n* layers of atoms, explain why the probability of an α particle being deflected by a given atom is therefore about 1 in 10 000*n*.

b Assuming this probability is equal to the ratio of the cross-sectional area of the nucleus to that of the atom, estimate the diameter of a nucleus for atoms of diameter 0.5 nm in a metal foil of thickness 10 μm.

c Calculate the least distance of approach of an α particle to a gold nucleus ($Z = 79$) if the initial kinetic energy of the α particle is 8.0×10^{-13} J.

$\varepsilon_0 = 8.85 \times 10^{-12}$ F m^{-1}, $e = 1.6 \times 10^{-19}$ C

8.2 The properties of α, β, and γ radiation

The discovery of radioactivity

In 1896, Henri Becquerel was investigating materials that glow when placed in an X-ray beam. He wanted to find out if strong sunlight could make uranium salts glow. He prepared a sample and placed it in a drawer on a wrapped photographic plate, ready to test the salts on the next sunny day. When he developed the film, he was amazed to see the image of a key. He had put the key on the plate in the drawer and then put the uranium salts on top of the key. He realised that uranium salts emit radiation which can penetrate paper and blacken a photographic film. The uranium salts were described as being radioactive. The task of investigating radioactivity was passed on by Becquerel to one of his students, Marie Curie. Within a few years, Marie Curie discovered other elements which are radioactive. One of these elements, radium, was found to be over a million times more radioactive than uranium.

Marie Curie established the nature of radioactive materials. She showed how radioactive compounds could be separated and identified. She and her husband, Pierre, and Henri Becquerel won the 1903 Nobel prize for physics for their discovery of radioactivity. After Pierre's death in 1906, she continued her painstaking research and was awarded a second Nobel prize in 1911 – an unprecedented honour.

Rutherford's investigations on radioactivity

Rutherford wanted to find out what the radiation emitted by radioactive substances was and what caused it. In 1899, he found that the radiation:

- ionised air, making it conduct electricity – he made a detector which could measure the radiation from its ionising effect
- was of two types. One type, which he called alpha (α) radiation, was easily absorbed. The other type, which he called beta (β) radiation, was more penetrating. A third type of radiation, called gamma (γ) radiation, even more penetrating than β radiation, was discovered a year later.

Further tests showed that a magnetic field deflects α and β radiation in opposite directions and has no effect on γ radiation. From the deflection direction, it was concluded that α radiation consists of positively charged particles and β radiation consists of negatively charged particles. Gamma radiation was later shown to consist of high-energy **photons**.

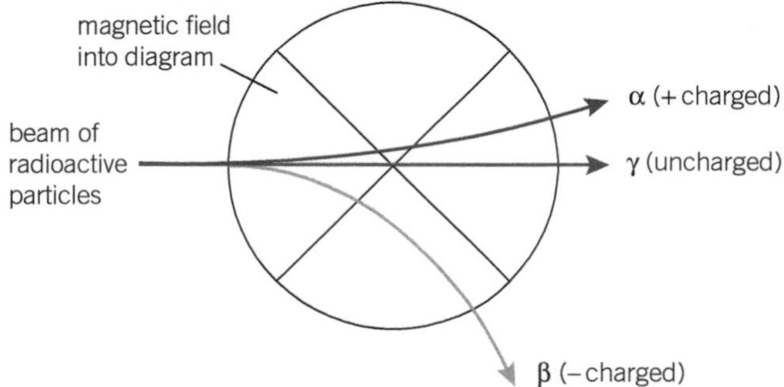

▲ **Figure 2** *Deflection by a magnetic field*

Learning objectives:

- → Define α, β, and γ radiation.
- → Explain why it is dangerous.
- → Describe the properties of α, β, and γ radiation.

Specification reference: 3.3.3

▲ **Figure 1** *Marie Curie (1867–1934)*

Study tip

Don't confuse the radiation with the source; the source is radioactive not the radiation.

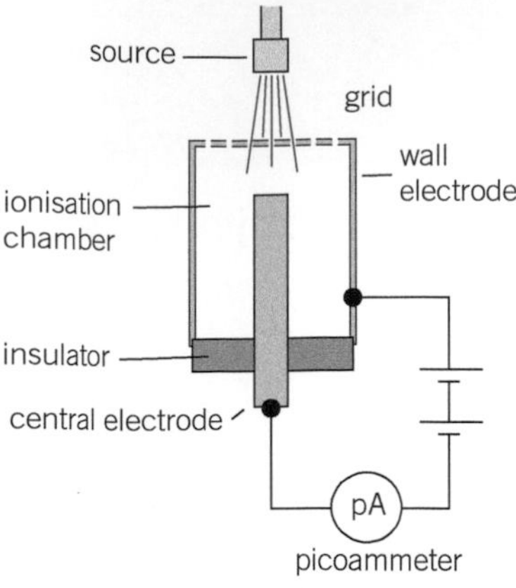

▲ **Figure 3** *Investigating ionisation*

Synoptic link

Alpha particles from a given source have the same range in air as each other, whereas β particles do not. The α particles from a given isotope are always emitted with the same kinetic energy. This is because each α particle and the nucleus that emits it move apart with equal and opposite amounts of momentum. This isn't the case with β particles because a neutrino or antineutrino is emitted as well. In β emission, the nucleus, the β particle, and a neutrino or antineutrino share the energy released in variable proportions. See Topic 7.2, Stable and unstable nuclei.

Radioactivity experiments

Ionisation

The ionising effect of each type of radiation can be investigated using an ionisation chamber and a picoammeter, as shown in Figure 3. The chamber contains air at atmospheric pressure. Ions created in the chamber are attracted to the oppositely charged electrode where they are discharged. Electrons pass through the picoammeter as a result of ionisation in the chamber. The current is proportional to the number of ions per second created in the chamber.

- Alpha radiation causes strong ionisation. However, if the source is moved away from the top of the chamber, ionisation ceases beyond a certain distance. This is because α radiation has a range in air of no more than a few centimetres.
- Beta radiation has a much weaker ionising effect than α radiation. Its range in air varies up to a metre or more. A β particle therefore produces fewer ions per millimetre along its path than an α particle does.
- Gamma radiation has a much weaker ionising effect than either α or β radiation. This is because photons carry no charge so they have less effect than α or β particles do.

Cloud chamber observations

A cloud chamber contains air saturated with a vapour at a very low temperature. Due to ionisation of the air, an α or a β particle passing through the cloud chamber leaves a visible track of minute condensed vapour droplets. This is because the air space is supersaturated. When an ionising particle passes through the supersaturated vapour, the ions produced trigger the formation of droplets.

- Alpha particles produce straight tracks that radiate from the source and are easily visible. The tracks from a given isotope are all of the same length, indicating that the α particles have the same range.
- Beta particles produce wispy tracks that are easily deflected as a result of collisions with air molecules. The tracks are not as easy to see as α particle tracks because β particles are less ionising than α particles.

a *Alpha particle tracks*

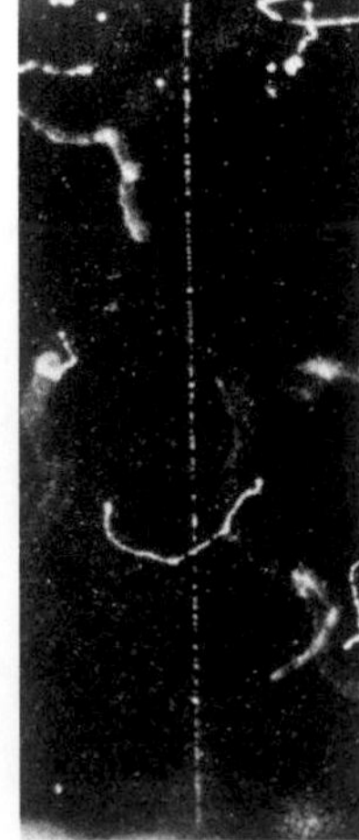

b *Beta particle tracks*

▲ **Figure 4** *Cloud chamber photographs. The yellow track in a is caused by an alpha particle colliding with the nucleus of a gas atom.*

Absorption tests

Figure 5 shows how a Geiger–Müller tube and a counter may be used to investigate absorption by different materials. Each particle of radiation that enters the tube is registered by the counter as a single count.

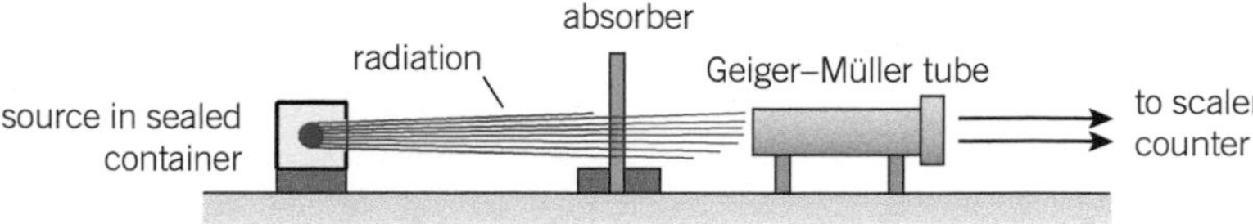

▲ **Figure 5** *Investigating absorption*

The number of counts in a given time is measured and used to work out the **count rate**, which is the number of counts divided by the time taken. Before the source is tested, the count rate due to background radioactivity must be measured. This is the count rate without the source present.

- The count rate is then measured with the source at a fixed distance from the tube without any absorber present. The background count rate is then subtracted from the count rate with the source present to give the corrected (i.e., true) count rate from the source.
- The count rate is then measured with the absorber in a fixed position between the source and the tube. The corrected count rates with and without the absorber present can then be compared.

By using absorbers of different thickness of the same material, the effect of the absorber thickness can be investigated. Figure 6 shows a typical set of measurements for the absorption of β radiation by aluminium. Note that the count rate scale is a logarithmic scale (see Topic 27.3).

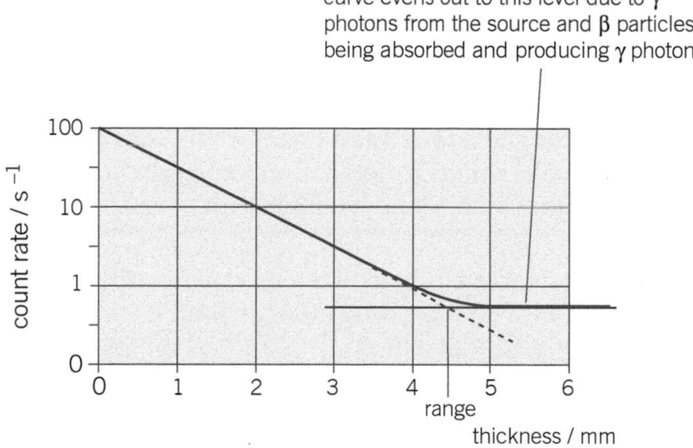

▲ **Figure 6** *Count rate against absorber thickness*

Absorption summary

- Alpha radiation is absorbed completely by paper and thin metal foil.
- Beta radiation is absorbed completely by about 5 mm of metal.
- Gamma radiation is absorbed almost completely by several centimetres of lead.

The Geiger–Müller tube

The Geiger–Müller tube is a sealed metal tube that contains argon gas at low pressure. The thin mica window at the end of the tube allows α and β particles to enter the tube. Gamma photons can enter the tube through the tube wall as well. A metal rod down the middle of the tube is at a positive potential, as shown in Figure 7. The tube wall is connected to the negative terminal of the power supply and is earthed.

Notes

The *dead time* of the tube, the time taken to regain its non-conducting state after an ionising particle enters it, is typically of the order of 0.2 ms. Another particle that enters the tube in this time will not cause a voltage pulse. Therefore, the count rate should be no greater than about $5000\,s^{-1}\left(=\frac{1}{0.2}\,ms\right)$.

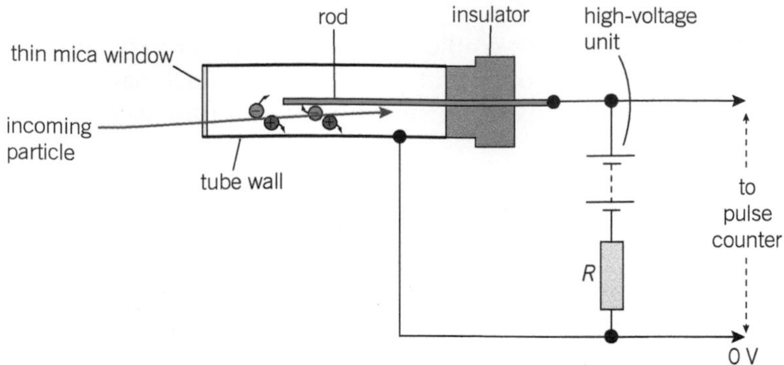

▲ **Figure 7** *A Geiger–Müller tube*

When a particle of **ionising radiation** enters the tube, the particle ionises the gas atoms along its track. The negative ions are attracted to the rod and the positive ions to the wall. The ions accelerate and collide with other gas atoms, producing more ions. These ions produce further ions in the same way so that, within a very short time, many ions are created and discharged at the electrodes. A pulse of charge passes round the circuit through resistor R, causing a voltage pulse across R which is recorded as a single count by the pulse counter.

Range in air

The arrangement in Figure 5 without the absorbers may be used to investigate the range of each type of radiation in air. The count rate is measured for different distances between the source and the tube, starting with the source close to the tube. The background count rate must also be measured in the absence of the source so the corrected count rate can be calculated for each distance.

- Alpha radiation has a range of only a few centimetres in air. The count rate decreases sharply once the tube is beyond the range of the α particles. This can be seen in Figure 4 as the tracks from the source are the same length, indicating that the particles from a given source have the same range and therefore the same initial kinetic energy. The range differs from one source to another, indicating that the initial kinetic energy differs from one source to another.
- Beta radiation has a range in air of up to about a metre. The count rate gradually decreases with increasing distance until it is the same as the background count rate at a distance of about 1 m. The reason for the gradual decrease of count rate as the distance increases is that the β particles from any given source have a range of initial kinetic energies up to a maximum. Faster β particles travel further in air than slower β particles as they have greater initial kinetic energy.
- Gamma radiation has an unlimited range in air. The count rate gradually decreases with increasing distance because the radiation spreads out in all directions, as shown in Figure 2. The proportion of the γ photons from the source entering the tube decreases according to the inverse square law. See Topic 8.3 for more on the properties of α, β, and γ radiation.

▼ **Table 1** *Nature and properties of α, β, and γ radiation*

	α radiation	β radiation	γ radiation
Nature	2 protons + 2 neutrons	β^- = electron (β^+ = positron)	photon of energy of the order of MeV
Range in air	fixed range, depends on energy, can be up to 100 mm	range up to about 1 m	follows the inverse square law
Deflection in a magnetic field	deflected	opposite direction to α particles, and more easily deflected	not deflected
Absorption	stopped by paper or thin metal foil	stopped by approx. 5 mm of aluminium or 2–3 mm of steel	stopped or significantly reduced by several centimetres of lead or about 5 cm of steel
Ionisation	produces about 10^4 ions per mm in air at standard pressure	produces about 100 ions per mm in air at standard pressure	very weak ionising effect
Energy of each particle/photon	constant for a given source	varies up to a maximum for a given source	constant for a given source

Summary questions

1 A beam of radiation from a radioactive substance passes through paper and is then stopped by an aluminium plate of thickness 5 mm.

- **a** What type of particles are in this beam?
- **b** Describe a further test you could do to check your answer in part **a**.

2 What type of radioactivity was responsible for the image of the key seen by Becquerel in the effect described at the beginning of this topic?

- **a** Explain why an image of the key was produced on the photographic plate.
- **b** Which type of radiation from a radioactive source is
  - least ionising?
  - most ionising?
- **c** When an α-emitting source above an ionisation chamber grid is moved gradually away from the grid, the ionisation current suddenly drops to zero. Explain why the current suddenly drops to zero.

3 In an absorption test as shown in Figure 5 using a β-emitting source and a Geiger–Müller counter, a count rate of 8.2 counts per second was obtained without the absorber present and a count rate of 3.7 counts per second was obtained with the absorber present. The background count rate was 0.4 counts per second. What percentage of the β particles hitting the absorber

- **a** pass through it?
- **b** are stopped by the absorber?

4 In an investigation to find out the type of radiation emitted by a radioactive source, a Geiger–Müller tube was placed near the source and its count rate was significantly more than the background count rate. When an aluminium plate was placed between the source and the tube, the count rate was reduced but it was still significantly more than the background count rate.

- **a** What can be concluded from these observations?
- **b** When the distance from the source to the tube was doubled with the aluminium plate still present, the corrected count rate decreased to about 25%. What conclusion can be drawn from this observation?

8.3 More about α, β, and γ radiation

Learning objectives:

- → Describe what happens to the nucleus in a radioactive change.
- → Describe how the intensity of γ radiation changes as it spreads out.
- → Explain how to represent the change in a nucleus when it emits α, β, or γ radiation.

Specification reference: 3.3.3

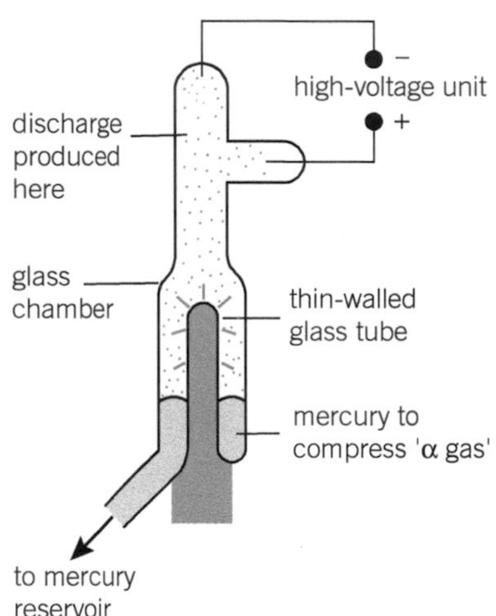

▲ **Figure 1** *Identifying α particles*

The nature of α, β, and γ radiation

Alpha radiation consists of positively charged particles. Each α particle is composed of two protons and two neutrons, the same as the nucleus of a helium atom. Rutherford devised an experiment in which α particles were collected as a gas in a glass tube fitted with two electrodes. When a voltage was applied to the electrodes, the gas conducted electricity and emitted light. Using a spectrometer, he proved that the spectrum of light from the tube was the same as from a tube filled with helium gas.

Rutherford made the discovery that neutralised α particles are the same as helium some years before his discovery that every atom contains a nucleus. After he established the nuclear model of the atom, it was realised that the nucleus of the hydrogen atom, the lightest known atom, was a single positively charged particle which became known as the **proton**. Rutherford realised that other nuclei contain protons, and he predicted the existence of neutral particles of similar mass, **neutrons**, in the nucleus. For example, the helium nucleus carries twice the charge of the hydrogen nucleus and therefore contains two protons. However, its mass is four times the mass of the hydrogen nucleus so Rutherford predicted that it contained two neutrons as well as two protons. The existence of the neutron was established in 1932 by James Chadwick, one of Rutherford's former students.

Beta radiation from naturally occurring radioactive substances consists of fast-moving electrons. This was proved by measuring the deflection of a beam of β particles using electric and magnetic fields. The measurements were used to work out the specific charge (which is the charge/mass) of the particles. This was shown to be the same as the specific charge of the electron. An electron is created and emitted from a nucleus with too many neutrons as a result of a neutron suddenly changing into a proton.

A nucleus with too many protons is also unstable and emits a **positron**, the antiparticle of the electron, when a proton changes to a neutron. Such unstable nuclei are not present in naturally occurring radioactive substances. They are created when high-energy protons collide with nuclei. As outlined in Chapter 7, the theory that for every type of particle there is a corresponding antiparticle was put forward by Paul Dirac in 1928. The first antiparticle to be discovered, the positron, was discovered by Carl Anderson four years later.

Gamma radiation consists of photons with a wavelength of the order of 10^{-11} m or less. This discovery was made by using a crystal to diffract a beam of γ radiation in a similar way to the diffraction of light by a diffraction grating.

Synoptic link

Gamma photons are emitted from 'excited' nuclei with energies of the order of MeV (where 1 MeV $= 1.6 \times 10^{-13}$ J), which means their wavelength is of the order of 10^{-11} m or less. See Topic 7.3, Photons.

The inverse square law for γ radiation

The **intensity of radiation, *I***, is the radiation energy per second passing normally through unit area.

- For a point source that emits n γ photons per second, each of energy hf, the radiation energy per second from the source = nhf.
- At distance r from the source, all the photons emitted from the source pass through a total area of $4\pi r^2$ (the surface area of a sphere of radius r).

So the intensity I of the radiation at this distance

$$= \frac{\text{radiation energy per second}}{\text{total area}} = \frac{nhf}{4\pi r^2}$$

Therefore $I = \frac{k}{r^2}$, where the constant $k = \frac{nhf}{4\pi}$.

Thus I varies with the inverse square of distance r.

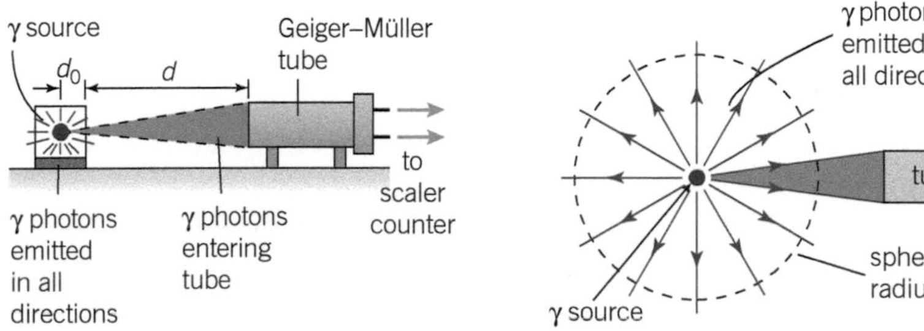

▲ **Figure 2** *Verifying the inverse square law for γ radiation*

The equations for radioactive change

A **nuclide** ${}^{A}_{Z}X$ contains Z protons and $A - Z$ neutrons.

Alpha emission

An α particle consists of two neutrons and two protons so is represented by the symbol ${}^{4}_{2}\alpha$. As explained in Topic 7.2, the equation below represents the change that takes place when a nuclide ${}^{A}_{Z}X$ emits an α particle to form a new nuclide ${}^{A-4}_{Z-2}Y$.

$${}^{A}_{Z}X \rightarrow {}^{4}_{2}\alpha + {}^{A-4}_{Z-2}Y$$

Electron (β^-) emission

A negative β particle (i.e., an electron) is represented by the symbol ${}^{0}_{-1}\beta$ (or β^-) as its charge = $-e$ and it is not a neutron or a proton.

The equation below represents the changes that take place when a nucleus ${}^{A}_{Z}X$ emits a β^- particle. In effect, a neutron in a neutron-rich nucleus changes into a proton, and an (electron) antineutrino $\overline{\nu}_e$ is emitted at the same time as the β^- particle is created.

$${}^{A}_{Z}X \rightarrow {}^{0}_{-1}\beta + {}^{A}_{Z+1}Y + \overline{\nu}_e$$

Positron (β^+) emission

A positive β particle (i.e., a positron) is represented by the symbol ${}^{0}_{+1}\beta$ (or β^+) as its charge = $+e$ and it is not a neutron or a proton.

The equation below represents the changes that take place when a nucleus ${}^{A}_{Z}X$ emits a positron. In effect, a proton in a proton-rich nucleus changes into a neutron and an electron neutrino ν_e is emitted at the same time as the β^+ particle is created.

$${}^{A}_{Z}X \rightarrow {}^{0}_{+1}\beta + {}^{A}_{Z-1}Y + \nu_e$$

Notes

1 If the intensity at distance r_0 is I_0, then:

- at distance $2r_0$, the intensity is $\frac{I_0}{4}$
- at distance $3r_0$, the intensity is $\frac{I_0}{9}$, etc.

Radiation workers use remote handling devices (and lead screens) to keep as far as possible from γ sources to reduce their exposure to the radiation.

2 To verify the inverse square law for a γ source, use a Geiger–Müller counter to measure the count rate C at different measured distances, d, from the tube and the background count rate, C_0, without the source present. The corrected count rate $C - C_0$ is proportional to the intensity of the radiation.

As shown in Figure 2 the source is at unknown distance, d_0, inside its sealed container. Using the inverse square law for γ radiation gives the corrected count rate

$$C - C_0 = \frac{\text{constant}}{(d + d_0)^2}$$

Therefore, a graph of d (on the y-axis) against $\frac{1}{(C - C_0)^{\frac{1}{2}}}$ should give a straight line with a negative intercept $-d_0$.

Electron capture

Some proton-rich nuclides can capture an inner-shell electron. This causes a proton in the nucleus to change into a neutron with the emission of an electron neutrino ν_e at the same time.

$$^{A}_{Z}\mathrm{X} + {}^{0}_{-1}\mathrm{e} \rightarrow {}^{A}_{Z-1}\mathrm{Y} + \nu_e$$

The inner-shell vacancy is filled by an outer-shell electron, as a result causing an X-ray photon to be emitted by the atom.

Gamma emission

No change occurs in the number of protons or neutrons of a nucleus when it emits a γ photon.

A γ photon is emitted if a nucleus has excess energy after it has emitted an α or a β^- particle.

Decay of free neutrons

A neutron outside the nucleus is referred to as a free neutron. A free neutron will decay into a proton, emitting a β^- particle and an antineutrino in the process. This change can be represented by the equation

$$^{1}_{0}\mathrm{n} \rightarrow {}^{1}_{1}\mathrm{p} + {}^{0}_{-1}\beta + \bar{\nu}$$

The proton is a stable particle as it does not decay.

Study tip

Remember for the inverse square law, doubling the distance reduces the intensity to a quarter and trebling the distance reduces it to a ninth!

Hint

1 atomic mass unit (abbreviated as u) $= \frac{1}{12} \times$ the mass of a $^{12}_{6}\mathrm{C}$ atom.

The mass of a proton is 1.00728 u and the mass of a neutron is 1.00867 u.

The difference between A and the mass of a nucleus in atomic mass units, although very small for most atoms, is important in nuclear energy calculations, which you will meet in Topic 24.3.

Summary questions

1 Copy and complete each of the following equations representing α emission:

a $^{238}_{92}\mathrm{U} \rightarrow {}_{90}\mathrm{Th} +$ **b** $_{90}\mathrm{Th} \rightarrow {}^{224}_{88}\mathrm{Ra} +$

2 Copy and complete each of the following equations representing β^- emission:

a $^{64}_{29}\mathrm{Cu} \rightarrow {}_{30}\mathrm{Zn} + {}^{0}\beta +$ **b** $_{15}\mathrm{P} \rightarrow {}^{32}\mathrm{S} + {}^{0}\beta +$

3 The bismuth isotope $^{213}_{83}\mathrm{Bi}$ decays by emitting a β^- particle to form an unstable isotope of polonium (Po) which then decays by emitting an α particle to form an unstable isotope of lead (Pb). This isotope then decays by emitting a β^- particle to form a stable isotope of bismuth.

a Write down the symbol for each of the three product nuclides in this sequence.

b Write down the number of protons and the number of neutrons in a nucleus of

i the bismuth isotope $^{213}_{83}\mathrm{Bi}$

ii the stable bismuth isotope.

4 A point source of γ radiation is placed 200 mm from the end of a Geiger–Müller tube. The corrected count rate was measured at 12.7 counts per second. Calculate:

a the corrected count rate if the source was moved to a distance of 400 mm from the tube

b the distance between the source and the tube for a corrected count rate of 20 counts per second.

8.4 The dangers of radioactivity

Learning objectives:

- → Explain why ionising radiation is harmful.
- → State the factors that determine whether α, β, or γ are the most dangerous.
- → Discuss how exposure to ionising radiation can be reduced.

Specification reference: 3.3.3

The hazards of ionising radiation

Ionising radiation is hazardous because it damages living cells. Such radiation includes **X-rays**, protons, and neutrons as well as α, β, and γ radiation. Ionising radiation affects living cells because:

- it can destroy cell membranes which causes cells to die, or
- it can damage vital molecules such as DNA directly or indirectly by creating 'free radical' ions which react with vital molecules. Normal cell division is affected and nuclei become damaged. Damaged DNA may cause cells to divide and grow uncontrollably, causing a tumour which may be cancerous. Damaged DNA in a sex cell (i.e., an egg or a sperm) may cause a mutation which may be passed on to future generations.

As a result of exposure to ionising radiation, living cells die or grow uncontrollably or mutate, affecting the health of the affected person (somatic effects) and possibly affecting future generations (genetic effects). High doses of ionising radiation kill living cells. Cell mutation and cancerous growth occur at low doses as well as at high doses. There is no evidence of the existence of a threshold level of ionising radiation below which living cells would not be damaged.

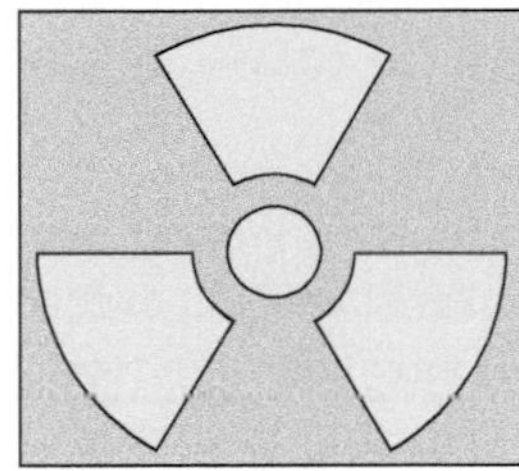

▲ **Figure 1** *A radioactive warning sign*

Radiation monitoring

Anyone using equipment that produces ionising radiation must wear a *film badge* to monitor his or her exposure to ionising radiation. The badge contains a strip of photographic film in a light-proof wrapper. Different areas of the film are covered by absorbers of different materials and different thicknesses. When the film is developed, the amount of exposure to each form of ionising radiation can be estimated from the blackening of the film. If the badge is overexposed, the wearer is not allowed to continue working with the equipment.

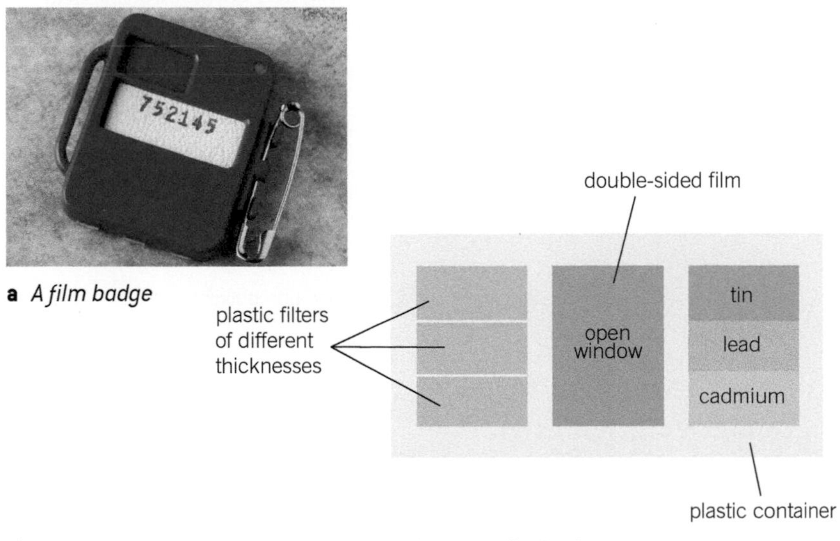

a *A film badge*

▲ **Figure 2** **b** *Inside a film badge*

The biological effect of ionising radiation depends on the dose received and the type of radiation. The dose is measured in terms of the energy absorbed per unit mass of matter from the radiation. The same dose of different types of ionising radiation has different effects. For example, α radiation produces far more ions per millimetre than γ radiation in the same substance so it is far more damaging. However, α radiation from a source outside the body cannot penetrate the skin's outer layer of dead cells so is much less damaging than if the source were inside the body.

Extension

Radiation dose limits

For any dose of ionising radiation, its **dose equivalent**, measured in sieverts (Sv), is the dose due to 250 kV X-rays that would have the same effect. For example, 1 millisievert (mSv) of α radiation has the same biological effect as 10 mSv of 250 kV X-rays.

Maximum permissible exposure limits are recommended safety limits for the annual dose equivalent which a radiation worker should not exceed. The recommended limit is 15 mSv per year, although the average dose due to occupation is much lower at 2 mSv per year. This is based on the death rate of survivors of the atomic bombs dropped on Hiroshima and Nagasaki, which is estimated at 3 deaths per 100 000 survivors for each millisievert of radiation. Thus the risk to a radiation worker exposed to 2 mSv per year for 5 years would be 3 in 10 000.

Scientists think there is no lower limit below which ionising radiation is harmless. They recommend that exposure to ionising radiation should be as low as reasonably practical or achievable (abbreviated as ALARP in the UK or ALARA in the USA). It would be unethical to carry out tests that expose people or animals to ionising radiation to find out how they react to it. So the evidence for 'no lower limit' comes from the study of groups of people such as Chernobyl survivors who have been exposed to ionising radiation.

Q: Look at Figure 3. State two sources of background radiation that affect everyone all the time.

Answer: Food and drink, cosmic rays.

Study tip

ALARA stands for 'as low as reasonably achievable'. Risks are always reduced by increasing the distance from sources and shortening the time exposure.

Background radiation

We are all subject to **background radiation** which occurs naturally due to cosmic radiation and from radioactive materials in rocks, soil, and in the air. Background radiation varies with location due to local geological features. For example, radon gas, which is radioactive, can accumulate in poorly ventilated areas of buildings in certain locations. Figure 3 shows the sources of background radiation in the UK.

Safe use of radioactive materials

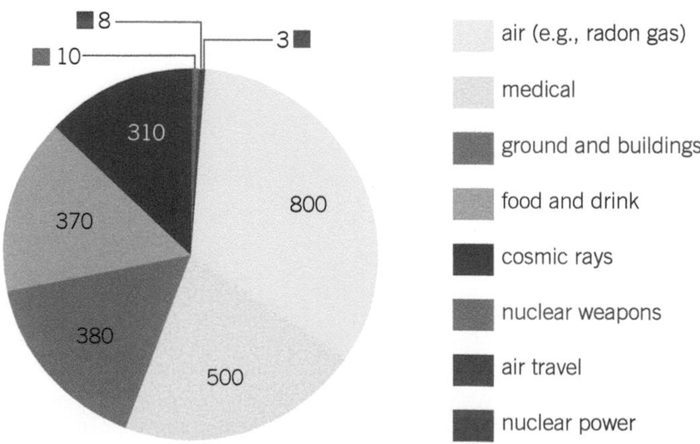

▲ **Figure 3** *Sources of background radiation in the UK in terms of the average radiation dose per person per year in microsieverts (μSv)*

Because radioactive materials produce ionising radiation, they must be stored and used with care. In addition, disposal of a radioactive substance must be carried out in accordance with specific regulations. Only approved institutions are allowed to use radioactive materials. Approval is subject to regular checks, and approved institutions are categorised according to purpose.

1 **Storage of radioactive materials** should be in lead-lined containers. Most radioactive sources produce γ radiation as well as α or β radiation so the lead lining of a container must be thick enough to reduce the γ radiation from the sources in the container to about the background level. In addition, regulations require that the containers are under lock and key, and that a record of the sources is kept.

2 **When using radioactive materials**, established rules and regulations must be followed. No source should be allowed to come into contact with the skin.

- Solid sources should be transferred using handling tools such as tongs or a glove-box or using robots. The handling tools ensure the material is as far from the user as practicable so the intensity of the γ radiation from the source at the user is as low as possible and the user is beyond the range of α or β radiation from the source.
- Liquid and gas sources and solids in powder form should be in sealed containers. This is to ensure radioactive gas cannot be breathed in and radioactive liquid cannot be splashed on the skin or drunk.
- Radioactive sources should not be used for longer than is necessary. The longer a person is exposed to ionising radiation, the greater the dose of radiation received.

Summary questions

1 What is meant by ionisation? Explain why a source of α radiation is not as dangerous as a source of β radiation provided the sources are outside the body.

2 a Discuss the reasons why ionising radiation is hazardous to a person exposed to the radiation.

b i What is the purpose of a film badge worn by a radiation worker?

ii With the aid of a diagram, describe what is in a film badge and how the film badge is tested.

3 a Explain why a radioactive source should be

i kept in a lead-lined storage box when not in use

ii transferred using a pair of tongs with long handles.

b Discuss the precautions you would take when carrying out an experiment using a source of γ radiation.

4 a State *two* sources of ionising radiation which are likely to affect people in certain occupations more than the general public.

b Radon gas is a source of α radiation that can seep into buildings from the ground. Explain why the presence of this gas in the air in a building is a serious health hazard.

Synoptic link

The intensity of a γ beam that passes through an absorber decreases exponentially with the thickness of the absorber. If a certain thickness of a material cuts the intensity of a γ beam to half, twice the thickness will cut the intensity to a quarter, and so on. See Topic 8.5, Radioactive decay.

Hint

A typical radioactive source used in a school might produce of the order of 10^5 particles per second, each typically of energy of the order of MeV ($1\ \text{MeV} = 1.6 \times 10^{-13}$ J). Show for yourself that the energy transfer per second from such a source is of the order of $10^{-8}\ \text{J s}^{-1}$. In 15 minutes or so, the source would transfer 10^{-5} J to its surroundings. If this amount of energy were to be absorbed by about 10 kg of living tissue, the dose would be about 10^{-6} Sv ($= 1\ \mu\text{Sv}$) which is not insignificant.

8.5 Radioactive decay

Learning objectives:

- → State what is meant by a decay curve.
- → Define the half-life of a radioactive isotope.
- → Discuss whether anything affects radioactive decay.

Specification reference: 3.3.3

Synoptic link

Although in theory a radioactive decay curve never falls to zero, in practice it eventually falls to a level which is indistinguishable from background radiation. See Topic 8.4, The dangers of radioactivity.

Study tip

Know how to use a calculator to raise 0.5 to a power.

Half-life

When a nucleus of a radioactive isotope emits an α or a β particle, it becomes a nucleus of a different element because its proton number changes. The number of nuclei of the initial radioactive isotope therefore decreases. The mass of the initial isotope decreases gradually as the number of nuclei of the isotope decreases. Figure 1 shows how the mass decreases with time. The curve is referred to as a **decay curve**. The mass of the isotope decreases with time at a slower and slower rate. Measurements show that the mass decreases exponentially, which means that the mass drops by a constant factor (e.g., × 0.8) in equal intervals of time. For example, if the initial mass of the radioactive isotope is 100 g and the mass decreases to a factor of × 0.8 every 1000 s, then

- after 1000 s, the mass remaining = 80 g (= 0.8 × 100 g),
- after 2000 s, the mass remaining = 64 g (= 0.8 × 0.8 × 100 g),
- after 3000 s, the mass remaining = 51 g (= 0.8 × 0.8 × 0.8 × 100 g).

A convenient measure for the rate of decrease is the time taken for a decrease by half. This is the half-life of the process.

The half-life, $T_{\frac{1}{2}}$, of a radioactive isotope is the time taken for the mass of the isotope to decrease to half the initial mass. This is the same as the time taken for the number of nuclei of the isotope to decrease to half the initial number.

Consider a sample of a radioactive isotope X which initially contains 100 g of the isotope.

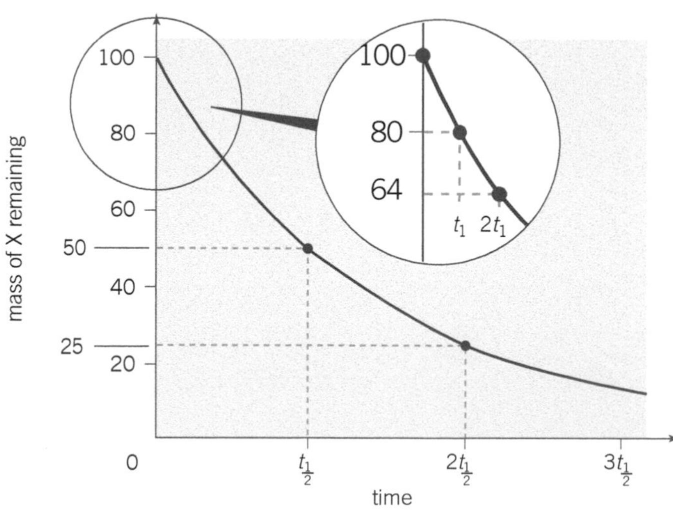

▲ **Figure 1** *A radioactive decay curve*

- After one half-life, the mass of X remaining = 0.5 × 100 = 50 g.
- After two half-lives from the start, the mass of X remaining = $0.5^2 \times 100$ = 25 g.
- After three half-lives from the start, the mass of X remaining = $0.5^3 \times 100$ = 12.5 g.
- After n half-lives from the start, **the mass of X remaining = $0.5^n m_0$,** where m_0 = the initial mass.

The mass of X decreases exponentially. This is because radioactive decay is a **random** process and the number of nuclei that decay in a certain time is in proportion to the number of nuclei of X remaining. To understand this idea, consider a game of dice where there are 1000 dice, each representing a nucleus of X. The throw of a dice is a random process in which each face has an equal chance of being uppermost.

- First throw – when the dice are all thrown, you would expect $\frac{1}{6}$ of the dice to show the same figure on the upper surface.

Let all the dice that show '1' uppermost represent nuclei that have disintegrated, an expected total of 167 $\left(= \frac{1000}{6}\right)$. If these are removed, then 833 dice remain.

- Second throw – the remaining dice are thrown to give $\frac{1}{6}$ of 833 as the expected number of 1's. So 694 dice $\left(= 833 - \frac{833}{6}\right)$ remain.
- Third throw – the remaining dice are thrown to give $\frac{1}{6}$ of 694 as the expected number of 1's. So 578 dice $\left(= 694 - \frac{694}{6}\right)$ remain.
- Fourth throw – the remaining dice are thrown to give $\frac{1}{6}$ of 578 as the expected number of 1's. So 482 dice $\left(= 578 - \frac{578}{6}\right)$ remain.

The analysis shows that four throws are needed to reduce the number of dice to less than half the initial number. Prove for yourself that a further four throws would reduce the number of dice to 25% of the initial number. Figure 2 shows how the number of dice remaining decreases with time. The curve has the same shape as Figure 1. The half-life of the process is 3.8 'throws'.

The number of nuclei N of a radioactive isotope remaining after n half-lives = $0.5^n \times$ the initial number N_0.

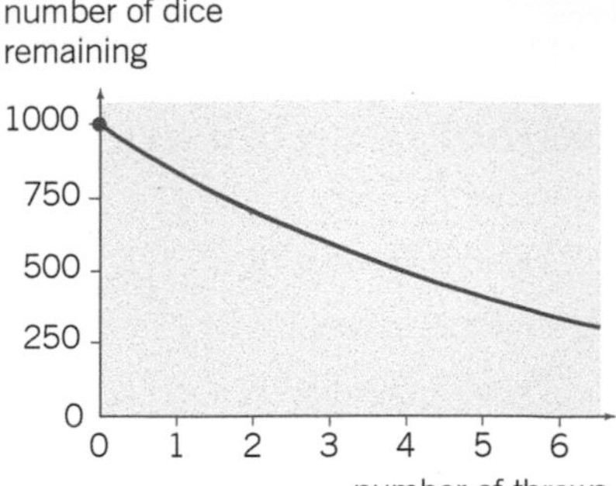

▲ **Figure 2** *Exponential decrease*

Synoptic link

To see how to model the dice model in a computer spreadsheet, see Topic 27.6, Graphical and computational modelling.

Summary questions

$N_A = 6.02 \times 10^{23}\,\text{mol}^{-1}$, $1\,\text{MeV} = 1.6 \times 10^{-13}\,\text{J}$

1 Figure 3 shows how the mass of a certain radioactive isotope decreases with time. Use the graph to work out:

a the half-life of this isotope

b the mass of the isotope remaining after 120 s.

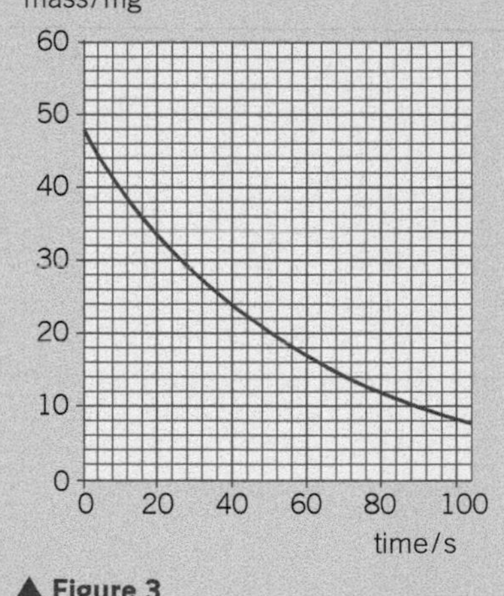

▲ **Figure 3**

2 A freshly prepared sample of a radioactive isotope X contains 1.8×10^{15} atoms of the isotope. The half-life of the isotope is 8.0 hours. Calculate:

a the number of atoms of this isotope remaining after

i 8 h **ii** 24 h

b the number of atoms of X that would have decayed after

i 8 h **ii** 24 h

c the energy transfer from the sample in 24 h if the isotope emits α particles of energy 5 MeV.

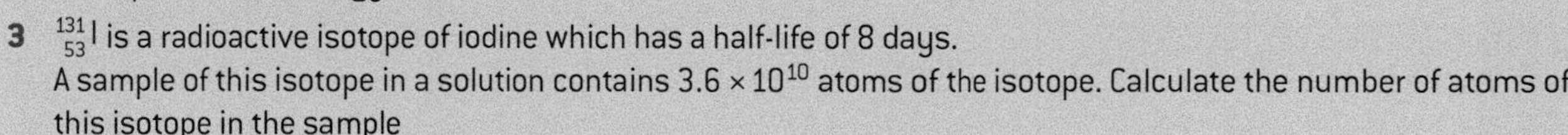

3 $^{131}_{53}\text{I}$ is a radioactive isotope of iodine which has a half-life of 8 days.
A sample of this isotope in a solution contains 3.6×10^{10} atoms of the isotope. Calculate the number of atoms of this isotope in the sample

a 8.0 days later **b** 64 days later.

4 $^{137}_{55}\text{Cs}$ is a radioactive isotope of caesium which has a half-life of 35 years. A sample of this isotope has a mass of 1.0×10^{-3} kg.
Calculate the mass of the isotope remaining in the sample after 70 years.

8.6 Radioactive isotopes in use

Learning objectives:

- → Describe how to do radioactive dating.
- → Define radioactive tracers.
- → Describe industrial uses of radioactivity.

Specification reference: 3.3.3

Radioactive isotopes are used for many purposes. The choice of an isotope for a particular purpose depends on its half-life and on the type of radiation it emits. For some uses, the choice also depends on how the isotope is obtained and on whether or not it produces a stable decay product. In addition, the toxicity and biochemical suitability of the pharmaceuticals to which it is attached need to be considered in medical and related applications. The following examples in this topic and in Topic 8.7 are intended to provide a wider awareness of important uses of radioactive substances and to set contexts in which knowledge and understanding of radioactivity is developed further.

Radioactive dating

Carbon dating

Living plants and trees contain a small percentage of the radioactive isotope of carbon, ${}^{14}_{6}C$, which is formed in the atmosphere as a result of cosmic rays knocking out neutrons from nuclei. These neutrons then collide with nitrogen nuclei to form carbon-14 nuclei.

$${}^{1}_{0}n + {}^{14}_{7}N \rightarrow {}^{14}_{6}C + {}^{1}_{1}P$$

Carbon dioxide from the atmosphere is taken up by living plants as a result of photosynthesis. So a small percentage of the carbon content of any plant is carbon-14. This isotope has a half-life of 5570 years so there is negligible decay during the lifetime of a plant. Once a tree has died, no further carbon is taken in so the proportion of carbon-14 in the dead tree decreases as the carbon-14 nuclei decay. The age of a sample of dead wood can be determined by measuring the amount of carbon-14 in a sample of dead wood and the amount of carbon-14 in the same mass of living wood. The ratio of the dead wood measurement to the living wood measurement gives 0.5^n where n is the number of half-lives of carbon-14. You can then calculate n.

Worked example

A certain sample of dead wood is found to contain 7.1×10^{10} atoms of carbon-14. An equal mass of living wood is found to contain 3.3×10^{11} atoms of carbon-14. Calculate the age of the sample. Give your answer in years to 3 significant figures.

The half-life of carbon-14 is 5570 years.

Solution

$N_0 = 3.3 \times 10^{11}$ atoms of carbon-14

$N = 7.1 \times 10^{10}$ atoms of carbon-14 after n half-lives

Using $\frac{N}{N_0} = 0.5^n$ gives $0.5^n = \frac{(7.1 \times 10^{10})}{(3.3 \times 10^{11})} = 0.215$

To find n,

either by trial and error: note that $n = 2$ gives $0.5^n = 0.25$, so try different values of n between 0.20 and 0.25 to find the value of n that gives $0.5^n = 0.215$

or by using the rule that $\ln 0.5^n = n \ln 0.5$ (see Topic 27.3), which gives $n \ln 0.5 = \ln 0.215$

Therefore, $n = \frac{\ln 0.215}{\ln 0.5} = 2.22$

So, the age of the sample = 2.22 half-lives = 2.22 × 5570 years = 12 300 years

Hint

A useful check is to estimate the number of half-lives needed for the number of atoms to decrease from 3.3×10^{11} to 7.1×10^{10}. You should find that just over two half-lives are needed, corresponding to about 11 000 years.

Argon dating

Ancient rocks contain trapped argon gas as a result of the decay of the radioactive isotope of potassium, ${}^{40}_{19}\text{K}$, into the argon isotope ${}^{40}_{18}\text{Ar}$. This happens when its nucleus captures an inner shell electron. As a result, a proton in the nucleus changes into a neutron, and a neutrino is emitted.

The equation for the change is ${}^{40}_{19}\text{K} + {}^{0}_{-1}\text{e} \rightarrow {}^{40}_{18}\text{Ar} + \nu_e$

The potassium isotope ${}^{40}_{19}\text{K}$ also decays by β^- emission to form the calcium isotope ${}^{40}_{20}\text{Ca}$. This process is eight times more probable than electron capture.

$${}^{40}_{19}\text{K} \rightarrow {}^{0}_{-1}\beta + {}^{40}_{20}\text{Ca} + \overline{\nu}_e$$

The effective half-life of the decay of ${}^{40}_{19}\text{K}$ is 1250 million years. The age of the rock (i.e., the time from when it solidified) can be calculated by measuring the proportion of argon-40 to potassium-40.

For every N potassium-40 atoms now present, if there is one argon-40 atom present, there must have originally been $N + 9$ potassium atoms. (i.e., one that decayed into argon-40 + eight that decayed into calcium-40 + N remaining). The equation $\frac{N}{N_0} = 0.5^n$ can then be used to find the age of the sample.

For example, suppose for every four potassium-40 atoms now present, a certain rock now has one argon-40 atom. Therefore, $N = 4$ and $N_0 = 13$. Substituting these values into the equation $\frac{N}{N_0} = 0.5^n$ gives $\frac{4}{13} = \frac{N}{N_0} = 0.5^n$.

Therefore, using either method described in the worked example opposite (to find n) gives $n = 1.70$. The age of the sample is therefore $1.70\ T_{1/2}$ or 1.70×1250 years which is 2130 million years.

Hint

A useful check is to estimate the number of half-lives needed for n to decrease from 13 to 4. You should find that between 1 and 2 half-lives are needed, corresponding to an age of between 1250 and 2500 million years.

Synoptic link

The metastable isotope of technetium, ${}^{99}_{43}\text{Tc}^{m}$, is widely used in medical diagnosis because it is a γ emitter with a half-life of 6 hours and it can be prepared on site. See Topic 8.7, Nuclear energy levels.

Radioactive tracers

A radioactive tracer is used to follow the path of a substance through a system. Table 1, on the next page, gives some examples. In general, the radioactive isotope(s) in the tracer should:

- have a half-life which is stable enough for the necessary measurements to be made and short enough to decay quickly after use
- emit β radiation or γ radiation so it can be detected outside the flow path.

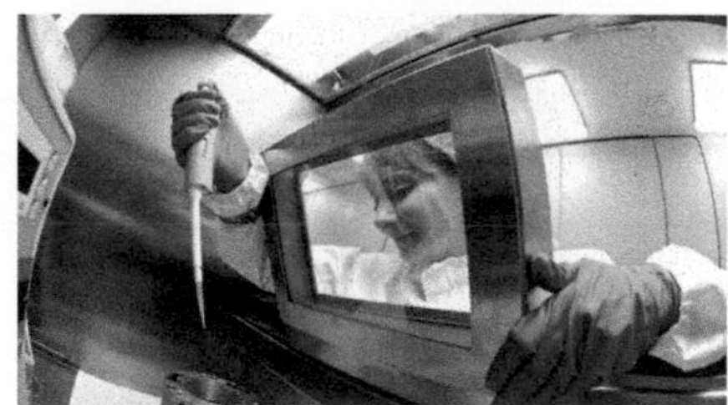

▲ **Figure 1** *Using tracers*

Industrial uses of radioactivity

The examples below are just three of a wide range of applications of radioactivity in industry and technology.

Engine wear

The rate of wear of a piston ring in an engine can be measured by fitting a ring that is radioactive. As the ring slides along the piston compartment, radioactive atoms transfer from the ring to the engine oil. By measuring the radioactivity of the oil, the mass of radioactive metal transferred from the ring can be determined and the rate of wear calculated. A metal ring can be made radioactive by exposing it to neutron radiation in a nuclear reactor. Each nucleus that absorbs a neutron becomes unstable and disintegrates by β^- emission.

▲ **Figure 2** *Measuring engine wear*

▼ **Table 1** *Examples of radioactive tracers*

Application	Method	Tracer
Detecting underground pipe leaks	Radioactive tracer injected into the flow. A detector on the surface above the pipeline is used to detect leakage.	Injected fluid contains a β emitter or a γ emitter (depending on factors such as depth or soil density) as α radiation would be absorbed by the pipes.
Modelling oil reservoirs mathematically to improve oil recovery	Water containing a radioactive tracer is injected into an oil reservoir at high pressure, forcing some of the oil out. Detectors at the production wells monitor breakthrough of the radioactive isotope.	*Tritiated* water ${}^{3}_{1}H_2O$, a β emitter with a half-life of 12 years.
Investigating the uptake of fertilisers by plants	Plant watered with a solution containing a fertiliser. By measuring the radioactivity of the leaves, the amount of fertiliser reaching them can be determined.	Fertiliser contains phosphorus, ${}^{32}_{15}P$, a β emitter with a half-life of 14 days.
Monitoring the uptake of iodine by the thyroid gland	Patient is given a solution containing sodium iodide which contains a small quantity of radioactive iodine, ${}^{133}_{53}I$. The activity of the patient's thyroid and the activity of an identical sample prepared at the same time is measured 24 h later.	Solution of sodium iodide contains iodine ${}^{131}_{53}I$, a β emitter with a half-life of 8 days.

Thickness monitoring

Metal foil is manufactured by using rollers to squeeze plate metal on a continuous production line. A detector measures the amount of radiation passing through the foil. If the foil is too thick, the detector reading drops. A signal from the detector is fed back to the control system to make the rollers move closer together and so make the foil thinner. The source used is a β^- emitter with a long half-life. Alpha radiation would be absorbed completely by the foil and γ radiation would pass straight through without absorption. A similar process is used in the process used to manufacture paper.

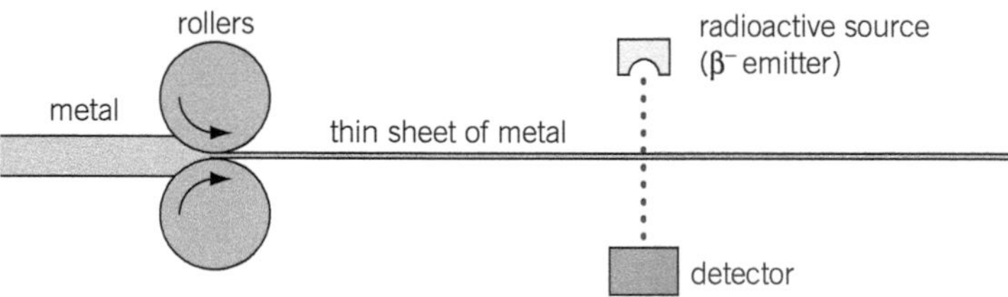

▲ **Figure 3** *The manufacture of metal foil*

Study tip

The choice of a radioactive isotope for a particular application is determined by:

- the half-life and the type of radiation needed
- the toxicity and biochemical suitability of the pharmaceuticals to which it is attached
- whether or not a stable product is needed.

Power sources for remote devices

Satellites, weather sensors, and other remote devices can be powered using a radioactive isotope in a thermally insulated sealed container which absorbs all the radiation emitted by the isotope. A thermocouple attached to the container produces electricity as a result of the container becoming warm through absorbing radiation.

For an initial mass *m* of the isotope, the shorter its half-life is, the smaller the mass of the isotope remaining after a certain time. The radioactive isotope in the source needs to have a reasonably long half-life, so it does not need to be replaced frequently, but a very long half-life may require too much mass in order to generate the necessary power.

Summary questions

$N_A = 6.02 \times 10^{23}\ \text{mol}^{-1}$, $1\ \text{MeV} = 1.6 \times 10^{-13}\ \text{J}$

1 **a** Explain why living wood is slightly radioactive.

b A sample of ancient wood of mass 0.50 g is found to contain 2.8×10^{11} atoms of carbon-14. A sample of living wood of the same mass contains 3.3×10^{11} atoms of carbon-14. Calculate the age of the sample of wood.

The half-life of radioactive carbon $^{14}_{6}\text{C}$ is 5570 years.

2 The radioactive isotope of iodine, $^{131}_{53}\text{I}$, is used for medical diagnosis of the kidneys. The isotope has a half life of 8 days. A sample of the isotope is to be given to a patient in a glass of water. The passage of the isotope through each kidney is then monitored using two detectors outside the body. The isotope is required to have 8.0×10^{11} atoms of iodine-131 at the time it is given to the patient.

a Calculate:

i the number of iodine-131 atoms in the sample 24 hours after it was given to the patient

ii the number of iodine-131 atoms in the sample when it was prepared 24 hours before it was given to the patient

iii the mass of $^{131}_{53}\text{I}$ in the sample when it was prepared.

b The reading from the detector near one of the patient's kidneys rises then falls. The reading from the other detector which is near the other kidney rises and does not fall. Discuss the conclusions that can be drawn from these observations.

3 **a** In the manufacture of metal foil, describe how the thickness of the foil is monitored using a radioactive source and a detector.

b Explain why the source needs to:

i be a β emitter, not an α emitter or a γ emitter

ii have a long half-life.

4 A cardiac pacemaker is a device used to ensure that a faulty heart beats at a suitable rate. The required electrical energy in one type of pacemaker is obtained from the energy released by a radioactive isotope. The radiation is absorbed inside the pacemaker. As a result, the absorbing material gains thermal energy and heats a thermocouple attached to the absorbing material. The voltage from the thermocouple provides the source of electrical energy for the pacemaker.

a **i** Discuss whether the radioactive source should be an α emitter, a β emitter, or a γ emitter.

ii The radioactive source needs to have a reasonably long half-life, otherwise it would need to be replaced frequently. Discuss the disadvantages of using a radioactive source with a very long half-life.

b The energy source for a remote weather station is the radioactive isotope of strontium, $^{90}_{38}\text{Sr}$, which has a half-life of 28 years. It emits β particles of energy 0.40 MeV. For an initial mass of 10 g of this isotope, calculate:

i the mass of the isotope remaining after 5.0 years

ii the average energy released per second over 5.0 years.

8.7 Nuclear energy levels

Learning objectives:

- → Discuss what you can tell about radioactive isotopes from their neutron-to-proton ratio.
- → Explain why naturally occurring isotopes don't emit β^+ radiation.
- → Describe what happens to an unstable nucleus that emits γ radiation.

Specification reference: 3.3.3

Study tip

The neutron–proton ratio is high for β^- and low for β^+ emitters.

There are no α emitters for $Z <$ about 60.

The largest stable nuclide is ${}^{209}_{83}Bi$.

Most α-emitting nuclides are larger than ${}^{209}_{83}Bi$.

Those α-emitting nuclides that are smaller than ${}^{209}_{83}Bi$ generally lie below the N–Z stability belt.

Synoptic link

β^- emitters can also be manufactured by bombarding stable isotopes with neutrons. β^+ emitters can only be produced by bombarding stable isotopes with protons. The protons need to have sufficient kinetic energy to overcome coulomb repulsion from the nucleus. See Topic 8.1, The discovery of the nucleus.

Radioactive series

Many radioactive isotopes decay to form another isotope which might itself be unstable. If the *daughter* nucleus is unstable, it will decay to form a nucleus of a different isotope (which may itself be stable or unstable) by either emitting:

- a further α particle, or
- a β^- particle if it has too many neutrons (i.e., its neutron-to-proton ratio becomes too large)
- a β^+ particle if it has too many protons (i.e., its neutron-to-proton ratio becomes too small).

Thus an unstable nucleus, before it becomes stable, may undergo a series of isotopic changes in which each change involves an emission of an α particle or a β particle. Naturally occurring radioactive isotopes decay through a series of such changes with one or more of the changes having a very long half-life – this explains the reason why such isotopes have not decayed completely.

For example, the uranium isotope ${}^{238}_{92}U$ has a half-life of 4500 million years and decays through a series of changes outlined below.

$${}^{238}_{92}U \xrightarrow{\alpha} {}^{234}_{90}Th \xrightarrow{\alpha} {}^{230}_{88}Ra \xrightarrow{\beta^-} {}^{230}_{89}Ac$$

$$\xrightarrow{\beta^-} {}^{230}_{90}Th \xrightarrow{\alpha} {}^{226}_{88}Ra \longrightarrow \ldots \longrightarrow {}^{206}_{82}Pb$$

Any radioactive series may be represented on an N–Z graph by a sequence of *decay arrows* as shown in Figure 2, which represents the first five changes above. Notice there are no β^+ emissions in the sequence because such an emission after an α emission would

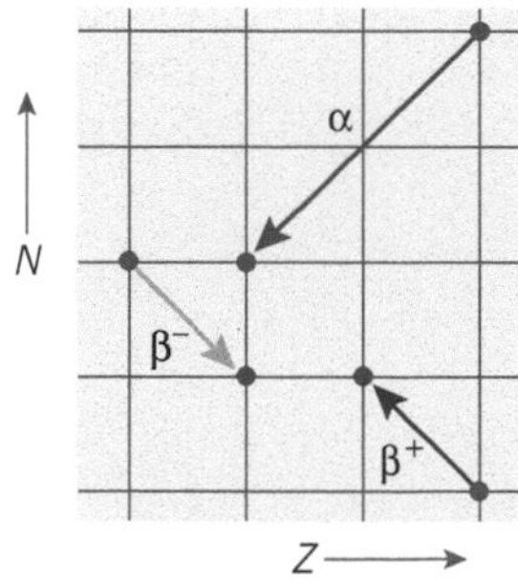

▲ **Figure 1** *N–Z changes*

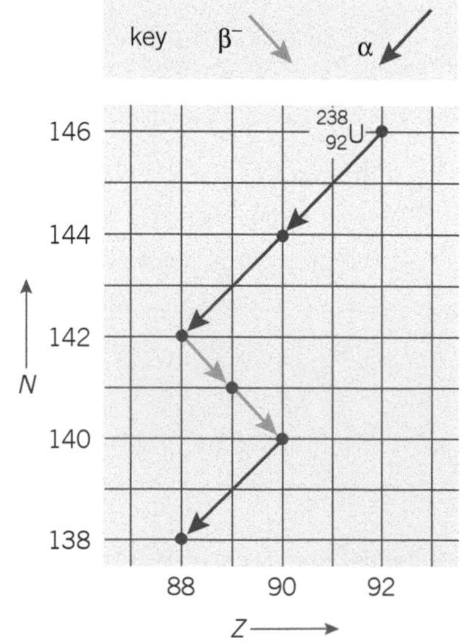

▲ **Figure 2** *Part of a radioactive decay series*

make the nucleus more neutron rich, so making it more unstable. A β^- emission after an α emission makes the nucleus less neutron-rich causing it to be more stable.

Nuclear energy levels

After an unstable nucleus emits an α or a β particle or undergoes electron capture, it might emit a γ photon. Emission of a γ photon does not change the number of protons or the number of neutrons in the nucleus but it does allow the nucleus to lose energy. This happens if the 'daughter' nucleus is formed in an **excited state** after it emits an α or a β particle or undergoes electron capture. The excited state is usually short lived and the nucleus moves to its lowest energy state, its **ground state**, either directly or via one or more lower-energy excited states. We can represent such changes by means of an energy level diagram, as shown by the example in Figure 3 in which:

- a magnesium $^{27}_{12}$Mg nucleus decays by β^- emission (shown as a blue arrow) to form an aluminium $^{27}_{13}$Al nucleus in an excited state 1.02 MeV above the ground state at zero energy
- the aluminium nucleus de-excites by emitting (as shown by the green arrows) either a 1.02 MeV γ photon (1), or a 0.19 MeV γ photon (2) followed by a 0.83 MeV γ photon (3).

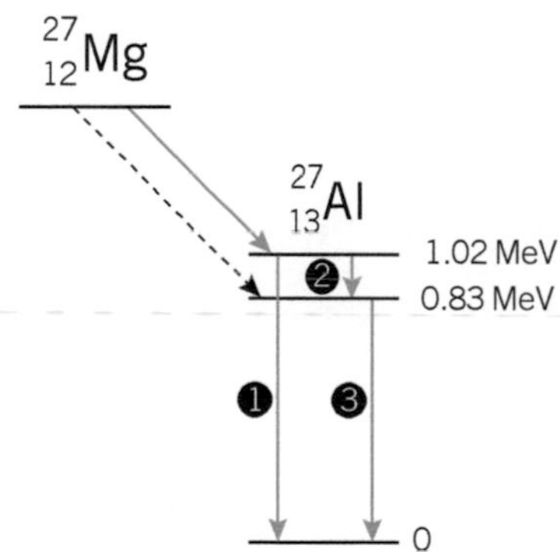

▲ **Figure 3** *Nuclear energy states*

The technetium generator

The technetium generator is used in hospitals to produce a source which emits γ radiation only. Some radioactive isotopes such as the technetium isotope $^{99}_{43}$Tc (technetium-99m) form in an excited state after an α emission or a β emission and stay in the excited state long enough to be separated from the parent isotope. Such a long-lived excited state is said to be a **metastable state**. Nuclei of the technetium isotope $^{99}_{43}$Tc form in a metastable state (indicated by the symbol $^{99}_{43}$Tc^{m}) after β^- emission from nuclei of the molybdenum isotope $^{99}_{42}$Mo which has a half-life of 67 hours. The metastable state $^{99}_{43}$Tc^{m} has a half-life of 6 hours and decays to the ground state by γ emission.

Technetium $^{99}_{43}$Tc in the ground state is a β^- emitter with a half-life of 500 000 years and it forms a stable product. Therefore, a sample of $^{99}_{43}$Tc^{m} with no molybdenum present effectively emits only γ photons. Such samples of technetium $^{99}_{43}$Tc^{m} are used in medical diagnosis applications, as outlined on the next page.

Notes

1. In Figure 3 the presence of two excited states is indicated by the fact that the two smaller γ energies add up to the largest γ energy.
2. In this example, the parent nucleus can also decay by β^- emission to the excited state of $^{27}_{13}$Al at 0.83 MeV then by emission of a γ photon of energy 0.83 MeV to the ground state. The β^- emission for this change is shown on Figure 3 by the dashed arrow.

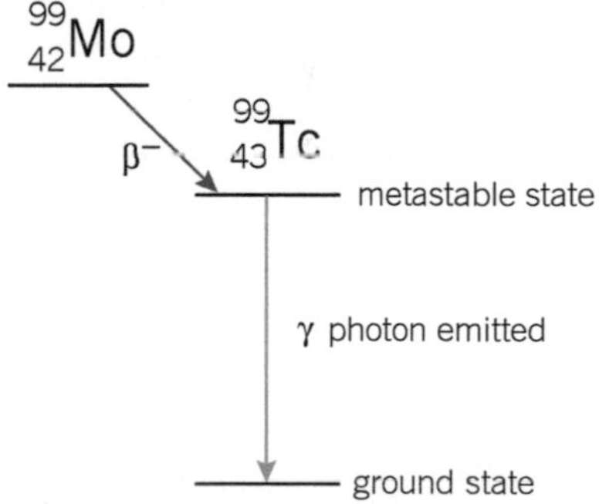

▲ **Figure 4** *The metastable state of technetium $^{99}_{43}$Tc*

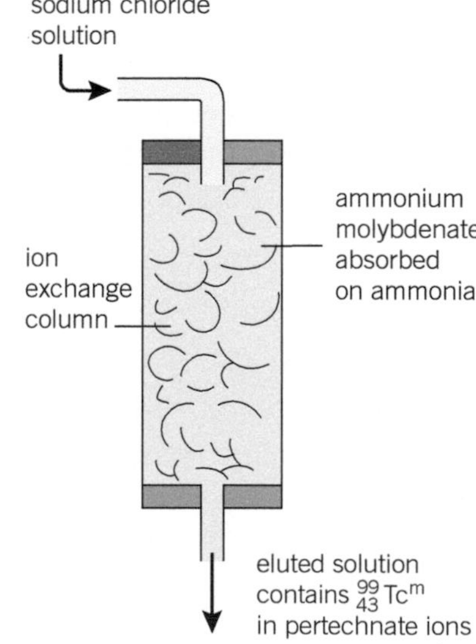

▲ **Figure 5** *The technetium generator*

The technetium generator consists of an ion exchange column containing ammonium molybdenate exposed to neutron radiation several days earlier to make a significant number of the molybdenum nuclei unstable. When a solution of sodium chloride is passed through the column, some of the chlorine ions exchange with pertechnate ions but not with molybdenate ions so the solution that emerges contains ${}^{99}_{43}Tc^m$ nuclei.

Application

Examples of diagnostic uses of ${}^{99}_{43}Tc^m$

- **Monitoring blood flow** through the brain using external detectors after a small quantity of sodium pertechnate solution is administered intravenously.
- **The γ camera** is designed to 'image' internal organs and bones by detecting γ radiation from sites in the body where a γ-emitting isotope such as ${}^{99}_{43}Tc^m$ nuclei is located. For example, bone deposits can be located using a phosphate tracer labelled with ${}^{99}_{43}Tc^m$. The γ camera itself consists of detectors called photomultiplier tubes in a lead shield behind a lead collimator grid, which ensures each tube only detects γ photons emitted from nuclei located at a well-defined spot directly in front of the tube.

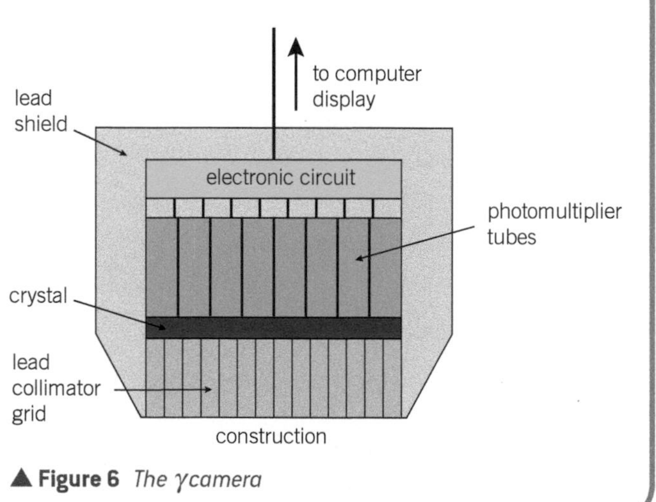

▲ **Figure 6** *The γ camera*

Summary questions

1 **a** What is meant by a metastable state of a radioactive isotope?

b The metastable isotope technetium ${}^{99}_{43}Tc^m$ decays by gamma emission with a half-life of 6 hours. Technetium ${}^{99}_{43}Tc$ in its ground state is a β^- emitter with a half-life of 500 000 years and it forms a stable product.

Explain why the radiation from a pure sample of the metastable isotope after 48 hours is less than 0.5% of the initial radiation.

2 A nucleus of the polonium isotope ${}^{216}_{84}Po$ decays to form a stable nucleus X by emitting in succession an α particle, a β^- particle, a further β^- particle, then another α particle.

a Determine the number of protons and the number of neutrons in X.

b A different isotope of polonium is formed in this decay series. Determine the mass number of this isotope.

3 **a** Explain what is meant by electron capture.

b State one similarity and one difference between electron capture and positron emission.

4 The germanium isotope ${}^{77}_{32}Ge$ has a metastable state which decays to the ground state by emission of a 0.16 MeV γ photon. The isotope decays by β^- emission to form the arsenic isotope ${}^{77}_{33}As$ in an excited state 0.48 MeV above the ground state of ${}^{77}_{33}As$.

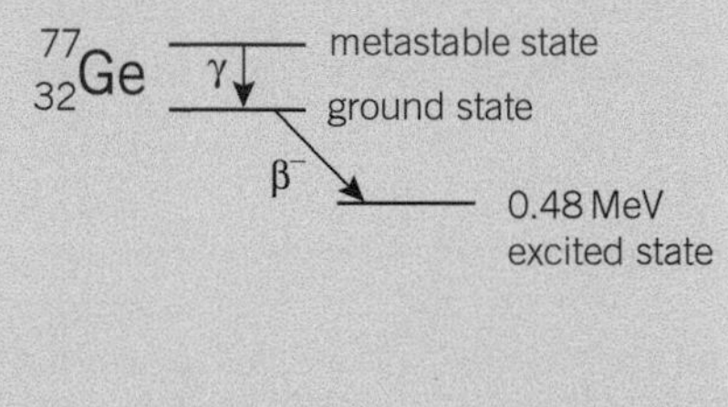

a Copy and complete the energy level diagram to show these changes.

b The excited state of ${}^{77}_{33}As$ also decays to the ground state via an excited state which is 0.27 MeV above the ground state. Calculate the energies of the γ photons emitted in this decay.

Practice questions: Chapter 8

1 **(a)** Copy and complete each of the following equations for radioactive disintegration

(i) $^{214}_{84}\text{Po} \rightarrow ^{?}_{?}\alpha + ^{?}_{?}\text{Pb}$

(ii) $^{210}_{83}\text{Bi} \rightarrow ^{?}_{?}\beta + ^{?}_{?}\text{Po}$

(b) Use the two equations in (a) to describe the changes in the nucleus that occur in each case.

2 **Figure 1** shows the apparatus in which α particles are directed at a metal foil in order to investigate the structure of the atom.

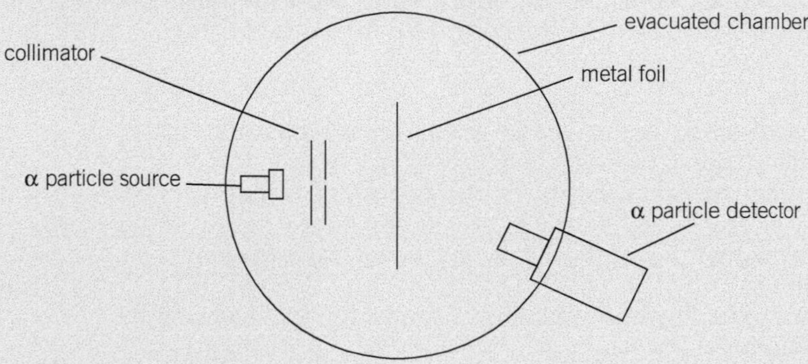

▲ **Figure 1**

(a) (i) Give *two* reasons why the metal foil should be thin.

(ii) Explain why the incident beam of α particles should be narrow. (*3 marks*)

(b) Describe and explain *one* feature of the distribution of the scattered α particles that suggests the nucleus contains most of the mass of an atom. (*2 marks*)

(c) **Figure 2** shows three α particles with the same constant velocity incident on an atom in the metal foil. They all approach the nucleus close enough to be deflected by at least 10°.

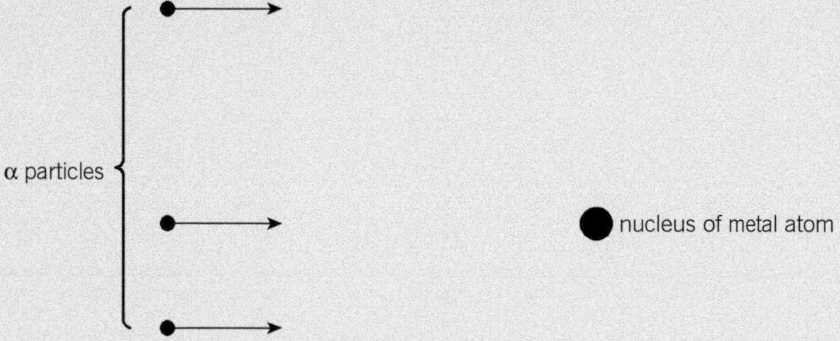

▲ **Figure 2**

Copy the diagram and draw the paths followed by the three α particles whose initial directions are shown by the arrows. (*3 marks*)

AQA, 2007

3 A radioactive nucleus decays with the emission of an alpha particle and a gamma-ray photon.

(a) Describe the changes that occur in the proton number and the nucleon number of the nucleus. (*2 marks*)

(b) Comment on the relative penetrating powers of the two types of ionising radiation. (*1 mark*)

(c) Gamma rays from a point source are travelling towards a detector. The distance from the source to the detector is changed from 1.0 m to 3.0 m. Calculate:

$$\frac{\text{intensity of radiation at 3.0 m}}{\text{intensity of radiation at 1.0 m}}$$

(*2 marks*)

AQA, 2006

4 **(a)** A radioactive source gives an initial count rate of 110 counts per second. After 10 minutes the count rate is 84 counts per second. background radiation = 3 counts per second

(i) Give *three* origins of the radiation that contributes to this background radiation.

(ii) Calculate the half-life of the radioactive source in s.

(iii) Estimate how long the count rate due to the source would take to decrease to less than 5 counts per second. (*7 marks*)

(b) Discuss the dangers of exposing the human body to a source of α radiation. In particular compare the dangers when the α source is held outside, but in contact with the body, with those when the source is placed inside the body. (*3 marks*)

AQA, 2004

5 Iodine-123 is a radioisotope used medically as a tracer to monitor thyroid and kidney functions. The decay of an iodine-123 nucleus produces a gamma ray which, when emitted from inside the body of a patient, can be detected externally.

(a) Why are gamma rays the most suitable type of nuclear radiation for this application? (*2 marks*)

(b) In a laboratory experiment on a sample of iodine-123 the following data were collected.

Time / h	0	4	8	12	16	20	24	28	32
Count rate / counts s^{-1}	512	401	338	279	217	191	143	119	91

Why was it unnecessary to correct these values for background radiation? (*2 marks*)

(c) Draw a graph of count rate against time. (*2 marks*)

(d) Use your graph to find an accurate value for the half-life of iodine-123. Show clearly the method you use. (*3 marks*)

(e) Give *two* reasons why radioisotopes with short half-lives are particularly suitable for use as a medical tracer. (*2 marks*)

AQA, 2004

9 Electric current
9.1 Current and charge

Electrical conduction

To make an electric current pass round a circuit, the circuit must be complete and there must be a source of potential difference, such as a battery, in the circuit. The electric current is the rate of flow of charge in the wire or component. The current is due to the passage of charged particles. These charged particles are referred to as **charge carriers**.

- In metals, the charge carriers are conduction electrons. They move about inside the metal, repeatedly colliding with each other and the fixed positive ions in the metal.
- In comparison, when an electric current is passed through a salt solution, the charge is carried by ions, which are charged atoms or molecules.

A simple test for conduction of electricity is shown in Figure 1. The meter shows a non-zero reading whenever any conducting material is connected into the circuit. The battery forces the charge carriers through the conducting material and causes them to pass through the battery and the meter. If the test material is a metal, the charge carriers in all parts of the circuit are electrons. These electrons enter the battery at its positive terminal after passing through the metal and the ammeter, and leave at the negative terminal to continue to cycle again.

The convention for the direction of current in a circuit is from positive (+) to negative (−), as shown in Figure 2. The convention was agreed long before the discovery of electrons. When it was set up, it was known that an electric current is a flow of charge one way round a circuit. However, it was not known whether the current was due to positive charge flowing round the circuit from + to −, or if it was due to negative charge flowing from − to +.

The unit of current is the *ampere* (A), which is defined in terms of the magnetic force between two parallel wires when they carry the same current. The symbol for current is I.

The unit of charge is the *coulomb* (C), equal to the charge flow in one second when the current is one ampere. The symbol for charge is Q.

For a current I, the charge flow ΔQ in time Δt is given by

$$\Delta Q = I \Delta t$$

For example, the charge flow for a current of:

- 1 A in 10 s is 10 C
- 5 A in 200 s is 1000 C
- 10 mA in 500 s is 5 C.

For charge flow ΔQ in a time interval Δt, the current I is given by

$$I = \frac{\Delta Q}{\Delta t}$$

The equation shows that a current of 1 A is due to a flow of charge of 1 coulomb per second. As the magnitude of the charge of the electron is 1.6×10^{-19} C, a current of 1 A along a wire must be due to 6.25×10^{18} electrons passing along the wire each second.

Learning objectives:

→ Define an electric current.

→ Calculate the charge flow in a circuit.

→ Define charge carriers.

Specification reference: 3.4.1

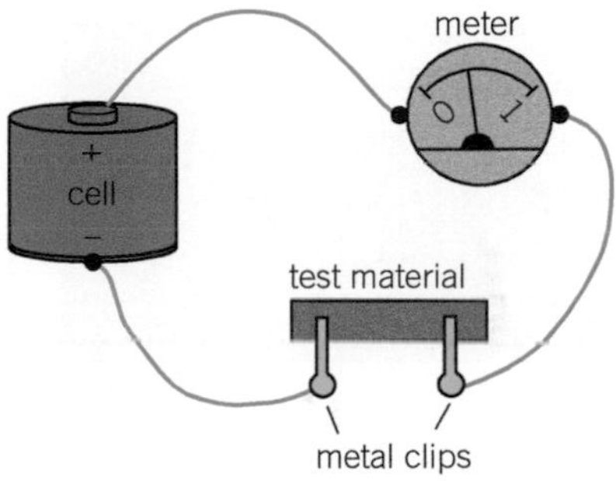

▲ **Figure 1** *Testing for conduction*

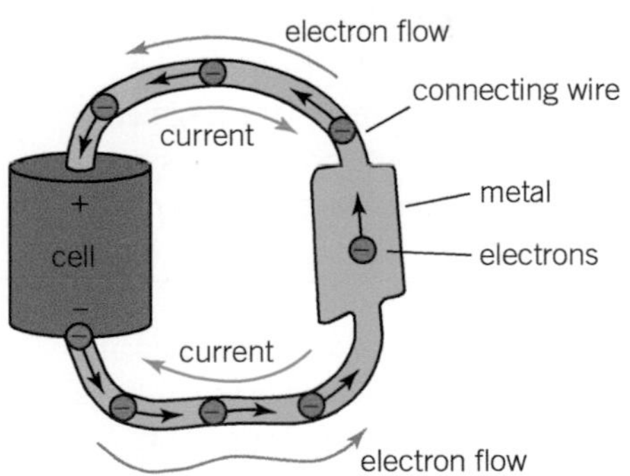

▲ **Figure 2** *Convention for current*

Hint

We use Δ (the Greek capital letter delta) where a change takes place. So $\frac{\Delta Q}{\Delta t}$ means 'change per second' of charge (i.e., rate of change of charge).

Application

Physics and the human genome project

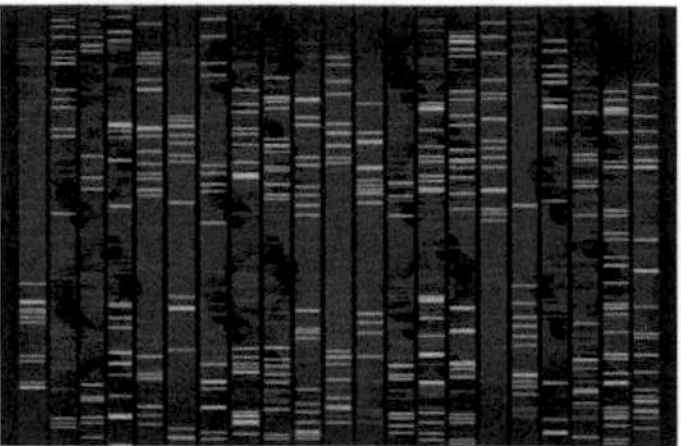

To map the human genome, fragments of DNA are tagged with amino acid bases. Each tagged fragment carries a negative charge. A voltage is applied across a strip of gel with a spot of liquid on it containing tagged fragments. The fragments are attracted to the positive electrode. The smaller the fragment, the faster it moves, so the fragments separate out according to size as they move to the positive electrode. The fragments pass through a spot of laser light, which causes a dye attached to each tag to fluoresce as it passes through the laser spot. Light sensors linked to a computer detect the glow from each tag. The computer is programmed to work out and display the sequence of bases in the DNA fragments.

Q: Why are the tagged fragments attracted to the positive electrode?

Answer: Because they carry a negative charge, and opposite charges attract.

More about charge carriers

Materials can be classified in electrical terms as conductors, insulators, or semiconductors.

- In an **insulator,** each electron is attached to an atom and cannot move away from the atom. When a voltage is applied across an insulator, no current passes through the insulator, because no electrons can move through the insulator.
- In a **metallic conductor,** most electrons are attached to atoms but some are delocalised – the delocalised electrons are the charge carriers in the metal. When a voltage is applied across the metal, these conduction electrons are attracted towards the positive terminal of the metal.
- In a **semiconductor,** the number of charge carriers increases with an increase of temperature. The resistance of a semiconductor therefore decreases as its temperature is raised. A pure semiconducting material is referred to as an intrinsic semiconductor because conduction is due to electrons that break free from the atoms of the semiconductor.

Application

Rechargeable batteries

A car battery is a 12 V rechargeable battery designed to supply a very large current to start the engine. The battery is recharged when the car engine is running. Smaller rechargeable batteries are used in portable electronic equipment, for example, in mobile phones. Such a battery supplies a much smaller current than a car battery. Disposable batteries can't be recharged. Once a disposable battery has run down, it is no longer of any use – not as environmentally friendly as a rechargeable battery!

Summary questions

$e = 1.6 \times 10^{-19}$ C

1. **a** The current in a certain wire is 0.35 A. Calculate the charge passing a point in the wire **i** in 10 s, **ii** in 10 min.
   b Calculate the average current in a wire through which a charge of 15 C passes in **i** 5 s, **ii** 100 s.
2. Calculate the number of electrons passing a point in the wire in 10 min when the current is **a** 1.0 μA, **b** 5.0 A.
3. In an electron beam experiment, the beam current is 1.2 mA. Calculate:
   a the charge flowing along the beam each minute
   b the number of electrons that pass along the beam each minute.
4. A certain type of rechargeable battery is capable of delivering a current of 0.2 A for 4000 s before its voltage drops and it needs to be recharged. Calculate the maximum time it could be used for without being recharged if the current through it was **a** 0.5 A , **b** 0.1 A.

9.2 Potential difference and power

Learning objectives:

- → Define potential difference.
- → Calculate electrical power.
- → Explain how energy transfers take place in electrical devices.

Specification reference: 3.4.1 and 3.4.4

Energy and potential difference

When a torch bulb is connected to a battery, electrons deliver energy from the battery to the torch bulb. Each electron moves around the circuit and takes a fixed amount of energy from the battery as it passes through it. The electrons then deliver energy to the bulb as they pass through it. After delivering energy to the bulb, each electron re-enters the battery via the positive terminal to be resupplied with more energy to deliver to the bulb.

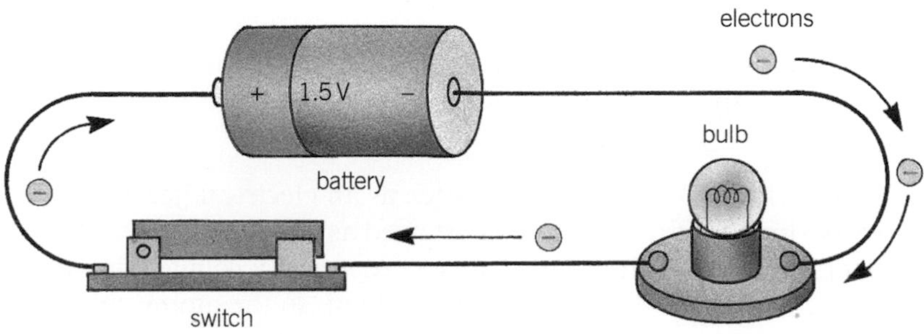

▲ **Figure 1** *Energy transfer by electrons*

A battery has the potential to transfer energy from its chemical store if the battery is not part of a complete circuit. When the battery is in a circuit, each electron passing through a circuit component does work to pass through the component and therefore transfers some or all of its energy. The work done by an electron is equal to its loss of energy. The work done per unit charge is defined as the **potential difference** (abbreviated as p.d.) or *voltage* across the component.

Potential difference is defined as the work done (or energy transfer) per unit charge. The unit of p.d. is the volt, which is equal to 1 joule per coulomb.

If work W is done when charge Q flows through the component, the p.d. across the component, V, is given by

$$V = \frac{W}{Q}$$

Rearranging this equation gives $W = QV$ for the work done or energy transfer when charge Q passes through a component which has a p.d. V across its terminals.

For example:

- If 30 J of work is done when 5 C of charge passes through a component, the p.d. across the component must be 6 V $\left(= \frac{30\,\text{J}}{5\,\text{C}}\right)$.
- If the p.d. across a component in a circuit is 12 V, then 3 C of charge passing through the component would transfer 36 J of energy from the battery to the component.

The e.m.f. of a source of electricity is defined as the electrical energy produced per unit charge passing through the source. The unit of e.m.f. is the volt, the same as the unit of p.d.

▲ **Figure 2** *Sources of e.m.f.*

Study tip

Remember 1 volt = 1 joule per coulomb.

Synoptic link

You will meet emf in more detail in Topic 10.3, Electromotive force and internal resistance.

For a source of e.m.f. ε in a circuit, the electrical energy produced when charge Q passes through the source = $Q\varepsilon$. This energy is transferred to other parts of the circuit, and some may be dissipated in the source itself due to the source's internal resistance.

Energy transfer in different devices

▲ **Figure 3** *Electrical devices*

An electric current has a heating effect when it passes through a component with resistance. It also has a magnetic effect, which is made use of in electric motors and loudspeakers.

1 In a device that has resistance, such as an electrical heater, the work done on the device is transferred as thermal energy. This happens because the charge carriers repeatedly collide with atoms in the device and transfer energy to them, so the atoms vibrate more and the resistor becomes hotter.
2 In an electric motor turning at a constant speed, the work done on the motor is equal to the energy transferred to the load and surroundings by the motor, so the kinetic energy of the motor remains constant. The charge carriers are electrons that need to be forced through the wires of the spinning motor coil against the opposing force on the electrons due to the motor's magnetic field.
3 For a loudspeaker, the work done on the loudspeaker is transferred as sound energy. Electrons need to be forced through the wires of the vibrating loudspeaker coil against the force on them due to the loudspeaker magnet.

Electrical power and current

Consider a component or device that has a potential difference V across its terminals and a current I passing through it. In time Δt:

- the charge flow through it, $Q = I\,\Delta t$
- the work done by the charge carriers, $W = QV = (I\,\Delta t)\,V = IV\,\Delta t$.

$$\textbf{work done } W = IV\Delta t$$

The energy transfer ΔE in the component or device is equal to the work done W.

Because power = $\frac{\text{energy}}{\text{time}}$, the electrical power P supplied to the device is

$$\frac{IV\,\Delta t}{\Delta t} = IV$$

$$\textbf{electrical power } P = IV$$

Notes:

1 This equation can be rearranged to give $I = \frac{P}{V}$ or $V = \frac{P}{I}$.
2 The unit of power is the watt (W). Therefore one volt is equal to one watt per ampere. For example, if the p.d. across a component is 4 V, then the power delivered to the component is 4 W per ampere of current.

Summary questions

1 Calculate the energy transfer in 1200 s in a component when the p.d. across it is 12 V and the current is
   a 2 A
   b 0.05 A.
2 A 6 V, 12 W light bulb is connected to a 6 V battery. Calculate:
   a the current through the light bulb
   b the energy transfer to the light bulb in 1800 s.
3 A 230 V electrical appliance has a power rating of 800 W. Calculate **i** the energy transfer in the appliance in 1 min, **ii** the current taken by the appliance.
4 A battery has an e.m.f. of 9 V and negligible internal resistance. It is capable of delivering a total charge of 1350 C. Calculate:
   a the maximum energy the battery could deliver
   b the power it would deliver to the components of a circuit if the current through it was 0.5 A
   c how long the battery would last for, if it were to supply power at the rate calculated in part **b**.

9.3 Resistance

Definitions and laws

The resistance of a component in a circuit is a measure of the difficulty of making current pass through the component. Resistance is caused by the repeated collisions between the charge carriers in the material with each other and with the fixed positive ions of the material.

The resistance of any component is defined as

$$\textbf{the p.d. across the component} \over \textbf{the current through it}$$

For a component which passes current I when the p.d. across it is V, its resistance R is given by the equation

$$R = \frac{V}{I}$$

The unit of resistance is the ohm (Ω), which is equal to 1 volt per ampere.

Rearranging the above equation gives $V = IR$ or $I = \frac{V}{R}$

Reminder about prefixes

▼ **Table 1** *Prefixes*

Prefix	nano	micro	milli	kilo	mega	giga
Symbol	n	μ	m	k	M	G
Value	10^{-9}	10^{-6}	10^{-3}	10^{3}	10^{6}	10^{9}

Worked example

The current through a component is 2.0 mA when the p.d. across it is 12 V. Calculate:

a its resistance at this current
b the p.d. across the component when the current is 50 μA, assuming its resistance is unchanged.

Solution

a $R = \frac{V}{I} = \frac{12}{2.0 \times 10^{-3}} = 6000\ \Omega$

b $V = IR = 50 \times 10^{-6} \times 6000 = 0.30\ \text{V}$

Measurement of resistance

A resistor is a component designed to have a certain resistance, which is the same regardless of the current through it. The resistance of a resistor can be measured using the circuit shown in Figure 1.

- The ammeter is used to measure the current through the resistor. The ammeter must be in series with the resistor, so the same current passes through both the resistor and the ammeter.
- The voltmeter is used to measure the p.d. across the resistor. The voltmeter must be in parallel with the resistor so that they

Learning objectives:

→ Describe what causes electrical resistance.

→ Discuss when Ohm's law can be used.

→ Explain what a superconductor is.

Specification reference: 3.4.2 and 3.4.3

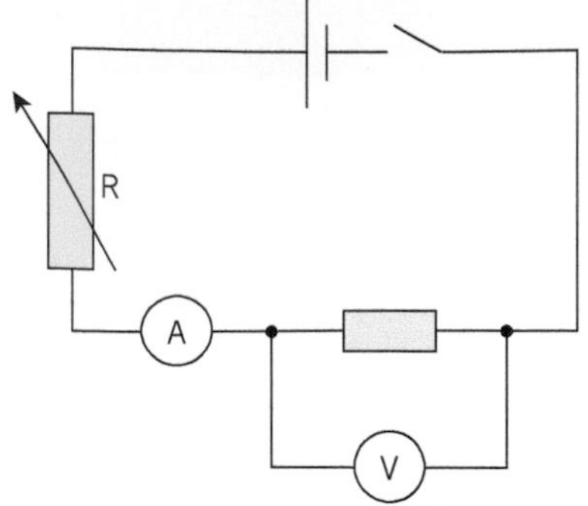

▲ **Figure 1** *Measuring resistance*

have the same p.d. Also, no current should pass through the voltmeter, otherwise the ammeter will not record the exact current through the resistor. In theory, the voltmeter should have infinite resistance. In practice, a voltmeter with a sufficiently high resistance would be satisfactory.

- The variable resistor is used to adjust the current and p.d. as necessary. To investigate the variation of current with p.d., the variable resistor is adjusted in steps. At each step, the current and p.d. are recorded from the ammeter and voltmeter, respectively. The measurements can then be plotted on a graph of p.d. against current, as shown in Figure 2 (or for current against p.d. – see Topic 9.4).

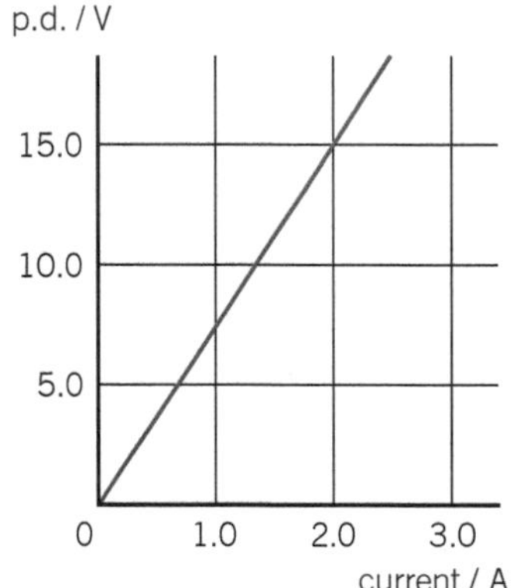

▲ **Figure 2** *Graph of p.d. versus current for a resistor*

The graph for a resistor is a straight line through the origin. The resistance is the same, regardless of the current. The resistance is equal to the gradient of the graph only because the gradient is constant and at any point is equal to the p.d. divided by the current. In other words, the p.d. across the resistor is proportional to the current. The discovery that the p.d. across a metal wire is proportional to the current through it was made by Georg Ohm in 1826, and is known as Ohm's law.

Ohm's law states that the p.d. across a metallic conductor is proportional to the current through it, provided the physical conditions do not change.

Notes:

1 Ohm's law is equivalent to the statement that the resistance of a metallic conductor under constant physical conditions (e.g., temperature) is constant.

2 For an ohmic conductor, $V = IR$, where R is constant. A resistor is a component designed to have a certain resistance.

3 If the current and p.d. measurements for an ohmic conductor are plotted with current on the y-axis and p.d. on the x-axis, the gradient of this graph gives $\frac{1}{R}$.

Resistivity

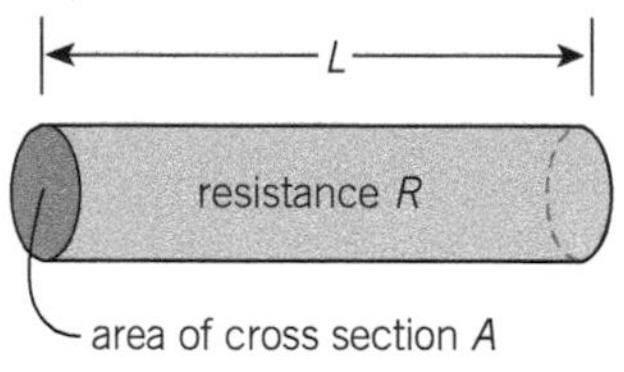

▲ **Figure 3** *Resistivity*

For a conductor of length L and uniform cross-sectional area A, as shown in Figure 3, its resistance R is:

- proportional to L
- inversely proportional to A.

Therefore, $R = \frac{\rho L}{A}$, where ρ is a constant for that material, known as its resistivity.

Rearranging this equation gives the following equation, which can be used to calculate the resistivity of a sample of material of length L and uniform cross-sectional area A:

$$\textbf{resistivity, } \rho = \frac{RA}{L}$$

Notes:

1 The unit of resistivity is the ohm metre (Ω m).

2 For a conductor with a circular cross section of diameter d, $A = \frac{\pi d^2}{4}$ (= πr^2 where radius $r = \frac{d}{2}$).

To determine the resistivity of a wire:

- Measure the diameter of the wire d using a micrometer at several different points along the wire, to give a mean value for d to calculate its cross-sectional area A.
- Measure the resistance R of different lengths L of wire to plot a graph of R against L (Figure 4).

The resistivity of the wire is given by the graph gradient × A.

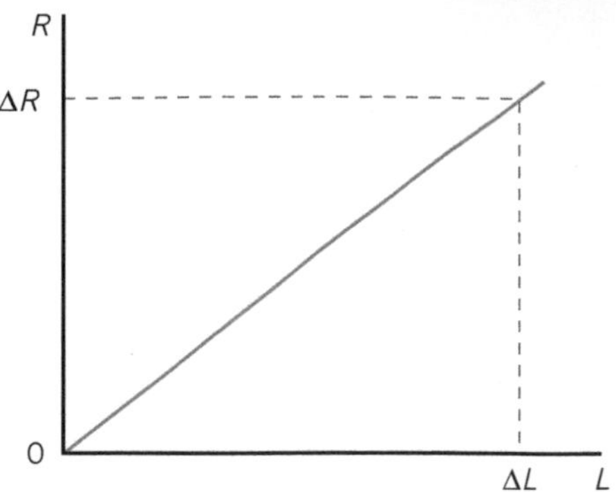

▲ **Figure 4** *Graph of resistance against length for a wire*

Superconductivity

A **superconductor** is a wire or a device made of material that has zero resistivity at and below a **critical temperature** that depends on the material. This property of the material is called **superconductivity**. The wire or device has zero resistance below the critical temperature of the material. When a current passes through it, there is no p.d. across it because its resistance is zero. So the current has no heating effect.

A superconductor material loses its superconductivity if its temperature is raised above its critical temperature. In 2014, the highest critical temperature was 150 K (−123 °C) for a compound containing mercury, barium, calcium, copper, and oxygen. Any material with a critical temperature above 77 K (−196 °C), the boiling point of liquid nitrogen, is referred to as a high-temperature superconductor.

Superconductors are used to make high-power electromagnets that generate very strong magnetic fields in devices such as MRI scanners and particle accelerators. These strong magnetic fields are also used in the development of new applications such as lightweight electric motors and power cables that transfer electrical energy without energy dissipation.

▼ **Table 2** *Resistivity values of different materials at room temperature*

Material	Resistivity / Ω m
Copper	1.7×10^{-8}
Constantan	5.0×10^{-7}
Carbon	3×10^{-5}
Silicon	2300
PVC	about 10^{14}

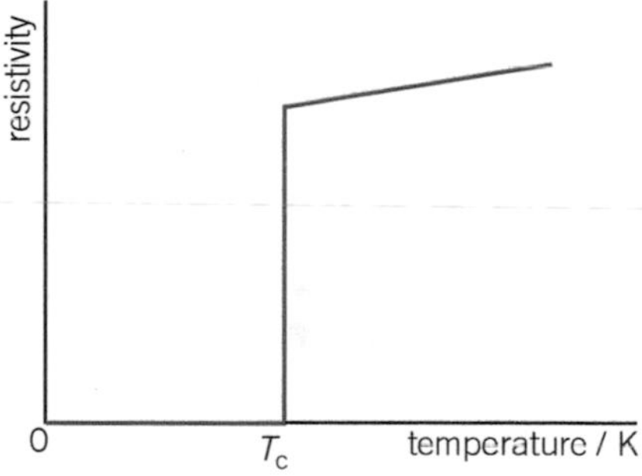

▲ **Figure 5** *Resistivity of a superconductor versus temperature near the critical temperature*

Summary questions

1 a Complete the table below by calculating the missing value for each resistor.

	1	2	3	4	5
Current / A	2.0	0.45		5.0×10^{-3}	
P.d. / V	12.0		5.0	0.80	5.0×10^{4}
Resistance / Ω		22	4.0×10^{4}		2.0×10^{7}

b Find the resistance of the resistor that gave the results shown in Figure 2.

2 Calculate the resistance of a uniform wire of diameter 0.32 mm and length 5.0 m. The resistivity of the material = 5.0×10^{-7} Ω m.

3 Calculate the resistance of a rectangular strip of copper of length 0.08 m, thickness 15 mm, and width 0.80 mm. The resistivity of copper = 1.7×10^{-8} Ω m.

4 A wire of uniform diameter 0.28 mm and length 1.50 m has a resistance of 45 Ω. Calculate:

a its resistivity

b the length of this wire that has a resistance of 1.0 Ω.

Synoptic link

For more information about MRI, see Topic 13.7, Wave–particle duality.

Study tip

Resistivity is a property of a material. Don't confuse resistivity and resistance.

9.4 Components and their characteristics

Learning objectives:

- → Describe how the current through a filament lamp varies with p.d.
- → State the characteristics of a diode.
- → Describe what we use a thermistor for.

Specification reference: 3.4.2

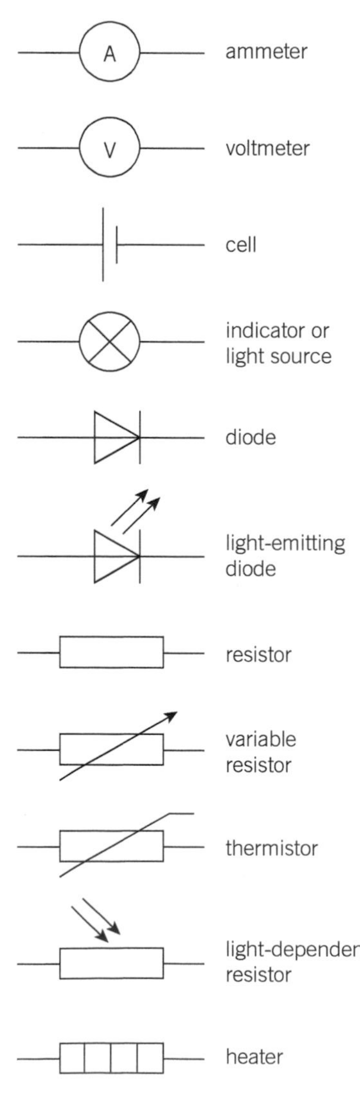

▲ **Figure 1** *Circuit components*

Circuit diagrams

Each type of component has its own symbol, which is used to represent the component in a circuit diagram. You need to recognise the symbols for different types of components to make progress – just like a motorist needs to know what different road signs mean. Note that on a circuit diagram, the direction of the current is always shown from + to – round the circuit.

You should be able to recognise the component symbols shown in Figure 1. Here are some notes about some of the components.

- A cell is a source of electrical energy. Note that a battery is a combination of cells.
- The symbol for an indicator is the same as that for a light source (including a filament lamp, but not a light-emitting diode).
- A diode allows current in one direction only. A light-emitting diode (or LED) emits light when it conducts. The direction in which the diode conducts is referred to as its forward direction. The opposite direction is referred to as its reverse direction. Examples of the use of diodes include the protection of dc circuits (in case the voltage supply is connected the wrong way round).
- A resistor is a component designed to have a certain resistance.
- The resistance of a thermistor decreases with increasing temperature, if the thermistor is an intrinsic semiconductor such as silicon. Such a thermistor is referred to as an ntc (negative temperature coefficient) thermistor.
- The resistance of a **light-dependent resistor** (LDR) decreases with increasing light intensity.

Investigating the characteristics of different components

To measure the variation of current with p.d. for a component, use either:

- a potential divider to vary the p.d. from zero (See Topic 10.5), or
- a variable resistor to vary the current to a minimum.

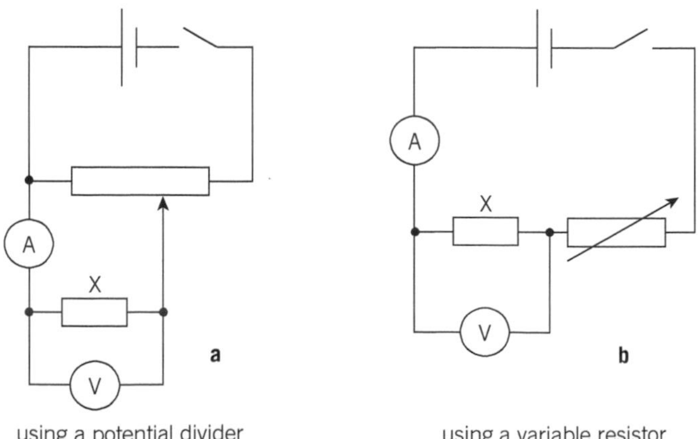

▲ **Figure 2** *Investigating component characteristics*

The advantage of using a potential divider is that the current through the component and the p.d. across it can be reduced to zero. This is not possible with a variable resistor circuit.

The measurements for each type of component are usually plotted as a graph of current (on the y-axis) against p.d. (on the x-axis). Typical graphs for a wire, a filament lamp, and a thermistor are shown in Figure 3. Note that the measurements are the same, regardless of which way the current passes through each of these components.

In both circuits, an ammeter sensor and a voltmeter sensor connected to a data logger could be used to capture data (i.e., to measure and record the readings) which could then be displayed directly on a computer.

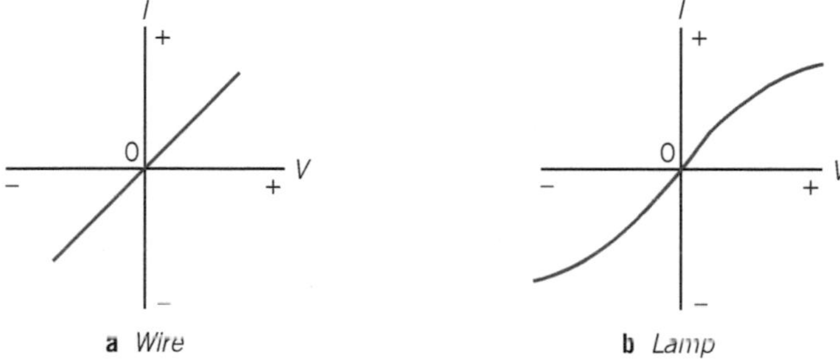

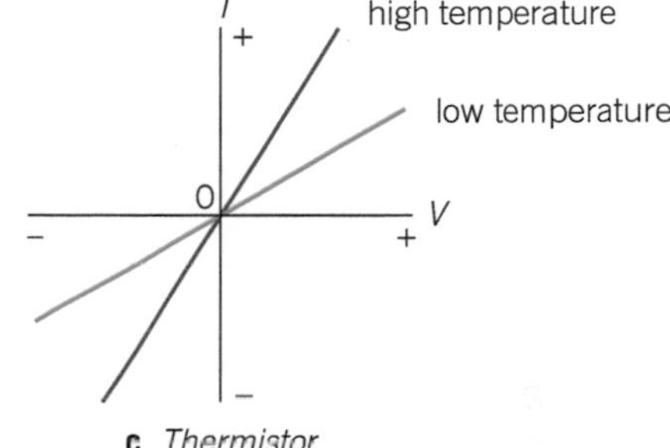

▲ **Figure 3** *Current versus p.d. for different components*

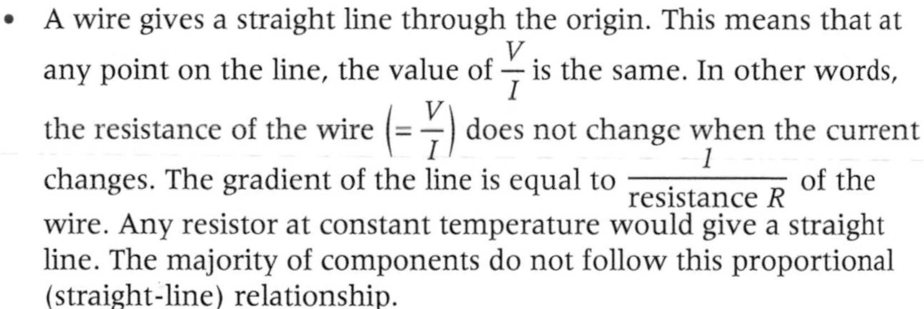

- A wire gives a straight line through the origin. This means that at any point on the line, the value of $\frac{V}{I}$ is the same. In other words, the resistance of the wire $\left(= \frac{V}{I}\right)$ does not change when the current changes. The gradient of the line is equal to $\frac{1}{\text{resistance } R}$ of the wire. Any resistor at constant temperature would give a straight line. The majority of components do not follow this proportional (straight-line) relationship.
- A filament bulb gives a curve with a decreasing gradient because its resistance increases as it becomes hotter.
- A thermistor at constant temperature gives a straight line. The higher the temperature, the greater the gradient of the line, as the resistance falls with increase of temperature. The same result is obtained for a light-dependent resistor in respect of light intensity.

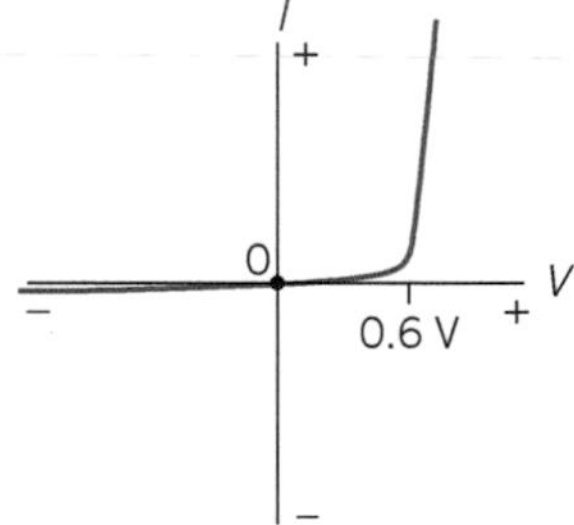

▲ **Figure 4** *Current versus p.d. for a diode*

The diode

To investigate the characteristics of the diode, one set of measurements is made with the diode in its forward direction (i.e., forward biased) and another set with it in its reverse direction (i.e., reverse biased). The current is very small when the diode is reverse biased and can only be measured using a milliammeter.

Typical results for a silicon diode are shown in Figure 4. A silicon diode conducts easily in its forward direction above a p.d. of about 0.6 V and hardly at all below 0.6 V or in the opposite direction.

Hint

Remember, a diode needs a certain p.d. to conduct.

Resistance and temperature

The resistance of a metal increases with increase of temperature. This is because the positive ions in the conductor vibrate more when its temperature is increased. The charge carriers (conduction electrons) therefore cannot pass through the metal as easily when a p.d. is applied across the conductor. A metal is said to have a **positive temperature coefficient** because its resistance increases with increase of temperature.

The resistance of an intrinsic semiconductor decreases with increase of temperature. This is because the number of charge carriers (conduction electrons) increases when the temperature is increased. A thermistor made from an intrinsic semiconductor therefore has a **negative temperature coefficient**. Its percentage change of resistance per kelvin change of temperature is much greater than for a metal. For this reason, thermistors are often used as the temperature-sensitive component in a temperature sensor.

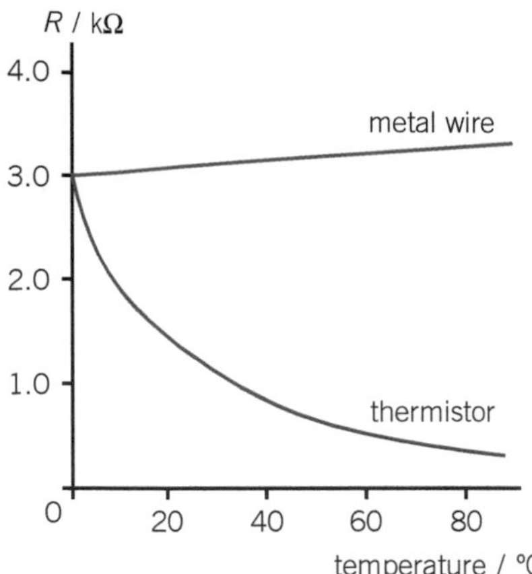

▲ **Figure 5** *Resistance variation with temperature for a thermistor and a metal wire*

Figure 5 shows how the resistance of a thermistor and a metal wire vary with temperature. Both components have the same resistance at 0 °C. The resistance of the thermistor decreases non-linearly with increase of temperature, whereas the resistance of the metal wire increases much less over the same temperature range.

Summary questions

1 A filament bulb is labelled '3.0 V, 0.75 W'.

- **a** Calculate its current and its resistance at 3.0 V.
- **b** State and explain what would happen to the filament bulb if the current was increased from the value in **a**.

2 A certain thermistor has a resistance of 50 000 Ω at 20 °C and a resistance of 4000 Ω at 60 °C.

It is connected in series with an ammeter and a 1.5 V cell. Calculate the ammeter reading when the thermistor is:

- **a** at 20 °C
- **b** at 60 °C.

3 A silicon diode is connected in series with a cell and a torch bulb.

- **a** Sketch the circuit diagram showing the diode in its forward direction.
- **b** Explain why the torch bulb would not light if the polarity of the cell were reversed in the circuit.

4 The resistance of a certain metal wire increased from 25.3 Ω at 0 °C to 35.5 Ω at 100 °C. Assuming the resistance over this range varies linearly with temperature, calculate:

- **a** the resistance at 50 °C
- **b** the temperature when the resistance is 30.0 Ω.

Practice questions: Chapter 9

1 The following measurements were made in an investigation to measure the resistivity of the material of a certain wire.

P.d. across the wire / V	0.0	2.0	4.0	6.0	8.0	10.0
Current through the wire / A	0.00	0.15	0.31	0.44	0.62	0.74

Length of wire = 1.60 m.
Diameter of wire = 0.28 mm.

(a) Plot a graph of the p.d. against the current. *(4 marks)*

(b) Show that the p.d., V, across the wire varies with the current I according to the equation

$$V = \frac{\rho L I}{A}$$

where ρ is the resistivity of the wire, L is its length, and A is its area of cross section. *(2 marks)*

(c) Use the graph to calculate the resistivity of the material of the wire. *(6 marks)*

2 A battery is connected across a uniform conductor. The current in the conductor is 40 mA.

(a) Calculate the total charge that flows past a point in the conductor in 3 minutes.

(b) Using data from page 497 calculate the number of electron charge carriers passing the same point in the conductor in this time.

(c) If 8.6 J of energy are transferred to the conductor in this time, calculate the potential difference across the conductor.

(d) Calculate the resistance of the conductor. *(6 marks)*

AQA, 2004

3 **(a)** **Figure 1** shows a graph of V against I for a filament lamp. Calculate the maximum resistance of the lamp over the range shown by the graph. *(3 marks)*

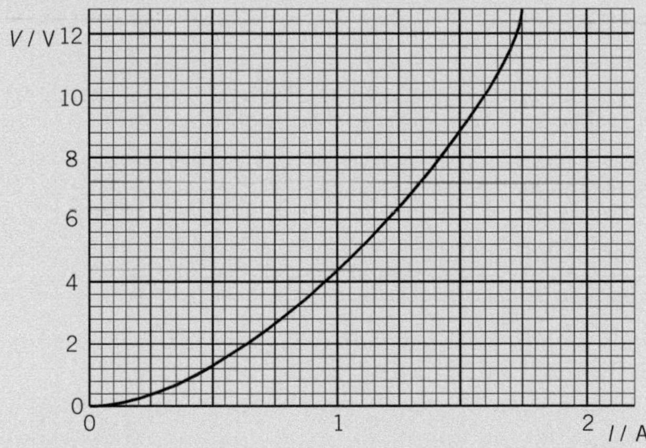

▲ **Figure 1**

(b) Sketch on a copy of the axes below a graph of current against potential difference for a diode. *(2 marks)*

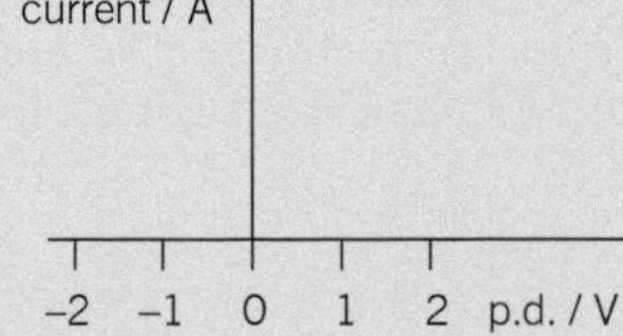

AQA, 2002

4 **Figure 2** shows the general shape of the current–time graph during the 2 seconds after a 12 V filament lamp is switched on.

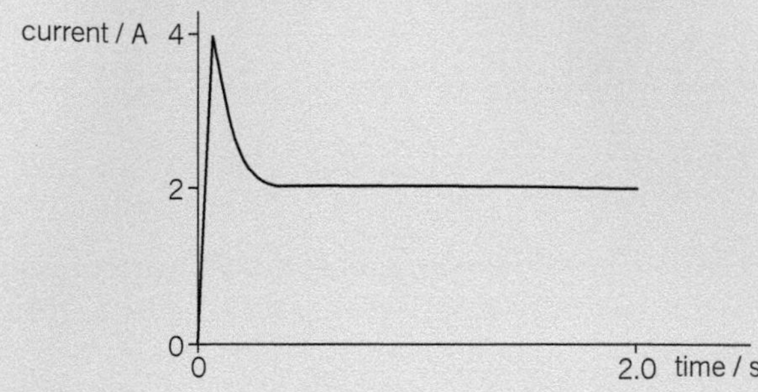

▲ **Figure 2**

(a) A student wishes to perform an experiment to obtain this graph.
 (i) Explain why sampling data using a sensor and a computer is a sensible option.
 (ii) Suggest a suitable sampling rate for such an experiment, giving a reason for your answer. *(3 marks)*

(b) Explain why the current rises to a high value before falling to a steady value and why a filament is more likely to fail when being switched on than at other times. *(6 marks)*

AQA, 2005

5 (a) A metal wire of length 1.4 m has a uniform cross-sectional area = $7.8 \times 10^{-7}\,m^2$. Calculate the resistance, *R*, of the wire.
resistivity of the metal = $1.7 \times 10^{-8}\,\Omega m$ *(2 marks)*

(b) The wire is now stretched to twice its original length by a process that keeps its volume constant.
If the resistivity of the metal of the wire remains constant, show that the resistance increases to 4*R*. *(2 marks)*

AQA, 2003

6 (a) The resistivity of a material in the form of a uniform resistance wire is to be measured. The area of cross section of the wire is known.
The apparatus available includes a battery, a switch, a variable resistor, an ammeter, and a voltmeter.
 (i) Draw a circuit diagram, using some or all of this apparatus, which would enable you to determine the resistivity of the material.
 (ii) Describe how you would make the necessary measurements, ensuring that you have a range of values.
 (iii) Show how a value of the resistivity is determined from your measurements. *(9 marks)*

(b) A sheet of carbon-reinforced plastic measuring 80 mm × 80 mm × 1.5 mm has its two large surfaces coated with highly conducting metal film. When a potential difference of 240 V is applied between the metal films, there is a current of 2.0 mA in the plastic. Calculate the resistivity of the plastic. *(3 marks)*

AQA, 2002

7 (a) (i) What is a *superconductor*?
 (ii) With the aid of a sketch graph, explain the term *transition temperature*. *(3 marks)*

(b) Explain why superconductors are very useful for applicatons which require very large electric currents and name *two* such applicatons. *(3 marks)*

AQA, 2006

Answers to the Practice Questions and Section Questions are available at
www.oxfordsecondary.com/oxfordaqaexams-alevel-physics

10 Direct current circuits
10.1 Circuit rules

Current rules

1 **At any junction in a circuit, the total current leaving the junction is equal to the total current entering the junction.**

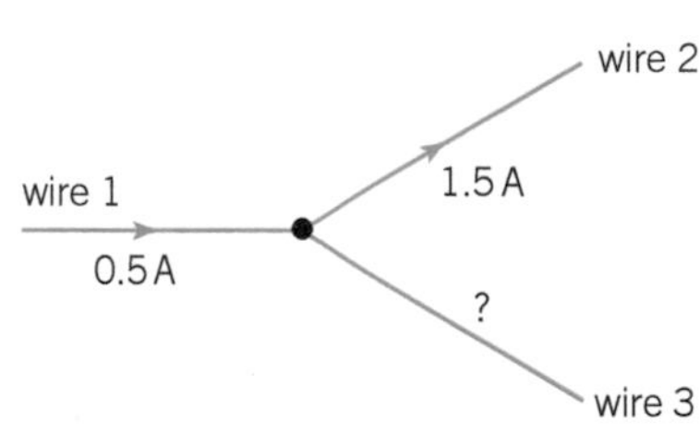

▲ **Figure 1** *At a junction*

For example, Figure 1 shows a junction of three wires where the current in two of the wires (wire 1 and wire 2) is given. The current in wire 3 must be 1.0 A *into* the junction, because the total current into the junction (= 1.0 A along wire 3 + 0.5 A along wire 1) is the same as the total current out of the junction (= 1.5 A along wire 2).

The junction rule holds because the rates of charge flowing into and out of a junction are always equal. The current along a wire is the charge flow per second. In Figure 1, the charge *entering* the junction each second is 0.5 C along wire 1 and 1.0 C along wire 3. The charge *leaving* the junction each second must therefore be 1.5 C as the junction does not retain charge.

2 **Components in series**

- **The current entering a component is the same as the current leaving the component**. In other words, components do not use up current. The charge per second entering a component is equal to the charge per second leaving it. In Figure 2, ammeters A_1 and A_2 show the same reading because they are measuring the same current.
- **The current passing through two or more components in series is the same through each component**. This is because the rate of flow of charge through each component is the same at any instant. The same amount of charge passing through any one component each second passes through every other component each second. In Figure 2, the ammeters read the same because the same amount of charge each second passes through each component.

Learning objectives:

- → State the rules for series and parallel circuits.
- → State the principles behind these rules.
- → Describe how we use the rules in circuits.

Specification reference: 3.4.4

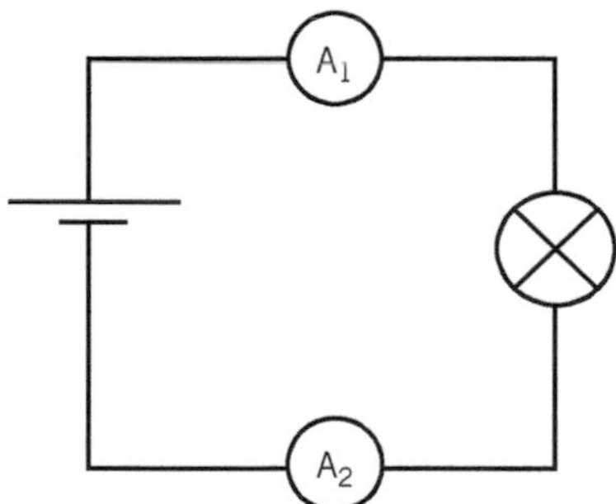

▲ **Figure 2** *Components in series*

Potential difference rules

The potential difference (abbreviated as p.d.), or voltage, between any two points in a circuit is defined as the energy transfer per coulomb of charge that flows from one point to the other.

- If the charge carriers lose energy, the potential difference is a potential drop.
- If the charge carriers gain energy, which happens when they pass through a battery or cell, the potential difference is a potential rise equal to the p.d. across the battery or cell's terminals.

Study tip

You should learn the rules in bold within this topic in preparation for the rest of this chapter.

The rules for potential differences are listed below with an explanation of each rule in energy terms.

1 **For two or more components in series, the total p.d. across all the components is equal to the sum of the potential differences across each component.**

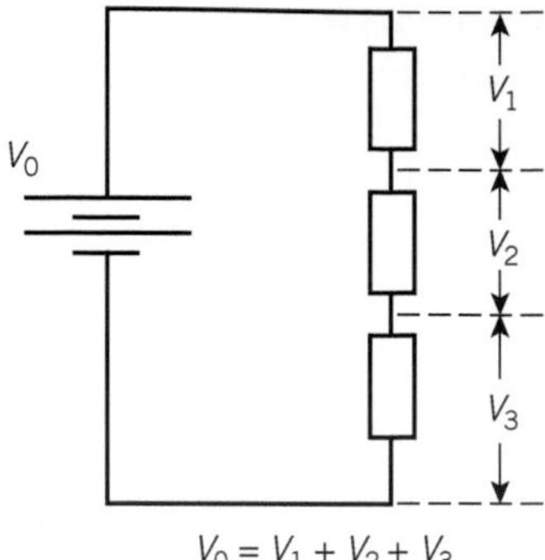

▲ **Figure 3** *Adding potential differences*

In Figure 3, the p.d. across the battery terminals is equal to the sum of the potential differences across the three resistors. This is because the p.d. across each resistor is the energy delivered per coulomb of charge to that resistor. So the sum of potential differences across the three resistors is the total energy delivered to the resistors per coulomb of charge passing through them, which is the p.d. across the battery terminals.

2 **The p.d. across components in parallel is the same.**

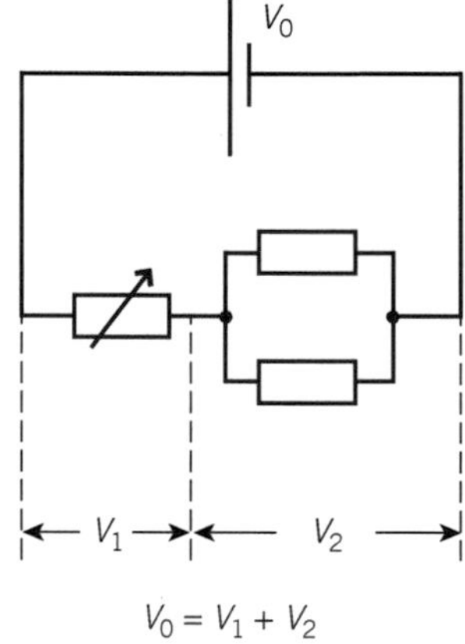

▲ **Figure 4** *Two components in parallel*

In Figure 4, charge carriers can pass through either of the two resistors in parallel. The same amount of energy is delivered by a charge carrier regardless of which of the two resistors it passes through.

Suppose the variable resistor is adjusted so the p.d. across it is 4 V. If the battery p.d. is 12 V, the p.d. across each of the two resistors in parallel is 8 V (= 12 V − 4 V). This is because each coulomb of charge leaves the battery with 12 J of electrical energy and uses 4 J on passing through the variable resistor. Therefore, each coulomb of charge has 8 J of electrical energy to deliver to either of the two parallel resistors.

3 **For any complete loop of a circuit, the sum of the e.m.f.s round the loop is equal to the sum of the potential drops around the loop.** This follows from the fact that the total e.m.f. in a loop is the total electrical energy per coulomb produced in the loop and the sum of the potential drops is the electrical energy per coulomb delivered round the loop. The above statement follows therefore from the conservation of energy.

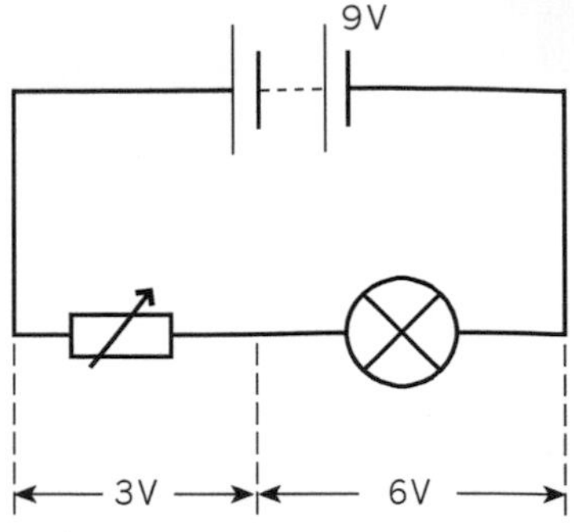

▲ **Figure 5** *The loop rule*

For example, in Figure 5, the battery has an e.m.f. of 9 V. If the variable resistor is adjusted so that the p.d. across the light bulb is 6 V, the p.d. across the variable resistor is 3 V (= 9 V – 6 V). The total e.m.f. in the circuit is 9 V due to the battery. This is equal to the sum of the potential differences round the circuit (= 3 V across the variable resistor + 6 V across the light bulb). In other words, every coulomb of charge leaves the battery with 9 J of electrical energy and supplies 3 J to the variable resistor and 6 J to the light bulb.

Each time a charge carrier goes round the circuit:

- a certain amount of energy E is transferred to it from the battery
- it transfers energy equal to $\frac{E}{3}$ to the variable resistor, and $\frac{2E}{3}$ to the light bulb.

Hint

Sources of e.m.f. usually possess some internal resistance (see Topic 10.3). This can be discounted if a question states that the internal resistance of a source of e.m.f. is negligible.

Summary questions

1 A battery which has an e.m.f. of 6 V and negligible internal resistance is connected to a 6 V 6 W light bulb in parallel with a 6 V 24 W light bulb, as shown in Figure 6.

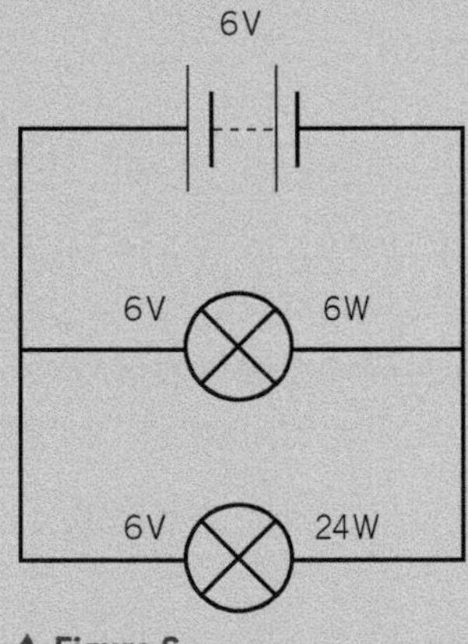

▲ **Figure 6**

Calculate **a** the current through each light bulb, **b** the current from the battery, **c** the power supplied by the battery.

2 A 4.5 V battery of negligible internal resistance is connected in series with a variable resistor and a 2.5 V 0.5 W torch bulb.

a Sketch the circuit diagram for this circuit.

b The variable resistor is adjusted so that the p.d. across the torch bulb is 2.5 V. Calculate **i** the p.d. across the variable resistor, **ii** the current through the torch bulb.

3 A 6.0 V battery of negligible internal resistance is connected in series with an ammeter, a 20 Ω resistor, and an unknown resistor R.

a Sketch the circuit diagram.

b The ammeter reads 0.20 A. Calculate **i** the p.d. across the 20 Ω resistor, **ii** the p.d. across R, **iii** the resistance of R.

4 In **3**, when the unknown resistor is replaced with a torch bulb, the ammeter reads 0.12 A. Calculate **a** the p.d. across the torch bulb, **b** the resistance of the torch bulb.

10.2 More about resistance

Learning objectives:

- → Calculate resistances in series and in parallel.
- → Define resistance heating.
- → Calculate the current and p.d.s for each component in a circuit.

Specification reference: 3.4.4

Resistors in series

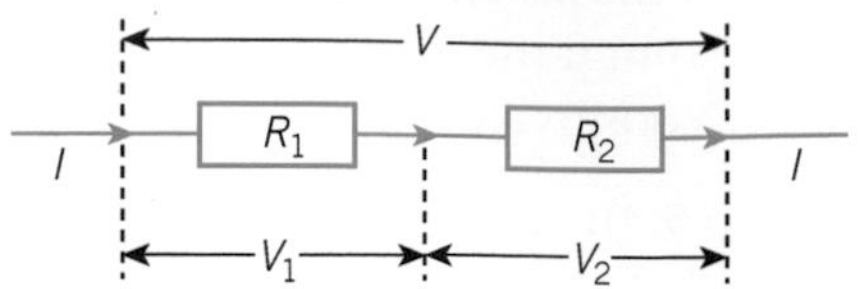

▲ **Figure 1** *Resistors in series*

Resistors in series pass the same current. The total p.d. is equal to the sum of the individual p.d.s.

- For two resistors R_1 and R_2 in series, as in Figure 1, when current I passes through the resistors,
  the p.d. across R_1, $V_1 = IR_1$, and
  the p.d. across R_2, $V_2 = IR_2$

The total p.d. across two resistors, $V = V_1 + V_2 = IR_1 + IR_2$

Therefore, the total resistance $R = \frac{V}{I} = \frac{IR_1 + IR_2}{I} = R_1 + R_2$

- For two or more resistors R_1, R_2, R_3, etc., in series, the theory can easily be extended to show that the **total resistance is equal to the sum of the individual resistances**.

$$\boldsymbol{R = R_1 + R_2 + R_3 + \ldots}$$

Resistors in parallel

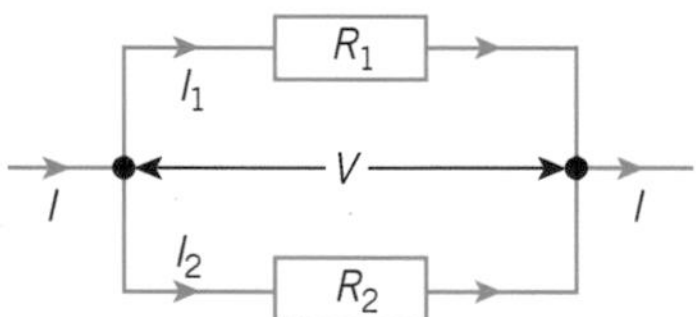

▲ **Figure 2** *Resistors in parallel*

Resistors in parallel have the same p.d. The current through a parallel combination of resistors is equal to the sum of the individual currents.

- For two resistors R_1 and R_2 in parallel, as in Figure 2, when the p.d. across the combination is V,
  the current through resistor R_1, $I_1 = \frac{V}{R_1}$
  the current through resistor R_2, $I_2 = \frac{V}{R_2}$

The total current through the combination, $I = I_1 + I_2 = \frac{V}{R_1} + \frac{V}{R_2}$

Since the total resistance $R = \frac{V}{I}$, then the total current $I = \frac{V}{R}$

Therefore $\frac{V}{R} = \frac{V}{R_1} + \frac{V}{R_2}$

Cancelling V from each term gives the following equation, which is used to calculate the total resistance R:

$$\frac{1}{R} = \frac{1}{R_1} + \frac{1}{R_2}$$

- For two or more resistors R_1, R_2, R_3, etc., in parallel, the theory can easily be extended to show that the total resistance R is given by

$$\frac{1}{R} = \frac{1}{R_1} + \frac{1}{R_2} + \frac{1}{R_3} + \ldots$$

Resistance heating

The heating effect of an electric current in any component is due to the resistance of the component. The charge carriers repeatedly collide with the positive ions of the conducting material. There is a net transfer of energy from the charge carriers to the positive ions as a result of these collisions. After a charge carrier loses kinetic energy in such a collision, the force due to the p.d. across the material accelerates it until it collides with another positive ion.

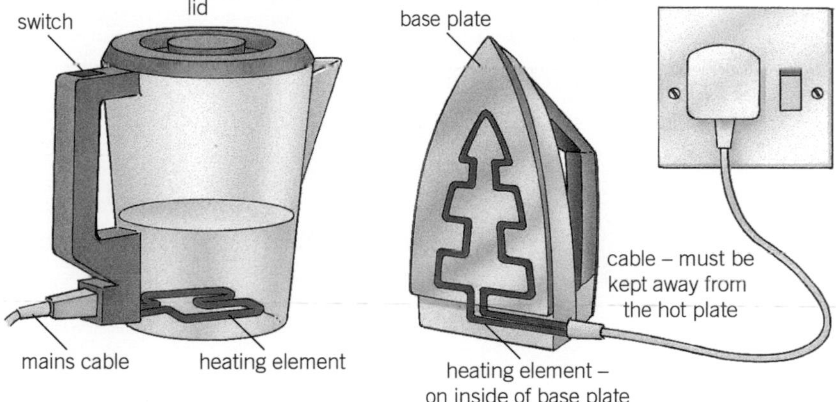

▲ **Figure 3** *Heating elements*

For a component of resistance R, when current I passes through it, the p.d. across the component, $V = IR$.

Therefore the power supplied to the component is given by

$P = IV = I^2R \left(= \frac{V^2}{R}\right)$.

Therefore the energy per second transferred to the component as thermal energy $= I^2R$.

If the component is at constant temperature, heat transfer to the surroundings takes place at the same rate. Therefore,

the rate of heat transfer = I^2R

- If the component heats up, its temperature rise depends on the power supplied to it (I^2R) less the rate of heat transfer from it to the surroundings.
- The energy transferred to the object by the electric current in time t = power × time = I^2Rt.
- The energy transfer per second to the component (i.e., the power supplied to it) does not depend on the direction of the current.

Study tip

You need to know how to use the resistor rules correctly.

Notice the resistance of two resistors in parallel is equal to their product divided by their sum. For example, the resistance of a 3 Ω resistor in parallel with a 6 Ω resistor is $\frac{3 \times 6}{(3+6)} = 2\,\Omega$.

Synoptic link

You learnt about resistance in Topic 9.3, Resistance and Topic 9.4, Components and their characteristics.

Study tip

The power of the resistor could also have been calculated in one step using the equation $P = \frac{V^2}{R}$. Prove to yourself that you can obtain the same answer using this method.

Worked example

The p.d. across a 1000 Ω resistor in a circuit was measured at 6.0 V. Calculate the electrical power supplied to the resistor.

Solution

Current $I = \frac{V}{R} = \frac{6.0\,\text{V}}{1000\,\Omega} = 6.0\,\text{mA}$

Power $P = I^2 R = (6.0 \times 10^{-3})^2 \times 1000 = 0.036\,\text{W}$

Summary questions

1 Calculate the total resistance of each of the resistor combinations in Figure 4.

a
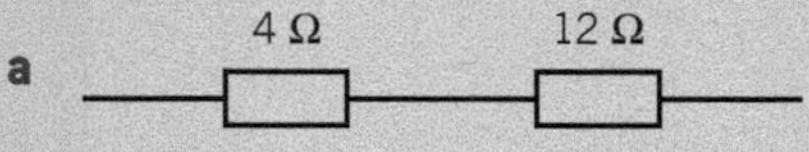

b
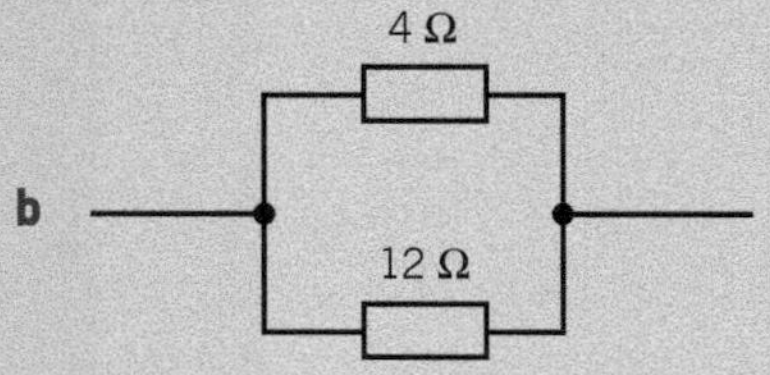

c
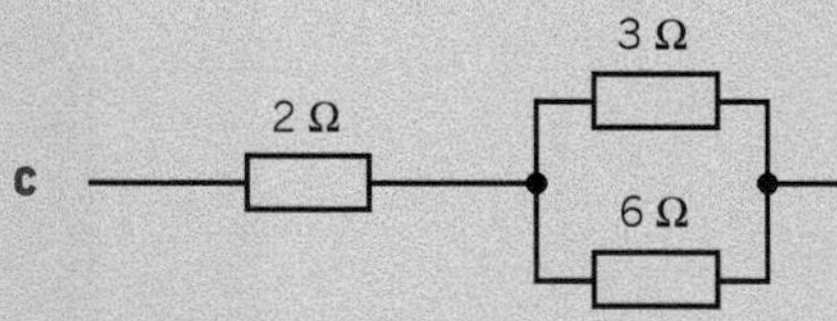

▲ **Figure 4**

2 A 3 Ω resistor and a 6 Ω resistor are connected in parallel with each other. The parallel combination is connected in series with a 6.0 V battery and a 4 Ω resistor, as shown in Figure 5. Assume the battery itself has negligible internal resistance.

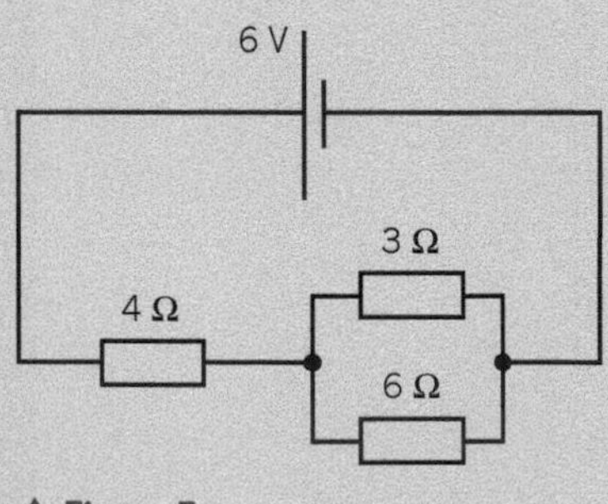

▲ **Figure 5**

Calculate:

a the combined resistance of the 3 Ω resistor and the 6 Ω resistor in parallel

b the total resistance of the circuit

c the battery current

d the power supplied to the 4 Ω resistor.

3 A 2 Ω resistor and a 4 Ω resistor are connected in series with each other. The series combination is connected in parallel with a 9 Ω resistor and a 3 V battery of negligible internal resistance, as shown in Figure 6.

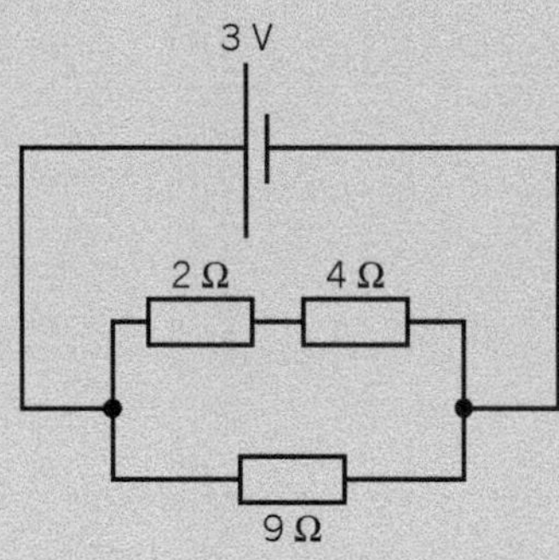

▲ **Figure 6**

Calculate:

a the total resistance of the circuit

b the battery current

c the power supplied to each resistor

d the power supplied by the battery.

4 Calculate:

a the power supplied to a 10 Ω resistor when the p.d. across it is 12 V

b the resistance of a heating element designed to operate at 60 W and 12 V.

10.3 Electromotive force and internal resistance

Learning objectives:

- → Explain why the p.d. of a battery (or cell) in use is less than its e.m.f.
- → Measure the internal resistance of a battery.
- → Describe how much power is wasted in a battery.

Specification reference: 3.4.6

Internal resistance

The **internal resistance** of a source of electricity is due to opposition to the flow of charge through the source. This causes electrical energy produced by the source to be dissipated inside the source when charge flows through it.

- The **electromotive force** (e.m.f., symbol ε) of the source is the electrical energy per unit charge produced by the source. If electrical energy E is given to a charge Q in the source,

$$\varepsilon = \frac{E}{Q}$$

- The **p.d. across the terminals** of the source is the electrical energy per unit charge delivered by the source when it is in a circuit. The terminal p.d. is less than the e.m.f. whenever current passes through the source. The difference is due to the internal resistance of the source.

The internal resistance of a source is the loss of potential difference per unit current in the source when current passes through the source.

In circuit diagrams, the internal resistance of a source may be shown as a resistor (labelled 'internal resistance') in series with the usual symbol for a cell or battery, as in Figure 1.

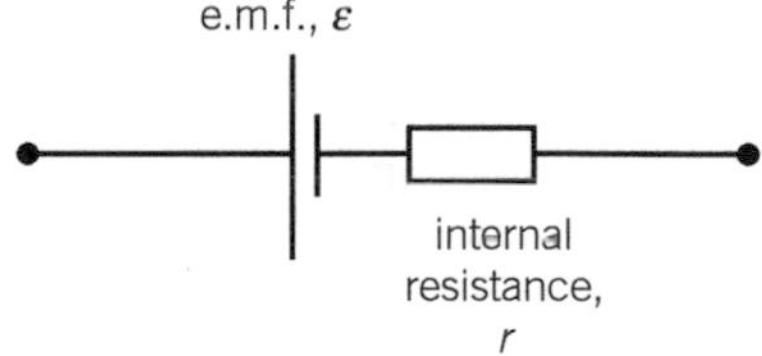

▲ **Figure 1** *Internal resistance*

When a cell of e.m.f. ε and internal resistance r is connected to an external resistor of resistance R, as shown in Figure 2, all the current through the cell passes through its internal resistance and the external resistor. So the two resistors are in series, which means that the total resistance of the circuit is $r + R$. Therefore, the current through the cell, $I = \dfrac{\varepsilon}{R + r}$.

In other words, the cell e.m.f. $\varepsilon = I(R + r) = IR + Ir$ = the terminal p.d. + the lost, or wasted, p.d.

$$\varepsilon = IR + Ir$$

The lost p.d. inside the cell (i.e., the p.d. across the internal resistance of the cell) is equal to the difference between the cell e.m.f. and the p.d. across its terminals. In energy terms, the lost p.d. is the energy per coulomb dissipated or wasted inside the cell due to its internal resistance.

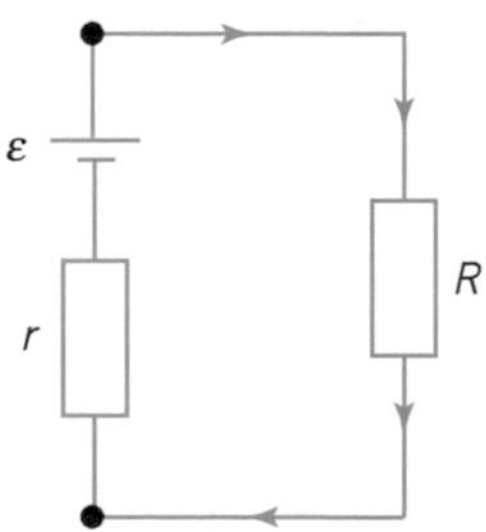

▲ **Figure 2** *Electromotive force and internal resistance*

Power

Multiplying each term of the above equation by the cell current I gives

$$\textbf{power supplied by the cell, } I\varepsilon = I^2R + I^2r$$

In other words, the power supplied by the cell = the power delivered to R + the power wasted in the cell due to its internal resistance.

The power delivered to $R = I^2R = \dfrac{\varepsilon^2}{(R + r)^2}R$ since $I = \dfrac{\varepsilon}{R + r}$.

Figure 3 shows how the power delivered to R varies with the value of R.

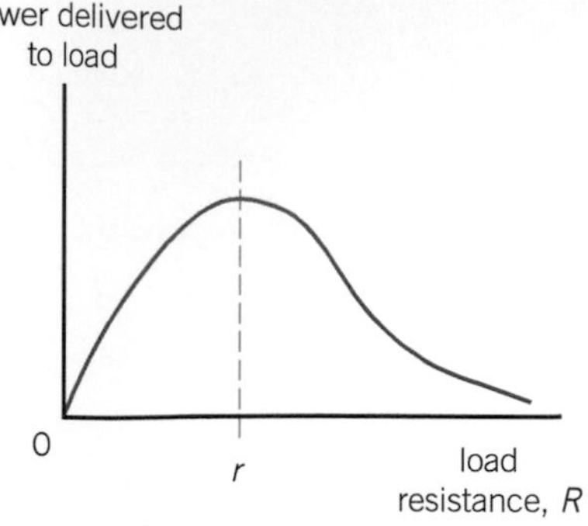

▲ **Figure 3** *Power delivered to a load versus load resistance*

It can be shown that the peak of this power curve is at $R = r$. In other words, when a source delivers power to a load, **maximum power is delivered to the load when the load resistance is equal to the internal resistance of the source**. The load is then said to be *matched* to the source. (Although you don't need to know this, you might find it useful if you ever need to replace an amplifier or a speaker.)

Measurement of internal resistance

The p.d. across the terminals of a cell, when the cell is in a circuit, can be measured by connecting a high-resistance voltmeter directly across the terminals of the cell. Figure 4 shows how the terminal p.d. can be measured for different values of current.

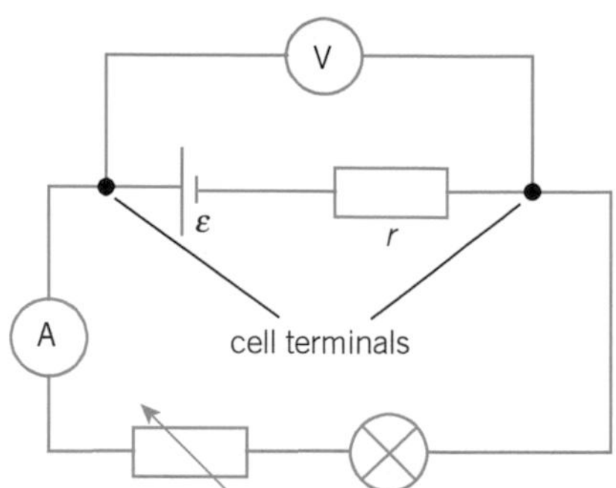

▲ **Figure 4** *Measuring internal resistance*

The current is changed by adjusting the variable resistor. The lamp (or a fixed resistor) limits the maximum current that can pass through the cell. The ammeter is used to measure the cell current. The measurements of terminal p.d. and current for a given cell may be plotted on a graph, as shown in Figure 5.

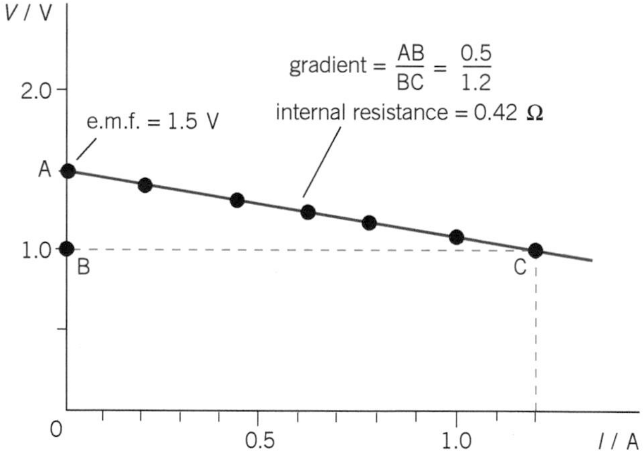

▲ **Figure 5** *A graph of terminal p.d. versus current*

The terminal p.d. decreases as the current increases. This is because the lost p.d. increases as the current increases.

- The terminal p.d. is equal to the cell e.m.f. at zero current. This is because the lost p.d. is zero at zero current.
- The graph is a straight line with a negative gradient. This can be seen by rearranging the equation $\varepsilon = IR + Ir$ to become $IR = \varepsilon - Ir$. Because IR represents the terminal p.d. V, then

$$V = \varepsilon - Ir$$

By comparison with the standard equation for a straight line, $y = mx + c$, a graph of V on the y-axis against I on the x-axis gives a straight line with a gradient $-r$ and a y-intercept ε.

Study tip

Ensure you can relate $V = \varepsilon - Ir$ to $y = mx + c$.

Figure 5 shows the gradient triangle ABC in which AB represents the lost p.d. and BC represents the current. So the gradient AB/BC = lost voltage ÷ current = internal resistance r.

Note:

The internal resistance and the e.m.f. of a cell can be calculated if the terminal p.d. is known for two different values of current.

- For current I_1, the terminal p.d. $V_1 = \varepsilon - I_1 r$.
- For current I_2, the terminal p.d. $V_2 = \varepsilon - I_2 r$.

Subtracting the first equation from the second gives:

$$V_1 - V_2 = (\varepsilon - I_1 r) - (\varepsilon - I_2 r) = I_2 r - I_1 r = (I_2 - I_1)r$$

Therefore, $r = \dfrac{V_1 - V_2}{I_2 - I_1}$.

So r can be calculated from the above equation and then substituted into either equation for the cell p.d. to enable ε to be calculated.

Worked example

A cell of unknown e.m.f. ε and internal resistance r is connected in series with an ammeter, a switch S, and a 10.0 Ω resistor P, as shown in Figure 6.

When the switch is closed, the ammeter reading is 0.40 A. When P is replaced by a 5.0 Ω resistor Q, the ammeter reading becomes 0.60 A. Calculate the internal resistance of the cell.

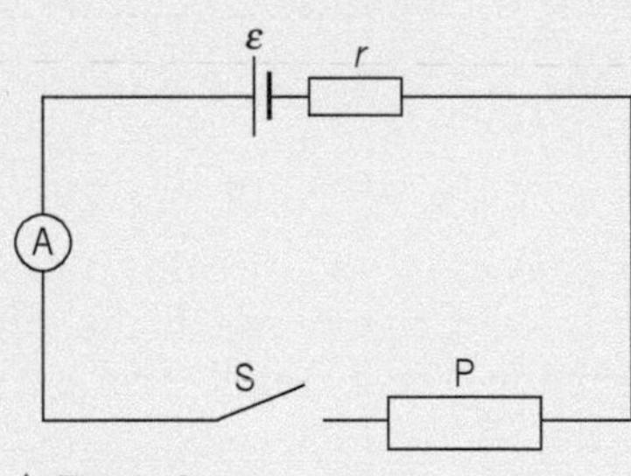

▲ Figure 6

Solution

Applying the equation $\varepsilon = IR + Ir$ to the circuit with P,
$\varepsilon = (0.40 \times 10.0) + (0.4r) = 4.0 + 0.4r$

Applying the equation $\varepsilon = IR + Ir$ to the circuit with Q,
$\varepsilon = (0.60 \times 5.0) + (0.6r) = 3.0 + 0.6r$

Therefore

$4.0 + 0.4r = 3.0 + 0.6r$

Rearranging this equation gives

$0.6r - 0.4r = 4.0 - 3.0$

$0.2r = 1.0$

$r = \dfrac{1.0}{0.2} = 5.0\,\Omega$

By substituting the value of r into either of the equations, $\varepsilon = 6.0$ V.

Summary questions

1 A battery of e.m.f. 12 V and internal resistance 1.5 Ω is connected to a 4.5 Ω resistor. Calculate:

- a the total resistance of the circuit
- b the current through the battery
- c the lost p.d.
- d the p.d. across the cell terminals.

2 A cell of e.m.f. 1.5 V and internal resistance 0.5 Ω is connected to a 2.5 Ω resistor. Calculate:

- a the current
- b the terminal p.d.
- c the power delivered to the 2.5 Ω resistor
- d the power wasted in the cell.

3 The p.d. across the terminals of a cell is 1.1 V when the current from the cell is 0.20 A, and 1.3 V when the current is 0.10 A. Calculate:

- a the internal resistance of the cell
- b the cell's e.m.f.

4 A battery of unknown e.m.f. ε and internal resistance r is connected in series with an ammeter and a resistance box R. The current is 2.0 A when $R = 4.0\,\Omega$ and 1.5 A when $R = 6.0\,\Omega$. Calculate ε and r.

10.4 More circuit calculations

Learning objectives:

→ Calculate currents in circuits with:

- resistors in series and parallel
- more than one cell
- diodes in the circuit.

Specification reference: 3.4.4 and 3.4.6

Study tip

Check the p.ds round a circuit add up to the battery p.d.

Circuits with a single cell and one or more resistors

Here are some rules:

1 Sketch the **circuit diagram** if it is not drawn.
2 To calculate the **current** passing through the cell, calculate the total circuit resistance using the resistor combination rules. Don't forget to add on the internal resistance of the cell if that is not negligible.

$$\textbf{cell current} = \frac{\textbf{cell e.m.f.}}{\textbf{total circuit resistance}}$$

3 To work out the current and p.d. for each resistor, start with the **resistors in series with the cell** which therefore pass the same current as the cell current.

P.d. across each resistor in series with the cell = current × the resistance of each resistor.

4 To work out the current through **parallel resistors**, work out the combined resistance and multiply by the cell current to give the p.d. across each resistor.

$$\textbf{Current through each resistor} = \frac{\textbf{p.d. across the parallel combination}}{\textbf{resistor's resistance}}$$

Circuits with cells in series

The same rules as above apply except the current through the cells is calculated by dividing the overall (net) e.m.f. by the total resistance.

- If the cells are connected in the same direction in the circuit, as in Figure 1a, the net e.m.f. is the sum of the individual e.m.fs. For example, in Figure 1a, the net e.m.f. is 3.5 V.
- If the cells are connected in opposite directions to each other in the circuit, as in Figure 1b, the net e.m.f. is the difference between the e.m.fs in each direction. For example, in Figure 1b, the net e.m.f. is 0.5 V in the direction of the 2.0 V cell.
- The total internal resistance is the sum of the individual internal resistances. This is because the cells, and therefore the internal resistances, are in series.

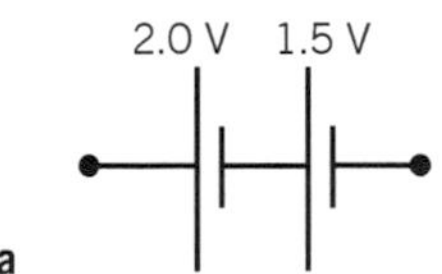

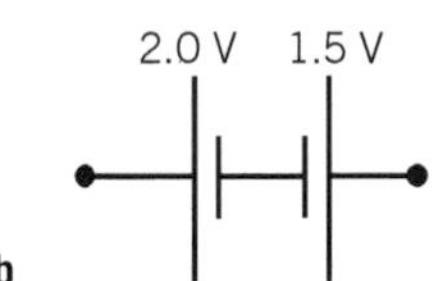

▲ **Figure 1** *Cells in series*

Worked example

A cell of e.m.f. 3.0 V and internal resistance 2.0 Ω and a cell of e.m.f. 2.0 V and internal resistance 1.0 Ω are connected in series with each other and with a 7.0 Ω resistor, as in Figure 2. Calculate the p.d. across the 7.0 Ω resistor.

Solution

The net e.m.f. of the two batteries = 3.0 + 2.0 = 5.0 V in the direction of the 3.0 V cell.

The total circuit resistance = 1.0 Ω + 2.0 Ω + 7.0 Ω = 10.0 Ω

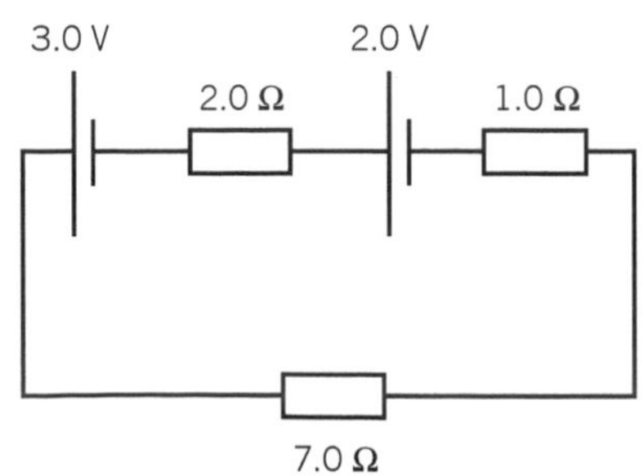

▲ **Figure 2**

Therefore the cell current = $\frac{\text{net e.m.f.}}{\text{total circuit resistance}} = \frac{5.0\,\text{V}}{10.0\,\Omega} = 0.50\,\text{A}$

The p.d. across the 7.0 Ω resistor = current × resistance
= 0.50 A × 7.0 Ω = 3.5 V

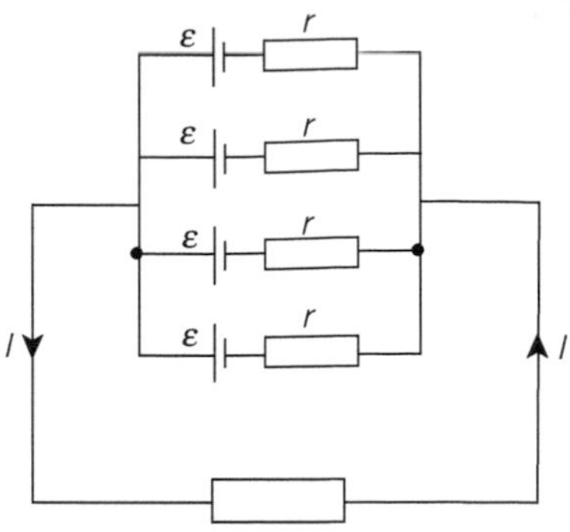

▲ **Figure 3** *Cells in parallel*

Circuits with identical cells in parallel

For a circuit with n identical cells in parallel, the current through each cell = $\frac{I}{n}$, where I is the total current supplied by the cells.

Therefore, the lost p.d. in each cell = $\frac{I}{n}r = \frac{Ir}{n}$, where r is the internal resistance of each cell.

Hence the terminal p.d. across each cell, $V = \varepsilon - \frac{Ir}{n}$.

Each time an electron passes through the cells, it travels through one of the cells only (as the cells are in parallel), therefore the cells act as a source of e.m.f. ε and internal resistance $\frac{r}{n}$.

Application

Solar panels

A solar panel consists of many parallel rows of identical solar cells in series. For example, using single solar cells with a maximum e.m.f. of 0.45 V and an internal resistance of 1.0 Ω, a row of 20 cells in series would have a maximum e.m.f. of 9.0 V and an internal resistance of 20 Ω. Forty such rows in parallel would still give a maximum e.m.f. of 9.0 V but would have an internal resistance of 0.5 Ω $\left(= \frac{20\,\Omega}{40}\right)$.

Synoptic link

For more about solar panels, see Chapter 26, Renewable energy.

Diodes in circuits

Assume that a silicon diode has:

- a forward p.d. of 0.6 V whenever a current passes through it
- infinite resistance in the reverse direction or at p.ds less than 0.6 V in the forward direction.

Therefore, in a circuit with one or more diodes:

- a p.d. of 0.6 V exists across a diode that is forward biased and passing a current
- a diode that is reverse biased has infinite resistance.

For example, suppose a diode is connected in its forward direction in series with a 1.5 V cell of negligible internal resistance and a 1.5 kΩ resistor, as in Figure 4.

The p.d. across the diode is 0.6 V because it is forward biased. Therefore, the p.d. across the resistor is 0.9 V (= 1.5 V − 0.6 V). The current through the resistor is therefore 6.0×10^{-4} A $\left(= \frac{0.9\,\text{V}}{1500\,\Omega}\right)$.

However, if the diode in Figure 4 were reversed, the circuit current would be zero, so the p.d. across the resistor would also be zero. The p.d. across the diode would therefore be 1.5 V (equal to the cell e.m.f.).

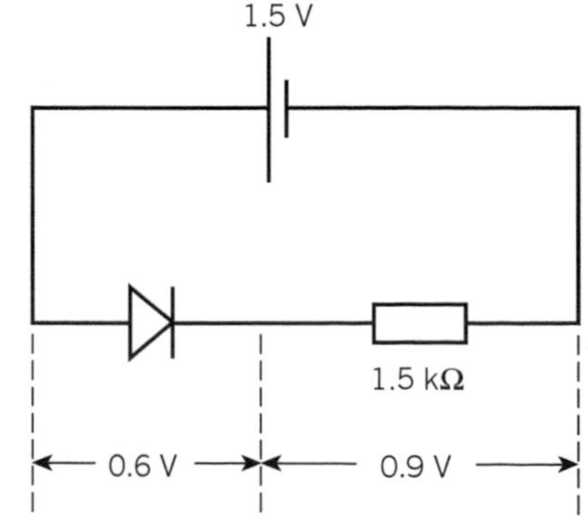

▲ **Figure 4** *Using a diode*

Extension

Kirchhoff's laws

The following two circuit rules, which are called Kirchhoff's laws, can be used to analyse any dc circuit, regardless of how many loops and cells are in the circuit. The example on the following page illustrates their use.

At any junction in a circuit, the total current entering the junction is equal to the total current leaving the junction.

For any complete loop in a circuit, the sum of the e.m.fs around the loop is equal to the sum of the potential drops around the loop.

Two cells, P and Q, and a resistor R of resistance 4.0 Ω are connected in parallel with each other, as shown in Figure 5. Cell P has an e.m.f. of 2.0 V and an internal resistance of 1.5 Ω. Cell Q has an e.m.f. of 1.5 V and an internal resistance of 2.0 Ω. Calculate the current in each cell.

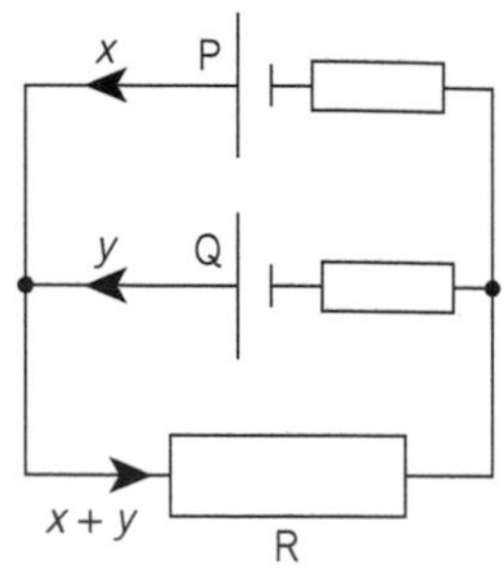

▲ Figure 5

Solution

Let the current in P be x and the current in Q be y.

Using Kirchhoff's first law, the current in R is therefore $x + y$.

In the complete loop consisting of P and R:

The sum of the e.m.fs in the loop is 2.0 V.

The p.d. across the internal resistance of P is $1.5x$.

The p.d. across R is $4.0(x + y)$.

Therefore, the sum of the p.ds around the loop = $1.5x + 4.0(x + y) = 5.5x + 4.0y$.

Using Kirchhoff's second law, $\mathbf{5.5x + 4.0y = 2.0}$. **(1)**

In the complete loop consisting of Q and R:

The sum of the e.m.fs in the loop is 1.5 V.

The p.d. across the internal resistance of Q is $2.0y$.

The p.d. across R is $4.0(x + y)$.

Therefore, the sum of the p.ds around the loop = $2.0y + 4.0(x + y) = 4.0x + 6.0y$.

Using Kirchhoff's second law, $\mathbf{4.0x + 6.0y = 1.5}$. **(2)**

Multiplying **(1)** by 3 gives $16.5x + 12.0y = 6.0$.

Multiplying **(2)** by 2 gives $8.0x + 12.0y = 3.0$.

Subtracting **(2)** from **(1)** gives $8.5x = 3.0$.

$x = \frac{3.0}{8.5} = 0.35$ A.

Prove for yourself that substituting this value into either **(1)** or **(2)** gives $y = -0.015$ A.

Note:

The negative value of y shows that the current in Q is in the reverse direction to that assumed in Figure 5.

Q: If cell Q had been reversed, the total e.m.f. in the loop of Q and R would have been −1.5 V. Calculate the current in each cell for this circuit.

Answer: x = 1.06 A, y = 0.96 A.

Summary questions

1 A cell of e.m.f. 3.0 V and negligible internal resistance is connected to a 4.0 Ω resistor in series with a parallel combination of a 24.0 Ω resistor and a 12.0 Ω resistor, as shown in Figure 6. Calculate:

a the total resistance of the circuit

b the cell current

c the current and p.d. for each resistor.

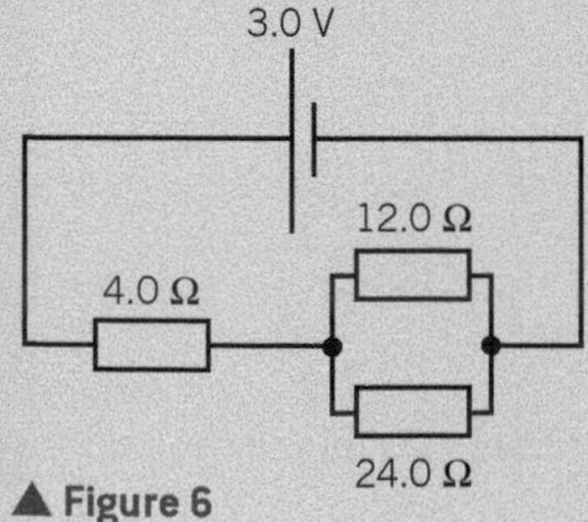

▲ Figure 6

2 A 15.0 Ω resistor, a battery of e.m.f. 12.0 V with an internal resistance of 3.0 Ω, and a battery of e.m.f. 9.0 V with an internal resistance of 2.0 Ω are connected in series. The batteries act in the same direction in the circuit. Sketch the circuit diagram and calculate:

a the total resistance of the circuit

b the cell current

c the current and p.d. across the 15 Ω resistor.

3 Two 8 Ω resistors and a battery of e.m.f. 12.0 V and internal resistance 8 Ω are connected in series with each other. Sketch the circuit diagram and calculate **i** the power delivered to each external resistor, **ii** the power wasted due to internal resistance.

10.5 The potential divider

The theory of the potential divider

A **potential divider** consists of two or more resistors in series with each other and with a source of fixed potential difference. The potential difference of the source is divided between the components in the circuit, as they are in series with each other. By making a suitable choice of components, a potential divider can be used:

- to supply a p.d. which is fixed at any value between zero and the source p.d.
- to supply a variable p.d.
- to supply a p.d. that varies with a physical condition such as temperature or pressure.

To supply a fixed p.d.

Consider two resistors R_1 and R_2 in series connected to a source of fixed p.d., V_0, as shown in Figure 1.

The total resistance of the combination $= R_1 + R_2$.

Therefore, the current I through the resistors is given by

$$I = \frac{\text{p.d. across the resistors}}{\text{total resistance}} = \frac{V_0}{R_1 + R_2}$$

So the p.d. V_1 across resistor R_1 is given by

$$V_1 = IR_1 = \frac{V_0 R_1}{R_1 + R_2}$$

and the p.d. V_2 across resistor R_2 is given by

$$V_2 = IR_2 = \frac{V_0 R_2}{R_1 + R_2}$$

These two equations show that the p.d. across each resistor as a proportion of the source p.d. is the same as the resistance of the resistor in proportion to the total resistance. In other words, if the resistances are 5 kΩ and 10 kΩ, respectively:

- the p.d. across the 5 kΩ resistor is $\frac{1}{3}\left(=\frac{5}{15}\right)$ of the source p.d.
- the p.d. across the 10 kΩ resistor is $\frac{2}{3}\left(=\frac{10}{15}\right)$ of the source p.d.

Also, dividing the equation for V_1 by the equation for V_2 gives

$$\frac{V_1}{V_2} = \frac{R_1}{R_2}$$

This equation shows that

the ratio of the p.ds across each resistor is equal to the resistance ratio of the two resistors.

To supply a variable p.d.

The source p.d. is connected to a fixed length of uniform resistance wire. A sliding contact on the wire can then be moved along the wire, as illustrated in Figure 2, giving a variable p.d. between the contact and one end of the wire. A uniform track of a suitable material may be used instead of resistance wire. The track may be linear or circular, as

Learning objectives:

→ Describe a potential divider.

→ Explain how we can supply a variable p.d. from a battery.

→ Explain how we can design sensor circuits.

Specification reference: 3.4.5

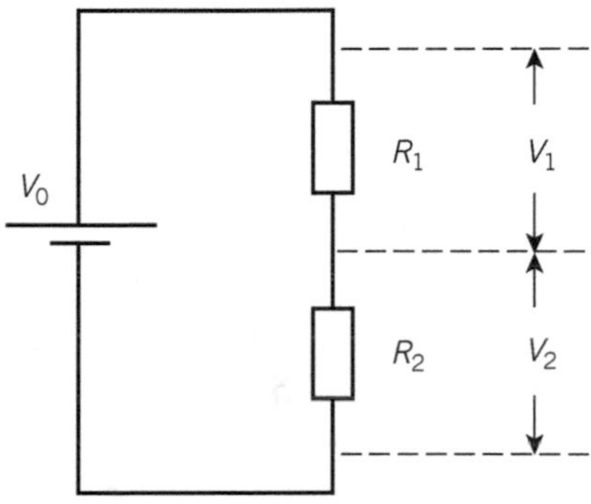

▲ **Figure 1** *A potential divider*

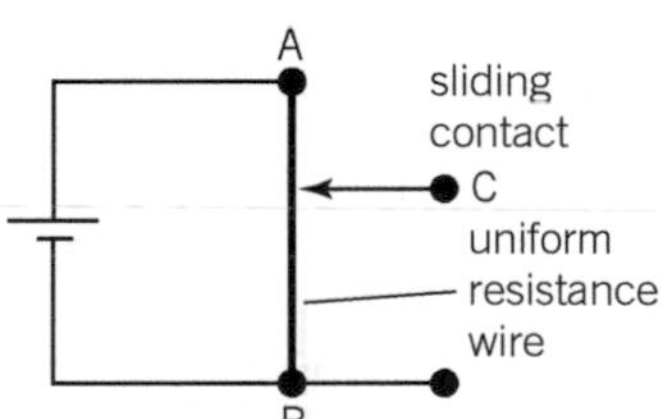

a *A linear track using resistance wire*

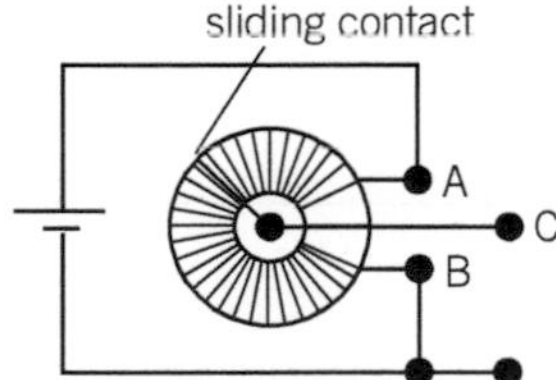

b *A circular track*

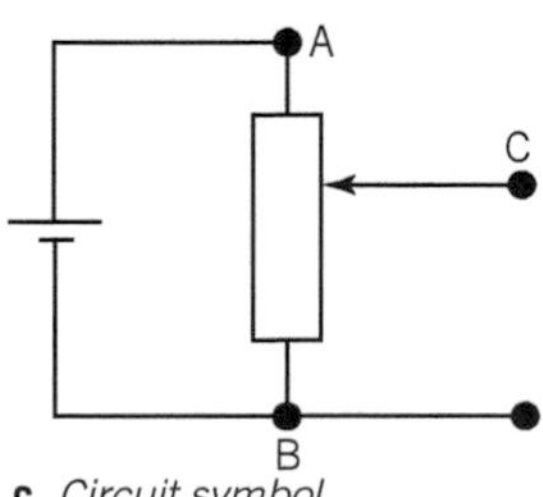

c *Circuit symbol*

▲ **Figure 2** *Potential dividers used to supply a variable p.d.*

Study tip

Ensure you understand the basic potential divider rule.

in Figures 2a and b. The circuit symbol for a variable potential divider is also shown in Figure 2c.

A variable potential divider can be used:

- As a simple audio volume control to change the loudness of the sound from a loudspeaker. The audio signal p.d. is supplied to the potential divider in place of a cell or battery. The variable output p.d. from the potential divider is supplied to the loudspeaker.
- To vary the brightness of a light bulb between zero and normal brightness (Figure 3). In contrast with using a variable resistor in series with the light bulb and the source p.d., the use of a potential divider enables the current through the light bulb to be reduced to zero. If a variable resistor in series with the light bulb had been used, there would still be a current through the light bulb when the variable resistor is at maximum resistance.

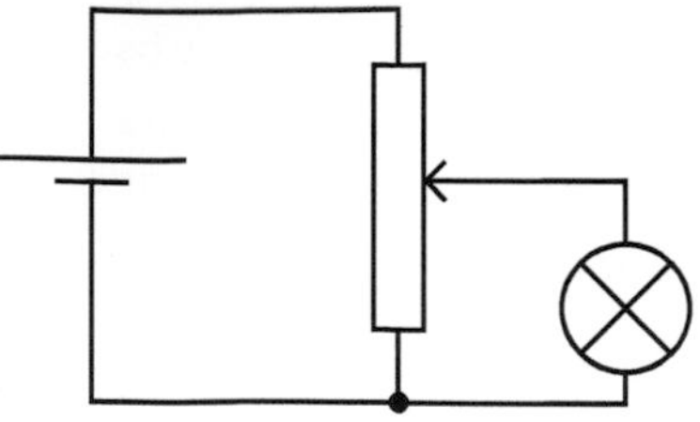

▲ **Figure 3** *Brightness control using a variable potential divider*

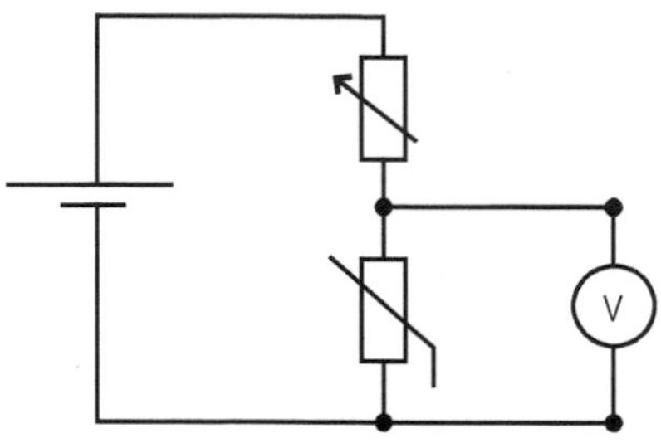

▲ **Figure 4** *A temperature sensor*

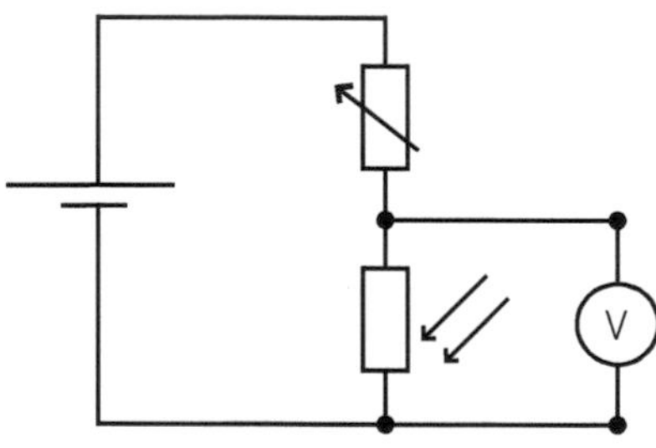

▲ **Figure 5** *A light sensor*

Sensor circuits

A **sensor circuit** produces an output p.d. which changes as a result of a change of a physical variable such as temperature or light intensity.

1 A **temperature sensor** consists of a potential divider made using a thermistor and a variable resistor, as in Figure 4.

With the temperature of the thermistor constant, the source p.d. is divided between the thermistor and the variable resistor. By adjusting the variable resistor, the p.d. across the thermistor can then be set at any desired value that triggers a response such as switching on a heater if the temperature decreases. When the temperature of the thermistor changes, its resistance changes so the p.d. across it changes. For example, suppose the variable resistor is adjusted so that the p.d. across the thermistor at 20 °C is exactly half the source p.d. If the temperature of the thermistor then decreases, its resistance increases, so the p.d. across it increases and switches a heater on.

2 A **light sensor** uses a light-dependent resistor (LDR) and a variable resistor, as in Figure 5. The p.d. across the LDR changes when the incident light intensity on the LDR changes. If the light intensity increases, the resistance of the LDR falls and the p.d. across the LDR falls.

Summary questions

1 A 12 V battery of negligible internal resistance is connected to the fixed terminals of a variable potential divider which has a maximum resistance of 50 Ω. A 12 V light bulb is connected between the sliding contact and the negative terminal of the potential divider. Sketch the circuit diagram and describe how the brightness of the light bulb changes when the sliding contact is moved from the negative to the positive terminal of the potential divider.

2 a A potential divider consists of an 8.0 Ω resistor in series with a 4.0 Ω resistor and a 6.0 V battery of negligible internal resistance. Calculate **i** the current, **ii** the p.d. across each resistor.

b In the circuit in **a**, the 4 Ω resistor is replaced by a thermistor with a resistance of 8 Ω at 20 °C and a resistance of 4 Ω at 100 °C. Calculate the p.d. across the fixed resistor at **i** 20 °C, **ii** 100 °C.

3 A light sensor consists of a 5.0 V cell, an LDR, and a 5.0 kΩ resistor in series with each other. A voltmeter is connected in parallel with the resistor. When the LDR is in darkness, the voltmeter reads 2.2 V.

a Calculate **i** the p.d. across the LDR, **ii** the resistance of the LDR when the voltmeter reads 2.2 V.

b Describe and explain how the voltmeter reading would change if the LDR were exposed to daylight with no adjustment made to the variable resistor.

Practice questions: Chapter 10

1 A student is given three resistors of resistance 3.0 Ω, 4.0 Ω, and 6.0 Ω, respectively.
(a) Draw the arrangement, using all three resistors, which will give the largest resistance.
(b) Calculate the resistance of the arrangement you have drawn.
(c) Draw the arrangement, using all three resistors, which will give the smallest resistance.
(d) Calculate the resistance of the arrangement you have drawn. *(5 marks)*
AQA, 2005

2 In the circuit shown in **Figure 1**, the battery has negligible internal resistance.

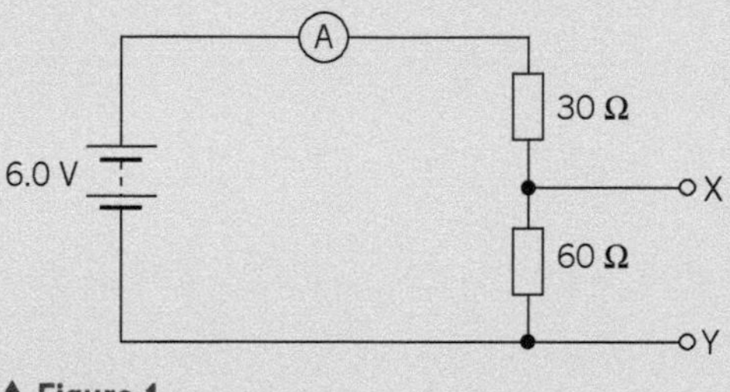

▲ **Figure 1**

Calculate the current in the ammeter when
(a) the terminals X and Y are short circuited, that is, connected together *(2 marks)*
(b) the terminals X and Y are connected to a 30 Ω resistor. *(4 marks)*
AQA, 2002

3 (a) Define the electrical resistance of a component. *(2 marks)*
(b) Calculate the total resistance of the arrangement of resistors in **Figure 2**. *(3 marks)*

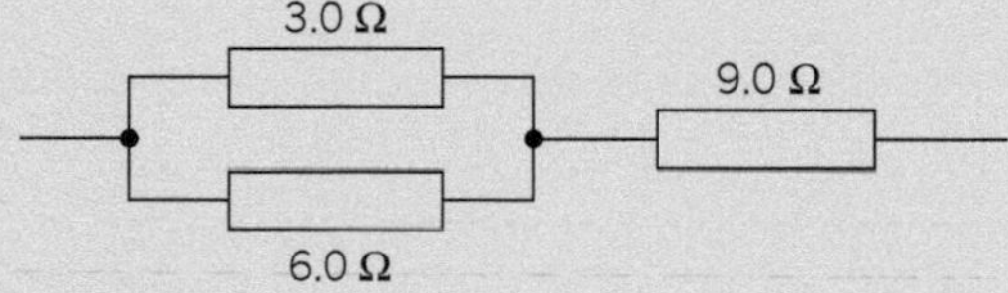

▲ **Figure 2**

(c) i Calculate the current in the 3.0 Ω resistor in **Figure 2** when the current in the 9.0 Ω resistor is 2.4 A.
ii Calculate the total power dissipated by the arrangement of resistors in **Figure 2** when the current in the 9.0 Ω resistor is 2.4 A. *(4 marks)*
AQA, 2006

4 In the circuit shown in **Figure 3** the resistor network between the points P and Q is connected in series to a resistor R, an ammeter, and a battery of negligible internal resistance.

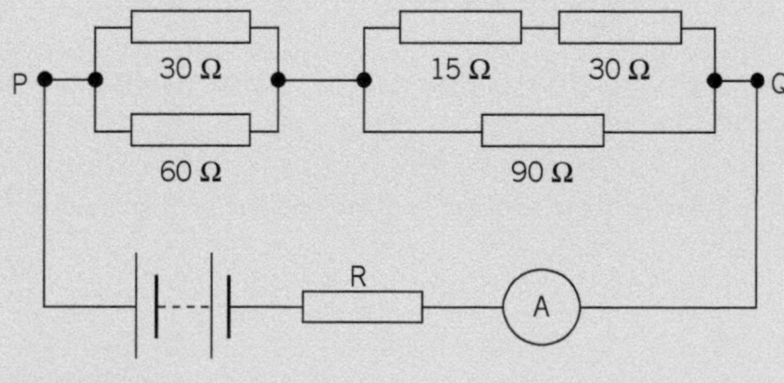

▲ **Figure 3**

(a) Determine the equivalent resistance of the network between the points P and Q. *(3 marks)*

(b) (i) If the current through the ammeter is 50 mA, calculate the total charge that flows through the resistor R in 4 minutes.
(ii) If 18 J of energy are transferred to the resistor R in this time, calculate the potential difference across R.
(iii) Calculate the resistance of R.
(iv) Calculate the e.m.f. of the battery. *(6 marks)*

AQA, 2007

5 In the circuit shown in **Figure 4**, the battery has an e.m.f. of 12 V and an internal resistance of 2.0 Ω. The resistors A and B each have a resistance of 30 Ω.

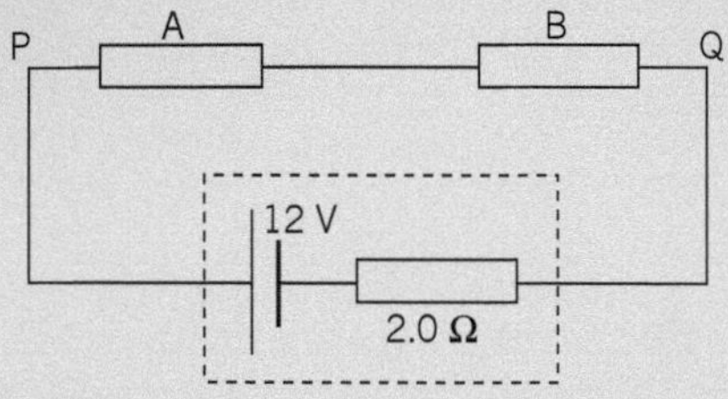

▲ **Figure 4**

Calculate:
(a) the total current in the circuit
(b) the voltage between the points P and Q
(c) the power dissipated in resistor A
(d) the energy dissipated by resistor A in 20 s. *(8 marks)*

AQA, 2003

6 **Figure 5** shows an extra high tension (EHT) supply of e.m.f. 5000 V and internal resistance 2 MΩ.

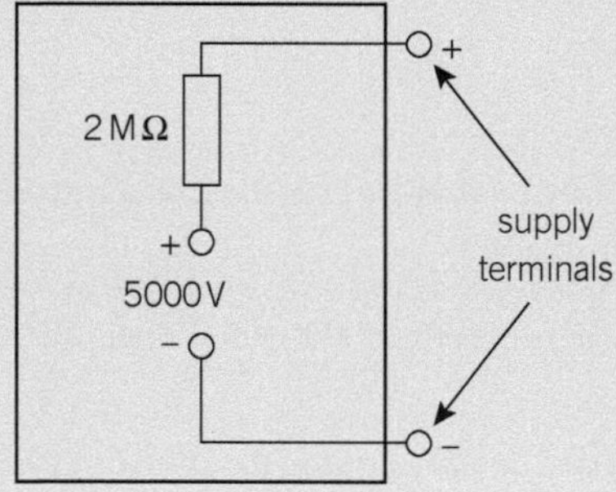

▲ **Figure 5**

(a) A lead of negligible resistance is connected between the supply terminals producing a short circuit.
(i) State the magnitude of the terminal potential difference between the supply terminals.
(ii) Calculate the current in the circuit.
(iii) Calculate the minimum power rating for the resistor used to provide the internal resistance. *(4 marks)*

(b) Explain briefly why the supply is designed with such a high internal resistance. *(1 mark)*

AQA, 2005

Answers to the Practice Questions and Section Questions are available at www.oxfordsecondary.com/oxfordaqaexams-alevel-physics

11 Oscillations and waves

11.1 Oscillations

Learning objectives:

- → Explain what is meant by one complete oscillation.
- → Define amplitude, frequency, and period.
- → Use equations for the time period of a simple pendulum and of an object oscillating on a spring.

Specification reference: 3.5.1

Measuring oscillations

There are many examples of oscillations in everyday life. A car that travels over a bump bounces up and down for a short time afterwards. Every microcomputer has an electronic oscillator to drive its internal clock. A child on a swing moves forwards then backwards repeatedly. In this simple example, one full cycle of motion is from maximum height at one side to maximum height on the other side and then back again. The lowest point is called the **equilibrium** position, because it is where the child eventually comes to a standstill. The child in motion is described as *oscillating* about equilibrium.

Other examples of oscillating motion include:

- an object on a spring moving up and down repeatedly
- a pendulum moving to and fro repeatedly
- a ball-bearing rolling from side to side, as in Figure 1
- a small boat rocking from side to side.

An oscillating object moves repeatedly one way then in the opposite direction through its equilibrium position. The **displacement** of the object (i.e., distance and direction) from equilibrium continually changes during the motion. In one full cycle after being released from a non-equilibrium position, the displacement of the object:

- decreases as it returns to equilibrium, then
- reverses and increases as it moves away from equilibrium in the opposite direction, then
- decreases as it returns to equilibrium, then
- increases as it moves away from equilibrium towards its starting position.

The **amplitude** of the oscillations is the maximum displacement of the oscillating object from equilibrium. If the amplitude is constant and no frictional forces are present, the oscillations are described as **free vibrations**. See Topic 11.2, Energy and oscillations.

The **time period**, T, of the oscillating motion is the time for one complete cycle of oscillation. One full cycle after passing through any position, the object passes through that same position in the same direction.

The **frequency** of oscillations is the number of cycles per second made by an oscillating object.

Synoptic link

The displacement of an object from a fixed point is its distance from that point in a particular direction. See Topic 2.1, Speed and velocity.

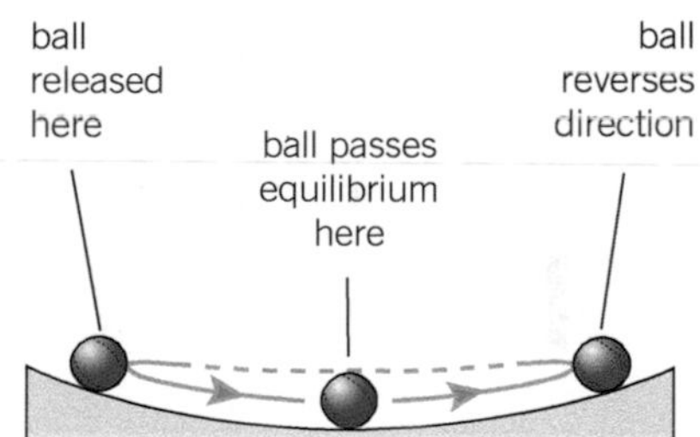

▲ **Figure 1** *Oscillating motion*

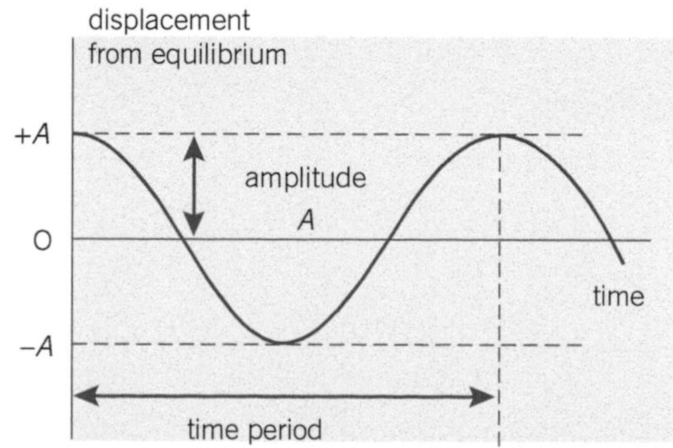

▲ **Figure 2** *Variation of displacement with time*

Study tip

Make sure you know how to apply these definitions.

The unit of frequency is the hertz (Hz), which is one cycle per second.

For oscillations of frequency f, the time period $T = \frac{1}{f}$

Note:

The **angular frequency** ω of the oscillating motion is defined as $\frac{2\pi}{T}$ ($= 2\pi f$). The unit of ω is the radian per second (rad s^{-1}).

Time period equations

1 **For small oscillations of a simple pendulum of length L,** as shown in Figure 3, the time period T is given by the equation

$$T = 2\pi\sqrt{\frac{L}{g}}$$

where g is the gravitational field strength where the pendulum is located.

Worked example

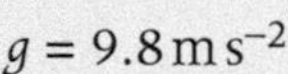

$g = 9.8\,m\,s^{-2}$

Calculate the time period of a simple pendulum of length 2.50 m.

Solution

$$T = 2\pi\sqrt{\frac{L}{g}} = 2\pi\sqrt{\frac{2.5}{9.8}} = 3.17\,s$$

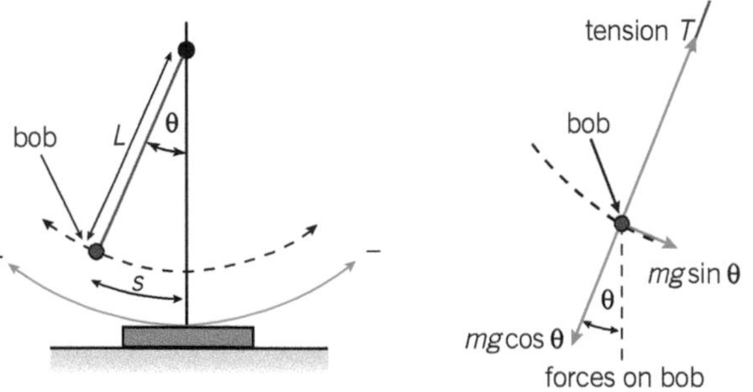

▲ **Figure 3** *The oscillations of a simple pendulum – the equation is valid provided θ does not exceed about 10°*

2 **For an object of mass m oscillating on a spring, as shown in Figure 4,** the time period T is given by the equation

$$T = 2\pi\sqrt{\frac{m}{k}}$$

where k is the spring constant of the spring.

Synoptic link

The derivation of these two equations is in Topic 16.3.

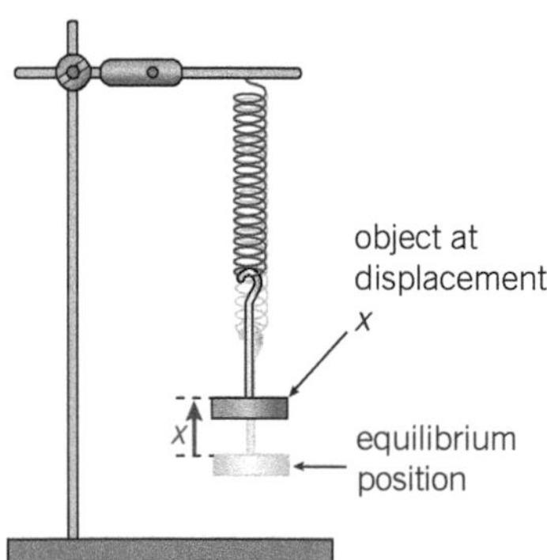

▲ **Figure 4** *The oscillations of a loaded spring*

Worked example

An object of mass 0.20 kg oscillates vertically on a loaded spring, as shown in Figure 4. A student timed ten complete oscillations three times:

5.6 s, 5.8 s, 5.9 s

a Calculate the time period of the oscillations.
b Calculate the spring constant of the oscillations.

Solution

a Mean time for ten oscillations $= \frac{(5.6 + 5.8 + 5.9)}{3} = 5.77\,\text{s}$

Time period $T = \frac{5.77}{10} = 0.577\,\text{s}$

b Rearranging $T = 2\pi\sqrt{\frac{m}{k}}$ gives

$$k = \frac{4\pi^2 m}{T^2} = \frac{4\pi^2 \times 0.20\,\text{kg}}{(0.577)^2} = 23.7\,\text{N m}^{-1}$$

Hints

1 The time period of a simple pendulum $T = 2\pi\sqrt{\frac{L}{g}}$

2 The time period of an object of mass m *oscillating* on a spring $T = 2\pi\sqrt{\frac{m}{k}}$

Summary questions

1 Describe how the velocity of a bungee jumper changes during a jump from the moment he jumps off the starting platform to the next time he reaches the highest point of his jump.

2 a Define free vibrations (free oscillations).

b A metre rule is clamped to a table so that part of its length projects at right angles from the edge of the table, as shown in Figure 5. A 100 g mass is attached to the free end of the rule. When the free end of the rule is depressed downwards then released, the mass oscillates. Describe how you would find out if the oscillations of the mass are free oscillations.

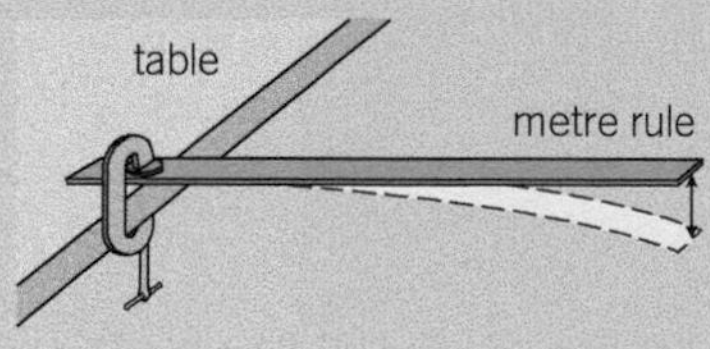

▲ **Figure 5**

3 An object suspended from the lower end of a vertical spring is displaced downwards from equilibrium. It takes 9.6 s to undergo 20 complete cycles of oscillation. Calculate:

a its time period

b its frequency of oscillation.

4 a i A simple pendulum undergoes 10 oscillations in 20.0 s. Calculate its length.

ii Estimate the uncertainty in the length of the pendulum if the timing of 20 s was accurate to within 0.2 s.

b An object of mass 0.36 kg suspended from the lower end of a spring as in Figure 4 oscillates vertically with a time period of 0.74 s. Calculate the spring constant of the spring.

11.2 Energy and oscillations

Learning objectives:

- → Describe how, in simple harmonic motion, kinetic energy and potential energy vary with displacement.
- → Describe how these energies vary with time if damping is negligible.
- → Describe the effects of damping on the characteristics of oscillations.

Specification reference: 3.5.1 and 3.5.2

Free oscillations

A freely oscillating object oscillates with a constant amplitude because there is no friction acting on it. The only forces acting on it combine to provide the restoring force. If friction were present, the amplitude of oscillations would gradually decrease, and the oscillations would eventually cease.

Observe the oscillations of a simple pendulum over many cycles, and you should find that the decrease of amplitude from one cycle to the next is scarcely measurable. Nevertheless, over many cycles, the amplitude does decrease noticeably. So friction is present, even if its effect is insignificant over a single cycle.

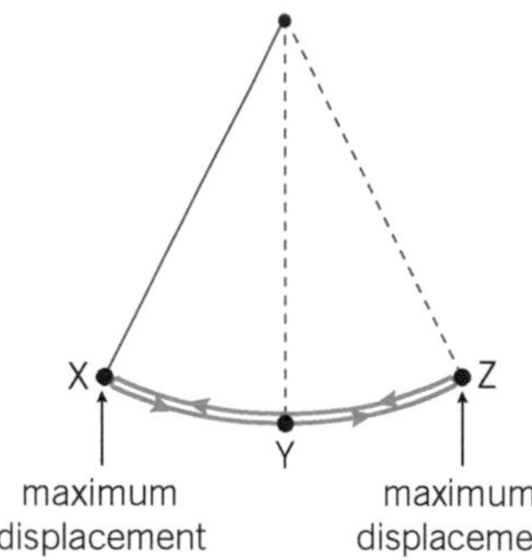

Position	E_P	E_K
X	E_{TOTAL}	0
Y	0	E_{TOTAL}
Z	E_{TOTAL}	0

▲ **Figure 1** *The energy changes of a simple pendulum*

Consider the example of a small object of mass m oscillating on a spring. The energy of the system changes from kinetic energy to potential energy and back again every half-cycle after passing though equilibrium. As long as friction is absent, the total energy of the system is constant and is equal to its maximum potential energy.

- The potential energy E_P changes with displacement x from equilibrium, in accordance with the equation $E_P = \frac{1}{2}kx^2$, where k is the spring constant of the spring.
- The total energy, E_T, of the system is therefore $\frac{1}{2}kA^2$, where A is the amplitude of the oscillations.
- Because the total energy is $E_T = E_K + E_P$, where E_K is the kinetic energy of the oscillating mass, then $E_K = E_T - E_P = \frac{1}{2}k(A^2 - x^2)$.

Synoptic link

The energy stored in a spring stretched by extension x from its equilibrium length is $\frac{1}{2}kx^2$, where k is the spring constant. See Topic 6.2, Springs.

Energy–displacement graphs

1 The potential energy curve is parabolic in shape, given by $E_P = \frac{1}{2}kx^2$

2 The kinetic energy curve is an inverted parabola, given by
$E_K = E_T - E_P = \frac{1}{2}k(A^2 - x^2)$

The sum of the kinetic energy and the potential energy is always equal to $\frac{1}{2}kA^2$, which is the potential energy at maximum displacement. This is the same as the kinetic energy at zero displacement. So the two curves add together to give a horizontal line for the total energy.

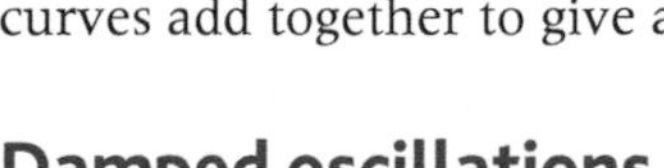

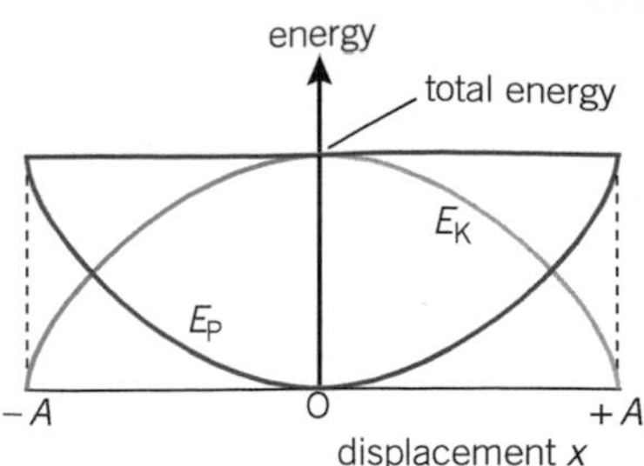

▲ **Figure 2** *Energy variation with displacement*

Damped oscillations

The oscillations of a simple pendulum gradually die away because air resistance gradually reduces the total energy of the system. In any oscillating system where friction or air resistance is present, the amplitude decreases. The forces causing the amplitude to decrease are described as *dissipative forces* because they dissipate the energy of the system to the surroundings as thermal energy. The motion is said to be **damped** if dissipative forces are present.

Study tip

E_K is equal to the total energy at $x = 0$ (i.e., $E_P = 0$ when $x = 0$).

- **Light damping** occurs when the time period is independent of the amplitude so each cycle takes the same length of time as the oscillations die away. Figure 3 shows how the displacement of a lightly damped oscillating system decreases with time. The amplitude gradually decreases, reducing by the same fraction each cycle.

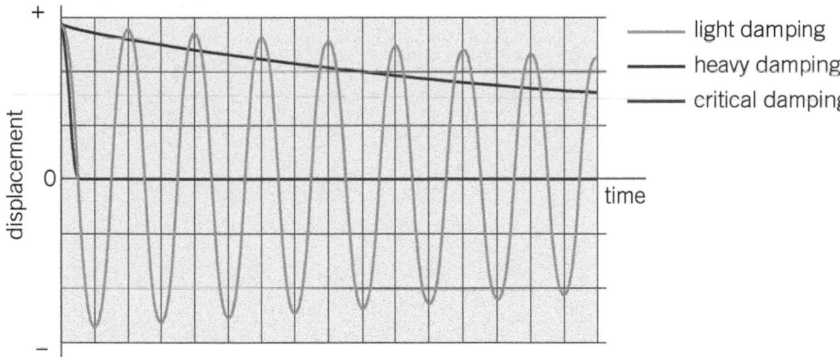

▲ **Figure 3** *Damping*

- **Critical damping** is just enough to stop the system oscillating after it has been displaced from equilibrium and released. The oscillating system returns to equilibrium in the shortest possible time without overshooting if the damping is critical. Such damping is important in mass–spring systems such as a vehicle suspension system where you would have an uncomfortable ride if the damping was too light or too heavy. Figure 3 also shows how the displacement changes with time when the damping is critical.
- **Heavy damping** occurs when the damping is so strong that the displaced object returns to equilibrium much more slowly than if the system is critically damped. No oscillating motion occurs. For example, a mass on a spring in thick oil would return to equilibrium very slowly after being displaced and released.

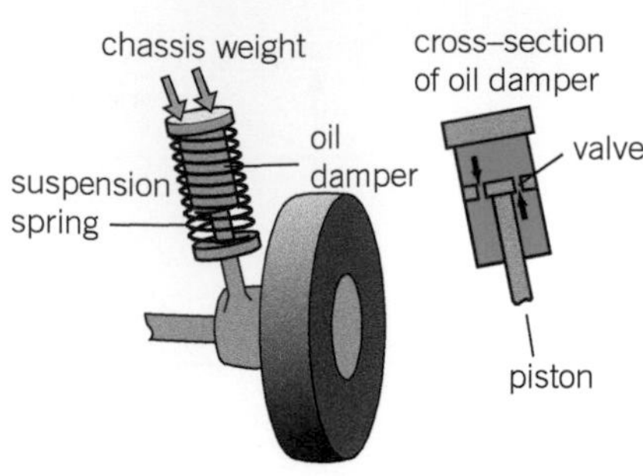

▲ **Figure 4** *Car suspension*

Application

A car suspension system

The suspension system of a car includes a coiled spring near each wheel between the wheel axle and the car chassis. When the wheel is jolted, for example, on a bumpy road, the spring smooths out the force of the jolts. An oil damper fitted with each spring prevents the chassis from bouncing up and down too much.

Without oil dampers, the occupants of the car would continue to be thrown up and down until the oscillations died away. The flow of oil through valves in the piston of each damper provides a frictional force that damps the oscillating motion of the chassis. The dampers are designed to ensure the chassis returns to its equilibrium position in the shortest possible time after each jolt with little or no oscillations. The suspension system is therefore at or close to critical damping.

Summary questions

$g = 9.81\ \text{m s}^{-2}$

1 A simple pendulum consists of a small metal sphere of mass 0.30 kg attached to a thread. The sphere is displaced through a height of 10 mm with the thread taut then released. It takes 15.0 s to make ten complete cycles of oscillation.

a Calculate:

i the time period of the pendulum

ii the length of the pendulum

iii the initial potential energy of the pendulum relative to its equilibrium position.

b Sketch graphs on the same axes to show how the potential energy and the kinetic energy of the pendulum vary with its displacement from equilibrium.

2 A glider of mass 0.45 kg on a frictionless air track is attached to two stretched springs at either end, as shown in Figure 5. A force of 3.0 N is needed to displace the glider from equilibrium and hold it at a displacement of 50 mm. The glider is then released, and it oscillates freely on the air track.

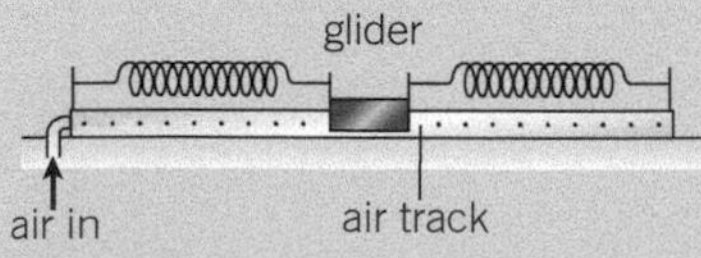

▲ **Figure 5**

Calculate:

a **i** the spring constant k for the system

ii the time period of the oscillations

b **i** the initial potential energy of the system when the glider is held at a displacement of 50 mm

ii the maximum kinetic energy of the glider

iii the speed of the glider at a displacement of 25 mm.

3 **a** State whether the damping in each of the following examples is light, critical, or heavy:

i a child on a swing displaced from equilibrium then released

ii oil in a U-shaped tube displaced from equilibrium then released.

b Discuss how effective a car suspension damper would be if the oil in the damper were replaced by oil that was much more viscous.

4 The amplitude of an oscillating mass on a spring decreases by 4% each cycle from an initial amplitude of 100 mm. Calculate the amplitude after:

a 5 cycles of oscillation

b 20 cycles of oscillation.

11.3 Forced vibrations and resonance

Learning objectives:

- → State the circumstances in which resonance occurs.
- → Distinguish between free vibrations and forced vibrations.
- → Explain why a resonant system reaches a maximum amplitude of vibration.

Specification reference: 3.5.2

Forced vibrations

Imagine pushing someone on a swing at regular intervals. If each push is timed suitably, the swing goes higher and higher. These pushes are a simple example of a **periodic force**, which is a force applied at regular intervals.

- When the system oscillates without a periodic force being applied to it, the system's frequency is called its *natural frequency*.
- When a periodic force is applied to an oscillating system, the response depends on the frequency of the periodic force. The system undergoes **forced vibrations**, or *forced oscillations*, when a periodic force is applied to it.

Figure 1 shows how a periodic force can be applied to an oscillating system consisting of a small object of fixed mass attached to two stretched springs.

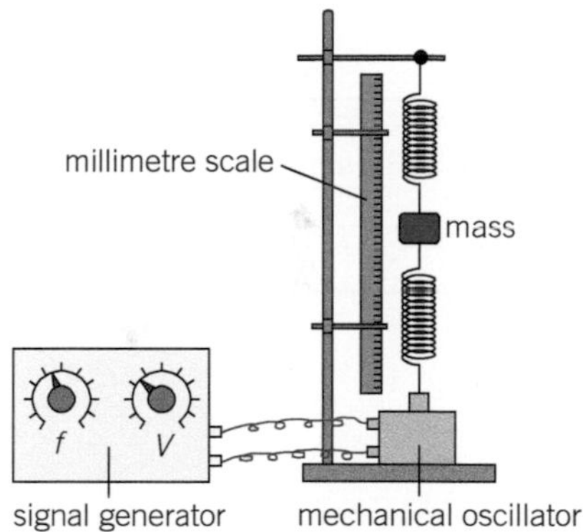

▲ **Figure 1** *Forced vibrations*

The bottom end of the lower spring is attached to a mechanical oscillator, which is connected to a signal generator. The top end of the upper spring is fixed. The mechanical oscillator pulls repeatedly on the lower spring at a frequency that can be changed by adjusting the signal generator. The frequency of the oscillator is the *applied frequency*. The response of the system is measured from the amplitude of oscillations of the system. The variation of the response with the applied frequency is shown in Figure 2.

Consider the effect of increasing the applied frequency from zero:

As the applied frequency increases, the amplitude of oscillations of the system increases until it reaches a maximum amplitude at a particular frequency, and then the amplitude decreases again.

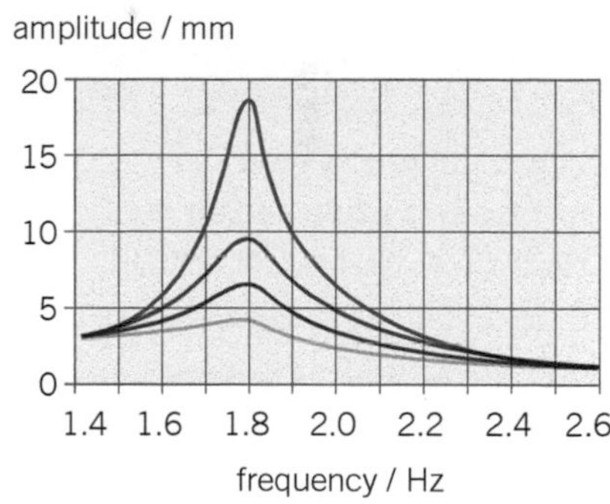

▲ **Figure 2** *Resonance curves*

Resonance

When the system is oscillating at the maximum amplitude, the periodic force is then exactly in phase with the velocity of the oscillating system, and the system is in **resonance**. The frequency at the maximum amplitude is called the **resonant frequency**.

The lighter the damping,

- the larger the maximum amplitude becomes at resonance, and
- the closer the resonant frequency is to the natural frequency of the system.

So, when the damping is lighter, the resonance curve that you see in Figure 2 is sharper.

As the applied frequency becomes increasingly larger than the resonant frequency of the mass–spring system, the amplitude of oscillations decreases more and more,

For an oscillating system with little or no damping, at resonance,

the applied frequency of the periodic force = the natural frequency of the system

More examples of resonance

Barton's pendulums

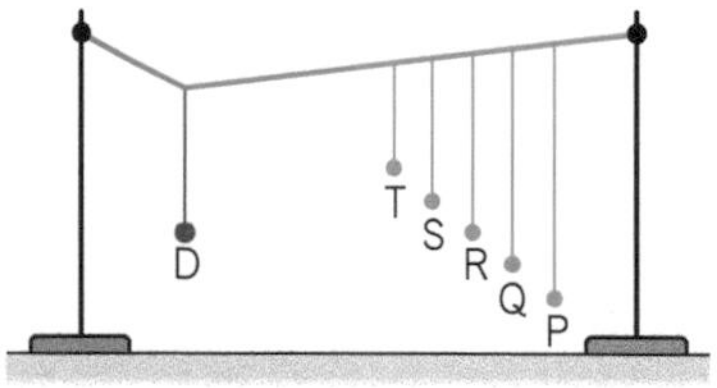

Figure 3 shows five simple pendulums, P, Q, R, S, and T, of different lengths hanging from a supporting thread that is stretched between two fixed points. A single driver pendulum D of the same length as one of the other pendulums is also tied to the thread.

▲ **Figure 3** *Barton's pendulums*

The driver pendulum D is displaced and released so that it oscillates in a plane perpendicular to the plane of the pendulums at rest. The effect of the oscillating motion of D is transmitted along the support thread, subjecting each of the other pendulums to forced oscillations. Pendulum R responds much more than any other pendulum. This is because it has the same length and therefore the same time period as D. So its natural frequency is the same as the natural frequency of D. Therefore, R oscillates in resonance with D because it is subjected to forced oscillations of the same frequency as its own natural frequency of oscillations. The response of each of the other pendulums depends on how close its length is to the length of D and whether it is shorter or longer than D.

Synoptic link

When **stationary waves** are formed on a stretched string, the string may be said to be in resonance. The string resonates and forms stationary wave patterns called harmonics. The stationary wave formed at the lowest frequency is called the first harmonic. Stationary waves are formed at higher frequencies that are multiples of the first-harmonic frequency. See Topic 11.8, Stationary and progressive waves, and Topic 11.9, More about stationary waves on strings.

Bridge oscillations

A bridge span can oscillate because of its springiness and its mass. If a bridge span is not fitted with dampers, it can be made to oscillate at resonance if the bridge span is subjected to a suitable periodic force.

1 A crosswind can cause a periodic force on the bridge span. If the wind speed is such that the periodic force is equal to the natural frequency of the bridge span, resonance can occur in the absence of damping. The dramatic collapse of the Tacoma Narrows Bridge in the United States in 1940 was due to such resonance.
2 A steady trail of people in step with each other walking across a footbridge can cause resonant oscillations of the bridge span if there is not enough damping. Soldiers marching in columns are taught to break out of step with each other when they cross a footbridge, to avoid causing resonance. Soon after it was opened, the Millennium Bridge in London had to be closed and fitted with a more suitable damping system because it swayed in resonance when people first walked across it.

Notes

1. At resonance, the periodic force acts on the system at the same point in each cycle, causing the amplitude to increase to a maximum value limited only by damping. At maximum amplitude, energy supplied by the periodic force is lost at the same rate because of the effects of damping.
2. The applied frequency at resonance (i.e., the resonant frequency) is equal to the natural frequency only when there is little or no damping. Resonance occurs at a slightly lower frequency than the natural frequency if the damping is not light. The lighter the damping, the closer the resonant frequency is to the natural frequency.

▲ **Figure 4** *The collapse of the Tacoma Narrows Bridge*

▲ **Figure 5** *The Millennium Bridge, London*

Summary questions

1 **a** A mass suspended on a vertical spring is made to oscillate by applying a periodic force of natural frequency f_0.

- **i** Define resonance.
- **ii** Explain why the frequency of the periodic force needs to be f_0 to cause resonance.

b With reference to the mass–spring system shown in Figure 1, state and explain what the effect would be on the resonant frequency of

- **i** increasing the mass
- **ii** replacing the springs with stiffer springs.

2 A 0.12 kg mass suspended on a vertical spring is made to resonate by applying a periodic force of frequency 2.4 Hz to it. Calculate:

- **a** the spring constant of the system
- **b** the frequency at which the system would resonate if the mass were doubled.

3 The panel of a washing machine vibrates loudly when the drum rotates at a particular frequency. Explain why this happens only when the drum rotates at this frequency.

4 A vehicle of mass 850 kg has a suspension system that is lightly damped. When it is driven without extra load by a driver of mass 50 kg over speed bumps spaced 15 m apart at a speed of 3.0 m s^{-1}, the vehicle resonates.

- **a** Explain why this effect happens.
- **b** Calculate the speed that resonance would occur at over the same speed bumps if the vehicle had also been carrying an extra load of 130 kg.

11.4 Waves and vibrations

Learning objectives:

- → Explain the differences between transverse and longitudinal waves.
- → Define a plane-polarised wave.
- → Describe a physics test that can distinguish transverse waves from longitudinal waves.

Specification reference: 3.5.4

Synoptic link

You have met the full spectrum of electromagnetic waves in more detail in Topic 7.3, Photons.

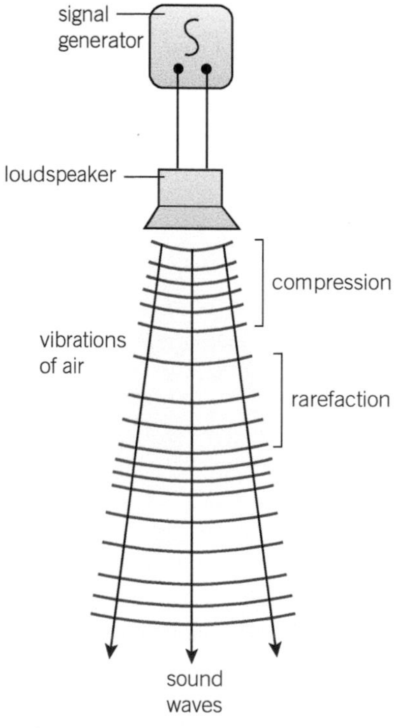

▲ **Figure 1** *Creating sound waves in air*

Study tip

Vibrations of the particles in a longitudinal wave are in the same direction as that along which the wave travels.

Types of waves

Waves that pass through a substance are vibrations which pass through that substance. For example, sound waves in air are created by making a surface vibrate so it sends compression waves through the surrounding air. Sound waves, seismic waves, and waves on strings are examples of waves that pass through a substance. These types of waves are often referred to as **mechanical waves**. When waves progress through a substance, the particles of the substance vibrate in a certain way which makes nearby particles vibrate in the same way, and so on.

Electromagnetic waves are oscillating electric and magnetic fields that progress through space without the need for a substance. The vibrating electric field generates a vibrating magnetic field, which generates a vibrating electric field further away, and so on. Electromagnetic waves include radio waves, microwaves, infrared radiation, light, ultraviolet radiation, X-rays, and gamma radiation.

Longitudinal and transverse waves

Longitudinal waves are waves in which the direction of vibration of the particles is **parallel** to (along) the direction in which the wave travels. Sound waves, primary seismic waves, and compression waves on a slinky toy are all longitudinal waves. Figure 2 shows how to send longitudinal waves along a slinky. When one end of the slinky is moved to and fro repeatedly, each 'forward' movement causes a compression wave to pass along the slinky as the coils push into each other. Each 'reverse' movement causes the coils to move apart so an expansion wave passes along the slinky.

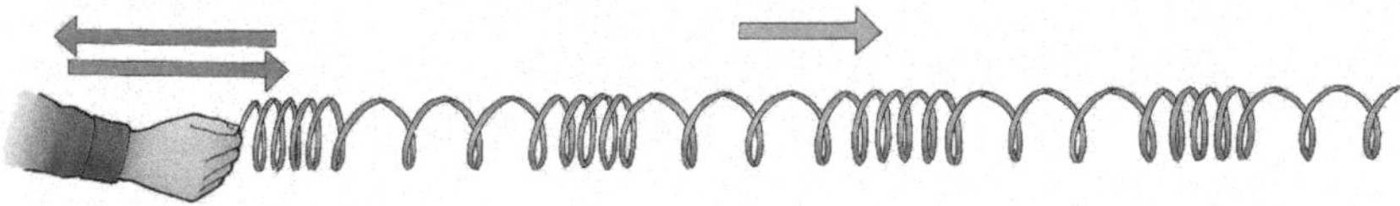

▲ **Figure 2** *Longitudinal waves on a slinky*

Transverse waves are waves in which the direction of vibration is **perpendicular** to the direction in which the wave travels. Electromagnetic waves, secondary seismic waves, and waves on a string or a wire are all transverse waves.

Figure 3 shows transverse waves travelling along a rope. When one end of the rope is moved from side to side repeatedly, these sideways movements travel along the rope, as each unaffected part of the rope is pulled sideways when the part next to it moves sideways.

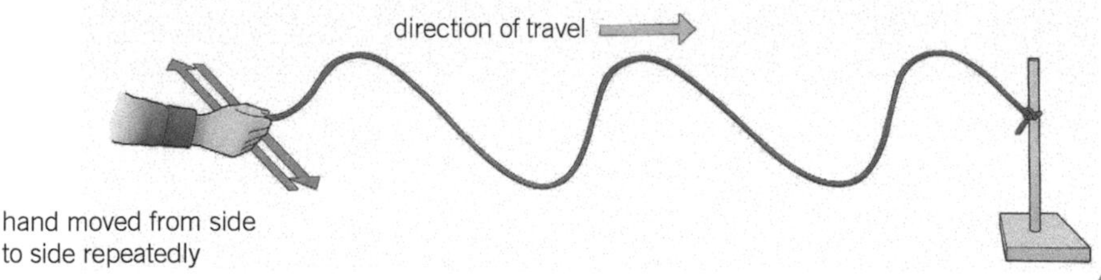

▲ **Figure 3** *Making rope waves*

Polarisation

Transverse waves are **plane-polarised waves** if the vibrations stay in one plane only. If the vibrations change from one plane to another, the waves are **unpolarised**. Longitudinal waves cannot be polarised.

Figure 4 shows unpolarised waves travelling on a rope. When they pass through a slit in a board, as in Figure 4, they become polarised because only the vibrations parallel to the slit pass through it.

Light from a filament lamp or a candle is unpolarised. If unpolarised light is passed through a polaroid filter, the transmitted light is polarised as the filter only allows through light which vibrates in a certain direction, according to the alignment of its molecules.

If unpolarised light is passed through two polaroid filters, the transmitted light intensity changes if one polaroid is turned relative to the other one. The filters are said to be crossed when the transmitted intensity is a minimum. At this position, the polarised light from the first filter cannot pass through the second filter, as the alignment of molecules in the second filter is at 90° to the alignment in the first filter. This is like passing rope waves through two letter boxes at right angles to each other, as shown in Figure 4.

Light is part of the spectrum of electromagnetic waves. The plane of polarisation of an electromagnetic wave is defined as the plane in which the electric field oscillates.

Polaroid sunglasses reduce the glare of light reflected by water or glass. The reflected light is polarised and the intensity is reduced when it passes through the polaroid sunglasses.

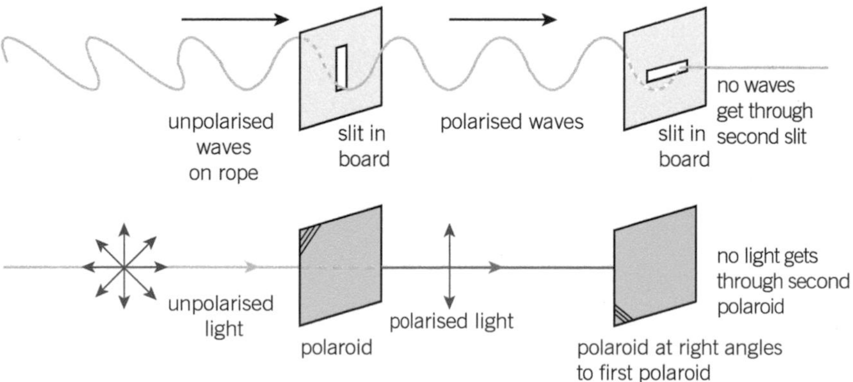

▲ **Figure 4** *Rope model and diagram to explain polarisation*

Application

Good reception

Radio waves from a transmitter are polarised. The aerial of a radio receiver needs to be aligned in the same plane as the radio waves to obtain the best reception.

Summary questions

1 Classify the following types of waves as either longitudinal or transverse: **a** radio waves, **b** microwaves, **c** sound waves, **d** secondary seismic waves.

2 Sketch a snapshot of a longitudinal wave travelling on a slinky coil, indicating the direction in which the wave is travelling and areas of high density (compression) and low density (rarefaction).

3 Sketch a snapshot of a transverse wave travelling along a rope, indicating the direction in which the wave is travelling and the direction of motion of the particles at the points of zero displacement.

4 **a** What is meant by a polarised wave?

b A light source is observed through two pieces of polaroid which are initially aligned parallel to each other. Describe and explain what you would expect to observe as one of the polaroids is rotated through 180°.

11.5 Measuring waves

Learning objectives:

- → Explain what is meant by the amplitude of a wave.
- → Explain what is meant by the wavelength of a wave.
- → Calculate the frequency of a wave from its period.

Specification reference: 3.5.3

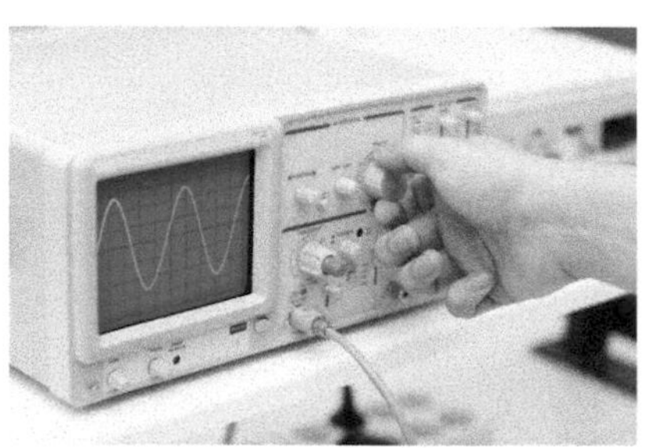

▲ **Figure 1** *Using a cathode ray oscilloscope (CRO) to give a voltage–time graph of a wave*

When we make a phone call, sound waves are converted to electrical waves. In an intercontinental phone call, these waves are carried by electromagnetic waves from ground transmitters to satellites in space and back to ground receivers, where they are converted back to electrical waves then back to sound waves. The electronic circuits ensure that these sound waves are very similar to the original sound waves. The engineers who design and maintain communications systems need to measure the different types of waves at different stages, to make sure the waves are not distorted.

Key terms

The following terms, some of which are illustrated in Figure 2, are used to describe waves.

- The **displacement** of a vibrating particle is its distance and direction from its equilibrium position.
- The **amplitude** of a wave is the maximum displacement of a vibrating particle. For a transverse wave, this is the height of a wave crest or the depth of a wave trough from its equilibrium position.
- The **wavelength** of a wave is the least distance between two adjacent vibrating particles with the same displacement and velocity at the same time (e.g., distance between adjacent crests).
- One complete **cycle** of a wave is from maximum displacement to the next maximum displacement (e.g., from one wave peak to the next).
- The **period** of a wave is the time for one complete wave to pass a fixed point.
- The **frequency** of a wave is the number of cycles of vibration of a particle per second, or the number of complete waves passing a point per second. The unit of frequency is the hertz (Hz).

For waves of frequency f, the period of the wave = $\frac{1}{f}$

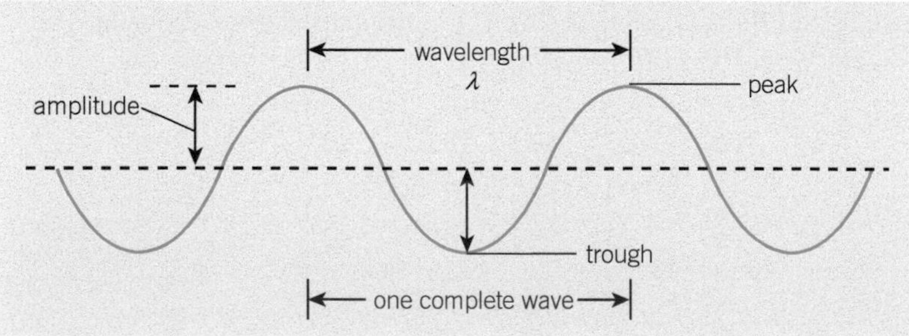

▲ **Figure 2** *Parts of a wave*

Study tip

Amplitude is measured from the equilibrium position to maximum positive or maximum negative, not from maximum positive to maximum negative.

Wave speed

The higher the frequency of a wave, the shorter its wavelength. For example, if waves are sent along a rope, the higher the frequency at which they are produced, the closer together the wave peaks are. The same effect can be seen in a ripple tank when straight waves are produced at a constant frequency. If the frequency is raised to a higher value, the waves are closer together.

Figure 3 represents the crests of straight waves travelling at a constant speed in a ripple tank.

- Each wave crest travels a distance equal to one wavelength (λ) in the time taken for one cycle.
- The time taken for one cycle $= \frac{1}{f}$, where f is the frequency of the waves.

Therefore, the speed of the waves, $c = \frac{\text{distance travelled in one cycle}}{\text{time taken for one cycle}}$

$$= \frac{\lambda}{\frac{1}{f}} = f\lambda$$

For waves of frequency f and wavelength λ

$$\textbf{wave speed } c = f\lambda$$

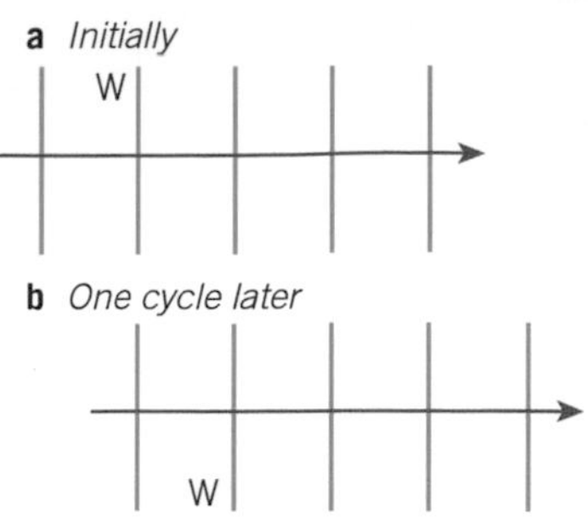

▲ **Figure 3** *Wave speed*

Phase difference

The **phase** of a vibrating particle at a certain time is the fraction of a cycle it has completed since the start of the cycle. The **phase difference** between two particles vibrating at the same frequency is the fraction of a cycle between the vibrations of the two particles, measured either in degrees or radians, where 1 cycle = 360° = 2π **radians**. For two points at distance d apart along a wave of wavelength λ

$$\textbf{the phase difference in radians} = \frac{2\pi d}{\lambda}$$

Study tip

Phase difference is measured as an angle (in radians or degrees), not in terms of wavelength.

Figure 4 shows three successive snapshots of the particles of a transverse wave progressing from left to right across the diagram. Particles O, P, Q, R, and S are spaced approximately $\frac{1}{4}$ of a wavelength apart. Table 1 shows the phase difference between O and each of the other particles.

▼ **Table 1** *Phase differences*

	P	Q	R	S
Distance from O in terms of wavelength, λ	$\frac{\lambda}{4}$	$\frac{\lambda}{2}$	$3\frac{\lambda}{4}$	λ
Phase difference relative to O / radians	$\frac{\pi}{2}$	π	$3\frac{\pi}{2}$	2π

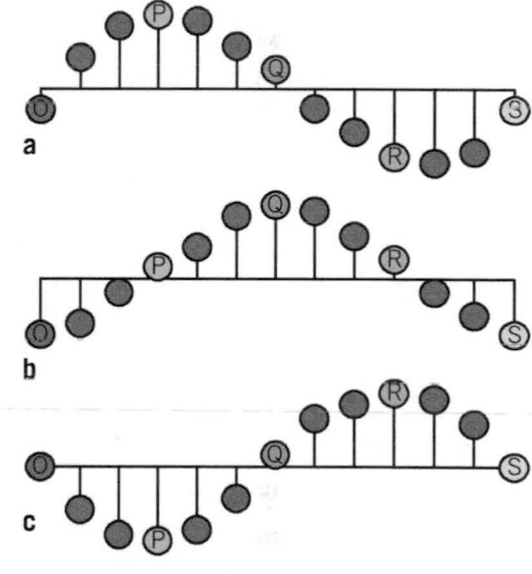

▲ **Figure 4** *Progressive waves*

Summary questions

1. Sound waves in air travel at a speed of 340 m s^{-1} at 20 °C. Calculate the wavelength of sound waves in air which have a frequency of **a** 3400 Hz, **b** 18 000 Hz.
2. Electromagnetic waves in air travel at a speed of 3.0×10^8 m s^{-1}. Calculate the frequency of light waves of wavelength **a** 0.030 m, **b** 600 nm.
3. Figure 5 shows a waveform on an oscilloscope screen when the y-sensitivity of the oscilloscope was 0.50 V cm^{-1} and the time base was set at 0.5 ms cm^{-1}. Determine the amplitude and the frequency of this waveform.
4. **a** For the waves in Figure 4, measure
   - **i** the amplitude and the wavelength
   - **ii** the phase difference between P and R
   - **iii** the phase difference between P and S.

   b What would be the displacement and direction of motion of particle Q three-quarters of a period after the last snapshot?

▲ **Figure 5**

11.6 Wave properties 1

Learning objectives:

- → Explain what causes waves to refract when they pass across a boundary.
- → Demonstrate the direction light waves bend when they travel out of glass and into air.
- → Explain what we mean by diffraction.

Specification reference: 3.5.8

Wave properties such as reflection, refraction, and diffraction occur with many different types of waves. A **ripple tank** may be used to study these wave properties. The tank is a shallow transparent tray of water with sloping sides. The slopes prevent waves reflecting off the sides of tank. If they did reflect, it would be difficult to see the waves.

- The waves observed in a ripple tank are referred to as **wavefronts**, which are lines of constant phase (e.g., crests).
- The direction in which a wave travels is at right angles to the wavefront.

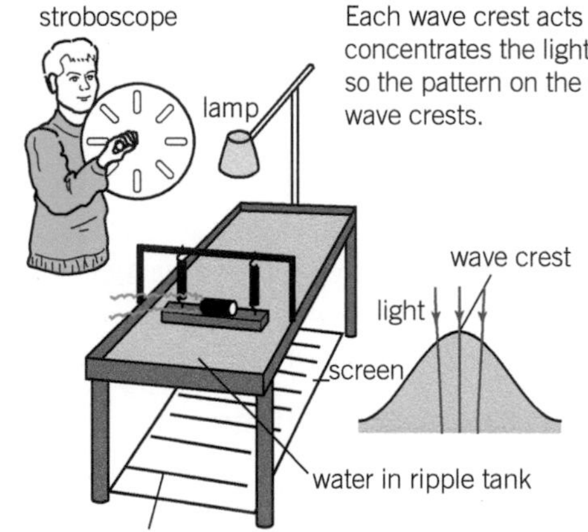

▲ **Figure 1** *The ripple tank*

▲ **Figure 2** *Reflection of plane waves*

Reflection

Straight waves directed at a certain angle to a hard flat surface (the reflector) reflect off at the same angle, as shown in Figure 2. The angle between the reflected wavefront and the surface is the same as the angle between the incident wavefront and the surface. Therefore the direction of the reflected wave is at the same angle to the reflector as the direction of the incident wave. This effect is observed when a light ray is directed at a plane mirror. The angle between the incident ray and the mirror is equal to the angle between the reflected ray and the mirror.

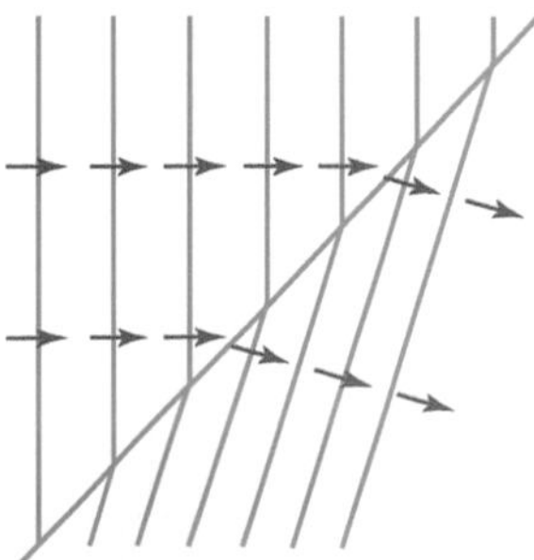

▲ **Figure 3** *Refraction*

Refraction

When waves pass across a boundary at which the wave speed changes, the wavelength also changes. If the wavefronts approach at an angle to the boundary, they change direction as well as changing speed. This effect is known as **refraction**.

Figure 3 shows the refraction of water waves in a ripple tank when they pass across a boundary from deep to shallow water at an angle to the boundary. Because they move more slowly in the shallow water, the wavelength is smaller in the shallow water and therefore they change direction.

Refraction of light is observed when a light ray is directed into a glass block at an angle (i.e., not along the normal). The light ray changes direction when it crosses the glass boundary. This happens because light waves travel more slowly in glass than in air.

Diffraction

Diffraction occurs when waves spread out after passing through a gap or round an obstacle. The effect can be seen in a ripple tank when straight waves are directed at a gap, as shown in Figure 4. Note that the wavelength doesn't change when the waves are diffracted.

- The narrower the gap, the more the waves spread out.
- The longer the wavelength, the more the waves spread out.

To explain why the waves are diffracted on passing through the gap, consider each point on a wavefront as a secondary emitter of wavelets. The wavelets from the points along a wavefront travel only in the direction in which the wave is travelling, not in the reverse direction, and they combine to form a new wavefront spreading beyond the gap.

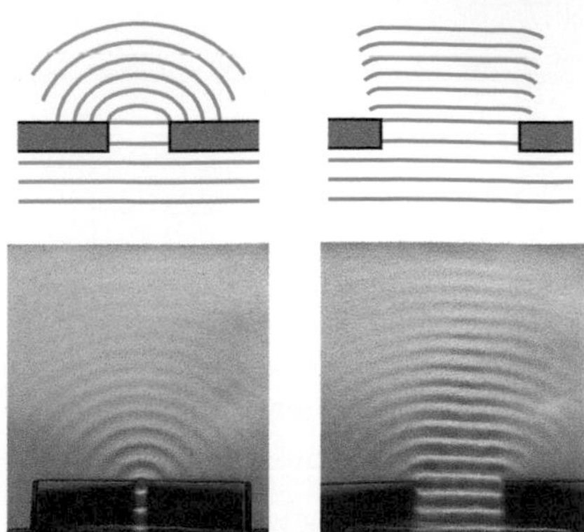

▲ **Figure 4** *The effect of the gap width*

Dish design

Satellite TV dishes in Europe need to point south, because the satellites orbit the Earth directly above the equator. The bigger the dish, the stronger the signal it can receive, because more radio waves are reflected by the dish onto the aerial. But a bigger dish reflects the radio waves to a smaller focus, because it diffracts the waves less. The dish therefore needs to be aligned more carefully than a smaller dish, otherwise it will not focus the radio waves onto the aerial.

Synoptic link

You will meet refraction and reflection of light in more detail in Topic 12.5, Refraction of light, and Topic 12.6, More about refraction.

Summary questions

1. Copy and complete the diagram in Figure 6 by showing the wavefront after it has reflected from the straight reflector. Also show the direction of the reflected wavefront.

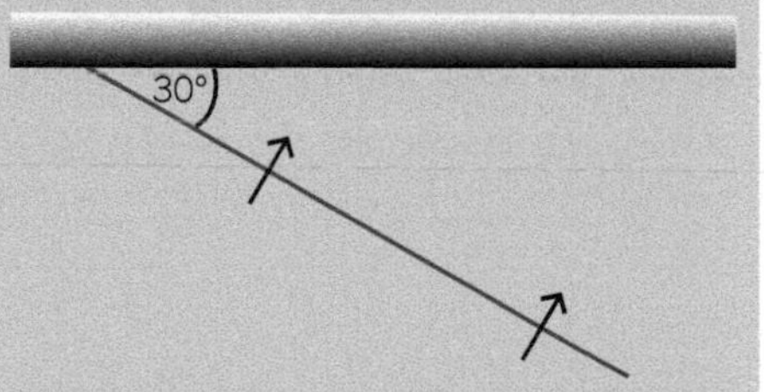

reflector

30°

▲ **Figure 6**

2. A circular wave spreads out from a point P on a water surface, which is 0.50 m from a flat reflecting wall. The wave travels at a speed of 0.20 m s^{-1}. Sketch the arrangement and show the position of the wavefront **a** 2.5 s, **b** 4.0 s after the wavefront was produced at P.

3. Copy and complete the diagram in Figure 7 by showing the wavefronts after they pass across the boundary and have been refracted. Also show the direction of the refracted waves.

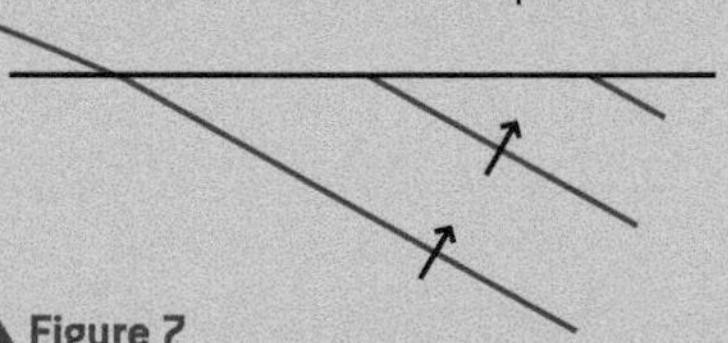

boundary

▲ **Figure 7**

4. Water waves are diffracted on passing through a gap. How is the amount of diffraction changed as a result of:
   - **a** widening the gap without changing the wavelength?
   - **b** increasing the wavelength of the water waves without changing the gap width?
   - **c** increasing the wavelength of the water waves and reducing the gap width?
   - **d** widening the gap and increasing the wavelength of the waves?

Study tip

Remember from your GCSE that the normal is an imaginary line perpendicular to a boundary between two materials or a surface.

Synoptic link

You will meet diffraction in more detail in Topic 12.3, Diffraction.

▲ **Figure 5** *A satellite TV dish*

11.7 Wave properties 2

Learning objectives:

- → Explain how two waves combine to produce reinforcement.
- → Describe the phase difference between two waves if they cancel each other out.
- → Explain why total cancellation is rarely achieved in practice.

Specification reference: 3.5.5 and 3.5.7

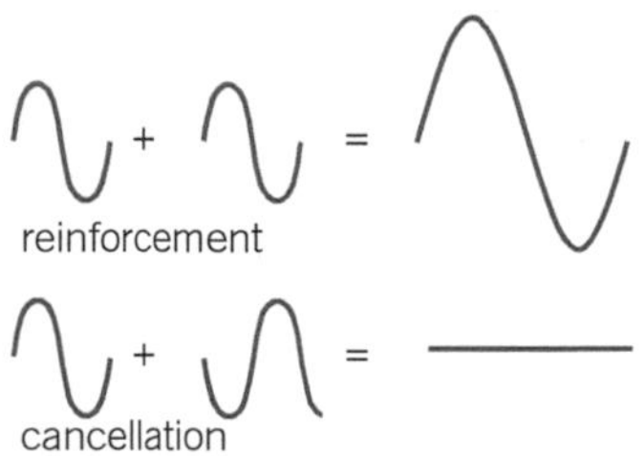

▲ **Figure 1** *Superposition*

▲ **Figure 2** *Making stationary waves*

Study tip

Be careful when using the term 'minimum'. When two troughs meet at the same point, they superpose to produce maximum negative displacement, not a minimum.

Synoptic link

You will meet the interference of light in more detail in Topic 12.1, Double slit interference.

The principle of superposition

When waves meet, they pass through each other. At the point where they meet, they combine for an instant before they move apart. This combining effect is known as **superposition**. Imagine a boat hit by two wave crests at the same time from different directions. Anyone on the boat would know it had been hit by a supercrest, the combined effect of two wave crests.

The principle of superposition states that when two waves meet, the total displacement at a point is equal to the sum of the individual displacements at that point.

- Where a crest meets a crest, a **supercrest** is created – the two waves reinforce each other.
- Where a trough meets a trough, a **supertrough** is created – the two waves reinforce each other.
- Where a crest meets a trough of the same amplitude, the resultant displacement is **zero**, and the two waves cancel each other out. If they are not the same amplitude, the resultant is called a minimum.

Further examples of superposition

1 Stationary waves on a rope

Stationary waves are formed on a rope if two people send waves continuously along a rope from either end, as shown in Figure 2. The two sets of waves are referred to as **progressive waves** to distinguish them from stationary waves. They combine at fixed points along the rope to form points of no displacement or **nodes** along the rope. At each node, the two sets of waves are always 180° out of phase, so they cancel each other out. Stationary waves are described in more detail in Topics 11.8 and 11.9.

2 Water waves in a ripple tank

A vibrating dipper on a water surface sends out circular waves. Figure 3 shows a snapshot of two sets of circular waves produced in this way in a ripple tank. The waves pass through each other continuously.

- Points of cancellation are created where a crest from one dipper meets a trough from the other dipper. These points of cancellation are seen as gaps in the wavefronts.
- Points of reinforcement are created where a crest from one dipper meets a crest from the other dipper, or where a trough from one dipper meets a trough from the other dipper.

As the waves are continuously passing through each other at constant frequency and at a constant phase difference, cancellation and reinforcement occur at fixed positions. This effect is known as **interference**. **Coherent** sources of waves produce an interference pattern where they overlap, because they vibrate at the same frequency with a constant phase difference. If the phase difference changed at random, the points of cancellation and reinforcement would move about at random, and no interference pattern would be seen.

Tests using microwaves

A microwave transmitter and receiver can be used to demonstrate reflection, refraction, diffraction, interference, and polarisation of microwaves. The transmitter produces microwaves of wavelength 3.0 cm. The receiver can be connected to a suitable meter, which gives a measure of the intensity of the microwaves at the receiver.

▲ **Figure 3** *Interference of water waves*

1 Place the receiver in the path of the microwave beam from the transmitter. Move the receiver gradually away from the transmitter and note that the receiver signal decreases with distance from the transmitter. This shows that the microwaves become weaker as they travel away from the transmitter.
2 Place a metal plate between the transmitter and the receiver to show that microwaves cannot pass through metal.
3 Use two metal plates to make a narrow slit and show that the receiver detects microwaves that have been diffracted as they pass through the slit. Show that if the slit is made wider, less diffraction occurs.
4 Use a narrow metal plate with the two plates from step 3 above to make a pair of slits, as in Figure 4. Direct the transmitter at the slits and use the receiver to find points of cancellation and reinforcement, where the microwaves from the two slits overlap.

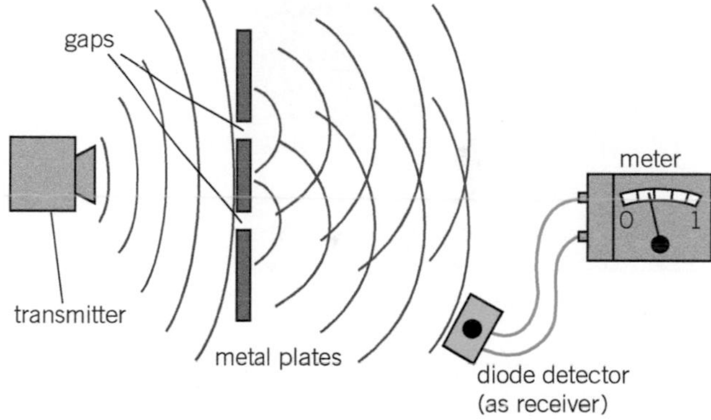

▲ **Figure 4** *Interference of microwaves*

Summary questions

1 Figure 5 shows two wave pulses on a rope travelling towards each other. Sketch a snapshot of the rope:

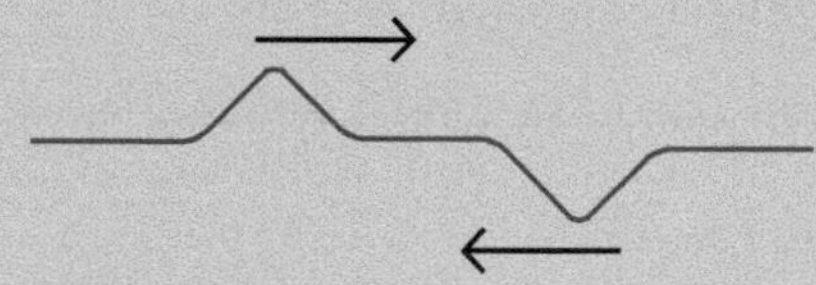

▲ **Figure 5**

  a when the two waves are passing through each other

  b when the two waves have passed through each other.

2 How would you expect the interference pattern in Figure 3 to change if:

  a the two dippers are moved further apart?

  b the frequency of the waves produced by the dippers is reduced?

3 Microwaves from a transmitter are directed at a narrow slit between two metal plates. A receiver is placed in the path of the diffracted microwaves, as shown in Figure 6.

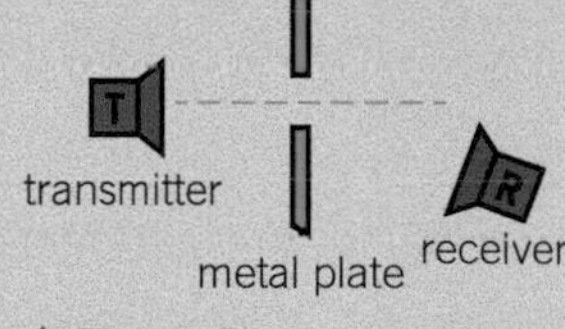

▲ **Figure 6**

How would you expect the receiver signal to change if **a** the receiver is moved towards the top of the figure in a line parallel to the metal plates, **b** the slit is then made narrower?

4 Microwaves from a transmitter are directed at two parallel slits in a metal plate (Figure 4). A receiver is placed on the other side of the metal plate on a line parallel to the plate. When the receiver is moved a short distance along the line, the receiver signal decreases then increases again.

  a Explain why the signal decreased when it was first moved.

  b Explain why the signal increased as it continued to move.

11.8 Stationary and progressive waves

Learning objectives:

- → Describe the necessary condition for the formation of a stationary wave.
- → Deduce whether a stationary wave is formed by superposition.
- → Explain why nodes are formed in fixed positions.

Specification reference: 3.5.5

▲ **Figure 1** *Vibrations of a guitar string*

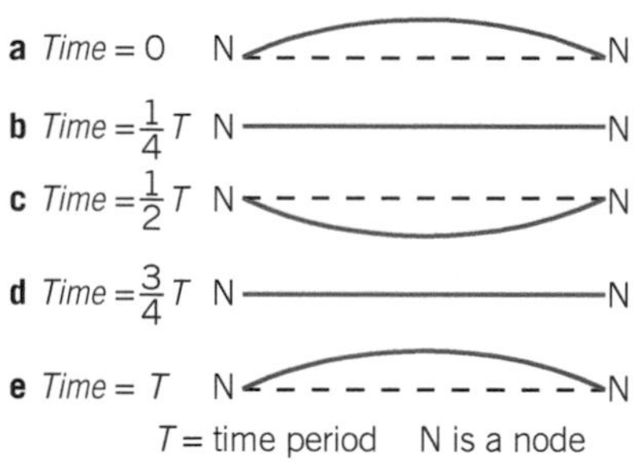

▲ **Figure 2** *First harmonic vibrations*

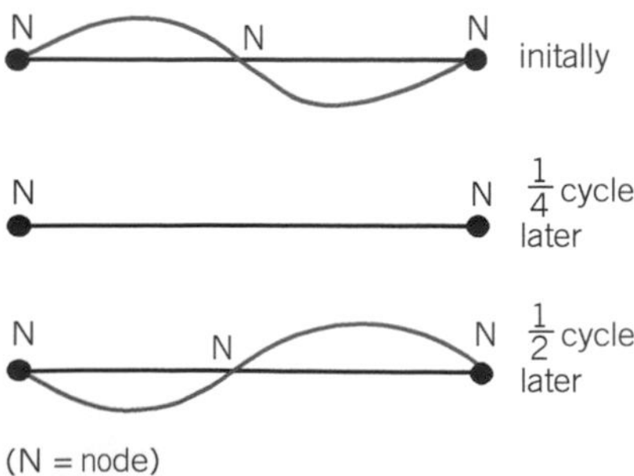

▲ **Figure 3** *A stationary wave of two loops*

Formation of stationary waves

When a guitar string is plucked, the sound produced depends on the way in which the string vibrates. If the string is plucked gently at its centre, a stationary wave of constant frequency is set up on the string. The sound produced therefore has a constant frequency. If the guitar string is plucked harshly, the string vibrates in a more complicated way, and the note produced contains other frequencies as well as the frequency produced when it is plucked gently.

As explained in Topic 11.7, a stationary wave is formed when two progressive waves pass through each other. This can be achieved on a string in tension by fixing both ends and making the middle part vibrate, so progressive waves travel towards each end, reflect at the ends, and then pass through each other.

The simplest stationary wave pattern on a string is shown in Figure 2. This is called the first harmonic of the string (sometimes referred to as its **fundamental mode of vibration**). It consists of a single loop that has a **node** (a point of no displacement) at either end. The string vibrates with maximum amplitude midway between the nodes. This position is referred to as an **antinode**. In effect, the string vibrates from side to side repeatedly. For this pattern to occur, the distance between the nodes at either end (i.e., the length of the string) must be equal to one half-wavelength of the waves on the string.

$$\textbf{Distance between adjacent nodes} = \frac{1}{2}\lambda$$

If the frequency of the waves sent along the rope from either end is raised steadily, the pattern in Figure 2 disappears and a new pattern is observed with two equal loops along the rope. This pattern, shown in Figure 3, has a node at the centre as well as at either end. It is formed when the frequency is twice as high as in Figure 2, corresponding to half the previous wavelength. Because the distance from one node to the next is equal to half a wavelength, the length of the rope is therefore equal to one full wavelength.

Stationary waves that vibrate freely do not transfer energy to their surroundings. The amplitude of vibration is zero at the nodes so there is no energy at the nodes. The amplitude of vibration is a maximum at the antinodes, so there is maximum energy at the antinodes. Because the nodes and antinodes are at fixed positions, no energy is transferred in a freely vibrating stationary wave pattern.

Explanation of stationary waves

Consider a snapshot of two progressive waves passing through each other.

- When they are in phase, they reinforce each other to produce a large wave, as shown in Figure 4a on the next page.
- A quarter of a cycle later, the two waves have each moved one-quarter of a wavelength in opposite directions. As shown in Figure 4b, they are now in antiphase so they cancel each other.
- After a further quarter cycle, the two waves are back in phase. The resultant is again a large wave as in Figure 4a, except reversed.

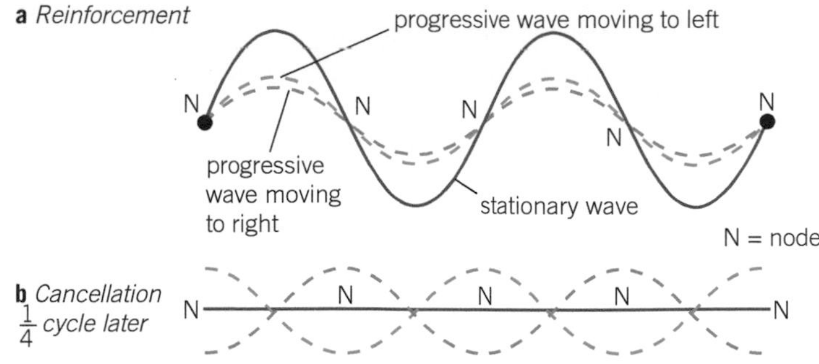

▲ **Figure 4** *Explaining stationary waves*

The points where there is no displacement (i.e., the nodes) are fixed in position throughout. Between these points, the stationary wave oscillates between the nodes. In general, in any stationary wave pattern:

1. The amplitude of a vibrating particle in a stationary wave pattern varies with position from zero at a node to maximum amplitude at an antinode.
2. The phase difference between two vibrating particles is
   - zero if the two particles are between adjacent nodes or separated by an even number of nodes
   - 180° (= π radians) if the two particles are separated by an odd number of nodes.

	Stationary waves	Progressive waves
Frequency	all particles except those at the nodes vibrate at the same frequency	all particles vibrate at the same frequency
Amplitude	the amplitude varies from zero at the nodes to a maximum at the antinodes	the amplitude is the same for all particles
Phase difference between two particles	equal to $m\pi$, where m is the number of nodes between the two particles	equal to $\frac{2\pi d}{\lambda}$, where d = distance apart of the two particles and λ is the wavelength

More examples of stationary waves

1 Sound in a pipe

Sound resonates at certain frequencies in an air-filled tube or pipe. In a pipe closed at one end, these resonant frequencies occur when there is an antinode at the open end and a node at the other end.

2 Using microwaves

Microwaves from a transmitter are directed normally at a metal plate, which reflects the microwaves back towards the transmitter. When a detector is moved along the line between the transmitter and the metal plate, the detector signal is found to be zero (or at a minimum) at equally spaced positions along the line. The reflected waves and the waves from the transmitter form a stationary wave pattern. The positions where no signal (or a minimum) is detected are where nodes occur. They are spaced at intervals of half a wavelength.

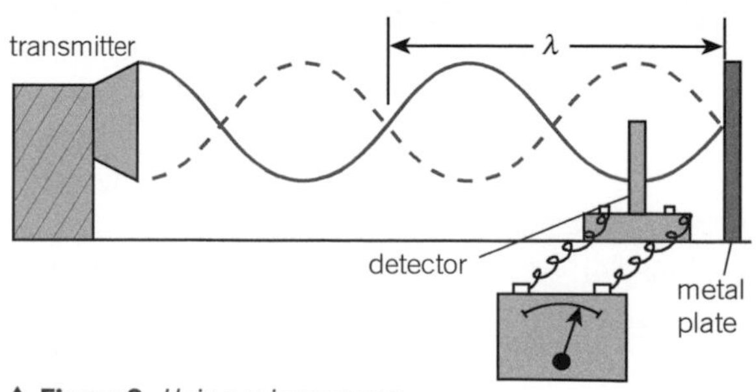

▲ **Figure 6** *Using microwaves*

Summary questions

1. a Sketch the stationary wave pattern seen on a rope when there is a node at either end and an antinode at its centre.
   b If the rope in part **a** is 4.0 m in length, calculate the wavelength of the waves on the rope.
2. The stationary wave pattern shown in Figure 5 is set up on a rope of length 3.0 m.

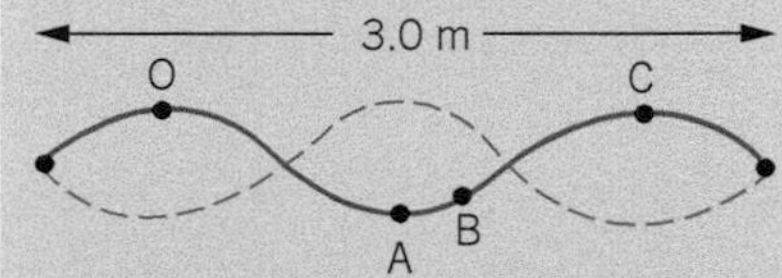

▲ **Figure 5**

   a Calculate the wavelength of these waves.
   b State the phase difference between the particle vibrating at O and the particle vibrating at
   i A ii B iii C.
3. State two differences between a stationary wave and a progressive wave in terms of the vibrations of the particles.
4. The detector in Figure 6 is moved along the line between the transmitter and the metal plate. The detector signal is zero at positions 15 mm apart.
   a Explain why the signal is at a minimum at certain positions.
   b Calculate the wavelength of the microwaves.

11.9 More about stationary waves on strings

Learning objectives:

- Explain what condition must be satisfied at both ends of the string.
- Describe the simplest possible stationary wave pattern that can be formed.
- Compare the frequencies of higher harmonics with the first harmonic frequency.

Specification reference: 3.5.5

Stationary waves on a vibrating string

A controlled arrangement for producing stationary waves is shown in Figure 1. A string or wire is tied at one end to a mechanical oscillator connected to a frequency generator. The other end of the string passes over a pulley and supports a weight, which keeps the tension in the string constant. As the frequency of the generator is increased from a very low value, different stationary wave patterns are seen on the string. In every case, the length of string between the pulley and the oscillator has a node at either end.

- The **first harmonic pattern of vibration** is seen at the lowest possible frequency that gives a pattern. This has an antinode at the middle as well as a node at either end. Because the length L of the vibrating section of the string is between adjacent nodes and the distance between adjacent nodes is $\frac{1}{2}\lambda_1$,
  the wavelength of the waves that form this pattern, the first harmonic wavelength, $\boldsymbol{\lambda_1 = 2L}$.
  Therefore, the first harmonic frequency $f_1 = \frac{c}{\lambda_1} = \frac{c}{2L}$, where c is the speed of the progressive waves on the wire.
- The next stationary wave pattern, the **second harmonic**, is where there is a node at the middle, so the string is in two loops. The wavelength of the waves that form this pattern $\lambda_2 = L$ because each loop has a length of half a wavelength.
  Therefore, the frequency of the second harmonic vibrations
  $$f_2 = \frac{c}{\lambda_2} = \frac{c}{L} = 2f_1.$$
- The next stationary wave pattern, the **third harmonic**, is where there are nodes at a distance of $\frac{1}{3}L$ from either end and an antinode at the middle. The wavelength of the waves that form this pattern $\lambda_3 = \frac{2}{3}L$ because each loop has a length of half a wavelength.
  Therefore, the frequency of the third harmonic vibrations
  $$f_3 = \frac{c}{\lambda_3} = \frac{3c}{2L} = 3f_1.$$

In general, stationary wave patterns occur at frequencies f_1, $2f_1$, $3f_1$, $4f_1$, and so on, where f is the first harmonic frequency of the fundamental vibrations. This is the case in any vibrating linear system that has a node at either end.

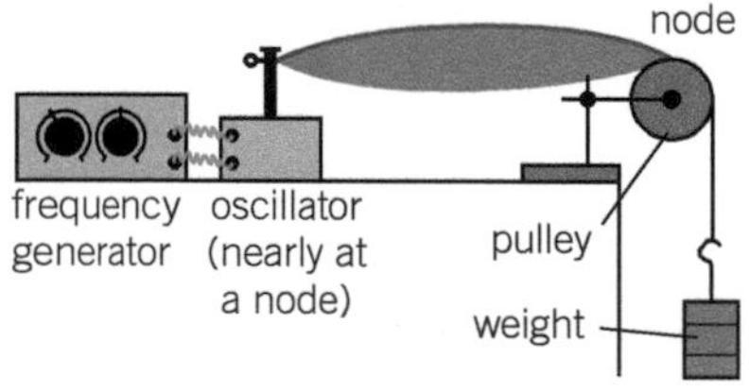

N = node A = antinode
(dotted line shows string half a cycle earlier)

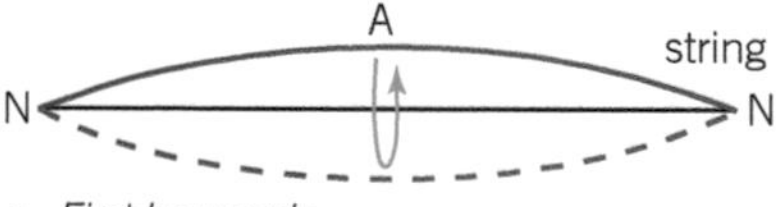

a *First harmonic*

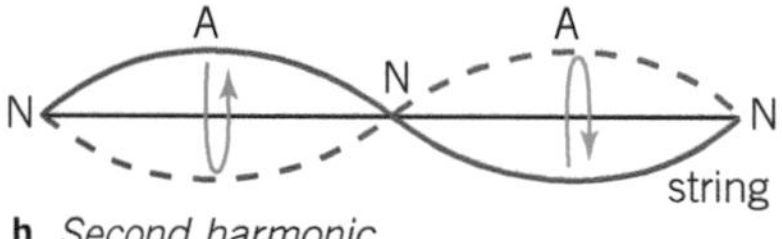

b *Second harmonic*

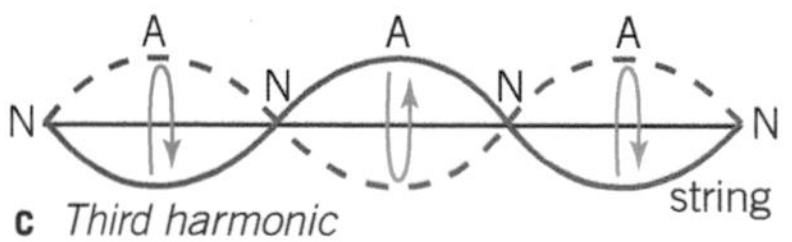

c *Third harmonic*

▲ **Figure 1** *Stationary waves on a string*

Explanation of the stationary wave patterns on a vibrating string

In the arrangement shown in Figure 1, consider what happens to a progressive wave sent out by the oscillator. The crest reverses its phase when it reflects at the fixed end and travels back along the string as a trough. When it reaches the oscillator, it reflects and reverses phase again, travelling away from the oscillator once more as a crest. If this crest is reinforced by a crest created by the oscillator, the amplitude of the wave is increased. This is how a stationary wave is formed. The key condition is that the time taken for a wave to travel along the

Study tip

Number of loops = number of nodes – 1.

string and back should be equal to the time taken for a whole number of cycles of the oscillator.

- The time taken for a wave to travel along the string and back is $t = \frac{2L}{c}$, where c is the speed of the waves on the string.
- The time taken for the oscillator to pass through a whole number of cycles $= \frac{m}{f}$, where f is the frequency of the oscillator and m is a whole number.

Therefore the key condition may be expressed as $\frac{2L}{c} = \frac{m}{f}$.

Rearranging this equation gives $f = \frac{mc}{2L} = mf_1$ and $\lambda = \frac{c}{f_1} = \frac{2L}{m}$.

In other words,

- stationary waves are formed at frequencies f_1, $2f_1$, $3f_1$, etc.
- the length of the vibrating section of the string $L = \frac{m\lambda}{2}$ = a whole number of half wavelengths.

Application

Making music

A guitar produces sound when its strings vibrate as a result of being plucked. When a stretched string or wire vibrates, its pattern of vibration is a mix of its first and higher harmonics. The sound produced is the same mix of frequencies which change with time as the pattern of vibration changes. A **spectrum analyser** can be used to show how the intensity of a sound varies with frequency and with time. Combined with a **sound synthesiser**, the original sound can be altered by amplifying or suppressing different frequency ranges.

The **pitch** of a note corresponds to frequency. This means that the pitch of the note from a stretched string can be altered by changing the tension of the string or by altering its length.

- Raising the tension or shortening the length increases the pitch.
- Lowering the tension or increasing the length lowers the pitch.

By changing the length or altering the tension, a vibrating string or wire can be tuned to the same pitch as a tuning fork. However, the sound from a vibrating string includes all the harmonic frequencies, whereas a tuning fork vibrates only at a single frequency. The wire is tuned when its first harmonic frequency is the same as the tuning fork frequency. It can be shown that the first harmonic frequency f_1 depends on the tension T in the wire, its length L, and its mass per unit length μ according to the equation $f_1 = \frac{1}{2L}\sqrt{\frac{T}{\mu}}$

Note:

A simple visual check when using a tuning fork to tune a wire is to balance a small piece of paper on the wire at its centre. Placing the base of the vibrating tuning fork on one end of the wire will cause the paper to fall off if the wire is tuned correctly.

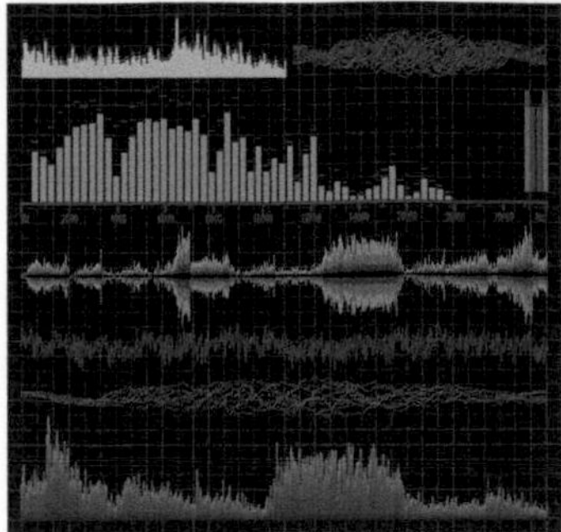

▲ **Figure 2** *A spectrum analyser*

Summary questions

1. A stretched wire of length 0.80 m vibrates at its first harmonic with a frequency of 256 Hz. Calculate **a** the wavelength of the progressive waves on the wire, **b** the speed of the progressive waves on the wire.
2. The first harmonic frequency of vibration of a stretched wire is inversely proportional to the length of the wire. For the wire in **Q1** at the same tension, calculate the length of the wire to produce a frequency of **a** 512 Hz, **b** 384 Hz.
3. The tension in the wire in **Q1** is 40 N. Calculate **a** the mass per unit length of the wire, **b** the diameter of the wire if its density (mass per unit volume) is 7800 kg m^{-3}.
4. The speed, c, of the progressive waves on a stretched wire varies with the tension T in the wire, in accordance with the equation $c = (\frac{T}{\mu})^{\frac{1}{2}}$, where μ is the mass per unit length of the wire. Use this formula to explain why a nylon wire and a steel wire of the same length, diameter, and tension produce notes of different pitch. State, with a reason, which wire would produce the higher pitch.

11.10 Using an oscilloscope

Learning objectives:

- → Describe how an oscilloscope can be used.
- → Interpret waveforms on an oscilloscope to give peak voltage and wavelength.

Specification reference: 3.10.5

An oscilloscope consists of a specially made electron tube and associated control circuits. An electron gun at one end of the glass tube emits electrons in a beam towards a fluorescent screen at the other end of the tube, as shown in Figure 1. Light is emitted from the spot on the screen where the beam hits the screen.

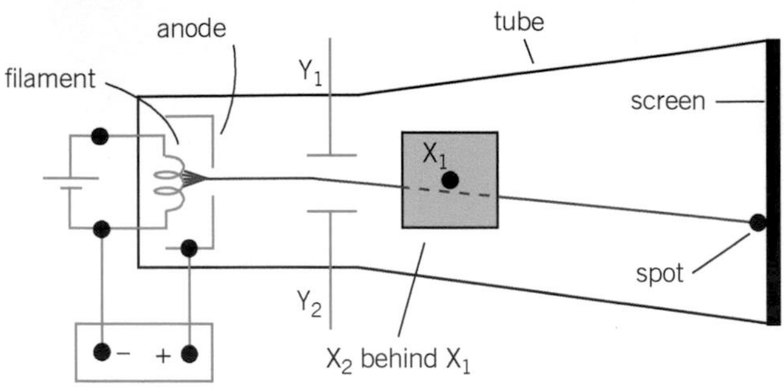

▲ **Figure 1** *An oscilloscope tube*

How to use an oscilloscope

The position of the spot of light on the screen is affected by the p.d. across either pair of deflecting plates. With no p.d. across either set of deflecting plates, the spot on the screen stays in the same position. If a p.d. is applied across the X-plates, the spot deflects horizontally. A p.d. across the Y-plates makes it deflect vertically. In both cases, **the displacement of the spot is proportional to the applied p.d.**

To display a waveform:

- The X-plates are connected to the oscilloscope's **time base circuit** which makes the spot move at constant speed left to right across the screen, then back again much faster. Because the spot moves at constant speed across the screen, the x-scale can be calibrated, usually in milliseconds or microseconds per centimetre.
- The p.d. to be displayed is connected to the Y-plates via the Y-input so the spot moves up and down as it moves left to right across the screen. As it does so, it traces out the waveform on the screen. Because the vertical displacement of the spot is proportional to the p.d. applied to the Y-plates, the Y-input is calibrated in volts per centimetre (or per division if the grid on the oscilloscope screen is not a centimetre grid). The calibration value is usually referred to as the Y-sensitivity or Y-gain of the oscilloscope.

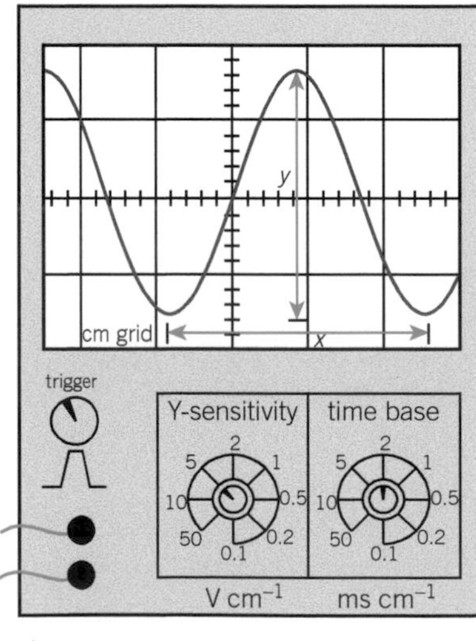

▲ **Figure 2** *Using an oscilloscope*

Figure 2 shows the trace produced when an alternating p.d. is applied to the Y-input. The screen is marked with a centimetre grid.

1. To measure the peak p.d., observe that the waveform height from the bottom to the top of the wave is 3.2 cm. The amplitude (i.e., peak height) of the wave is therefore 1.6 cm. As the Y-gain is set at 5.0 V cm^{-1}, the peak p.d. is therefore 8.0 V (= 5.0 V cm^{-1} × 1.6 cm).
2. To measure the frequency of the alternating p.d., we need to measure the time period T (the time for one full cycle) of the waveform. Then we can calculate the frequency f as $f = \frac{1}{T}$.

3 We can see from the waveform that one full cycle corresponds to a distance of 3.3 cm across the screen horizontally. As the time base control is set at 2 ms cm^{-1}, the time period T is therefore 6.6 ms (= 2 ms cm^{-1} × 3.3 cm). Therefore, the frequency of the alternating p.d. is 152 Hz.

Measuring the speed of ultrasound

The time base circuit of an oscilloscope can be used to trigger an ultrasound transmitter so it sends out a short pulse of ultrasound waves. An ultrasound receiver can be used to detect the transmitted pulse. If the receiver signal is applied to the Y-input of the oscilloscope, the waveform of the received pulse can be seen on the oscilloscope screen, as shown in Figure 3.

Because the pulse takes time to travel from the transmitter to the receiver, it is displayed on the screen at the point reached by the spot as it sweeps across from left to right. By measuring the horizontal distance on the screen from the leading edge of the pulse to the start of the sweep of the spot, the travel time of the pulse from the transmitter to the receiver can be determined. For example, if the pulse is 3.5 cm from the start of the sweep of the spot and the time base control is set at 0.2 ms cm^{-1}, the travel time of the pulse must be 0.7 ms (= 0.2 ms cm^{-1} × 3.5 cm). If the distance from the transmitter to the receiver is known, the speed of ultrasound can be calculated (from distance / travel time).

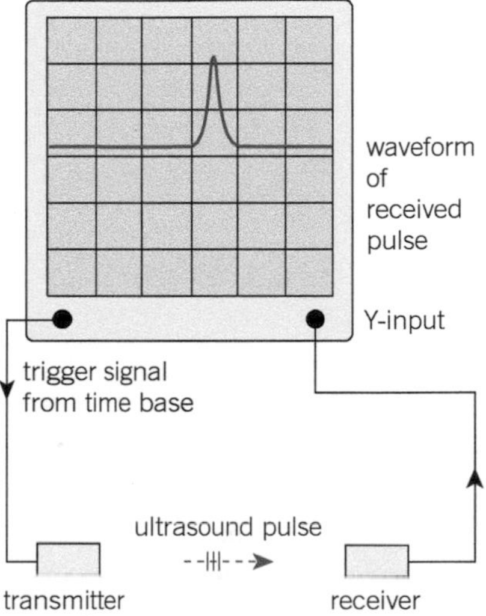

▲ **Figure 3** *Measuring the speed of ultrasound*

Summary questions

1 The trace of an oscilloscope is displaced vertically by 0.9 cm when a p.d. of 4.5 V is applied to the Y-input.

a Calculate **i** the Y-gain of the oscilloscope, **ii** the displacement of the spot when a p.d. of 12 V is applied to the Y-input.

b An alternating p.d. is applied to the Y-input instead. The height of the waveform from the bottom to the top is 6.5 cm when the Y-gain is 0.5 V cm^{-1}. Calculate the peak value of the alternating p.d.

2 The time base control of an oscilloscope is set at 10 ms cm^{-1} and an alternating p.d. is applied to the Y-input. The horizontal distance across two complete cycles is observed to be 4.4 cm. Calculate:

a the time period of the alternating p.d.

b its frequency.

3 Figure 4 shows the waveform on an oscilloscope screen when an alternating p.d. was applied to the Y-plates.

a The Y-gain of the oscilloscope is 5.0 V cm^{-1}. Calculate the peak value of the alternating p.d.

b The time base setting of the oscilloscope is 5 ms cm^{-1}. Calculate the time period and the frequency of the alternating p.d. (Assume each square represents 1 cm^2.)

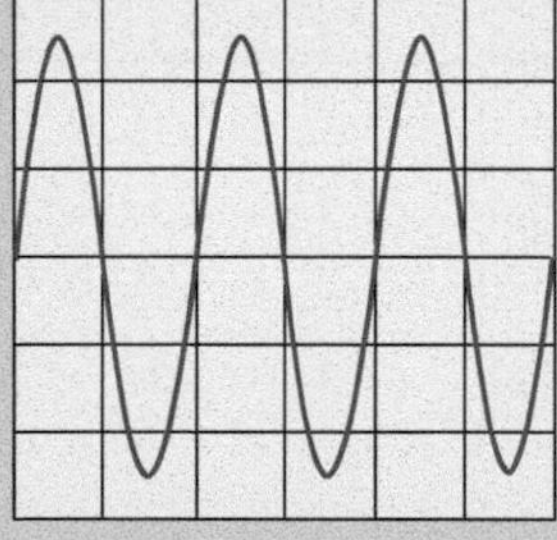

▲ **Figure 4**

4 Copy the grid of Figure 4 and sketch the trace you would observe if a constant p.d. of 10.0 V were applied to the Y-input.

11.11 Ultrasound imaging

Learning objectives:

→ Explain the principles of the generation and detection of ultrasound waves using piezo-electric transducers.

→ Explain the main principles behind the use of ultrasound to obtain diagnostic information about internal structures.

→ Define specific acoustic impedance and explain the importance in relation to the intensity reflection coefficient at a boundary.

→ Recall and solve problems by using the equation $I = I_0 e^{-\mu x}$ for the attenuation of ultrasound in matter.

Specification reference: 3.5.4

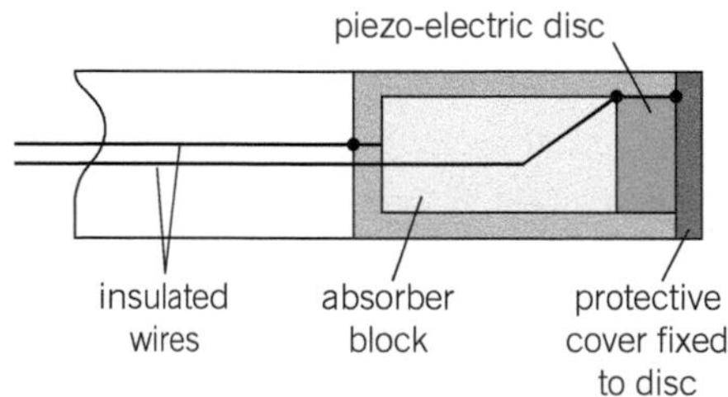

a *Probe construction*

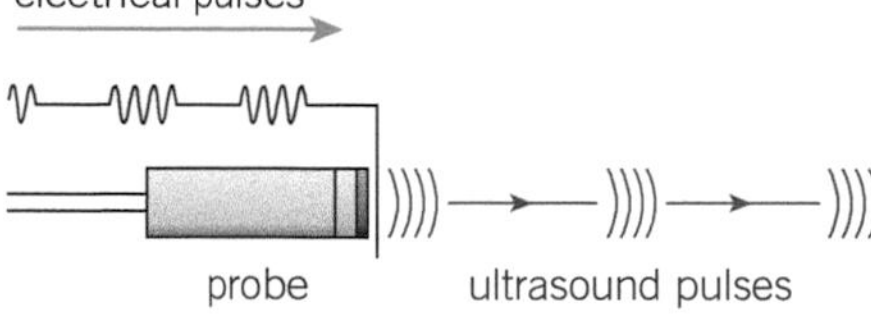

b *Ultrasound pulses*

▲ **Figure 1** *An ultrasound probe*

Producing ultrasound waves

Ultrasound consists of sound waves at frequencies of more than about 20 kHz (ie. above the range of the human ear). Unlike X-rays, ultrasound radiation is non-ionising radiation and therefore does not damage living tissue. For medical imaging, ultrasounds at frequencies between about 1 and 10 MHz are used, as diffraction would be significant at lower frequencies and reduced intensity due to absorption would be significant at higher frequencies.

An ultrasound probe used in ultrasound scanning is a hand-held device placed in contact with the body surface to direct pulses of ultrasound into the body. Each emitted pulse is partially reflected by internal boundaries in the body. The reflected pulses are then detected by the probe before the next pulse is emitted.

Figure 1 shows the construction of an ultrasound probe.

- The probe contains a piezo-electric **transducer** in the shape of a disc. Piezo-electricity is a property of certain solids whereby a p.d. applied between opposite faces causes a change of distance between the two faces. When an alternating p.d. is applied between the faces of the disc, the disc vibrates due to its changing thickness.
- By applying an alternating p.d. of frequency equal to the resonant frequency of vibration of the disc, the disc vibrates at resonance and creates ultrasound waves in the surrounding medium. The thickness of the disc determines its resonant frequency.
- An absorber pad of backing material behind the disc prevents ultrasound waves created at the two surfaces of the disc from cancelling each other out. The pad also damps the vibrations of the disc rapidly after each pulse is emitted.

More about piezo-electricity

A **transducer** is any device that is designed to convert energy from one form to another. A piezo-electric transducer generates a p.d. when it is squeezed. The piezo-electric material contains positive and negative ions, which are held together by the electrostatic forces they exert on each other. The centre of the negative charge of each molecule is in the same position as the centre of positive charge. When pressure is applied to opposite surfaces of the material, the centres of charge of the positive and negative ions are displaced slightly in opposite directions, causing a potential difference between the two surfaces. The effect is known as the **piezo-electric effect** and is displayed by crystals such as quartz.

To apply a p.d. across a quartz crystal, opposite surfaces to which the pressure is applied must be coated with metal so an electrical connection can be made to each surface.

The piezo-electric effect is reversible in that the application of a p.d. across a piezo-electric material causes the distance across the material to increase or decrease according to the polarity of the p.d.

Absorption and reflection of ultrasound waves

Ultrasound waves can be reflected and refracted just like sound waves. When ultrasound waves reach a boundary between two substances, some of the wave energy is reflected and some is transmitted, as shown in Figure 3.

Considering the energy reaching the boundary in 1 second, as energy cannot be created or destroyed, the sum of the reflected energy and the transmitted energy is equal to the incident energy. Therefore the incident intensity I = the reflected intensity I_R + the transmitted intensity I_T.

The fraction of the incident energy that is reflected or transmitted depends on:

- the angle of incidence θ of the incident waves
- the densities ρ_1 and ρ_2 of the two substances
- the wave speeds c_1 and c_2 of ultrasound waves in the two substances.

The **acoustic impedance** of a substance, Z, is defined by the equation

$$Z = \rho c$$

The unit of Z is given by the product of the unit of density (i.e. $kg\,m^{-3}$) and the unit of speed (i.e. $m\,s^{-1}$). Hence the unit of Z is $kg\,m^{-2}\,s^{-1}$.

Specific values for different substances are given in Table 1. Notice the very small value for air and the large values for quartz and bone compared with the other substances listed in the table.

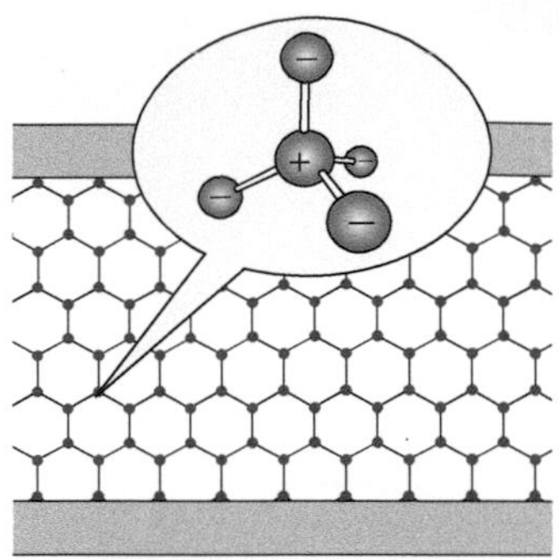

a *Unstressed*

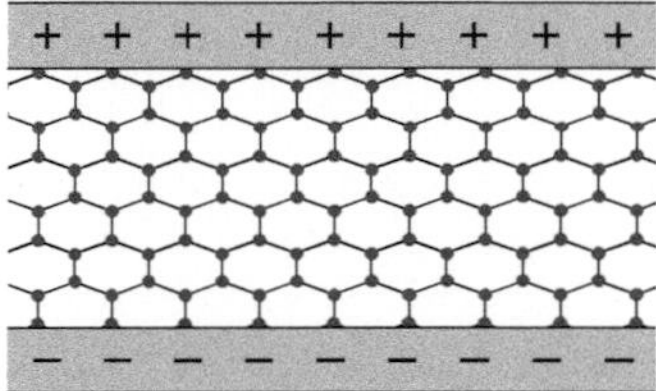

b *Compressed*

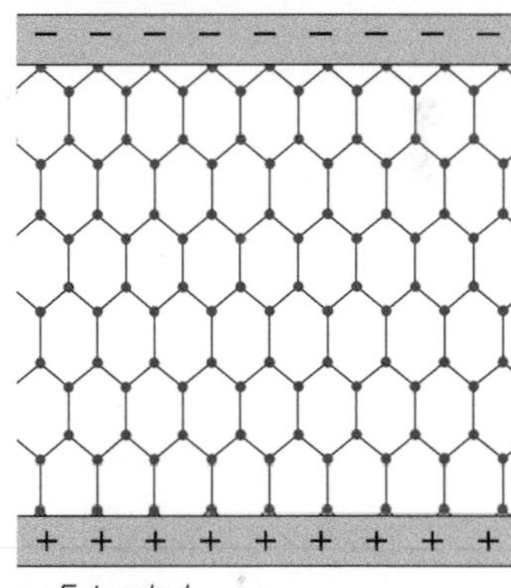

c *Extended*

▲ **Figure 2** *Piezo-electricity*

▼ **Table 1**

Substance	Acoustic impedance Z / $kg\,m^{-2}\,s^{-1}$
Air	430
Blood	1.59×10^6
Bone	6.80×10^6
Fat	1.38×10^6
Muscle	1.70×10^6
Quartz	1.52×10^7
Soft tissue	1.63×10^6
Water	1.50×10^6

When ultrasound waves are incident on a boundary between two substances with acoustic impedances Z_1 and Z_2, the ratio of the reflected intensity to the incident intensity, $\frac{I_R}{I}$, the reflection coefficient, is given by the equation

$$\frac{I_R}{I} = \frac{(Z_2 - Z_1)^2}{(Z_2 + Z_1)^2}$$

- If Z_1 and Z_2 are almost equal, the ratio is close to zero and the reflected intensity is very small compared with the incident intensity. In other words, most of the wave energy is transmitted.
- If Z_1 and Z_2 are very different, the ratio is close to 1 so most of the wave energy is reflected. The transmitted intensity is very small compared with the incident intensity. In other words, most of the wave energy is reflected.

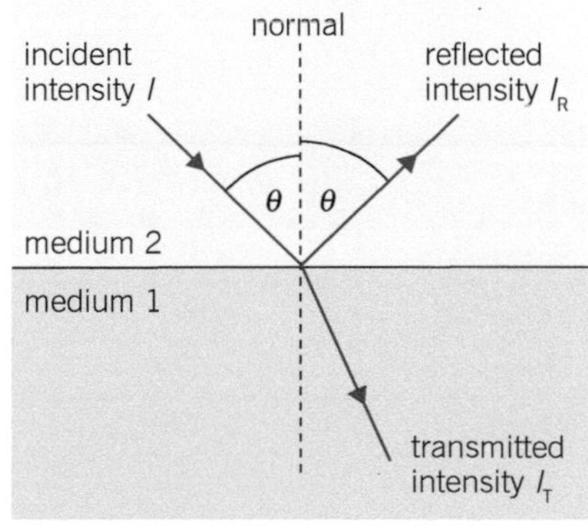

▲ **Figure 3** *Reflection and transmission at a boundary*

The intensity of the reflected pulses from the ultrasound probe when they return to it depends on:

1 the ratio of the reflected intensity to the incident intensity at each boundary, and
2 the absorption of the ultrasound waves by each substance they pass through.

When ultrasound is directed from a probe into the body, any air trapped between the probe and the body will cause most of the ultrasound energy to be reflected from the body. This is because the acoustic impedances of air and soft tissue are very different so the ratio $\frac{I_R}{I}$ is close to 1. To eliminate such trapped air, a **coupling medium** such as a gel is applied between the probe and the body surface. Such substances have similar acoustic impedances to soft tissue so the ratio $\frac{I_R}{I}$ is close to zero. In other words, most of the wave energy is transmitted into the body.

Reflection at tissue boundaries is significant and can't be avoided. For example, the ratio $\frac{I_R}{I}$ for a boundary between soft tissue and fat is 6.9×10^{-3} (using values from Table 1).

- The reflected pulses are therefore much weaker than the pulses leaving the probe.
- Also, the further a boundary is from the probe, the further the pulses reflected from that boundary travel and the weaker the reflected pulses from it will be due to absorption (see below).

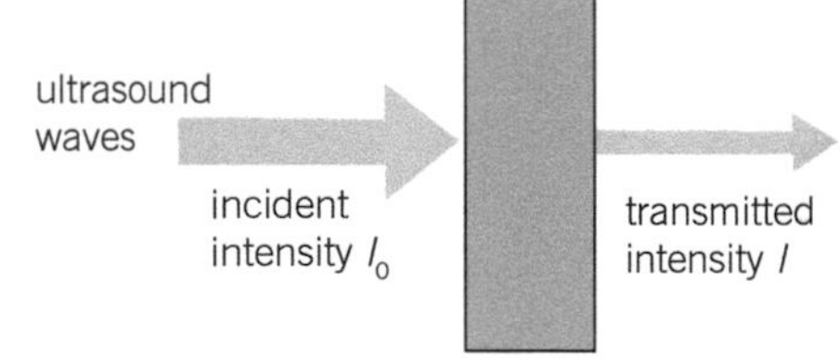

▲ **Figure 4** *Absorption*

▼ **Table 2**

Substance	Absorption (relative to air)
Soft tissue	0.3
Bone	4
Muscle	0.7
Water	0.001

Absorption of ultrasound waves depends on the substances the waves pass through and the distance travelled through each substance, the intensity of the waves decreases exponentially with distance. The energy absorbed by the substance causes the temperature of the substance to increase.

When a parallel beam of ultrasound waves travels through a substance, which means that the intensity decreases by a constant factor in equal distances. For example, if the intensity decreases from I_0 to $0.8 I_0$ in a distance of 10 mm, then

- after 20 mm, the intensity = $0.8 \times 0.8\, I_0 = 0.64\, I_0$
- after 30 mm, the intensity = $0.8 \times 0.64\, I_0 = 0.51\, I_0$
- after 40 mm, the intensity = $0.8 \times 0.51\, I_0 = 0.41\, I_0$

Notes:

1 The absorption of ultrasound waves varies according to the substance and the frequency of the waves.
2 Table 1 gives the absorption of ultrasound at 1MHz for the same distance in different substances relative to water.

Ultrasound scans

An **ultrasound scanner** consists of an ultrasound probe connected to a control unit and a visual display unit. In the simplest scan system, referred to as the **A-scan system**, a pulse generator is used to supply electrical pulses to the probe and to trigger the oscilloscope time base each time a pulse is generated. Figure 5 shows an A-scan of an eye.

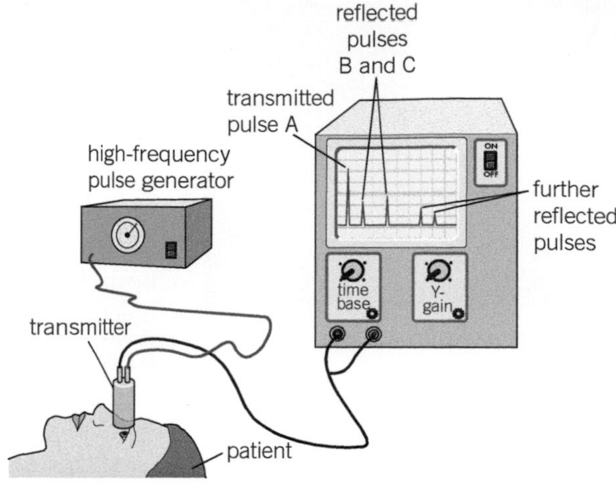

▲ **Figure 5** *The A-scan system*

- In each scan, a pulse is generated by the probe, and, before the next pulse is generated, the probe detects reflected pulses from the boundaries in the path of the pulse. The reflected pulses, or echoes, detected by the probe are amplified and displayed on the oscilloscope.
- Each time the time base is triggered, the oscilloscope beam sweeps across the screen from left to right. The time base of the oscilloscope is adjusted to display on the screen all the reflected pulses for each transmitted pulse. As a result, each pulse on the screen will be very narrow as the duration of each pulse is much shorter than the time for each sweep of the oscilloscope screen.

1 The position of each reflected pulse on the screen depends on the transit time of the ultrasound pulse (i.e., the time taken by the ultrasound pulse to travel from the probe to the internal boundary that reflected the pulse and back).

2 The further a reflected pulse on the screen appears from the transmitted pulse:
   - the longer the transit time of the pulse in the body and the further away the boundary is from the probe
   - the smaller the pulse height will be because the ultrasound pulse is partially reflected by the boundaries it passes through, and the substances it passes through absorb some of its energy.

3 The transit time is proportional to the distance from the probe to the boundary. Therefore, the greater the distance from the probe to the boundary, the further the pulse appears across the screen. The oscilloscope can be used to measure the transit time, t, of a pulse. The distance travelled by the pulse $s = vt$, where v is the speed of ultrasound waves in the body. Therefore, the distance from the probe to the internal boundary causing the pulse $= \frac{1}{2}vt$. Using this equation, the screen could be calibrated in terms of distance from the probe so the distance between a boundary and any other boundary or the probe can be measured directly.

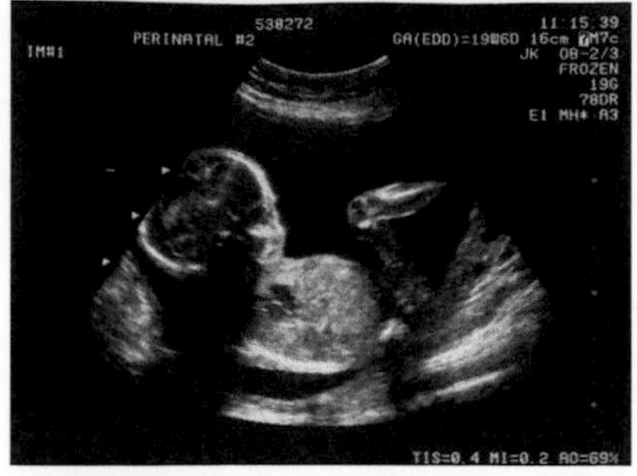

▲ **Figure 6** *A B-scan image of an unborn baby*

In the **B-scan system**, the probe has a number of ultrasound transducers side by side, each one sending out ultrasound pulses in a slightly different direction to the others. The signals from the transducers due to the reflected pulses are processed by a computer such that each reflected pulse is displayed as a bright spot on the screen in the correct direction and at the correct distance from the probe. As the probe is moved over the body surface, the bright spots on the screen build up a two-dimensional image of the reflecting boundaries scanned. The image may be enhanced and stored electronically.

Comparison with X-rays

- B-scans are used for pre-natal scans (i.e., to observe unborn babies in the womb) rather than X-ray CT scans. This is because ultrasound waves are non-ionising and, at the intensities used in scanning, do not damage human tissue.
- Ultrasound reflects at bone/tissue boundaries as well as at internal boundaries between soft substances such as fat, muscle, and tissue because the acoustic impedance of such substances differs. Therefore, ultrasound images show such 'soft' boundaries whereas X-ray images do not (because X-rays always pass through such boundaries without any reflection).

Summary questions

Use the data in Table 1 where necessary.

1 An ultrasound probe generates ultrasound waves at a frequency of 2.5 MHz. The speed of ultrasound in air = 350 m s^{-1} and in soft tissue = 1550 m s^{-1}.

a Calculate the wavelength of the ultrasound waves from this probe in **i** air, **ii** soft tissue.

b Explain why ultrasound waves of much lower frequency are unsuitable for medical imaging.

2 **a** **i** With the aid of a diagram, describe the construction of an ultrasound probe and how it produces ultrasound waves.

ii Explain the function of the backing block in an ultrasound probe.

b In the A-scan arrangement shown in Figure 5, the furthest boundary from the probe is the retina.

i Explain the presence of each pulse on the screen in terms of the cross-section of the patient's eye.

ii Calculate the distance between the boundary responsible for pulse B and the retina in Figure 5 if the distance from the probe to the furthest boundary is 24 mm.

3 **a** Use the data in Table 1 to calculate the reflection coefficient of the boundary between **i** air and skin, **ii** water and skin. Assume skin is soft tissue.

b Use the results of your calculation to explain why a gel must be applied between an ultrasound probe and the skin when the probe is used.

c A body organ has a density of 1040 kg m^{-3} and the speed of sound through it is 1580 m s^{-1}.

i Calculate the acoustic impedance of the organ tissue.

ii Use the data in Table 1 to calculate the reflection coefficient of the boundary between the organ and the surrounding soft tissue.

4 **a** State the main differences between an A-scan and a B-scan.

b Ultrasound waves and X-rays are both used for medical imaging. Explain why an ultrasound scan rather than an X-ray scan is used for scanning a baby in the womb.

Practice questions: Chapter 11

1 A student adjusted the tension of a stretched metal wire of length 820 mm so that when it vibrated it emitted sound at the same frequency as a 256 Hz tuning fork.

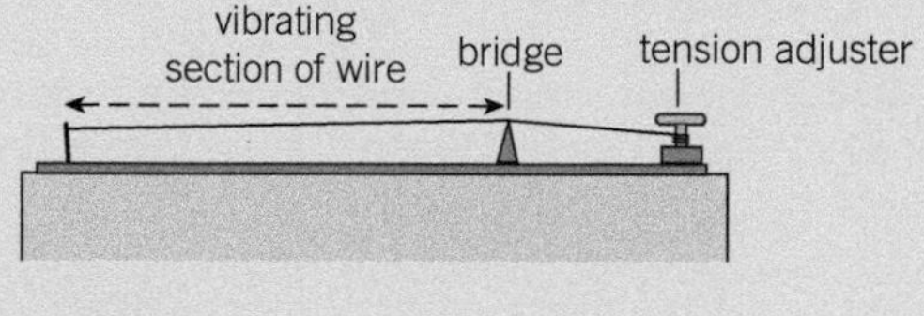

▲ **Figure 1**

She then altered the length L of the vibrating section of the wire as shown in **Figure 1**, until it emitted sound at the same frequency as a 512 Hz tuning fork. She repeated the test several times and obtained the following measurements for the length L:

425 mm   407 mm   396 mm   415 mm   402 mm

(a) (i) Calculate the mean length L at 512 Hz.

(ii) Estimate the uncertainty in this length measurement. *(2 marks)*

(b) The student thought the measurements showed that the frequency of sound f emitted by the wire is inversely proportional to the length L of the vibrating section.

(i) Discuss whether or not the measurements support this hypothesis.

(ii) In order to test the hypothesis further, state what further measurements the student could make and show how these measurements should be used. Assume further calibrated tuning forks are available. *(9 marks)*

2 An ultrasound signal from a ship travels vertically downwards through the water. The wavelength of the waves is 5.3×10^{-2} m and the frequency of the waves is 29 kHz.

(a) Calculate the speed of the sound through the water. *(3 marks)*

(b) The sound is reflected from the sea bed and is received back at the ship 0.23 s after it is transmitted. Calculate the depth of the water. *(2 marks)*

AQA, 2007

3 **(a)** State the characteristic features of

(i) longitudinal waves

(ii) transverse waves. *(3 marks)*

(b) Daylight passes horizontally through a fixed polarising filter P. An observer views the light emerging through a second polarising filter Q, which may be rotated in a vertical plane about point X as shown in **Figure 2**.

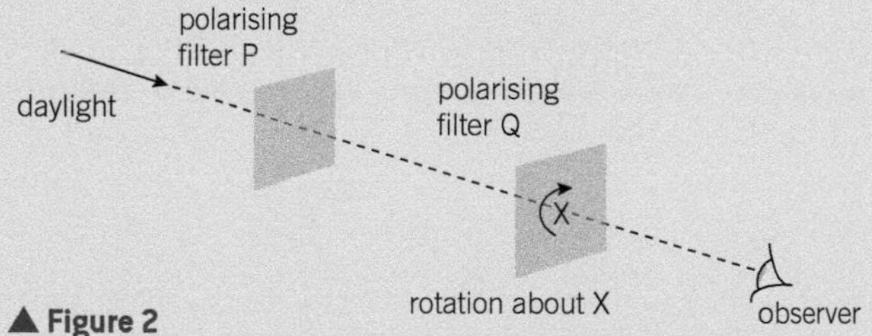

▲ **Figure 2**

Describe what the observer would see as Q is rotated slowly through 360°. *(2 marks)*

AQA, 2005

4 Polarisation is a property of one type of wave.

(a) There are two general classes of wave, longitudinal and transverse. Which class of wave can be polarised? *(1 mark)*

(b) Give *one* example of the type of wave that can be polarised. *(1 mark)*

(c) Explain why some waves can be polarised but others cannot. *(3 marks)*

AQA, 2002

5 **Figure 3** shows three particles in a medium that is transmitting a sound wave. Particles A and C are separated by one wavelength, and particle B is halfway between them when no sound is being transmitted.

(a) Name the type of wave that is involved in the transmission of this sound. *(1 mark)*

(b) At one instant particle A is displaced to the point A′ indicated by the tip of the arrow in **Figure 3**. Show on a copy of **Figure 3** the displacements of particles B and C at the same instant. Label the positions B′ and C′, respectively. *(2 marks)*

AQA, 2005

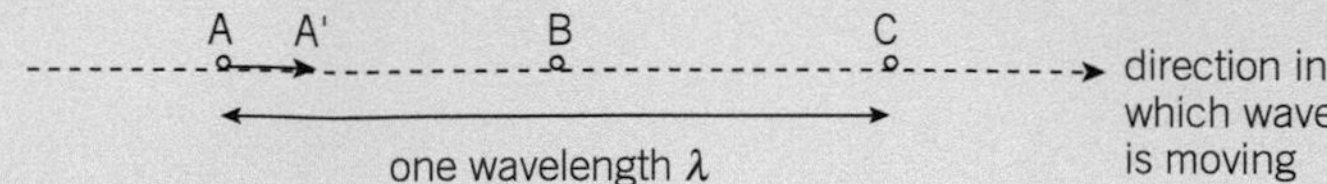

▲ **Figure 3**

6 **Figure 4** represents a stationary wave on a stretched string. The continuous line shows the position of the string at a particular instant when the displacement is a maximum. P and S are the fixed ends of the string. Q and R are the positions of the nodes. The speed of the waves on the string is $200\,\mathrm{m\,s^{-1}}$.

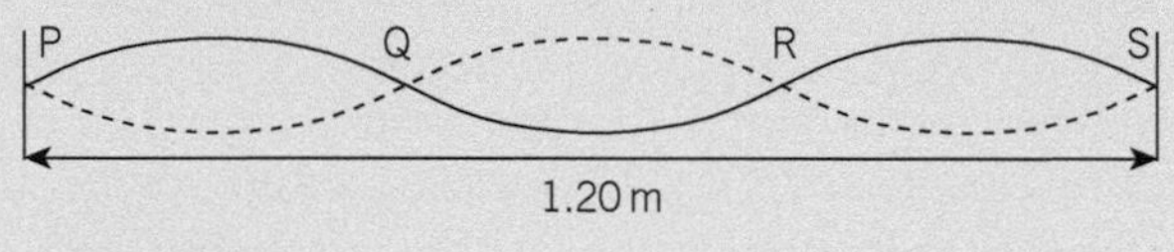

▲ **Figure 4**

(a) State the wavelength of the waves on the string.

(b) Calculate the frequency of vibration.

(c) Draw on a copy of the diagram the position of the string 3.0 ms later than the position shown.
Explain how you arrive at your answer. *(5 marks)*

AQA, 2004

7 Short pulses of sound are reflected from the wall of a building 18 m away from the sound source. The reflected pulses return to the source after 0.11 s.

(a) Calculate the speed of sound. *(3 marks)*

(b) The sound source now emits a continuous tone at a constant frequency. An observer, walking at a constant speed from the source to the wall, hears a regular rise and fall in the intensity of the sound. Explain how the *minima* of intensity occur. *(3 marks)*

AQA, 2002

8 A microwave transmitter directs waves towards a metal plate, as shown in **Figure 5**. When a microwave detector is moved along a line normal to the transmitter and the plate, it passes through a sequence of equally spaced maxima and minima of intensity.

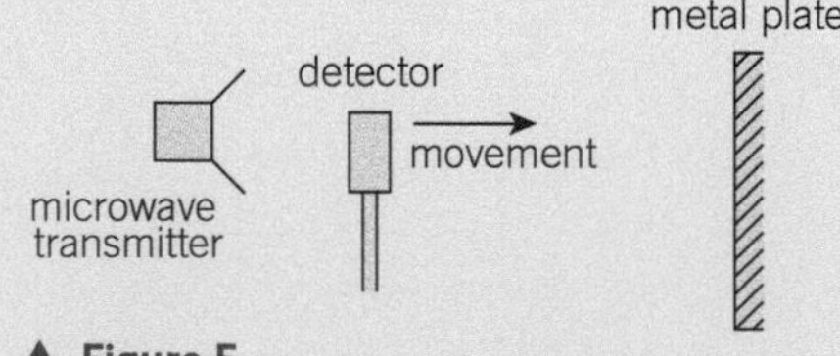

▲ **Figure 5**

(a) Explain how these maxima and minima are formed. *(4 marks)*

(b) The detector is placed at a position where the intensity is a minimum. When it is moved a distance of 144 mm it passes through nine maxima and reaches the ninth minimum from the starting point. Calculate:

(i) the wavelength of the microwaves

(ii) the frequency of the microwave transmitter. *(3 marks)*

AQA, 2003

12 Optics

12.1 Double slit interference

Learning objectives:

- → State the general condition for the formation of a bright fringe.
- → Describe Young's double slit experiment.
- → Describe what factors could be (i) increased or (ii) decreased to increase the fringe spacing.

Specification reference: 3.5.6

Young's double slit experiment

The wave nature of light was first suggested by Christiaan Huygens in the 17th century but it was rejected at the time in favour of Sir Isaac Newton's corpuscular theory of light. Newton considered that light was composed of tiny particles, which he referred to as corpuscles, and he was able to explain reflection and refraction using his theory. Huygens was also able to explain reflection and refraction using his wave theory. However, because of Newton's much stronger scientific reputation, Newton's theory of light remained unchallenged for over a century until 1803, when Thomas Young demonstrated interference of light.

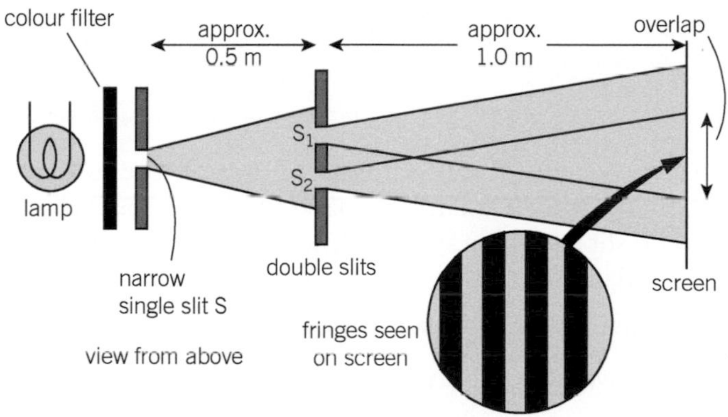

▲ **Figure 2** *Young's double slit experiment*

▲ **Figure 1** *Thomas Young 1773–1829*

To observe interference of light, we can illuminate two closely spaced parallel slits (double slits) using a suitable light source, as described below. The two slits act as **coherent** sources of waves, which means that they emit light waves with a constant phase difference and the same frequency.

1 An arrangement like the one used by Young is shown in Figure 2. Young would have used a candle instead of a light bulb to illuminate a narrow single slit S. The double slit arrangement S_1 and S_2 is illuminated by light from the single slit S. Alternate bright and dark fringes, referred to as **Young's fringes**, can be seen on a white screen placed where the diffracted light from the double slits overlaps. The fringes are evenly spaced and parallel to the double slits.

Note:

If the single slit is too wide, each part of it produces a fringe pattern, which is displaced slightly from the pattern due to adjacent parts of the single slit. As a result, the dark fringes of the double slit pattern become narrower than the bright fringes, and contrast is lost between the dark and the bright fringes.

2 A laser beam from a low-power laser could be used instead of the light bulb and the single slit. Figure 3 shows the arrangement. The fringes **must** be displayed on a screen, as a beam of laser light will damage the retina if it enters the eye. See Topic 12.2 for more information about laser light.

▲ **Figure 3** *Using a laser to demonstrate interference*

Practical link

Warning: Never look along a laser beam, even after reflection.

Synoptic link

You have met the principle of superposition in more detail in Topic 11.7, Wave properties 2.

The fringes are formed due to **interference of light** from the two slits:

- **Where a bright fringe is formed**, the light from one slit reinforces the light from the other slit. In other words, the light waves from each slit arrive in phase with each other.
- **Where a dark fringe is formed**, the light from one slit cancels the light from the other slit. In other words, the light waves from the two slits arrive **180° out of phase**.

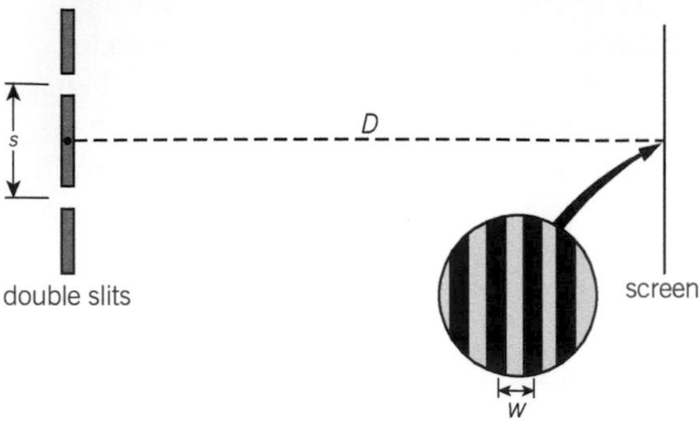

▲ **Figure 4** *Diagram to show w, D, and s for a Young's double slit experiment*

Study tip

Convert all distances into metres when substituting into the fringe spacing equation. The wavelength of visible light is usually quoted in nanometres (1 nm = 10^{-9} m).

The distance from the centre of a bright (or dark) fringe to the centre of the next bright fringe (or dark) is called the fringe separation. This depends on the slit spacing s and the distance D from the slits to the screen, in accordance with the equation

fringe separation, $w = \frac{\lambda D}{s}$**, where λ is the wavelength of light**

The equation shows that the fringes become more widely spaced if:

- the distance D from the slits to the screen is increased
- the wavelength λ of the light used is increased
- the slit spacing, s, is reduced. Note that the slit spacing is the distance between the centres of the slits.

The theory of the double slit equation

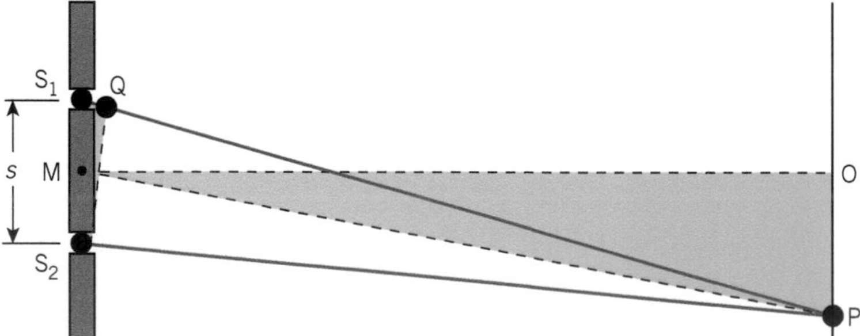

▲ **Figure 5** *The theory of the double slit experiment*

Consider the two slits S_1 and S_2 shown in Figure 5. At a point P on the screen where the fringes are observed, light emitted from S_1 arrives later than light from S_2 emitted at the same time. This is because the distance S_1P is greater than the distance S_2P. The difference between distances S_1P and S_2P is referred to as the path difference.

- **For reinforcement at P, the path difference $S_1P - S_2P = m\lambda$,** where $m = 0, 1, 2$, etc.

Light emitted simultaneously from S_1 and S_2 arrives in phase at P so reinforcement occurs at this point.

- **For cancellation at P, the path difference $S_1P - S_2P = (m + \frac{1}{2})\lambda$,** where $m = 0, 1, 2,$ etc.

Light emitted simultaneously from S_1 and S_2 arrives at P out of phase by 180°, so cancellation occurs at P.

In Figure 5, a point Q along line S_1P has been marked such that $QP = S_2P$.

Therefore, the path difference $S_1P - S_2P$ is represented by the distance S_1Q.

Consider triangles S_1S_2Q and MOP, where M is the midpoint between the two slits and O is the midpoint of the central bright fringe of the pattern. The two triangles are very nearly similar in shape, as angles $S_1\widehat{S_2}Q$ and $P\widehat{M}O$ are almost equal and the long sides of each triangle are of almost equal length. Therefore

$$\frac{S_1Q}{S_1S_2} = \frac{OP}{OM}$$

If P is the mth bright fringe from the centre (where $m = 0, 1, 2,$ etc.), then $S_1Q = m\lambda$ and $OP = mw$, where w is the distance between the centres of adjacent bright fringes. Also, OM = distance D and S_1S_2 = slit spacing s.

Therefore,

$$\frac{m\lambda}{s} = \frac{mw}{D}$$

Rearranging this equation gives

$$\lambda = \frac{sw}{D}$$

or

$$w = \frac{\lambda D}{s}$$

By measuring the slit spacing s, the fringe separation w, and the slit–screen distance D, the wavelength λ of the light used can be calculated. The formula is valid only if the fringe separation, w, is much less than the distance D from the slits to the screen. This condition is to ensure that the triangles S_1S_2Q and MOP are very nearly similar in shape.

Notes:

1 To measure the fringe separation, w, measure across several fringes from the centre of a dark fringe to the centre of another dark fringe, because the centres of the dark fringes are easier to locate than the centres of the bright fringes. Obtain w by dividing your measurement by the number of fringes you measured across. This means that the derivation above can also be done using the distance between the centres of adjacent dark fringes.

2 Two loudspeakers connected to the *same* signal generator can be used to demonstrate interference as they are coherent sources of sound waves. You can detect points of cancellation and reinforcement by ear as you move across in front of the speakers.

The Young's slit equation can then be used to estimate the wavelength of the sound waves (provided the wavelength is small compared with the distance between the speakers).

Summary questions

1 In a double slit experiment using red light, a fringe pattern is observed on a screen at a fixed distance from the double slits. How would the fringe pattern change if:

- **a** the screen is moved closer to the slits?
- **b** one of the double slits is blocked completely?

2 The following measurements were made in a double slit experiment:

slit spacing $s = 0.4$ mm, fringe separation $w = 1.1$ mm, slit–screen distance $D = 0.80$ m.

Calculate the wavelength of light used.

3 In **Q2**, the double slit arrangement was replaced by a pair of slits with a slit spacing of 0.5 mm. Calculate the fringe separation for the same slit–screen distance and wavelength.

4 The following measurements were made in a double slit experiment:

slit spacing $s = 0.4$ mm, fringe separation $w = 1.1$ mm, wavelength of light used = 590 nm. Calculate the distance from the slits to the screen.

12.2 More about interference

Learning objectives:

- Identify coherent sources.
- Explain why slits, rather than two separate light sources, are used in Young's double slit experiment.
- Describe the roles of diffraction and interference when producing Young's fringes.

Specification reference: 3.5.6

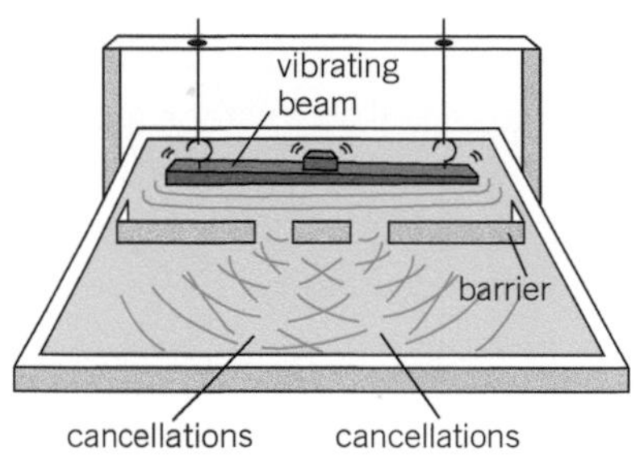

▲ **Figure 1** *Interference of water waves*

Study tip

It isn't necessary for two sources to be in phase for them to be coherent – a constant phase difference between them is sufficient. However, the waves must be of the same frequency.

▲ **Figure 3** *The double slit fringe pattern*

Coherence

The double slits are described as **coherent sources** because they emit light waves of the same frequency with a constant phase difference, provided we illuminate the double slits with laser light, or with light from a narrow single slit if we are using non-laser light. Each wave crest or wave trough from the single slit always passes through one of the double slits a fixed time after it passes through the other slit. The double slits therefore emit wavefronts with a constant phase difference.

The double slit arrangement is like the ripple tank demonstration in Figure 1. Straight waves from the beam vibrating on the water surface diffract after passing through the two gaps in the barrier, and produce an interference pattern where the diffracted waves overlap. If one gap is closer to the beam than the other, each wavefront from the beam passes through the nearer gap first. However, the time interval between the same wavefront passing through the two gaps is always the same so the waves emerge from the gaps with a constant phase difference.

Light from two nearby lamp bulbs could not form an interference pattern because the two light sources emit light waves at random. The points of cancellation and reinforcement would change at random, so no interference pattern is possible.

Wavelength and colour

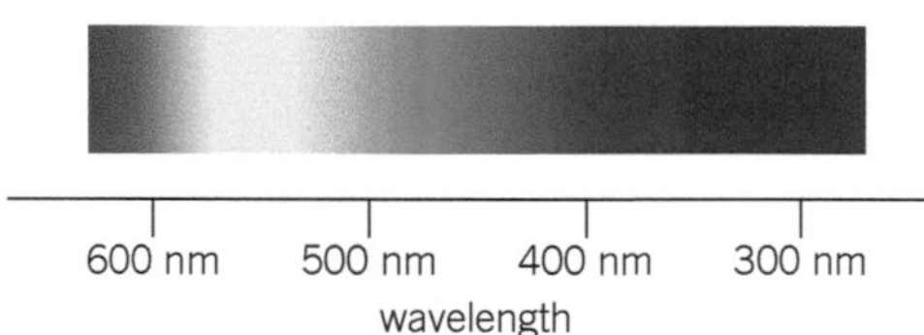

▲ **Figure 2** *Wavelength and colour*

In the double slit experiment, the fringe separation depends on the colour of light used. White light is composed of a continuous spectrum of colours, corresponding to a continuous range of wavelengths from about 350 nm for violet light to about 650 nm for red light. Each colour of light has its own wavelength, as shown in Figure 2.

The fringe pattern photographs in Figure 3 show that the fringe separation is greater for red light than for blue light. This is because red light has a longer wavelength than blue light. The fringe spacing, w, depends on the wavelength λ of the light, according to the formula

$$w = \frac{\lambda D}{s}$$

as explained in Topic 12.1, so the longer the wavelength of the light used, the greater the fringe separation.

Light sources

Vapour lamps and discharge tubes produce light with a dominant colour. For example, the sodium vapour lamp produces a yellow/orange glow, which is due to light of wavelength 590 nm. Other wavelengths of light are also emitted from a sodium vapour lamp but the colour due to light of wavelength 590 nm is much more intense than any other colour. A sodium vapour lamp is in effect a monochromatic light source, because its spectrum is dominated by light of a certain colour.

Light from a filament lamp or from the Sun is composed of the colours of the spectrum and therefore covers a continuous range of wavelengths from about 350 nm to about 650 nm. If a beam of white light is directed at a colour filter, the light from the filter is a particular colour because it contains a much narrower range of wavelengths than white light does.

Light from a laser differs from non-laser light in two main ways:

- Laser light is highly **monochromatic**, which means we can specify its wavelength to within a nanometre. The wavelength depends on the type of laser that produces it. For example, a helium–neon laser produces red light of wavelength 635 nm. Because a laser beam is almost perfectly parallel and monochromatic, a convex lens can focus it to a very fine spot. The beam power is then concentrated in a very small area. This is why a laser beam is very dangerous if it enters the eye. The eye lens would focus the beam on a tiny spot on the retina and the intense concentration of light at that spot would destroy the retina.

  Never look along a laser beam, even after reflection.

- **A laser is a convenient source of coherent light.** When we use a laser to demonstrate double slit interference, we can illuminate the double slits directly. We do not need to make the light pass through a narrow single slit first as we do with light from a non-laser light source. This is because a laser is a source of coherent light. A light source emits light as a result of electrons inside its atoms moving to lower energy levels inside the atoms. Each such electron emits a photon, which is a packet of electromagnetic waves of constant frequency. Inside a laser, each emitted photon causes more photons to be emitted as it passes through the light-emitting substance. These stimulated photons are in phase with the photon that caused them. As a result, the photons in a laser beam are in phase with each other. So the laser is a coherent source of light. In comparison, in a non-laser light source, the atoms emit photons at random so the photons in such a beam have random phase differences.

Synoptic link

You have met photons in more detail in Topic 7.3, Photons.

Application

DVDs

Laser light is used to read DVDs in computers and DVD players. The track of a DVD is very narrow and is encoded with pits representing digital information. Laser light is reflected from the track. The pits cause the laser light to change in intensity. This change is then converted by a detector into an electronic signal to produce the output.

Study tip

You need to know that a laser is a coherent source of light and a non-laser source is not a coherent source.

White light fringes

Figure 3 shows the fringe patterns observed with blue light and with red light. As explained above, the blue light fringes are closer together than the red light fringes. However, the central fringe of each pattern is in the same position on the screen. The fringe pattern produced by white light is shown in Figure 4. Each component colour of white light produces its own fringe pattern, and each pattern is centred on the screen at the same position. As a result:

- the central fringe is white because every colour contributes at the centre of the pattern.
- the inner fringes are tinged with blue on the inner side and red on the outer side. This is because the red fringes are more spaced out than the blue fringes and the two fringe patterns do not overlap exactly.
- the outer fringes merge into an indistinct background of white light, becoming fainter with increasing distance from the centre. This is because, where the fringes merge, different colours reinforce and therefore overlap.

▲ **Figure 4** *White light fringes*

Summary questions

1 a Sketch an arrangement that may be used to observe the fringe pattern when light from a narrow slit illuminated by a sodium vapour lamp is passed through a double slit.

b Describe the fringe pattern you would expect to observe in part **a**.

2 Describe how the fringe pattern would change in **Q1a** if the narrow single slit is replaced by a wider slit.

3 Double slit interference fringes are observed using light of wavelength 590 nm and a double slit of slit spacing 0.50 mm. The fringes are observed on a screen at a distance of 0.90 m from the double slits. Calculate the fringe separation of these fringes.

4 Describe and explain the fringe pattern that would be observed in **Q3** if the light source were replaced by a white light source.

12.3 Diffraction

Observing diffraction

Diffraction is the spreading of waves when they pass through a gap or by an edge. This general property of all waves is very important in the design of optical instruments, such as cameras, microscopes, and telescopes. For example, when a telescope is used to observe planets, we can often see features that are not evident when observed directly. This is partly because less diffraction occurs when waves pass through a wide gap than through a narrow gap. Therefore, because a telescope is much wider than the eye pupil, much less diffraction occurs when using a telescope than when observing with the unaided eye.

Diffraction of water waves through a gap can be observed using a ripple tank. This arrangement shows that the diffracted waves spread out more if:

- the gap is made narrower, or
- the wavelength is made larger.

In addition, close examination of the diffracted waves should reveal that each diffracted wavefront has breaks either side of the centre. These breaks are due to waves diffracted by adjacent sections on the gap being out of phase and cancelling each other out in certain directions.

Diffraction of light by a single slit can be demonstrated by directing a parallel beam of light at the slit. The diffracted light forms a pattern that can be observed on a white screen. The pattern shows a central fringe with further fringes either side of the central fringe, as shown in Figure 2. The intensity of the fringes is greatest at the centre of the central fringe. Figure 2 also shows the variation of the intensity of the diffracted light with the distance from the centre of the fringe pattern.

Learning objectives:

- → Explain why diffraction of light is important in the design of optical instruments.
- → Compare the single slit diffraction pattern with the pattern of Young's fringes.
- → Describe the effect of the single slit pattern on the brightness of Young's fringes.

Specification reference: 3.5.7

Synoptic link

You have met the study of water waves using a ripple tank in more detail in Topic 11.6, Wave properties 1.

Hint

The central bright fringe of the single slit pattern is twice as wide as the others, and much brighter.

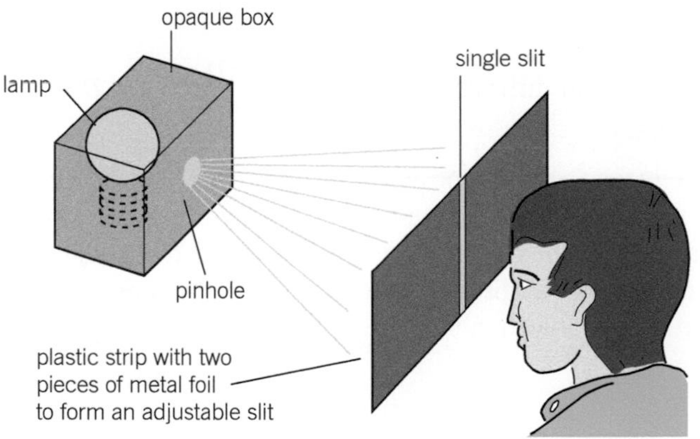

▲ **Figure 1**

▲ **Figure 2** *Single slit diffraction*

Notes:

- The central fringe is twice as wide as each of the outer fringes (measured from minimum to minimum intensity).
- Each of the outer fringes is the same width.
- The outer fringes are much less intense than the central fringe.
- For white light, the central fringe would be white and each of the other fringes would be tinged with blue nearer the central fringe and red on the other side.

Application

Microscopes and diffraction

Microscopes are often fitted with a blue filter as greater detail (better resolution) can be seen in microscope images observed with blue light than with white light. In an electron microscope, the resolution is increased when a higher voltage setting is used. This is because the higher the voltage, the greater the speed of the electrons in the microscope beam and therefore the smaller the de Broglie wavelength of the electrons. Hence the electrons are diffracted less as they pass through the microscope's magnetic lenses. See page 257.

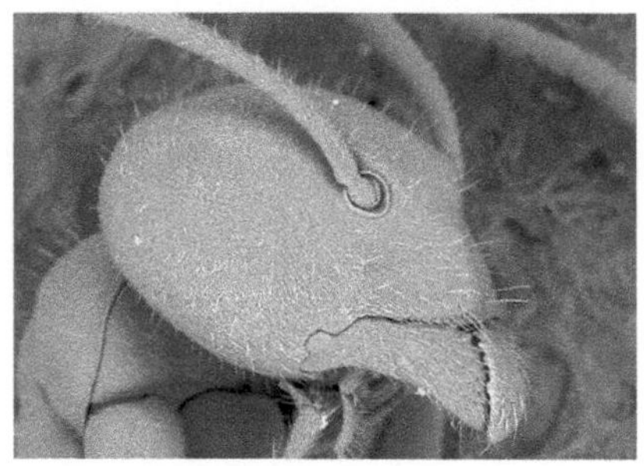

▲ **Figure 3** *An electron microscope image*

Synoptic link

You will meet the de Broglie wavelength in Topic 13.7, Wave–particle duality.

More about single slit diffraction

If the single slit pattern is observed:

- Using different sources of monochromatic light in turn, the observations show that the greater the wavelength, the wider the fringes.
- Using an adjustable slit, the observations show that making the slit narrower makes the fringes wider.

It can be shown theoretically that the width W of the central fringe observed on a screen at distance D from the slit is given by

$$W = \frac{\text{the wavelength of the light } (\lambda)}{\text{the width of the single slit } (a)} \times 2D$$

Therefore, the width of each fringe is proportional to $\frac{\lambda}{a}$. For this reason, the fringes are narrower using blue light than if red light had been used.

Single slit diffraction and Young's fringes

In the double slit experiment in Topic 12.1, light passes through the two slits of the double slit arrangement and produces an interference pattern. However, if the slits are too wide and too far apart, no interference pattern is observed. This is because interference can only occur if the light from the two slits overlaps. For this to be the case:

- each slit must be narrow enough to make the light passing through it diffract sufficiently
- the two slits must be close enough so the diffracted waves overlap on the screen.

In general, for monochromatic light of wavelength λ, incident on two slits of aperture width a at slit separation s (from centre to centre),

- the fringe spacing of the interference fringes, $w = \frac{\lambda D}{s}$
- the width of the central diffraction fringe, $W = \frac{2\lambda D}{a}$, where D is the slit–screen distance.

Figure 4 shows the intensity variation with distance across the screen in terms of distance from the centre of the pattern. Using the expressions above, you should be able to see that only a few interference fringes will be observed in the central diffraction fringe if the slit separation s is small compared to the slit width a.

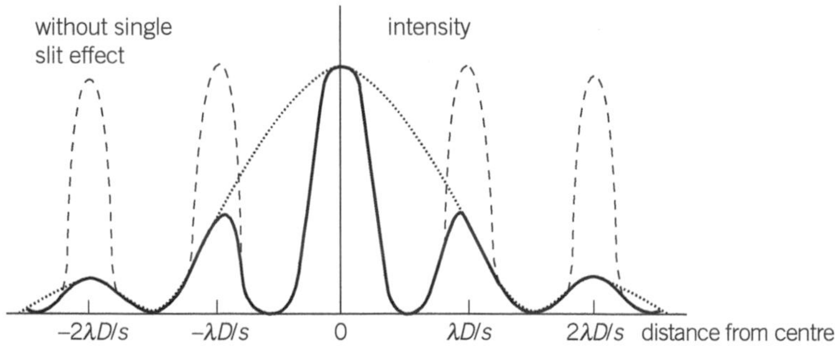

▲ **Figure 4** *Intensity distribution for Young's fringes*

Extension

Explaining single slit diffraction

In Figure 2, the intensity minima occur at evenly spaced positions on the screen. To explain this, consider wavelets emitted simultaneously by equally spaced point sources across the slit each time a wavefront arrives at the slit. At a certain position, P, on the screen (Figure 5), the wavelet from each point source arrives a short time after the wavelet from the adjacent point source nearer P. Therefore, at any given time, the phase difference ϕ between the contributions from any two adjacent point sources is constant.

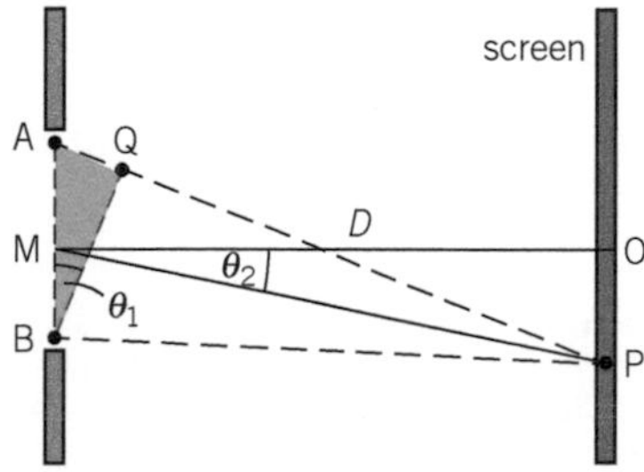

▲ **Figure 5**

The contributions at P from each point source can be represented by phased vectors or **phasors** as shown in Figure 6. Each phasor is at angle ϕ to each adjacent phasor. In Figure 6a, the resultant is the vector sum of four phasors, which is non-zero. In Figure 6b, the resultant is zero because $\phi \times$ the number of phasors is equal to 2π.

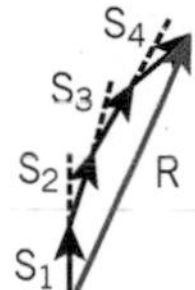

a *Resultant R = the vector sum of four phasors*

b *Resultant R = 0*

▲ **Figure 6**

More generally, intensity minima occur at positions P so that the phase difference between contributions from two point sources A and B at opposite sides of the slit is equal to $2\pi m$, where m is a whole number (see Figure 5), so the path difference AP – BP = $m\lambda$.

In Figure 5, point Q on the line AP is such that QP = BP. Therefore AQ = AP – QP = AP – BP = $m\lambda$.

Because $\widehat{AQB}$ is almost 90°, $\sin\theta_1 = \frac{AQ}{AB} = \frac{m\lambda}{a}$, where the aperture width a = AB, and angle $\theta_1 = \widehat{ABQ}$.

Consider triangles AQB and MOP, where M is the midpoint of the slit, and O is the point on the screen where the intensity is greatest. The two triangles are almost identical. This means that $\frac{OP}{OM} = \frac{AQ}{AB} = \frac{m\lambda}{a}$.

Intensity minima occurs at evenly spaced positions P such that

$$OP = \frac{m\lambda D}{a} \quad \text{where } D = \text{distance OM.}$$

Question

1 Light of wavelength 620 mm is used to form a single slit diffraction pattern on a screen which is 800 mm from a slit of width 0.24 mm. Calculate the width of the central maximum.

Answer: 4.1 mm.

Summary questions

1 Red light from a laser is directed normally at a slit that can be adjusted in width. The diffracted light from the slit forms a pattern of diffraction fringes on a white screen. Describe how the appearance of the fringes changes if the slit is made wider.

2 A narrow beam of white light is directed normally at a single slit. The diffracted light forms a pattern on a white screen.

  a A blue filter is placed in the path of the beam before it reaches the slit. The distance across five fringes including the central fringe is 18 mm. Calculate the width of the central fringe.

  b The blue filter is replaced by a red filter. Compare the red fringe pattern with the blue fringe pattern.

3 Figure 3 in Topic 12.2 shows the interference fringes observed in a Young's fringes experiment.

  a Explain why the fringes at the outer edges are dimmer than the fringes nearer the centre.

  b Describe how the appearance of the fringes would change if the slits were made wider without changing their centre-to-centre separation.

  c Sketch a graph to show how the intensity of the fringes varies with distance from the centre of the central fringe.

12.4 The diffraction grating

Learning objectives:

- → Explain why a diffraction grating diffracts monochromatic light in certain directions only.
- → If a coarser grating is used, explain the effect on the number of diffracted beams produced and on the spread of each diffracted beam.
- → Determine the grating spacing for any given grating, if it is not known.

Specification reference: 3.5.7

Testing a diffraction grating

A **diffraction grating** consists of a plate with many closely spaced parallel slits ruled on it. When a parallel beam of monochromatic light is directed normally at a diffraction grating, light is transmitted by the grating in certain directions only. This is because:

- the light passing through each slit is diffracted
- the diffracted light waves from adjacent slits reinforce each other in certain directions only, including the incident light direction, and cancel out in all other directions.

Figure 1 shows one way to observe the effect of a diffraction grating on the incident light beam.

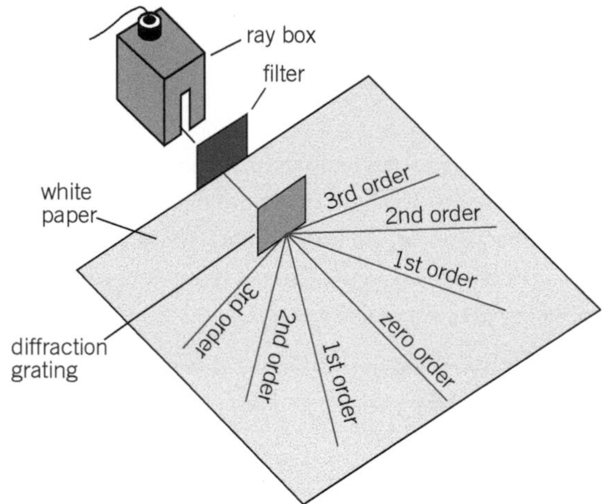

▲ **Figure 1** *The diffraction grating*

The central beam, referred to as the zero-order beam, is in the same direction as the incident beam. The other transmitted beams are numbered outwards from the zero-order beam. The angle of diffraction between each transmitted beam and the central beam increases if:

- light of a longer wavelength is used (e.g., by replacing a blue filter with a red filter)
- a grating with closer slits is used.

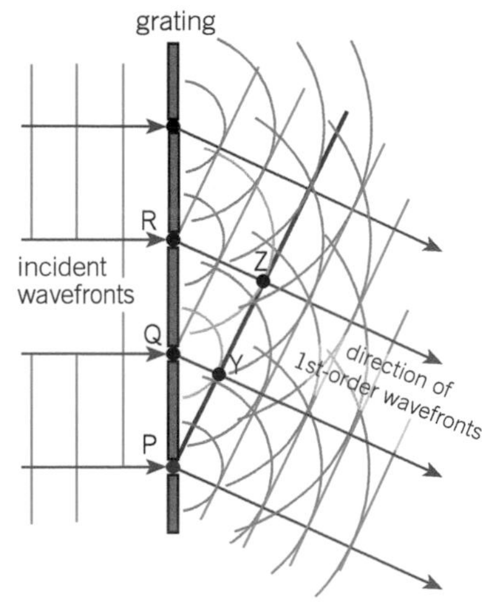

▲ **Figure 2** *Formation of the first-order wavefront*

The diffraction grating equation

Consider a magnified view of part of a diffraction grating and a snapshot of the diffracted waves, as shown in Figure 2. Each slit diffracts the light waves that pass through it. As each diffracted wavefront emerges from a slit, it reinforces a wavefront from a slit adjacent to it. In this example, the wavefront emerging at P reinforces the wavefront emitted from Q one cycle earlier, which reinforces the wavefront emitted from R one cycle earlier, and so on. The effect is to form a new wavefront PYZ.

Figure 3 shows the formation of a wavefront of the nth-order beam. The wavefront emerging from slit P reinforces a wavefront emitted n cycles earlier by the adjacent slit Q. This earlier wavefront therefore must have travelled a distance of n wavelengths from the slit. Thus the perpendicular distance QY from the slit to the wavefront is equal to $n\lambda$, where λ is the wavelength of the light waves.

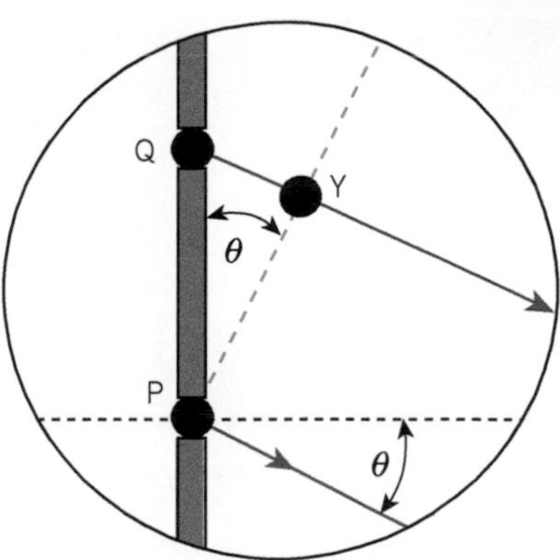

▲ **Figure 3** *The nth-order wavefront*

Since the angle of diffraction of the beam, θ, is equal to the angle between the wavefront and the plane of the slits, it follows that $\sin\theta = \frac{QY}{QP}$, where QP is the grating spacing d.

Substituting d for QP and $n\lambda$ for QY into the equation above gives $\sin\theta = \frac{n\lambda}{d}$. Rearranging this equation gives the diffraction grating equation for the angle of diffraction of the nth-order beam

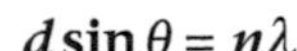

$$d\sin\theta = n\lambda$$

Notes:

1 The number of slits per metre on the grating $N = \frac{1}{d}$, where d is the grating spacing.

2 For a given order and wavelength, the smaller the value of d, the greater the angle of diffraction. In other words, the larger the number of slits per metre, the bigger the angle of diffraction.

3 Fractions of a degree are usually expressed either as a decimal or in minutes (abbreviated as ′) where $1° = 60′$.

4 To find the maximum number of orders produced, substitute $\theta = 90°$ ($\sin\theta = 1$) in the grating equation and calculate n using $n = \frac{d}{\lambda}$.
The maximum number of orders is given by the value of $\frac{d}{\lambda}$ rounded down to the nearest whole number.

Hint

A minimum in the single slit diffraction will suppress a diffracted order from the grating, and there will be a missing order in the pattern as a result.

Study tip

The number of maxima observed is $2n + 1$, where n is the greatest order.

Application

Diffraction gratings in action

Diffraction gratings can be made by cutting parallel grooves very close together on a smooth glass plate. Each groove transmits some incident light and reflects or scatters some. So the grooves act as coherent emitters of waves just as if they were slits. However, the effective slit width needs to be much smaller than the grating spacing. This is so that the diffracted waves from each groove spread out widely. If they did not, the higher-order beams would be much less intense than the lower-order beams.

We can use a diffraction grating in a **spectrometer** to study the spectrum of light from any light source including stars and to measure light wavelengths very accurately. A spectrometer is designed to measure angles to within 1 arc minute, which is one-sixtieth of a degree. The angle of diffraction of a diffracted beam can be measured very accurately. The grating spacing of a diffraction grating can therefore be measured very accurately using light of a known wavelength. The grating can then be used to measure light of any wavelength.

Most industrial and research laboratories now use a **spectrum analyser**, which is an electronic spectrometer linked to a computer that gives a visual display of the variation of intensity with wavelength. You do not need to know how spectrometers or spectrum analysers work or are used.

▲ **Figure 4** *A spectrometer fitted with a diffraction grating*

Synoptic link

You will meet line spectra in more detail in Topic 13.5, Energy levels and spectra.

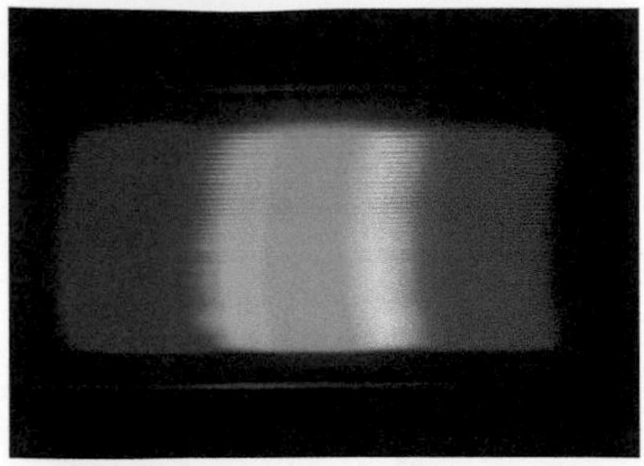
▲ **Figure 5** *A continuous spectrum*

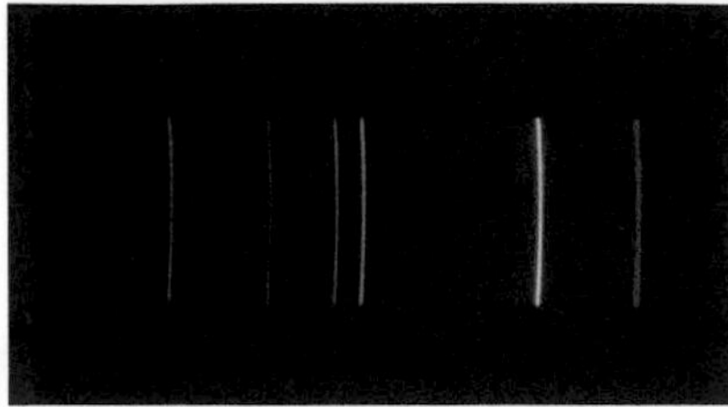
▲ **Figure 6** *A line emission spectrum*

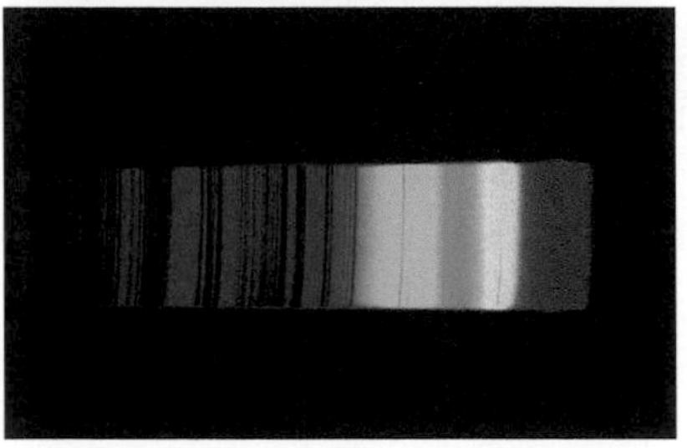
▲ **Figure 7** *A line absorption spectrum*

Types of spectra

Continuous spectra

The spectrum of light from a filament lamp is a continuous spectrum of colour from deep violet at about 350 nm to deep red at about 650 nm, as shown in Figure 5. The most intense part of the spectrum depends on the temperature of the light source. The hotter the light source, the shorter the wavelength of the brightest part of the spectrum. By measuring the wavelength of the brightest part of a continuous spectrum, we can therefore measure the temperature of the light source.

Line emission spectra

A glowing gas in a vapour lamp or a discharge tube emits light at specific wavelengths so its spectrum consists of narrow vertical lines of different colours, as shown in Figure 6. The wavelengths of the lines are characteristic of the chemical element that produced the light. If a glowing gas contains more than one element, the elements in the gas can be identified by observing its line spectrum.

Line absorption spectra

A line absorption spectrum is a continuous spectrum with narrow dark lines at certain wavelengths. For example, if the spectrum of light from a filament lamp is observed after passing it through a glowing gas, thin dark vertical lines are observed superimposed on the continuous spectrum, as shown in Figure 7. The pattern of the dark lines is due to the elements in the glowing gas. These elements absorb light of the same wavelengths they can emit at so the transmitted light is missing these wavelengths. The atoms of the glowing gas that absorb light then emit the light subsequently but not necessarily in the same direction as the transmitted light.

Summary questions

1 A laser beam of wavelength 630 nm is directed normally at a diffraction grating with 300 lines per millimetre. Calculate:

- **a** the angle of diffraction of each of the first two orders
- **b** the number of diffracted orders produced.

2 Light directed normally at a diffraction grating contains wavelengths of 580 and 586 nm only. The grating has 600 lines per mm.

- **a** How many diffracted orders are observed in the transmitted light?
- **b** For the highest order, calculate the angle between the two diffracted beams.

3 Light of wavelength 430 nm is directed normally at a diffraction grating. The first-order transmitted beams are at 28° to the zero-order beam. Calculate:

- **a** the number of slits per millimetre on the grating
- **b** the angle of diffraction for each of the other diffracted orders of the transmitted light.

4 A diffraction grating is designed with a slit width of 0.83 μm. When used in a spectrometer to view light of wavelength 430 nm, diffracted beams are observed at angles of 14° 55′ and 50° 40′ to the zero-order beam. (Reminder: 1 degree = 60 minutes.)

- **a** Assuming the low-angle diffracted beam is the first-order beam, calculate the number of lines per mm on the grating.
- **b** Explain why there is no diffracted beam between the two observed beams. What is the order number for the beam at 50° 40′?

12.5 Refraction of light

The wave theory of light can be used to explain reflection and refraction of light. However, when we consider the effect of lenses or mirrors on the path of light, we usually prefer to draw diagrams using light rays and normals. Light rays represent the direction of travel of wavefronts. The **normal** is an imaginary line perpendicular to a boundary between two materials or a surface.

Refraction is the change of direction that occurs when light passes at an angle across a boundary between two transparent substances. Figure 2 shows the change of direction of a light ray when it enters and when it leaves a rectangular glass block in air. The light ray bends:

- towards the normal when it passes from air into glass
- away from the normal when it passes from glass into air.

No refraction takes place if the incident light ray is along the normal.

At a boundary between two transparent substances, the light ray bends:

- towards the normal if it passes into a more dense substance
- away from the normal if it passes into a less dense substance.

Investigating the refraction of light by glass

Use a ray box to direct a light ray into a rectangular glass block at different angles of incidence near the midpoint P of one of the longer sides, as shown in Figure 2. Note that the angle of incidence is the angle between the incident light ray and the normal at the point of incidence.

For each angle of incidence at P, mark the point Q where the light ray leaves the block. The angle of refraction is the angle between the normal at P and the line PQ. Measurements of the angles of incidence and refraction for different incident rays show that:

- the angle of refraction, r, at P is always less than the angle of incidence, i
- the ratio of $\frac{\sin i}{\sin r}$ is the same for each light ray. This is known as Snell's law. The ratio is referred to as the **refractive index**, n, of glass.

For a light ray travelling from air into a transparent substance,

$$\textbf{the refractive index of the substance } n = \frac{\sin i}{\sin r}$$

Notice that **partial reflection** also occurs when a light ray in air enters glass (or any other refractive substance).

Worked example

A light ray is directed into a glass block of refractive index 1.5 at an angle of incidence of 40°. Calculate the angle of refraction of this light ray.

Learning objectives:

- → Explain what we mean by rays.
- → State Snell's law.
- → Comment on whether refraction is different for a light ray travelling from a transparent substance into air.

Specification reference: 3.5.8

Study tip

Remember from your GCSE that the normal is an imaginary line perpendicular to a boundary between two materials or a surface.

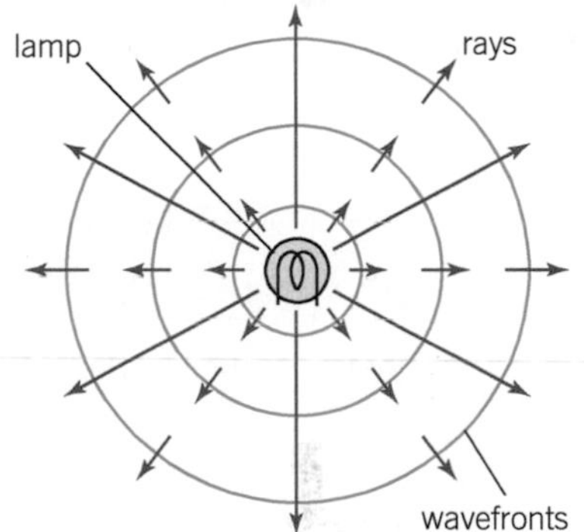

▲ **Figure 1** *Rays and waves*

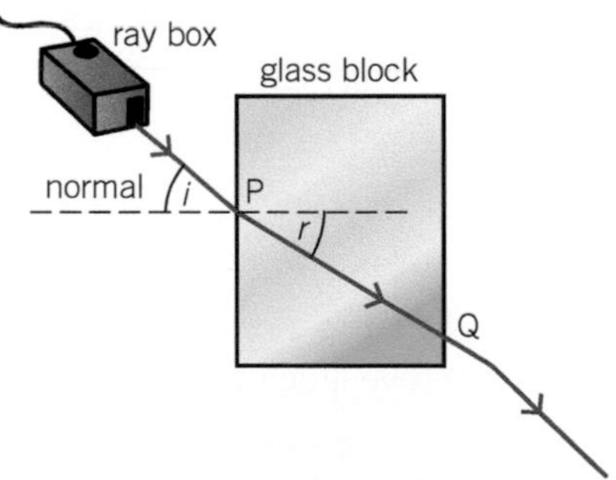

▲ **Figure 2** *Investigating refraction*

Study tip

Set your calculator mode to degrees when using Snell's law.

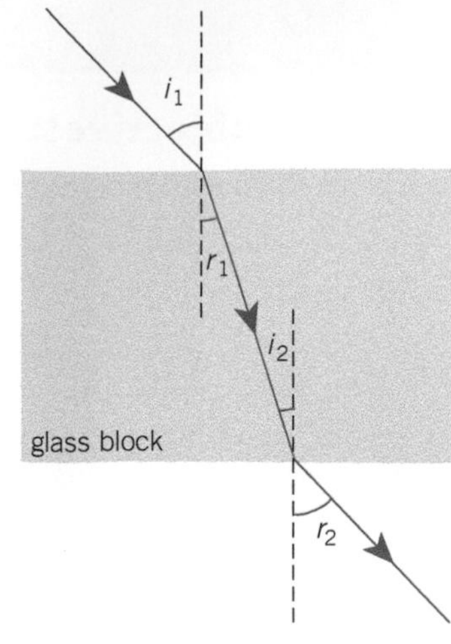

▲ **Figure 3** *Comparing glass to air refraction with air to glass refraction*

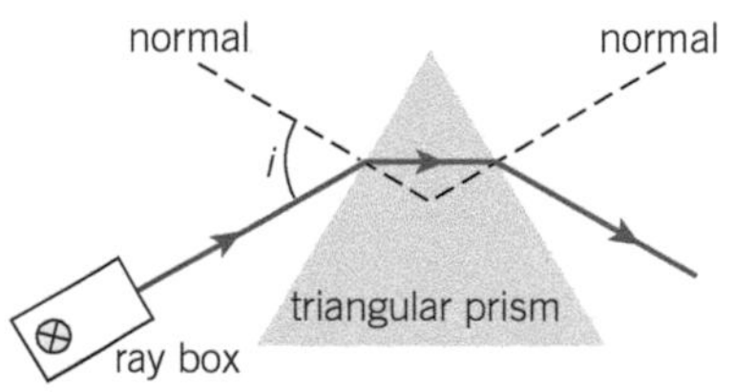

▲ **Figure 4** *Refraction by a glass prism*

Solution

$i = 40°$, $n = 1.5$

Rearranging: $\frac{\sin i}{\sin r} = n$ gives $\sin r = \frac{\sin i}{n} = \frac{\sin 40}{1.5} = \frac{0.643}{1.5} = 0.429$

Therefore $r = 25°$.

Comparing glass to air refraction with air to glass refraction

In Figure 3, notice that the angle of refraction of the light ray emerging from the block is the same as the angle of incidence of the light ray entering the block. This is because the two sides of the block at which refraction occurs are parallel to each other.

- If i_1 and r_1 are the angles of incidence and refraction at the point where the light ray enters the block, then the refractive index of the glass $n = \frac{\sin i_1}{\sin r_1}$.
- At the point where the light ray leaves the block, $i_2 = r_1$ and $r_2 = i_1$, so $\frac{\sin i_2}{\sin r_2} = \frac{1}{n}$.

Refraction of a light ray by a triangular prism

Figure 4 shows the path of a monochromatic light ray through a triangular prism. The light ray refracts towards the normal where it enters the glass prism then refracts away from the normal where it leaves the prism.

Summary questions

1 **a** A light ray in air is directed into water from air. The refractive index of water is 1.33. Calculate the angle of refraction of the light ray in the water for an angle of incidence of

i 20° **ii** 40° **iii** 60°.

b A light ray in water is directed at the water surface at an angle of incidence of 40°.

i Calculate the angle of refraction of the light ray at this surface.

ii Sketch the path of the light ray showing the normal and the angles of incidence and refraction.

2 A light ray is directed from air into a glass block of refractive index 1.5. Calculate:

a the angle of refraction at the point of incidence if the angle of incidence is **i** 30° **ii** 60°

b the angle of incidence if the angle of refraction at the point of incidence is **i** 35° **ii** 40°.

3 A light ray in air is directed at the flat side of a semicircular glass block at an angle of incidence of 50°, as shown in Figure 5. The angle of refraction at the point of incidence is 30°.

a Calculate the refractive index of the glass.

b Calculate the angle of refraction if the angle of incidence is changed to 60°.

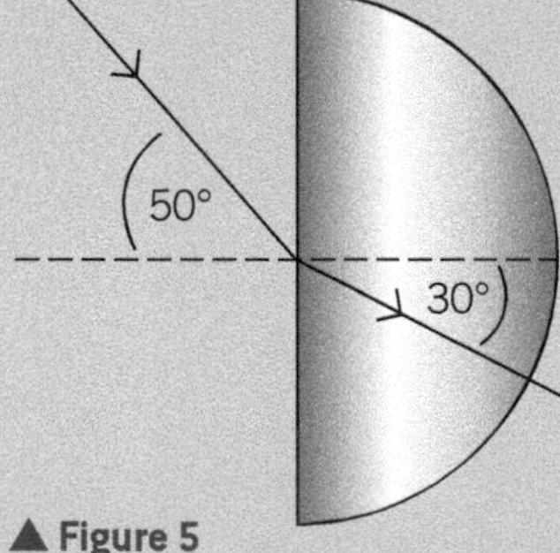

▲ **Figure 5**

4 A light ray enters an equilateral glass prism of refractive index 1.55 at the midpoint of one side of the prism at an angle of incidence of 35°.

a Sketch this arrangement and show that the angle of refraction of the light ray in the glass is 22°.

b **i** Show that the angle of incidence where the light ray leaves the glass prism is 38°.

ii Calculate the angle of refraction of the light ray where it leaves the prism.

12.6 More about refraction

Learning objectives:

- → Explain what happens to the speed of light waves when they enter a material such as water.
- → Relate refractive index to the speed of light waves.
- → Explain why a glass prism splits white light into the colours of a spectrum.

Specification reference: 3.5.8

Explaining refraction

Refraction occurs because the speed of the light waves is different in each substance. The amount of refraction that takes place depends on the speed of the waves in each substance.

Consider a wavefront of a light wave when it passes across a straight boundary from a vacuum (or air) into a transparent substance, as shown in Figure 1. Suppose the wavefront moves from XY to X′Y′ in time t, crossing the boundary between X and Y′. In this time, the wavefront moves

- a distance ct at speed c in a vacuum from Y to Y′
- a distance $c_s t$ at speed c_s in the substance from X to X′.

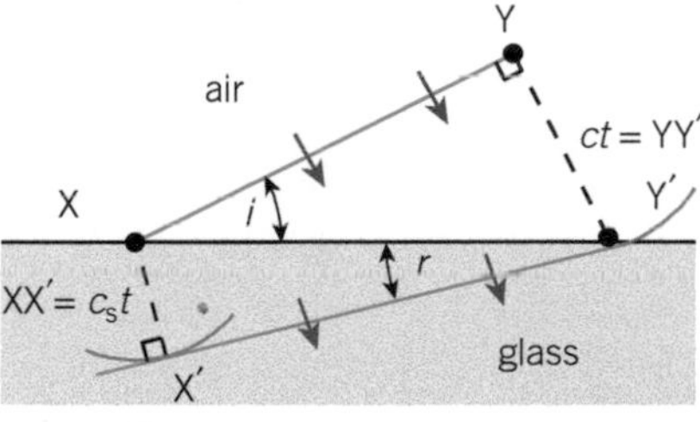

▲ **Figure 1** *Explaining refraction*

Considering triangle XYY′, since YY′ is the direction of the wavefront in the vacuum and is therefore perpendicular to XY, then YY′ = XY′ sin$i$, where i = angle YXY′.

$$ct = XY' \sin i$$

Considering triangle XX′Y′, since XX′ is the direction of the wavefront in the substance and is therefore perpendicular to X′Y′, then XX′ = XY′ sin$r$, where r = angle XY′X′.

$$c_s t = XY' \sin r$$

Combining these two equations therefore gives

$$\frac{\sin i}{\sin r} = \frac{c}{c_s}$$

Therefore the refractive index of the substance $n_s = \dfrac{c}{c_s}$

This equation shows that the smaller the speed of light is in a substance, the greater is the refractive index of the substance.

Note:

The frequency f of the waves does not change when refraction occurs. As $c = f\lambda$ and $c_s = \lambda_s f$, where λ and λ_s are the wavelengths of the waves in a vacuum and in the substance, respectively, then it follows that the refractive index of the substance,

$$n_s = \frac{c}{c_s} = \frac{\lambda}{\lambda_s}$$

Worked example

The speed of light is $3.00 \times 10^8\,\mathrm{m\,s^{-1}}$ in a vacuum. A certain type of glass has a refractive index of 1.62. Calculate the speed of light in the glass.

Solution

Rearranging $n_s = \dfrac{c}{c_s}$ gives $c_s = \dfrac{c}{n_s} = \dfrac{3.0 \times 10^8}{1.62} = 1.85 \times 10^8\,\mathrm{m\,s^{-1}}$

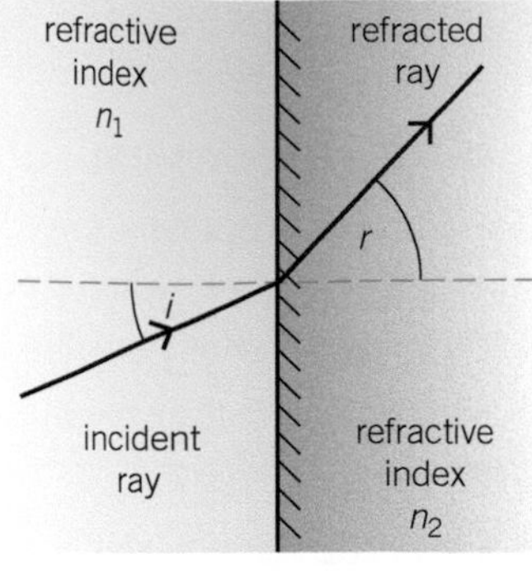

▲ **Figure 2** *The n sin i rule*

Refraction at a boundary between two transparent substances

Consider a light ray crossing a boundary from a substance in which the speed of light is c_1 to a substance in which the speed of light is c_2. Using the same theory as on the previous page gives

$$\frac{\sin i}{\sin r} = \frac{c_1}{c_2}$$

where i = the angle between the incident ray and the normal and r = the angle between the refracted ray and the normal.

This equation may be rearranged as $\frac{1}{c_1}\sin i = \frac{1}{c_2}\sin r$.

Multiplying both sides of this equation by c, the speed of light in a vacuum, gives

$$\frac{c}{c_1}\sin i = \frac{c}{c_2}\sin r$$

Substituting n_1 for $\frac{c}{c_1}$, where n_1 is the refractive index of substance 1, and n_2 for $\frac{c}{c_2}$, where n_2 is the refractive index of substance, gives Snell's law,

$$n_1 \sin\theta_1 = n_2 \sin\theta_2$$

where $\theta_1 = i$ and $\theta_2 = r$.

Study tip

Always use the equation $n_1 \sin\theta_1 = n_2 \sin\theta_2$ in calculations if you know three of its four quantities.

Worked example

A light ray crosses the boundary between water of refractive index 1.33 and glass of refractive index 1.50 at an angle of incidence of 40°. Calculate the angle of refraction of this light ray.

Solution

$n_1 = 1.33$, $n_2 = 1.50$, $\theta_1 = 40°$

Using $n_1 \sin\theta_1 = n_2 \sin\theta_2$ gives $1.33 \sin 40 = 1.50 \sin\theta_2$

$$\therefore \sin\theta_2 = \frac{1.33 \sin 40}{1.5} = 0.57$$

$$\theta_2 = 35°$$

Study tip

During refraction, the speed and wavelength both change, but the frequency stays constant.

Note:

When a light ray passes from a vacuum into a transparent substance of refractive index n,

$$\frac{\sin\theta_1}{\sin\theta_2} = n$$

where θ_1 = the angle between the incident ray and the normal

θ_2 = the angle between the refracted ray and the normal

$$n = \frac{\text{the speed of light in a vacuum}}{\text{the speed of light in the transparent substance}}$$

Study tip

Remember that the refractive index of air is approximately 1. This is very similar to the refractive index of a vacuum.

The speed of light in air at atmospheric pressure is 99.97% of the speed of light in a vacuum.

Therefore, the refractive index of air is 1.0003. For most purposes, the refractive index of air may be assumed to be 1.

The white light spectrum

We can use a prism to split a beam of white light from a filament lamp (or sunlight) into the colours of the spectrum by a glass prism, as shown in Figure 3. This happens because white light is composed of light with a continuous range of wavelengths, from red at about 650 nm to violet at about 350 nm. The glass prism refracts light by different amounts, depending on its wavelength. The shorter the wavelength in air, the greater the amount of refraction. So each colour in the white light beam is refracted by a different amount. This dispersive effect occurs because the speed of light in glass depends on wavelength. Violet light travels more slowly than red light in glass so the refractive index of violet light is greater than the refractive index of red light.

▲ **Figure 3** *The incident beam of white light enters the prism at its right-hand side*

Summary questions

1 Water waves of frequency 4.0 Hz travelling at a speed of 0.16 m s^{-1} travel across a boundary from deep to shallow water where the speed is 0.12 m s^{-1}.

a Calculate the wavelength of these waves
i in the deep water **ii** in the shallow water.

b The incident wavefronts cross the boundary at an angle of 25° to the boundary. Calculate the angle of the refracted wavefronts to the boundary.

2 **a** The speed of light in a vacuum is 3.00×10^{8} m s^{-1}. Calculate the speed of light in
i glass of refractive index 1.52 **ii** water.
The refractive index of water = 1.33.

b A light ray passes across a plane boundary from water into glass at an angle of incidence of 55°. Use the refractive index values from part **a** to show that the angle of refraction of this light ray is 46°.

3 Calculate the angle of refraction for a light ray entering glass of refractive index 1.50 at an angle of incidence of 40° in:

a air **b** water of refractive index 1.33.

4 A white light ray is directed through the curved side of a semicircular glass block at the midpoint of the flat side, as shown in Figure 4. The angle of incidence of the light ray at the flat side is 30°.

The refractive index of the glass for red light is 1.52 and for blue light is 1.55.

a Calculate the angle of refraction at the midpoint of
i the red component, and
ii the blue component of the light ray.

b Hence show that the angle between the red and blue components of the refracted light ray is 1.3°.

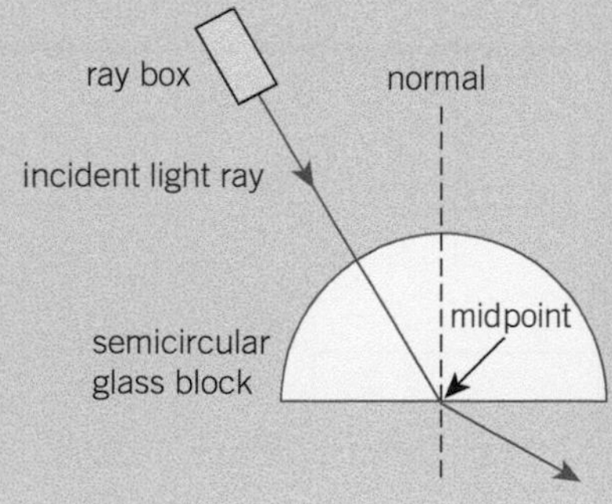

▲ **Figure 4**

12.7 Total internal reflection

Learning objectives:

- → State the conditions for total internal reflection.
- → Relate the critical angle to refractive index.
- → Explain why diamonds sparkle.

Specification reference: 3.5.8

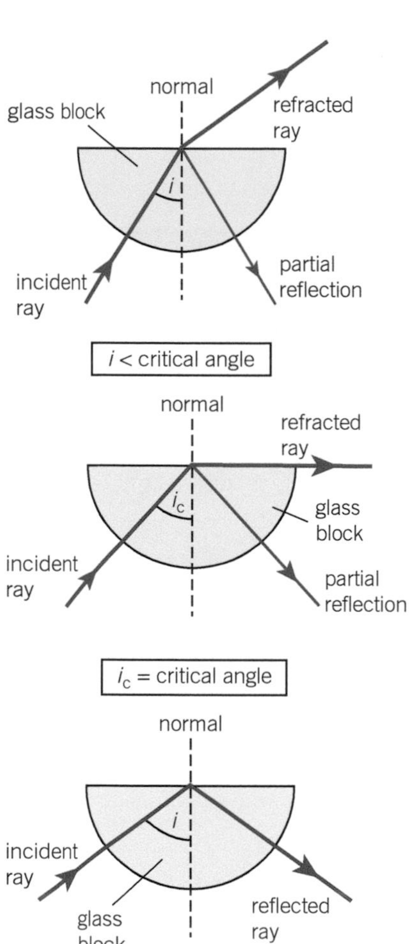

▲ **Figure 1** *Total internal reflection*

Study tip

Partial internal reflection always occurs at a boundary when the angle of incidence is less than or equal to the critical angle.

Investigating total internal reflection

When a light ray travels from glass into air, it refracts away from the normal. If the angle of incidence is increased to a certain value known as the **critical angle**, the light ray refracts along the boundary. Figure 1 shows this effect. If the angle of incidence is increased further, the light ray undergoes **total internal reflection** at the boundary, the same as if the boundary were replaced by a plane mirror.

In general, total internal reflection can only take place if:

1 the incident substance has a **larger refractive index** than the other substance

2 the angle of incidence **exceeds the critical angle**.

At the critical angle i_c, the angle of refraction is 90° because the light ray emerges along the boundary. Therefore, $n_1 \sin i_c = n_2 \sin 90$ where n_1 is the refractive index of the incident substance and n_2 is the refractive index of the other substance. Since $\sin 90 = 1$, then

$$\sin \theta_c = \frac{n_2}{n_1}$$

Prove for yourself that the critical angle for the boundary between glass of refractive index 1.5 and air (refractive index = 1) is 42°. This means that if the angle of incidence of a light ray in the glass is greater than 42°, the light ray undergoes total internal reflection back into the glass at the boundary.

Why do diamonds sparkle when white light is directed at them?

When white light enters a diamond, it is split into the colours of the spectrum. Diamond has a very high refractive index of 2.417 so it separates the colours more than any other substance does. In addition, the high refractive index gives diamond a critical angle of 24.4°. So a light ray in a diamond may be totally internally reflected many times before it emerges, which means its colours spread out more and more. So the diamond sparkles with different colours.

Optical fibres

Optical fibres are used in medical **endoscopes** to see inside the body, and in communications to carry light signals. Figure 2 (on the next page) shows the path of a light ray along an optical fibre. The light ray is totally internally reflected each time it reaches the fibre boundary, even where the fibre bends, unless the radius of the bend is too small. At each point where the light ray reaches the boundary, the angle of incidence exceeds the critical angle of the fibre.

A communications optical fibre allows pulses of light that enter at one end, from a transmitter, to reach a receiver at the other end. Such fibres need to be highly transparent to minimise absorption of light, which would otherwise reduce the amplitude of the pulses progressively the further they travel in the fibre. Each fibre consists of a core surrounded by a layer of cladding of lower refractive index to reduce light loss from the core. Light loss would also reduce the amplitude of the pulses.

- Total internal reflection takes place at the core–cladding boundary. At any point where two fibres are in direct contact, light would cross from one fibre to the other if there were no cladding. Such crossover would mean that the signals would not be secure, as they would reach the wrong destination.
- The core must be very narrow to prevent **modal** (i.e., multipath) **dispersion**. This occurs in a wide core because light travelling along the axis of the core travels a shorter distance per metre of fibre than light that repeatedly undergoes total internal reflection. A pulse of light sent along a wide core would become longer than it ought to be. If it was too long, it would merge with the next pulse.

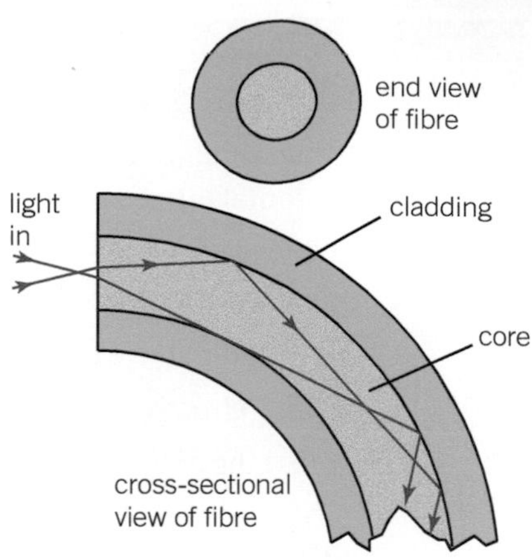

▲ **Figure 2** *Fibre optics*

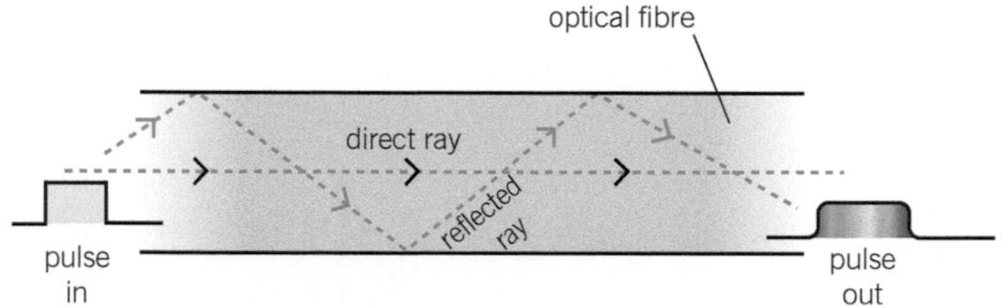

▲ **Figure 3** *Modal dispersion*

Maths link

Q: Make an order-of-magnitude estimate of the number of fibres of diameter 10 μm in a bundle of fibres of diameter 10 mm.

Answer: 10^6.

Note:

Pulse broadening also occurs if white light is used instead of monochromatic light (light of a single wavelength). This **material dispersion** (sometimes referred to as spectral dispersion) is because the speed of light in the glass of the optical fibre depends on the wavelength of light travelling through it. Violet light travels more slowly than red light in glass. The difference in speed would cause white light pulses in optical fibres to become longer, as the violet component falls behind the faster red component of each pulse. So the light (or infrared radiation) used must be monochromatic to prevent pulse merging.

The **medical endoscope** contains two bundles of fibres. The endoscope is inserted into a body cavity, which is then illuminated using light sent through one of the fibre bundles. A lens over the end of the other fibre bundle is used to form an image of the body cavity on the end of the fibre bundle. The light that forms this image travels along the fibres to the other end of the fibre bundle where the image can be observed. This fibre bundle needs to be a **coherent bundle**, which means that the fibre ends at each end are in the same relative positions.

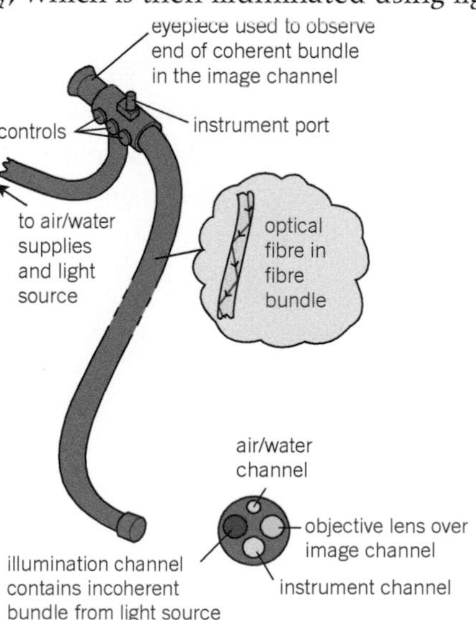

a *Endoscope design*

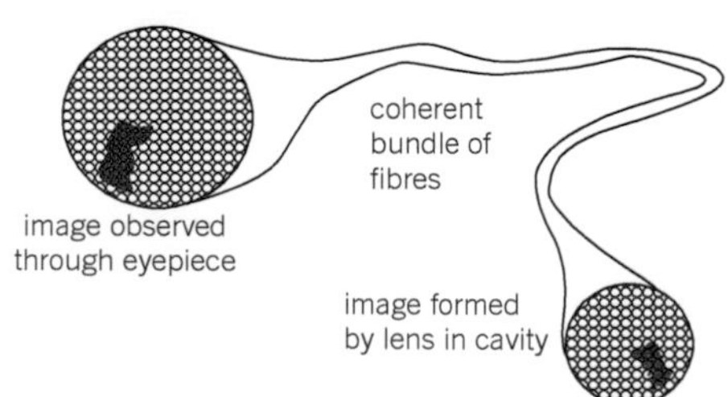

b *A coherent bundle*

▲ **Figure 4** *The endoscope*

Summary questions

1 a State two conditions for a light ray to undergo total internal reflection at a boundary between two transparent substances.

b Calculate the critical angle for

i glass of refractive index 1.52 and air

ii water (refractive index = 1.33) and air.

2 a Show that the critical angle at a boundary between glass of refractive index 1.52 and water (refractive index = 1.33) is 61°.

b Figure 5 shows the path of a light ray in water of refractive index 1.33 directed at an angle of incidence of 40° at a thick glass plate of refractive index 1.52.

i Calculate the angle of refraction of the light ray at P.

ii State the angle of incidence of the light ray at Q.

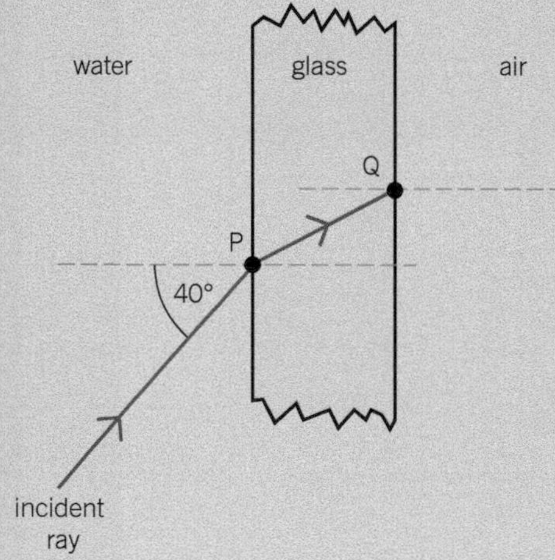

▲ **Figure 5**

c Sketch the path of the light ray beyond Q.

3 A window pane made of glass of refractive index 1.55 is covered on one side only with a transparent film of refractive index 1.40.

a Calculate the critical angle of the film–glass boundary,

b A light ray in air is directed at the film at an angle of incidence of 45° as shown in Figure 6. Calculate:

i the angle of refraction in the film

ii the angle of refraction of the ray where it leaves the window pane.

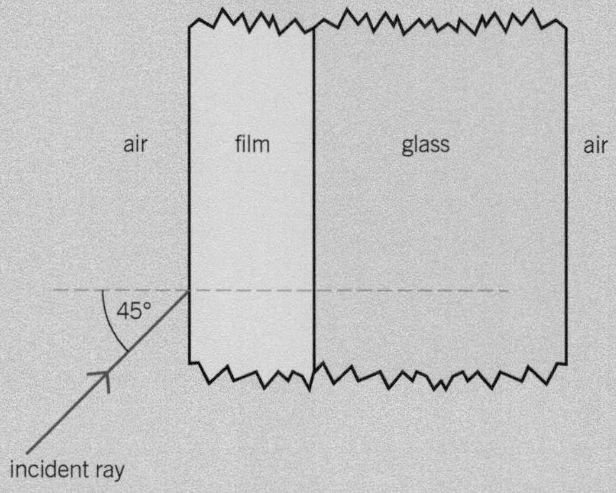

▲ **Figure 6**

4 a In a medical endoscope, the fibre bundle used to view the image is coherent.

i What is meant by a coherent fibre bundle?

ii Explain why this fibre bundle needs to be coherent.

b i Why is an optical fibre used in communication composed of a core surrounded by a layer of cladding of lower refractive index?

ii Why is it necessary for the core of an optical communications fibre to be narrow?

Practice questions: Chapter 12

1 In a secure communications system, an optical fibre is used to transmit digital signals in the form of infrared pulses. The fibre has a thin core, which is surrounded by cladding that has a lower refractive index than the core.

(a) **Figure 1** shows a cross section of a straight length of the fibre.

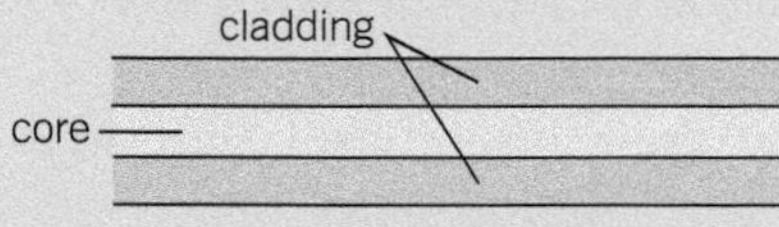

▲ **Figure 1**

(i) The core has a refractive index of 1.52 and the cladding has a refractive index of 1.35. Show that the critical angle at the core–cladding boundary is 62.6°.

(ii) Sketch the path of a light ray in the core that is totally internally reflected when it reaches the core–cladding boundary. *(6 marks)*

(b) (i) Explain why the cladding is necessary.

(ii) An optical fibre is used to transmit digital images from a security camera to a computer where the images are stored. A prominent notice near the camera informs people that the camera is in use. Discuss the benefits and drawbacks associated with making and storing images of people in this situation. *(7 marks)*

2 **Figure 2** shows a cube of glass. A ray of light, incident at the centre of a face of the cube, at an angle of incidence θ, goes on to meet another face at an angle of incidence of 50°, as shown in **Figure 2**.
Critical angle at the glass–air boundary = 45°

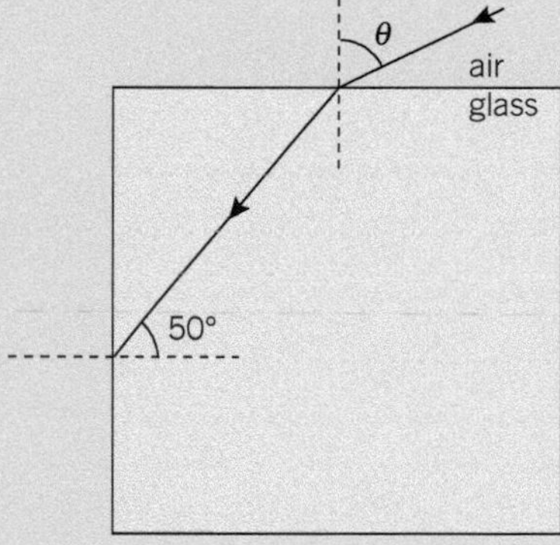

▲ **Figure 2**

(a) Copy the diagram and draw the continuation of the path of the ray, showing it passing through the glass and out into the air. *(3 marks)*

(b) Show that the refractive index of the glass is 1.41. *(2 marks)*

(c) Calculate the angle of incidence, θ. *(3 marks)*

AQA, 2005

3 **Figure 3** shows a ray of monochromatic light, in the plane of the paper, incident on the end face of an optical fibre.

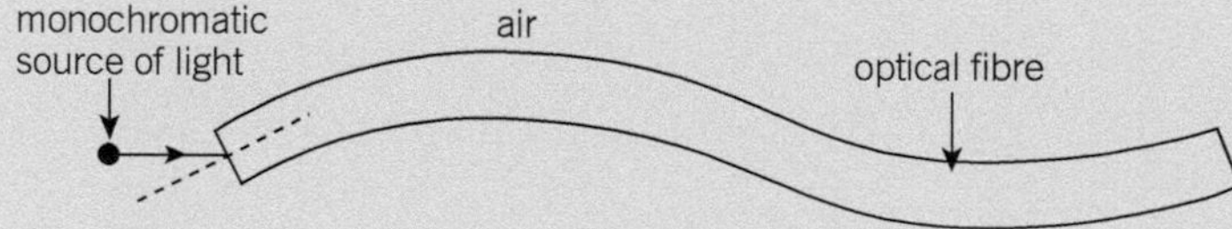

▲ **Figure 3**

(a) (i) Draw on a copy of the diagram the complete path followed by the incident ray, showing it entering into the fibre and emerging from the fibre at the far end.

(ii) State any changes that occur in the speed of the ray as it follows this path from the source.
Calculations are not required. *(4 marks)*

(b) (i) Calculate the critical angle for the optical fibre at the air boundary.
refractive index of the optical fibre glass = 1.57

(ii) The optical fibre is now surrounded by cladding of refractive index 1.47. Calculate the critical angle at the core–cladding boundary.

(iii) State *one* advantage of cladding an optical fibre. *(6 marks)*

AQA, 2004

4 **(a)** State what is meant by coherent sources of light. *(2 marks)*

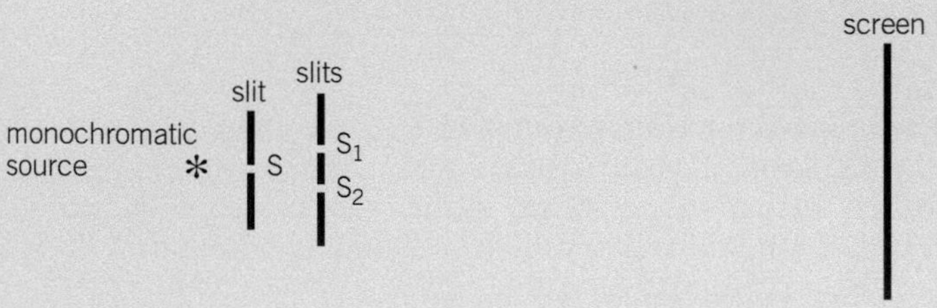

screen
slits
slit
monochromatic source
S
S_1
S_2

▲ **Figure 4**

(b) Young's fringes are produced on the screen from the monochromatic source by the arrangement shown in **Figure 4**.

(i) Explain why slit S should be narrow.

(ii) Why do slits S_1 and S_2 act as coherent sources? *(4 marks)*

(c) The pattern on the screen may be represented as a graph of intensity against position on the screen. The central fringe is shown on the graph in **Figure 5**. Copy and complete this graph to represent the rest of the pattern. *(2 marks)*

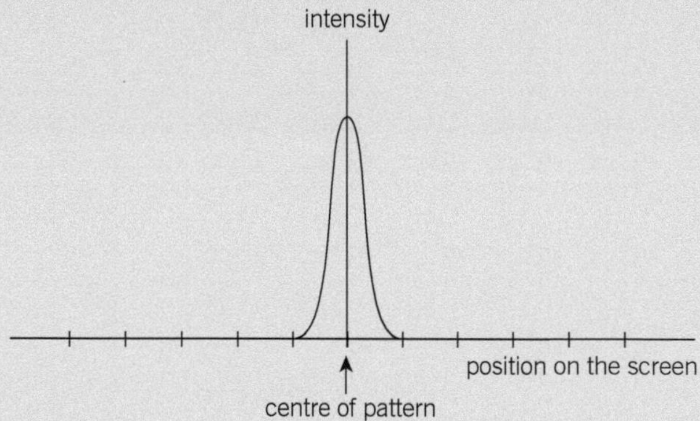

▲ **Figure 5**

AQA, 2005

5 A diffraction grating has 940 lines per mm.

(a) Calculate the distance between adjacent lines on the grating. *(1 mark)*

(b) Monochromatic light is incident on the grating and a second-order spectral line is formed at an angle of 55° from the normal to the grating. Calculate the wavelength of the light. *(3 marks)*

AQA, 2006

13 Quantum phenomena

13.1 Photoelectricity

The discovery of the photoelectric effect

A metal contains conduction electrons, which move about freely inside the metal. These electrons collide with each other and with the positive ions of the metal. When Heinrich Hertz discovered how to produce and detect radio waves, he found that the sparks produced in his spark gap detector when radio waves were being transmitted were stronger when ultraviolet radiation was directed at the spark gap. Further investigations on the effect of electromagnetic radiation on metals showed that electrons are emitted from the surface of a metal when electromagnetic radiation above a certain frequency was directed at the metal. This effect is known as the **photoelectric effect**.

Puzzling problems

The following observations were made about the photoelectric effect after Hertz's discovery. These observations were a major problem because they could not be explained using the idea that light is a wave.

1 Photoelectric emission of electrons from a metal surface does *not* take place if the frequency of the incident electromagnetic radiation is below a certain value known as the **threshold frequency**. This minimum frequency depends on the type of metal. This means that the **wavelength** of the incident light must be less than a *maximum* value equal to the speed of light divided by the threshold frequency since $\lambda = \frac{c}{f}$.

2 The number of electrons emitted per second is proportional to the intensity of the incident radiation, provided the frequency is greater than the threshold frequency. However, if the frequency of the incident radiation is less than the threshold frequency, no photoelectric emission from that metal surface can take place, no matter how intense the incident radiation is.

3 Photoelectric emission occurs without delay as soon as the incident radiation is directed at the surface, provided the frequency of the radiation exceeds the threshold frequency, and regardless of intensity.

The wave theory of light cannot explain either the existence of a threshold frequency or why photoelectric emission occurs without delay. According to wave theory, each conduction electron at the surface of a metal should gain some energy from the incoming waves, regardless of how many waves arrive each second.

Einstein's explanation of the photoelectric effect

The photon theory of light was put forward by Einstein in 1905 to explain the photoelectric effect. Einstein assumed that light is composed of wavepackets or **photons**, each of energy equal to hf, where f is the frequency of the light and h is the Planck constant. The accepted value for h is 6.63×10^{-34} J s.

$$\textbf{Energy of a photon} = hf$$

For electromagnetic waves of wavelength λ, the energy of each photon $E = hf = \frac{hc}{\lambda}$, where c is the speed of the electromagnetic waves.

Learning objectives:

- → Explain the photoelectric effect.
- → Define a photon.
- → Discuss how the photon model was established.

Specification reference: 3.5.10

Practical link

Observing the photoelectric effect

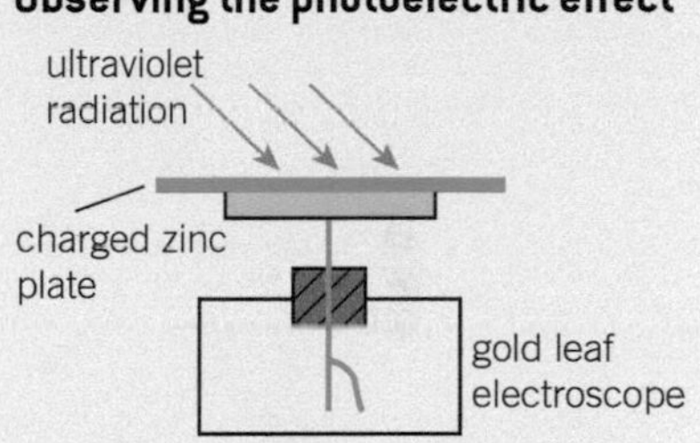

Ultraviolet radiation from a UV lamp is directed at the surface of a zinc plate placed on the cap of a gold leaf electroscope. This device is a very sensitive detector of charge. When it is charged, the thin gold leaf of the electroscope rises – it is repelled from the metal stem, because they both have the same type of charge.

If the electroscope is charged negatively, the leaf rises and stays in position. However, if ultraviolet light is directed at the zinc plate, the leaf gradually falls. The leaf falls because conduction electrons at the zinc surface leave the zinc surface when ultraviolet light is directed at it. The emitted electrons are referred to as **photoelectrons**.

If the electroscope is charged positively, the leaf rises and stays in position, regardless of whether or not ultraviolet light is directed at the zinc plate.

Synoptic link

You have met Einstein's ideas about photons in Topic 7.3, Photons.

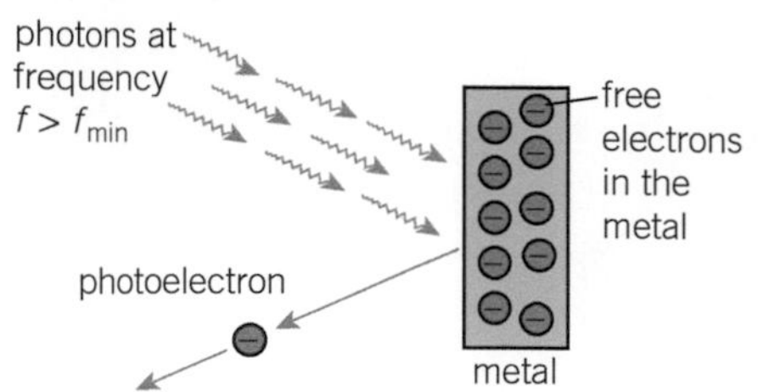

▲ **Figure 1** *Explaining the photoelectric effect – one electron absorbs one photon*

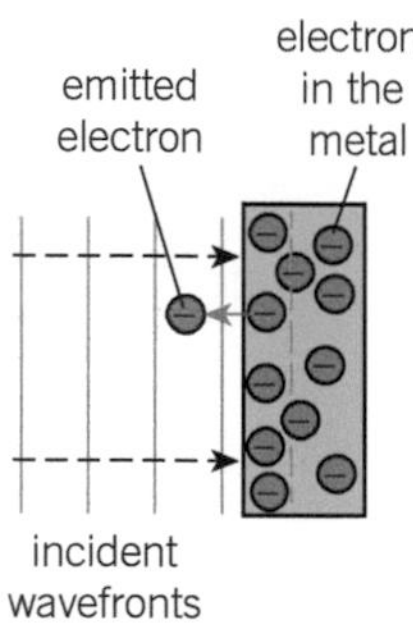

▲ **Figure 2** *Each electron in the metal can only absorb some of energy from each incident wavefront if light is a wave, so emission would take longer – this model is incorrect*

To explain the photoelectric effect, Einstein said that:

- When light is incident on a metal surface, an electron at the surface absorbs a *single* photon from the incident light and therefore gains energy equal to hf, where hf is the energy of a light photon.
- An electron can leave the metal surface if the energy gained from a single photon exceeds the **work function of the metal, ϕ**. This is the minimum energy needed by an electron to escape from the metal surface. Excess energy gained by the photoelectron becomes its kinetic energy.

The maximum kinetic energy of an emitted electron is therefore

$$E_{K(max)} = hf - \phi$$

Rearranging this equation gives

$$hf = E_{K(max)} + \phi$$

Emission can take place from a metal surface provided $E_{K(max)} > 0$ or $hf > \phi$, so the threshold frequency of the metal is $f_{min} = \frac{\phi}{h}$

Stopping potential

Electrons that escape from the metal plate can be attracted back to it by giving the plate a sufficient positive charge. The minimum potential needed to stop photoelectric emission is called the **stopping potential V_s**. At this potential, the maximum kinetic energy of the emitted electron is reduced to zero because each emitted electron must do extra work equal to $e \times V_S$ so they lose kinetic energy as they move away from the metal surface. Hence its maximum kinetic energy is equal to $e \times V_S$.

Conclusive experimental evidence for Einstein's photon theory was obtained by Robert Millikan. Millikan measured the stopping potential for a range of metals using light of different frequencies. His results fitted Einstein's photoelectric equation very closely. After these results were checked and evaluated by other physicists through peer review, the scientific community accepted that light consists of photons.

Summary questions

$h = 6.63 \times 10^{-34}$ J s, $c = 3.00 \times 10^{8}$ m s^{-1}

1 a What is meant by photoelectric emission from a metal surface?

b Explain why photoelectric emission from a metal surface only takes place if the frequency of the incident radiation is greater than a certain value.

2 a Calculate the frequency and energy of a photon of wavelength

i 450 nm

ii 1500 nm.

b A metal surface at zero potential emits electrons from its surface if light of wavelength 450 nm is directed at it. However, electrons are not emitted when light of wavelength 650 nm is used. Explain these observations.

3 The work function of a certain metal plate is 1.1×10^{-19} J. Calculate:

a the threshold frequency of incident radiation for this metal

b the maximum kinetic energy of photoelectrons emitted from this plate when light of wavelength 520 nm is directed at the metal surface.

4 Light of wavelength 635 nm is directed at a metal plate at zero potential. Electrons are emitted from the plate with a maximum kinetic energy of 1.5×10^{-19} J. Calculate:

a the energy of a photon of this wavelength

b the work function of the metal

c the threshold frequency of electromagnetic radiation incident on this metal.

13.2 More about photoelectricity

Into the quantum world

At the end of the 19th century, a physicist named Max Planck suggested that the energy of each vibrating atom is **quantised** – only certain levels of energy are allowed. He said the energy could only be in multiples of a basic amount, or quantum, hf, where f is the frequency of vibration of the atom and h is a constant, which became known as the Planck constant. He imagined the energy levels to be like the rungs of a ladder, with each atom absorbing or emitting radiation when it moved up or down a level.

There was a problem that the physicists of the time couldn't solve – the photoelectric effect. They couldn't fully explain it using the wave theory of radiation. Einstein did so by inventing a new theory of radiation – the photon model. His key idea was that electromagnetic radiation consists of photons, or wavepackets, of energy hf, where f is the frequency of the radiation and h is the Planck constant. Einstein's photon model and Planck's theory of vibrating atoms showed that energy is quantised – a completely new way of thinking about energy.

More about conduction electrons

The conduction electrons in a metal move about at random, like the molecules of a gas. The average kinetic energy of a conduction electron depends on the temperature of the metal.

The work function of a metal is the *minimum* energy needed by a conduction electron to escape from the metal surface when the metal is at zero potential. The work function of a metal is of the order of 10^{-19} J, which is about 20 times greater than the average kinetic energy of a conduction electron in a metal at 300 K.

When a conduction electron absorbs a photon, its kinetic energy increases by an amount equal to the energy of the photon. If the energy of the photon exceeds the work function of the metal, the conduction electron can leave the metal. If the electron does not leave the metal, it collides repeatedly with other electrons and positive ions, and it quickly loses its extra kinetic energy.

The vacuum photocell

A vacuum photocell is a glass tube that contains a metal plate, referred to as the photocathode, and a smaller metal electrode referred to as the anode. Figure 2 shows a vacuum photocell in a circuit. When light of a frequency greater than the threshold frequency for the metal is directed at the photocathode, electrons are emitted from the cathode and are attracted to the anode. The microammeter in the circuit can be used to measure the photoelectric current. This is proportional to the number of electrons per second that transfer from the cathode to the anode.

Learning objectives:

- → Explain why Einstein's photon model was revolutionary.
- → Define a quantum.
- → Explain why an electron can't absorb several photons to escape from a metal.

Specification reference: 3.5.10

▲ **Figure 1** *Albert Einstein 1879–1955*

Study tip

Remember that the work function is characteristic of the metal and the threshold frequency relates the work function to the incident radiation.

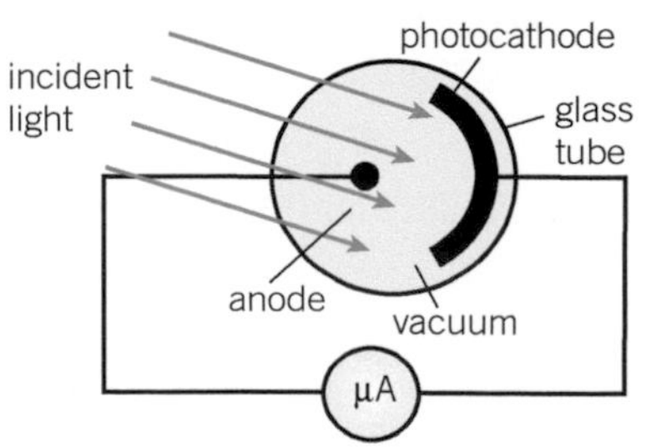

▲ **Figure 2** *Using a vacuum photocell*

Synoptic link

The equation
charge, Q (C) = current, I (A) × time, t (s)
was covered in Topic 9.1, Current and charge.

- For a photoelectric current I, the number of photoelectrons per second that transfer from the cathode to the anode = $\frac{I}{e}$, where e is the charge of the electron.
- The photoelectric current is proportional to the intensity of the light incident on the cathode. The light intensity is a measure of the energy per second carried by the incident light, which is proportional to the number of photons per second incident on the cathode. Because each photoelectron must have absorbed one photon to escape from the metal surface, the number of photoelectrons emitted per second (i.e., the photoelectric current) is therefore proportional to the intensity of the incident light.
- The intensity of the incident light does *not* affect the maximum kinetic energy of a photoelectron. No matter how intense the incident light is, the energy gained by a photoelectron is due to the absorption of one photon only. Therefore, the maximum kinetic energy of a photoelectron is still given by $E_{Kmax} = hf - \phi$.
- The maximum kinetic energy of the photoelectrons emitted for a given frequency of light can be measured using a photocell.
- If the measurements for different frequencies are plotted as a graph of E_{Kmax} against f, a straight line of the form $y = mx + c$ is obtained. This is in accordance with the above equation as $y = E_K$ and $x = f$. Note that the gradient of the line $m = h$ and the y-intercept, $c = -\phi$. The x-intercept is equal to the threshold frequency.

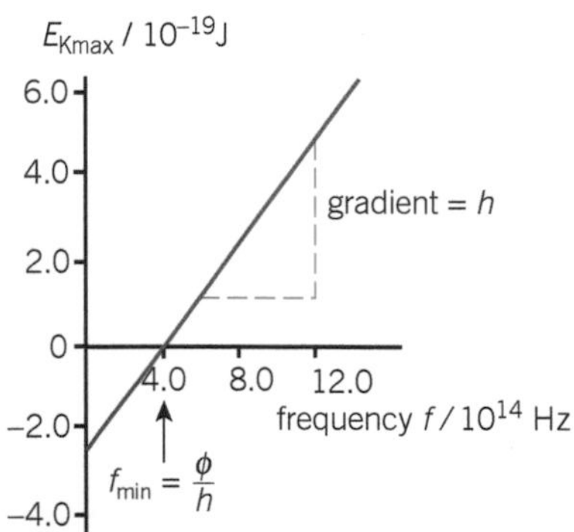

▲ **Figure 3** *A graph of E_{Kmax} against frequency*

Summary questions

$h = 6.63 \times 10^{-34}$ J s, $c = 3.00 \times 10^{8}$ m s^{-1}, $e = 1.6 \times 10^{-19}$ C

1 A vacuum photocell is connected to a microammeter. Explain the following observations.
  a When the cathode was illuminated with blue light of low intensity, the microammeter showed a non-zero reading.
  b When the cathode was illuminated with an intense red light, the microammeter reading was zero.

2 A vacuum photocell is connected to a microammeter. When light is directed at the photocell, the microammeter reads 0.25 μA.
  a Calculate the number of photoelectrons emitted per second by the photocathode of the photocell.
  b Explain why the microammeter reading is doubled if the intensity of the incident light is doubled.

3 A narrow beam of light of wavelength 590 nm and of power 0.5 mW is directed at the photocathode of a vacuum photocell, which is connected to a microammeter which reads 0.4 μA. Calculate:
  a the energy of a single light photon of this wavelength
  b the number of photons incident on the photocathode per second
  c the number of electrons emitted per second from the photocathode.

4 a Use Figure 3 to estimate **i** the threshold frequency, **ii** the work function of the metal.
  b A metal surface has a work function of 1.9×10^{-19} J. Light of wavelength 435 nm is directed at the metal surface. Calculate the maximum kinetic energy of the photoelectrons emitted from this metal surface.

13.3 Collisions of electrons with atoms

Ionisation

An **ion** is a charged atom. The number of electrons in an ion is not equal to the number of protons. An ion is formed from an uncharged atom by adding or removing electrons from the atom. Adding electrons makes the atom into a negative ion. Removing electrons makes the atom into a positive ion.

Any process of creating ions is called **ionisation**. For example:

- Alpha, beta, and gamma radiation create ions when they pass through substances and collide with the atoms of the substance.
- Electrons passing through a fluorescent tube create ions when they collide with the atoms of the gas or vapour in the tube.

Learning objectives:

→ Explain what is meant by ionisation of an atom.

→ Explain what is meant by excitation of an atom.

→ Explain what happens inside an atom when it becomes excited.

Specification reference: 3.5.9

Extension

Measuring ionisation energy

We can measure the energy needed to ionise a gas atom by making electrons collide at increasing speed with the gas atoms in a sealed tube. The electrons are emitted from a heated filament in the tube and are attracted to a positive metal plate, the anode, at the other end of the tube. The gas needs to be at sufficiently low pressure, otherwise there are too many atoms in the tube and the electrons cannot reach the anode.

The potential difference (p.d.) between the anode and the filament is increased so as to increase the speed of the electrons. The circuit is shown in Figure 2. The ammeter records a very small current due to electrons from the filament reaching the anode. No ionisation occurs until the electrons from the filament reach a certain speed. At this speed, each electron arrives near the anode with just enough kinetic energy to ionise a gas atom, by knocking an electron out of the atom. Ionisation near the anode causes a much greater current to pass through the ammeter.

By measuring the p.d. between the filament and the anode, when the current starts to increase, we can calculate the ionisation energy of a gas atom, as this is equal to the work done W on each electron from the filament (and the work done is transformed to kinetic energy). The work done on each electron from the filament is given by its charge e × the tube potential difference V. Therefore, the **ionisation energy of a gas atom = eV.**

Q: Why does the anode need to be positive relative to the filament?

Answer: To attract electrons from the filament.

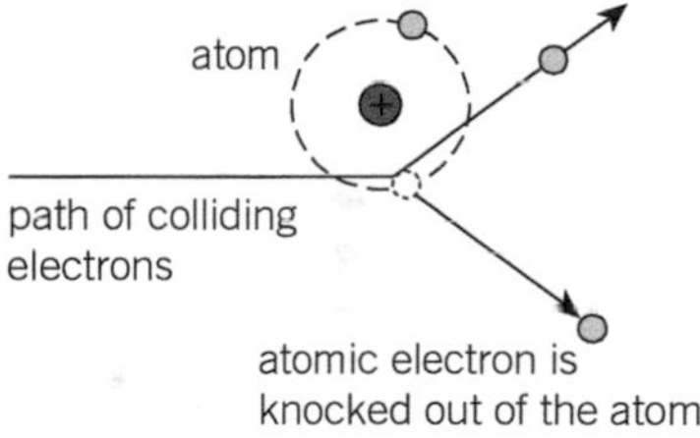

▲ **Figure 1** *Ionisation by collision*

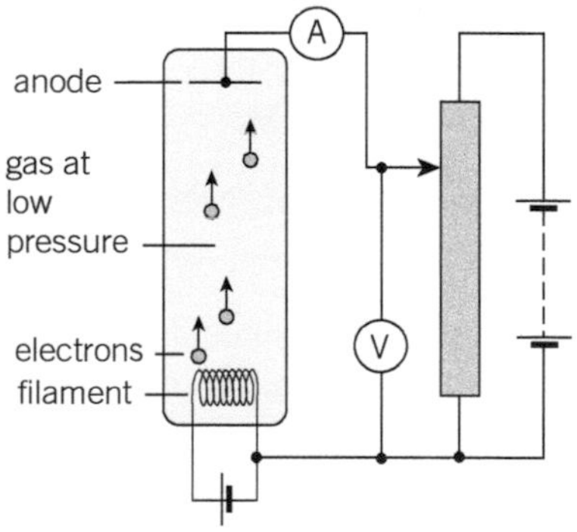

▲ **Figure 2** *Measuring ionisation energy*

The electron volt

The **electron volt** is a unit of energy equal to the work done when an electron is moved through a p.d. of 1 V. For a charge q moved through a p.d. V, the work done = qV. Therefore, the work done when an electron moves through a potential difference of 1 V is equal to 1.6×10^{-19} J (= 1.6×10^{-19} C × 1 V). This amount of energy is defined as 1 electron volt (eV).

Synoptic link

You have studied potential difference in detail in Topic 9.2, Potential difference and power.

Study tip

Don't confuse atomic electrons with electrons that hit the atom.

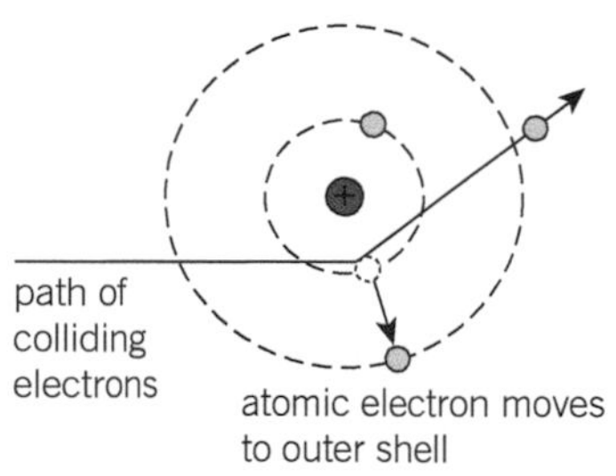

▲ **Figure 3** *A simple model of excitation by collision*

For example, the work done on

- an electron when it moves through a potential difference of 1000 V is 1000 eV
- an ion of charge $+2e$ when it moves through a potential difference of 10 V is 20 eV.

Excitation by collision

Using gas-filled tubes with a metal grid between the filament and the anode, we can show that gas atoms can absorb energy from colliding electrons without being ionised. This process, known as **excitation**, happens at certain energies, which are characteristic of the atoms of the gas. If a colliding electron loses all its kinetic energy when it causes excitation, the current due to the flow of electrons through the gas is reduced. If the colliding electron does not have enough kinetic energy to cause excitation, it is deflected by the atom, with no overall loss of kinetic energy.

The energy values at which an atom absorbs energy are known as its **excitation energies**. We can determine the excitation energies of the atoms in the gas-filled tube by increasing the potential difference between the filament and the anode and measuring the p.d. when the anode current falls. For example, two prominent excitation energies of a mercury atom are 4.9 eV and 5.7 eV. This would mean that the current would fall at 4.9 V and 5.7 V.

When excitation occurs, the colliding electron makes an electron inside the atom move from an inner shell to an outer shell. Energy is needed for this process, because the atomic electron moves away from the nucleus of the atom. The excitation energy is always less than the ionisation energy of the atom, because the atomic electron is not removed completely from the atom when excitation occurs.

Summary questions

$e = 1.6 \times 10^{-19}$ C

1 State one difference and one similarity between ionisation and excitation.

2 a The mercury atom has an ionisation energy of 10.4 eV. Calculate this ionisation energy in joules.

b An electron with 12.0 eV of kinetic energy collides with a mercury atom and ionises it. Calculate the kinetic energy, in eV, of the electron after the collision.

3 Complete the sentences below.

a The ionisation energy of neon is much greater than that of sodium. Therefore, ______ energy is needed to remove an electron from a sodium atom than from a neon atom.

b An electron that causes excitation of an atom by colliding with it has ________ kinetic energy before the collision than after the collision. The atom's internal energy _____ and the electron's kinetic energy ______.

4 a Describe what happens to a gas atom when an electron collides with it and causes it to absorb energy from the electron without being ionised.

b Explain why a gas atom cannot absorb energy from a slow-moving electron that collides with it.

13.4 Energy levels in atoms

Electrons in atoms

The electrons in an atom are trapped by the electrostatic force of attraction of the nucleus. They move about the nucleus in allowed orbits, or shells, surrounding the nucleus. The energy of an electron in a shell is constant. An electron in a shell near the nucleus has less energy than an electron in a shell further away from the nucleus. Each shell can only hold a certain number of electrons. For example, the innermost shell (i.e., the shell nearest to the nucleus) can only hold two electrons and the next nearest shell can only hold eight electrons.

Each type of atom has a certain number of electrons. For example, a helium atom has two electrons. Thus, in its lowest energy state, a helium atom has both electrons in the innermost shell.

The lowest energy state of an atom is called its **ground state**. When an atom in the ground state absorbs energy, one of its electrons moves to a shell at higher energy, so the atom is now in an **excited state**. We can use the excitation energy measurements to construct an **energy level** diagram for the atom, as shown in Figure 1. This shows the allowed energy values of the atom. Each allowed energy corresponds to a certain electron configuration in the atom. Note that the ionisation level may be considered as the zero reference level for energy, instead of the ground state level. The energy levels below the ionisation level would then need to be shown as negative values, as shown on the right-hand side of Figure 1.

Learning objectives:

- → Explain what energy levels are.
- → Explain what happens when excited atoms de-excite.
- → Explain how a fluorescent tube works.

Specification reference: 3.5.9

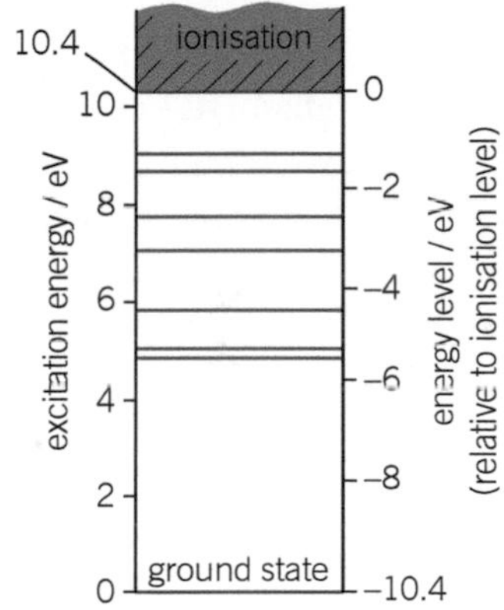

▲ **Figure 1** *The energy levels of the mercury atom*

De-excitation

Did you know that gases at low pressure emit light when they are made to conduct electricity? For example, a neon tube emits red-orange light when it conducts. The gas-filled tube used to measure excitation energies in Topic 13.3 emits light when excitation occurs. This happens because the atoms absorb energy as a result of excitation by collision, but they do not retain the absorbed energy permanently.

The electron configuration in an excited atom is unstable because an electron that moves to an outer shell leaves a vacancy in the shell it moves from. Sooner or later, the vacancy is filled by an electron from an outer shell transferring to it. When this happens, the electron emits a photon. The atom therefore moves to a lower energy level (the process of **de-excitation**).

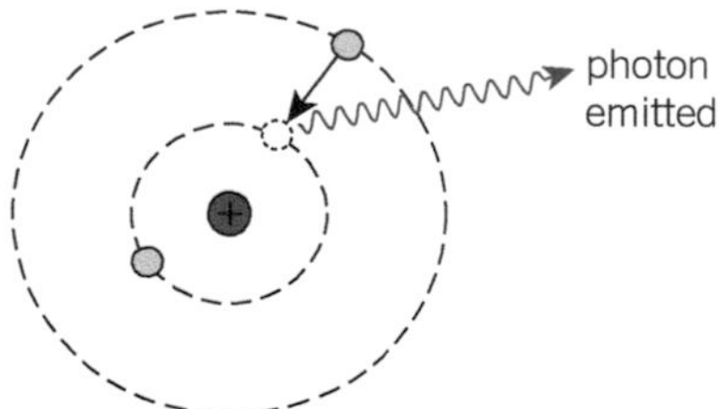

▲ **Figure 2** *De-excitation by photon emission*

The energy of the photon is equal to the energy lost by the electron and therefore by the atom. For example, when a mercury atom at an excitation energy level of 4.9 eV de-excites to the ground state, it emits a photon of energy 4.9 eV. The mercury atom could also de-excite indirectly via several energy levels if intermediate energy levels are present. However, since there are no intermediate levels between 4.9 eV and the ground state, the only way that the excited electron in this example can de-excite is to release a photon of energy 4.9 eV.

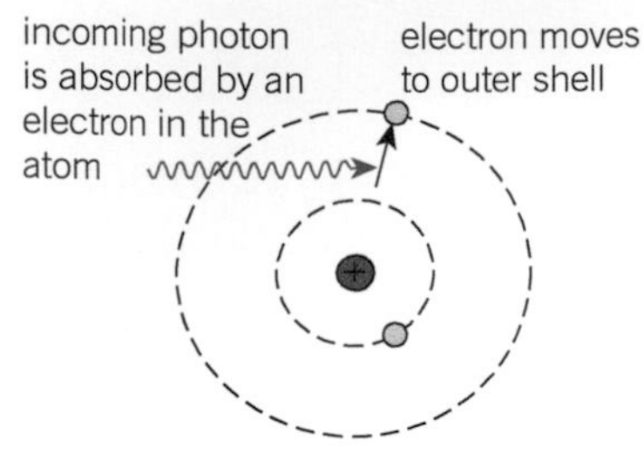

▲ **Figure 3** *Excitation by photon absorption*

In general, when an electron moves from energy level E_1 to a lower energy level E_2,

the energy of the emitted photon $hf = E_1 - E_2$.

Excitation using photons

An electron in an atom can absorb a photon and move to an outer shell where a vacancy exists – but only if the energy of the photon is exactly equal to the gain in the electron's energy (Figure 3). In other words, the photon energy must be exactly equal to the difference between the final and initial energy levels of the atom. If the photon's energy is smaller or larger than the difference between the two energy levels, it will not be absorbed by the electron.

Hint

A photon can be absorbed and cause ionisation if its energy is greater than or equal to the difference between the ionisation level and the ground state. However, excitation requires photons of specific energies.

Fluorescence

An atom in an excited state can de-excite directly or indirectly to the ground state, regardless of how the excitation took place. An atom can absorb photons of certain energies and then emit photons of the same or lesser energies. For example, a mercury atom in the ground state could

- be excited to its 5.7 eV energy level by absorbing a photon of energy 5.7 eV, then
- de-excite to its 4.9 eV energy level by emitting a photon of energy 0.8 eV, then
- de-excite to the ground state by emitting a photon of energy 4.9 eV.

Figure 4 represents these changes on an energy level diagram.

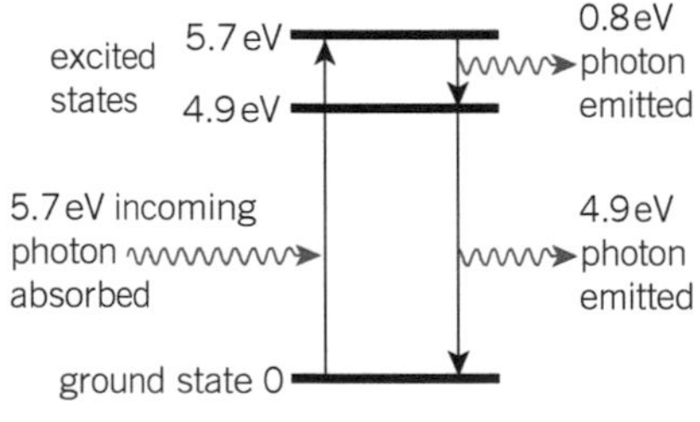

▲ **Figure 4** *Fluorescence*

This overall process explains why certain substances **fluoresce** or glow with visible light when they absorb ultraviolet radiation. Atoms in the substance absorb ultraviolet photons and become excited. When the atoms de-excite, they emit visible photons. When the source of ultraviolet radiation is removed, the substance stops glowing.

The **fluorescent tube** is a glass tube with a fluorescent coating on its inner surface. The tube contains mercury vapour at low pressure. When the tube is on, it emits visible light because:

- ionisation and excitation of the mercury atoms occur as they collide with each other and with electrons in the tube
- the mercury atoms emit ultraviolet photons, as well as visible photons and photons of much less energy, when they de-excite
- the ultraviolet photons are absorbed by the atoms of the fluorescent coating, causing excitation of the atoms
- the coating atoms de-excite in steps and emit visible photons.

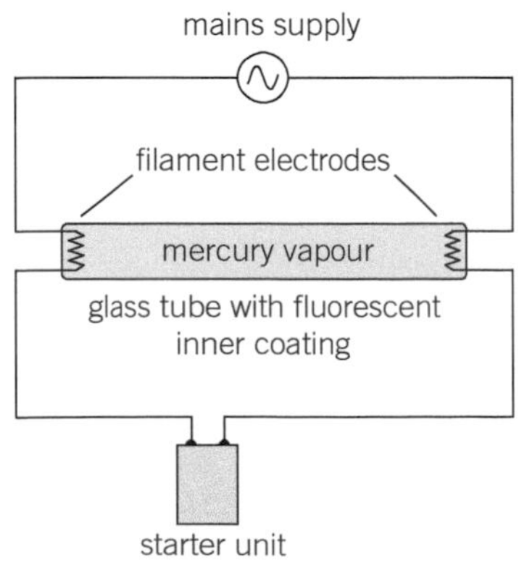

▲ **Figure 5** *The fluorescent tube*

Figure 5 shows the circuit for a fluorescent tube. The tube is much more efficient than a filament lamp. A typical 100 W filament lamp releases about 10–15 W of light energy. The rest of the energy supplied to it is wasted as heat. In contrast, a fluorescent tube can produce the same light output with no more than a few watts of power wasted as heat.

Application

Starting a fluorescent tube

A fluorescent tube has a filament electrode at each end. A starter unit is necessary because the mains voltage is too small to ionise the vapour in the tube when the electrodes are cold. When the tube is first switched on, the gas (argon) in the starter switch unit conducts and heats a bimetallic strip, making it bend, so the switch closes. The current through the starter unit increases enough to heat the filament electrodes. When the bimetallic switch closes, the gas in the starter unit stops conducting, and so the bimetallic strip cools and the switch opens. The mains voltage now acts between the two electrodes, which are now hot enough for ionisation of the gas to occur.

Low-energy light bulbs

Did you know that the UK government has stopped most sales of filament light bulbs? We now need to use low-energy light bulbs instead. Such a light bulb uses much less power than a filament light bulb which has the same light output. This is because the light is produced by a folded-up fluorescent tube instead of a glowing filament so less energy is wasted as heat. A 100 W filament bulb emits about 15 W of light and wastes the rest. In contrast, a low-energy light bulb with the same light output of 15 W wastes only about 5 W. Prove for yourself that the low-energy light bulb is five times more efficient than the filament light bulb. Using them at home would cut your electricity bill considerably and help to cut carbon emissions.

▲ **Figure 6** *A low-energy light bulb and a filament light bulb*

Worked example

Estimate the energy wasted in kWh by a 60 W filament lamp in its lifetime of 1100 h. Assume its efficiency is 10%.

Solution

Energy supplied = power × time = 0.06 kW × 1100 h = 66 kWh

Energy wasted = 90% of 66 kWh ≈ 60 kWh

Summary questions

$e = 1.6 \times 10^{-19}$ C

1 Figure 1 shows some of the energy levels of the mercury atom.

a Estimate the energy needed to excite the atom from the ground state to the highest excitation level shown in the diagram.

b Mercury atoms in an excited state at 5.7 eV can de-excite directly or indirectly to the ground state. Show that the photons released could have six different energies.

2 a In terms of electrons, state two differences between excitation and de-excitation.

b A certain type of atom has excitation energies of 1.8 eV and 4.6 eV.

i Sketch an energy level diagram for the atom using these energy values.

ii Calculate the possible photon energies from the atom when it de-excites from the 4.6 eV level. Use a downward arrow to indicate on your diagram the energy change responsible for each photon energy.

3 An atom absorbs a photon of energy 3.8 eV and subsequently emits photons of energy 0.6 eV and 3.2 eV.

a Sketch an energy level diagram representing these changes.

b In terms of electrons in the atom, describe how the above changes take place.

4 Explain why the atoms in a fluorescent tube stop emitting light when the electricity supply to it is switched off.

13.5 Energy levels and spectra

Learning objectives:

- → Define a line spectrum.
- → Explain why atoms emit characteristic line spectra.
- → Calculate the wavelength of light for a given electron transition.

Specification reference: 3.5.9

▲ **Figure 1** *A line spectrum*

A colourful spectrum

A rainbow is a natural display of the colours of the spectrum of sunlight. Raindrops split sunlight into a continuous spectrum of colours. Figure 3 in Topic 12.6 shows how we can use a prism to split a beam of white light from a filament lamp into a continuous spectrum. The wavelength of the light photons that produce the spectrum increases across the spectrum from deep violet at less than 400 nm to deep red at about 650 nm.

If we use a tube of glowing gas as the light source instead of a filament lamp, we see a spectrum of discrete lines of different colours, as shown in Figure 1, instead of a continuous spectrum.

The wavelengths of the lines of a line spectrum of an element are characteristic of the atoms of that element. By measuring the wavelengths of a line spectrum, we can therefore identify the element that produced the light. No other element produces the same pattern of light wavelengths. This is because the energy levels of each type of atom are unique to that atom. So the photons emitted are characteristic of the atom.

- Each line in a line spectrum is due to light of a certain colour and therefore a certain wavelength.
- The photons that produce each line all have the same energy, which is different from the energy of the photons that produce any other line.
- Each photon is emitted when an atom de-excites due to one of its electrons moving to an inner shell.
- If the electron moves from energy level E_1 to a lower energy level E_2

the energy of the emitted photon $hf = E_1 - E_2$

For each wavelength λ, we can calculate the energy of a photon of that wavelength as its frequency $f = \frac{c}{\lambda}$, where c is the speed of light. Given the energy level diagram for the atom, we can therefore identify on the diagram the transition that causes a photon of that wavelength to be emitted.

Synoptic link

You have met energy levels in more detail in Topic 13.4, Energy levels in atoms.

Synoptic link

You have seen how to measure the wavelengths in a line spectrum in Topic 12.4, The diffraction grating.

Worked example:

$c = 3.0 \times 10^8\,\mathrm{m\,s^{-1}}$, $e = 1.6 \times 10^{-19}\,\mathrm{C}$, $h = 6.63 \times 10^{-34}\,\mathrm{J\,s}$

A mercury atom de-excites from its 4.9 eV energy level to the ground state. Calculate the wavelength of the photon released.

Solution

$E_1 - E_2 = 4.9 - 0 = 4.9\,\mathrm{eV} = 4.9 \times 1.6 \times 10^{-19}\,\mathrm{J} = 7.84 \times 10^{-19}\,\mathrm{J}$

Therefore, $f = \frac{E_1 - E_2}{h} = \frac{7.84 \times 10^{-19}}{6.63 \times 10^{-34}} = 1.18 \times 10^{15}\,\mathrm{Hz}$

$\lambda = \frac{c}{f} = \frac{3.0 \times 10^8}{1.18 \times 10^{15}} = 2.54 \times 10^{-7}\,\mathrm{m} = 254\,\mathrm{nm}$

Extension

The Bohr atom

The hydrogen atom is the simplest type of atom – just one proton as its nucleus and one electron. The energy levels of the hydrogen atom, relative to the ionisation level, are given by the general formula

$$E = -\frac{13.6\text{ eV}}{n^2}$$

where $n = 1$ for the ground state, $n = 2$ for the next excited state, etc.

Therefore, when a hydrogen atom de-excites from energy level n_1 to a lower energy level n_2, the energy of the emitted photon is given by

$$E = \left(\frac{1}{n_2^{\,2}} - \frac{1}{n_1^{\,2}}\right) \times 13.6\text{ eV}$$

Each energy level corresponds to the electron in a particular shell. Thus the above formula gives the energy of a photon released when an electron in the hydrogen atom moves from one shell to a shell at lower energy.

Figure 2 shows an example of a transition that can take place in an excited hydrogen atom. The energy level formula for hydrogen was first deduced from the measurements of the wavelengths of the lines. Later, the Danish physicist Niels Bohr applied the quantum theory to the motion of the electron in the hydrogen atom, and so produced the first theoretical explanation of the energy level formula for hydrogen.

Q: Which transition between adjacent levels in an excited hydrogen atom gives the highest energy of the released photon?

Answer: From $n = 2$ to $n = 1$.

Application

The discovery of helium

Measurements of the wavelength of light are important in branches of science such as astronomy and forensic science, as they enable us to identify the chemical elements in the light source. Helium was discovered from the spectrum of sunlight. A pattern of lines in the spectrum was observed at wavelengths that had never been observed from any known gas, and were therefore due to the presence of a previously unknown element in the Sun. Helium is produced as a result of the nuclear fusion of hydrogen nuclei in the Sun, and was given the name helium from *helios*, the Greek word for sun. Helium is also present in the Earth, produced as alpha particles from the radioactive decay of elements such as uranium. It can be collected at oil wells and stored for use in fusion reactors, helium–neon lasers, and for very low temperature devices. In the liquid state, below a temperature of 2.17 K, it becomes a superfluid. This is a fluid with no resistance to flow, which escapes from an open container by creeping – as a thin film – up and over the sides of the container.

Study tip

Don't forget that lines in a spectrum are always due to energy levels.

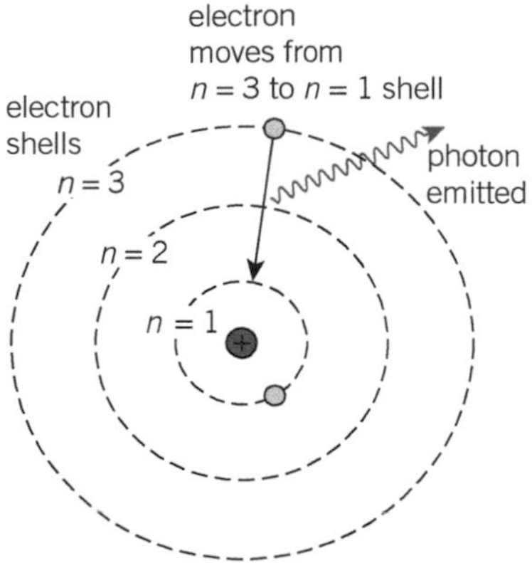

▲ **Figure 2** *An electron transition in the hydrogen atom*

Summary questions

$h = 6.63 \times 10^{-34}$ J s

$c = 3.00 \times 10^{8}$ m s^{-1}

$e = 1.6 \times 10^{-19}$ C

1. State two differences between a continuous spectrum and a line spectrum.
2. A mercury atom de-excites from 5.7 eV to 4.9 eV. For the photon emitted, calculate:
   - **a** its energy in J
   - **b** its wavelength.
3. A line spectrum has a line at a wavelength of 620 nm. Calculate:
   - **a** the energy, in J, of a photon of this wavelength
   - **b** the energy loss, in eV, of an atom that emits a photon of this wavelength.
4. Explain why the line spectrum of an element is unique to that element and can be used to identify it.

13.6 X-rays

Learning objectives:

- → Explain the principles of the production of X-rays by electron bombardment of a metal target.
- → Describe the main features of a modern X-ray tube, including control of the intensity and hardness of the X-ray beam.
- → Show an understanding of the use of X-rays in imaging internal body structures, including a simple analysis of the causes of sharpness and contrast in X-ray imaging.
- → Recall and solve problems by using the equation $I = I_0e^{-\mu x}$ for the attenuation of X-rays in matter.

Specification reference: 3.5.9

The production and properties of X-rays

X-ray imaging in medicine is an example of a diagnostic technique that is non-invasive. X-rays are electromagnetic waves of wavelength of the order of 0.1 nm or less. Figure 1 shows how a diagnostic X-ray tube works. The current through the filament wire heats the wire, which causes electrons to be emitted from the wire. These electrons are attracted from the filament, or **cathode**, to the **anode** when the anode is positive relative to the filament, typically 20–100 kV for X-ray imaging. The electrons are stopped when they collide with the anode and they emit X-rays in the process.

For an anode potential V, the maximum energy of an X-ray photon = eV so the maximum frequency $f_{max} = \frac{eV}{h}$.
Therefore the minimum wavelength λ_{min} is given by

$$\lambda_{min} = \frac{hc}{eV}$$

Note:

X-ray tubes used for therapy to destroy tumours are designed differently because they need to produce photons at higher energies. Such tubes operate at voltages from 250 kV to an upper limit (due to insulation breakdown) of 300 kV. X-ray photons from such tubes can destroy tumours no deeper than 5 cm beneath the skin. For deeper tumours, gamma photons with energies of the order of 1 MeV from radioactive isotopes are used.

The spectrum of photon energies from an *X*-ray tube

An X-ray tube produces a continuous spectrum of photon energies up to the maximum value of eV, where V is the maximum tube voltage, as shown in Figure 2. Raising the tube voltage increases the intensity at all photon energies up to the maximum photon energy as well as increasing the maximum photon energy.

In addition, intensity spikes are produced, which are characteristic of the atoms of the anode and do not change position when the tube voltage is altered. However, if the tube voltage is reduced sufficiently, each spike will disappear when the maximum photon energy is less than the energy of the photons at the spike.

The spikes are caused by the excitation of atoms in the anode when electrons from the filament collide with them. As a result, electrons in the atoms move temporarily from the innermost shells of the atom to higher energy levels. When these electrons return to their original levels, they emit X-ray photons at energies which are characteristic of the anode atoms. These emitted X-rays form patterns of line spectra, each pattern corresponding to

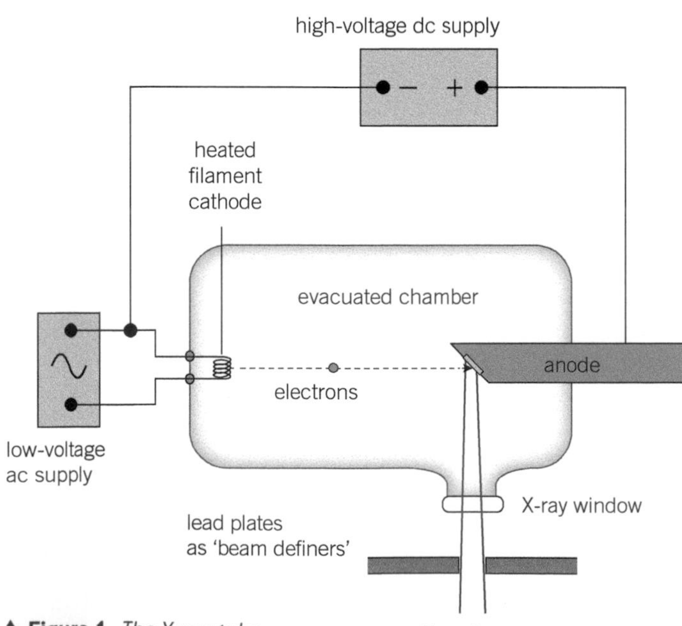

▲ **Figure 1** *The X-ray tube*

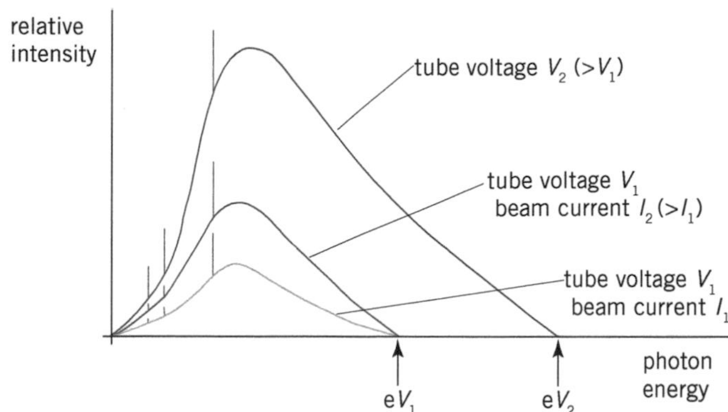

▲ **Figure 2** *The energy spectrum of an X-ray tube*

electrons returning from the outer energy levels to a particular electron shell. The pattern for electrons returning to:

- the innermost energy shell (n = 1) is referred to as the K-series
- the second energy shell (n = 2) is referred to as the L-series
- the third shell (n = 3) is referred to as the M-series.

Notes:

1 The energy of an X-ray photon is often expressed in electronvolts (eV) where $1\,\text{eV} = 1.6 \times 10^{-19}\,\text{J}$.

2 The power supplied to an X-ray tube = IV, where I is the beam current. The % efficiency of an X-ray tube is the percentage of the power supplied emitted as X-radiation. A typical X-ray tube has an efficiency of about 1%. The wasted energy is dissipated as heat at the anode.

X-ray imaging

When an X-ray picture is made, X-rays from the X-ray tube are directed for a specified time at the relevant area of the body with a film cassette on the other side of the body. Bones, teeth, and other dense matter in the path of the X-rays absorb X-rays much more than muscle and body tissue does. When the film is developed, the areas of the film exposed to X-rays are darker than the unexposed areas, so a negative image of the bones, teeth, etc., is formed on the developed film.

For any given application, the following factors must be taken into account before the X-ray tube is used.

1 **The penetrating power or 'hardness' of the X-ray beam** is increased by increasing the tube voltage. The higher the energy of the X-ray photon, the further it can travel through matter. The X-rays used to give an image of the bones of a broken arm do not need to penetrate as far as the X-rays used to give an image of an organ in the body. Increasing the tube voltage increases the maximum energy of the photons emitted by the tube so the beam is more penetrating. Low-energy photons are easily absorbed. An **aluminium filter** placed between the X-ray tube and the patient is used to absorb such photons which would otherwise be absorbed by the body and cause unnecessary exposure of the body to X-rays.

2 **The intensity of the X-ray beam** is the radiation energy per second passing through unit area at right angles to the area. The darkening of an X-ray film depends on the intensity of the X-radiation as well as on the

duration of exposure. The greater the intensity or the longer the duration of exposure, the darker the exposed parts of the film will be. An organ that moves would need a shorter duration of exposure and therefore greater intensity than a bone in the arm which can be held still. The intensity depends on the number of electrons per second reaching the anode and therefore on the tube current (since the tube current is a measure of the number of electrons per second reaching the anode). The tube current is controlled by the current through the filament wire. If the filament current is increased, the intensity of the X-ray beam is increased because:

- the filament becomes hotter and emits more electrons per second
- more electrons per second hit the anode so more X-ray photons are released each second.

3 **The width of the beam** is set using 'beam definer' lead plates to ensure only the part of the patient to be X-rayed is exposed to X-rays. Lead plates surrounding the X-ray tube are used to prevent people other than the patient being exposed to X-rays (see Figure 3).

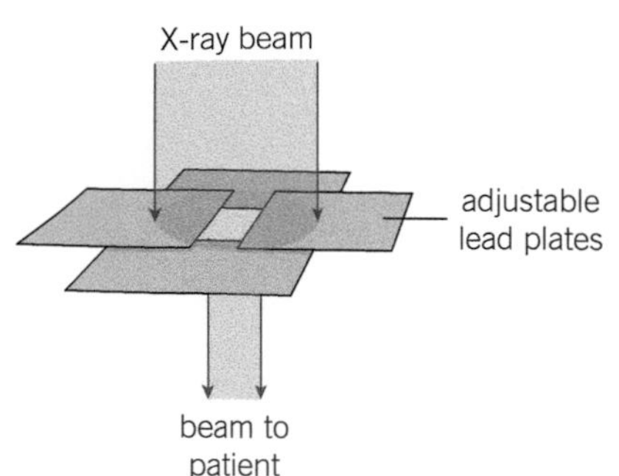

▲ **Figure 3** *Beam definers*

Image quality

Image sharpness

The sharpness of an X-ray image is determined by how clearly the edges of structures in the image are defined. A sharp image is one in which such edges can clearly be seen.

To form a sharp image on the film, the X-rays need to originate from a small area of the anode, and X-rays scattered by body organs and tissues need to be stopped from reaching the film.

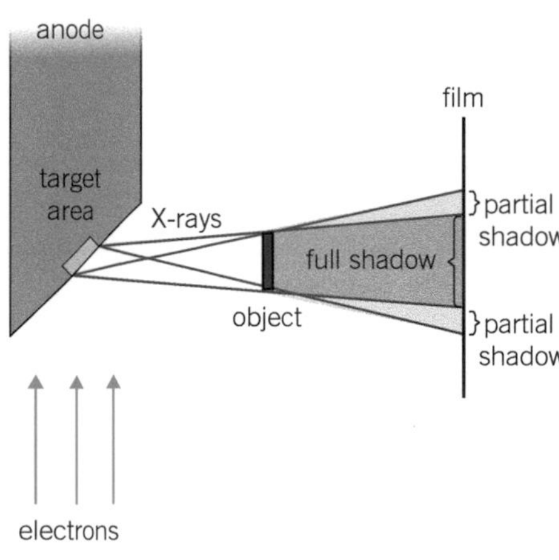

▲ **Figure 4** *Sharpness*

If the area of the anode is too large, the images will be blurred at the edges by large partial shadows, as shown in Figure 4. However, if the area is too small, the intense concentration of electrons in this area of the target area will damage the anode. To prevent overheating of the anode, it usually consists of a tungsten metal block set in a copper cylinder which is kept cool by pumping water or oil through it. Tungsten is chosen as it has a high melting point and copper is chosen as it is an excellent conductor of heat.

- X-ray photons may be scattered by atoms in the body tissues which they pass through. Some scattered X-ray photons may be scattered into the shadow areas on the film of bones or body organs. This would lessen the contrast between the images on the film of bones and body organs and the surrounding tissues. To eliminate scattered X-rays, a lead grid is placed between the patient and the film, as shown in Figure 5. Lead is used because it is a very effective absorber of X-rays.

The grid holes are aligned with the direction in which the unscattered X-rays are travelling so unscattered X-rays that enter the holes of the grid pass straight through it. However, scattered X-rays are absorbed by the grid because they travel mostly through lead after reaching the grid.

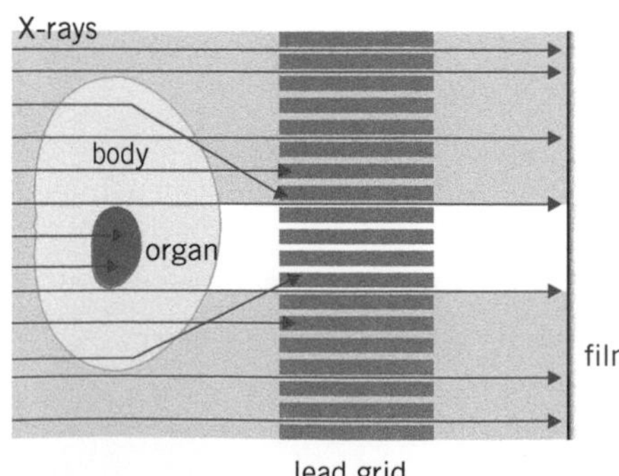

▲ **Figure 5** *Using a lead grid*

Contrast

An X-ray image with good contrast has areas where the film is very dark due to exposure to X-rays and other areas that are hardly darkened by X-rays. Bones and teeth are good absorbers of X-rays so they give images that stand out in good contrast with the surrounding tissue.

Body organs such as the stomach are not as effective at absorbing X-rays as bones and teeth. In order to obtain good X-ray images of an organ, a **contrast medium** is used. For example, a patient about to undergo a stomach X-ray is given a drink containing barium sulfate. Because barium is a good absorber of X-rays, X-rays that would otherwise pass through the stomach are absorbed so the contrast between the image of the stomach and its surroundings is vastly improved. A contrast medium is also used to obtain X-ray images of blood vessels where the contrast medium is injected into the bloodstream.

Contrast is lessened if:

- **The duration of exposure is too long**. The light areas and the dark areas of the film both become darker but the increase of darkness is greater in the light areas of the film. So the difference in darkness between the light and dark areas is reduced.
- **The X-rays are too penetrating**. Increasing the energy of the photons by increasing the tube voltage would increase their penetrating power. So more X-rays would pass through the organ and reach the shadow area of the film.
- **Too much scattering of X-ray photons** occurs when they pass through the tissue surrounding the organ.

Contrast can be improved if the film in its cassette is covered with a sheet of fluorescent substance. X-rays directed at a fluorescent substance cause the atoms of the substance to emit light photons. Each X-ray photon might cause many light photons to be emitted, darkening the film more in the areas which are exposed to X-rays. In addition to improving the contrast, the exposure of the patient to X-rays can be reduced as the film is more sensitive to darkening.

Summary questions

1 An X-ray tube operates at an anode potential of 50 kV.

a **i** Show that the minimum wavelength of the X-rays from this tube is 0.025 nm.

ii When the beam current at 50 kV is 0.2 mA, the tube operates at an efficiency of 1.5%. Calculate the radiation energy per second produced by the tube.

b State and explain what change in the X-ray beam occurs if

i the tube current is increased

ii the anode voltage is increased.

2 **a** Describe how the intensity of the X-rays from an X-ray tube operating at a constant potential difference varies with the wavelength of the X-rays.

b Explain why X-rays are emitted at certain wavelengths which are characteristic of the anode atoms.

3 **a** What is meant by

i the sharpness

ii the contrast of an X-ray image?

b State the function of a scattering grid and explain why it is necessary when an X-ray image is obtained.

4 Explain why a contrast medium is used when an X-ray picture of the stomach is made.

13.7 Wave–particle duality

Learning objectives:

- → Explain why we say photons have a dual nature.
- → Describe how we know that matter particles have a dual nature.
- → Discuss why we can change the wavelength of a matter particle but not that of a photon.

Specification reference: 3.5.11

The dual nature of light

Light is part of the electromagnetic spectrum of waves. The theory of electromagnetic waves predicted the existence of electromagnetic waves beyond the visible spectrum. The subsequent discovery of X-rays and radio waves confirmed these predictions and seemed to show that the nature of light had been settled. Many scientists in the late 19th century reckoned that all aspects of physics could be explained using Newton's laws of motion and the theory of electromagnetic waves. They thought that the few minor problem areas, such as the photoelectric effect, would be explained sooner or later using Newton's laws of motion and Maxwell's theory of electromagnetic waves. However, as explained in Topic 13.1, the photoelectric effect was not explained until Einstein put forward the radical theory that light consists of photons, which are particle-like packets of electromagnetic waves. Light has a dual nature, in that it can behave as a wave or as a particle, according to circumstances.

- The **wave-like nature** is observed when **diffraction** of light takes place. This happens, for example, when light passes through a narrow slit. The light emerging from the slit spreads out in the same way as water waves spread out after passing through a gap. The narrower the gap or the longer the wavelength, the greater the amount of diffraction.

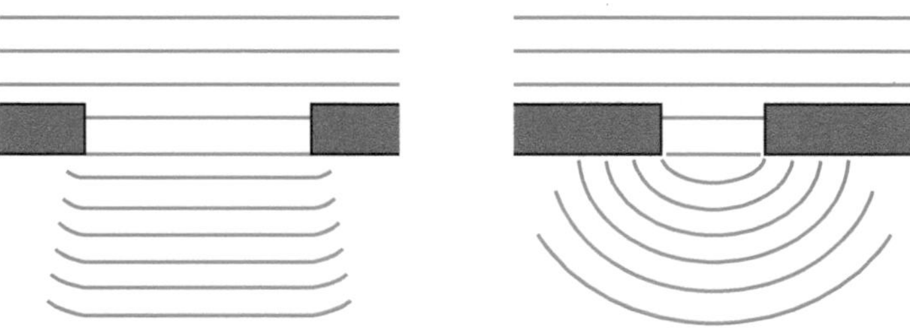

▲ **Figure 1** *Diffraction*

- The **particle-like nature** is observed, for example, in the photoelectric effect. When light is directed at a metal surface and an electron at the surface absorbs a photon of frequency f, the kinetic energy of the electron is increased from a negligible value by hf. The electron can escape if the energy it gains from a photon exceeds the work function of the metal.

Matter waves

If light has a dual wave–particle nature, perhaps particles of matter also have a dual wave–particle nature. Electrons in a beam can be deflected by a magnetic field. This is evidence that electrons have a particle-like nature. The idea that matter particles also have a wave-like nature was first considered by Louis de Broglie in 1923 (**de Broglie hypothesis**).

By extending the ideas of duality from photons to matter particles, de Broglie put forward the hypothesis that:

- matter particles have a dual wave–particle nature
- the wave-like behaviour of a matter particle is characterised by a wavelength, its **de Broglie wavelength**, λ, which is related to the momentum, p, of the particle by means of the equation

$$\lambda = \frac{h}{p}$$

Since the momentum of a particle is defined as its mass × its velocity, according to de Broglie's hypothesis, a particle of mass m moving at velocity v has a de Broglie wavelength given by

$$\lambda = \frac{h}{mv}$$

Note:

The de Broglie wavelength of a particle can be altered by changing the velocity of the particle.

Study tip

Don't mix up matter waves and electromagnetic waves, and don't confuse their equations.

Evidence for de Broglie's hypothesis

The wave-like nature of electrons was discovered when, three years after de Broglie put forward his hypothesis, it was demonstrated that a beam of electrons can be diffracted. Figure 2 shows in outline how this is done. After this discovery, further experimental evidence, using other types of particles, confirmed the correctness of de Broglie's theory.

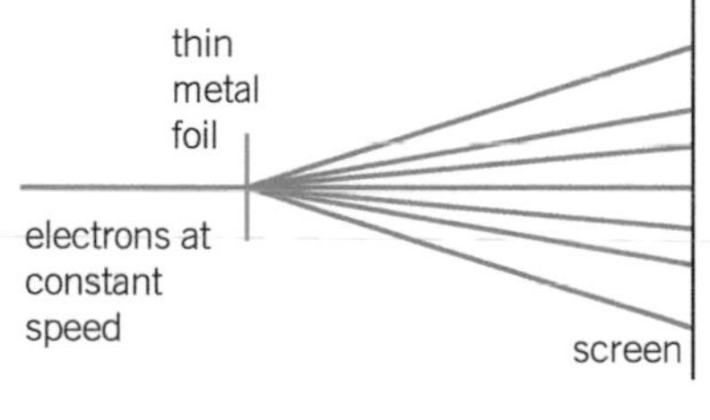

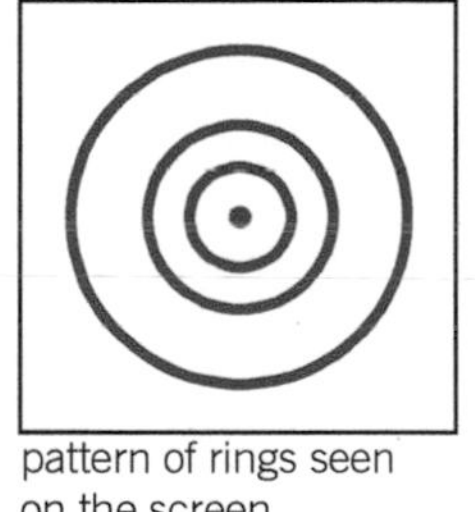

▲ **Figure 2** *Diffraction of electrons*

- A narrow beam of electrons in a vacuum tube is directed at a thin metal foil. A metal is composed of many tiny crystalline regions. Each region, or grain, consists of positive ions arranged in fixed positions in rows in a regular pattern. The rows of atoms cause the electrons in the beam to be diffracted, just as a beam of light is diffracted when it passes through a slit.
- The electrons in the beam pass through the metal foil and are diffracted in certain directions only, as shown in Figure 2. They form a pattern of rings on a fluorescent screen at the end of the tube. Each ring is due to electrons diffracted by the same amount from grains of different orientations, at the same angle to the incident beam.
- The beam of electrons is produced by attracting electrons from a heated filament wire to a positively charged metal plate, which has a small hole at its centre. Electrons that pass through the hole form the beam. The speed of these electrons can be increased by increasing the potential difference between the filament and the metal plate. This makes the diffraction rings smaller, because the increase of speed makes the de Broglie wavelength smaller. So less diffraction occurs and the rings become smaller.

Synoptic link

You have met the PET scanner in Topic 7.4, Particles and antiparticles.

Extension

Energy levels explained

An electron in an atom has a fixed amount of energy that depends on the shell it occupies. Its de Broglie wavelength has to fit the shape and size of the shell. This is why its energy depends on the shell it occupies.

For example, an electron in a spherical shell moves round the nucleus in a circular orbit. The circumference of its orbit must be equal to a whole number of de Broglie wavelengths (circumference = $n\lambda$, where n = 1 or 2 or 3, etc.). This condition can be used to derive the energy level formula for the hydrogen atom – and it gives you a deeper insight into quantum physics.

Q: Does the de Broglie wavelength of an electron increase or decrease when it moves to an orbit where it travels faster?

Answer: The wavelength becomes smaller because its momentum becomes greater.

Application

Quantum technology

The PET scanner is an example of quantum physics in use. Some further applications of quantum technology include

- The STM (scanning tunneling microscope) is used to map atoms on solid surfaces. The wave nature of electrons allows them to tunnel between the surface and a metal tip a few nanometres above the surface as the tip scans across the surface.
- The TEM (transmission electron microscope) is used to obtain very detailed images of objects and surface features too small to see with optical microscopes. Electrons are accelerated in a TEM to high speed so their de Broglie wavelength is so small that they can give very detailed images.
- The MR (magnetic resonance) body scanner used in hospitals detects radio waves emitted when hydrogen atoms in a patient in a strong magnetic field flip between energy levels.
- SQUIDs (superconducting quantum interference devices) are used to detect very very weak magnetic fields, for example SQUIDs are used to detect magnetic fields produced by electrical activity in the brain.

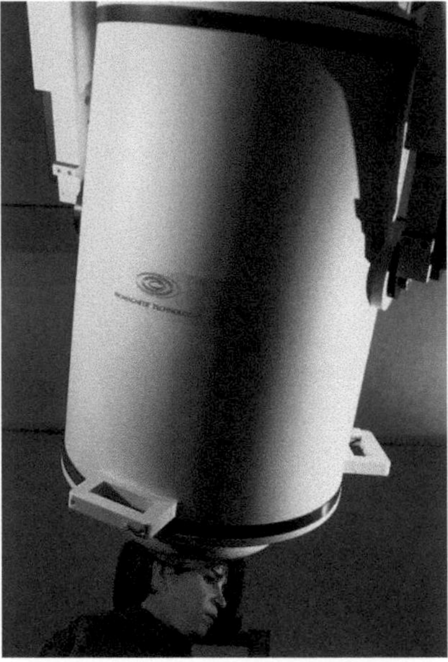

▲ **Figure 3** *An example of a SQUID*

Summary questions

$h = 6.6 \times 10^{-34}$ J s
the mass of an electron = 9.1×10^{-31} kg
the mass of a proton = 1.7×10^{-27} kg

1 With the aid of an example in each case, explain what is meant by the dual wave–particle nature of
 a light
 b matter particles, for example, electrons.

2 State whether each of the following experiments demonstrates the wave nature or the particle nature of matter or of light:
 a the photoelectric effect
 b electron diffraction.

3 Calculate the de Broglie wavelength of
 a an electron moving at a speed of 2.0×10^{7} m s^{-1}
 b a proton moving at the same speed.

4 Calculate the momentum and speed of
 a an electron that has a de Broglie wavelength of 500 nm
 b a proton that has the same de Broglie wavelength.

Practice questions: Chapter 13

1 When light at sufficiently high frequency, f, is incident on a metal surface, the maximum kinetic energy, E_{Kmax}, of a photoelectron is given by

$E_{Kmax} = hf - \phi$, where ϕ is the work function of the metal.

(a) State what is meant by the work function ϕ.

(b) The following results were obtained in an experiment to measure E_{Kmax} for different frequencies, f: *(1 mark)*

f / 10^{14} Hz	5.6	6.2	6.8	7.3	8.3	8.9
E_{Kmax} / 10^{-19} J	0.8	1.2	1.7	2.1	2.9	3.4

(i) Use these results to plot a graph of E_{Kmax} against f.
(ii) Explain why your graph confirms the equation above.
(iii) Use your graph to determine the value of h and to calculate the work function of the metal. *(13 marks)*

(c) Measurements like those above were first made to test the correctness of Einstein's explanation of the photoelectric effect. The equation above was a prediction by Einstein using the photon theory.
Why is it important to test any new theory by testing its predictions experimentally? *(2 marks)*

2 **(a)** One quantity in the photoelectric equation is a characteristic property of the metal that emits photoelectrons. Name and define this quantity. *(2 marks)*

(b) A metal is illuminated with monochromatic light. Explain why the kinetic energy of the photoelectrons emitted has a range of values up to a certain maximum. *(3 marks)*

(c) A gold surface is illuminated with monochromatic ultraviolet light of frequency 1.8×10^{15} Hz. The maximum kinetic energy of the emitted photoelectrons is 4.2×10^{-19} J. Calculate, for gold:
(i) the work function, in J
(ii) the threshold frequency. *(5 marks)*

AQA, 2006

3 **Figure 1** shows how the maximum kinetic energy of electrons emitted from the cathode of a photoelectric cell varies with the frequency of the incident radiation.

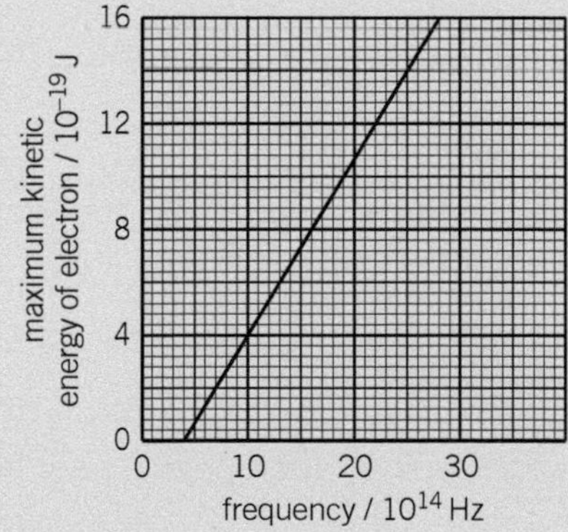

▲ **Figure 1**

(a) Calculate the maximum wavelength of electromagnetic radiation that can release photoelectrons from the cathode surface.

speed of electromagnetic radiation in a vacuum = $3.0 \times 10^8\,\text{m}\,\text{s}^{-1}$ *(3 marks)*

(b) Another photoelectric cell uses a different metal for the photocathode. This metal requires twice the minimum energy for electron release compared to the metal in the first cell.

(i) If drawn on the same axes, how would the graph line obtained for this second cell compare with the one for the first cell?

(ii) Explain your answer with reference to the Einstein photoelectric equation. *(3 marks)*

AQA, 2003

4 A fluorescent light tube contains mercury vapour at low pressure. The tube is coated on the inside, and contains two electrodes.

(a) Explain why the mercury vapour is at a low pressure. *(1 mark)*

(b) Explain the purpose of the coating on the inside of the tube. *(3 marks)*

AQA, 2003

5 The lowest energy levels of a mercury atom are shown below. The diagram is *not* to scale.

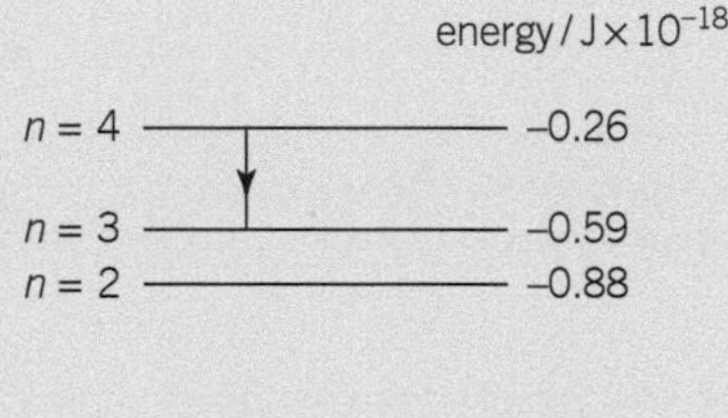

(a) Calculate the frequency of an emitted photon due to a transition, shown by an arrow, from level $n = 4$ to level $n = 3$. *(2 marks)*

(b) Which transition would cause the emission of a photon of a longer wavelength than that emitted in the transition from level $n = 4$ to level $n = 3$? *(1 mark)*

AQA, 2002

6 **(a)** State what is meant by the duality of electrons. Give *one* example of each type of behaviour. *(3 marks)*

(b) (i) Calculate the speed of an electron which has a de Broglie wavelength of 1.3×10^{-10} m.

(ii) A particle when travelling at the speed calculated in (b)(i) has a de Broglie wavelength of 8.6×10^{-14} m.
Calculate the mass of the particle. *(4 marks)*

AQA, 2007

7 Electrons travelling at a speed of $5.00 \times 10^5\,\text{m}\,\text{s}^{-1}$ exhibit wave properties.

(a) What phenomenon can be used to demonstrate the wave properties of electrons? Details of any apparatus used are not required. *(1 mark)*

(b) Calculate the wavelength of these electrons. *(2 marks)*

(c) Calculate the speed of muons with the same wavelength as these electrons.
mass of muon = 207 × mass of electron *(3 marks)*

(d) Both electrons and muons were accelerated from rest by the same potential difference. Explain why they have different wavelengths. *(2 marks)*

AQA, 2003

Further practice questions: multiple choice

1 Which one of the following alternatives **A–D** gives the ratio $\dfrac{\text{specific charge of a carbon } ^{12}_{6}\text{C nucleus}}{\text{specific charge of the proton}}$?

A $\frac{1}{3}$ **B** $\frac{1}{2}$ **C** 1 **D** 2

2 An unstable nucleus X emits an α particle then a β^- particle to form a stable nucleus Y. The atomic mass of X is P and its atomic number is Q. Which one of the following alternatives **A–D** gives the atomic mass and atomic number of Q?

	A	B	C	D
atomic mass	P-4	P-4	P-2	P-2
atomic number	Q-2	Q-1	Q-2	Q-1

3 A radioactive isotope R has a half-life of 8 days and it decays to form a stable isotope S. A pure sample of R is prepared. After 24 days, which one of the following alternatives **A–D** gives the ratio

$$\frac{\text{the number of atoms of S}}{\text{the number of atoms of R}}$$

A 1 **B** 3 **C** 4 **D** 7

4 A conduction electron at the surface of a metal escapes from the surface after absorbing a photon. The work function of the metal is ϕ and the metal is at zero potential. Which one of the following inequalities about wavelength λ of the photon is true?

A $\lambda < \dfrac{hc}{\phi}$ **B** $\lambda < \dfrac{h\phi}{c}$ **C** $\lambda < \dfrac{\phi}{hc}$ **D** $\lambda < \dfrac{c}{h\phi}$

5 Which one of the following statements about polarisation is true?

A Sound waves can be polarised.
B Some electromagnetic waves cannot be polarised.
C The vibrations of a polarised wave are always in the same plane.
D Light waves are always polarised.

6 Which of the following is correct for the first two harmonics of a stationary wave on a string?

A The frequency of the first harmonic is twice the frequency of the second harmonic.
B Both harmonics have a node at the midpoint of the string.
C Between two nodes all parts of the wave vibrate in phase.
D Adjacent nodes of the second harmonic are twice as far apart as adjacent nodes of the first harmonic.

7 A parallel beam of monochromatic light is directed at normal incidence at two narrow parallel slits spaced 0.7 mm apart. Interference fringes are formed on a screen which is perpendicular to the direction of the incident beam at a distance of 0.900 m from the slits. The distance across five fringe spaces is measured at 4.2 mm.
Which one of the following alternatives **A–D** gives the wavelength of the light?

A 130 nm **B** 450 nm **C** 550 nm **D** 650 nm

8 Two particles X and Y at the same initial position accelerate uniformly from rest along a straight line. After 1.0 s, X is 0.20 m ahead of Y. The separation of X and Y after 2.0 s from the start is

A 0.40 m **B** 0.80 m **C** 1.20 m **D** 1.60 m

9 Two trolleys, P and Q, travelling in opposite directions at the same speed, collide and move together after the collision in the direction in which P was originally travelling. Which *one* of the following statements about the collision is true?

A The collision is elastic.
B The mass of P is greater than the mass of Q.
C The force exerted by P on Q is greater than the force exerted by Q on P.
D The change of momentum of P is greater than the change of momentum of Q.

10 A box of mass m slides down a slope at constant velocity as shown in **Figure 1**. The slope is inclined at angle θ to the horizontal. The diagram shows the frictional force F and the normal reaction force N of the surface on the block.

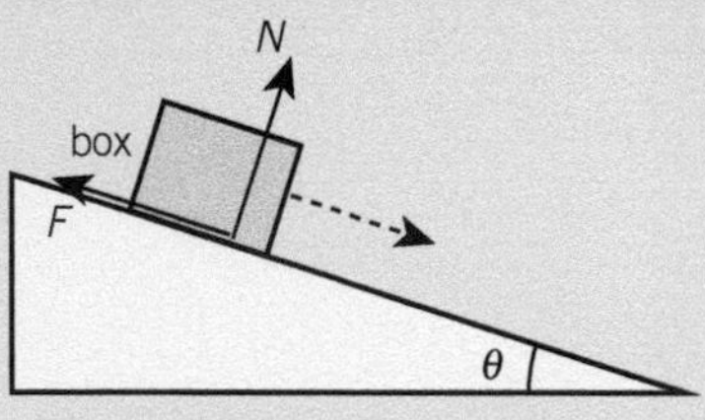

▲ **Figure 1**

Which one of the following statements about forces F and N is true?

A $F = mg\cos\theta$ **B** $F = N\sin\theta$ **C** $F = N + mg$ **D** $F^2 = (mg)^2 - N^2$

11 Two wires P and Q of the same material have lengths L and $2L$, and different diameters d_p and d_q respectively.
When the same force is applied to each wire, the extension of wire P is four times the extension of wire Q. Which one of the following alternatives **A–D** gives the ratio $\frac{d_p}{d_q}$?

A $\frac{1}{2\sqrt{2}}$ **B** $\frac{1}{\sqrt{2}}$ **C** $\sqrt{2}$ **D** $2\sqrt{2}$

12 Four resistors of resistances 1.0 Ω, 2.0 Ω, 3.0 Ω, and 4.0 Ω are connected together as shown in **Figure 2**. A battery is connected across the 1.0 Ω resistor.

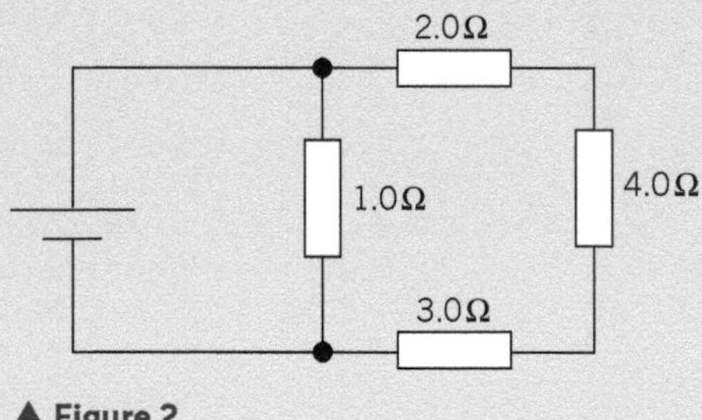

▲ **Figure 2**

If I_0 is the battery current and I_1 is the current in the 4.0 Ω resistor, which one of the following alternatives **A–D** gives the ratio $\frac{I_0}{I_1}$?

A 0.1 **B** 0.5 **C** 1.5 **D** 10

Further practice questions

1 **(a)** An electron is trapped in a solid between a group of atoms where the potential is +2.8 V.
The de Broglie wavelength of this electron is 1.2 nm. Calculate
(i) its speed
(ii) its kinetic energy
(iii) the sum of its kinetic energy and its potential energy. *(5 marks)*

(b) (i) Calculate the energy of a photon of wavelength 650 nm.
(ii) State and explain whether or not the electron in part (a) can escape from this group of atoms as a result of absorbing this photon. *(3 marks)*

AQA, 2005

2 **(a)** (i) State what is meant by the *wave–particle duality* of electromagnetic radiation.
(ii) Which aspect of the dual nature of electromagnetic radiation is demonstrated by the photoelectric effect? *(2 marks)*

(b) A metal plate is illuminated with ultraviolet radiation of frequency 1.67×10^{15} Hz. The maximum kinetic energy of the liberated electrons is 3.0×10^{-19} J.
(i) Calculate the work function of the metal.
(ii) The radiation is maintained at the same frequency but the intensity is doubled. State what changes, if any, occur to the number of electrons released per second and to the maximum kinetic energy of these electrons.
(iii) The metal plate is replaced by another metal plate of different material. When illuminated by radiation of the same frequency no electrons are liberated. Explain why this happens and what can be deduced about the work function of the new metal. *(7 marks)*

AQA, 2001

3 **Figure 3** shows the forces acting on a stationary kite. The force F is the force that the air exerts on the kite.

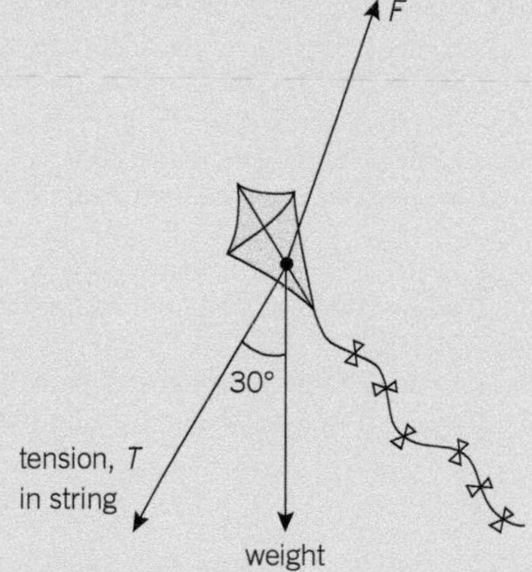

▲ **Figure 3**

(a) Show on the diagram how force F can be resolved into horizontal and vertical components. *(1 mark)*

(b) The magnitude of the tension, T, is 25 N.
Calculate:
(i) the horizontal component of the tension
(ii) the vertical component of the tension. *(2 marks)*

(c) (i) Calculate the magnitude of the vertical component of F when the weight of the kite is 2.5 N.

(ii) State the magnitude of the horizontal component of F.

(iii) Hence calculate the magnitude of F. *(5 marks)*

AQA, 2002

4 A sprinter is shown before a race, stationary in the *set* position, as shown in **Figure 4**. Force F is the resultant force on the sprinter's finger tips. The reaction force Y on her forward foot is 180 N, and her weight W is 520 N. X is the vertical reaction force on her back foot.

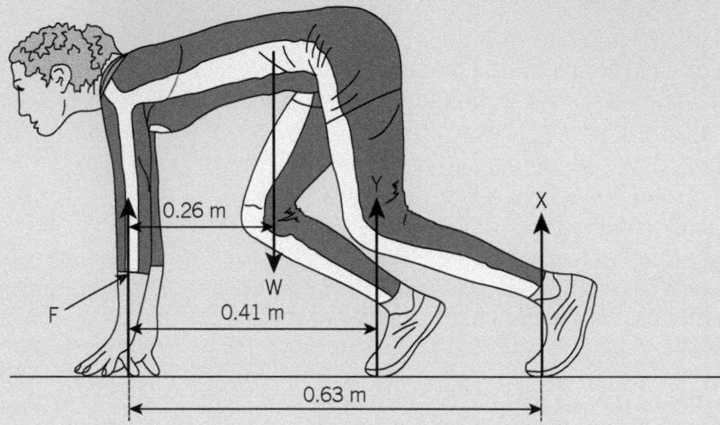

▲ **Figure 4**

(a) (i) Calculate the moment of the sprinter's weight W about her fingertips. Give an appropriate unit.

(ii) By taking moments about her finger tips, calculate the force on her back foot X.

(iii) Calculate the force F. *(6 marks)*

(b) The sprinter starts running and reaches a horizontal velocity of $9.3\,\mathrm{m\,s^{-1}}$ in a distance of 35 m.

(i) Calculate her average acceleration over this distance.

(ii) Calculate the resultant force necessary to produce this acceleration. *(4 marks)*

AQA, 2012

5 In a vehicle impact, a car runs into the back of a lorry. The car driver sustains serious injuries, which would have been much less had the car been fitted with a driver's air bag.

(a) Explain why the effect of the impact on the driver would have been much less if an air bag had been fitted and had inflated in the crash. *(4 marks)*

(b) Calculate the deceleration of the car if it was travelling at a speed of $18\,\mathrm{m\,s^{-1}}$ when the impact occurred and was brought to rest in a distance of 2.5 m. *(2 marks)*

AQA, 2004

6 The silver isotope $^{106}_{47}\mathrm{Ag}$ decays by emitting a β^{+} particle and another particle to form a stable palladium (Pa) isotope.

(a) The equation below represents the decay of a $^{106}_{47}\mathrm{Ag}$ nucleus.

$$^{106}_{47}\mathrm{Ag} \rightarrow {}^{a}_{b}\mathrm{Pa} + {}^{c}_{d}\beta + \mathrm{X}$$

(i) Copy the equation showing the correct values of a, b, c, and d. *(2 marks)*

(ii) Identify particle X and state its charge. *(1 mark)*

(iii) Describe how the number of protons and the number of neutrons in the nucleus changes when a silver $^{106}_{47}\mathrm{Ag}$ nucleus decays. *(1 mark)*

(b) A positron is the antiparticle of an electron.

(i) State what is meant by an *antiparticle*. (*2 marks*)
(ii) Describe what happens when a positron collides with an electron. (*2 marks*)

(c) **Figure 5** shows how the number of silver $^{106}_{47}$Ag nuclei in a sample of the isotope changes as the silver nuclei decay.

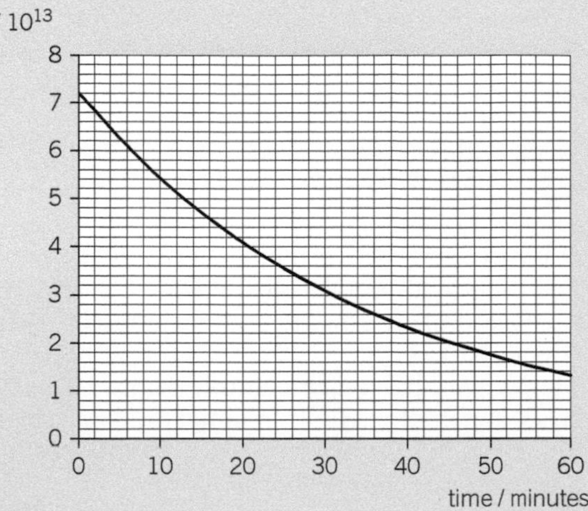

▲ **Figure 5**

(i) Estimate the half-life of the isotope. Explain how you obtained your estimate. (*2 marks*)
(ii) Calculate the number of $^{106}_{47}$Ag nuclei remaining after 180 minutes. (*3 marks*)
(iii) Each silver nucleus that decays releases 2.04 MeV of energy. Determine the initial rate of release of energy in J s^{-1} from the sample. (*3 marks*)

7 Sail systems are being developed to reduce the running costs of cargo ships. The sail and ship's engines work together to power the ship. One of these sails is shown in **Figure 6** pulling at an angle of 40° to the horizontal.

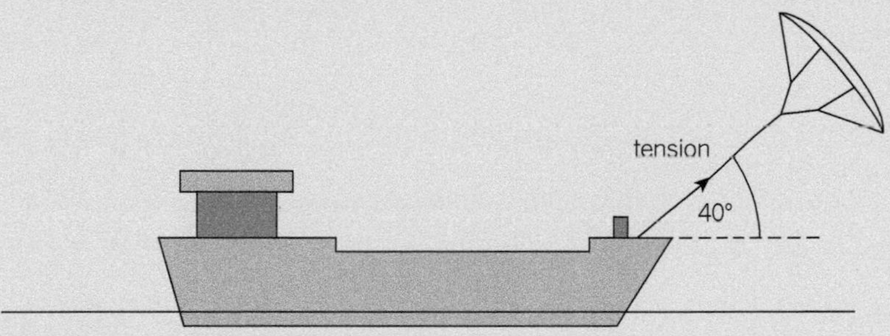

▲ **Figure 6**

(a) The average tension in the cable is 170 kN. Show that, when the ship travels 1.0 km, the work done by the sail on the ship is 1.3×10^8 J. (*2 marks*)
(b) With the sail and the engines operating, the ship is travelling at a steady speed of 7.0 m s^{-1}.
(i) Calculate the power developed by the sail.
(ii) Calculate the percentage of the ship's power requirement that is provided by the wind when the ship is travelling at this speed. The power output of the engines is 2.1 MW. (*4 marks*)
(c) The angle of the cable to the horizontal is one of the factors that affects the horizontal force exerted by the sail on the ship. State *two* other factors that would affect this force. (*2 marks*)

AQA, 2012

8 An aerial system consists of a horizontal copper wire of length 38 m supported between two masts, as shown in **Figure 7**. The wire transmits electromagnetic waves when an alternating potential is applied to it at one end.

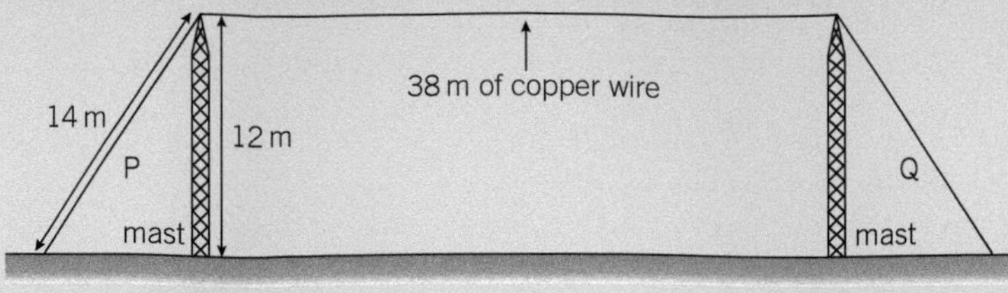

▲ Figure 7

(a) The wavelength of the radiation transmitted from the wire is twice the length of the copper wire. Calculate the frequency of the transmitted radiation. *(1 mark)*

(b) The ends of the copper wire are fixed to masts of height 12.0 m. The masts are held in a vertical position by cables, labelled P and Q, as shown in **Figure 7**.

(i) P has a length of 14.0 m and the tension in it is 110 N. Calculate the tension in the copper wire.

(ii) The copper wire has a diameter of 4.0 mm. Calculate the stress in the copper wire.

(iii) Discuss whether the wire is in danger of breaking if it is stretched further due to movement of the top of the masts in strong winds.
breaking stress of copper = 3.0×10^8 Pa *(7 marks)*

AQA, 2006

9 The filament of a 60 W, 230 V mains lamp is a coil of thin tungsten wire.

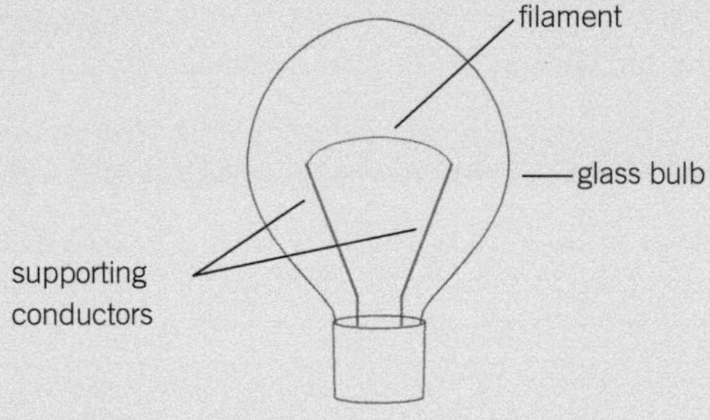

▲ Figure 8

(a) When the lamp is new, the filament wire has a radius of 80 μm and operates at a temperature of 2500 K.

(i) Calculate the resistance of the filament at 2500 K.

(ii) Show that the length of the wire in the filament is 0.25 m.
resistivity of tungsten = $7.0 \times 10^{-5}\,\Omega$m *(4 marks)*

(b) Near the end of its working life, the radius of the filament has decreased to 70 μm and its working temperature is 2300 K. At this temperature, the resistivity of tungsten is $6.4 \times 10^{-5}\,\Omega$m. Discuss whether the lamp consumes electrical energy at a rate of 60 W near the end of its working life. *(3 marks)*

AQA, 2006

10 **Figure 9** shows how the resistance R of three electrical components varies with temperature θ in °C.

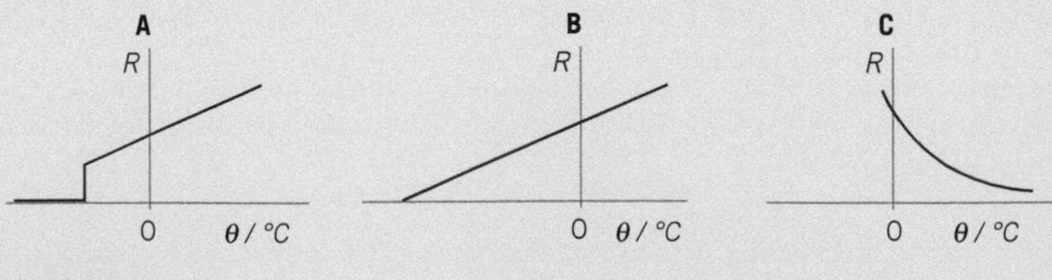

▲ Figure 9

(a) Indicate below which one of **A**, **B**, or **C** shows the correct graph for
(i) a wire-wound resistor
(ii) a thermistor
(iii) a superconductor. *(2 marks)*

(b) The metal wire used to manufacture the wire-wound resistor has a resistance per metre of 26 Ω and a diameter of 0.23 mm. Calculate the resistivity of the material from which the wire is made. *(4 marks)*

AQA, 2007

11 A manufacturer asks you to design the heating element in a car rear-window demister. The design brief calls for an output of 48 W at a potential difference of 12 V.
Figure 10 shows where the eight elements will be on the car window before electrical connections are made to them.

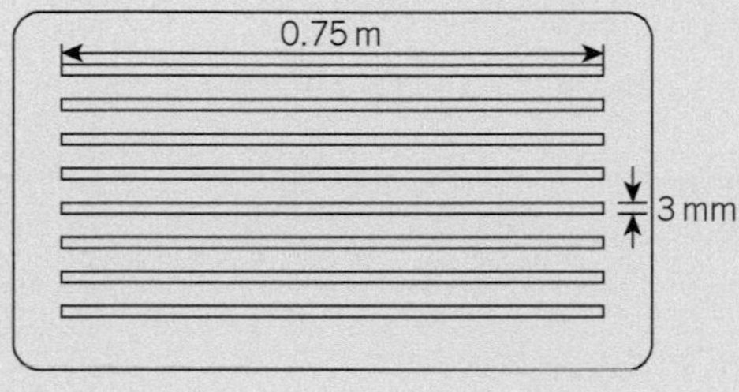

▲ **Figure 10**

(a) Calculate the current supplied by the power supply. *(1 mark)*

(b) One design possibility is for the eight elements to be connected in parallel.
(i) Calculate the current in each element in this parallel arrangement.
(ii) Calculate the resistance required for each element. *(3 marks)*

(c) Another design possibility is to have the eight elements connected in series.
(i) Calculate the current in each element in this series arrangement.
(ii) Calculate the resistance required for each element. *(4 marks)*

(d) State *one* disadvantage of the series design compared to the parallel arrangement. *(1 mark)*

(e) The series design is adopted. Each element is to have a rectangular cross section of 0.12 mm by 3.0 mm. The length of each element is to be 0.75 m.
(i) State the unit of resistivity.
(ii) Calculate the resistivity of the material from which the element must be made. *(3 marks)*

AQA, 2002

12 A copper connecting wire is 0.75 m long and has a cross-sectional area of $1.3 \times 10^{-7}\,m^2$.

(a) Calculate the resistance of the wire. (Resistivity of copper = $1.7 \times 10^{-7}\,\Omega m$.) *(2 marks)*

(b) A 12 V 25 W lamp is connected to a power supply of negligible internal resistance using two of the connecting wires. The lamp is operating at its rated power.
(i) Calculate the current flowing in the lamp.
(ii) Calculate the p.d. across each of the wires.
(iii) Calculate the e.m.f. (electromotive force) of the power supply. *(4 marks)*

(c) The lamp used in (b) is connected by the same two wires to a power supply of the same e.m.f., but whose internal resistance is not negligible.
State and explain what happens to the brightness of the lamp when compared to its brightness in (b). *(2 marks)*

AQA, 2013

13 When a note is played on a string instrument such as a violin, the sound it produces consists of the first harmonic and many higher harmonics. **Figure 11** shows the shape of the string for a stationary wave that corresponds to a higher harmonic. The positions of maximum and zero displacement for this harmonic are shown. The ends A and B of the string are fixed, and P, Q, and R are points on the string.

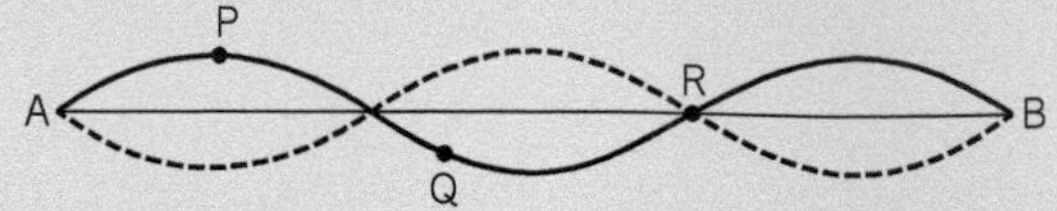

▲ **Figure 11**

(a) (i) Describe and compare the motion of points P and Q on the string. *(3 marks)*

(ii) What can you say about the motion of point R on the string? *(1 mark)*

(iii) What is the phase relationship between point Q on the string and the midpoint of the string? *(1 mark)*

(b) The string has a length of 0.60 m and is vibrating at a frequency of 510 Hz.

(i) Calculate the wavelength and the speed of the progressive waves on the wire. *(2 marks)*

(ii) The tension in the wire is 10 N. Calculate the mass per unit length of the wire. *(3 marks)*

14 (a) A laser emits *monochromatic light*. Explain the meaning of the term monochromatic light. *(1 mark)*

Figure 12 shows a laser emitting blue light directed at a single slit, where the slit width is greater than the wavelength of the light. The intensity graph for the diffracted blue light is shown.

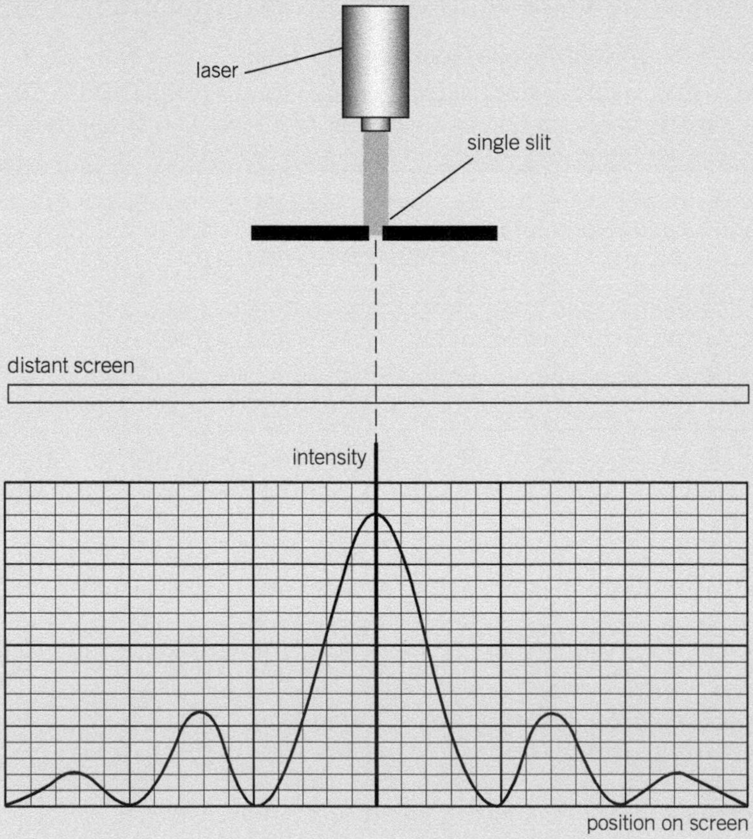

laser

single slit

distant screen

intensity

position on screen

▲ **Figure 12**

(b) On the axes shown on a copy of **Figure 12**, sketch the intensity graph for a laser emitting red light. *(2 marks)*

(c) State and explain *one* precaution that should be taken when using laser light. *(2 marks)*

(d) The red laser light is replaced by a non-laser source emitting white light. Describe how the appearance of the pattern would change. *(3 marks)*

AQA, 2013

Answers to the Practice Questions and Section Questions are available at www.oxfordsecondary.com/oxfordaqaexams-alevel-physics

14 More on mathematical skills

14.1 Data handling

This chapter covers the mathematical skills you will require for this course.

Scientific units

Scientists use a single system of units to avoid unnecessary effort and time converting between different units of the same quantity. This system, the **Système International** (or **SI system**) is based on a defined unit for certain physical quantities including those listed in Table 1. Units of all other quantities are derived from the SI **base units**.

▼ **Table 1** *SI base units*

Physical quantity	Unit
Mass	kilogram (kg)
Length	metre (m)
Time	second (s)
Electric current	ampere (A)
Temperature	kelvin (K)

The following examples show how the units of all other physical quantities are derived from the base units.

- The unit of density is the kilogram per cubic metre ($kg\,m^{-3}$).
- The unit of force is the newton (N) which in terms of base units is $kg\,m\,s^{-2}$ (ie. the unit of mass × the unit of acceleration)

More about using a calculator

1 '**Exp**', '**EE**', or '**× 10^x**', is the calculator button you press to key in a **power of ten**. To key in a number in standard form (e.g., 3.0×10^8), the steps are as follows:
  - Step 1 Key in the number between 1 and 10 (e.g., 3.0).
  - Step 2 Press the calculator button marked 'Exp' (or 'EE' on some calculators).
  - Step 3 Key in the power of ten (e.g., 8).

  If the display reads '3.0 08' this should be read as 3.0×10^8 (not 3.0^8 which means 3.0 multiplied by itself eight times). If the power of ten is a negative number (e.g., 10^{-8} not 10^8), press the calculator button marked '+/–' after step 3 (or before, if you are using a graphic calculator) to change the sign of the power of ten.

2 '**inv**' is the button you press if you want the calculator to give the value of the inverse of a function. For example, if you want to find out the angle which has a sine of 0.5, you key in 0.5 on the display then press 'inv' then 'sin' to obtain the answer of 30°. Some calculators have a 'second function' or '**shift**' button that you press instead of the 'inv' button.

3 '**log$_{10}$**', '**log**',' or '**lg**' is the button you press to find out what a number is as a power of ten. For example, press 'log' then key in 100 and the display will show 2, because $100 = 10^2$. Logarithmic scales have equal intervals for each power of ten.

4 'x^y', 'x^', or 'x^y' allows you to raise a number to any power. For example, if you want to work out the value of 2^8, key in 2 onto the display then x^y then 8, and press =. The display should then show 256 as the decimal value of 2^8. Raising a number N to the power $\frac{1}{n}$ gives the nth root of N. For example, the cube root of 29791 is $(29791)^{\frac{1}{3}}$ which equals 31.

▲ **Figure 1** *Displaying powers of ten*

Hint

Do not confuse $\log_{10}$ (base 10) with ln or $\log_e$ (base e). Log_{10} allows you to find a number as a power of ten, while ln or $\log_e$ allows you to find a number as a power of e, the natural log.

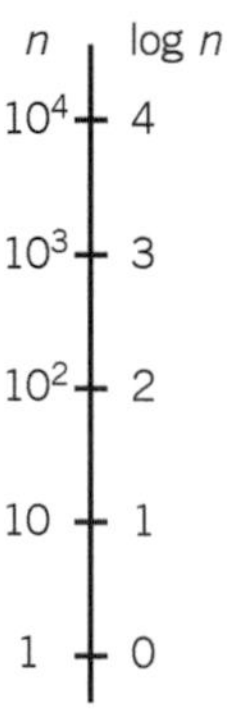

▲ **Figure 2** *A logarithmic scale*

Worked example

Calculate the cube root of 2.9×10^6.

Solution

Step 1 Key in 2.9×10^6 as explained on Page 259.

Step 2 Press the y^x button.

Step 3 Key in $(1 \div 3)$.

Step 4 Press =.

The display should show '1.426×10^2' so the answer is 142.6.

Significant figures

A calculator display shows a large number of digits. When you use a calculator, you should always round up or round down the final answer of a calculation to the same number of significant figures as the data given. Sometimes, a numerical answer to one part of a question has to be used in a subsequent calculation, in which case the numerical answer to the first part should be carried forward without rounding it up or down. For example, if you need to calculate the value of $d \sin 65.0°$, where $d = 1.64$, the calculator will show $9.063077870 \times 10^{-1}$ for the sine of 65.0°. Multiplying this answer by 1.64 then gives 1.486344771, which should then be rounded off to 1.49 so it has the same number of significant figures as 1.64 (i.e., to 3 significant figures).

Summary questions

Write your answers to each of the following questions in standard form, where appropriate, and to the same number of significant figures as the data.

1 Copy and complete the following conversions.

a i 500 mm = _____ m
ii 3.2 m = _____ cm
iii 9560 cm = _____ m

b i 0.45 kg = _____ g
ii 1997 g = _____ kg
iii 54 000 kg = _____ g

c i 20 cm^2 = _____ m^2
ii 55 mm^2 = _____ m^2
iii 0.050 cm^2 = _____ m^2.

2 **a** Write the following values in standard form:

i 150 million km in metres
ii 365 days in seconds
iii 630 nm in metres
iv 25.7 μg in kilograms
v 150 m in millimetres
vi 1.245 μm in metres.

b Write the following values with a prefix instead of in standard form.

i 3.5×10^4 m = _____ km
ii 6.5×10^{-7} m = _____ nm
iii 3.4×10^6 g = _____ kg
iv 8.7×10^8 W = _____ MW = _____ GW.

3 **a** Use the equation average speed $= \dfrac{\text{distance}}{\text{time}}$ to calculate the average speed in m s^{-1} of:

i a vehicle that travels a distance of 9000 m in 450 s
ii a vehicle that travels a distance of 144 km in 2 h
iii a particle that travels a distance of 0.30 nm in a time of 2.0×10^{-18} s
iv the Earth on its orbit of radius 1.5×10^{11} m, given the time taken per orbit is 365.25 days.

b Use the equation resistance $= \dfrac{\text{potential difference}}{\text{current}}$ to calculate the resistance of a component for the following values of current I and p.d. V:

i $V = 15$ V, $I = 2.5$ mA
ii $V = 80$ mV, $I = 16$ mA
iii $V = 5.2$ kV, $I = 3.0$ mA
iv $V = 250$ V, $I = 0.51$ μA
v $V = 160$ mV, $I = 53$ mA.

4 **a** Calculate each of the following:

i 6.7^3
ii $(5.3 \times 10^4)^2$
iii $(2.1 \times 10^{-6})^4$
iv $(0.035)^2$
v $(4.2 \times 10^8)^{\frac{1}{2}}$
vi $(3.8 \times 10^{-5})^{\frac{1}{4}}$.

b Calculate each of the following:

i $\dfrac{2.4^2}{3.5 \times 10^3}$
ii $\dfrac{3.6 \times 10^{-3}}{6.2 \times 10^2}$
iii $\dfrac{8.1 \times 10^4 + 6.5 \times 10^3}{5.3 \times 10^4}$
iv $7.2 \times 10^{-3} + \dfrac{6.2 \times 10^4}{2.6 \times 10^6}$.

14.2 Trigonometry

Angles and arcs

- Angles are measured in degrees or radians. The scale for conversion is $360° = 2\pi$ radians. The symbol for the radian is rad, so 1 rad = $\frac{360}{2\pi} = 57.3°$ (to 3 significant figures).
- The circumference of a circle of radius $r = 2\pi r$. So the circumference can be written as the angle in radians (2π) round the circle × r.
- For a segment of a circle, the length of the arc of the segment is in proportion to the angle θ which the arc makes to the centre of the circle. This is shown in Figure 1. Because the arc length is $2\pi r$ (the circumference) for an angle of 360° (= 2π radians), then

$$\frac{\textbf{arc length, } s}{2\pi r} = \frac{\theta \textbf{ in degrees}}{360°}$$

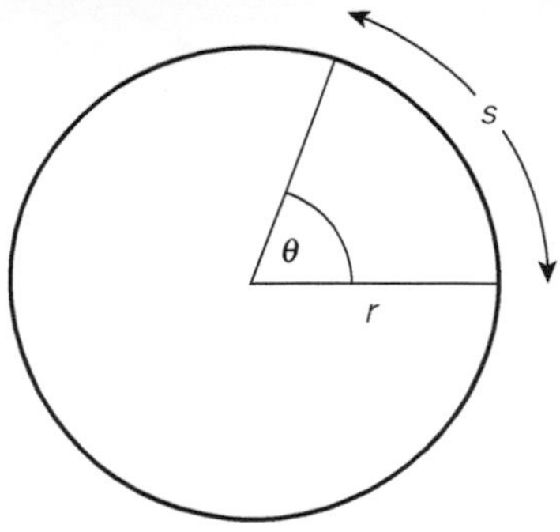

▲ **Figure 1** *Arcs and segments*

Triangles and trigonometry

Area rule

As shown in Figure 2,

the area of any triangle = $\frac{1}{2}$ × its height × its base

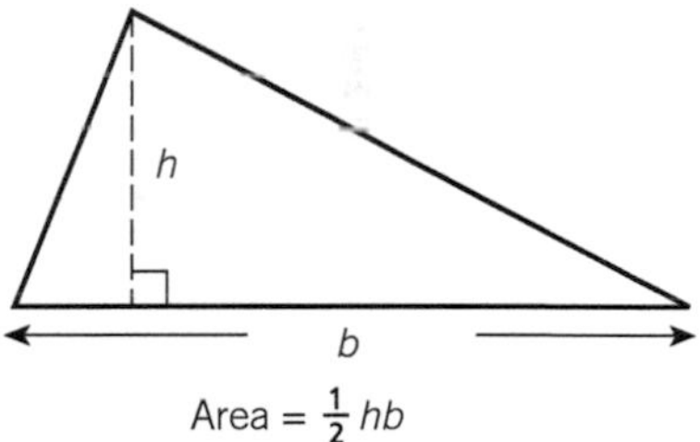

▲ **Figure 2** *The area of a triangle*

Trigonometry calculations using a calculator

A scientific calculator has a button you can press to use either degrees or radians. Make sure you know how to switch your calculator from one of these two modes to the other. For example,

- $\sin 30° = 0.50$, whereas $\sin 30$ rad $= -0.99$
- inv $\sin 0.17$ in degree mode = 9.79°, whereas inv $\sin 0.17$ in rad mode = 0.171.

Also, watch out when you calculate the sine, cosine, or tangent of the product of a number and an angle. For example, sin (2 × 30°) comes out as 1.05 if you forget the brackets, instead of the correct answer of 0.867. The reason for the error is that, unless you insert the brackets, the calculator is programmed to work out sin 2° then multiply the answer by 30.

Hint

Angle in degrees = $\frac{360}{2\pi}$ × angle in radians

Trigonometry functions

Consider again the definitions of the sine, cosine, and tangent of an angle, as applied to the right-angled triangle in Figure 3.

$\sin\theta = \frac{o}{h}$ **where** o **= the length of the side opposite angle** θ

$\cos\theta = \frac{a}{h}$ h **= the length of the hypotenuse**

$\tan\theta = \frac{o}{a}$ a **= the length of the side adjacent to angle** θ

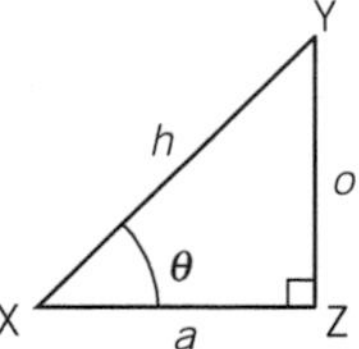

▲ **Figure 3** *A right-angled triangle*

Pythagoras's theorem and trigonometry

Pythagoras's theorem states that, for any right-angled triangle, the square of the hypotenuse = the sum of the squares of the other two sides.

Applying Pythagoras's theorem to the right-angled triangle in Figure 3 gives

$$h^2 = o^2 + a^2$$

Synoptic link

You met vectors and how to resolve them in Topic 1.1, Vectors and scalars.

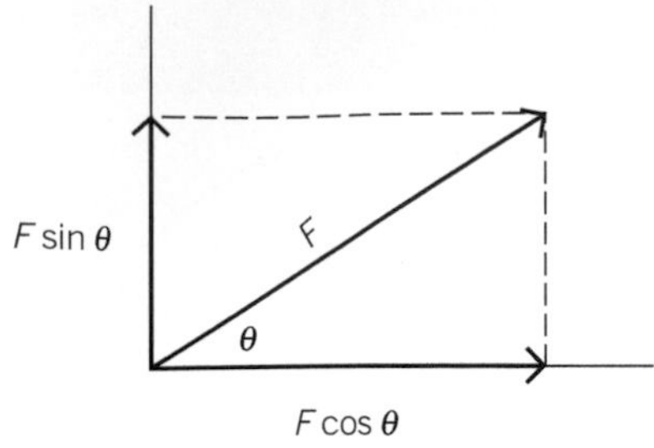

▲ **Figure 4** *Resolving a vector*

Since $o = h \sin\theta$ and $a = h \cos\theta$, then the above equation may be written:

$$h^2 = h^2 \sin^2\theta + h^2 \cos^2\theta$$

Cancelling h^2 therefore gives the following useful link between $\sin\theta$ and $\cos\theta$:

$$1 = \sin^2\theta + \cos^2\theta$$

Vector rules

Resolving a vector

Any vector can be resolved into two perpendicular components in the same plane as the vector, as shown by Figure 4. The force vector F is resolved into a horizontal component $F\cos\theta$ and a vertical component $F\sin\theta$, where θ is the angle between the line of action of the force and the horizontal line.

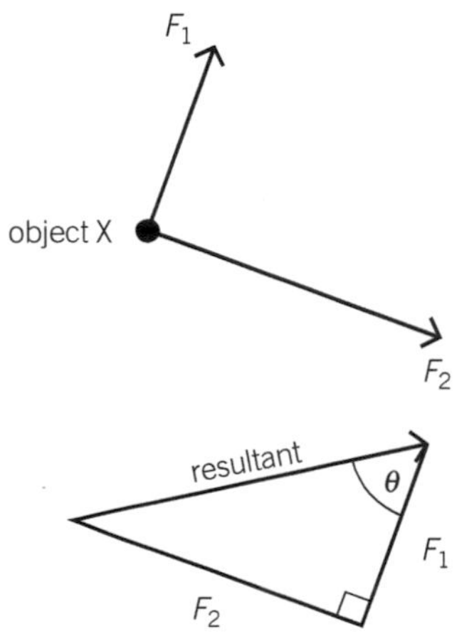

▲ **Figure 5** *Adding two perpendicular vectors*

Adding two perpendicular vectors

Figure 5 shows two perpendicular forces F_1 and F_2 acting on a point object X. The combined effect of these two forces, the resultant force, is given by the vector triangle in Figure 5. This is a right-angled triangle, where the resultant force is represented by the hypotenuse.

- Applying Pythagoras's theorem to the triangle gives $F^2 = F_1^{\,2} + F_2^{\,2}$, where F is the magnitude of the resultant force.
  Therefore $F = (F_1^{\,2} + F_2^{\,2})^{\frac{1}{2}}$.
- Applying the trigonometry equation $\tan\theta = \frac{o}{a}$, the angle between the resultant force and force F_1 is given by $\tan\theta = \frac{F_2}{F_1}$.

Summary questions

1 a Calculate the circumference of a circle of radius 0.250 m.

b Calculate the length of the arc of a circle of radius 0.250 m for the following angles between the arc and the centre of the circle:

i 360° **ii** 240° **iii** 60°.

2 For the right-angled triangle XYZ in Figure 6, calculate:

a angle YXZ (= θ) if XY = 80 mm and

i XZ = 30 mm

ii XZ = 60 mm

iii YZ = 30 mm

iv YZ = 70 mm.

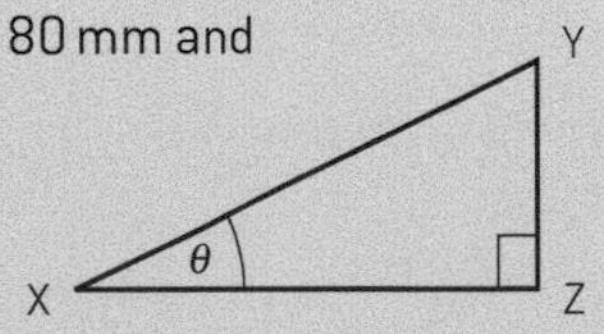

Y
X θ Z

▲ **Figure 6**

b XZ if

i XY = 20 cm and θ = 30°

ii XY = 22 m and θ = 45°

iii YZ = 18 mm and θ = 75°

iv YZ = 47 cm and θ = 25°.

3 a A right-angled triangle XYZ has a hypoteneuse XY of length 55 mm and side XZ of length 25 mm. Calculate the length of the other side.

b An aircraft travels a distance of 30 km due north from an airport P to an airport Q. It then travels due east for a distance of 18 km to an airport R. Calculate

i the distance from P to R

ii the angle QPR.

4 a Calculate the horizontal component A and the vertical component B of:

i a 6.0 N force at 40° to the vertical

ii a 10.0 N force at 20° to the vertical

iii a 7.5 N force at 50° to the horizontal.

b Calculate the magnitude and direction of the resultant of a 2.0 N force acting due north and a 3.5 N force acting due east.

14.3 More about algebra

Signs and symbols

Every physical quantity in physics has a symbol which is used to represent it in equations. When an equation is used to calculate a physical quantity, the result of the calculation is expressed as a numerical value and the correct unit for the quantity. An answer giving a numerical value only is insufficient because the correct unit should also be given.

Most of the symbols used in physics each represent a single physical quantity. However, certain symbols are used to represent more than one physical quantity. For example,

- the capital letter I is used to represent electric current or the intensity of radiation or of sound
- the capital letter E is used to represent energy or the Young modulus or electric field strength (which you will meet in Chapter 18).

So the context in which such symbols are used needs to be considered to establish what the symbol represents in that context. As you move further through your A Level Physics course, take care when you meet symbols that represent more than one quantity.

Signs you need to recognise

- Inequality signs are often used in physics. You need to be able to recognise the meaning of the signs in Table 1. For example, the inequality $I \geq 3$ A means that the current is greater than or equal to 3 A. This is the same as saying that the current is not less than 3 A.

▼ **Table 1** *Signs*

Sign	Meaning	Sign	Meaning	Sign	Meaning
$>$	greater than	$>>$	much greater than	$<x^2>$	mean square value
$<$	less than	$<<$	much less than	$\propto$	is proportional to
$\geq$	greater than or equal to	$\approx$	approximately equals to	Δ	change of
$\leq$	less than or equal to	$<x>$	mean value	$\surd$	square root

- The approximation sign is used where an estimate or an order-of-magnitude calculation is made, rather than a precise calculation. For an order-of-magnitude calculation, the final value is written with one significant figure only, or even rounded up or down to the nearest power of ten. Order-of-magnitude calculations are useful as a quick check after using a calculator. For example, if you are asked to calculate the density of a 1.0 kg metal cylinder of height 0.100 m and diameter 0.071 m, you ought to obtain a value of 2530 kg m^{-3} using a calculator. Now let's check the value quickly:

$$\text{volume} = \pi(\text{radius})^2 \times \text{height}$$

$$\approx 3 \times (0.04)^2 \times 0.1 \approx 48 \times 10^{-5}\,\text{m}^3$$

$$\text{density} = \frac{\text{mass}}{\text{volume}}$$

$$\approx \frac{1.0}{50 \times 10^{-5}} \approx 2000\,\text{kg}\,\text{m}^{-3}$$

This confirms our 'precise' calculation.

- Proportionality is represented by the $\propto$ sign. A simple example of its use in physics is for Hooke's law – the tension in a spring is directly proportional to its extension:

$$\textbf{tension } T \propto \textbf{extension } \Delta L$$

By introducing a constant of proportionality k, the link above can be made into an equation:

$$T = k\Delta L$$

where k is defined as the spring constant. With any proportionality relationship, if one of the variables is increased by a given factor (e.g., ×3), the other variable is increased by the same factor. So in the above example, if T is trebled, then extension ΔL is also trebled. A graph of tension T on the y-axis against extension ΔL on the x-axis would give a straight line through the origin.

Synoptic link

You met Hooke's law and the spring constant in Topic 6.2, Springs.

More about equations

Rearranging an equation with several terms

The equation $v = u + at$ is an example of an equation with two terms on the right-hand side. These terms are u and at. To make t the subject of the equation:

1. Isolate the term containing t on one side by subtracting u from both sides to give $v - u = at$.
2. Isolate t by dividing both sides of the equation $v - u = at$ by a to give
   $\frac{(v-u)}{a} = \frac{at}{a} = t$
   Note that a cancels out in the expression $\frac{at}{a}$
3. The rearranged equation may now be written
   $t = \frac{(v-u)}{a}$

Rearranging an equation containing powers

Suppose a quantity is raised to a power in a term in an equation, and that quantity is to be made the subject of the equation. For example, consider the equation $V = \frac{4}{3}\pi r^3$ where r is to be made the subject of the equation.

1. Isolate r^3 from the other factors in the equation by dividing both sides by 4π, then multiplying both sides by 3 to give $\frac{3V}{4\pi} = r^3$.
2. Take the cube root of both sides to give $\left(\frac{3V}{4\pi}\right)^{\frac{1}{3}} = r$.
3. Rewrite the equation with r on the left-hand side if necessary.

More about powers

1. Powers add for identical quantities when two terms are multiplied together. For example, if
   $y = ax^n$ and $z = bx^m$, then $yz = ax^m\, bx^n = abx^{m+n}$.
2. An equation of the form $y = \frac{k}{z^n}$ may be written in the form $y = kz^{-n}$.

3 The nth root of an expression is written as the power $\frac{1}{n}$. For example, the square root of x is $x^{\frac{1}{2}}$. Therefore, rearranging $y = x^n$ to make x the subject gives $x = y^{\frac{1}{n}}$.

Summary questions

1 Complete each of the following statements:

a if $x > 5$, then $\frac{1}{x} <$

b if $4 < x < 10$, then ____ $< \frac{1}{x} <$

c if x is positive and $x^2 > 100$, then $\frac{1}{x}$ ____.

2 a Make t the subject of each of the following equations:

i $v = u + at$ iii $y = k(t - t_0)$

ii $s = \frac{1}{2}at^2$ iv $F = \frac{mv}{t}$.

b Solve each of the following equations:

i $2z + 6 = 10$ iii $\frac{2}{z-4} = 8$

ii $2(z + 6) = 10$ iv $\frac{4}{z^2} = 36$.

3 a Make x the subject of each of the following equations:

i $y = 2x^{\frac{1}{2}}$ iii $yx^{\frac{1}{3}} = 1$

ii $2y = x^{-\frac{1}{2}}$ iv $y = \frac{k}{x^2}$.

b Solve each of the following equations:

i $x^{-\frac{1}{2}} = 2$ iii $\frac{8}{x^2} = 32$

ii $3x^2 = 24$ iv $2(x^{\frac{1}{2}} + 4) = 12$.

4 Use the data given with each equation below to calculate:

a the volume V of a wire of radius $r = 0.34$ mm and length $L = 0.840$ m, using the equation $V = \pi r^2 L$

b the radius r of a sphere of volume $V = 1.00 \times 10^{-6}\,\text{m}^3$, using the equation $V = \frac{4}{3}\pi r^3$

c the time period T of a simple pendulum of length $L = 1.50$ m, using the equation $T = 2\pi\left(\frac{L}{g}\right)^{\frac{1}{2}}$, where $g = 9.8\,\text{m}\,\text{s}^{-2}$

d the speed v of an object of mass $m = 0.20$ kg and kinetic energy $E_k = 28$ J, using the equation $E_k = \frac{1}{2}mv^2$.

14.4 Straight-line graphs

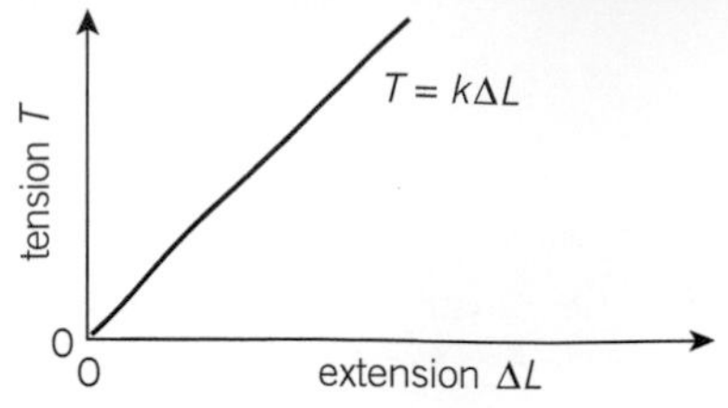

▲ **Figure 1** *Straight-line graph*

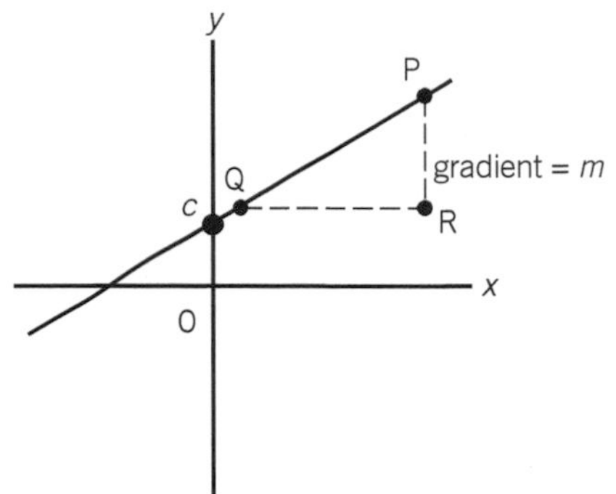

▲ **Figure 2** *Graph of $y = mx + c$*

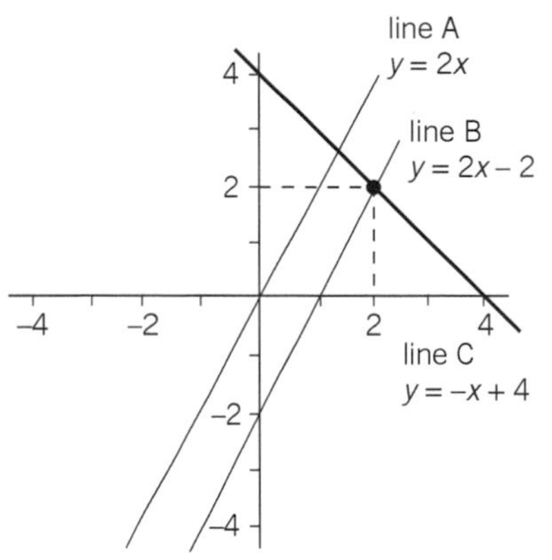

▲ **Figure 3** *Straight-line graphs*

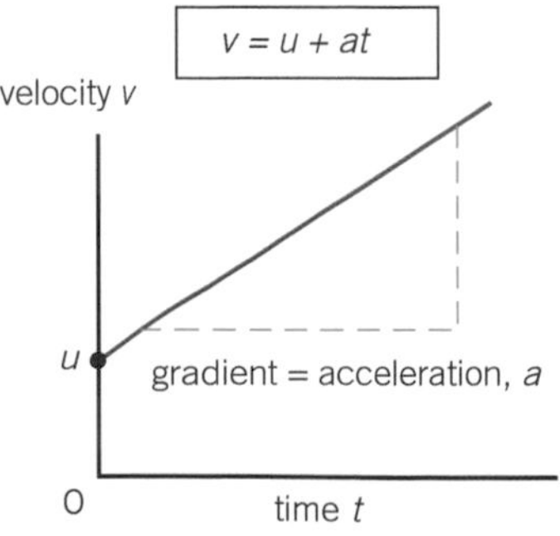

▲ **Figure 4** *Graph of motion at constant acceleration*

The general equation for a straight-line graph

Links between two physical quantities can be established most easily by plotting a graph. One of the physical quantities is represented by the vertical scale (the ordinate, often called the y-axis) and the other quantity by the horizontal scale (the abscissa, often called the x-axis). The coordinates of a point on a graph are the x- and y-values, usually written (x, y), of the point.

The simplest link between two physical variables is where the plotted points define a straight line. For example, Figure 1 shows the link between the tension in a spring and the extension of the spring – the gradient of the line is constant and the line passes through the origin. Any situation where the y-variable is directly proportional to the x-variable gives a straight line through the origin. For Figure 1, the gradient of the line is the spring constant k. The relationship between the tension T and the extension ΔL may therefore be written as $T = k\,\Delta L$.

The general equation for a straight-line graph is usually written in the form $\boldsymbol{y = mx + c}$, where m = the gradient of the line and c = the y-intercept.

- The gradient m can be measured by marking two points P and Q far apart on the line. The triangle PQR, as shown in Figure 2, is then used to find the gradient. If (x_P, y_P) and (x_Q, y_Q) represent the x- and y-coordinates of points P and Q, respectively, then

$$\textbf{gradient } m = \frac{y_P - y_Q}{x_P - x_Q}$$

- The y-intercept, c, is the point at $x = 0$ where the line crosses the y-axis. To find the y-intercept of a line on a graph that does not show $x = 0$, measure the gradient as above, then use the coordinates of any point on the line with the equation $y = mx + c$ to calculate c. For example, rearranging $y = mx + c$ gives $c = y - mx$. Therefore, using the coordinates of point Q in Figure 2, the y-intercept $c = y_Q - mx_Q$.

Examples of straight-line graphs

- Line A $c = 0$, so the line passes through the origin. Its equation is $y = 2x$.
- Line B $m > 0$, so the line has a positive gradient. Its equation is $y = 2x - 2$.
- Line C $m < 0$, so the line has a negative gradient. Its equation is $y = -x + 4$.

Straight-line graphs and physics equations

You need to be able to work out gradients and intercepts for equations you meet in physics that generate straight-line graphs. Some further examples in addition to Figure 1 are described below.

1 **The velocity v of an object moving at constant acceleration a at time t** is given by the equation $v = u + at$, where u is its velocity at time $t = 0$. Figure 4 shows the corresponding graph of velocity v on the y-axis against time t on the x-axis.

Rearranging the equation as $v = at + u$ and comparing this with $y = mx + c$ shows that

- the gradient m = acceleration a
- the y-intercept c = the initial velocity u.

2 **The p.d., V, across the terminals of a battery of e.m.f. ε and internal resistance r** varies with current in accordance with the equation $V = \varepsilon - Ir$. Figure 5 shows the corresponding graph of p.d., V, on the y-axis against current, I, on the x-axis.
Rearranging the equation as $V = -rI + \varepsilon$ and comparing this with $y = mx + c$ shows that
   - the gradient $m = -r$
   - the y-intercept $c = \varepsilon$ so the intercept on the y-axis gives the e.m.f. ε of the battery.

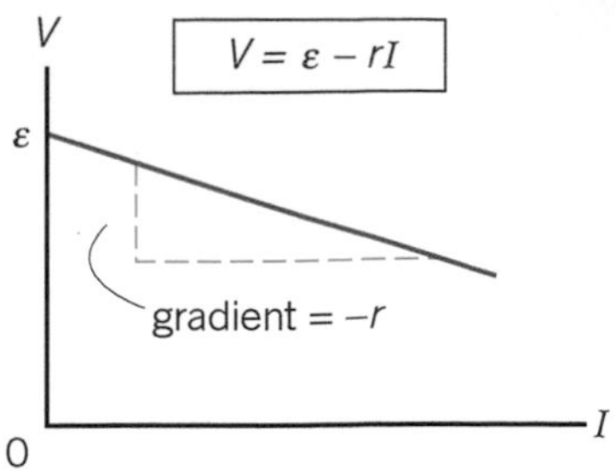

▲ **Figure 5** *Graph of p.d. versus I for a battery*

3 **The maximum kinetic energy E_{Kmax} of a photoelectron** emitted from a metal surface of work function ϕ varies with frequency f of the incident radiation, in accordance with the equation $E_{Kmax} = hf - \phi$. Figure 6 shows the corresponding graph of E_{Kmax} on the y-axis against f on the x-axis.

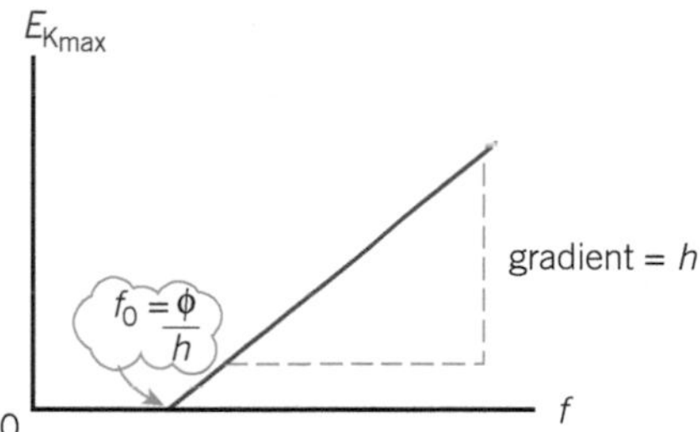

▲ **Figure 6** *Graph of photoelectric emission*

Comparing the equation $E_{Kmax} = hf - \phi$ with $y = mx + c$ shows that
   - the gradient $m = h$
   - the y-intercept $c = -\phi$.

Note that the x-intercept is where $y = 0$ on the line. Let the coordinates of the x-intercept be $(x_0, 0)$. Therefore $mx_0 + c = 0$ so $x_0 = -\frac{c}{m}$. Applied to Figure 6, the x-intercept is therefore $\frac{\phi}{h}$. Since the x-intercept is the threshold frequency f_0, then $f_0 = \frac{\phi}{h}$.

Simultaneous equations

In physics, simultaneous equations can be solved graphically by plotting the line for each equation. The solution of the equations is given by the coordinates of the points where the lines meet. For example, lines B and C in Figure 3 meet at the point (2, 2) so $x = 2$, $y = 2$ are the only values of x and y that fit both equations.

Solving simultaneous equations doesn't require graph plotting if the equations can be arranged to fit one of the variables. Start by rearranging to make y the subject of each equation, if necessary. Considering the example above:

equation of line B is $y = 2x - 2$

equation of line C is $y = -x + 4$.

At the point where they meet, their coordinates are the same, so solving $2x - 2 = -x + 4$ gives the x-coordinate. Rearranging this equation gives $3x = 6$ so $x = 2$.

Since $y = 2x - 2$, then $y = (2 \times 2) - 2 = 2$.

Summary questions

1 For each of the following equations, which represent straight-line graphs, write down **i** the gradient, **ii** the y-intercept, **iii** the x-intercept:

a $y = 3x - 3$ **c** $y + x = 5$

b $y = -4x + 8$ **d** $2y + 3x = 6$.

2 **a** A straight line on a graph has a gradient $m = 2$ and passes through the point (2, −4). Work out **i** the equation for this line, **ii** its y-intercept.

b The velocity v (in m s^{-1}) of an object varies with time t (in s) in accordance with the equation $v = 5 + 3t$. Determine **i** the acceleration of the object, **ii** the initial velocity of the object.

3 **a** Plot the equations $y = x + 3$ and $y = -2x + 6$ over the range from $x = -3$ to $x = +3$. Write down the coordinates of the point P where the two lines cross.

b Write down the equation for the line OP, where O is the origin of the graph.

4 Solve the following pairs of simultaneous equations:

a $y = 2x - 4$, $y = -x + 2$

b $y = 3x - 4$, $x + y = 8$

c $2x + 3y = 4$, $x + 2y = 2$.

14.5 More on graphs

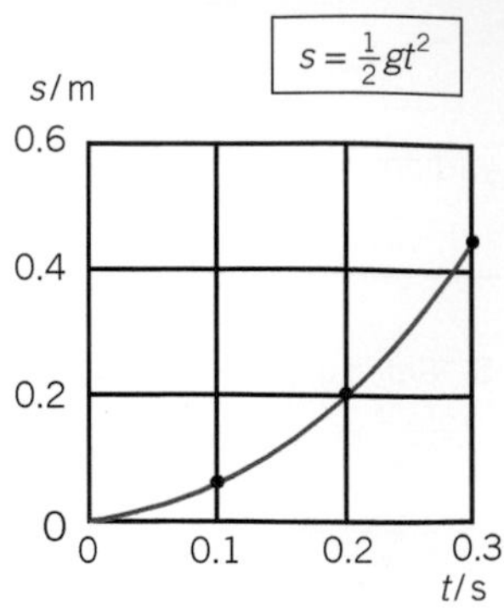

▲ **Figure 1** *Graph of s against t*

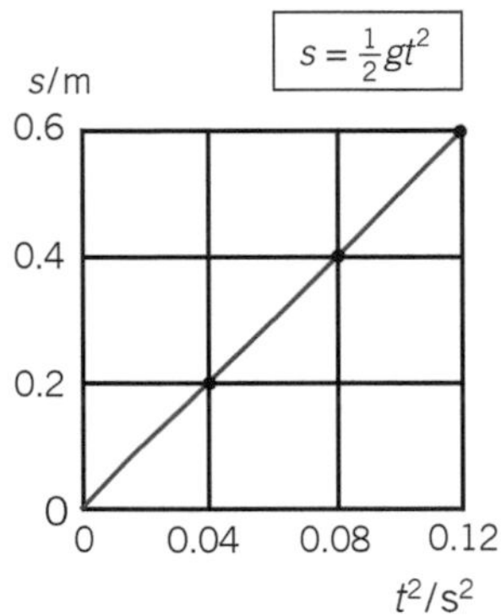

▲ **Figure 2** *Graph of s against t^2*

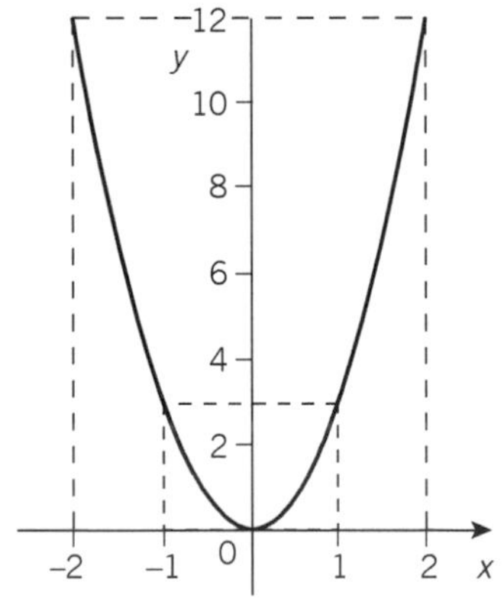

▲ **Figure 3** *Graph of $y = 3x^2$*

Synoptic link

You met projectiles in Topic 2.7, Projectile motion 1, and Topic 2.8, Projectile motion 2.

Curves and equations

Graphs with curves occur in physics in two ways.

1. In practical work, where one physical variable is plotted against another and the plotted points do not lie along a straight line. For example, a graph of p.d. on the y-axis against current on the x-axis for a filament lamp is a curve that passes through the origin.
2. In theory work, where an equation containing two physical variables is not in the form of the equation for a straight line ($y = mx + c$). For example, for an object released from rest, a graph of distance fallen, s, on the y-axis against time, t, on the x-axis is a curve, because $s = \frac{1}{2}gt^2$. Figure 1 shows this equation represented on a graph.

Knowledge of the general equations for some common curves is an essential part of physics. When a curve is produced as a result of plotting a set of measurements in a practical experiment, few conclusions can be drawn as the exact shape of a curve is difficult to test. In comparison, if the measurements produce a straight line, it can then be concluded that the two physical variables plotted on the axes are related by an equation of the form $y = mx + c$.

If a set of measurements produces a curve rather than a straight line, knowledge of the theory could enable the measurements to be processed in order to give a straight-line graph, which would then be a confirmation of the theory. For example, the distance and time measurements that produced the curve in Figure 1 could be plotted as distance fallen, s, on the y-axis against t^2 on the x-axis (where t is the time taken). Figure 2 shows the idea.

If a graph of s against t^2 gives a straight line, this would confirm that the relationship between s and t is of the form $s = kt^2$, where k is a constant. Because theory gives $s = \frac{1}{2}gt^2$, it can then be concluded that $k = \frac{1}{2}g$.

From curves to straight lines

Parabolic curves

These curves describe the flight paths of projectiles or other objects acted on by a constant force that is not in the same direction as the initial velocity of the object. In addition, parabolic curves occur where the energy of an object depends on some physical variable.

The general equation for a parabola is $y = kx^2$. Figure 3 shows the shape of the parabola $y = 3x^2$. Equations of the form $y = kx^2$ pass through the origin and they are symmetrical about the y-axis. This is because equal positive and negative values of x always give the same y-value.

The **flight path for a projectile** projected horizontally at speed u has coordinates $x = ut$, $y = \frac{1}{2}gt^2$, where x = horizontal distance travelled, y = vertical distance fallen, and t is the time from initial projection. Combining these equations gives the flight path equation $y = \frac{gx^2}{2u^2}$

which is the same as the parabola equation $y = kx^2$ where $\frac{g}{2u^2}$ is represented by k in the equation.

A set of measurements plotted as a graph of vertical distance fallen, y, against horizontal distance travelled, x, would be a parabolic curve, as shown in Figure 3. However, a graph of y against x^2 should give a straight line (of gradient k) through the origin (Figure 4), because $y = kx^2$.

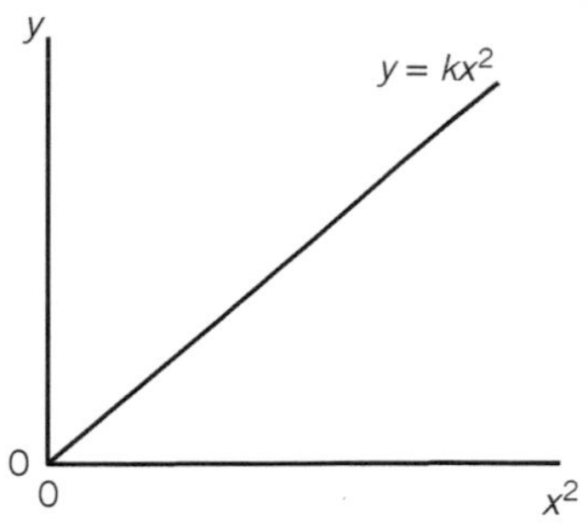

▲ **Figure 4** *Graph of y against x^2*

Inverse curves

An inverse relationship between two variables x and y is of the form $y = \frac{k}{x}$, where k is a constant. The variable y is said to be inversely proportional to variable x. For example, if x is doubled, y is halved.

Figure 5 shows the curve for $y = \frac{10}{x}$.

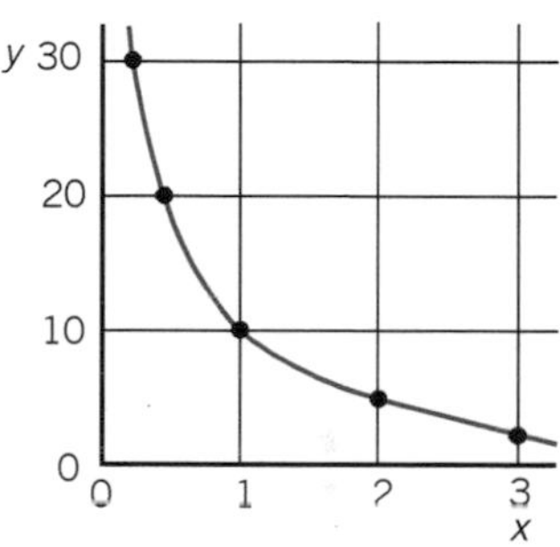

▲ **Figure 5** *Graph of $y = \frac{10}{x}$*

The curve tends towards either axis but never actually meets the axes. The correct mathematical word for 'tending towards but never meeting' is asymptotic. Consider the following example.

The resistance R of a wire of constant length L varies with the wire's area of cross section, A, in accordance with the equation $R = \frac{\rho L}{A}$, where ρ is the resistivity of the wire.

Therefore, R is inversely proportional to A. A graph of R (on the vertical axis) against A would therefore be a curve like Figure 5. However, a graph of R (on the vertical axis) against $\frac{1}{A}$ is a straight line through the origin (Figure 6). The gradient of this straight line is ρL.

Summary questions

1 The potential energy, E_p, stored in a stretched spring varies with the extension ΔL of the spring, in accordance with the equation $E_p = \frac{1}{2}k\,\Delta L^2$. Sketch a graph of E_p against **a** ΔL, **b** ΔL^2.

2 The energy E_{ph} of a photon varies with its wavelength λ in accordance with the equation $E_{ph} = \frac{hc}{\lambda}$, where h is the Planck constant and c is the speed of light. Sketch a graph of E_{ph} against **a** λ, **b** $\frac{1}{\lambda}$.

3 The current I through a wire of resistivity ρ varies with the length L, area of cross section A, and p.d. V, in accordance with the equation $I = \frac{VA}{\rho L}$.

a Sketch a graph of I against **i** V, **ii** L, **iii** $\frac{1}{L}$.

b Explain how you would determine the resistivity from the graph of I against **i** V, **ii** $\frac{1}{L}$.

4 An object released from rest falls at constant acceleration a and passes through a horizontal beam at speed u. The distance it falls in time t after passing through the light beam is given by the equation $s = ut + \frac{1}{2}at^2$.

a Show that $\frac{s}{t} = u + \frac{1}{2}at$.

b **i** Sketch a graph of $\frac{s}{t}$ on the vertical axis against t on the horizontal axis.

ii Explain how u and a can be determined from the graph.

Synoptic link

You met resistivity in Topic 9.3, Resistance.

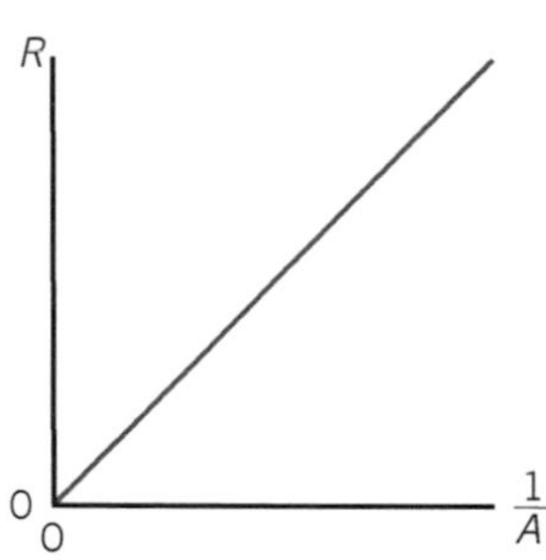

▲ **Figure 6** *Graph of R against $\frac{1}{A}$*

14.6 Graphs, gradients, and areas

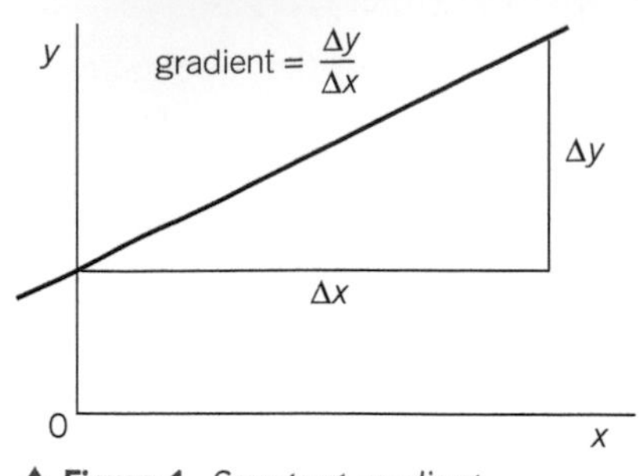

▲ **Figure 1** *Constant gradient*

Gradients

1 The gradient of a straight line = $\frac{\Delta y}{\Delta x}$, where Δy is the change of the quantity plotted on the y-axis and Δx is the change of the quantity plotted on the x-axis. As shown in Figure 1, the gradient of a straight line is obtained by drawing as large a gradient triangle as possible and measuring the height Δy and the base Δx of this triangle, using the scale on each axis.

Note:

As a rule, when you plot a straight-line graph, always choose a scale for each axis that covers at least half the length of each axis. This will enable you to draw the line of best fit as accurately as possible. The measurement of the gradient of the line will therefore be more accurate. If the y-intercept is required and it cannot be read directly from the graph, or extrapolating (ie. drawing) the line back to the y-axis, it can be calculated by substituting the value of the gradient and the coordinates of a point on the line into the equation $y = mx + c$.

Synoptic link

You have met lines of best fit in the Evaluating your results section on Page 9.

2 The gradient at a point on a curve = the gradient of the tangent to the curve at that point.

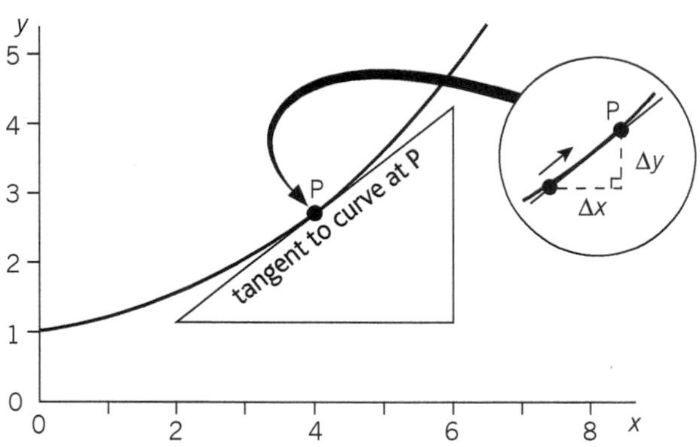

▲ **Figure 2** *Tangent to a curve at a point*

The tangent to the curve at a point is a straight line that touches the curve at that point, without cutting across it. To see why, mark any two points on a curve and join them by a straight line. The gradient of the line is $\frac{\Delta y}{\Delta x}$, where Δy is the vertical separation of the two points and Δx is the horizontal separation. Now repeat with one of the points closer to the other – the straight line is now closer in direction to the curve. If the points are very close, the straight line between them is almost along the curve. The gradient of the line is then virtually the same as the gradient of the curve at that position. Figure 2 shows this idea. In other words, the gradient of the straight line $\frac{\Delta y}{\Delta x}$ becomes equal to the gradient of the curve as $\Delta x \rightarrow 0$. The curve gradient is written as $\frac{dy}{dx}$ where $\frac{d}{dx}$ means rate of change.

The gradient of the tangent is a straight line and is obtained as explained above. Drawing the tangent to a curve requires practice. This skill is often needed in practical work. The **normal** at the point where the tangent touches the curve is the straight line perpendicular to the tangent at that point. An accurate technique for drawing the normal to a curve using a plane mirror is shown in Figure 3. At the point where the normal intersects the curve, the curve and its mirror image should join smoothly, without an abrupt change of gradient where they join. After positioning the mirror surface correctly, the normal can then be drawn and then used to draw the tangent to the curve.

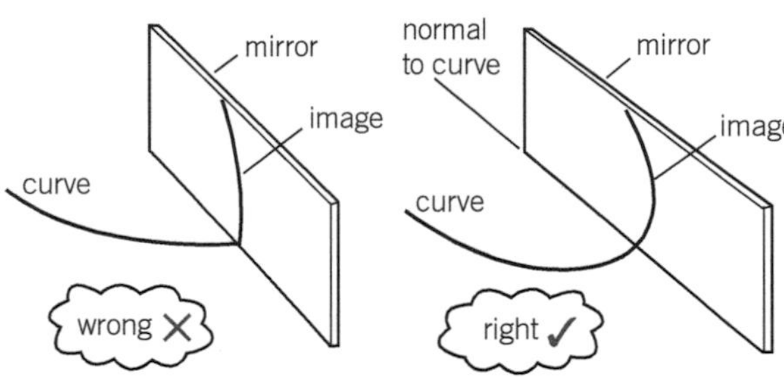

▲ **Figure 3** *Drawing the normal to a curve*

Hint

To estimate the area under a curve, use the graph grid to count how many grid squares are under the curve. Where the curve crosses a grid square, count it as a whole square if at least half of the grid square is under the curve otherwise ignore it.

Turning points

A turning point on a curve is where the gradient of the curve is zero. This happens where a curve reaches a peak with a fall either side (i.e., a maximum) or where it reaches a trough with a rise either side (i.e., a minimum). Where the gradient represents a physical quantity, a turning point is where that physical quantity is zero. Figure 4 shows an example of a curve with a turning point. This is a graph of the vertical height y against time for a projectile that reaches a maximum height, then descends as it travels horizontally. The gradient represents the vertical component of velocity. At maximum height, the gradient of the curve is zero, so the vertical component of velocity is zero at that point.

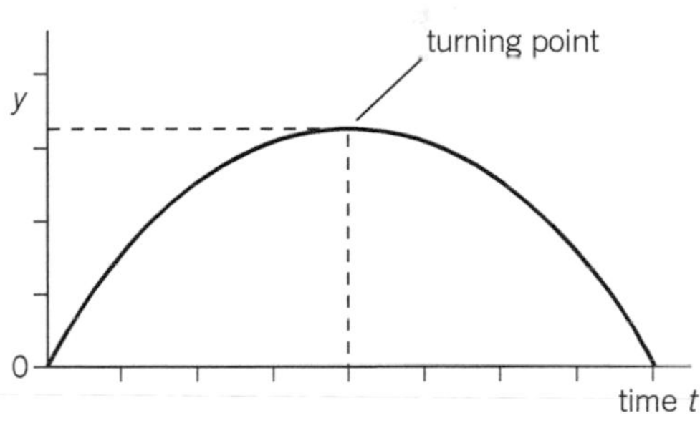

▲ **Figure 4** *Turning points*

Note:

If the equation of a curve is known, the gradient can be determined by the process of **differentiation**. This mathematical process is not needed for AS Level physics. The essential feature of the process is that, for a function of the form $y = kx^n$, the gradient (written as $\frac{dy}{dx}$) $= nkx^{n-1}$.

For example, if $y = \frac{1}{2}gt^2$, then $\frac{dy}{dt} = gt$.

Areas and graphs

The area under a line on a graph can give useful information if the product of the y-variable and the x-variable represents another physical variable. For example, consider Figure 5, which is a graph of the tension in a spring against its extension. Since tension × extension is force × distance, which equals work done, then the area under the line represents the work done to stretch the spring.

Figure 5b shows a tension against extension graph for a rubber band. Unlike Figure 5a, the area under the curve is not a triangle, but it still represents work done, in this case the work done to stretch the rubber band.

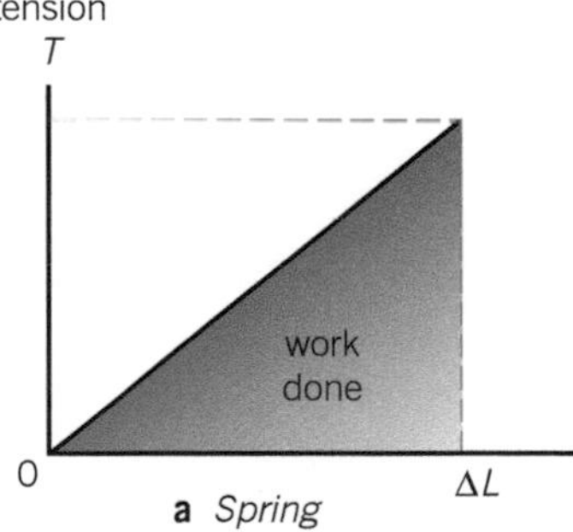

a *Spring*

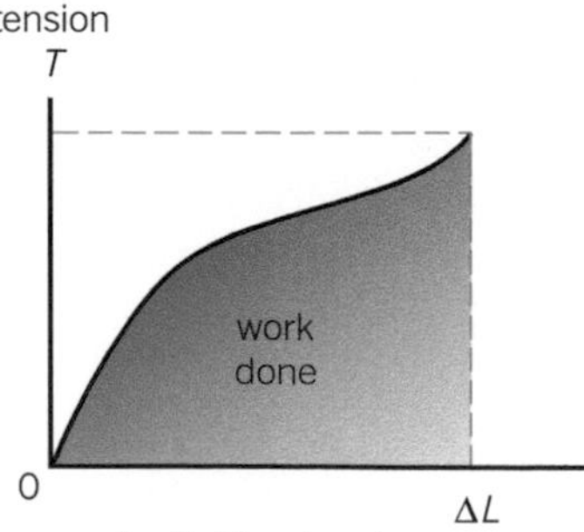

b *Rubber band*

▲ **Figure 5** *Tension versus extension*

Synoptic link

You have met work done as the area under a graph in Topic 6.2, Springs.

Worked example

Figure 6 shows how the tension in a rubber band varies with its extension. Use the graph to estimate the work done to stretch the rubber band to an extension of 220 mm.

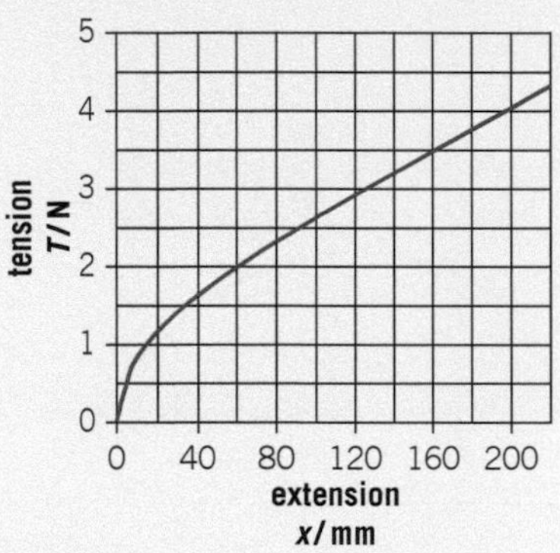

▲ **Figure 2** *Tension against extension for a rubber band*

Solution

Step 1 The number of grid squares under the line (including those more than half below the line) = sum of the number of grid squares in each 20 mm col = 2 + 3 + 4 + 4 + 5 + 6 + 6 + 7 + 7 + 8 + 8 = 60

Step 2 Work done represented by each square = height × width measured from the scale = 0. 5 N × 20 mm = 10 mJ = 0.010 J

Step 3 Total work done = number of squares × work done for 1 square = 60 × 0.010 J = 0.60 J

The product of the y-variable and the x-variable must represent a physical variable with a physical meaning if the area is to be of use. A graph of mass against volume for different sizes of the same material gives a straight line through the origin. The mass is directly proportional to the volume, and the gradient gives the density. But the area under the line has no physical significance since mass × volume does not represent a physical variable.

Note that even where the area does represent a physical variable, it may not have any physical meaning. For example, for a graph of p.d. against current, the product of p.d. and current represents power, but this physical quantity has no meaning in this situation.

More examples of curves where the area is useful include:

- velocity against time, where the area between the line and the time axis represents displacement
- acceleration against time, where the area between the line and the time axis represents change of velocity
- power against time, where the area between the curve and the time axis represents energy
- potential difference against charge, where the area between the curve and the charge axis represents energy.

Summary questions

1 **a** Sketch a velocity against time graph (with time on the x-axis) to represent the equation $v = u + at$, where v is the velocity at time t.

b What feature of the graph represents **i** the acceleration, **ii** the displacement?

2 **a** Sketch a graph of current (on the y-axis) against p.d. (on the x-axis) to show how the current through an ohmic conductor varies with p.d.

b How can the resistance of the conductor be determined from the graph?

3 An electric motor is supplied with energy at a constant rate.

a Sketch a graph to show how the energy supplied to the motor increases with time.

b Explain how the power supplied to the motor can be determined from the graph.

4 A steel ball bearing is released in a tube of oil and falls to the bottom of the tube.

a Sketch graphs to show how **i** the velocity, **ii** the acceleration of the ball changes with time from the instant of release to the point of impact at the bottom of the tube.

b What is represented on graph **a i** by **i** the gradient, **ii** the area under the line?

c What is represented on graph **a ii** by the area under the line?

15 Motion in a circle

15.1 Uniform circular motion

Learning objectives:

- → Recognise uniform motion in a circle.
- → Describe what you need to measure to find the speed of an object moving in uniform circular motion.
- → Define angular displacement and angular speed.

Specification reference: 3.6.3

In a cycle race, the cyclists pedal furiously at top speed. The speed of the perimeter of each wheel is the same as the cyclist's speed, provided the wheels do not slip on the ground. If the cyclist's speed is constant, the wheels must turn at a steady rate. An object rotating at a steady rate is said to be in **uniform circular motion**.

Consider a point on the perimeter of a wheel of radius r rotating at a steady speed.

- The circumference of the wheel = $2\pi r$.
- The frequency of rotation $f = \frac{1}{T}$, where T is the time for one rotation.

The speed v of a point on the perimeter = $\frac{\text{circumference}}{\text{time for one rotation}}$ $= \frac{2\pi r}{T} = 2\pi r f$

$$v = \frac{2\pi r}{T}$$

▲ **Figure 1** *In uniform circular motion*

Worked example

A cyclist is travelling at a speed of $25\,\text{m}\,\text{s}^{-1}$ on a bicycle that has wheels of radius 750 mm. Calculate:

a the time for one rotation of the wheel

b **i** the frequency of rotation of the wheel

ii the number of rotations of the wheel in one minute.

Solution

a Rearranging speed $v = \frac{2\pi r}{T}$ gives the time for one rotation, $T = \frac{2\pi r}{v}$

Therefore, $T = \frac{2\pi \times 0.75}{25} = 0.19\,\text{s}$

b **i** Frequency $f = \frac{1}{T} = \frac{1}{0.19} = 5.3\,\text{Hz}$

ii Number of rotations in 1 min = $60 \times 5.3 = 320$ (2 s.f.)

Angular displacement and angular speed

The London Eye is a very popular tourist attraction. The wheel has a diameter of 130 m and takes passengers high above the surrounding buildings, giving a glorious view on a clear day. Each full rotation of the wheel takes 30 minutes. So each capsule takes its passengers through an angle of 0.2° per second ($= \frac{\pi}{900}$ radians per second). This means that each capsule turns through an angle of

- 2° in 10 s,
- 20° ($= \frac{2\pi}{18}$ radians) in 100 s,
- 90° ($= \frac{\pi}{2}$ radians) in 450 s.

▲ **Figure 2** *The London Eye*

Hint

Remember that 360° = 2π radians.
180° = π radians
90° = $\frac{\pi}{2}$ radians.
Prove for yourself that 1 rad = 57.3°.

For any object in uniform circular motion, the object turns through an angle of $\frac{2\pi}{T}$ radians per second, where T is the time taken, in seconds,

Hint

1 In time t, an object in uniform circular motion at speed v moves along the arc of the circle through a distance s of

$$s = vt = \frac{2\pi rt}{T} = \theta r$$

2 Speed $v = \frac{2\pi r}{T} = \omega r$

(because $\omega = \frac{2\pi}{T}$)

for one complete rotation. In other words, the angular displacement of the object per second is $\frac{2\pi}{T}$.

The **angular displacement**, θ (i.e., the angle in radians), of the object in time t is $\frac{2\pi t}{T} = 2\pi ft$, where T is the time for one rotation and $f\ (= \frac{1}{T})$ is the frequency of rotation.

The **angular speed**, ω, (omega), is defined as the angle it turns through per second.

So, $\omega = \frac{\text{angular displacement}}{\text{time taken}} = \frac{\theta}{t} = \frac{2\pi}{T} = 2\pi f$

The unit of ω is the *radian per second* (rad s^{-1}).

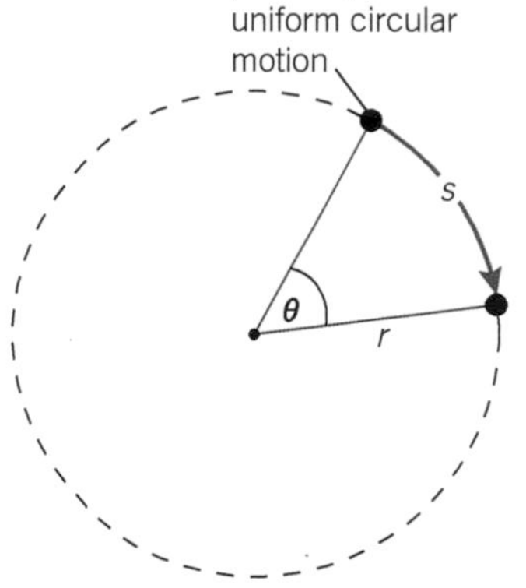

▲ **Figure 3** *Arcs and angles*

Study tip

The unit symbol for the radian is 'rad'. Make sure you know how to convert angles from degrees to radians, and remember to use rad for θ and rad s^{-1} for ω.

Worked example

A cyclist travels at a speed of $12\,\text{m s}^{-1}$ on a bicycle that has wheels of radius 0.40 m. Calculate:

a the frequency of rotation of each wheel
b the angular speed of each wheel
c the angle the wheel turns through in 0.10 s in
i radians, ii degrees.

Solution

a Circumference of wheel $= 2\pi r = 2\pi \times 0.4 = 2.5\,\text{m}$

Time for one wheel rotation, $T = \frac{\text{circumference}}{\text{speed}} = \frac{2.5}{12} = 0.21\,\text{s}$

Frequency $f = \frac{1}{T} = \frac{1}{0.21} = 4.8\,\text{Hz}$

b Angular speed $\omega = \frac{2\pi}{T} = 30\,\text{rad s}^{-1}$

c i Angle the wheel turns through in 0.10 s,

$$\theta = \frac{2\pi t}{T} = \frac{2\pi \times 0.10}{0.21} = 3.0\,\text{rad}$$

ii $\theta = 3.0 \times \frac{360}{2\pi} = 170°$ (2 s.f.)

Summary questions

1 Calculate the angular displacement in radians of the tip of the minute hand of a clock in:
   a 1 second
   b 1 minute
   c 1 hour.

2 An electric motor turns at a frequency of 50 Hz. Calculate:
   a its time period
   b the angle it turns through in radians in
   i 1 ms, ii 1 s.

3 The Earth takes exactly 24 h for one full rotation. Calculate:
   a the speed of rotation of a point on the equator
   b the angle the Earth turns through in 1 s in
   i degrees, ii radians.
   The radius of the Earth = 6400 km.

4 A satellite in a circular orbit of radius 8000 km takes 120 minutes per orbit. Calculate:
   a its speed
   b its angular displacement in 1.0 s in
   i degrees, ii radians.

15.2 Centripetal acceleration

When an object is moving in a circle at a constant speed, the **velocity** of the object is continually changing direction. Because its velocity is changing, the object is *accelerating*. If this seems odd because the object's speed always stays constant (uniform), remember that acceleration is the change of velocity per second. Passengers on the London Eye might not notice they are being accelerated, but if the wheel rotated at a bigger speed than usual, they certainly would!

The velocity of an object in uniform circular motion at any point along its path is along the tangent to the circle at that point (see Figure 1). The direction of the velocity changes continually as the object moves along its circular path. This change in the direction of the velocity is towards the centre of the circle. So, the acceleration of the object is towards the centre of the circle and is called **centripetal acceleration**. Centripetal means *towards the centre of the circle*.

For an object moving at constant speed v in a circle of radius r,

its centripetal acceleration $a = \dfrac{v^2}{r}$

Proof of $a = \dfrac{v^2}{r}$

- Consider an object in uniform circular motion at speed v moving in a short time interval δt from position A to position B along the perimeter of a circle of radius r. The distance AB along the circle is $\delta s = v\delta t$. You can see this in Figure 2.
- The line from the object to the centre of the circle at C turns through angle $\delta\theta$ when the object moves from A to B. The velocity direction of the object turns through the same angle $\delta\theta$, as you can see in Figure 2.
- The change of velocity is δv = velocity at B – velocity at A. You can see this in Figure 2 in the velocity vector triangle.
- The triangles ABC and the velocity vector triangle have the same shape because they both have two sides of equal length with the same angle, $\delta\theta$, between the two sides.

If $\delta\theta$ is small, then $\dfrac{\delta v}{v} = \dfrac{\delta s}{r}$

Because $\delta s = v\delta t$, then $\dfrac{\delta v}{v} = \dfrac{v\delta t}{r}$

So, acceleration $a = \dfrac{\text{change of velocity}}{\text{time taken}} = \dfrac{\delta v}{\delta t} = \dfrac{v^2}{r}$ towards the centre

Centripetal force

To make an object move around on a circular path, it must be acted on by a resultant force that changes its direction of motion.

The resultant force on an object moving around a circle at constant speed is called the **centripetal force** because it acts in the same direction as the centripetal acceleration, which is towards the centre of the circle.

- For an object whirling around rapidly on the end of a string, the tension in the string provides most of the centripetal force.
- For a planet moving around the Sun, the force of gravity on the planet due to the Sun is the centripetal force.

Learning objectives:

- → Explain why velocity is not constant when an object is travelling uniformly in a circle.
- → Determine the direction of the acceleration.
- → Calculate the centripetal force.

Specification reference: 3.6.3

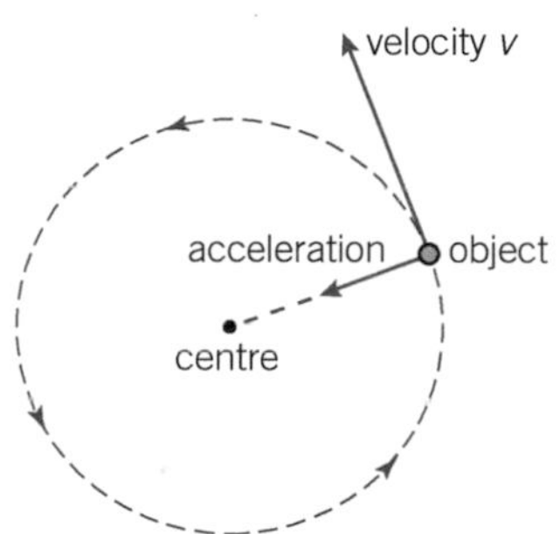

▲ **Figure 1** *Centripetal acceleration*

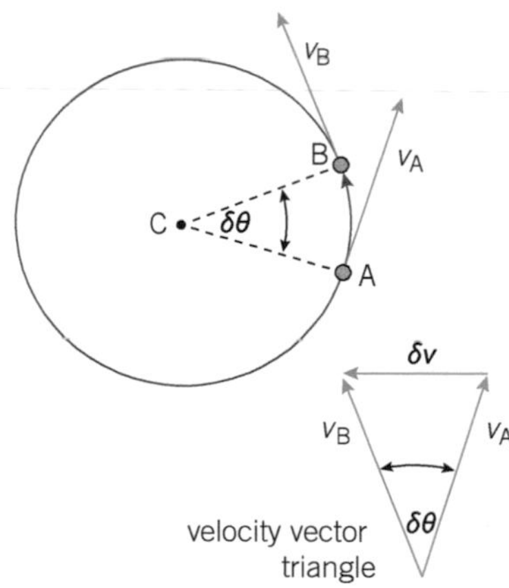

▲ **Figure 2** *Proving* $a = \dfrac{v^2}{r}$

Synoptic link

Remember that acceleration and velocity are vectors and that the direction of acceleration of an object is always the same as the direction of the resultant force on the object. See Topic 3.1, Force and acceleration.

Study tip

You *don't* need to know how to prove the equation for centripetal acceleration at A Level. The proof is given here just to give you a better understanding of the idea of centripetal acceleration.

Because speed $v = \omega r$, then

$$a = \frac{v^2}{r} = \frac{(\omega r)^2}{r} = \omega^2 r$$

The equation for centripetal acceleration can then be written as $a = \omega^2 r$

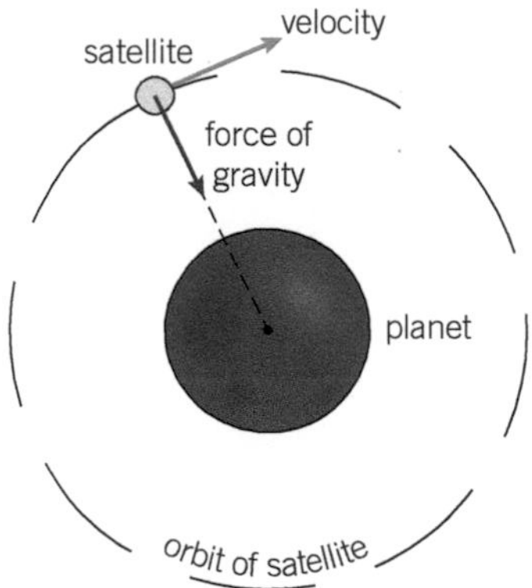

▲ **Figure 3** *A satellite in uniform circular motion*

Hint

1. If the object is acted on by a single force only (e.g., a satellite in orbit around the Earth), that force is the centripetal force and causes the centripetal acceleration.
2. The centripetal force is at right angles to the direction of the object's velocity. Therefore, no work is done by the centripetal force on the object because there is no displacement in the direction of the force. The kinetic energy of the object is therefore constant, so its speed is unchanged.

- For a satellite moving around the Earth, the force of gravity between the satellite and the Earth is the centripetal force.
- For a capsule on the London Eye, the centripetal force is the resultant of the support force on the capsule and the force of gravity on it.
- In Chapter 20, you will meet the use of a magnetic field to bend a beam of charged particles (e.g., electrons) in a circular path. The magnetic force on the moving charged particles is the centripetal force.

Any object that moves in circular motion is acted on by a resultant force that always acts towards the centre of the circle. The resultant force is the centripetal force and so causes centripetal acceleration.

Equation for centripetal force

For an object moving at constant speed v along a circular path of radius r, its centripetal acceleration $a = \frac{v^2}{r} = \omega^2 r$ (where $\omega = \frac{v}{r}$).

Therefore, applying Newton's second law for constant mass in the form $F = ma$ gives

$$\textbf{centripetal force } F = \frac{mv^2}{r} = m\omega^2 r$$

Summary questions

1. The wheel of the London Eye has a diameter of 130 m and takes 30 minutes for one full rotation. Calculate:
   - **a** the speed of a capsule
   - **b** **i** the centripetal acceleration of a capsule
     ii the centripetal force on a person of mass 65 kg in a capsule.
2. An object of mass 0.15 kg moves around a circular path of radius 0.42 m at a steady rate once every 5.0 s. Calculate:
   - **a** the speed and acceleration of the object
   - **b** the centripetal force on the object.
3. **a** The Earth moves around the Sun on a circular orbit of radius 1.5×10^{11} m, taking 365.25 days for each complete orbit. Calculate:
     i the speed
     ii the centripetal acceleration of the Earth on its orbit around the Sun.
   b A satellite is in orbit just above the surface of a spherical planet which has the same radius as the Earth and the same acceleration of free fall at its surface. Calculate:
     i the speed, **ii** the time for one complete orbit of this satellite.
   radius of the Earth = 6400 km   acceleration of free fall = 9.8 m s^{-2}
4. A hammer thrower spins a 2.0 kg hammer on the end of a rope in a circle of radius 0.80 m. The hammer takes 0.60 s to make one full rotation just before it is released. Calculate:
   - **a** the speed of the hammer just before it is released
   - **b** its centripetal acceleration
   - **c** the centripetal force on the hammer just before it is released.

15.3 On the road

Even on a very short journey, the effects of circular motion can be important. For example, a vehicle that turns a corner too fast could skid or topple over. A vehicle that goes over a curved bridge too fast might even lose contact briefly with the road surface. To make any object move on a circular path, the object must be acted on by a resultant force that is always towards the centre of curvature of its path.

Learning objectives:

- → Explain why a passenger in a car seems to be thrown outwards if the car goes round a bend too quickly.
- → Describe what happens to the force between the passenger and his seat when travelling over a curved bridge.
- → Identify the forces that provide the centripetal force on a banked track.

Specification reference: 3.6.3

Over the top of a hill

Consider a vehicle of mass m moving at speed v along a road that passes over the top of a hill or over the top of a curved bridge.

At the top of the hill, the support force S from the road on the vehicle is directly upwards in the opposite direction to its weight, mg. The resultant force on the vehicle is the difference between the weight and the support force. This difference acts towards the centre of curvature of the hill as the centripetal force. In other words,

$$mg - S = \frac{mv^2}{r}$$

where r is the radius of curvature of the hill.

The vehicle would lose contact with the road if its speed is equal to or greater than a particular speed, v_0. If this happens, then the support force is zero (i.e., $S = 0$), so $mg = \frac{mv_0^2}{r}$.

Therefore, the vehicle speed should not exceed v_0, where $v_0^2 = gr$, otherwise the vehicle will lose contact with the road surface at the top of the hill. Prove for yourself that a vehicle travelling over a curved bridge of radius of curvature 5 m would lose contact with the road surface if its speed exceeded $7\,\mathrm{m\,s^{-1}}$.

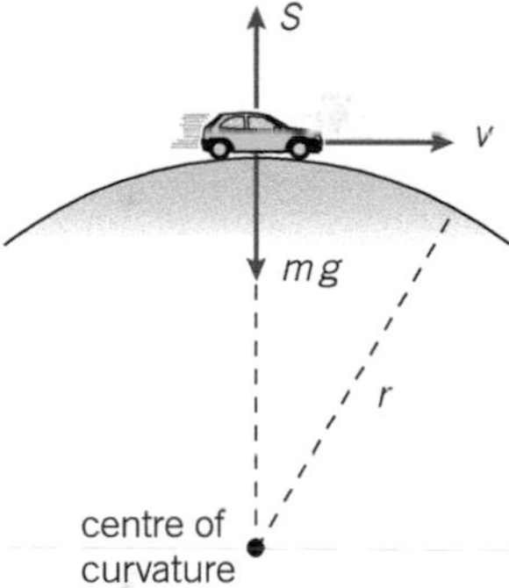

▲ **Figure 1** *Over the top*

On a roundabout

Consider a vehicle of mass m moving at speed v in a circle of radius r as it moves around a roundabout on a level road. The centripetal force is provided by the sideways force of friction between the vehicle's tyres and the road surface. In other words,

$$\textbf{force of friction, } F = \frac{mv^2}{r}$$

For no skidding or slipping to occur, the force of friction between the tyres and the road surface must be less than a limiting value, F_0, that is proportional to the vehicle's weight.

Therefore, for no slipping to occur, the speed of the vehicle must be less than a particular value v_0 that is given in the equation

$$\textbf{limiting force of friction, } F_0 = \frac{mv_0^2}{r}$$

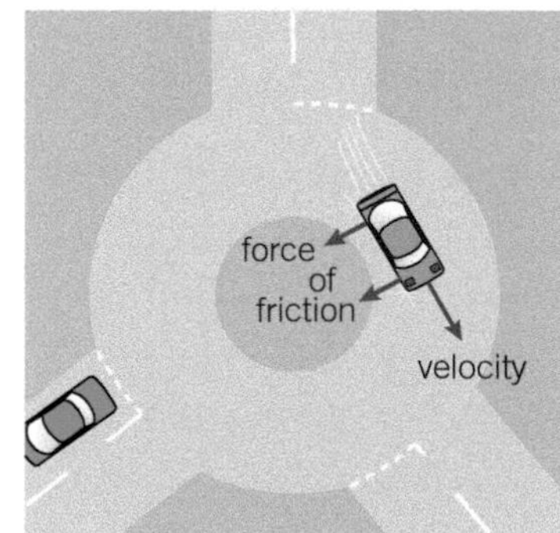

▲ **Figure 2** *On a roundabout*

Note:

Because F_0 is proportional to the vehicle's weight, then $F_0 = \mu mg$, where μ is the coefficient of friction.

Therefore, $\mu mg = \frac{mv_0^2}{r}$

The maximum speed for no slipping $v_0 = (\mu gr)^{\frac{1}{2}}$

Study tip

μ is not needed for A Level Physics, but you might meet it if you are studying A Level Mathematics.

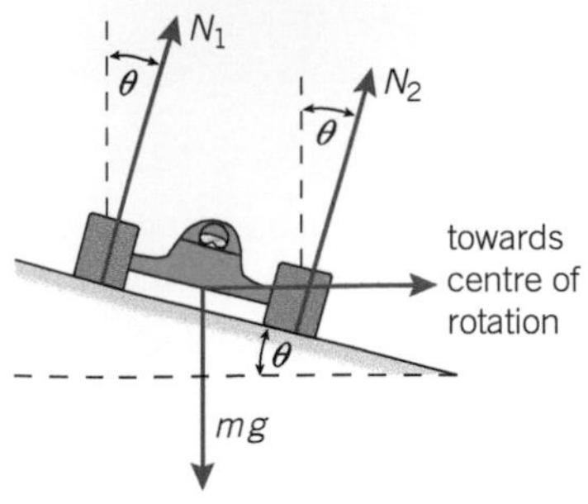

▲ **Figure 3** *A racing car taking a bend*

On a banked track

A race track is often banked where it curves. Motorway slip roads often bend in a tight curve. Such a road is usually banked to enable vehicles to drive around without any sideways friction acting on the tyres. Rail tracks on curves are usually banked to enable trains to move around the curve without slowing down too much. Imagine you are an engineer and you are asked to design a banked track for a horizontal motorway curve.

- Without any banking, the centripetal force on a road vehicle is provided only by sideways friction between the vehicle wheels and the road surface. As you saw in the example of a roundabout, a vehicle on a bend slips outwards if its speed is too high.
- On a banked track, the speed can be higher. To understand why, consider Figure 3, which represents the front view of a racing car of mass m on a banked track, where θ = the angle of the track to the horizontal. For there to be no sideways friction on the tyres from the road, the horizontal component of the support forces N_1 and N_2 must act as the centripetal force.

Resolving these forces into a horizontal component ($= (N_1 + N_2)\sin\theta$) and a vertical component ($= (N_1 + N_2)\cos\theta$), then

- because $(N_1 + N_2)\sin\theta$ acts as the centripetal force, $(N_1 + N_2)\sin\theta = \frac{mv^2}{r}$
- because $(N_1 + N_2)\cos\theta$ balances the weight (mg), $(N_1 + N_2)\cos\theta = mg$.

Therefore, $\tan\theta = \frac{(N_1 + N_2)\sin\theta}{(N_1 + N_2)\cos\theta} = \frac{mv^2}{mgr}$

Simplifying this equation gives the condition for no sideways friction:

$$\tan\theta = \frac{v^2}{gr}$$

In other words, there is no sideways friction if the speed v is such that

$$v^2 = gr\tan\theta$$

Synoptic link

The horizontal and vertical components of a force F that is at an angle θ to the vertical must be $F\sin\theta$ and $F\cos\theta$, respectively. See Topic 1.1, Vectors and scalars.

Study tip

Prove for yourself that if the banking angle θ is not to exceed 5° and the radius of curvature is 360 m, the speed for zero sideways friction is 18 m s^{-1}.

Summary questions

$g = 9.8$ m s^{-2}

1 A vehicle of mass 1200 kg passes over a bridge of radius of curvature 15 m at a speed of 10 m s^{-1}. Calculate:

  a the centripetal acceleration of the vehicle on the bridge

  b the support force on the vehicle when it was at the top.

2 The maximum speed for no skidding of a vehicle of mass 750 kg on a roundabout of radius 20 m is 9.0 m s^{-1}. Calculate:

  a the centripetal acceleration

  b the centripetal force on the vehicle when moving at this speed.

3 Explain why a circular athletics track is banked for sprinters but not for marathon runners.

4 At a racing car circuit, the track is banked at an angle of 25° to the horizontal on a bend that has a radius of curvature of 350 m.

  a Use the equation $v^2 = gr\tan\theta$ to calculate the speed of a vehicle on the bend if there is to be no sideways friction on its tyres.

  b Discuss and explain what could happen to a vehicle that took the bend too fast.

15.4 At the fairground

Learning objectives:
- → Describe when the contact force on a passenger on a Big Dipper ride is the greatest.
- → Describe the condition that applies when a passenger just fails to keep in contact with her seat.

Specification reference: 3.6.3

Many of the rides at a fairground or amusement park take people around in circles. Some examples are analysed below. It is worth remembering that centripetal acceleration values of more than 2 to 3g can be dangerous to the average person.

The Big Dipper

A ride that takes you at high speed through a big dip pushes you into your seat as you pass through the dip. The difference between the support force on you (acting upwards) and your weight acts as the centripetal force.

At the bottom of the dip, the support force S on you is vertically upwards, as shown in Figure 1.

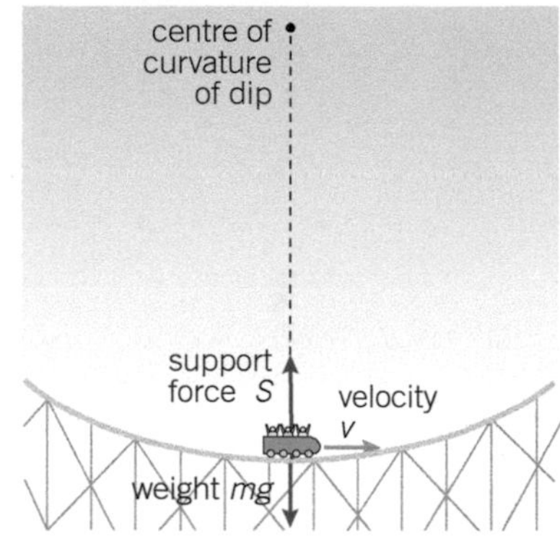

▲ **Figure 1** *In a dip*

Therefore, for a speed v at the bottom of a dip of radius of curvature r,

$$S - mg = \frac{mv^2}{r}$$

So the support force $S = mg + \frac{mv^2}{r}$.

The extra force you experience due to circular motion is therefore $\frac{mv^2}{r}$.

The very long swing

In another ride, a person of mass m on a very long swing of length L is released from height h above the equilibrium position. The maximum speed occurs when the swing passes through the lowest point. You can work this out by equating the gain of kinetic energy to the loss of potential energy:

$$\frac{1}{2}mv^2 = mgh$$

where v is the swing's speed as it passes through the lowest point.

$$\text{Therefore, } v^2 = 2gh$$

The person on the swing is on a circular path of radius L. At the lowest point, the support force S on the person from the rope is in the opposite direction to the person's weight, mg. The difference, $S - mg$, acts towards the centre of the circular path and provides the centripetal force. Therefore,

$$S - mg = \frac{mv^2}{L}$$

Because $v^2 = 2gh$, then $S - mg = \frac{2mgh}{L}$

In other words, $\frac{2mgh}{L}$ represents the extra support force the person experiences because of circular motion. Prove for yourself that for $h = L$ (i.e., a 90° swing), the extra support force is equal to twice the person's weight.

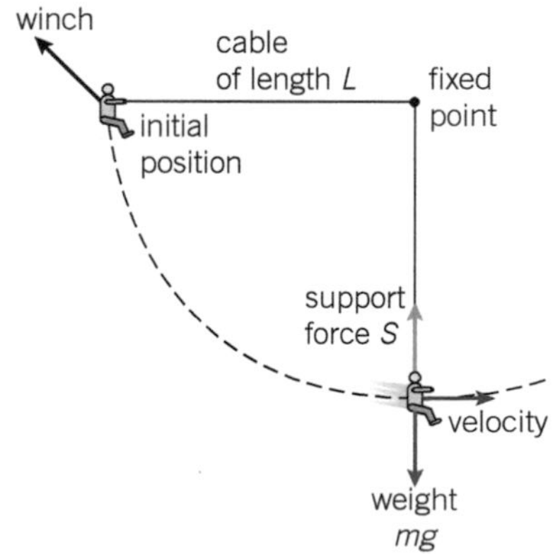

▲ **Figure 2** *The very long swing*

The Big Wheel

This ride takes its passengers around in a vertical circle on the inside of the circumference of a very large wheel. The wheel turns fast enough to stop the passengers falling out as they pass through the highest position.

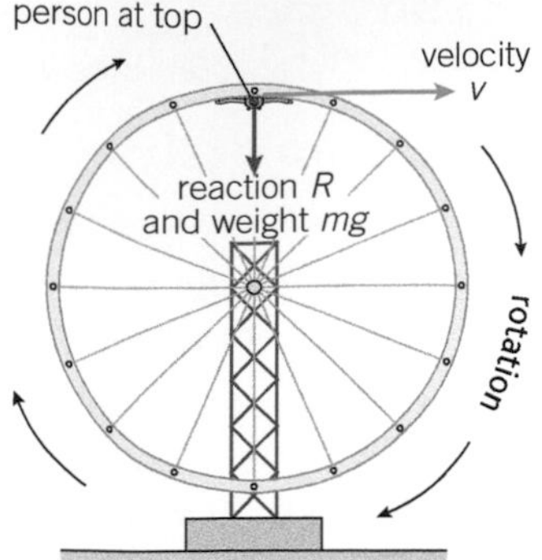

▲ **Figure 3** *The Big Wheel*

▲ **Figure 4** *g-forces at work*

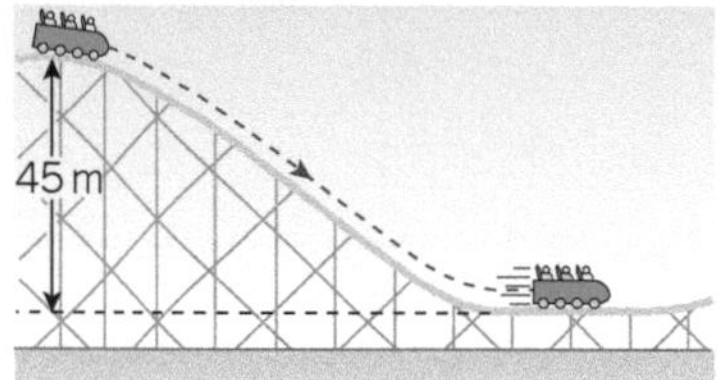

▲ **Figure 5**

At maximum height, the reaction R from the wheel on each person acts downwards. So, the resultant force at this position is equal to $mg + R$. This reaction force and the weight provide the centripetal force. Therefore, at the highest position, if the wheel speed is v,

$$mg + R = \frac{mv^2}{r}, \text{ where } r \text{ is the radius of the wheel}$$

$$\text{therefore, } R = \frac{mv^2}{r} - mg$$

At a particular speed v_0 such that $v_0^2 = gr$, then $R = 0$, so there would be no force on the person due to the wheel.

Application

Safe rides

Amusement rides are checked regularly to ensure they are safe. Accidents on such rides in the United Kingdom have to be investigated by the Health and Safety Executive (HSE). A passenger on a ride may experience 'g-forces' in different directions that may be back and forth, side to side, or normal to the track. The HSE have found that the g-forces in past accidents were within acceptable limits for amusement rides, and that other factors such as passenger behaviour and passenger height in relation to compartment design were more likely to have caused the accidents. In particular, passengers should be able to sit back in their seats with their feet on the foot rests or floor and be able to reach the hand holds comfortably.

Q: Calculate the maximum speed of the train in Figure 5 at the top of the loop if it is to stay in contact with the rails at the top. Assume that the loop radius is 10 m.

Answer: 10 m s^{-1}.

Summary questions

$g = 9.8\,\text{m}\,\text{s}^{-2}$

1 A train on a fairground ride is initially stationary before it descends through a height of 45 m into a dip that has a radius of curvature of 78 m, as shown in Figure 5.

- **a** Calculate the speed of the train at the bottom of the dip, assuming air resistance and friction are negligible.
- **b** Calculate:
  - **i** the centripetal acceleration of the train at the bottom of the dip
  - **ii** the extra support force on a person of weight 600 N in the train.

2 A very long swing at a fairground is 32 m in length. A person of mass 69 kg on the swing descends from a position when the swing is horizontal. Calculate:

- **a** the speed of the person at the lowest point
- **b** the centripetal acceleration at the lowest point
- **c** the support force on the person at the lowest point.

3 The Big Wheel at a fairground has a radius of 12.0 m and rotates once every 6.0 s. Calculate:

- **a** the speed of rotation of the perimeter of the wheel
- **b** the centripetal acceleration of a person on the perimeter
- **c** the support force on a person of mass 72 kg at the highest point.

4 The wheel of the London Eye has a diameter of 130 m and takes 30 minutes to complete one revolution. Calculate the change due to rotation of the wheel of the support force on a person of weight 500 N in a capsule at the top of the wheel.

Practice questions: Chapter 15

1 **Figure 1** shows a cross section of an automatic brake fitted to a rotating shaft. The brake pads are held on the shaft by springs.

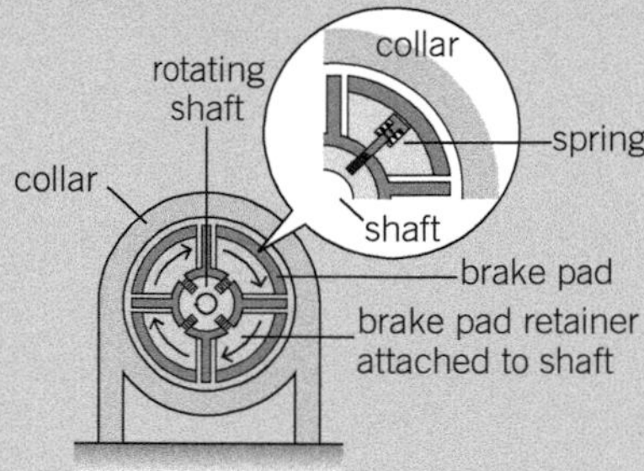

▲ **Figure 1**

(a) Explain why the brake pads press against the inner surface of the stationary collar if the shaft rotates too fast. *(3 marks)*

(b) Each brake pad and its retainer has a mass of 0.30 kg and its centre of mass is 60 mm from the centre of the shaft. The tension in the spring attached to each pad is 250 N. Calculate the maximum frequency of rotation of the shaft for no braking. *(4 marks)*

(c) Automatic brakes of the type described above are used on ships to prevent lifeboats falling freely when they are lowered on cables onto the water. Discuss how the performance of the brake would be affected if the springs gradually became weaker. *(2 marks)*

2 **(a)** A particle that moves uniformly in a circular path is accelerating yet moving at a constant speed.

Explain this statement by reference to the physical principles involved. *(3 marks)*

(b) A 0.10 kg mass is to be placed on a horizontal turntable that is then rotated at a fixed rate of 78 revolutions per minute. The mass may be placed on the table at any distance, *r*, from the axis of rotation, as shown in **Figure 2**.

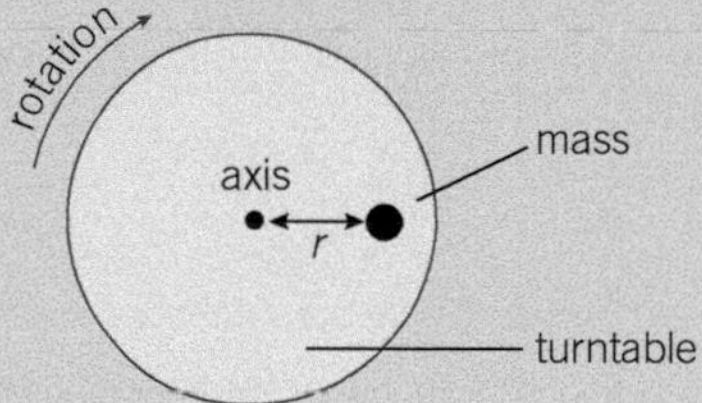

▲ **Figure 2**

If the maximum frictional force between the mass and the turntable is 0.50 N, calculate the maximum value of the distance *r* at which the mass would stay on the turntable at this rate of rotation. *(4 marks)*

AQA, 2007

3 **Figure 3** shows a dust particle at position D on a rotating vinyl disc. A combination of electrostatic and frictional forces act on the dust particle to keep it in the same position.

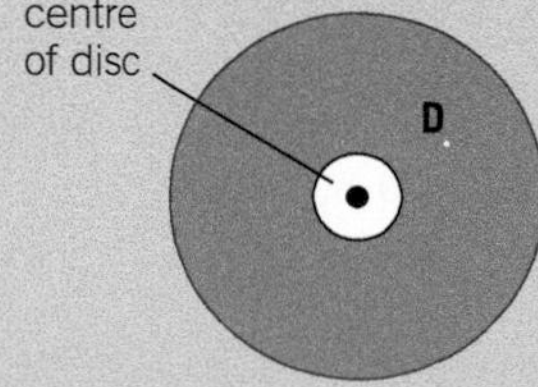

▲ **Figure 3**

The dust particle is at a distance of 0.125 m from the centre of the disc. The disc rotates at 45 revolutions per minute.

(a) Calculate the linear speed of the dust particle at D. *(3 marks)*

(b) (i) Copy the diagram and mark an arrow to show the direction of the resultant horizontal force on the dust particle.

(ii) Calculate the centripetal acceleration at position D. *(3 marks)*

(c) On looking closely at the rotating disc, it can be seen that there is more dust concentrated on the inner part of the disc than the outer part. Suggest why this should be so. *(3 marks)*

AQA, 2002

4 A strimmer is a tool for cutting long grass. A strimmer head such as that shown in **Figure 4a** is driven by a motor. This makes the plastic line rotate, causing it to cut the grass. To simplify analysis, the strimmer line is modelled as the arrangement shown in **Figure 4b**. In this model the effective mass of the line is considered to rotate at the end of the line.

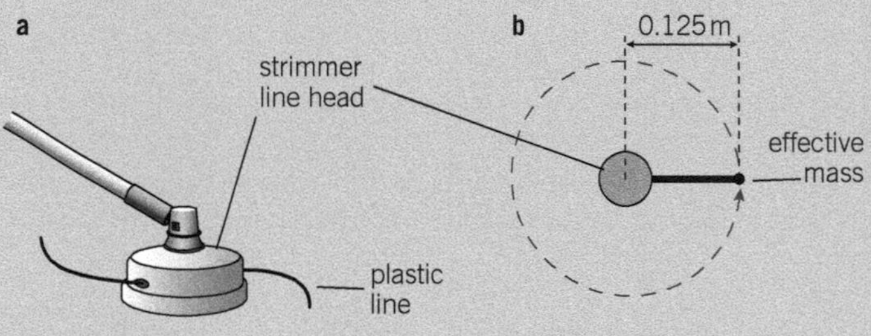

▲ **Figure 4**

In one strimmer, the effective mass of 0.80 g rotates in a circle of radius 0.125 m at 9000 revolutions per minute.

(a) Show that the angular speed of the line is approximately 9.4×10^2 rad s^{-1}. *(2 marks)*

(b) (i) Explain how the centripetal force is applied to the effective mass.

(ii) Calculate the centripetal force acting on the effective mass. *(4 marks)*

(c) The line strikes a pebble of mass 1.2 g, making contact for a time of 0.68 ms. This causes the pebble to fly off at a speed of 15 m s^{-1}. Calculate the average force applied to the pebble. *(3 marks)*

AQA, 2007

5 **Figure 5** shows a toy engine moving with a constant speed on a circular track of constant radius.

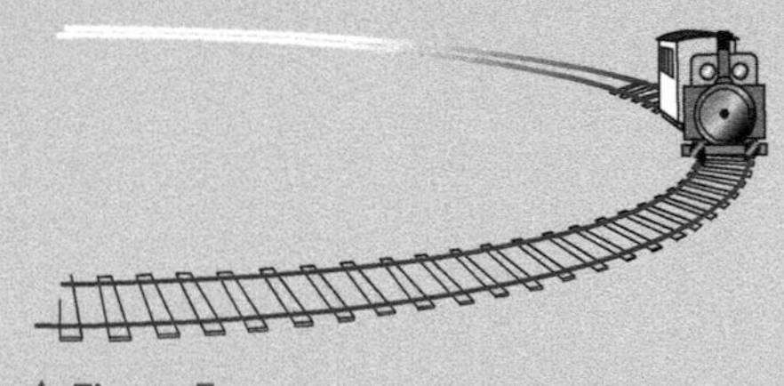

▲ **Figure 5**

(a) (i) Explain why the engine is accelerating even though its speed remains constant.

(ii) Mark on a copy of **Figure 5** the direction of the centripetal force acting on the engine. *(3 marks)*

(b) The total mass of the toy engine is 0.14 kg and it travels with a speed of 0.17 m s^{-1}. The radius of the track is 0.80 m. Calculate the centripetal force acting on the engine. *(2 marks)*

Figure 6 shows a close up of a pair of wheels as the engine moves towards you in the forward direction shown in **Figure 5**.

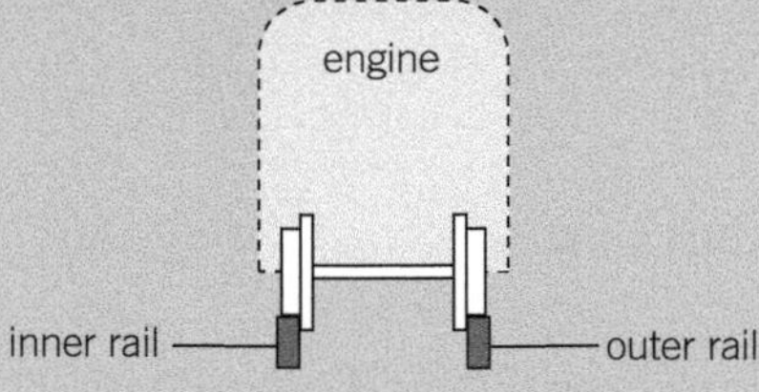

▲ **Figure 6**

(c) (i) State and explain on which wheel the centripetal force acts at the instant shown. You may use **Figure 6** to help your explanation.

(ii) For the toy engine going round a curved track, state and explain the two factors which determine the stress on each wheel. *(5 marks)*

AQA, 2004

6 A mass of 30 g is attached to a thread and spun in a circle of radius 45 cm. The circle is in a horizontal plane. The tension in the thread is 0.35 N.

(a) Calculate:

(i) the speed of the mass

(ii) the period of rotation of the mass. *(4 marks)*

(b) The mass M is now spun in a circle in a vertical plane as shown in **Figure 7**.

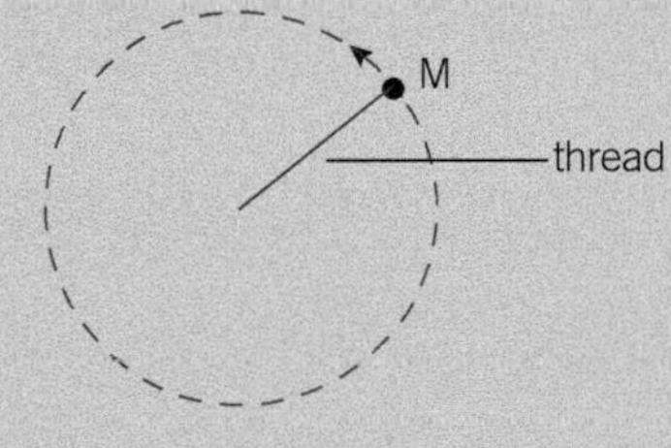

▲ **Figure 7**

(i) On a copy of **Figure 7**, label the forces acting on the mass, and use arrows to show their direction.

(ii) Without performing calculations, state and explain the difference between the tension in the thread when M is at the top of the circle and when it is at the bottom. *(6 marks)*

AQA, 2007

7 **Figure 8** shows the initial path taken by an electron when it is produced as a result of a collision in a cloud chamber.

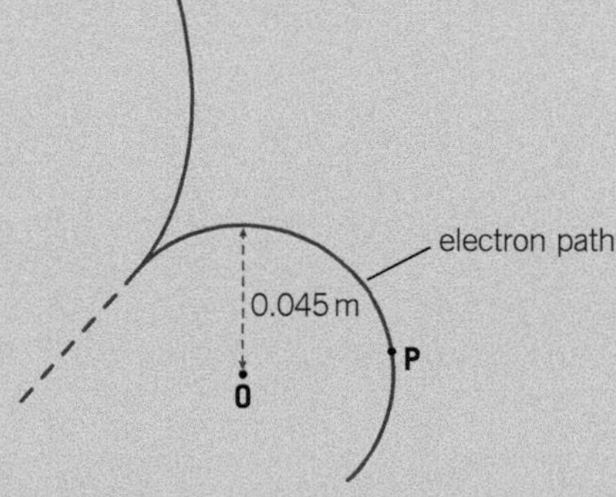

▲ **Figure 8**

The path is the arc of a circle of radius 0.045 m with centre O.
The speed of the electron is $4.2 \times 10^7\,\text{m}\,\text{s}^{-1}$. The mass of an electron is $9.1 \times 10^{-31}\,\text{kg}$.

(a) Calculate the momentum of the electron. *(2 marks)*

(b) Calculate the magnitude of the force acting on the electron that makes it follow the curved path. *(2 marks)*

(c) Show on a copy of **Figure 8** the direction of this force when an electron is at point P. *(1 mark)*

AQA, 2002

8 **Figure 9** shows a simple accelerometer designed to measure the centripetal acceleration of a car going around a bend following a circular path.

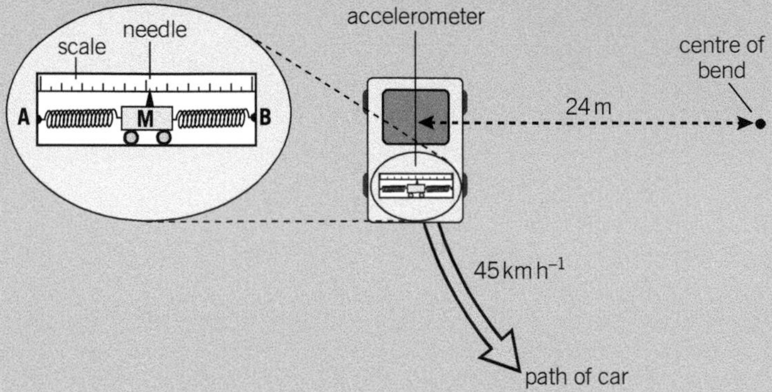

▲ **Figure 9**

The two ends A and B are fixed to the car. The mass M is free to move between the two springs.

The needle attached to the mass moves along a scale to indicate the acceleration.

In one instant a car travels around a bend of radius 24 m in the direction shown in **Figure 9**. The speed of the car is 45 km h^{-1}.

(a) State and explain the direction in which the pointer moves from its equilibrium position. *(3 marks)*

(b) (i) Calculate the acceleration that would be recorded by the accelerometer.

(ii) The mass M between the springs in the accelerometer is 0.35 kg. A test shows that a force of 0.75 N moves the pointer 27 mm. Calculate the displacement of the needle from the equilibrium position when the car is travelling with the acceleration in part (i). *(4 marks)*

AQA, 2003

Answers to the Practice Questions and Section Questions are available at www.oxfordsecondary.com/oxfordaqaexams-alevel-physics

16 Simple harmonic motion

16.1 The principles of simple harmonic motion

Learning objectives:

- → State the two fundamental conditions about acceleration that apply to simple harmonic motion.
- → Describe how displacement, velocity, and acceleration vary with time.
- → Describe the phase difference between displacement and (i) velocity, (ii) acceleration.

Specification reference: 3.6.4

An oscillating object speeds up as it returns to equilibrium and it slows down as it moves away from equilibrium. Figure 1 shows one way to record the displacement of an oscillating pendulum.

The variation of displacement with time is shown in Figure 2a. As long as friction is negligible, the amplitude of the oscillations is constant.

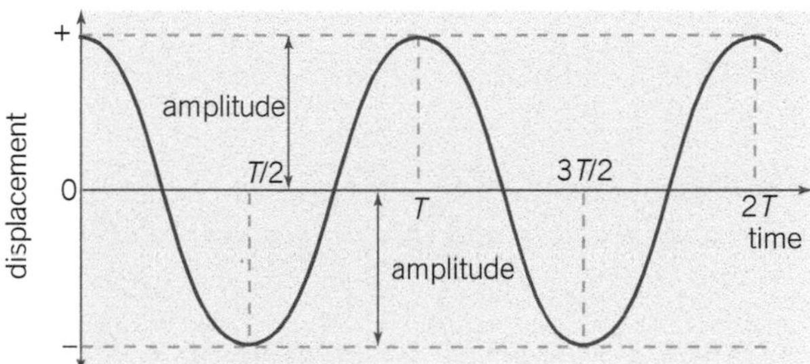

▲ **Figure 2a** *Displacement against time*

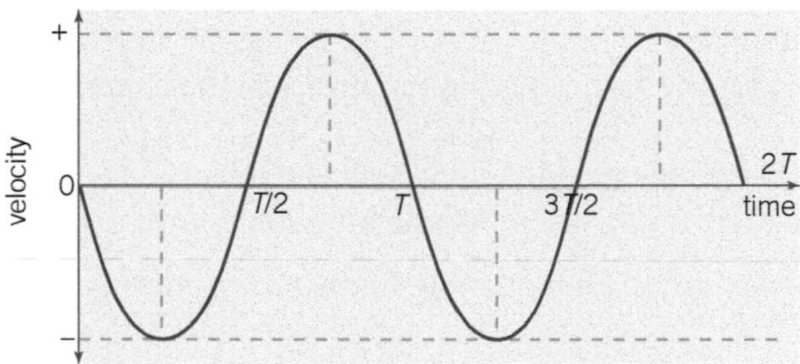

▲ **Figure 2b** *Velocity against time*

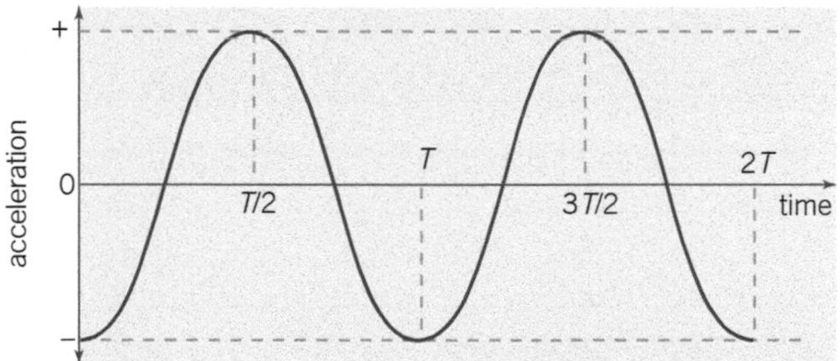

▲ **Figure 2c** *Acceleration against time*

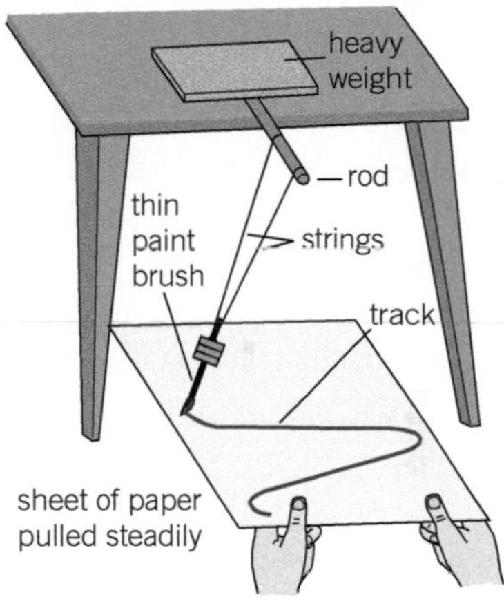

▲ **Figure 1** *Investigating oscillations*

The variation of velocity with time is given by the gradient of the displacement–time graph, as shown by Figure 2b.

- The magnitude of the velocity is greatest when the gradient of the displacement–time graph is greatest (i.e., at zero displacement when the object passes through equilibrium).
- The velocity is zero when the gradient of the displacement–time graph is zero (i.e., at maximum displacement in either direction).

The variation of acceleration with time is given by the gradient of the velocity–time graph, as shown by Figure 2c.

Synoptic link

The gradient of a displacement–time graph gives velocity. The gradient of a velocity–time graph gives acceleration. See Topic 2.5, Motion graphs.

Hint

1. The time period is independent of the amplitude of the oscillations.
2. Maximum displacement $x_{max} = \pm A$, where A is the amplitude of the oscillations. Therefore,
   - when $x_{max} = +A$, the acceleration $a = -\omega^2 A$, and
   - when $x_{max} = -A$, the acceleration $a = +\omega^2 A$.
3. The acceleration equation may also be written as $a = -(2\pi f)^2 x$, where frequency $f = \frac{1}{T}$.

Hint

The symbol ω represents angular speed in circular motion and angular frequency in simple harmonic motion. In both cases, $\omega = \frac{2\pi}{T}$ where T is the time period

- The acceleration is greatest when the gradient of the velocity–time graph is greatest. This is when the velocity is zero and occurs at maximum displacement in the opposite direction.
- The acceleration is zero when the gradient of the velocity–time graph is zero. This is when the displacement is zero and the velocity is a maximum.

By comparing Figures 2a and 2c directly, it can be seen that

the acceleration is always in the opposite direction to the displacement

In other words, if you call one direction the positive direction and the other direction the negative direction, the acceleration direction is always the opposite sign to the displacement direction.

Simple harmonic motion is defined as oscillating motion in which the acceleration is

1. proportional to the displacement, and
2. always in the opposite direction to the displacement.

$$a \propto -x$$

In other words, acceleration a = −constant × displacement x.

The minus sign in both equations tells you that the acceleration is in the opposite direction to the displacement. The constant of proportionality depends on the time period T of the oscillations. The shorter the time period, the faster the oscillations, which means the larger the acceleration at any given displacement. So the constant is greater, the shorter the time period. As you will see in Topic 16.3, the constant in this equation is ω^2, where ω, the angular frequency $= \frac{2\pi}{T}$.

Therefore, the defining equation for simple harmonic motion is

$$\textbf{acceleration } a = -\omega^2 x$$

where x = displacement and ω = angular frequency.

Summary questions

1 A small object attached to the end of a vertical spring oscillates with an amplitude of 25 mm and a time period of 2.0 s. The object passes through equilibrium moving upwards at time $t = 0$. What is the displacement and direction of motion of the object:

a $\frac{1}{4}$ cycle later? b $\frac{1}{2}$ cycle later?

c $\frac{3}{4}$ cycle later? d 1 cycle later?

2 For the oscillations in **Q1**, calculate:

a the frequency

b the acceleration of the object when its displacement is

i +25 mm ii 0 iii −25 mm.

3 A simple pendulum consists of a small weight on the end of a thread. The weight is displaced from equilibrium and released. It oscillates with an amplitude of 32 mm, taking 20 s to execute ten oscillations. Calculate:

a its frequency b its initial acceleration.

4 For the oscillations in **Q3**, the object is released at time $t = 0$. State the displacement and calculate the acceleration when

a $t = 1.0$ s b $t = 1.5$ s.

16.2 More about sine waves

Circles and waves

Consider a small object P in uniform circular motion, as shown in Figure 1. Measured from the centre of the circle at O, the coordinates of P are therefore $x = r\cos\theta$ and $y = r\sin\theta$, where θ is the angle between the x-axis and the radial line OP. The graph shows how the x-coordinate changes as angle θ changes. The curve is a cosine wave. It has the same shape as the simple harmonic motion curves in Figure 2 in Topic 16.1.

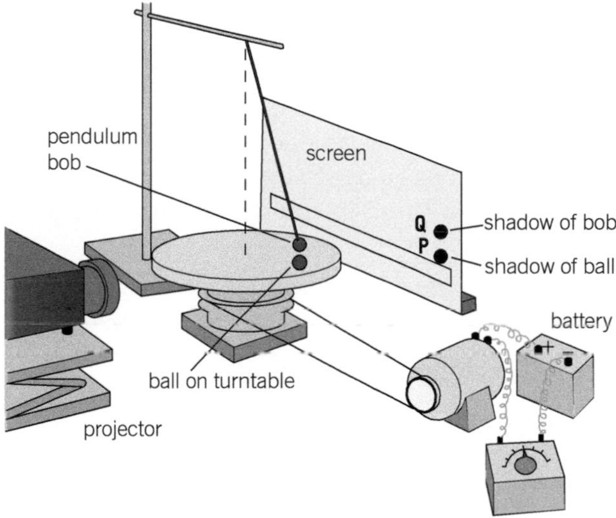

▲ **Figure 2** *Comparing simple harmonic motion with circular motion*

To see directly the link between simple harmonic motion and sine curves, consider the motion of the ball and the pendulum bob in Figure 2. A projector is used to cast a shadow of the ball (P) in uniform circular motion onto a screen alongside the shadow of the oscillating bob (Q). The two shadows keep up with each other exactly when their time periods are matched. In other words, P and Q at any instant have the same horizontal motion. So the acceleration of Q is the same as the acceleration of P's shadow on the screen.

- Because the ball (P) is in uniform circular motion, its acceleration $a = -\frac{v^2}{r}$, where v is its speed and r is the radius of the circle. Note that the minus sign indicates that its direction is towards the centre.

Because speed $v = \omega r$ (see Topic 15.1), then $a = -\omega^2 r$.

- The component of acceleration of the ball parallel to the screen is $a_x = a\cos\theta$, so the acceleration of the ball's shadow is $a_x = -\omega^2 r\cos\theta = -\omega^2 x$, where $x = r\cos\theta$ is the displacement of the shadow from the midpoint of the oscillations.
- Because the bob's motion is the same as the motion of the ball's shadow,

the acceleration of the bob (Q) $a_x = -\omega^2 x$

This is the defining equation for simple harmonic motion, and it shows why the constant of proportionality is ω^2.

Learning objectives:

- → State the equation that relates displacement to time for a body moving with simple harmonic motion.
- → State the point at which the oscillations must start for this equation to apply.
- → Calculate the velocity for a given displacement.

Specification reference: 3.6.4

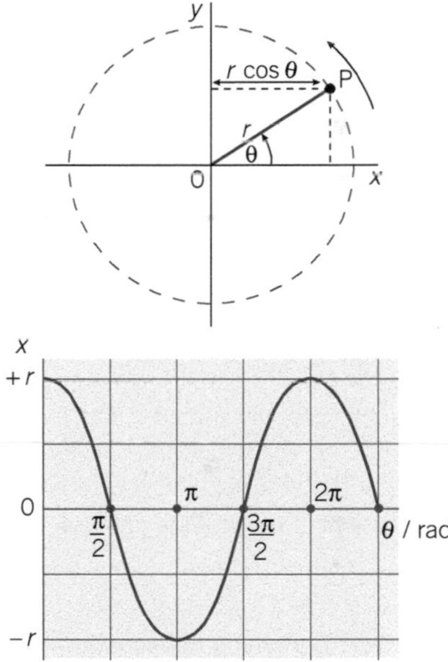

▲ **Figure 1** *Circles and waves*

Phase difference

The phase difference between the ball and the bob is constant but *not* zero if the shadow of the bob is *not* directly above the shadow of the bob. If the time interval is Δt between when they reach maximum displacement in the same direction, their motion is out of phase by a fraction of a cycle equal to $\frac{\Delta t}{T}$ (where T is their time period). So their phase difference, in radians, $= \frac{2\pi\Delta t}{T}$

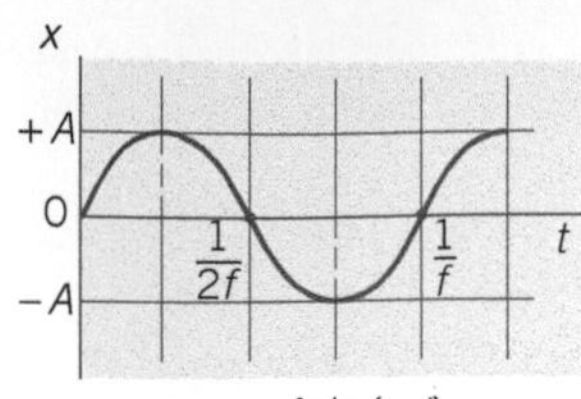

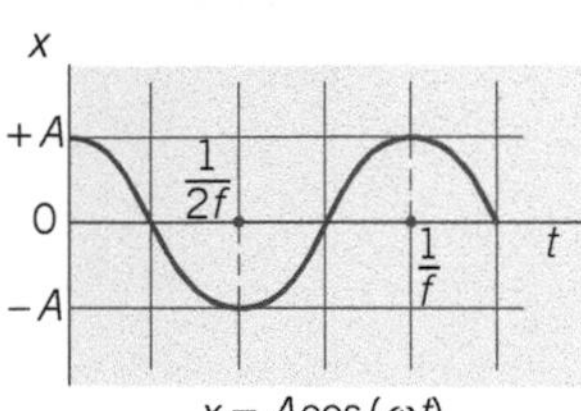

▲ **Figure 3** *Graphical solutions*

More on sine waves

1 The shape of the curves in Figure 3, described as sinusoidal curves, is the same as the shape of the simple harmonic motion curves in Figure 2 in Topic 16.1.

2 The general solution of $a = -\omega^2 x$ is $x = A \sin(\omega t + \phi)$, where ϕ is the phase difference between the instants when $t = 0$ and when $x = 0$. If timing were to start at the centre with oscillations moving positive, the x against t equation would become $x = A \sin \omega t$ as $\phi = 0$. You *don't* need to know this general solution.

3 The time period T does *not* depend on the amplitude of the oscillating motion. For example, the time period of an object oscillating on a spring is the same, regardless of whether the amplitude is large or small.

Study tip

Remember that ωt must be in radians. So make sure your calculator is in radian mode when you use the equation $x = A \cos(\omega t)$.

Sine wave solutions

For any object oscillating at frequency f in simple harmonic motion, its acceleration a at displacement x is given by

$$a = -\omega^2 x$$

where $\omega = 2\pi f$. The variation of displacement with time depends on the initial displacement and the initial velocity (i.e., the displacement and velocity at time $t = 0$).

As shown by the bottom graph in Figure 3, if $x = +A$ when $t = 0$ and the object has zero velocity at that instant, then its displacement at a later time t is given by

$$x = A \cos(\omega t)$$

Notes:

1 In the above equation, x is the displacement of the bob in Figure 2 from its equilibrium position. Its value changes from $-r$ to $+r$ and back again as the bob oscillates. Therefore, the amplitude of oscillation of the bob is $A = r$.

2 The displacement of the bob from equilibrium is $x = A\cos\theta$, where θ is the angle the ball moves through from its position when $x = A$.

At time t after the ball passes through this position,

$$\theta \text{ (in radians)} = \frac{2\pi t}{T} = \omega t$$

Therefore, the displacement of the bob at time t is given by

$$x = A\cos(\omega t)$$

Summary questions

1 An object oscillates in simple harmonic motion with a time period of 3.0 s and an amplitude of 58 mm. Calculate:

a its frequency   b its maximum acceleration.

2 The displacement of an object oscillating in simple harmonic motion varies with time according to the equation x (mm) = 12 cos 10t, where t is the time in seconds after the object's displacement was at its maximum positive value.

a Determine:

i the amplitude   ii the time period.

b Calculate the displacement of the object at $t = 0.1$ s.

3 An object on a spring oscillates with a time period of 0.48 s and a maximum acceleration of 9.8 m s^{-2}. Calculate:

a its frequency   b its amplitude.

4 An object oscillates in simple harmonic motion with an amplitude of 12 mm and a time period of 0.27 s. Calculate:

a its frequency

b its displacement and its direction of motion

i 0.10 s   ii 0.20 s after its displacement was +12 mm.

16.3 Applications of simple harmonic motion

Learning objectives:

- → State the conditions that must be satisfied for a mass–spring system or simple pendulum to oscillate with simple harmonic motion.
- → Describe how the period of a mass–spring system depends on the mass.
- → Describe how the period of a simple pendulum depends on its length.

Specification reference: 3.6.4

For any oscillating object, the resultant force acting on the object acts towards the equilibrium position. The resultant force is described as a *restoring force* because it always acts towards equilibrium. As long as the restoring force is proportional to the displacement from equilibrium, the acceleration is proportional to the displacement and always acts towards equilibrium. Therefore, the object oscillates with simple harmonic motion.

The oscillations of a mass–spring system

Use two stretched springs and a trolley, as shown in Figure 1. When the trolley is displaced then released, it oscillates backwards and forwards.

- The first half-cycle of the trolley's motion can be recorded using a length of ticker tape attached at one end to the trolley. When the trolley is released, the tape is pulled through a ticker timer that prints dots on the tape at a rate of 50 dots per second.

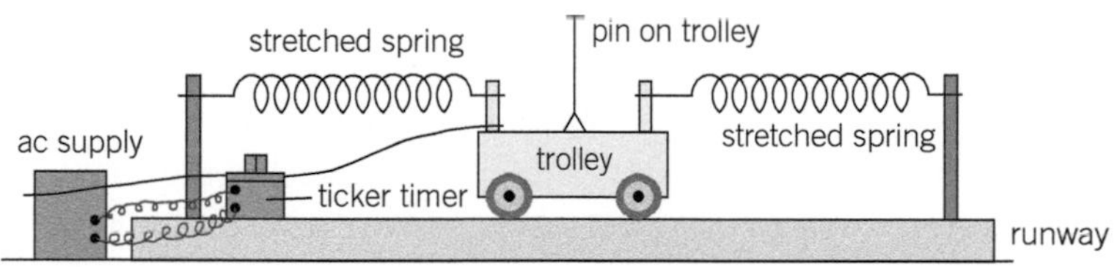

▲ **Figure 1** *Investigating oscillations*

- A graph of displacement against time for the first half-cycle can be drawn using the tape, as shown in Figure 2. The graph can be used to measure the time period, which can be checked (see the Hint box on the next page) if the trolley mass m and the combined spring constant k are known.
- A motion sensor linked to a computer can also be used to record the oscillating motion of the trolley. See Topic 4.3.

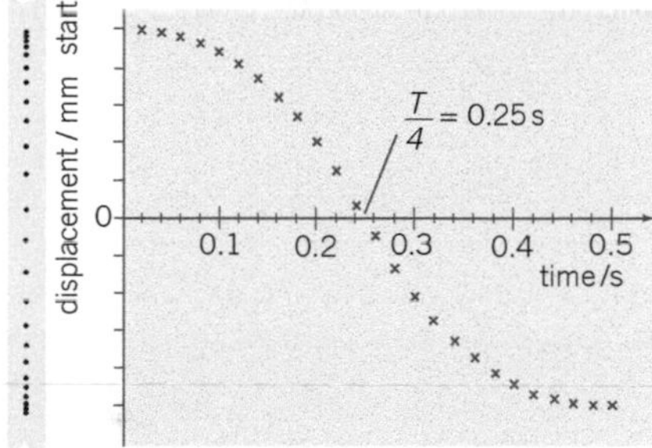

▲ **Figure 2** *Displacement–time curve from a ticker tape*

What determines the frequency of oscillation of a loaded spring?

In the above investigation, the frequency of oscillation of the trolley can be changed by loading the trolley with extra mass or by replacing the springs with springs of different stiffness. The frequency is reduced by:

1. **Adding extra mass.** This is because the extra mass increases the inertia of the system. At a given displacement, the trolley would therefore be slower than if the extra mass had not been added. Each cycle of oscillation would therefore take longer.
2. **Using weaker springs.** The restoring force on the trolley at any given displacement would be less, so the trolley's acceleration and speed at any given displacement would be less. Each cycle of oscillation would therefore take longer.

Study tip

When timing oscillations, use a fiduciary marker at the centre of oscillations and start counting with 0 as the oscillating object passes the mark in a particular direction.

To see exactly how the mass and the spring constant affect the frequency, consider a small object of mass m attached to a spring.

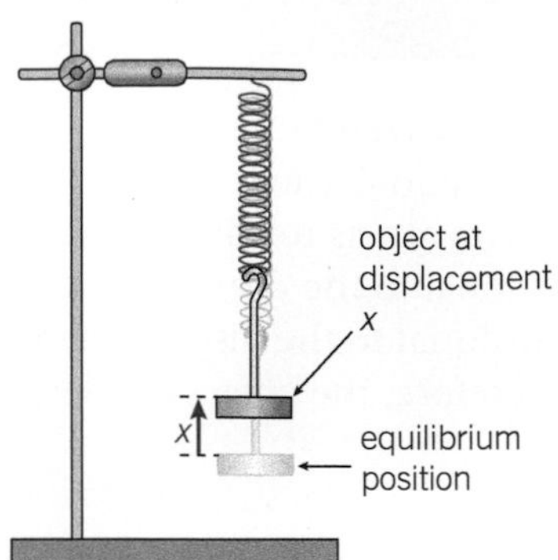

▲ **Figure 3** *The oscillations of a loaded spring*

Practical link

Investigating the oscillations of a loaded spring

Measure the oscillations of a loaded spring for different masses and verify the equation $T = 2\pi\sqrt{\frac{m}{k}}$.

Synoptic link

Because $T^2 = \frac{4\pi^2 m}{k}$, plotting a graph of T^2 on the y-axis against m on the x-axis should give a straight line through the origin with a gradient of $\frac{4\pi^2}{k}$ in accordance with the general equation for a straight-line graph $y = mx + c$.

See Topic 14.4, Straight-line graphs.

- Assuming the spring obeys Hooke's law, the tension T_s in the spring is proportional to its extension ΔL from its unstretched length. This relationship can be expressed using the equation $T_s = k\Delta L$, where k is the spring constant.
- When the object is oscillating and is at displacement x from its equilibrium position, the change of tension in the spring provides the restoring force on the object. Using the equation $T_s = k\Delta L$, the change of tension ΔT_s from equilibrium is therefore given by $\Delta T_s = -kx$, where the minus sign represents the fact that the change of tension always tries to restore the object to its equilibrium position.
- So the restoring force on the object is equal to $-kx$
- Therefore, the acceleration $a = \frac{\text{restoring force}}{\text{mass}} = \frac{-kx}{m}$

This equation may be written in the form $a = -\omega^2 x$, where $\omega^2 = \frac{k}{m}$.

The object therefore oscillates in simple harmonic motion because its acceleration $a = -\omega^2 x$.

Notes

1 Because $\omega = 2\pi f$, then $\omega^2 = \frac{k}{m}$ can be written as $(2\pi f)^2 = \frac{k}{m}$. Rearranging this equation gives $f = \frac{1}{2\pi}\sqrt{\frac{k}{m}}$. You can use this equation to calculate f if k and m are known. The equation also shows that the frequency is increased if k is increased or if m is reduced.

2 The time period of the oscillations $T = \frac{1}{f} = 2\pi\sqrt{\frac{m}{k}}$.
The time period does *not* depend on g. A mass–spring system on the Moon would have the same time period as it would on Earth.

3 The tension in the spring varies from $mg + kA$ to $mg - kA$, where A = amplitude.
- Maximum tension is when the spring is stretched as much as possible (i.e., $x = -A$ = maximum displacement downwards)
- Minimum tension is when the spring is stretched as little as possible (i.e., $x = +A$ = maximum displacement upwards)

Worked example

$g = 9.8\,\text{m s}^{-2}$

A spring of natural length 300 mm hangs vertically with its upper end attached to a fixed point. When a small object of mass 0.20 kg is suspended from the lower end of the spring in equilibrium, the spring is stretched to a length of 379 mm. Calculate:

a **i** the extension of the spring at equilibrium
ii the spring constant

b the time period of oscillations that the mass on the spring would have if the mass was to be displaced downwards slightly then released.

Solution

a **i** Extension of spring at equilibrium is $\Delta L_0 = 79\,\text{mm} = 0.079\,\text{m}$

ii Spring constant $k = \dfrac{mg}{\Delta L_0} = \dfrac{0.20 \times 9.8}{0.079} = 25\,\text{N m}^{-1}$

b $T = 2\pi\sqrt{\dfrac{0.20}{25}} = 0.56\,\text{s}$

The theory of the simple pendulum

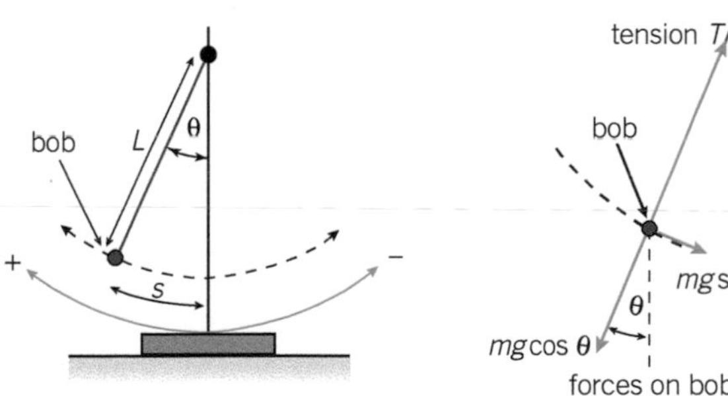

▲ **Figure 4** *The simple pendulum*

Consider a simple pendulum that consists of a bob of mass m attached to a thread of length L, as shown in Figure 4. If the bob is displaced from equilibrium then released, it oscillates about the lowest point. At displacement s from the lowest point, when the thread is at angle θ to the vertical, the weight mg has components:

- $mg\cos\theta$ perpendicular to the path of the bob, and
- $mg\sin\theta$ along the path towards the equilibrium position.

The restoring force $F = -mg\sin\theta$, so the acceleration

$$a = \frac{F}{m} = \frac{-mg\sin\theta}{m} = -g\sin\theta$$

As long as θ does not exceed approximately 10° then $\sin\theta = \dfrac{s}{L}$, therefore the acceleration $a = -\dfrac{g}{L}s = -\omega^2 s$, where $\omega^2 = \dfrac{g}{L}$. See Topic 27.1 for more about the small angle approximation.

So the object oscillates with simple harmonic motion because its acceleration is proportional to the displacement from equilibrium and always acts towards equilibrium.

Notes

1 Because $\omega = 2\pi f$, then $\omega^2 = \dfrac{g}{L}$ can be written as $(2\pi f)^2 = \dfrac{g}{L}$. Rearranging this equation gives $f = \dfrac{1}{2\pi}\sqrt{\dfrac{g}{L}}$.

So the time period

$$T = \frac{1}{f} = 2\pi\sqrt{\frac{L}{g}}$$

as long as the angle of the thread to the vertical does not exceed about 10°.

2 The time period T can be increased by increasing the length L of the pendulum. The length of the pendulum is the distance from the point of support to the centre of the bob.

3 As the bob passes through equilibrium, the tension T_s acts directly upwards. Therefore, the resultant force on the bob at this instant

$$T_s - mg = \frac{mv^2}{L}$$

where v is the speed as it passes through equilibrium.

Practical link

Investigating the simple pendulum

Measure the oscillations of a simple pendulum for different lengths and verify the equation $T = 2\pi\sqrt{\dfrac{L}{g}}$.

Synoptic link

Plotting a graph of T^2 on the y-axis against L on the x-axis should give a straight line through the origin with a gradient of $\dfrac{4\pi^2}{g}$.

See Topic 14.4, Straight-line graphs.

Synoptic link

Look back at Topic 11.2 Figure 3 to remind yourself how the kinetic energy, the potential energy and the total energy of a freely oscillating object varies with its displacement.

As explained in Topic 11.2, for an amplitude A, its kinetic energy E_K at displacement $x = \frac{1}{2}kA^2 - \frac{1}{2}kx^2$ where $\frac{1}{2}kA^2$ is its total energy E and $\frac{1}{2}kx^2$ is its potential energy E_P.

Therefore $\frac{1}{2}mv^2 = \frac{1}{2}kA^2 - \frac{1}{2}kx^2$ where v is its speed and m is its mass.

From Topic 16.3, $k/m = \omega^2$ where ω is the angular frequency of the object. Substituting $k = m\,\omega^2$ into the equation above gives $\frac{1}{2}mv^2 = \frac{1}{2}m\,\omega^2 A^2 - \frac{1}{2}m\,\omega^2 x^2$ which simplifies to give

$$\mathbf{v = \pm\sqrt{(A^2 - x^2)}}$$

where the ± sign indicates the two possible directions of the velocity.

Note

1. The maximum speed v_{max} is at $x = 0$. Substituting $x = 0$ into the above velocity equation therefore gives $v_{max} = \omega A$.
2. The maximum acceleration is at $x = \pm A$. Since the acceleration a at displacement x is given by the SHM equation $a = -\omega^2 x$, the maximum acceleration $a_{max} = \omega^2 A$

Summary questions

$g = 9.8\,\text{m}\,\text{s}^{-2}$

1 An object is suspended from the end of a vertical spring and set into oscillating motion along a vertical line. The amplitude of its oscillations is 20 mm, and it takes 6.5 s to perform 20 oscillations. Calculate:

- **a** **i** its time period **ii** its frequency
- **b** its acceleration and its speed when its displacement is
  - **i** 0 mm **ii** 10 mm **iii** 20 mm.

2 In the arrangement described in **Q1**, the object is replaced by an object of different mass. When the second object oscillates vertically, its acceleration a at displacement x is given by $a = -360x$.

- **a** Calculate:
  - **i** the frequency
  - **ii** the time period of the oscillations.
- **b** By comparing the frequency of the oscillations of the second object with that of the first, discuss whether the mass of the second object is greater than or less than the mass of the first object.

3 The upper end of a vertical spring of natural length 250 mm is attached to a fixed point. When a small object of mass 0.15 kg is attached to the lower end of the spring, the spring stretches to an equilibrium length of 320 mm.

- **a** Calculate:
  - **i** the extension of the spring at equilibrium
  - **ii** the spring constant.
- **b** The object is displaced vertically from its equilibrium position and released. Show that it oscillates at a frequency of 1.9 Hz and calculate its period of oscillation.

4 A mass of 0.50 kg is attached to the lower end of a vertical spring that has a spring constant of $25\,\text{N}\,\text{m}^{-1}$. The mass is displaced downwards by a distance of 50 mm then released.

- **a** Calculate:
  - **i** the force on the object at a displacement of 50 mm
  - **ii** the acceleration of the object at the instant it is released.
- **b** **i** Show that the acceleration a at displacement x is given by $a = -50x$.
  - **ii** Calculate the frequency of the oscillations and the displacement of the mass 0.050 s after it is released.

5 Calculate the time period of a simple pendulum:

- **a** of length
  - **i** 1.0 m **ii** 0.25 m
- **b** of length 1.0 m on the surface of the Moon, where $g = 1.6\,\text{m}\,\text{s}^{-2}$.

6 A simple pendulum and a mass suspended on a vertical spring have equal time periods on the Earth. Discuss whether or not they would have the same time periods on the surface of the Moon, where $g = 1.6\,\text{m}\,\text{s}^{-2}$.

Practice questions: Chapter 16

1 In an investigation, a small object is suspended from the lower end of a vertical steel spring which is fixed at its upper end, as shown in **Figure 1**.

A horizontal marker pin P is attached to the object. The vertical position, x, of the pin is measured against the millimetre scale of a metre rule clamped vertically in a fixed position. The measurement is made three times without, then with, the small object suspended from the spring.

(a) The readings obtained are shown in Table 1.

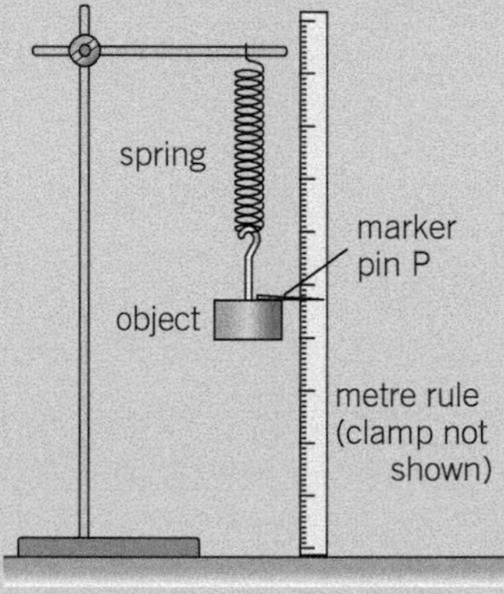

▲ **Figure 1**

▼ **Table 1**

	x / mm			Mean x / mm	Extension e / mm
Without the object on the spring	2	2	2	2.0	0
With the object on the spring	71	72	73		

(ii) Copy and complete Table 1 by calculating the mean vertical position of P and the extension of the spring when the object is placed on it.

(iii) The readings are taken to a precision of 0.5 mm using a millimetre ruler. Estimate the percentage 'uncertainty' in the extension. *(2 marks)*

(b) The time period, T, of small vertical oscillations of the object on the spring is also measured by timing 20 oscillations three times.
The timing readings for 20 oscillations are 10.98 s, 11.11 s, and 10.97 s.

(i) Calculate the time period T.

(ii) Use the readings to estimate the percentage 'uncertainty' in T. *(2 marks)*

(c) (i) Give an expression for the extension e of the spring in terms of the mass m of the object and the spring constant k of the spring.

(ii) Therefore show that $T = 2\pi\sqrt{\frac{e}{g}}$. *(3 marks)*

(d) The experiment is repeated with objects of different mass suspended from the spring. The measurements obtained are given in Table 2.

▼ **Table 2**

Object	e / mm	T / s
1	70	0.551
2	139	0.761
3	205	0.923
4	271	1.062
5	341	1.187
6	409	1.291

Plot a suitable graph using the above measurements to confirm the equation and to determine g. *(9 marks)*

(e) Discuss the accuracy of your determination of g. *(4 marks)*

2 The tuning fork shown in **Figure 2** is labelled 512 Hz and has the tip of each of its two prongs vibrating with simple harmonic motion of amplitude 0.85 mm.

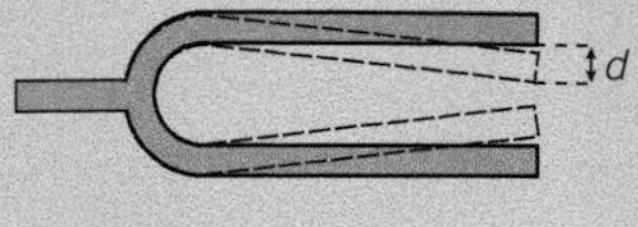

▲ Figure 2

(a) (i) **Figure 2** shows the extreme positions of the prongs. How is the distance marked d related to the amplitude of the prongs?

(ii) Sketch a graph to show how the displacement of one tip of the tuning fork changes with time. Mark each axis with an appropriate scale. *(4 marks)*

(b) (i) Calculate the maximum speed of the tip of a prong.

(ii) Calculate the maximum acceleration of the tip of a prong. *(4 marks)*

AQA, 2007

3 A simple pendulum consists of a 25 g mass tied to the end of a light string 800 mm long. The mass is drawn to one side until it is 20 mm above its rest position, as shown in **Figure 3**. When released it swings with simple harmonic motion.

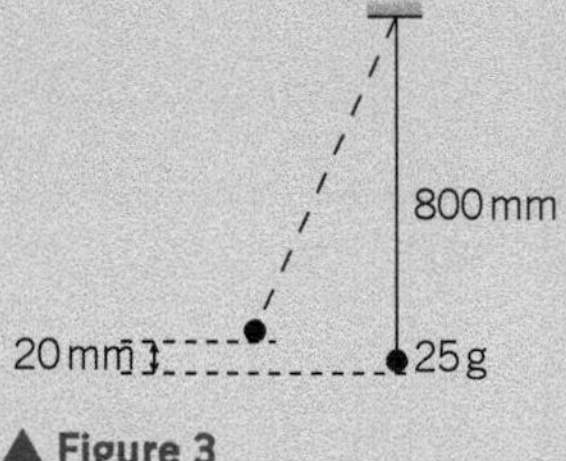

▲ Figure 3

(a) Calculate the period of the pendulum. *(2 marks)*

(b) Show that the initial amplitude of the oscillations is approximately 0.18 m, and that the maximum speed of the mass during the first oscillation is about 0.63 m s^{-1}. *(4 marks)*

(c) Calculate the magnitude of the tension in the string when the mass passes through the lowest point of the first swing. *(2 marks)*

AQA, 2003

4 **(a)** **Figure 4a** shows a demonstration used in teaching simple harmonic motion. A sphere rotates in a horizontal plane on a turntable. A lamp produces a shadow of the sphere. This shadow moves with approximate simple harmonic motion on the vertical screen.

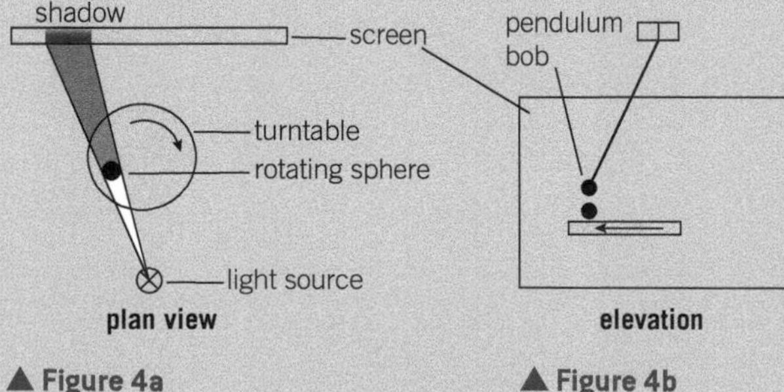

▲ Figure 4a

▲ Figure 4b

(i) The turntable has a radius of 0.13 m, and the teacher wishes the time taken for one cycle of the motion to be 2.2 s. The mass of the sphere is 0.050 kg.
Calculate the magnitude of the horizontal force acting on the sphere.
(ii) State the direction in which the force acts. *(3 marks)*

(b) **Figure 4b** shows how the demonstration might be extended. A simple pendulum is mounted above the turntable so that the shadows of the sphere and the pendulum bob can be seen to move in a similar way and with the same period.
(i) Calculate the required length of the pendulum.
(ii) Calculate the maximum acceleration of the pendulum bob when its motion has an amplitude of 0.13 m. *(3 marks)*

(c) **Figure 5** is a graph of displacement against time for the pendulum.

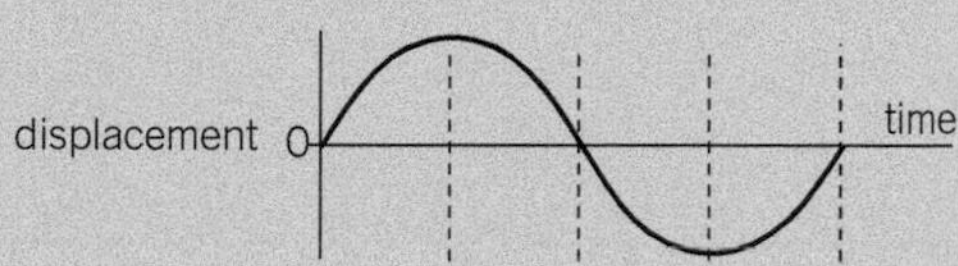

▲ **Figure 5**

Sketch, for the same interval, graphs of:
(i) acceleration against time for the bob, and
(ii) kinetic energy against time for the bob. *(4 marks)*

AQA, 2005

5 **(a)** Simple harmonic motion may be represented by the equation

$$a = -\omega^2 x$$

(i) Explain the significance of the minus sign in this equation.
(ii) Copy **Figure 6** and sketch the corresponding v–t graph to show how the phase of velocity v relates to that of the acceleration a. *(2 marks)*

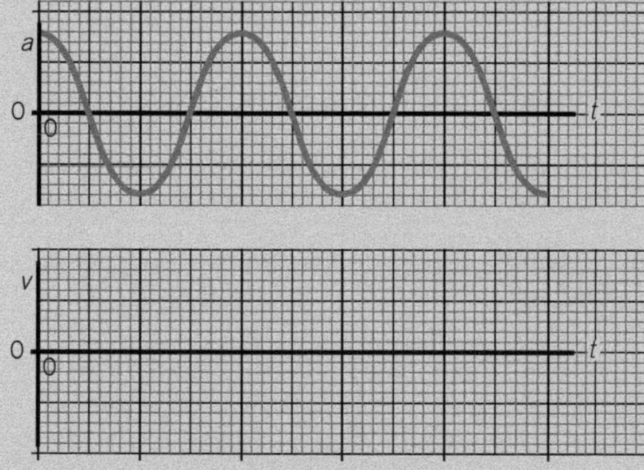

▲ **Figure 6**

(b) (i) A mass of 24 kg is attached to the end of a spring of spring constant 60 N m^{-1}. The mass is displaced 0.035 m vertically from its equilibrium position and released. Show that the maximum kinetic energy of the mass is about 40 mJ.
(ii) When the mass on the spring is quite heavily damped its amplitude halves by the end of each complete cycle. Sketch a graph to show how the kinetic energy, E_K (mJ), of the mass on the spring varies with time, t (s), over a single period.
Start at time, $t = 0$, with your maximum kinetic energy.
You should include suitable values on each of your scales. *(8 marks)*

AQA, 2004

6 To celebrate the Millennium in the year 2000, a footbridge was constructed across the River Thames in London. After the bridge was opened to the public it was discovered that the structure could easily be set into oscillation when large numbers of people were walking across it.

(a) What name is given to this kind of physical phenomenon, when caused by a periodic driving force? *(1 mark)*

(b) Under what condition would this phenomenon become particularly hazardous? Explain your answer. *(4 marks)*

(c) Suggest *two* measures which engineers might adopt to reduce the size of the oscillations of a bridge. *(2 marks)*

AQA, 2002

7 **Figure 7** shows how the displacement of the bob of a simple pendulum varies with time.

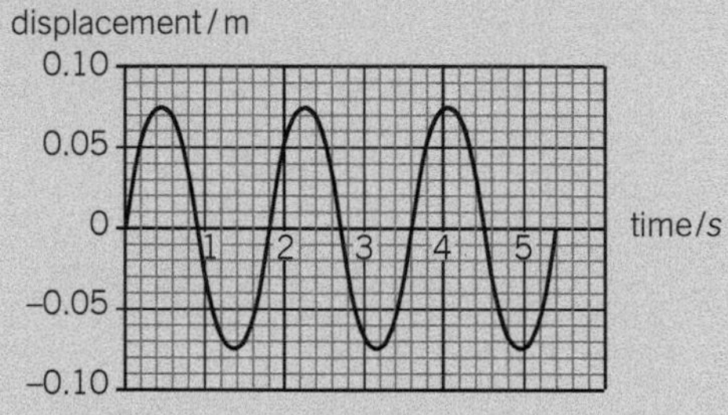

▲ **Figure 7**

(a) (i) Calculate the frequency of the oscillation.

(ii) State the magnitude of the amplitude of the oscillation.

(iii) State how the frequency and amplitude of a simple pendulum are affected by increased damping. *(5 marks)*

(b) Draw on a copy of **Figure 7** the displacement–time graph for a pendulum that has the same period and amplitude but oscillates 90° ($\frac{\pi}{2}$ radian) out of phase with the one shown. *(2 marks)*

(c) The pendulum bob has a mass of 8.0×10^{-3} kg. Calculate:

(i) the maximum acceleration of the bob during the oscillation

(ii) the total energy of the oscillations. *(5 marks)*

AQA, 2006

8 (a) A spring, which hangs from a fixed support, extends by 40 mm when a mass of 0.25 kg is suspended from it.

(i) Calculate the spring constant of the spring.

(ii) An additional mass of 0.44 kg is then placed on the spring and the system is set into vertical oscillation. Show that the oscillation frequency is 1.5 Hz. *(4 marks)*

(b) With both masses still in place, the spring is now suspended from a horizontal support rod that can be made to oscillate vertically, as shown in **Figure 8**, with amplitude 30 mm at several different frequencies.

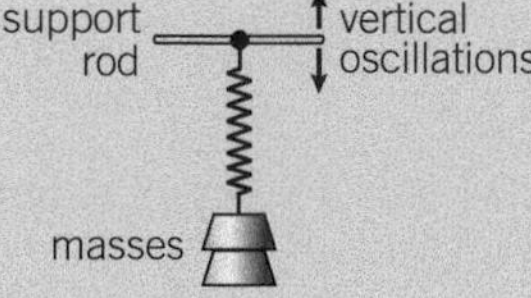

▲ **Figure 8**

Describe fully, with reference to amplitude, frequency, and phase, the motion of the masses suspended from the spring in each of the following cases.

(i) The support rod oscillates at a frequency of 0.2 Hz.

(ii) The support rod oscillates at a frequency of 1.5 Hz.

(iii) The support rod oscillates at a frequency of 10 Hz. *(6 marks)*

AQA, 2006

Answers to the Practice Questions and Section Questions are available at www.oxfordsecondary.com/oxfordaqaexams-alevel-physics

17 Gravitational fields

17.1 Newton's law of gravitation

We owe our understanding of gravitation to Isaac Newton. 'The notion of gravity was occasioned by the fall of an apple!', said Newton when asked what made him develop the idea of gravity. Newton's theory of gravitation was an enormous leap forward because it explains events from the down-to-earth falling apple to the motion of the planets. Like any good theory, it can be used to make predictions. For example, the return of a comet and its exact path can be calculated using Newton's theory of gravitation. So successful was his theory that it became known as Newton's law of gravitation.

Learning objectives:

- → Describe how gravitational attraction varies with distance.
- → Explain what is meant by an inverse-square law.
- → Discuss whether spherical objects, for example, planets, can be treated as point masses.

Specification reference: 3.7.2

How Newton established his theory of gravitation

Newton realised that gravity is universal. Any two masses exert a force of attraction on each other. He knew about the careful measurements of planetary motion made by astronomers like Johannes Kepler. Forty or more years before Newton established the theory of gravitation, Kepler had shown that the motion of the planets was governed by a set of laws. Kepler used his own and other astronomers' measurements of motion of each planet to show that each planet orbits the Sun. The measurements that he made for each planet were its time period T (i.e., the time for one complete orbit of the Sun) and the average radius r of its orbit. He used his measurements to show that the value of $\frac{r^3}{T^2}$ was the same for all the planets. This is called **Kepler's third law**.

▼ **Table 1** *Kepler's third law*

	Mercury	Venus	Earth	Mars	Jupiter	Saturn
Average radius r of orbit / 10^{10} m	6	11	15	23	78	143
Time T for one orbit / 10^7 s	0.8	1.95	3.2	5.9	37.4	93.0
$\frac{r^3}{T^2}$ / 10^{16} m^3 s^{-2}	337	350	330	349	340	338

To explain Kepler's third law, Newton started by assuming that the planets and the Sun were point masses. A scale model of the Solar System with the Sun represented by a 1p piece would put the Earth about a metre away, represented by a grain of sand! Newton realised that there is a force of gravitational attraction between any planet and the Sun, and that it caused the planet to orbit the Sun. He assumed that the force of gravitation between a planet and the Sun varies inversely with the square of their distance apart. In other words, if the force is F at distance d apart, then:

- at distance $2d$ apart, the force is $\frac{F}{4}$
- at distance $3d$ apart, the force is $\frac{F}{9}$
- at distance $4d$ apart, the force is $\frac{F}{16}$.

Notes

1 Work out for yourself the gravitational force between two point masses, each of mass 10 kg at 0.1 m apart. The values of m_1, m_2, and r must be put into the equation in units of kilograms and metres. The force of gravitational attraction works out at 6.7×10^{-7} N, which is far too small to notice except with extremely sensitive equipment. Only if one of the masses is very large does the force become noticeable, unless special techniques are used, as you will see later.

2 The equation for Newton's law of gravitation can be applied to any two objects provided r is the distance between their centres of mass.

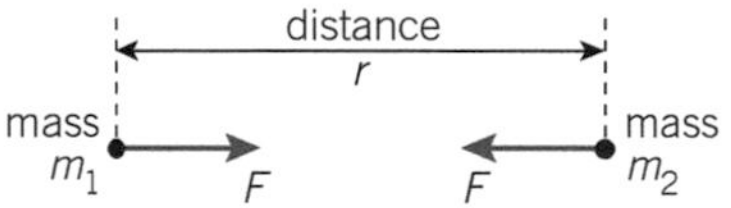

▲ **Figure 1** *Newton's law of gravitation*

Study tip

Think carefully before writing about the 'separation' of two objects. Do you mean distance between centres, or distance between surfaces?

Using this inverse-square law of force, Newton was able to prove that $\frac{r^3}{T^2}$ was the same for all of the planets. Newton then went on to use the inverse-square law of force to explain and make predictions for many other events involving gravity.

Newton's law of gravitation assumes that the gravitational force between any two **point** objects is:

- always an attractive force
- proportional to the mass of each object
- proportional to $\frac{1}{r^2}$, where r is their distance apart.

These last two requirements can be summarised as

$$\textbf{gravitational force } F = \frac{Gm_1m_2}{r^2}$$

where m_1 and m_2 are masses of the two objects (see Figure 1).

The constant of proportionality, G, in the above equation, is called the **universal constant of gravitation**. The unit of G can be worked out from the equation above.

Rearranged, the equation gives $G = \frac{Fr^2}{m_1\, m_2}$.

So G can be given units of $\text{N m}^2\text{kg}^{-2}$. The value of G is $6.67 \times 10^{-11}\,\text{N m}^2\text{kg}^{-2}$.

Worked example

$G = 6.67 \times 10^{-11}\,\text{N m}^2\text{kg}^{-2}$.

The distance from the centre of the Sun to the centre of the Earth is 1.5×10^{11} m. The mass of the Sun is 2.0×10^{30} kg, and the mass of the Earth is 6.0×10^{24} kg.

a The Earth has a diameter of 1.3×10^7 m. The Sun has a diameter of about 1.4×10^9 m. Explain why it is reasonable to consider the Sun and the Earth at a distance of 1.5×10^{11} m apart as point masses on this distance scale.

b Calculate the force of gravitational attraction between the Sun and the Earth.

Solution

a On a scale model where the centre of the Sun was 1 m away from the centre of the Earth, the Sun would be a sphere of diameter about 1 cm, and the Earth would be a sphere of diameter about 0.1 mm, no larger than a dot. The distance from the Earth to any part of the Sun is therefore the same to within 1%. Therefore this shows that they can be treated as point objects.

b $F = \frac{6.67 \times 10^{-11} \times 2.0 \times 10^{30} \times 6.0 \times 10^{24}}{(1.5 \times 10^{11})^2} = 3.6 \times 10^{22}\,\text{N}$

Extension

Cavendish's measurement of *G*

The first accurate measurement of G was made by Henry Cavendish in 1798. He devised a torsion balance made of two small lead balls at either end of a rod. The rod was suspended horizontally by a torsion wire, as in Figure 2. The wire was calibrated by measuring the couple required to twist it per degree. Then, with the rod at rest in equilibrium, two massive lead balls were brought near the torsion balance to make the wire twist. By measuring the angle it twisted through, the force of attraction between each massive lead ball and the small ball nearest to it was calculated. The distance between the centres of the small and large masses was also measured. Then G was calculated using the equation for the law of gravitation.

Q: Explain why the forces on the small balls make the torsion balance turn.

Answer: The two forces have a turning effect in the same direction on the balance.

Summary questions

$G = 6.67 \times 10^{-11}\,\text{N}\,\text{m}^2\,\text{kg}^{-2}$

1 **a** Calculate the force of gravitational attraction between two point objects of masses 60 kg and 80 kg at a distance of 0.5 m apart.

b Calculate the distance between two identical point objects, each of mass 0.20 kg, which exert a force of 9.0×10^{-8} N on each other.

2 **a** Calculate the force of gravitational attraction between the Earth and an object of mass 80 kg on the surface of the Earth, where $g = 9.8\,\text{N}\,\text{kg}^{-1}$.

b Use the result of your calculation in **a** to estimate the mass of the Earth. Assume that the mass of the Earth is concentrated at its centre.

radius of the Earth = 6.4×10^6 m

3 The Sun exerts a force of 6.0 N on a 1000 kg comet when it is at a distance of 1.5×10^{11} m from the Sun. Calculate the force due to the Sun on the comet when it is at a distance of:

a 0.5×10^{11} m from the Sun

b 7.5×10^{11} m from the Sun.

4 A space rocket of mass 1500 kg travels from the Earth to the Moon, a distance of 3.8×10^8 m.

a When the space rocket is mid-way between the Earth and the Moon, calculate the force of gravitational attraction on it:

i due to the Earth

ii due to the Moon.

b Calculate the magnitude and direction of the force of gravity of the Earth and the Moon on the space rocket when it is mid-way between the Earth and the Moon.

mass of the Earth = 6.0×10^{24} kg
mass of the Moon = 7.4×10^{22} kg

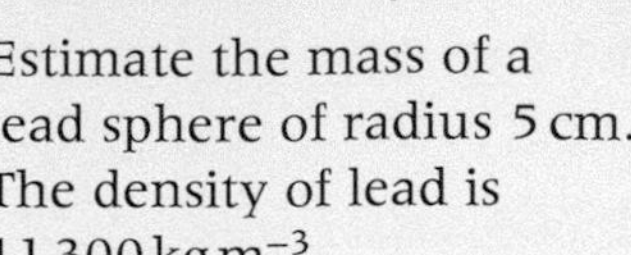

Worked example

Estimate the mass of a lead sphere of radius 5 cm. The density of lead is $11\,300\,\text{kg}\,\text{m}^{-3}$.

Solution

Volume ≈ (diameter)^3 = $(2 \times 0.05\,\text{m})^3 = 10^{-3}\,\text{m}^3$

Mass = volume × density = $10^{-3}\,\text{m}^3 \times 11\,300\,\text{kg}\,\text{m}^{-3} \approx 10\,\text{kg}$.

▲ **Figure 2** *Cavendish's measurement of G*

Synoptic link

Newton's law of gravitation and Coulomb's law of force between point charges are both inverse-square laws. For example, doubling the separation of two point masses or two point charges causes the force to reduce to a quarter. See Topic 18.4.

17.2 Gravitational field strength

Learning objectives:

- → Illustrate a gravitational field.
- → Explain what is meant by the strength of a gravitational field.
- → Define a radial field and a uniform field.

Specification reference: 3.7.3

▲ **Figure 1** *Earth in space*

Study tip

Remember that g is a vector, even though it is called strength.

What goes up must come down – or must it? Throw a ball into the air and it returns to you because of the Earth's gravity. The force of gravity on the ball pulls it back to Earth. The force of attraction between the ball and the Earth is an example of gravitational attraction, which exists between any two masses. It isn't obvious that there is a force of attraction between you and any object near you, but it is true. Any two masses exert a gravitational pull on each other. But the force is usually too weak to be noticed unless at least one of the masses is very large.

The mass of an object creates a force field around itself. Any other mass placed in the field is attracted towards the object. The second mass also has a force field around itself and this pulls on the first object with an equal force in the opposite direction. The force field round a mass is called a **gravitational field**.

If a small test mass is placed close to a massive body, the small mass and the body attract each other with equal and opposite forces. However, this force is too small to move the massive body noticeably. The small mass, assuming it is free to move, is pulled by the force towards the massive body. The path which the smaller mass would follow is called a **field line** or sometimes a **line of force**. Figure 2 shows the field lines near a planet. The lines are directed to the centre of the planet because a small object released near the planet would fall towards its centre.

The gravitational field strength at a point in a field, g, is the force per unit mass on a small test mass placed in the field at that point.

The test mass needs to be small, otherwise it might pull so much on the other object that it changes its position and alters the field. In general, the force on a small mass in a gravitational field varies from one position to another. If a small test mass, m, is at a particular position in a gravitational field where it is acted on by a gravitational force F, the gravitational field strength at that position is given by

$$g = \frac{F}{m}$$

The unit of gravitational field strength is the newton per kilogram ($\mathrm{N\,kg^{-1}}$). For example, the gravitational field strength of the Earth at the surface of the Earth is $9.8\,\mathrm{N\,kg^{-1}}$.

Free fall in a gravitational field

The weight of an object is the force of gravity on it. If an object of mass m is in a gravitational field, the gravitational force on the object is $F = mg$, where g is the gravitational field strength at the object's position. If the object is not acted on by any other force, it accelerates with

$$\textbf{acceleration, } a = \frac{\textbf{force}}{\textbf{mass}} = \frac{mg}{m} = g$$

The object therefore falls freely with acceleration g. So, g may also be described as the acceleration of a freely falling object.

An object that falls freely is unsupported. Although the object in this situation is commonly described as being weightless, it is better to describe it as unsupported because it is acted on by the force of gravity alone. The unit of gravitational field strength is $\mathrm{N\,kg^{-1}}$, and the unit of the acceleration of free fall is $\mathrm{m\,s^{-2}}$.

Field patterns

1. A **radial field** is where the field lines are like the spokes of a wheel, always directed to the centre. Figure 2 shows an example of a radial field. The force of gravity on a small mass near a much larger spherical mass is always directed to the centre of the larger mass. For example, the force on a small object near a spherical planet always acts towards the centre of the planet, regardless of the position of the object. The magnitude of g in a radial field decreases with increasing distance from the massive body.
2. A **uniform field** is where the gravitational field strength is the same in magnitude and direction throughout the field. The field lines are therefore parallel to one another and equally spaced.

Synoptic link

See Topic 2.4, Free fall, for more about the measurement of g.

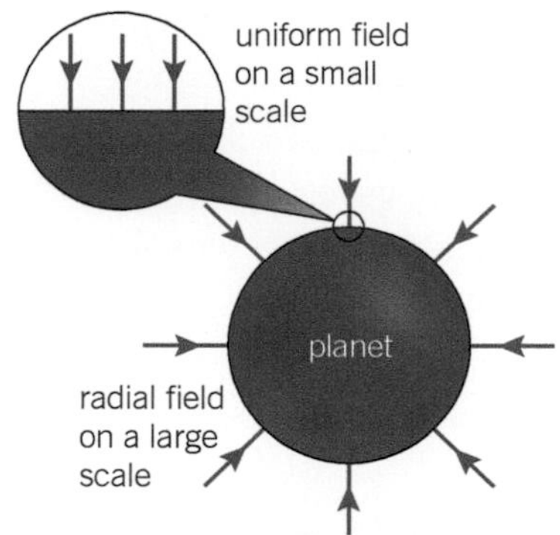

▲ **Figure 2** *Field lines*

Application

Is the Earth's gravitational field uniform or radial?

The force of gravity due to the Earth on a small mass decreases with distance from the Earth. So the gravitational field strength of the Earth falls with increasing distance from the Earth. The field is therefore radial. However, over small distances much less than the Earth's radius, the change of gravitational field strength is insignificant. For example, the measured value of g has the same magnitude ($= 9.8\,\text{N kg}^{-1}$) and direction (downwards) 100 m above the Earth as it is on the surface. In theory, g is smaller higher up, but the difference is too small to be noticeable – provided we don't go too high! Only over distances that are small compared with the Earth's radius can the Earth's field be considered uniform.

Summary questions

1 **a** Explain what is meant by a field line or a line of force of a gravitational field.

b With the aid of a diagram in each case, explain what is meant by:

i a radial field **ii** a uniform field.

2 **a** Calculate the gravitational force on:

i an object of mass 3.5 kg in a gravitational field at a position where $g = 9.5\,\text{N kg}^{-1}$

ii an object of mass 100 kg in a gravitational field at a position where $g = 1.6\,\text{N kg}^{-1}$.

b Calculate the gravitational field strength at a position in a gravitational field where:

i an object of mass 2.5 kg experiences a force of 40 N

ii an object of mass 18 kg experiences a force of 72 N.

3 Demonstrate that the acceleration of an object falling freely in a gravitational field is equal to g, where g is the gravitational field strength at that position.

4 Figure 3 represents a small part of the Earth's surface. Sketch the lines of force near this part of the Earth's surface:

a if the density of the Earth in this part is uniform

b if there is a large mass of dense matter under this part of the surface.

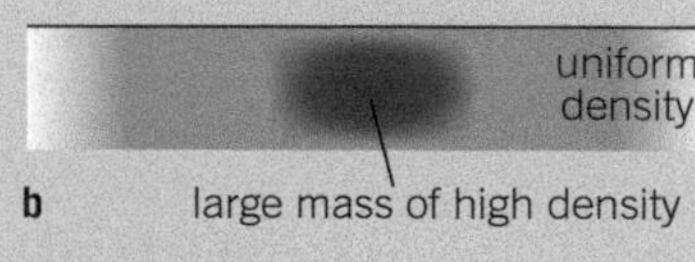

▲ **Figure 3**

17.3 Gravitational potential

Learning objectives:

- → Define gravitational potential.
- → Calculate the gravitational potential difference between two points.
- → Explain where an object would have to be placed for its gravitational potential energy to be zero.

Specification reference: 3.7.4

Imagine you are in a space rocket, about to blast off from the surface of a planet. The planet's gravitational field extends far into space although it becomes weaker with increased distance from the planet. To escape from the planet's pull due to gravity, the rocket must do work against the force of gravity on it due to the planet. If the rocket fuel doesn't provide enough energy to escape, you are doomed to return! The planet's gravitational field is like a trap which the rocket must climb out from to escape. As the rocket rises, its gravitational potential energy (g.p.e.) increases. If it falls back, its g.p.e. decreases. If the rocket carried a g.p.e. meter which was set to zero when it was far away from the planet, the meter reading would be negative when the rocket is on the surface of the planet and would increase towards zero as the rocket moves away from the planet.

▲ **Figure 1** *Into space*

Gravitational potential energy is the energy of an object due to its position in a gravitational field. The position for zero g.p.e. is at infinity – in other words, the object would be so far away that the gravitational force on it is negligible. Your rocket climbing out of the planet's field needs to increase its g.p.e. to zero to escape completely. At the surface, its g.p.e. is negative, so it needs to do work to escape from the field completely.

The **gravitational potential** at a point in a gravitational field is the g.p.e. per unit mass of a small test mass. This is equal to the work done per unit mass to move a small object from infinity to that point as the gravitational potential at infinity is zero. So, you can define gravitational potential at a point as *the work done per unit mass to move a small object from infinity to that point.*

The gravitational potential, *V*, at a point is the work done per unit mass to move a small object from infinity to that point.

Therefore, for a small object of mass m at a position where the gravitational potential is V, the work W that must be done to enable it to escape completely is given by $W = mV$. Rearranging this gives

$$V = \frac{W}{m}$$

The unit of gravitational potential is J kg^{-1}.

Suppose our rocket has a payload mass of 1000 kg and the gravitational potential at the surface of the planet is $-100\,\text{MJ kg}^{-1}$. Assume that the fuel is used quickly to boost the rocket to high speed. For the rocket to escape completely, the g.p.e. of the 1000 kg payload must increase from -100×1000 MJ to zero. So the work done on the payload must be at least 100 000 MJ to escape. If the rocket payload is only given 40 000 MJ of kinetic energy from the fuel, then it can only increase its g.p.e. by 40 000 MJ. So it can only reach a position in the field where the gravitational potential is $-60\,\text{MJ kg}^{-1}$.

Because the work done ΔW to move it from V_1 to V_2 is equal to the change of its gravitational potential energy, then $\Delta W = m\Delta V$.

Synoptic link

Gravitational fields act on objects due to their mass. Gravitational potential is defined in terms of work done per unit mass. Electric fields act on charged objects. Electric potential is defined in terms of work done per unit positive charge. See Topic 18.3, Electric potential.

Note

In general, if a small object of mass m is moved from gravitational potential V_1 to gravitational potential V_2, its change of g.p.e. $\Delta E_P = m(V_2 - V_1) = m\Delta V$, where $\Delta V = (V_2 - V_1)$.

Potential gradients

Equipotentials are surfaces of constant potential. Because of this, no work needs to be done to move along an equipotential surface. Hillwalkers know all about equipotentials because the contour lines on a map are lines of constant potential. A contour line joins points of equal height above sea level, so a hillwalker following a contour line has constant potential energy. Sensible hillwalkers take great care where the contour lines are very close to each other. One slip and their gravitational potential energy might fall dramatically!

The equipotentials near the Earth are circles, as shown in Figure 2. At increasing distance from the surface, the gravitational field becomes weaker. So the gain of gravitational potential energy per metre of height gain becomes less. In other words, away from the Earth's surface, the equipotentials for equal increases of potential are spaced further apart.

However, near the surface over a small region, the equipotentials are horizontal (i.e., parallel to the ground) as shown in Figure 3. This is because the gravitational field over a small region is uniform. A 1 kg mass raised from the surface by 1 m gains 9.8 J of gravitational potential energy. If it is raised another 1 m, it gains another 9.8 J. So its g.p.e. rises by 9.8 J for every metre of height it gains above the surface.

The **potential gradient** at a point in a gravitational field is the change of potential per metre at that point.

Near the Earth's surface, the potential changes by $9.8\,\text{J}\,\text{kg}^{-1}$ for every metre of height gained. So the potential gradient near the surface of the Earth is constant and equal to $9.8\,\text{J}\,\text{kg}^{-1}\,\text{m}^{-1}$. However, further from the Earth's surface, the potential gradient becomes less and less.

In general, for a change of potential ΔV over a small distance Δr, the potential gradient = $\frac{\Delta V}{\Delta r}$.

Consider a test mass m being moved away from a planet, as shown in Figure 3. To move m a small distance Δr in the opposite direction to the gravitational force F_{grav} on it, its gravitational potential energy must be increased

- by an equal and opposite force F acting through the distance Δr
- by an amount of energy equal to the work done by F, that is, $\Delta W = F\Delta r$.

For the test mass, the change of potential ΔV of the test mass $= \frac{\Delta W}{m}$, then $\Delta V = \frac{F\Delta r}{m}$, so $F = \frac{m\Delta V}{\Delta r}$. So, $F_{grav} = -\frac{m\Delta V}{\Delta r}$ as $F_{grav} = -F$.

Gravitational field strength $g = \frac{F_{grav}}{m} = -\frac{\Delta V}{\Delta r}$

Gravitational field strength g is the negative of the potential gradient.

$$g = -\frac{\Delta V}{\Delta r}$$

where the minus sign here shows you that g acts in the opposite direction to the potential gradient.

Study tip

$\Delta E_P = mg\Delta h$ can only be applied for values of Δh that are very small compared with the Earth's radius. $\Delta E_P = m\Delta V$ can *always* be applied. Remember that V is a scalar. You can use the equation $\Delta E_P = mg\Delta h$ to estimate how much potential energy is gained by a mountaineer who climbs to the top of any mountain on Earth. Even the height of Mount Everest, which is the highest mountain on Earth and about 10 km high, is much less than the Earth's radius, which is about 6360 km.

Q: A 70 kg mountaineer climbs a vertical distance 10 km. Estimate how many 100 g chocolate bars she would need to eat to release the same amount of energy as her gain of potential energy. Assume the energy released per gram of chocolate is 30 kJ.

Answer: 2 bars.

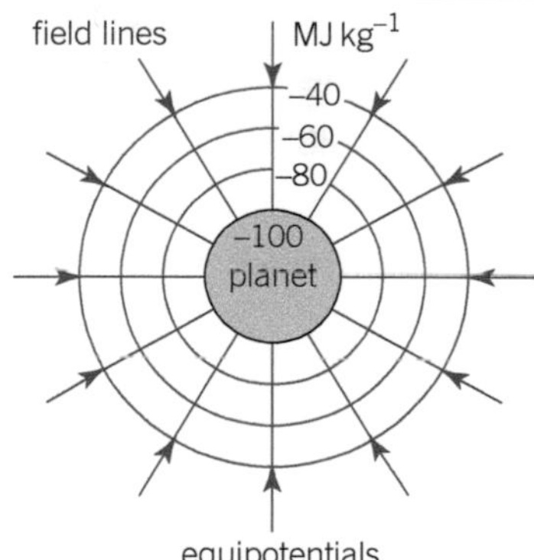

▲ **Figure 2** *Equipotentials near a planet*

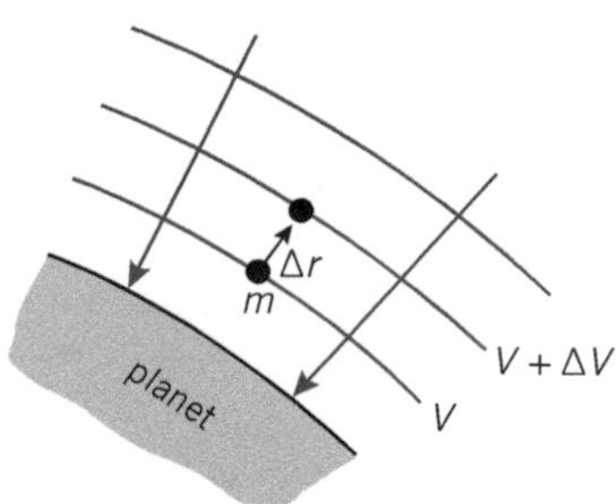

▲ **Figure 3** *Potential gradients*

Potential gradients are like contour gradients on a map. The closer the contours are on a map, the steeper the hill. Likewise, the closer the equipotentials are, the greater the potential gradient is and the stronger the field is. Where the equipotentials show equal changes of potential for equal changes of spacing, the potential gradient is constant. So the gravitational field strength is constant, and the field is uniform. Notice also that the gradient is always at right angles to the equipotentials, so the lines of force of the gravitational field are always perpendicular to the equipotentials.

Summary questions

$g = 9.8\,\text{N kg}^{-1}$

1 **a** Calculate the gain of gravitational potential energy of an object of mass 12 kg when its centre of mass is raised through a height of 2.0 m.

b Demonstrate that the gravitational potential difference between the Earth's surface and a point 2.0 m above the surface is $19.6\,\text{J kg}^{-1}$.

2 A rocket of mass 35 kg launched from the Earth's surface gains 70 MJ of gravitational potential energy when it reaches its maximum height.

a Calculate the gravitational potential difference between the Earth's surface and the highest point reached by the rocket.

b The gravitational potential of the Earth's gravitational field at the surface of the Earth is $-63\,\text{MJ kg}^{-1}$. Calculate:

i the gravitational potential at the highest point reached by the rocket

ii the work that would need to have been done by the rocket to escape from the Earth's gravitational field.

3 Figure 4 shows the equipotentials near a non-spherical object.

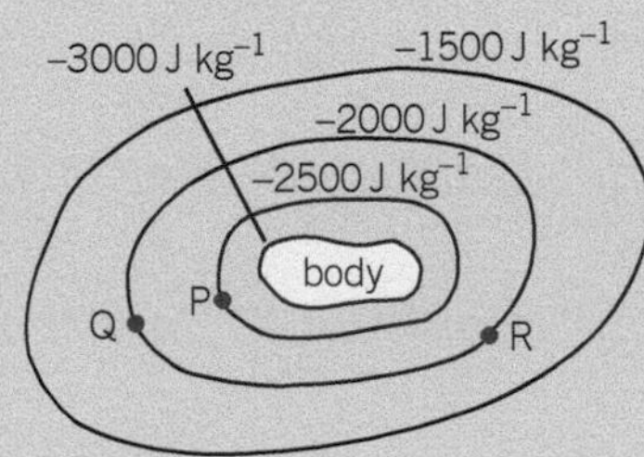

▲ **Figure 4**

a Calculate the gravitational potential energy of a 0.1 kg object at:

i P

ii Q

iii R.

b Calculate how much work must be done on the object to move it from:

i P to Q

ii Q to R.

4 Figure 5 shows equipotentials at a spacing of 1.0 km near a planet. The point labelled X is on the $-500\,\text{kJ kg}^{-1}$ equipotential.

a Demonstrate that the potential gradient at X is $5.0\,\text{J kg}^{-1}\,\text{m}^{-1}$.

b Using your answer from part **a** calculate the gravitational field strength at X.

c Calculate the work that would need to be done to remove an object of mass 50 kg from X to infinity.

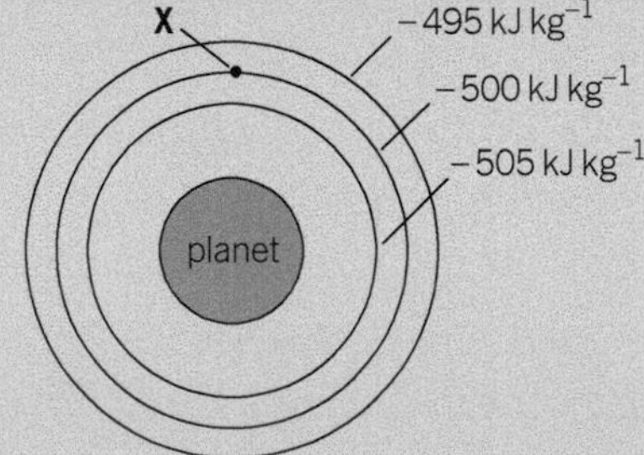

▲ **Figure 5**

17.4 Planetary fields

Gravitational field strength

The law of gravitation can be used to determine the **gravitational field strength** at any point in the field of a planet or any other spherical mass. Newton showed that the field of a spherical mass is the same as if the mass were concentrated at its centre. The field lines of a spherical mass are always directed towards the centre, so the field pattern is just the same as for a point mass, as shown in Figure 1.

- **For a point mass M**, the magnitude of the force of attraction on a test mass m (where $m << M$) at distance r from M is given by Newton's law of gravitation $F = \frac{GMm}{r^2}$.
  Therefore, the magnitude of the gravitational field strength at distance r is given by
  $$g = \frac{F}{m} = \frac{GM}{r^2}$$
- **For a spherical mass M of radius R**, the force of attraction on a test mass m at distance r **from the centre of M** is the same as if the mass M were concentrated at its centre. Therefore, the force of attraction between m and M is
  $$F = \frac{GMm}{r^2}$$
  So, the magnitude of the gravitational field strength at distance r is given by $g = \frac{F}{m} = \frac{GM}{r^2}$, only if distance r is greater than or equal to the radius R of the sphere.

The **magnitude of the gravitational field strength is $g = \frac{GM}{r^2}$** at distance r from a point object or from the centre of a sphere of mass M.

Note:

The gravitational field strength at the surface of a sphere of radius R and mass M is given by $g_s = \frac{GM}{R^2}$ because the distance from the centre, r, is equal to R at the surface.

Learning objectives:

- → Describe the shape of a graph of g against r for points outside the surface of a planet.
- → Compare this graph with the graph of V against r.
- → Explain the significance of the gradient of the V–r graph.

Specification reference: 3.7.3 and 3.7.4

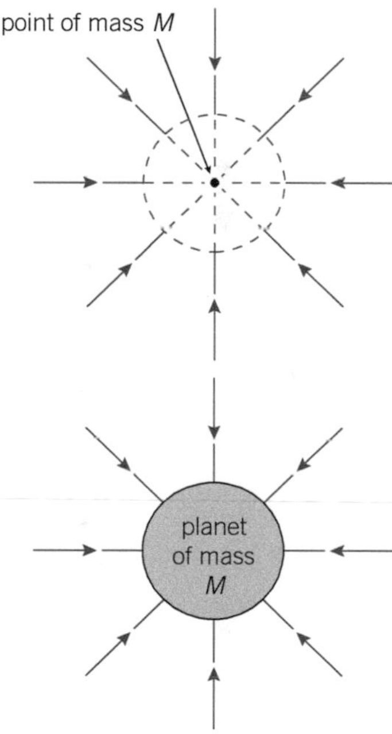

▲ **Figure 1** *Comparing fields*

Worked example

$G = 6.67 \times 10^{-11}\,\mathrm{N\,m^2\,kg^{-2}}$

The gravitational field strength at the surface of the Earth is $9.8\,\mathrm{N\,kg^{-1}}$. Calculate:

a the mass of the Earth

b the gravitational field strength of the Earth at a height of 1000 km above the surface.

The radius of the Earth = 6400 km.

Solution

a Rearranging $g_s = \frac{GM}{R^2}$ gives $M = \frac{g_s R^2}{G}$

$$= \frac{9.8 \times (6400 \times 10^3)^2}{6.67 \times 10^{-11}}$$

$$= 6.0 \times 10^{24}\,\text{kg}$$

b At height $h = 1000\,\text{km}$, $r = R + h = 7400\,\text{km}$

$$\therefore\ g = \frac{GM}{r^2} = \frac{6.67 \times 10^{-11} \times 6.0 \times 10^{24}}{(7400 \times 10^3)^2} = 7.3\,\text{N}\,\text{kg}^{-1}$$

The variation of *g* with distance from the centre of a spherical planet (or star)

Consider the gravitational field beyond the surface of a spherical planet of mass M and radius R. As you saw on the previous page, for any position at or beyond the surface, the magnitude of the gravitational field strength is given by $g = \frac{GM}{r^2}$, where r is the distance from the position to the centre of the sphere.

Because the surface gravitational field strength $g_s = \frac{GM}{R^2}$, then $GM = g_s R^2$.

Therefore, $g = \frac{g_s R^2}{r^2}$

The equation shows how g changes with increase of distance r from the centre of the planet.

- At distance $r = 2R$, $g = \frac{g_s R^2}{(2R)^2} = \frac{g_s}{4}$
- At distance $r = 3R$, $g = \frac{g_s R^2}{(3R)^2} = \frac{g_s}{9}$
- At distance $r = 4R$, $g = \frac{g_s R^2}{(4R)^2} = \frac{g_s}{16}$

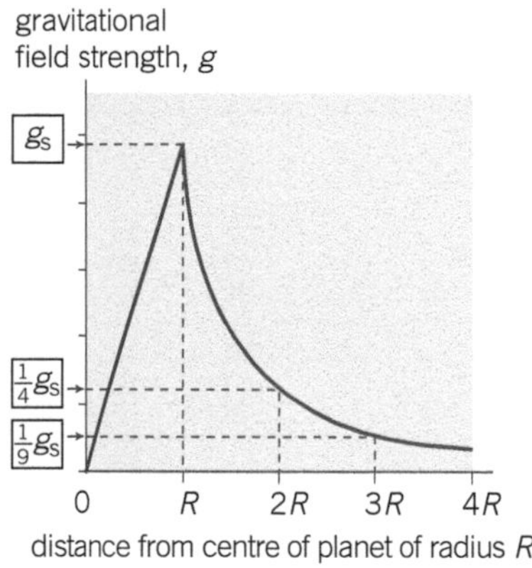

▲ **Figure 2** *Gravitational field strength*

Figure 2 shows how g varies with distance r. The shape of the curve beyond $r = R$ is an *inverse-square law* curve because g decreases in inverse proportion to r^2.

Application

Inside a planet

From the equation $g = \frac{GM}{r^2}$, you might think that the magnitude of g inside the Earth becomes ever larger and larger as r becomes smaller and smaller. However, inside the planet, only the mass in the sphere of radius r contributes to g. The remainder of the mass outside r up to the surface gives no resultant force. So as r becomes smaller, the mass M that contributes to g becomes smaller too. At the centre, the mass that contributes to g is zero. So g is zero at the centre. Figure 2 also shows how g inside the planet varies with distance from the centre. For a spherical planet of density ρ, you can demonstrate that $g = \frac{4\pi G\rho r}{3}$ inside the planet.

Gravitational potential near a spherical planet

At or beyond the surface of a spherical planet, the gravitational potential V at distance r from the centre of the planet of mass M is given by

$$V = -\frac{GM}{r}$$

Applying this equation to the surface of the Earth with $M = 6.0 \times 10^{24}$ kg and $r = 6.4 \times 10^6$ m gives a value of $-63\,\text{MJ}\,\text{kg}^{-1}$. This means that 63 MJ of work needs to be done to remove a 1 kg mass from the surface of the Earth to infinity.

The **escape velocity** from a planet is the minimum velocity an object must be given to escape from the planet when projected vertically from the surface.

- At the surface of a planet of radius R, the potential $V = -\frac{GM}{R}$. At infinity, the potential is, by definition, zero. Therefore, to move an object of mass m from the surface to infinity, the work that must be done is $\Delta W = m\Delta V = \frac{GMm}{R}$.
- If the object is projected at speed v, then for it to be able to escape from the planet, its initial kinetic energy $\frac{1}{2}mv^2 \geq \Delta W$.

So, $\frac{1}{2}mv^2 \geq \Delta W$, which gives $\frac{1}{2}mv^2 \geq \frac{GMm}{R}$

Therefore, $v^2 \geq \frac{2GM}{R}$

So, the escape velocity, $v_{esc} = \sqrt{\frac{2GM}{R}}$

Because $g = \frac{GM}{R^2}$, substituting gR^2 for GM therefore gives

$$v_{esc} = \sqrt{2gR}$$

Figure 3 shows how the force of gravity on a 1 kg mass (i.e., g) varies with distance from the surface of a planet. As you have just learnt, the equation for this curve is $g = \frac{g_s R^2}{r^2}$.

The area under the curve also represents the work done to move the 1 kg mass from infinity to the surface. Therefore the area under the curve in Figure 3 gives the value of the gravitational potential at the surface.

- Each grid square in Figure 3 represents a 1 N force acting for a distance of 2.5×10^6 m. So each grid square represents 2.5 MJ ($= 1\,\text{N} \times 2.5 \times 10^6\,\text{m}$) of work done.
- Therefore, the work done to move the 1 kg mass from infinity to the surface can be estimated by counting the number of grid squares under the curve and multiplying this number by 2.5 MJ.

By counting part-filled squares that are half filled or more as wholly filled squares, and neglecting part-filled squares that are less than half filled, show for yourself that this method gives an estimate for the work done that is very close to the value of 63 MJ determined above.

In effect, the area method used above is an application of *work done = force × distance moved*, with a variable force $F = \frac{GMm}{r^2}$ and $m = 1$ kg.

Consider one small step in moving the 1 kg mass from infinity to the surface. Suppose the distance from the centre of the Earth decreases from r_1 to r_2 in making this small step of distance Δr.

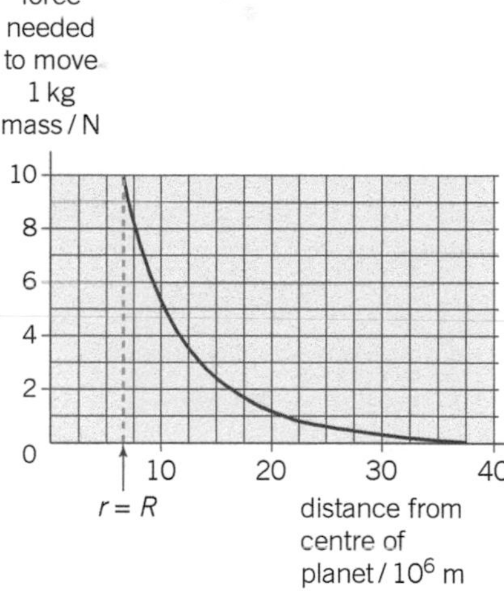

▲ **Figure 3** *Work done*

Hint

Figure 3 shows how g changes with distance from the centre of a planet. The area under any section of the curve gives the corresponding change of potential for this section. This is because the change of potential between any two points on the curve is equal to the work done to move a 1 kg mass from one point to the other.

The work done ΔW to make this small step is given by

$$\Delta W = F\Delta r = \frac{GMm}{r^2}\Delta r$$

In Figure 3, the work done ΔW in making each small step $\Delta r = 2.5 \times 10^6$ m is represented by each column of grid squares under the curve. So the total work done to move from infinity to the surface is given by the total area under the curve.

Extension

Where does the equation for gravitational potential $V = -\frac{GM}{r}$ come from?

You *don't* need to know the proof below. It is given here just to give you some further insight into the use of maths in physics.

The change of potential ΔV for the small step (from $r_1 = r$ to $r_2 = r - \Delta r$) $= V_2 - V_1$, where V_2 is the potential at r_2 and V_1 is the potential at r_1

$$\Delta V = \frac{\Delta W}{m} = F\frac{\Delta r}{m} = -\frac{GM}{r^2}\Delta r$$

where the minus sign indicates a decrease in r.

Because $\frac{1}{r_1} - \frac{1}{r_2} = \frac{r_2 - r_1}{r_1 r_2} = \frac{(r - \Delta r) - r}{r(r - \Delta r)} = \frac{-\Delta r}{r^2}$, only if $\Delta r << r$,

then $\Delta V = V_2 - V_1 = GM\left(\frac{1}{r_1} - \frac{1}{r_2}\right)$.

So, $V_2 = -\frac{GM}{r_2}$ and $V_1 = -\frac{GM}{r_1}$

Therefore, in general, the potential V at distance r from the centre of a planet is given by

$$V = -\frac{GM}{r}$$

Synoptic link

The gravitational potential at distance r from the centre of a spherical mass is proportional to $\frac{1}{r}$; the electric potential at distance r from a point charge is also proportional to $\frac{1}{r}$. See Topic 17.3.

Note

The gradient of the potential curve at any point is equal to $-g$, where g is the gravitational field strength at the point. This is because $g = -$ the potential gradient, as you learnt in Topic 17.3. The potential curve becomes less steep as the distance from the centre decreases. At any point on the curve, the gradient of the curve can be found by drawing a tangent to the curve at that point and measuring the gradient of the tangent.

Potential gradients near a spherical planet

The gravitational potential V near a spherical planet is inversely proportional to the distance r from the centre of the planet because $V = -\frac{GM}{r}$. Figure 4 shows how the gravitational potential of the Earth varies with distance. Note the potential curve is a $\frac{1}{r}$ curve, not an inverse-square (i.e., $\frac{1}{r^2}$) curve like the field strength curve in Figure 2. So the potential:

- at distance $2R$ from the centre is $0.50 \times$ the potential at distance R from the centre
- at distance $3R$ from the centre is $0.33 \times$ the potential at distance R from the centre
- at distance $4R$ from the centre is $0.25 \times$ the potential at distance R from the centre, and so on.

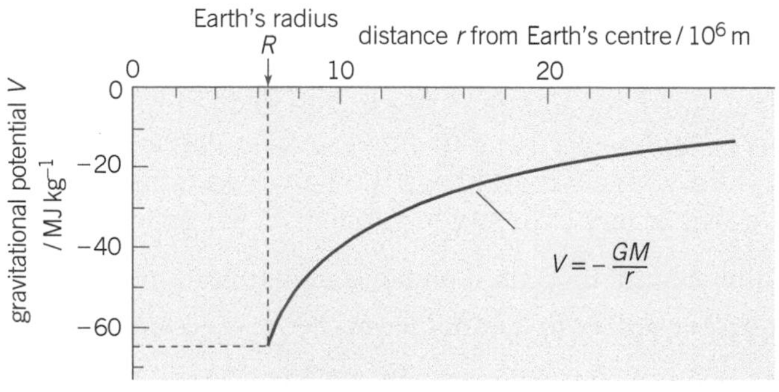

▲ **Figure 4** *Gravitational potential near the Earth*

Application

Multistage rockets

The Russian physicist Konstantin Tsiolovsky predicted in 1895 that space rockets would need to consist of parts in many stages, that is, be multistage. The reason is that rocket fuel releases no more than about 30 MJ kg^{-1} when it burns, and this is not enough to enable a single-stage rocket to climb into space from the Earth's surface, as the surface potential is about −63 MJ kg^{-1}. A multistage rocket jettisons the lowest stage when the fuel contained in that part has all been burnt, and only the top stage (the payload) escapes from the Earth's gravitational pull.

▲ **Figure 5** *A rocket during takeoff*

Summary questions

$G = 6.67 \times 10^{-11}\ \mathrm{N\,m^2\,kg^{-2}}$

1 The Moon has a radius of 1740 km, and its surface gravitational field strength is 1.62 N kg^{-1} to three significant figures.

  a Calculate the mass of the Moon.

  b The Moon's gravitational pull on the Earth causes the ocean tides. Demonstrate that the gravitational pull of the Moon on the Earth's oceans is approximately three-millionths of the gravitational pull of the Earth on its oceans. Assume that the distance from the Earth to the Moon is 380 000 km.

2 The Sun has a mass of 2.0×10^{30} kg and a mean diameter of 1.4×10^{9} m. Calculate:

  a its gravitational field strength at

    i its surface

    ii the Earth's orbit, which is at a distance of 1.5×10^{11} m from the Sun.

  b The Earth has a mass of 6.0×10^{24} kg. Show that the gravitational field strength of the Earth is equal and opposite to the gravitational field strength of the Sun at a distance of 260 000 km from the centre of the Earth.

3 The tip of the tallest mountain on the Earth, Mount Everest, is 9 km above sea level. The mean radius of the Earth to the nearest kilometre is 6378 km.

  a Calculate the difference between the gravitational field strength of the Earth at sea level and the top of Mount Everest.

  b Discuss whether it is reasonable to assume that the Earth's gravitational field is uniform between the surface and a height of 10 km above the surface.

  c Calculate the gain of potential energy of a mountaineer of mass 80 kg who travels to the top of the mountain from sea level.

4 Use the data in **Q1** to calculate the gravitational potential at the surface of the Moon and therefore calculate the work done to launch a 500 kg rocket from the surface so that it escapes from the Moon's gravitational field.

Study tip

Know how to draw $(\frac{1}{r^2})$ and $-(\frac{1}{r})$ graphs, and note how g changes much more sharply with distance than V does.

17.5 Satellite motion

Learning objectives:

- → State the condition needed for a satellite to be in a stable orbit.
- → Describe what happens to the speed of a satellite if it moves closer to the Earth.
- → Discuss why a geostationary satellite must be in an orbit above the equator.

Specification reference: 3.7.5

On any clear night you should be able to see satellites passing overhead in the night sky. Although they are pinpoints of light, they are noticeable because they move steadily through the constellations. You can find information on the Internet to help you identify some of them from their directions. However, satellite motion is not confined to artificial satellites orbiting the Earth. Any small mass that orbits a larger mass is a satellite. The Moon is the Earth's only natural satellite. Mars has two moons, Phobos and Deimos. Jupiter has at least 14 satellites including the four satellites Io, Callisto, Ganymede, and Europa, first observed by Galileo four centuries ago.

Newton knew that the time period T of a planet orbiting the Sun depends on the mean radius r of the orbit in accordance with Kepler's third law, $T^2 \propto r^3$.

Newton realised that the force of gravitational attraction between each planet and the Sun is the centripetal force that keeps the planet on its orbit. By assuming that the gravitational force is given by $\frac{GMm}{r^2}$, the gravitational field strength $\frac{GM}{r^2}$ is therefore equal to the centripetal acceleration $\frac{v^2}{r}$, where M is the mass of the Sun, m is the mass of the planet, r is the radius of the orbit, and v is the speed of the planet.

Therefore, the speed of the planet is given by $v^2 = \frac{GM}{r}$.

Because speed $v = \frac{\text{circumference of the orbit}}{\text{time period}} = \frac{2\pi r}{T}$,

then $\frac{(2\pi r)^2}{T^2} = \frac{GM}{r}$.

Rearranging this equation gives $\frac{r^3}{T^2} = \frac{GM}{4\pi^2}$.

Because $\frac{GM}{4\pi^2}$ is the same for all of the planets, then $\frac{r^3}{T^2}$ is the same for all of the planets.

So, by assuming that the force of attraction F varies with distance according to the inverse-square law (i.e., $F \propto \frac{1}{r^2}$), Newton was able to prove Kepler's third law.

▲ **Figure 1** *International Space Station in orbit*

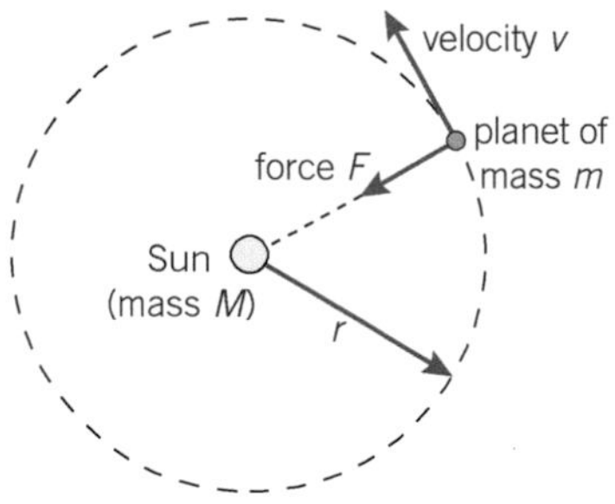

▲ **Figure 2** *Explanation of planetary motion*

Extension

The mass of the Sun

Newton's theory not only explains Kepler's laws, but it also allows the mass M to be calculated if the value of G is known. The Earth orbits the Sun once per year on a circular orbit of radius 1.5×10^{11} m. So you can prove for yourself that the value of $\frac{r^3}{T^2}$ for any planet is $3.4 \times 10^{18}\,\text{m}^3\,\text{s}^{-2}$.

Q: Given $G = 6.67 \times 10^{-11}\,\text{N m}^2\,\text{kg}^{-2}$, demonstrate that the mass of the Sun is 2.0×10^{30} kg.

Geostationary satellites

A geostationary satellite orbits the Earth directly above the equator and has a time period of exactly 24 h. It therefore remains in a fixed position above the equator because it has exactly the same time period as the Earth's rotation. The radius of orbit of a geostationary satellite can be calculated using the equation $\frac{r^3}{T^2} = \frac{GM}{4\pi^2}$.

So, $T = 24\,\text{h} = 24 \times 3600\,\text{s} = 86\,400\,\text{s}$

$$r^3 = \frac{GMT^2}{4\pi^2} = \frac{6.67 \times 10^{-11} \times 6.0 \times 10^{24} \times (86\,400)^2}{4\pi^2} = 7.6 \times 10^{22}\,\text{m}^3$$

$r = 4.2 \times 10^7\,\text{m} = 42\,000\,\text{km}$.

The radius of the Earth is 6400 km. Therefore, the height of a geostationary satellite above the Earth is 36 000 km (= 42 000 − 6400 km).

Note:

That a geosynchronous orbit is a 24 hour orbit inclined to the equator.

Application

SATNAV at work

▲ **Figure 3** *SATNAV at work*

Vehicle satellite navigation (SATNAV) units receive time signals from a system of global positioning system (GPS) of satellites that orbit the Earth at a height of just over 22 000 km. The signals are synchronised, and the receiver in each unit is programmed to measure the time differences between the signals and use the differences to pinpoint its position. The software in the unit is designed to display the local road map showing the vehicle's position and route ahead to the destination supplied by the driver. But beware – SATNAVs have been known to send heavy goods vehicles down unsuitable narrow roads!

Q: Estimate how far a vehicle moving at 30 m s^{-1} could travel in the time it takes for a radio signal to travel 44 000 km.

Answer: About 4 m.

Application

Vehicle tracking

Vehicles on the road in many countries are tracked by a network of thousands of cameras that automatically read and record vehicle number plates. Millions of number plates are recorded on a central database every day. GPS satellites pinpoint the location of each camera as well as the time and date of each record.

The benefits of nationwide vehicle tracking include the following.

- It ensures that vehicle registration and motor test certificates and motor insurance policies are up to date.
- It eliminates the need for paper tax discs on windscreens.
- It identifies stolen vehicles and vehicles suspected to have been used in criminal activities.

Concerns expressed by many people include the following.

- The movements of millions of people may be stored on the database for many years.
- The data might be misused, so that innocent people suffer further direct surveillance.

The benefits are clear, but the concerns raise issues about the way we use science. The concerns need to be addressed to ensure people have confidence in the system of vehicle tracking.

Synoptic link

By assuming that the satellite is in uniform circular motion, you can equate $\frac{mv^2}{r}$ to the gravitational force on the satellite. See Topic 15.2, Centripetal acceleration.

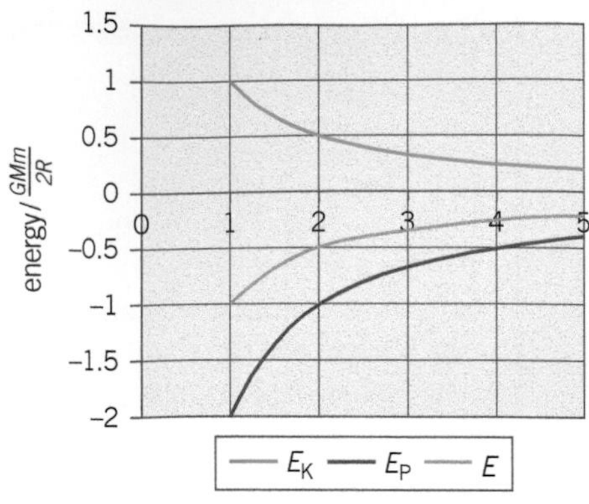

▲ **Figure 4** *Satellite energy*

The energy of an orbiting satellite

Consider a satellite of mass m in a circular orbit of radius r about a spherical planet or star of mass M. The speed of the satellite v is given by $v^2 = \frac{GM}{r}$, as explained earlier in this topic.

The kinetic energy E_K of the satellite $= \frac{1}{2}mv^2 = \frac{1}{2}m \times \frac{GM}{r} = \frac{GMm}{2r}$

The satellite is at distance r from the centre of the planet.

At this distance, the gravitational potential $V = -\frac{GM}{r}$.

So the satellite's gravitational potential energy $E_P = mV = -\frac{GMm}{r}$

Therefore the total energy of the satellite

$$E = E_P + E_K = -\frac{GMm}{r} + \frac{GMm}{2r} = -\frac{GMm}{2r}$$

For a satellite in a circular orbit of radius r, its total energy

$$\mathbf{E = -\frac{GMm}{2r}}$$

Figure 4 shows how E, E_P, and E_K vary with r. Notice that all three curves are $\frac{1}{r}$ curves and that the total energy is always negative.

Summary questions

$G = 6.67 \times 10^{-11}\,\text{N m}^2\,\text{kg}^{-2}$

radius of the Earth = 6400 km

$g = 9.8\,\text{N kg}^{-1}$ at the Earth's surface

1 **a** Two satellites X and Y are seen from the ground crossing the night sky at the same time. Satellite X crosses the sky faster than Y. State with a reason which satellite is higher.

b Explain why satellite TV dishes must be aligned carefully so they always point to the same position above the equator.

2 A space probe moving at a speed of $3.2\,\text{km s}^{-1}$ is in a circular orbit about a spherical planet. The time period of the satellite is 110 min. Calculate:

a the radius of the orbit

b the centripetal acceleration of the satellite

c the mass of the planet.

3 **a** A satellite moves at speed v in a circular orbit of radius r.

i Write down an expression for the centripetal acceleration of the satellite.

ii Demonstrate that the speed of the satellite is given by the equation $v^2 = gr$, where g is the gravitational field strength at the orbit.

b A satellite orbits the Earth in a circular orbit at a height of 100 km. Calculate:

i the gravitational field strength of the Earth at this distance

ii the speed of the satellite

iii the time period of the satellite.

4 **a** Demonstrate that the speed v of a satellite in a circular orbit of radius r about a planet of mass M is given by the equation $v^2 = \frac{GM}{r}$.

b A weather satellite is in a polar orbit above the Earth at a height of 1600 km.

i Demonstrate that its speed is $7.1\,\text{km s}^{-1}$.

ii Calculate its time period.

iii Explain why such a satellite can survey global weather patterns every day.

Practice questions: Chapter 17

Black holes

A black hole is an astronomical body that is so massive that nothing can escape from it. Anything that falls into a black hole would never be seen again. The radius of a black hole is defined by its *event horizon*, an imaginary spherical surface surrounding the body. Nothing inside the event horizon can escape, not even light. Radiation can be detected from matter drawn into a black hole before it is trapped inside the event horizon. Astronomers reckon there might be a black hole at the centre of our own galaxy, the Milky Way.

▲ **Figure 1** *The Milky Way*

The idea of a black hole was first thought up by John Michell in 1783, although the term itself was first used much later by the American physicist John Wheeler. Michell's idea was not tested until after Einstein published his *General Theory of Relativity* in 1916 in which he predicted mathematically that a strong gravitational field distorts space and time and bends light. He calculated that light grazing the Sun from a star would be deflected by 0.0005° due to the Sun's gravity. The prediction was confirmed by Sir Arthur Eddington in 1919, who led a scientific expedition to South America to test the prediction by photographing stars close to the Sun during a total solar eclipse. The eclipse photographs revealed that the star images were displaced relative to each other, just as Einstein had predicted.

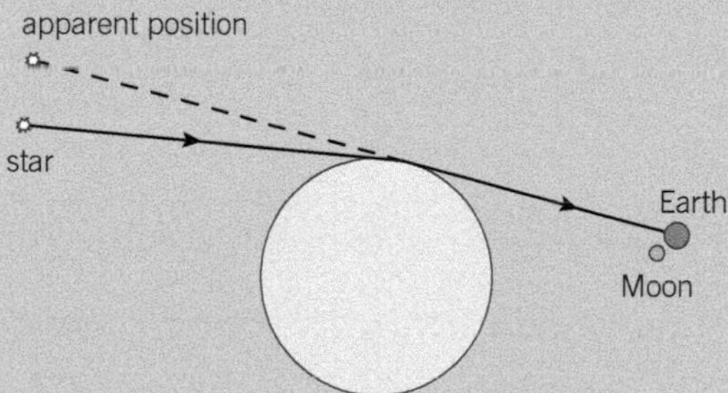

▲ **Figure 2** *Bending starlight*

The modern theory of black holes was started by Karl Schwarzschild, who used Einstein's theory to prove light could not escape from an object with a sufficiently strong gravitational field. He showed that such an object is surrounded by an event horizon, and that the radius, R, of the event horizon is given by $R = \frac{2GM}{c^2}$, where M is the mass of the black hole and c is the speed of light in free space.

Evidence for black holes has been obtained by astronomers. The central region of the M87 galaxy is rotating so fast that there must be a massive black hole at its centre. The X-ray source Cygnus X1 is a binary system consisting of a supergiant star accompanied by a very dense invisible star thought to be a black hole pulling matter from its companion.

$c = 3.0 \times 10^8\,\text{m s}^{-1}$, $G = 6.7 \times 10^{-11}\,\text{N m}^2\,\text{kg}^{-2}$

1 **(a)** (i) The mass of the Earth is 6.0×10^{24} kg. Show that the radius of the event horizon of a black hole that has the same mass as the Earth would be 8.9 mm.

(ii) Calculate the density of an object with the same mass as the Earth and a radius of 9.0 mm. The mass of the Earth = 6.0×10^{24} kg. *(5 marks)*

(b) (i) Use the scientific examples described above to explain what is meant by a scientific hypothesis.

(ii) What key discovery was made by Eddington in 1919 and what was the significance of this discovery? *(5 marks)*

(c) The Hubble Space Telescope has led to many new astronomical discoveries. Suppose your government commits itself to establishing a manned space observatory on the Moon by 2025. Some people take the view this project could lead to many new discoveries. Other people take the view that the money for the project would be better spent helping to improve living conditions in poorer countries. Discuss *one* argument in support of each of these views, and use your arguments to decide whether or not you would welcome such a project. *(5 marks)*

2 (a) (i) Explain what is meant by the *gravitational field strength* at a point in a gravitational field.

(ii) State the SI unit of gravitational field strength. *(2 marks)*

(b) Planet P has mass M and radius R. Planet Q has a radius $3R$. The values of the gravitational field strengths at the surfaces of P and Q are the same.

(i) Determine the mass of Q in terms of M.

(ii) **Figure 3** shows how the gravitational field strength above the surface of planet P varies with distance from its centre.

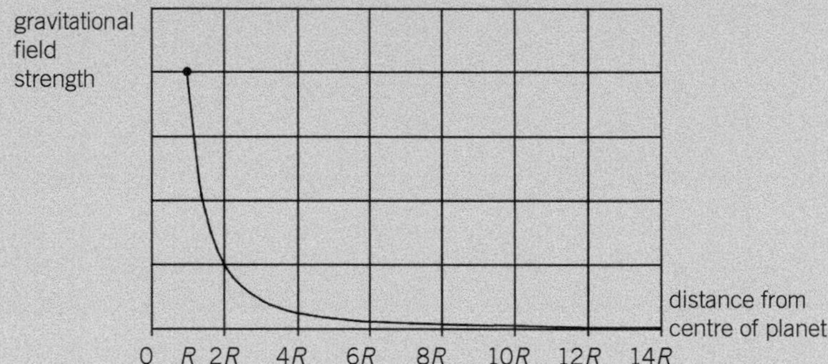

▲ **Figure 3**

Copy the diagram and draw the variation of the gravitational field strength above the surface of Q over the range shown. *(6 marks)*

AQA, 2006

3 (a) Artificial satellites are used to monitor weather conditions on Earth, for surveillance, and for communications. Such satellites may be placed in a *geostationary* orbit or in a low polar orbit. Describe the properties of the geostationary orbit and the advantages it offers when a satellite is used for communications. *(3 marks)*

(b) A satellite of mass m travels at angular speed ω in a circular orbit at a height h above the surface of a planet of mass M and radius R.

(i) Using these symbols, give an equation that relates the gravitational force on the satellite to the centripetal force.

(ii) Use your equation from part (b)(i) to show that the orbital period, T, of the satellite is given by

$$T^2 = \frac{4\pi^2(R+h)^3}{GM}$$

(iii) Explain why the period of a satellite in orbit around the Earth cannot be less than 85 minutes. Your answer should include a calculation to justify this value. *(6 marks)*

(c) Describe and explain what happens to the speed of a satellite when it moves to an orbit that is closer to the Earth. *(2 marks)*

AQA, 2006

4 (a) (i) Show that the gravitational field strength of the Earth at height h above the surface is given by

$$g = g_s \frac{R}{(R+h)^2}$$

where g_s is the gravitational field strength at the surface and R is the radius of the Earth.

(ii) Calculate the gravitational field strength of the Earth at a height of 200 km above its surface. *(5 marks)*

(b) An astronaut floats in a spacecraft which is in a circular orbit around the Earth. Discuss whether or not the astronaut is weightless in this situation. *(3 marks)*

AQA, 2007

5 NASA wishes to recover a satellite, at present stranded on the Moon's surface, and to place it in orbit around the Moon.

(a) (i) **Figure 4** shows a graph of how gravitational field strength due to the Moon varies with distance from the centre of the Moon. Copy the graph and mark the area that corresponds to the energy needed to move 1 kg from the surface of the Moon to a vertical height of 4000 km above the surface.

radius of the Moon = 1700 km

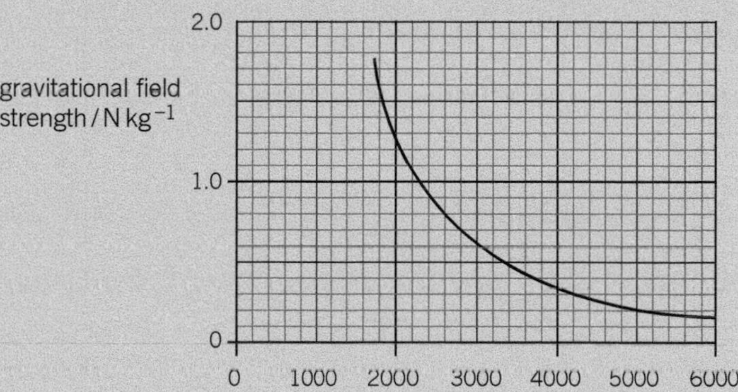

▲ **Figure 4**

(ii) The satellite has a mass of 450 kg. Estimate the change in gravitational potential energy of the satellite when it is moved from the surface of the Moon to a vertical height of 4000 km above the surface. *(6 marks)*

(b) NASA now decides to bring the satellite back to Earth. Explain why the amount of fuel required to return the satellite to Earth will be *much* less than the amount required to send it to the Moon originally. *(5 marks)*

AQA, 2004

6 (a) Explain what is meant by the *gravitational potential* at a point in a gravitational field. *(2 marks)*

(b) Use the following data to calculate the gravitational potential at the surface of the Moon.

mass of Earth = 81 × mass of Moon

radius of Earth = 3.7 × radius of Moon

gravitational potential at surface of the Earth = $-63\,\text{MJ}\,\text{kg}^{-1}$ *(3 marks)*

(c) Sketch a graph using axes as in **Figure 5** to indicate how the gravitational potential due to the Moon varies with distance along a line outwards from the surface of the Earth to the surface of the Moon. *(3 marks)*

AQA, 2005

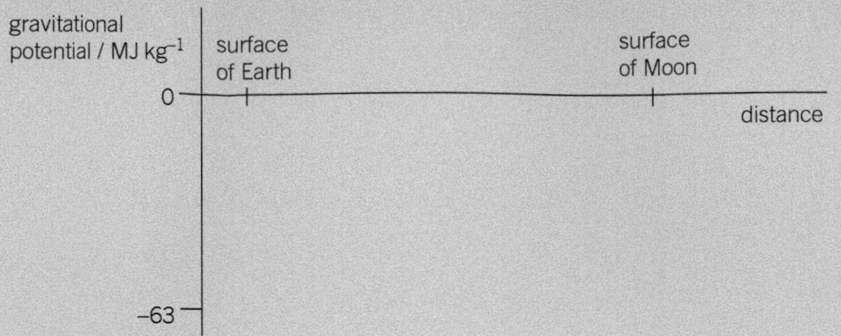

▲ **Figure 5**

7 (a) State the law that governs the magnitude of the force between two point masses. *(2 marks)*

(b) Table 1 shows how the gravitational potential varies for three points above the centre of the Sun.

▼ **Table 1**

Distance from centre of Sun / 10^8 m	Gravitational potential / 10^{10} J kg^{-1}
7.0 (surface of Sun)	−19
16	−8.3
35	−3.8

(i) Show that the data suggest that the potential is inversely proportional to the distance from the centre of the Sun. *(2 marks)*

(ii) Use the data to determine the gravitational field strength near the surface of the Sun. *(3 marks)*

(iii) Calculate the change in gravitational potential energy needed for the Earth to escape from the gravitational attraction of the Sun.
mass of the Earth = 6.0×10^{24} kg
distance of Earth from centre of Sun = 1.5×10^{11} m *(3 marks)*

(iv) Calculate the kinetic energy of the Earth due to its orbital speed around the Sun and therefore find the minimum energy that would be needed for the Earth to escape from its orbit. Assume that the Earth moves in a circular orbit. *(3 marks)*

AQA, 2005

8 For an object, such as a space rocket, to escape from the gravitational attraction of the Earth it must be given an amount of energy equal to the gravitational potential energy that it has on the Earth's surface. The minimum initial vertical velocity at the surface of the Earth that it requires to achieve this is known as the escape velocity.

(a) (i) Write down the equation for the gravitational potential energy of a rocket when it is on the Earth's surface. Take the mass of the Earth to be *M*, that of the rocket to be *m*, and the radius of the Earth to be *R*.

(ii) Show that the escape velocity, *v*, of the rocket is given by the equation

$$v = \sqrt{\frac{2GM}{R}}$$

(3 marks)

(b) The nominal escape velocity from the Earth is 11.2 km s^{-1}. Calculate a value for the escape velocity from a planet of mass four times that of the Earth and radius twice that of the Earth. *(2 marks)*

(c) Explain why the actual escape velocity from the Earth would be greater than the nominal value calculated from the equation given in part (a)(ii). *(2 marks)*

AQA, 2004

Answers to the Practice Questions and Section Questions are available at www.oxfordsecondary.com/oxfordaqaexams-alevel-physics

18 Electric fields
18.1 Field patterns

Static electricity

Most plastic materials can be charged quite easily by rubbing with a dry cloth. When charged, they can usually pick up small bits of paper. The bits of paper are attracted to the charged piece of plastic. Do charged pieces of plastic material attract each other? Figure 1 shows an arrangement to test for attraction. A charged perspex ruler will attract a charged polythene comb, but two charged rods of the same material always repel one another.

Like charges repel; unlike charges attract.

Electrons are responsible for charging in most situations. An uncharged atom contains equal numbers of protons and electrons. Add one or more electrons to an uncharged atom and it becomes negatively charged. Remove one or more electrons from an uncharged atom and it becomes positively charged. An uncharged solid contains equal numbers of electrons and protons. To make it negatively charged, electrons must be added to it. To make it positively charged, electrons must be removed from it. When an uncharged perspex rod is rubbed with an uncharged dry cloth, electrons transfer from the rod to the cloth. So the rod becomes positively charged, and the cloth becomes negatively charged.

- **Electrical conductors** such as metals contain lots of **free electrons**. These are electrons which move about inside the metal and are not attached to any one atom. To charge a metal, it must be first isolated from the Earth. Otherwise, any charge given to it is neutralised by electrons transferring between the conductor and the Earth. Then the isolated conductor can be charged by direct contact with any charged object. If an isolated conductor is charged positively and then earthed, electrons transfer from the Earth to the conductor to neutralise or discharge it (see Figure 2).
- **Electrically insulating materials** do not contain free electrons. All the electrons in an insulator are attached to individual atoms. Some insulators, such as perspex or polythene, are easy to charge because their surface atoms easily gain or lose electrons.

The shuttling ball experiment shows that an electric current is a flow of charge. A conducting ball is suspended by an insulating thread between two vertical plates, as in Figure 3. When a high voltage is applied across the two plates, the ball bounces back and forth between the two plates. Each time it touches the negative plate, the ball gains some electrons and becomes negatively charged. It is then repelled by the negative plate and is pulled across to the positive plate. When contact is made, electrons on the ball transfer to the positive plate so the ball now becomes positively charged and is repelled back to the negative plate to repeat the cycle. Therefore, the electrons from the high-voltage supply pass along the wire to the negative plate. There, they are ferried across to the other plate by the ball. Then they pass along the wire back to the supply. A microammeter in series with the plates shows that the shuttling ball causes a current round the circuit.

Learning objectives:

- → Explain how to charge a metal object.
- → Describe what the direction of an electric field line shows concerning a test charge.
- → Illustrate the strength of an electric field by using field lines.

Specification reference: 3.8.2

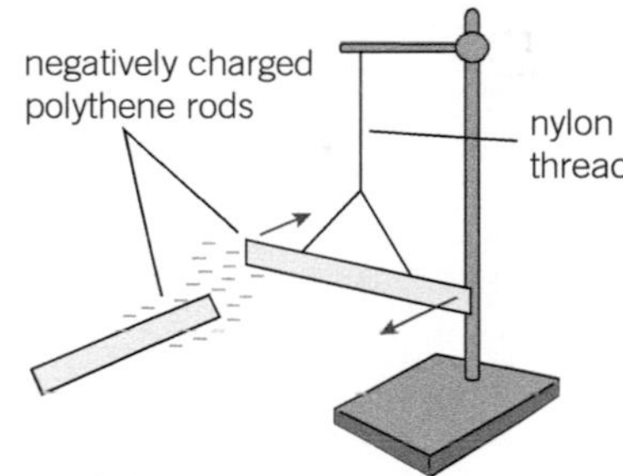

▲ **Figure 1** *Electrostatic forces*

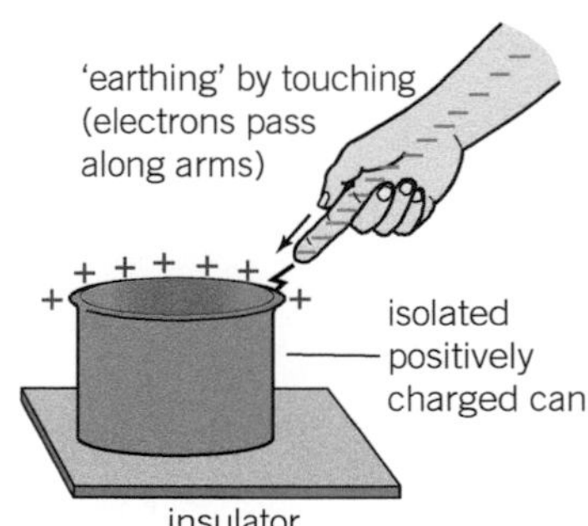

▲ **Figure 2** *Discharge to Earth*

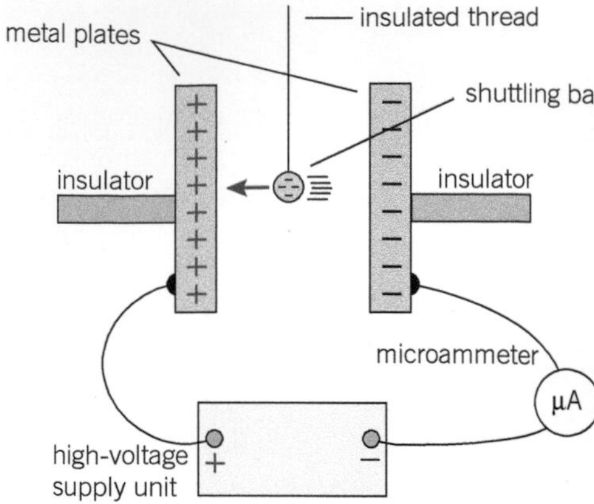

▲ **Figure 3** *The shuttling ball experiment*

If the plates are brought closer together, the ball shuttles back and forth even more rapidly. As a result, the microammeter reading increases because charge is ferried across at a faster rate.

Suppose the ball shuttles back and forth at frequency f. The time taken for each cycle is therefore $\frac{1}{f}$. The amount and type of charge on the ball depends on the voltage of the plate it last made contact with. Therefore, if the charge ferried across the gap each cycle is Q, the average current round the circuit is

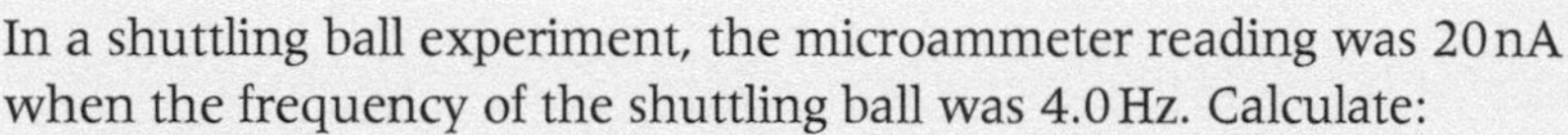

$$I = Qf \left(= \frac{\textbf{charge } Q}{\textbf{time for one cycle}}\right)$$

▲ **Figure 4** *The charged gold leaf electroscope*

Worked example

$e = 1.6 \times 10^{-19}\,\text{C}$

In a shuttling ball experiment, the microammeter reading was 20 nA when the frequency of the shuttling ball was 4.0 Hz. Calculate:

a the charge carried by the ball

b the number of electrons needed for the charge calculated in **a**.

Solution

a $Q = \frac{I}{f} = \frac{20 \times 10^{-9}}{4.0} = 5.0 \times 10^{-9}\,\text{C}$

b number of electrons $= \frac{Q}{e} = \frac{5.0 \times 10^{-9}}{1.6 \times 10^{-19}} = 3.1 \times 10^{10}$

The **gold leaf electroscope** is used to detect charge. If a charged object is in contact with the metal cap of the electroscope, some of the charge on the object transfers to the electroscope. As a result, the gold leaf and the metal stem which is attached to the cap gain the same type of charge and the leaf rises because it is repelled by the stem. See Topic 13.1, The photoelectric effect, for more about the gold leaf electroscope.

If another object with the same type of charge is brought near the electroscope, the leaf rises further because the object forces some charge on the cap to transfer to the leaf and stem.

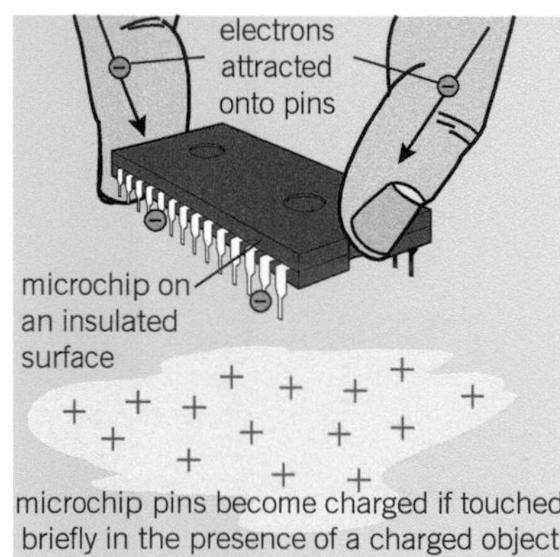

▲ **Figure 5** *Microchip damage*

Application

Chips and charge

People handling microchips wear antistatic clothing and they work in rooms fitted with antistatic floors. This is because a tiny amount of charge on the pins of an electronic chip can be enough to destroy circuits inside the chip. This can happen if the pins are touched in the presence of a charged body. The pins are earthed when they are touched. As a result, electrons transfer between the pins and Earth. If the connection to Earth is removed, the pins remain charged when the charged body is removed. Microchips are stored in antistatic packets and handled with special tools. Antistatic materials allow charge to flow across the surface.

Field lines and patterns

Any two charged objects exert equal and opposite forces on each other without being directly in contact. An electric field is said to surround each charge. Suppose a small positive charge is placed as a test charge near a body with a much bigger charge which is also positive. If the test charge is free to move, it will follow a path away from the body with the bigger charge. The path a free positive test charge follows is called a **line of force** or a **field line**.

The field lines of an electric field are the lines which positive test charges follow. The direction of an electric field line is the direction a positive test charge would move along. Figure 6 shows the patterns of fields around different charged objects. Each pattern is produced by semolina grains sprinkled on oil. An electric field is set up across the surface of the oil by connecting two metal conductors in the oil to the output terminals of a high-voltage supply unit. The grains line up along the field lines, like plotting compasses in a magnetic field.

- Oppositely charged objects create a field as shown in Figure 6a. The field lines become concentrated at the points. A positive test charge released from an off-centre position would follow a curved path to the negative point charge.
- A point object near an oppositely charged flat plate produces a field as shown in Figure 6b. The field lines are concentrated at the point object, but they are at right angles to the plate where they meet. The field is strongest where the lines are most concentrated.
- Two oppositely charged plates create a field as shown in Figure 6c. The field lines run parallel from one plate to the other, meeting the plates at right angles, except near the edges of the field. The field is *uniform* between the plates because the field lines are parallel to each other.

a *Oppositely charged points*

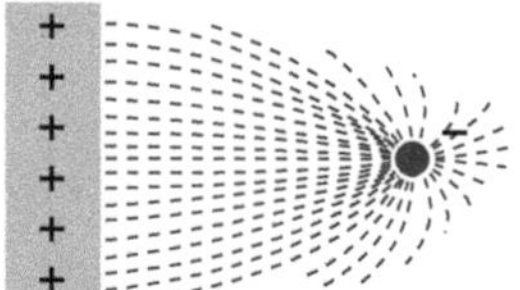

b *A point near a plate*

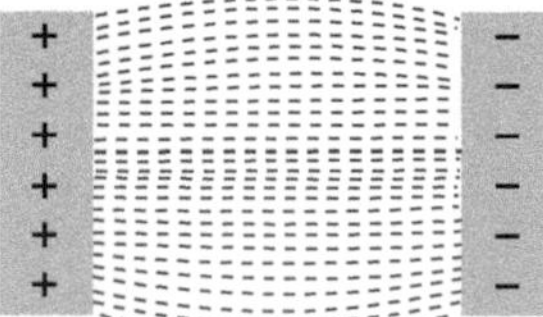

c *Oppositely charged plates*

▲ **Figure 6** *Electric field patterns*

Study tip

Note that an earthed conductor will become charged if a charged object is placed near it.

Synoptic link

The words *field* or *force field* are usually used where objects exert forces on each other without being in direct contact with each other.

- Objects exert gravitational forces of attraction on each other because they interact due to their mass. See Topic 17.2, Gravitational field strength.
- Charged objects exert an electric force on each other because they interact due to their charge. The force between two charged objects is attractive if the objects are oppositely charged. The force is repulsive if both objects have the same type of charge (both positive or both negative).
- Charged particles in motion exert magnetic forces on each other depending on their direction of motion and their relative speed. You will learn much more about magnetic fields in Chapter 20, Magnetic fields.

You can represent a field by lines of force and/or vectors that represent the magnitude and direction of the force on a suitable small object in the field.

A line of force (or field line) of:

- a gravitational field is a line along which a free point mass would move
- an electric field is a line along which a free positive charge would move.

You will see a more detailed comparison of electric fields and gravitational fields in Topic 18.6, Comparing electric fields and gravitational fields.

Summary questions

$e = 1.6 \times 10^{-19}$ C

1 Explain each of the following observations in terms of transfer of electrons:

  a An insulated metal can is given a positive charge by touching it with a positively charged rod.

  b A negatively charged metal sphere suspended on a thread is discharged by connecting it to the ground using a wire.

2 a In the shuttling ball experiment, explain why the ball shuttles faster if:

  i the potential difference between the plates is increased

  ii the plates are brought closer together.

  b A ball shuttles between two oppositely charged metal plates at a frequency of 2.5 Hz. The ball carries a charge of 30 nC each time it shuttles from one plate to the other. Calculate:

  i the average current in the circuit

  ii the number of electrons transferred each time the ball makes contact with a metal plate.

3 An insulated metal conductor is earthed before a negatively charged object is brought near to it.

  a Explain why the free electrons in the conductor move as far away from the charged object as they can.

  b The conductor is then briefly earthed. The charged object is then removed from the vicinity of the conductor. Explain why the conductor is left with an overall positive charge.

4 a A positively charged point object is placed near an earthed metal plate, as shown in Figure 7.

  i Explain why electrons gather at the surface of the metal plate near the object.

  ii Explain why there is a force of attraction between the object and the metal plate.

  b Sketch the pattern of the field lines of the electric field:

  i between two oppositely charged parallel plates

  ii between a positively charged point object and an earthed metal plate.

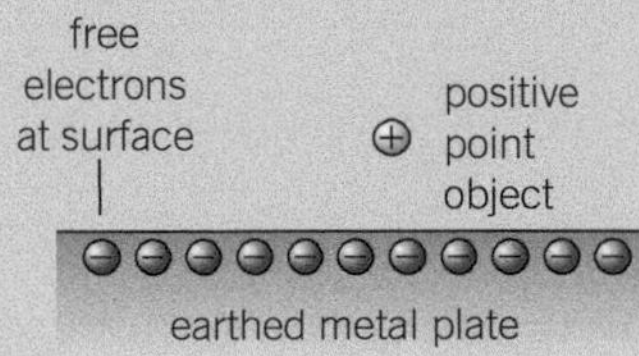

▲ **Figure 7**

18.2 Electric field strength

A charged object in an electric field experiences a force due to the field. Provided the object's size and charge are both sufficiently small, the object may be used as a 'test' charge to measure the strength of the field at any position in the field.

The electric field strength, E, at a point in the field is defined as the force per unit charge on a positive test charge placed at that point.

The unit of E is the newton per coulomb ($N\,C^{-1}$).

If a positive test charge Q at a certain point in an electric field is acted on by force F due to the electric field, the electric field strength E at that point is given by the equation

$$E = \frac{F}{Q}$$

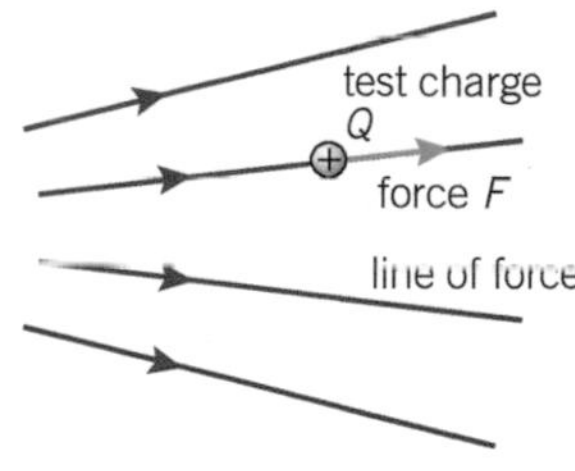

▲ **Figure 1** *Electric field strength*

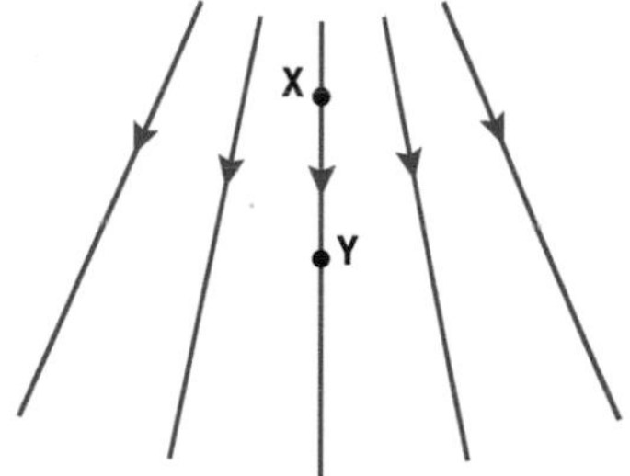

▲ **Figure 2** *A non-uniform field*

Worked example

$e = 1.6 \times 10^{-19}\,C$

Figure 2 shows the field lines of an electric field.

a Calculate the magnitude of electric field strength at X if a $+3.5\,\mu C$ test charge at X experiences a force of 70 mN.

b At a second position Y in the field, the electric field strength is $15\,000\,N\,C^{-1}$ in a direction downwards. Calculate the force on an electron at Y and state the direction of the force.

Solution

a $F = 70 \times 10^{-3}\,N$, $Q = 3.5 \times 10^{-6}\,C$

$$E = \frac{F}{Q} = \frac{70 \times 10^{-3}}{3.5 \times 10^{-6}} = 2.0 \times 10^{4}\,N\,C^{-1}$$

b $F = QE = 1.6 \times 10^{-19} \times 15\,000 = 2.4 \times 10^{-15}\,N$

The direction of the force on an electron at Y is directly upwards because the field line at Y is directly downwards and the charge on an electron is negative.

Learning objectives:

→ Describe how to measure, in principle, the strength of an electric field.

→ Discuss whether electric field strength E is a scalar or a vector, and describe how this affects the sign of a test charge that you should use.

→ Explain why E should be described as the force per unit charge instead of the force that acts on one coulomb of charge.

Specification reference: 3.8.2

Hint

1 Rearranging the equation gives $F = QE$ for the force F on a test charge Q at a point in the electric field, where the electric field strength is E.

2 Electric field strength is a **vector** in the same direction as the force on a positive test charge. In other words, the direction of a field line at any point is the direction of the electric field strength at that point. The force on a small charge in an electric field is:

- in the same direction as the electric field if the charge is positive
- in the opposite direction to the electric field if the charge is negative.

3 The charge of a test charge needs to be very much less than one coulomb because this amount of charge would affect the charges that cause the field, and so it would alter the electric field strength.

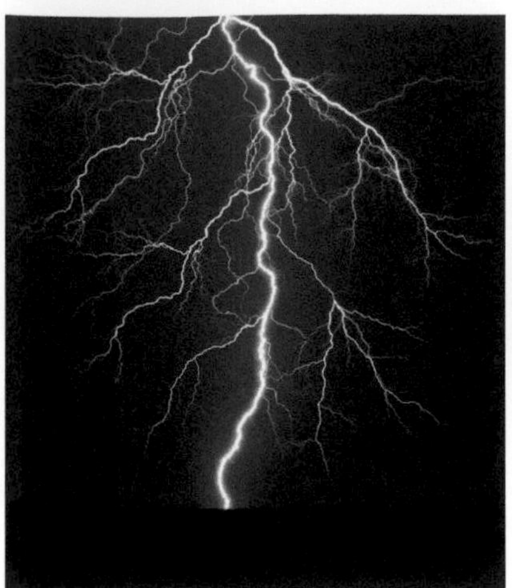

▲ **Figure 3** *Lightning*

Extension

The lightning conductor

Air is an insulator provided it is not subjected to an electric field that is too strong. Such a field ionises the air molecules by pulling electrons out of the molecules. In a thunderstorm, a lightning strike to the ground occurs when a cloud becomes more and more charged and the electric field in the air becomes stronger and stronger. The insulating property of air suddenly breaks down and a massive discharge of electric charge occurs between the cloud and the ground. A lightning conductor is a metal rod at the top of a tall building. The rod is connected to the ground by means of a thick metal conductor. When a charged cloud is overhead, it creates a very strong electric field near the tip of the conductor. Air molecules near the tip are ionised by this very strong field. The ions discharge the thundercloud making a lightning strike less likely.

Q: A lightning strike transfers 30 000 C through a p.d. of 50 000 V. Estimate how much energy is released.

Answer: 1500 MJ.

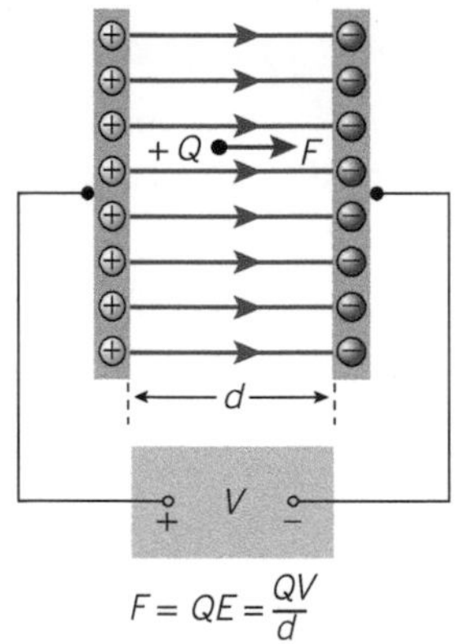

▲ **Figure 4** *The electric field strength between two parallel plates*

The electric field between two parallel plates

Figure 6c in Topic 18.1 shows that the field lines between two oppositely charged flat conductors are parallel to each other and at right angles to the plates. The field pattern for two oppositely charged flat plates is similar, as shown in Figure 4. The field lines are:

- parallel to each other
- at right angles to the plates
- from the positive plate to the negative plate.

The field between the plates is uniform. This is because the electric field strength has the same magnitude and direction everywhere between the plates. The electric field strength E can be calculated from the potential difference V between the plates and their separation d using the equation

$$E = \frac{V}{d}$$

To prove this equation, consider a small charge Q between the plates, as in Figure 4.

1. The force F on a small charge Q in the field is given by $F = QE$, where E is the electric field strength between the plates.
2. If the charge is moved from the positive to the negative plate, the work done W by the field on Q is given by W = force F × distance moved d, so $W = QEd$.
3. By definition, the potential difference between the plates, V, is the work done per unit charge when a small charge is moved through potential difference V.

   Therefore, $V = \frac{W}{Q} = \frac{QEd}{Q} = Ed$

   Rearranging $V = Ed$ gives $E = \frac{V}{d}$

Hint

The unit of E can be written either as the newton per coulomb ($N\,C^{-1}$) or as the volt per metre ($V\,m^{-1}$).

The link between the two can be seen by combining $F = QE$ and $E = \frac{V}{d}$ to give $F = \frac{QV}{d}$.

Rearranging this equation gives

$$\frac{F}{Q} = \frac{V}{d}\ (= E)$$

Therefore, the newton per coulomb and the volt per metre are both acceptable as the unit of E.

Worked example

A pair of parallel plates at a separation of 80 mm are connected to a high-voltage supply unit which maintains a constant p.d. of 6000 V between the plates. Calculate:

a the electric field strength between the plates

b the magnitude and direction of the electrostatic force on an ion of charge 4.8×10^{-19} C when it is between the plates.

Solution

a $E = \frac{V}{d} = \frac{6000}{80 \times 10^{-3}} = 7.5 \times 10^{4}\,\text{V m}^{-1}$

b $F = QE = 4.8 \times 10^{-19} \times 7.5 \times 10^{4} = 3.6 \times 10^{-14}\,\text{N}$
The force on the ion is directly towards the negative plate.

Synoptic link

E is the same everywhere in a uniform electric field, and *g* is the same everywhere in a uniform gravitational field. See Topic 17.3, Gravitational potential.

A beam of charged particles can be deflected by an electric field, as shown in Figure 5. By adjusting the strength of the field, the extent of the deflection can be controlled. The electric field used to deflect the beam is produced by applying a constant p.d. between the metal deflecting plates A and B.

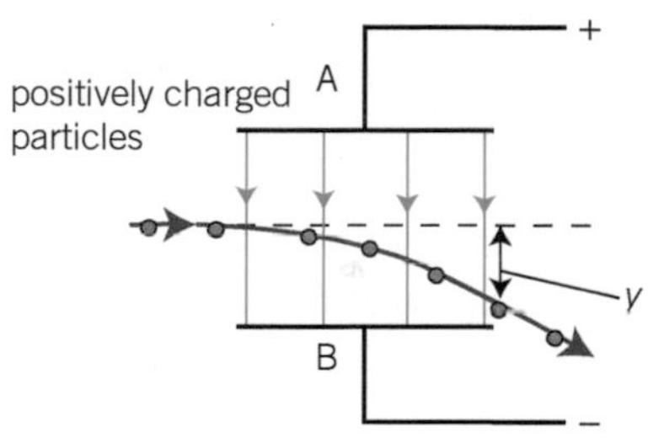

▲ **Figure 5** *Deflection of a beam of charged particles in an electric field*

The beam in Figure 5 consists of identical positively charged particles which are deflected towards the negative plate. If the particles had been negatively charged, they would have been deflected towards the positive plate. The force on each charged particle in the field is constant in magnitude and direction because the field is uniform so the beam curves in a parabolic path (just as a projectile projected horizontally does). The screen in the tube enables the path of the beam to be observed and measured.

If the p.d. between plates A and B is V_P, the force F on each charged particle is given by the equation

$$F = qE = \frac{qV_P}{d}$$

where d is the perpendicular distance between plates A and B, q is the charge of each charge particle, and $E = \left(\frac{V_P}{d}\right)$ is the electric field strength between the plates.

Study tip

Field lines should have arrows (+ to –) and, when not straight, they are smooth curves.

Notes:

1 The acceleration of each charged particle towards the negative plate $a = \frac{F}{m} = \frac{qV_P}{md}$, where m is the mass of the particle.

2 The time taken, t, by each charged particle to cross the field $= \frac{L}{v}$, where L is the length of each plate and v is the initial speed of the particle on entry to the field. This is the horizontal component of the particle's velocity in the field as it has no horizontal acceleration.

3 The deflection, y, of the charged particle on leaving the field is given by the equation $y = \frac{1}{2}at^2$. Using the above equations, it can be shown that y is directly proportional to the plate p.d., V_P.

Field factors 1

An electric field exists near any charged body. The greater the charge on the body, the stronger the electric field. For a charged metal conductor, the charge on it is spread across its surface. The more concentrated the charge on the surface, the greater the strength of the electric field above the surface.

Figure 6 shows the electric field pattern between a V-shaped conductor opposite a flat plate when a constant p.d. is applied between the plate and the conductor. The field lines are concentrated at the tip of the V because that is where charge on the V-shaped conductor is most concentrated.

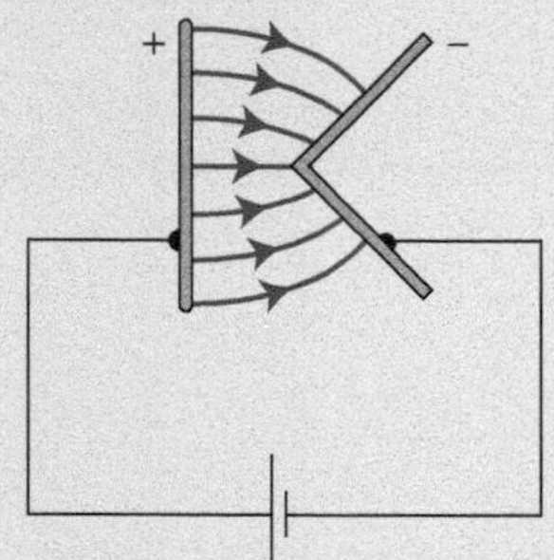

▲ **Figure 6** *The electric field near a metal tip*

Summary questions

$e = 1.6 \times 10^{-19}$ C

1 A +40 nC point charge Q_1 is placed in an electric field.

a Calculate the magnitude of the force on Q_1 if the electric field strength where Q_1 is placed is $3.5 \times 10^4\,\mathrm{V\,m^{-1}}$.

b Q_1 is moved to a different position in the electric field. The force on Q_1 at this position is 1.6×10^{-3} N. Calculate the magnitude of the electric field strength at this position.

2 Figure 7 shows the path of a charged dust particle in an electric field.

a The electric field strength at X is $65\,\mathrm{kV\,m^{-1}}$. The force due to the field on the particle when it is at X is 8.2×10^{-3} N towards the metal surface.

i Describe the type of charge carried by the particle.

ii Calculate the charge carried by the particle.

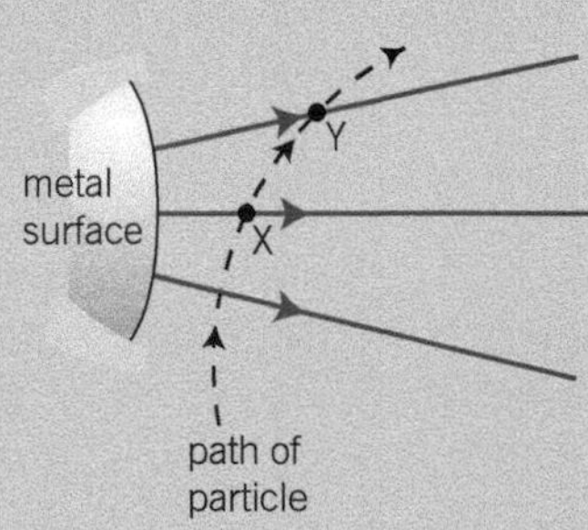

▲ **Figure 7**

b i Calculate the magnitude of the force on the particle when it is at Y, where the electric field strength is $58\,\mathrm{kV\,m^{-1}}$.

ii State the direction of the force on the particle when it is at Y.

3 A high-voltage supply unit is connected across a pair of parallel plates which are at a separation of 50 mm.

a The voltage is adjusted to 4.5 kV. Calculate:

i the electric field strength between the plates

ii the electrostatic force on a droplet in the field that carries a charge of 8.0×10^{-19} C.

b The separation between the plates is altered without changing the p.d. between the plates. The droplet in **a** is now acted on by a force of 4.5×10^{-14} N. Calculate the new separation between the plates.

4 A certain gas in a tube is subjected to an electric field of increasing strength. The gas becomes conducting when the electric field reaches a strength of $35\,\mathrm{kV\,m^{-1}}$.

a The electrodes in the tube are at a spacing of 84 mm. Assuming the field between the electrodes is uniform before the gas conducts, calculate the potential difference between the electrodes that is necessary to produce an electric field of strength $35\,\mathrm{kV\,m^{-1}}$ in the tube.

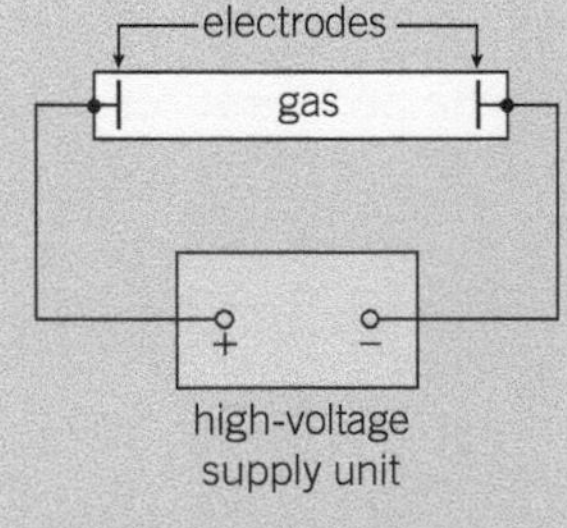

▲ **Figure 8**

b i Calculate the force on an electron in the tube when the electric field strength is $35\,\mathrm{kV\,m^{-1}}$.

ii Explain why the gas becomes conducting only when the electric field strength in the tube reaches a certain value.

18.3 Electric potential

The Van de Graaff generator

A Van de Graaff generator can easily produce sparks in air several centimetres in length. Figure 1 shows how a Van de Graaff generator works. Charge created when the rubber belt rubs against a pad is carried by the belt up to the metal dome of the generator. As charge gathers on the dome, the potential difference between the dome and Earth increases until sparking occurs.

A spark suddenly transfers energy from the dome. Work must be done to charge the dome because a force is needed to move the charge on the belt up to the dome. So the electric potential energy of the dome increases as it charges up. Some or all of this energy is transferred from the dome when a spark is created.

In general, work must be done to move a charged object X towards another object Y that has the same type of charge. Their electric potential energy increases as X moves towards Y.

The electric potential energy of X increases from zero if it is moved from infinity towards Y. The electric field of Y causes a force of repulsion to act on X. This force must be overcome to move X closer to Y.

The electric potential at a certain position in any electric field is defined as the work done per unit positive charge on a positive test charge (i.e., a small positively charged object) when it is moved from infinity to that position. By definition, the position of zero potential energy is infinity. Thus the electric potential at a certain position is the potential energy per unit positive charge of a positive test charge at that position.

The unit of electric potential is the volt (V), equal to $1\,\mathrm{J\,C^{-1}}$.

For a positive test charge Q placed at a position in an electric field where its electric potential energy is E_P, the electric potential V at this position is given by

$$V = \frac{E_P}{Q}$$

Note that rearranging this equation gives $E_P = QV$.

Suppose a $+1\,\mu\mathrm{C}$ test charge moves in an electric field from infinity to reach a certain position P, where the electric potential is $+1000\,\mathrm{V}$. The electric potential energy of the test charge at P is therefore $+1.0 \times 10^{-3}\,\mathrm{J}$ ($= QV = 1.0 \times 10^{-6}\,\mathrm{C} \times +1000\,\mathrm{V}$). In other words, $1.0 \times 10^{-3}\,\mathrm{J}$ of work is done on the test charge when it moves from infinity to P.

Note:

If a test charge $+Q$ is moved in an electric field from one position where the electric potential is V_1 to another where the electric potential is V_2, then the work done ΔW on it is given by $\Delta W = Q(V_2 - V_1)$.

Learning objectives:

- → Explain why potential is defined in terms of the work done per unit + positive charge.
- → Calculate the electric potential difference between two points.
- → Describe how to find the change in electric potential energy from p.d.
- → Explain why potential (and p.d.) is measured in V.

Specification reference: 3.8.3

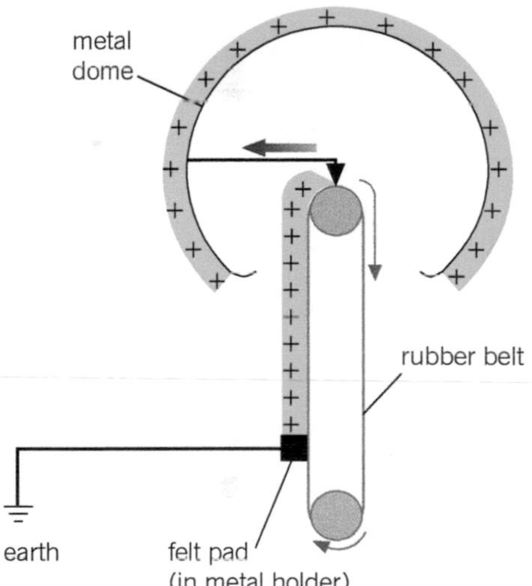

▲ **Figure 1** *The Van de Graaff generator*

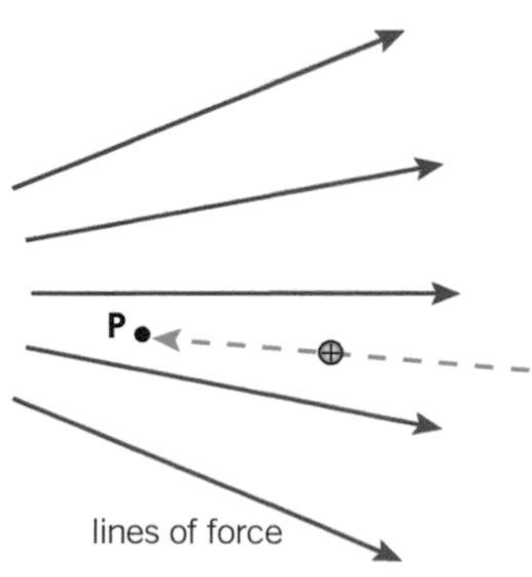

▲ **Figure 2**

Synoptic link

Equipotentials in an electric field are like map contours that are lines of constant height and are therefore lines of gravitational equipotential. See Topic 17.3, Gravitational potential.

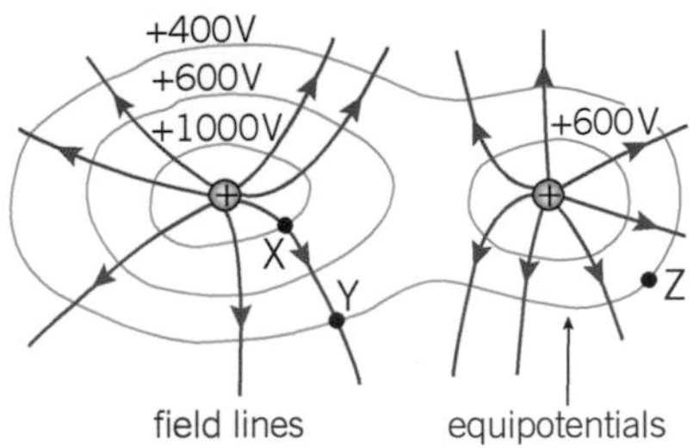

▲ **Figure 3** *Equipotentials*

Study tip

Remember that V is a scalar. Equipotentials always meet field lines at right angles.

Potential gradients

Equipotentials are surfaces of constant potential. A test charge moving along a line on an equipotential surface has constant potential energy. No work is done by the electric field on the test charge because the force due to the field is at right angles to the equipotential. In other words, the lines of force of the electric field cross an equipotential surface at right angles (ie. along the normal).

The equipotentials for an electric field are like equipotentials for a gravitational field. Both are lines of constant potential energy for the appropriate test object, in one case a test charge, and in the other case a test mass.

Figure 3 shows the equipotentials of the electric field due to two positively charged objects.

Suppose a +2.0 μC test charge is moved from X to Y.

The potential at X, V_X, is +1000 V, so the test charge at X has potential energy equal to $+2.0 \times 10^{-3}$ J ($= QV_X = +2.0 \times 10^{-6}\,\text{C} \times +1000\,\text{V}$).

The potential at Y, V_Y, is +400 V, so the test charge at Y has potential energy equal to $+8.0 \times 10^{-4}$ J ($= QV_Y = +2.0 \times 10^{-6}\,\text{C} \times +400\,\text{V}$).

Therefore, moving the test charge from X to Y lowers its potential energy by 1.2×10^{-3} J.

Note that if the test charge is moved from Y to Z along the +400 V equipotential, its potential energy stays constant at $+8.0 \times 10^{-4}$ J.

The **potential gradient** at any position in an electric field is the change of potential per unit change of distance in a given direction.

1 If the field is non-uniform, as in Figure 3, the potential gradient varies according to position and direction. The closer the equipotentials are, the greater the potential gradient is at right angles to the equipotentials.
2 If the field is uniform, such as the field between the two oppositely charged parallel plates shown in Figure 4, the equipotentials *between the plates* are equally spaced lines parallel to the plates. Figure 4 also shows how the potential relative to the negative plate changes with perpendicular distance x from the negative plate.

The graph shows that the potential relative to the negative plate is proportional to distance x. In other words, the potential gradient is:

- constant
- such that the potential increases in the opposite direction to the electric field
- equal to $\frac{V}{d}$.

As you learnt in Topic 18.2, the electric field strength E between the plates is equal to $\frac{V}{d}$ and is directed from the + to the − plate. In other words,

the electric field strength is equal to the negative of the potential gradient.

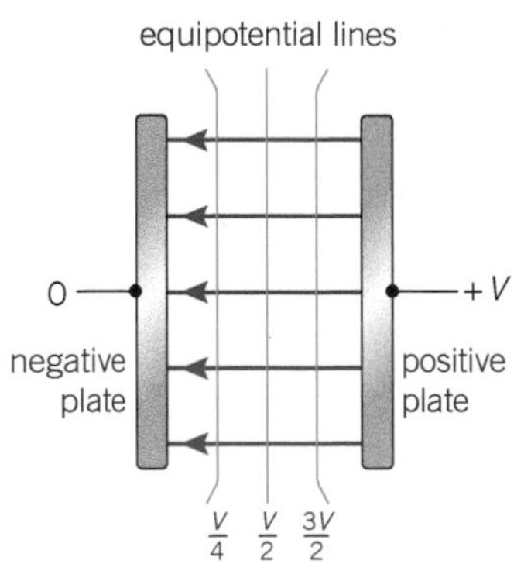

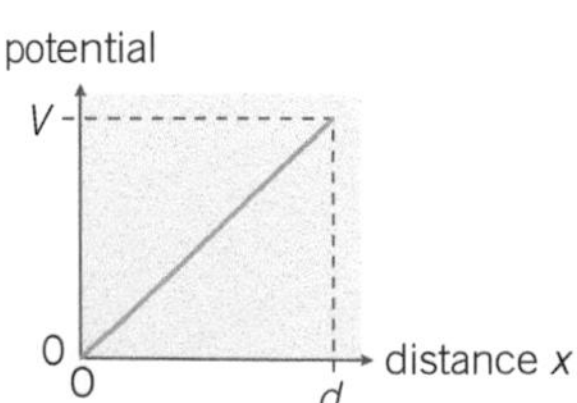

▲ **Figure 4** *A uniform potential gradient*

If the field is non-uniform, this statement still holds. However, the potential gradient at any position in an electric field is given by $\frac{\Delta V}{\Delta x}$ (instead of $\frac{V}{d}$), where ΔV is the change of potential between two closely spaced points at distance Δx apart.

Because the potential gradient is in the opposite direction to the lines of force of the electric field, then, in general,

$$\textbf{electric field strength } E = -\frac{\Delta V}{\Delta x}$$

Summary questions

$e = 1.6 \times 10^{-19}\,\text{C}$

1 An electron in a beam is accelerated from a potential of −50 V to a potential of +450 V. Calculate:

a the potential energy of the electron at

i −50 V

ii +450 V

b the change of potential energy of the electron.

2 In Figure 3, a test charge q is moved from X to Z. Calculate the change of potential energy of the test charge

a if $q = +3.0\,\mu\text{C}$

b if $q = -2.0\,\mu\text{C}$.

3 An oil droplet carrying a charge of $+2e$ is in air between two parallel metal plates separated by a distance of 20 mm. The p.d. between the plates is 5.0 V.

a Calculate:

i the potential gradient between the two plates

ii the force on the droplet.

b Calculate the change of electric potential energy of the oil droplet if it moves from the midpoint of the plates to the negative plate.

4 a Define electric potential and state its unit.

b Two parallel horizontal metal plates are placed one above the other at a separation of 20 mm. A potential difference of +60 V is applied between the plates with the top plate positive.

i Calculate the electric field strength between the plates.

ii Sketch a graph to show how the electric potential V between the plates varies with height h above the lower plate.

Field factors 2

The electric field between two oppositely charged parallel plates depends on the concentration of charge on the surface of the plates. The charge on each plate is spread evenly across the surface of the plate facing the other plate. Measurements show that the electric field strength between the plates is proportional to the charge per unit area on the facing surfaces.

Therefore, for charge Q on a plate of surface area A, the electric field strength E between the plates is proportional to $\frac{Q}{A}$.

Introducing a constant of proportionality, ε_0 (called epsilon nought), into this equation gives $\frac{Q}{A} = \varepsilon_0 E$. The value of ε_0 is 8.85×10^{-12} farads per metre (F m^{-1}), where the farad (the unit of capacitance – see Topic 19.1) is 1 coulomb per volt. It is called the permittivity of free space. It represents that charge per unit area on a surface in a vacuum that produces an electric field strength of one volt per metre between the plates. You *don't* need to know the equation $\frac{Q}{A} = \varepsilon_0 E$, but you *do* need to know about ε_0, which you will meet in Topic 18.4.

18.4 Coulomb's law

Learning objectives:

- → Describe how the force between two point charges depends on distance.
- → Calculate the force between two charged objects.
- → Explain what the sign of the force (+ or −) indicates.

Specification reference: 3.8.1

Coulomb's experiment

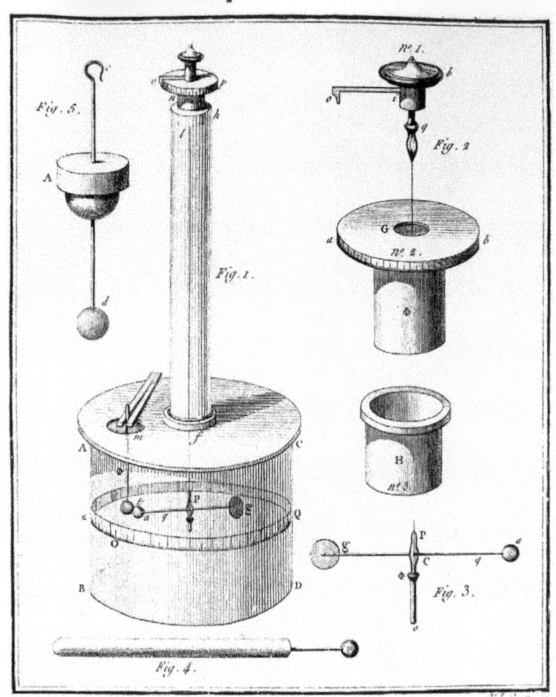

▲ **Figure 1** *Coulomb's torsion balance*

Like charges repel, and unlike charges attract. The force between two charged objects depends on how close they are to each other. The exact link was first established by Charles Coulomb in France in 1784. He devised a very sensitive torsion balance to measure the force between charged pith balls. Figure 1 shows the arrangement. A needle with a ball made of pith (a substance obtained from plants) at one end and a counterweight at the other end was suspended horizontally by a vertical wire. Another pith ball on the end of a thin vertical rod could be placed in contact with the first ball.

The pith balls were small enough to be considered as point objects. The ball on the rod was charged and then placed in contact with the other ball on the needle. The contact between them charged the second ball, which was then repelled by the ball fixed on the rod. This caused the wire to twist until the electrical repulsion was balanced by the twist built up in the wire. By turning the torsion head at the top of the wire, the distance between the two balls could be set at any required value. The amount of turning needed to achieve that distance gave the force. Some of Coulomb's many measurements are in Table 1.

The measurements fit the link that the force F is proportional to $\frac{1}{r^2}$. All of the other measurements made by Coulomb fitted the same link. Because the force is also proportional to the charge on each ball, Coulomb deduced the following equation, called **Coulomb's law of forces**, for the force F between two point charges Q_1 and Q_2

$$F = \frac{kQ_1Q_2}{r^2}$$

where r is their distance apart.

Note

▼ **Table 1** *Some of Coulomb's results*

Distance r	36	18	$8\frac{1}{2}$
Force F	36	144	567

Measurements for both variables were actually made in degrees, so the above values are in relative units. Can you make out a pattern for these measurements? Halving the distance from 36 to 18 makes the force increase by a factor of 4. Halving the distance from 18 to $8\frac{1}{2}$ (near enough 9) increases the force again by a factor of about 4.

The constant of proportionality, k, can be shown to be equal to $\frac{1}{4\pi\varepsilon_0}$, where ε_0 is the permittivity of free space. You first encountered this in Topic 18.3.

Coulomb's law is therefore written as

$$F = \frac{1}{4\pi\varepsilon_0}\frac{Q_1Q_2}{r^2}$$

where r = distance between the two point charges Q_1 and Q_2.

As you saw in Topic 18.3, $\varepsilon_0 = 8.85 \times 10^{-12}\,\mathrm{F\,m^{-1}}$

so $\frac{1}{4\pi\varepsilon_0} = 9.0 \times 10^{9}\,\mathrm{m\,F^{-1}}$

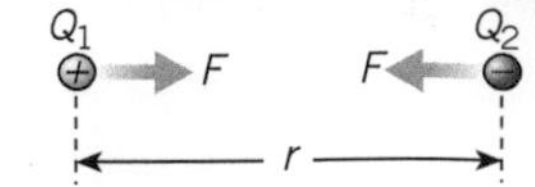

a *Unlike charges attract*

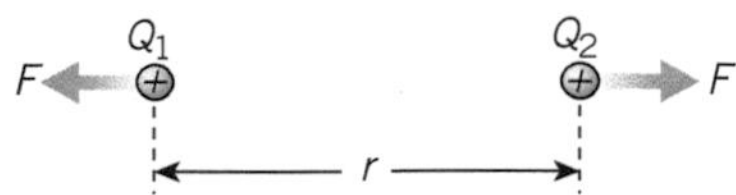

b *Like charges repel*

$$F = \frac{1}{4\pi\varepsilon_0}\frac{Q_1Q_2}{r^2}$$

▲ **Figure 2** *Coulomb's law*

Worked example

$e = 1.60 \times 10^{-19}\,\mathrm{C}$, $\varepsilon_0 = 8.85 \times 10^{-12}\,\mathrm{F\,m^{-1}}$

Calculate the magnitude of the force between a proton and an electron at a separation of $3.00 \times 10^{-10}\,\mathrm{m}$.

Solution

$$F = \frac{1}{4\pi\varepsilon_0}\frac{Q_1Q_2}{r^2} = \frac{1.60 \times 10^{-19} \times 1.60 \times 10^{-19}}{4\pi \times 8.85 \times 10^{-12} \times (3.00 \times 10^{-10})^2} = 2.56 \times 10^{-9}\,\mathrm{N}$$

Synoptic link

Coulomb's law of force between point charges and Newton's law of gravitation are both inverse-square laws. For example, doubling the separation of two point masses or two point charges causes the force to reduce to a quarter. See Topic 17.1, Newton's law of gravitation.

Extension

More on $k = \frac{1}{4\pi\varepsilon_0}$

In Topic 18.3, ε_0 was introduced as the constant of proportionality in the equation $\frac{Q}{A} = \varepsilon_0 E$ linking the electric field strength E at the surface of a flat conductor where the charge Q is evenly distributed over a surface area A.

As you will see in Topic 18.5, if Coulomb's law is applied to the force on a test charge q at distance r from a point charge Q, the force on the test charge $F = \frac{kQq}{r^2}$, so the electric field strength at distance r is given by

$$E = \frac{F}{Q} = \frac{kQ}{r^2}$$

By introducing $\frac{1}{4\pi\varepsilon_0}$ as k, the equation $E = \frac{kQ}{r^2}$ may be written as

$$\frac{Q}{4\pi r^2} = \varepsilon_0 E \text{ or } \frac{Q}{A} = \varepsilon_0 E$$

where $A = 4\pi r^2$ = the surface area of a sphere of radius r.

So, the general equation $\frac{Q}{A} = \varepsilon_0 E$ gives the surface charge density $\frac{Q}{A}$ needed on the surface of a conducting sphere in air to produce an electric field of strength E at the surface.

Q: Estimate the charge per unit area $\left(\frac{Q}{A}\right)$ on a surface at which the electric field strength is $1\,\mathrm{V\,m^{-1}}$.

Answer: $8.85 \times 10^{-12}\,\mathrm{C\,m^{-2}}$.

a

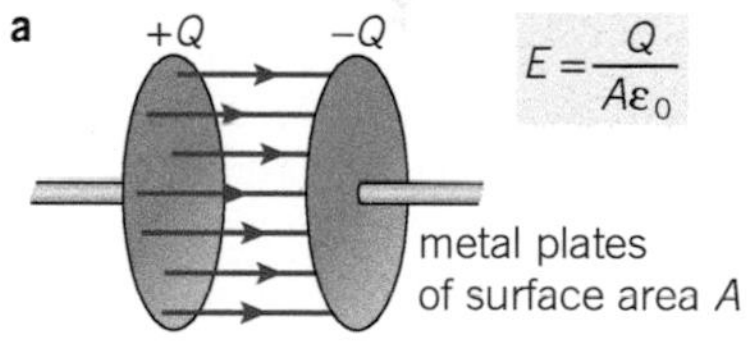

b

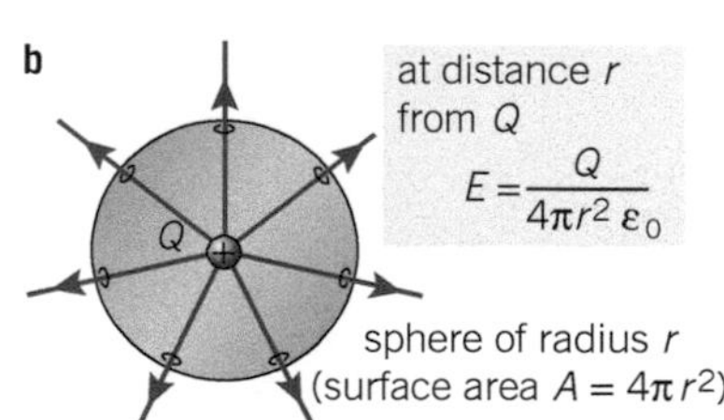

▲ **Figure 3** *Comparison of surface electric field strengths*

Synoptic link

You can use Newton's law of gravitation to compare the electrostatic force and the gravitational force between a proton (mass = 1.67×10^{-27} kg) and an electron (mass = 9.11×10^{-31} kg) at the same distance. If you need to, see Topic 17.4, Planetary fields. You should find that the electrostatic force is almost 2.3×10^{39} times stronger than the gravitational force.

Study tip

Remember to square r when calculating F.

When substituting charges, get the powers of 10 correct: $\mu = 10^{-6}$, $\text{n} = 10^{-9}$, $\text{p} = 10^{-12}$.

You can quickly check the solution given in the example on the previous page by approximating the calculation

$$\frac{1.60 \times 10^{-19} \times 1.60 \times 10^{-19}}{4\pi \times 8.85 \times 10^{-12} \times (3.0 \times 10^{-10})^2}$$

to $\dfrac{2 \times 10^{-38}}{100 \times 10^{-31}}$ to give 2×10^{-9} N.

This shows that there are no errors to the power of 10 in the calculation.

Application

Why do salt crystals dissolve in water?

A salt crystal is an ionic crystal. The sodium ions and the chlorine ions in a salt crystal are oppositely charged and the electrostatic forces between them hold them together. Salt crystals dissolve in water because the water weakens the electrostatic forces between the ions at the surface of the crystal. So the ions break free from the surface, and the crystal gradually dissolves. In fact, the force between two ions in water is about 80 times weaker than the force would be in air for the same distance apart. Note that Coulomb's law as stated above applies strictly to a vacuum. In terms of the force between two point charges in air, the force is effectively the same as it would be in a vacuum.

Summary questions

$\varepsilon_0 = 8.85 \times 10^{-12}\,\text{F m}^{-1}$

$e = 1.6 \times 10^{-19}$ C

1 Calculate the force between an electron and

  a a proton at a distance of 2.5×10^{-9} m

  b a nucleus of a nitrogen atom (charge $+7e$) at a distance of 2.5×10^{-9} m.

2 a Two point charges $Q_1 = +6.3$ nC and $Q_2 = -2.7$ nC exert a force of 3.2×10^{-5} N on each other when they are at a certain distance, d, apart. Calculate:

  i the distance d between the two charges

  ii the force between the two charges if they are moved to distance $3d$ apart.

  b A charge of +4.0 nC is added to each charge in part **a**. Calculate the force between Q_1 and Q_2 when they are at separation d.

3 A +30 nC point charge is at a fixed distance of 6.2 mm from a point charge Q. The charges attract each other with a force of 4.3×10^{-2} N.

  a Calculate the magnitude of charge Q and state whether Q is a positive or a negative charge.

  b The two charges are moved 2.5 mm further apart. Calculate the force between them in this new position.

4 Two point objects, X and Y, carry equal and opposite amounts of charge at a fixed separation of 3.6×10^{-2} m. The two objects exert a force on each other of 5.1×10^{-5} N.

  a Calculate the magnitude, Q, of each charge, and state whether the charges attract or repel each other.

  b The charge of each object is increased by adding a positive charge of $+2Q$ to each object. Calculate the separation at which the two objects would exert a force of 5.1×10^{-5} N on each other, and state whether the objects attract or repel each other.

18.5 Point charges

A point charge is a convenient expression for a charged object in a situation where distances under consideration are much greater than the size of the object. The same idea applies to a distant star which is considered as a point object because its diameter is much smaller than the distance to it from the Earth. A test charge in an electric field is a point charge that does not alter the electric field in which it is placed. Such an alteration would happen if an object with a sufficiently large charge is placed in an electric field and it causes a change in the distribution of charge that creates the field.

Consider the electric field due to a point charge $+Q$, as shown in Figure 1. The field lines radiate from Q because a test charge $+q$ in the field would experience a force directly away from Q wherever the test charge was placed. Coulomb's law gives the force F on the test charge q at distance r from Q as

$$F = \frac{1}{4\pi\varepsilon_0}\frac{Qq}{r^2}$$

Therefore, because, by definition, electric field strength $E = \frac{F}{q}$, the electric field strength at distance r from Q is given by

$$E = \frac{Q}{4\pi\varepsilon_0 r^2}$$

Note that if Q is negative, the above equation gives a negative value of E corresponding to the field lines pointing inwards towards Q.

Learning objectives:

→ State the equation that gives the electric field strength near a point charge.

→ State the equation that gives the potential associated with a point charge.

→ Explain why E is equal to zero inside a charged sphere.

Specification reference: 3.8.2 and 3.8.3

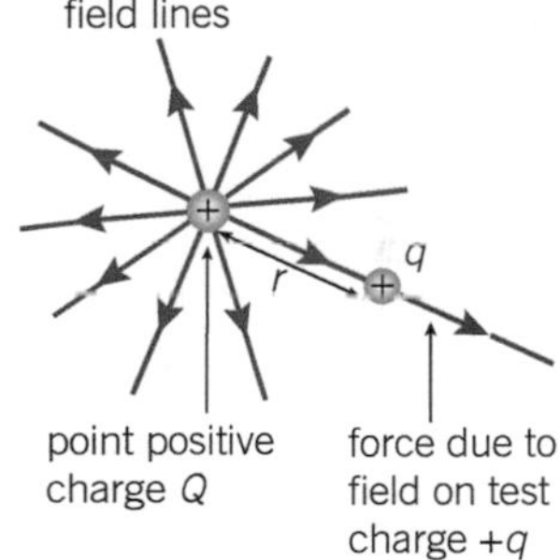

▲ **Figure 1** *Force near a point charge Q*

Worked example

$\varepsilon_0 = 8.85 \times 10^{-12}\,\text{F m}^{-1}$, $e = 1.6 \times 10^{-19}\,\text{C}$

Calculate the electric field strength due to a nucleus of charge $+82e$ at a distance of 0.35 nm.

Solution

$$E = \frac{Q}{4\pi\varepsilon_0 r^2} = \frac{+82 \times 1.6 \times 10^{-19}}{4\pi \times 8.85 \times 10^{-12} \times (0.35 \times 10^{-9})^2} = 9.6 \times 10^{11}\,\text{V m}^{-1}$$

Electric field strength as a vector

If a test charge is in an electric field due to several point charges, each charge exerts a force on the test charge. The resultant force per unit charge $\frac{F}{q}$ on the test charge gives the resultant electric field strength at the position of the test charge. Consider the following situations:

- **Forces in the same direction:** Figure 2a shows a test charge $+q$ on the line between a negative point charge Q_1 and a positive point charge Q_2. The test charge experiences a force $F_1 = qE_1$, where E_1 is the electric field strength due to Q_1, and a force $F_2 = qE_2$, where E_2 is the electric field strength due to Q_2. The two forces act in the same direction because Q_1 attracts q and Q_2 repels q. So the resultant force $F = F_1 + F_2 = qE_1 + qE_2$.

  Therefore, the resultant electric field strength

  $$E = \frac{F}{q} = \frac{qE_1 + qE_2}{q} = E_1 + E_2$$

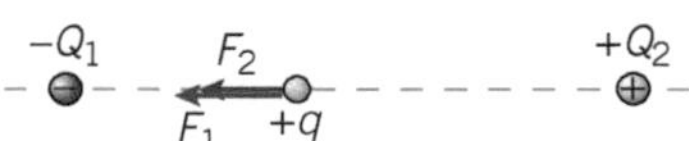

a *Forces in same direction*

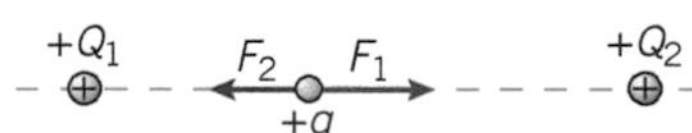

b *Forces in opposite direction*

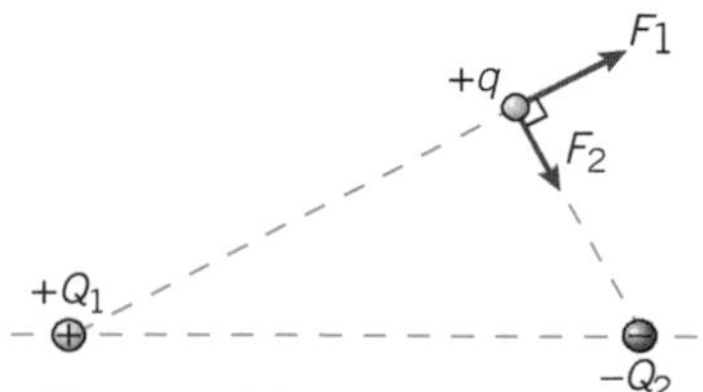

c *Forces at right angles*

▲ **Figure 2** *Combined electric fields*

Synoptic link

Remember how to calculate the resultant force of two perpendicular forces. See Topic 1.1, Vectors and scalars.

- **Forces in opposite directions:** Figure 2b shows a test charge $+q$ on the line between two positive point charges Q_1 and Q_2. The forces on the test charge are the same in magnitude as in Figure 2a but opposite in direction because Q_1 repels q and Q_2 repels q. So the resultant force $F = F_1 - F_2 = qE_1 - qE_2$.

  Therefore, the resultant electric field strength

$$E = \frac{F}{q} = \frac{qE_1 - qE_2}{q} = E_1 - E_2$$

- **Forces at right angles to each other:** Figure 2c shows a test charge $+q$ on perpendicular lines from two positive point charges Q_1 and Q_2. The forces on the test charge are smaller in magnitude than in Figures 2a and 2b (because the distances to Q_1 and Q_2 are larger) and are perpendicular to each other. The magnitude of the resultant force F is given by Pythagoras's equation $F^2 = F_1^2 + F_2^2$.

  As the resultant electric field strength $E = \frac{F}{q}$, then $E^2 = E_1^2 + E_2^2$ can be used to calculate the resultant electric field strength.

In general, the resultant electric field strength is the vector sum of the individual electric field strengths.

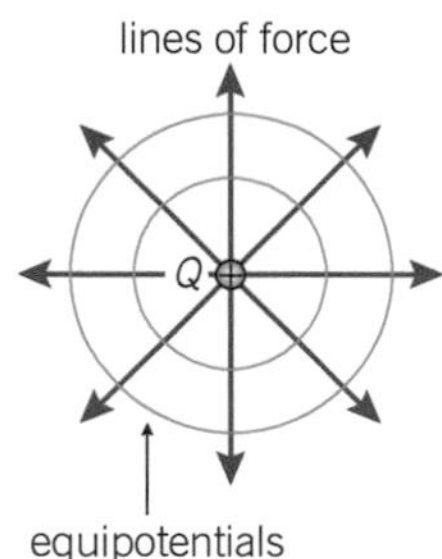

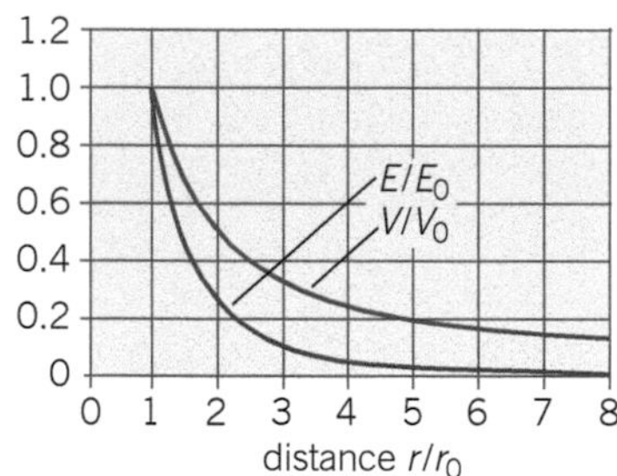

▲ **Figure 3** *The electric field and potential near a point charge*

More about radial fields

The electric field lines of force surrounding a point charge Q are radial. The equipotentials are therefore concentric circles centred on Q. For a charged sphere, you can say that the charge is at the centre of the sphere.

At distance r from Q, the electric field strength $E = \frac{Q}{4\pi\varepsilon_0 r^2}$

The equation was derived at the start of this topic. It shows that the electric field strength E is inversely proportional to the square of the distance r. Figure 3 shows how E varies with distance r from Q from a position that is distance r_0 from Q. The curve is an *inverse-square law* curve because E is proportional to $\frac{1}{r^2}$.

Notice that the equation is of the same form as the gravitational field strength equation $g = \frac{GM}{r^2}$ for the gravitational field strength at distance r from the centre of a spherical planet. See Topic 17.2 if you need to. The field strength equations are both inverse-square relationships because both the force between two point charges (Coulomb's law) and the force between two point masses (Newton's law of gravitation) vary with distance according to the inverse-square law.

The electric potential $V = \frac{Q}{4\pi\varepsilon_0 r}$ at distance r from Q.

Because both Coulomb's law $F = \frac{1}{4\pi\varepsilon_0}\frac{Q_1Q_2}{r^2}$ and Newton's law $F = \frac{Gm_1m_2}{r^2}$ are inverse-square relationships, the forces vary with distance in the same way. Therefore, the equation for electric potential near a point charge $V = \frac{Q}{4\pi\varepsilon_0 r}$ is of the same form as the gravitational potential near a point mass (or spherical mass). That is, $V = -\frac{GM}{r}$, which you derived in Topic 17.4. The equation shows that the electric potential V is inversely proportional to the distance r.

Study tip

A negative E indicates a field that acts towards a negative charge. But a negative V indicates a value less than zero: E is a vector, whilst V is a scalar. Note that E varies with distance more sharply than V does.

Figure 3 also shows how V varies with distance r from Q. The curve is *not* an inverse-square law curve, because V is proportional to $\frac{1}{r}$.

However, the gravitational potential in a gravitational field is always negative, because the force is always attractive. But the electric potential in the electric field near a point charge Q can be positive or negative according to whether Q is a positive or a negative charge.

The relationship between electric field strength and electric potential

1 Electric field strength = – the gradient of a potential against distance graph.
In Topic 18.3, you saw that at any position in an electric field, the electric field strength $E = -\frac{\Delta V}{\Delta x}$, where $\frac{\Delta V}{\Delta x}$ is the potential gradient at that position.

2 Change of potential = area under an electric field strength against distance graph.

Because electric field strength is the force per unit charge on a small positive test charge, a graph of electric field strength against distance shows how the force per unit charge on a positive test charge varies with distance. So the area under any section of the graph gives the work done per unit charge (i.e., change of potential) when a positive test charge is moved through the distance represented by that section.

Sparks and shocks

When a spherical metal conductor insulated from the ground is charged, the charge spreads out across the surface with the greatest concentration where the surface is most curved. This is why charge gathers at the tip of a lightning conductor when a charged cloud is overhead. As you saw in Topic 18.2, the electric field at the tip then becomes so strong that air molecules near the tip become ionised and the air conducts. This effect also explains why a fatal accident can happen if a conducting rod is held near an overhead high-voltage cable. An angler walking along a footpath near a railway line was electrocuted because the end of his carbon fibre fishing rod was inadvertently too close to the overhead cable along the track.

Summary questions

$\varepsilon_0 = 8.85 \times 10^{-12}\,\text{F m}^{-1}$

1 **a** Calculate the electric field strength at a distance of 3.2 mm from a +6.0 nC point charge.

b Calculate the distance from the point charge in **a** at which the electric field strength is $5.4 \times 10^5\,\text{V m}^{-1}$.

2 A +25 μC point charge Q_1 is at a distance of 60 mm from a +100 μC charge Q_2.

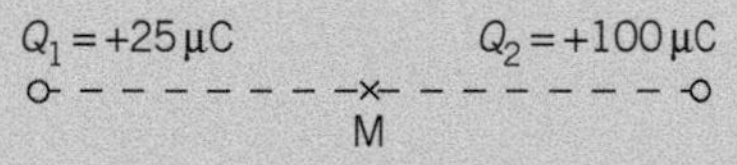

$Q_1 = +25\,\mu\text{C}$ $Q_2 = +100\,\mu\text{C}$
M

▲ **Figure 4**

a A +15 pC charge q is placed at M, 25 mm from Q_1 and 35 mm from Q_2. Calculate:

i the resultant electric field strength at M

ii the magnitude and direction of the force on q.

b Show that the electric field strength due to Q_1 and Q_2 is zero at the point which is 20 mm from Q_1 and 40 mm from Q_2.

3 A +15 μC point charge Q_1 is at a distance of 20 mm from a +10 μC charge Q_2.

a Calculate the resultant electric field strength:

i at M, the midpoint between the two charges

ii at the point P along the line between Q_1 and Q_2 which is 25 mm from Q_1 and 45 mm from Q_2.

b **i** Explain why there is a point along the line between the two charges at which the electric field strength is zero.

ii Calculate the distance from this point to Q_1 and to Q_2.

4 A +15 μC point charge Q_1 is at a distance of 30 mm from a –30 μC charge Q_2.

a Calculate the electric potential at the midpoint between the two charges.

b **i** Show that the electric potential is zero at a point between the two charges which is 10 mm from Q_1 and 20 mm from Q_2.

ii Calculate the electric field strength at this position and state its direction.

18.6 Comparing electric fields and gravitational fields

Learning objectives:

- → State which electrical quantity is analogous to mass.
- → State the main similarities between electric and gravitational fields.
- → State the principal differences between electric and gravitational fields.

Specification reference: 3.8.1 and 3.8.3

The similarities and differences between the two types of fields are listed in Table 1. In the mid 19th century, James Clerk Maxwell showed that electric and magnetic forces are different manifestations of the electromagnetic force. About two decades ago, physicists proved that the electromagnetic force and the nuclear force responsible for radioactive decay are different manifestations of a more fundamental force, the electroweak force. As yet, the gravitational force has still not been incorporated into this theoretical framework, despite repeated attempts by physicists to establish a unified theory. The fundamental nature of the gravitational force remains mysterious, even though we use it in everyday situations more than any other force.

Table 1 summarises the conceptual links between electric and gravitational fields.

▼ **Table 1** *Similarities and differences between gravitational and electric fields*

	Gravitational fields	Electrostatic fields
Similarities		
Line of force or a field line	path of a free test mass in the field	path of a free positive test charge in the field
Inverse-square law of force	Newton's law of gravitation $F = \frac{Gm_1m_2}{r^2}$	Coulomb's law of force $F = \frac{Q_1Q_2}{4\pi\varepsilon_0 r^2}$
Field strength	force per unit mass, $g = \frac{F}{m}$	force per unit + charge, $E = \frac{F}{q}$
Unit of field strength	N kg^{-1} or m s^{-2}	N C^{-1} or V m^{-1}
Uniform fields	g is the same everywhere, field lines are parallel and equally spaced	E is the same everywhere, field lines are parallel and equally spaced
Potential	gravitational potential energy per unit mass	electric potential energy per unit + charge
Unit of potential	J kg^{-1}	V (= J C^{-1})
Potential energy of two point masses or charges	$E_P = \frac{-Gm_1m_2}{r}$	$E_P = \frac{Q_1Q_2}{4\pi\varepsilon_0 r}$
Radial fields	due to a point mass or a uniform spherical mass M, $g = \frac{GM}{r^2}$ $V = \frac{-GM}{r}$	due to a point charge Q, $E = \frac{Q}{4\pi\varepsilon_0 r^2}$ $V = \frac{Q}{4\pi\varepsilon_0 r}$
Differences		
Action at a distance	between any two masses	between any two charged objects
Force	attracts only	unlike charges attract and like charges repel
Constant of proportionality in force law	G	$\frac{1}{4\pi\varepsilon_0}$

Practice questions: Chapter 18

1 **(a)** (i) Define the electric field strength, E, at a point in an electric field.
(ii) State whether E is a scalar or a vector quantity. *(3 marks)*

(b) Point charges of +4.0 nC and −8.0 nC are placed 80 mm apart, as shown in **Figure 1**.

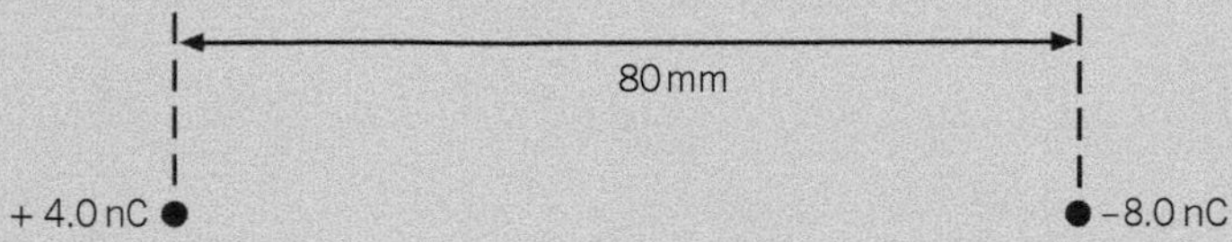

▲ **Figure 1**

(i) Calculate the magnitude of the force exerted on the +4.0 nC charge by the −8.0 nC charge.
(ii) Determine the distance from the +4.0 nC charge to the point, along the straight line between the charges, where the electric potential is zero. *(4 marks)*

(c) Point P in the diagram is equidistant from the two charges.
(i) On a copy of the diagram, draw two arrows at P to represent the directions and relative magnitudes of the components of the electric field at P due to each of the charges.
(ii) Draw an arrow, labelled R, on your diagram at P to represent the direction of the resultant electric field at P. *(3 marks)*

AQA, 2006

2 A small charged sphere of mass 2.1×10^{-4} kg, suspended from a thread of insulating material, was placed between two vertical parallel plates 60 mm apart. When a potential difference of 4200 V was applied to the plates, the sphere moved until the thread made an angle of 6.0° to the vertical, as shown in **Figure 2**.

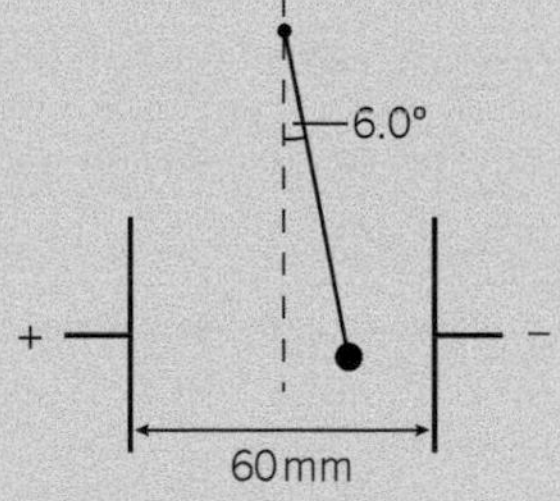

▲ **Figure 2**

(a) Show that the electrostatic force F on the sphere is given by $F = mg \tan 6.0°$ where m is the mass of the sphere. *(3 marks)*

(b) Calculate:
(i) the electric field strength between the plates
(ii) the charge on the sphere. *(3 marks)*

AQA, 2003

3 **Figure 3** shows some of the equipotential lines that are associated with a point negative charge Q.

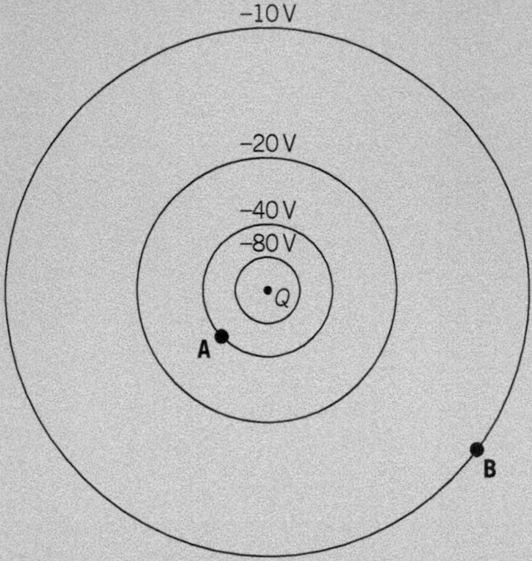

▲ **Figure 3**

(a) (i) Explain why the potentials have a negative sign.
(ii) Draw on a copy of the diagram three electric field lines. Use arrows to show the direction of the field. *(4 marks)*

(b) (i) Use data from the diagram, which is full size, to show that the charge Q is about -4.5×10^{-11} C.
(ii) Calculate the electric field strength at B. *(4 marks)*

(c) (i) Calculate the energy, in J, transferred when an electron moves from A to B in the field.
(ii) State and explain
- why the kinetic energy of the electron increases as it moves from A to B
- how the de Broglie wavelength of the electron changes as it moves from A to B.

(6 marks)

AQA, 2003

4 **Figure 4** shows the parallel deflecting plates with some dimensions of the ink-jet cartridge. In order to land in the centre of the gutter, the ink droplet must leave the plates at an angle of 35°. On entering the electric field the ink droplet carries a charge of -2×10^{-10} C and travels with a horizontal velocity of 20 m s^{-1}.

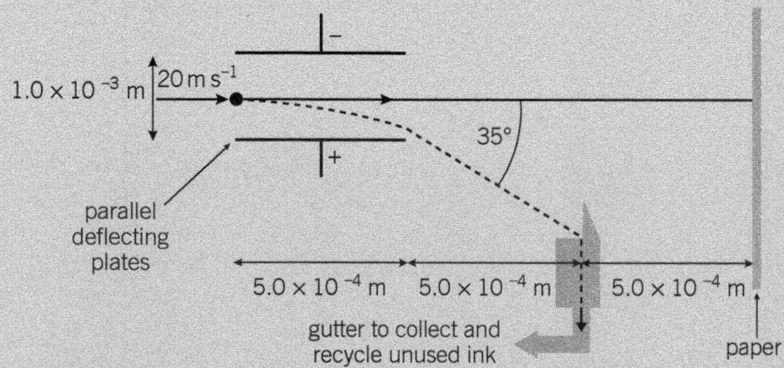

▲ **Figure 4**

(a) (i) Draw a vector diagram to show the components and the resultant of the velocity of the charged ink droplet as it leaves the deflecting field. Determine the size of the vertical component.

(ii) Find the time for which the ink droplet is between the deflecting plates and therefore calculate its vertical acceleration during this time.
(iii) For an ink droplet of mass 2.9×10^{-10} kg, calculate the electric force acting on the ink droplet whilst it is between the deflecting plates.
(iv) Calculate the electric field strength between the deflecting plates.
(v) Calculate the potential difference between the deflecting plates. *(12 marks)*

(b) The uncharged, undeflected ink droplets travel beyond the deflecting plates towards the paper. With the aid of a suitable calculation, discuss whether or not the printer manufacturer needs to take into consideration the droplet falling under gravity. *(4 marks)*

AQA, 2003

5 Dry air ceases to be an insulator if it is subjected to an electric field strength of $3.3\,\text{kV}\,\text{mm}^{-1}$ or more.

(a) (i) Show that the electric field strength *E* and the potential *V* at the surface of a charged sphere of radius *R* are related by

$$E = \frac{V}{R}$$

(ii) The dome of a Van de Graaff generator has a radius of 0.20 m. Calculate the maximum potential of this dome in dry air. *(5 marks)*

(b) Two high-voltage conductors are joined together using a small sphere, as shown in **Figure 5**.

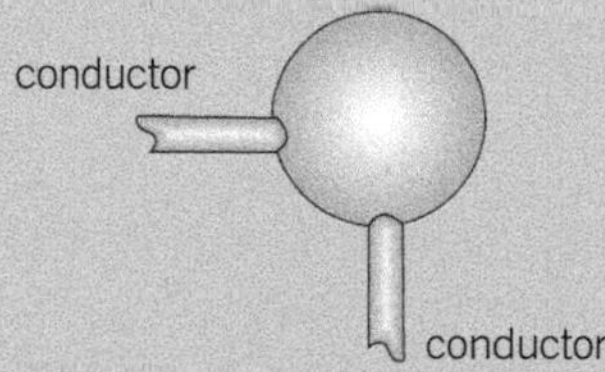

▲ **Figure 5**

The conductors are used to transmit alternating current at a peak potential of 140 kV. Calculate the minimum diameter of the sphere necessary to ensure the surrounding air does not conduct. *(2 marks)*

AQA, 2004

6 **(a)** **Figure 6** shows a small charged metal sphere carrying a charge *Q*. The potential of the sphere is 8000 V. Copy Figure 6.

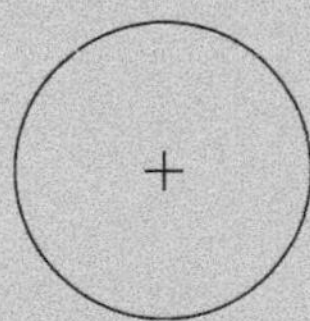

▲ **Figure 6**

(i) Draw on your diagram at least six lines to show the direction of the electric field in the region around the charged sphere.
(ii) Draw on your diagram the equipotential lines for potentials of 4000 V and 2000 V.
(iii) The equation for the field strength at a distance *r* from the sphere is $\dfrac{Q}{4\pi\varepsilon_0 r^2}$. State the name of the quantity represented by ε_0. *(5 marks)*

(b) **Figure 7** shows an arrangement for determining the charge on a small sphere.

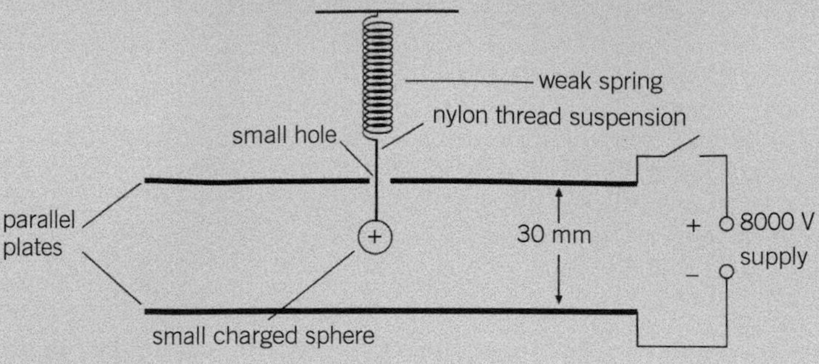

▲ **Figure 7**

The sphere is suspended from a spring of spring constant $0.18\,\mathrm{N\,m^{-1}}$. It hangs between two parallel plates which can be connected to a high-voltage supply.

(i) Explain why nylon thread is used for the suspension.

(ii) Calculate the extension of the spring when a sphere of mass 1.5 g is suspended from it.

(iii) Calculate the magnitude of the electric field strength between the plates when the 8000 V supply is switched on.

(iv) When the 8000 V supply is switched on, the sphere moves down a further 4.5 mm. Calculate the charge on the sphere. *(8 marks)*

(c) One problem with this arrangement is the oscillations of the sphere that occur when the switch is closed.

(i) Show that the period of the oscillations produced is about 0.6 s.

(ii) In practice the oscillations are damped. Sketch a graph showing how the amplitude of the oscillations changes with time for the damped oscillation.

(5 marks)

AQA, 2007

7 The Earth has an electric charge. The electric field strength outside the Earth varies in the same way as if this charge were concentrated at the centre of the Earth. The axes in **Figure 8** represent the electric field strength, E, and the distance from the centre of the Earth, r. The electric field strength at A has been plotted.

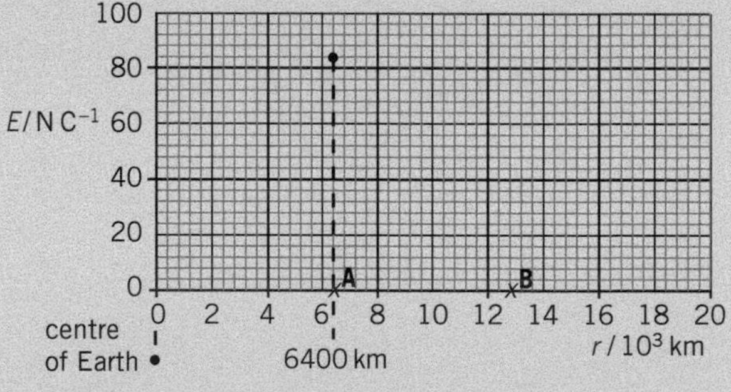

▲ **Figure 8**

(a) (i) Determine the electric field strength at B and then complete a copy of the graph to show how the electric field strength varies with distance from the centre of the Earth for distances greater than 6400 km.

(ii) State how you would use the graph to find the electric potential difference between the points A and B. *(4 marks)*

(b) (i) Calculate the total charge on the Earth.

(ii) The charge is distributed uniformly over the Earth's surface. Calculate the charge per square metre on the Earth's surface. *(4 marks)*

AQA, 2002

19 Capacitors

19.1 Capacitance

Learning objectives:

- → Describe in terms of electron flow what is happening when a capacitor charges up.
- → Relate the potential difference (p.d.) across the plates of a capacitor to the charge on its plate.
- → Discuss what capacitors are used for.

Specification reference: 3.8.4

A capacitor is a device designed to store charge. Two parallel metal plates placed near each other form a capacitor. When the plates are connected to a battery, electrons move through the battery and are forced from its negative terminal of the battery onto one of the plates. An equal number of electrons leave the other plate to return to the battery via its positive terminal. So each plate gains an equal and opposite charge.

A capacitor consists of two conductors insulated from each other. The symbol for a capacitor is shown in Figure 1b. As explained above, when a capacitor is connected to a battery, one of the two conductors gains electrons from the battery, and the other conductor loses electrons to the battery. When we say that the charge stored by the capacitor is Q, we mean that one conductor stores charge $+Q$ and the other conductor stores charge $-Q$.

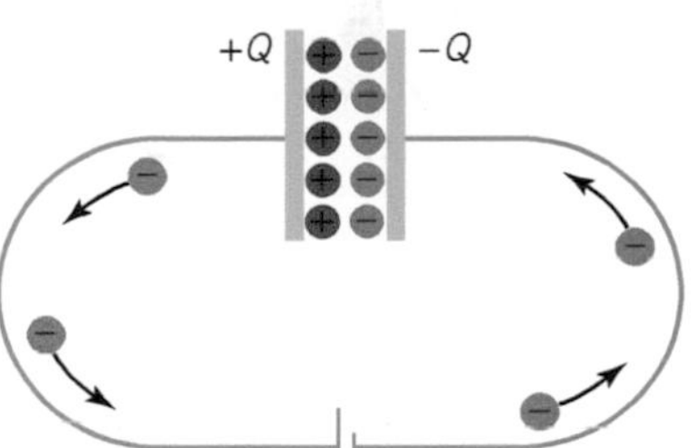

▲ **Figure 1a** *Storing charge on a capacitor*

▲ **Figure 1b** *The capacitor circuit symbol*

Charging a capacitor at constant current

Figure 2 shows how this can be achieved using a variable resistor, a switch, a microammeter, and a cell in series with the capacitor. When the switch is closed, the variable resistor is continually adjusted to keep the microammeter reading constant. At any given time t after the switch is closed, the charge Q on the capacitor can be calculated using the equation $Q = It$, where I is the current.

By using a high-resistance voltmeter connected in parallel with the capacitor, you can measure the capacitor potential difference (p.d.). To investigate how the capacitor p.d. changes with time for a constant current, use the variable resistor to keep the current constant, and either:

- use a stopwatch and measure the voltmeter reading at measured times, or
- use a data logger as shown in Figure 2.

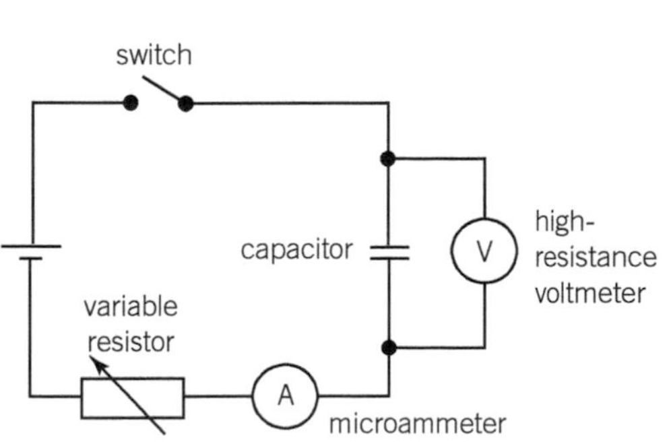

a *Circuit diagram*

A
C
V
to computer
data logger

b *Using a data logger*

▲ **Figure 2** *Investigating capacitors*

Typical readings for a current of 15 μA are shown in Table 1. The charge Q has been calculated using $Q = It$.

Study tip

The farad is a very large unit. In practice, **capacitance** is measured in µF, nF, or pF.

When you use $C = \frac{Q}{V}$, remember that charge in µC and p.d. in V gives capacitance in µF.

Summary questions

1 Complete the following table.

	(a)	(b)	(c)	(d)
Charge / µC	60	330		6.30
P.d. / V	12		9.0	4.5
Capacitance / µF		150	1100	

2 A 22 µF capacitor is charged by using a constant current of 2.5 µA to a p.d. of 12.0 V. Calculate:

- **a** the charge stored on the capacitor at 12.0 V
- **b** the time taken.

3 A capacitor is charged by using a constant current of 0.5 µA to a p.d. of 5.0 V in 55 s. Calculate:

- **a** the charge stored
- **b** the capacitance of the capacitor.

4 A capacitor is charged by using a constant current of 24 µA to a p.d. of 4.2 V in 38 s. The capacitor is then charged from 4.2 V by using a current of 14 µA in 50 s. Calculate:

- **a** charge stored at a p.d. of 4.2 V
- **b** the capacitance of the capacitor
- **c** the extra charge stored at a current of 14 µA
- **d** the p.d. after the extra charge was stored.

▼ **Table 1** *Current = 15 µA*

Time t / s	0	20	40	60	80	100
P.d. V / volts	0	0.29	0.62	0.90	1.22	1.50
Charge Q / µC	0	300	600	900	1200	1500

The graph of charge stored, Q, against p.d., V, for these measurements is shown in Figure 3. The measurements give a straight line passing through the origin. Therefore, the charge stored, Q, is proportional to the p.d., V. In other words, the charge stored per volt is constant.
$= Q_0(1 - e^{-t/RC})$ where $Q_0 = CV_0$

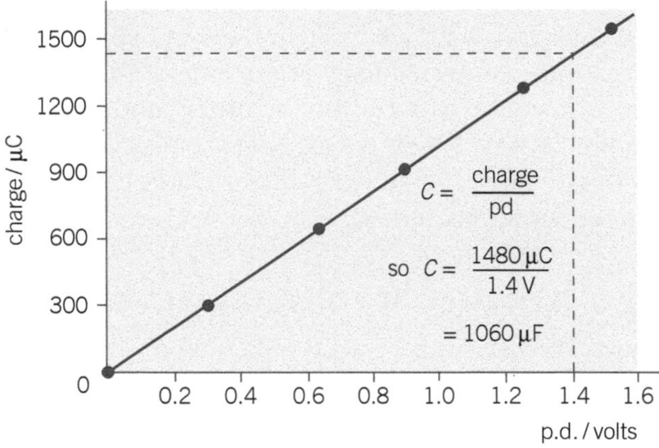

▲ **Figure 3** *Graph of results*

The capacitance C of a capacitor is defined as the charge stored per unit p.d.

The unit of capacitance is the farad (F), which is equal to one coulomb per volt. Note that $1.0\,\mu F = 1.0 \times 10^{-6}\,F$.

For a capacitor that stores charge Q at p.d. V, its capacitance can be calculated using the equation

$$C = \frac{Q}{V}$$

Note:
Rearranging this equation gives $Q = CV$ or $V = \frac{Q}{C}$.

Application

Capacitor uses

Capacitors are used in:

- smoothing circuits (i.e., circuits that smooth out unwanted variations in voltage)
- back-up power supplies (i.e., circuits that take over when the mains supply is interrupted)
- timing circuits, (i.e., circuits that switch on or off automatically after a preset delay)
- pulse-producing circuits (i.e., circuits that switch on and off repeatedly)
- tuning circuits (i.e., circuits that are used to select radio stations and TV channels)
- filter circuits (i.e., circuits that remove unwanted frequencies).

19.2 Energy stored in a charged capacitor

Learning objectives:

- → Explain why a capacitor stores energy as it is being charged.
- → Describe the form of energy that is stored by a capacitor.
- → Describe what happens to the amount of energy stored if the charge stored is doubled.

Specification reference: 3.8.4

When a capacitor is charged, energy is stored in it because electrons are forced onto one of its plates and taken off the other plate. The energy is stored in the capacitor as electric potential energy. A charged capacitor discharged across a torch bulb will release its energy in a brief flash of light from the bulb, as long as the capacitor has first been charged to the operating p.d. of the bulb. Charge flow is rapid enough to give a large enough current to light the bulb, but only for a brief time. You could replace the bulb with a miniature electric motor. The motor would briefly spin when the capacitor is discharged through it.

How much energy is stored in a charged capacitor? The charge is forced onto the plates by the battery. In the charging process, the p.d. across the plates increases in proportion to the charge stored, as shown in Figure 2.

Consider one step in the process of charging a capacitor of capacitance C when the charge on the plates increases by a small amount Δq from q to $q + \Delta q$. The energy stored ΔE in the capacitor is equal to the work done to force the extra charge Δq onto the plates and is given by $\Delta E = v\Delta q$, where v is the average p.d. during this step. In Figure 2, $v\Delta q$ is represented by the area of the vertical strip of width Δq and height v under the line. Therefore, the area of this strip represents the work done ΔE in this small step.

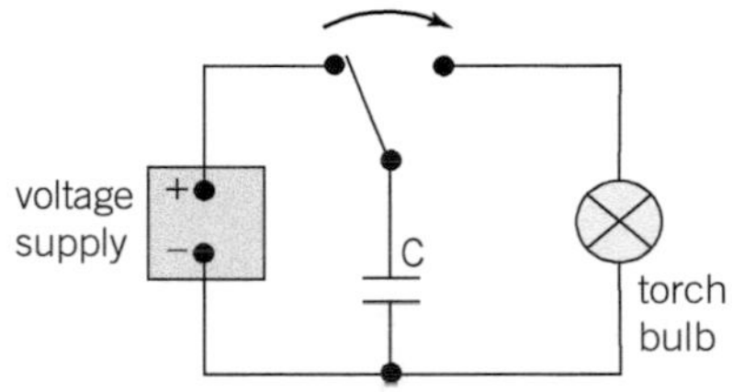

▲ **Figure 1** *Releasing stored energy*

Now consider all the small steps from zero p.d. to the final p.d. V. The total energy stored E is obtained by adding up the energy stored in each small step. In other words, E is represented by the total area under the line from zero p.d. to p.d. V. Because this area is a triangle of height V and base length Q ($= CV$), the total energy stored E = triangle area $= \frac{1}{2} \times$ height $\times$ base $= \frac{1}{2}VQ$.

$$\textbf{Energy stored by the capacitor } E = \frac{1}{2}QV$$

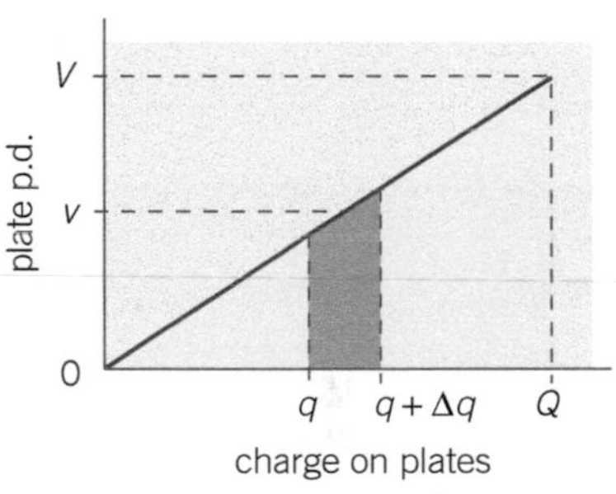

▲ **Figure 2** *Energy stored in a capacitor*

Notes:

1. Using $Q = CV$ or $V = \frac{Q}{C}$, you can write the above equation as $E = \frac{1}{2}CV^2$ or $E = \frac{1}{2}\frac{Q^2}{C}$.
2. In the charging process, the battery forces charge Q through p.d. V round the circuit and therefore transfers energy QV to the circuit. Thus, 50% of the energy supplied by the battery ($= \frac{1}{2}QV$) is stored in the capacitor. The other 50% is wasted due to resistance in the circuit as it is transferred to the surroundings when the charge flows in the circuit.

Measuring the energy stored in a charged capacitor

A joulemeter can measure the energy transfer from a charged capacitor to a light bulb when the capacitor discharges (Figure 3). Before the discharge starts, the capacitor p.d. V is measured and the joulemeter reading recorded. When the capacitor has discharged, the joulemeter reading is recorded again. The difference between these two joulemeter readings is the energy transferred from the capacitor during the discharge process. This is the total energy stored in the capacitor before it discharged. You can compare this with the calculation of the energy stored by using the equation $E = \frac{1}{2}CV^2$.

Study tip

Don't forget that doubling the charge also doubles the voltage.

You can see the effect of this on energy if you use the equation $E = \frac{1}{2}CV^2$ instead of $E = \frac{1}{2}QV$.

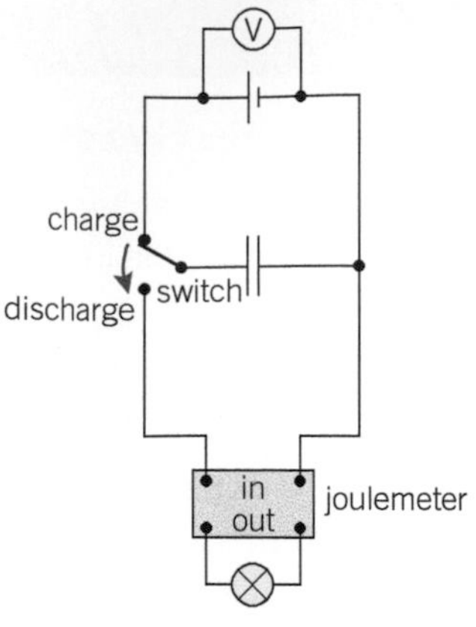

▲ **Figure 3** *Measuring energy stored*

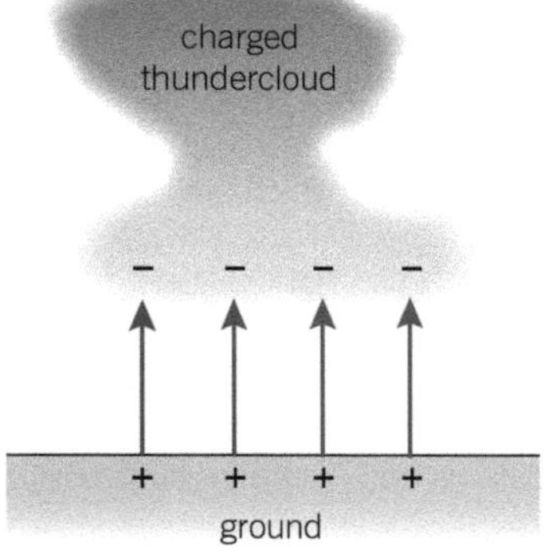

▲ **Figure 4** *Energy in a thundercloud*

The energy stored in a thundercloud

Imagine that a thundercloud and the Earth below it are like a pair of charged parallel plates. Because the thundercloud is charged, a strong electric field exists between the thundercloud and the ground. The potential difference between the thundercloud and the ground is $V = Ed$, where E is the electric field strength and d is the height of the thundercloud above the ground.

- For a thundercloud carrying a constant charge Q, the energy stored $= \frac{1}{2}QV = \frac{1}{2}QEd$.
- If the thundercloud is forced by winds to rise up to a new height d', then the energy stored $= \frac{1}{2}QEd'$.
- Because the electric field strength is unchanged (because it depends on the charge per unit area; see Topic 18.2), then the increase in the energy stored $= \frac{1}{2}QEd' - \frac{1}{2}QEd = \frac{1}{2}QE\Delta d$, where $\Delta d = d' - d$.

The energy stored increases because work is done by the force of the wind. It has to overcome the electrical attraction between the thundercloud and the ground to make the charged thundercloud move away from the ground.

The insulating property of air breaks down if it is subjected to an electric field stronger than about $300\,\text{kV}\,\text{m}^{-1}$. Prove for yourself that, for every metre rise of the thundercloud carrying a charge of 20 C, the energy stored would increase by 3 MJ. At a height of 500 m, the energy stored would be 1500 MJ.

Summary questions

1. Calculate the charge and energy stored in a $10\,\mu\text{F}$ capacitor charged to a p.d. of
   - **a** 3.0 V
   - **b** 6.0 V.
2. A $50\,000\,\mu\text{F}$ capacitor is charged from a 9 V battery then discharged through a light bulb in a flash of light lasting 0.2 s. Calculate:
   - **a** the charge and energy stored in the capacitor before discharge
   - **b** the average power supplied to the light bulb.
3. An uncharged $2.2\,\mu\text{F}$ capacitor is connected to a 3.0 V battery. Calculate:
   - **a** the charge and energy
     - **i** stored in the capacitor
     - **ii** supplied by the battery.
   - **b** Explain the difference between the energy supplied by the battery and the energy stored in the capacitor.
4. In Figure 5, a $4.7\,\mu\text{F}$ capacitor is charged from a 12.0 V battery by connecting the switch to X. The switch is then reconnected to Y to charge a $2.2\,\mu\text{F}$ capacitor from the first capacitor, causing the charge to be shared in proportion to the capacitance of each capacitor.

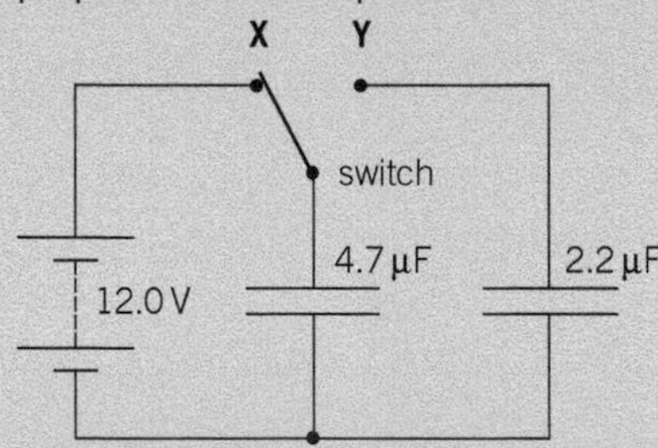

▲ **Figure 5**

Calculate:
   - **a** the initial charge and energy stored in the $4.7\,\mu\text{F}$ capacitor
   - **b** **i** the final charge stored by each capacitor
     ii the final p.d. across the two capacitors
   - **c** the final energy stored in each capacitor.

   Explain the loss of energy stored.

19.3 Dielectrics

Dielectric action

A capacitor is a device designed to store charge. The simplest type of capacitor is made of two parallel metal plates opposite each other as shown in Figure 1a in Topic 19.1. When a battery is connected to the plates, electrons from the negative terminal of the battery move onto the plate connected to that terminal. At the same time, electrons move from the other plate to the positive terminal of the battery, leaving this plate with a positive charge. The two plates store equal and opposite amounts of charge.

The charge stored on the plates can be increased by inserting a **dielectric** between the plates. Dielectrics are electrically insulating materials that increase the ability of a parallel-plate capacitor to store charge when a dielectric is placed between the plates of the capacitor. Polythene and waxed paper are examples of dielectrics.

Consider what happens when a dielectric is placed between two oppositely charged parallel plates connected to a battery. Each molecule of the dielectric becomes **polarised**. This means that its electrons are pulled slightly towards the positive plate as shown in Figure 1. So the surface of the dielectric facing the positive plate gains negative charge at the expense of the other side of the dielectric that faces the negative plate. The other surface of the dielectric loses negative charge, so some positive charge is left on its surface.

In some dielectric substances, the molecules are already polarised, but they lie in random directions. These molecules, called polar molecules, turn (ie rotate) when the dielectric is placed between the charged plates because their electrons are attracted slightly to the positive plate. The effect is the same as with non-polar molecules – the surface of the dielectric near the positive plate gains negative charge, and the other surface gains positive charge.

As a result, more charge is stored on the plates because:

- the positive side of the dielectric attracts more electrons from the battery onto the negative plate
- the negative side of the dielectric pushes electrons back to the battery from the positive plate.

The effect of a dielectric is to increase the charge stored in a capacitor for any given p.d. across the capacitor terminals. In other words, its effect is to increase the capacitance of the capacitor.

Relative permittivity

For a fixed p.d. across a parallel-plate capacitor with an empty space between its plates, the charge stored is increased by inserting a dielectric substance between the plates. The ratio of the charge stored with the dielectric to the charge stored without the dielectric may be defined as the **relative permittivity**, ε_r, of the dielectric substance.

$$\textbf{relative permittivity } \varepsilon_r = \frac{Q}{Q_0}$$

where Q = charge stored by a parallel-plate capacitor when the space between the plates of the capacitor is completely filled with the

Learning objectives:

- → Explain how a dielectric affects a capacitor.
- → Define relative permittivity and dielectric constant.
- → Describe the action of a simple polar molecule rotating in an electric field.

Specification reference: 3.8.4

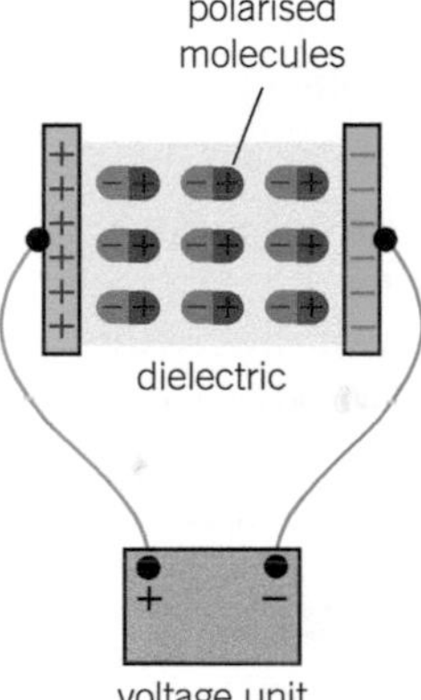

▲ **Figure 1** *Dielectric action*

Extension

Capacitor factors

In Topic 18.3, you saw that for an 'empty' parallel-plate capacitor, the charge stored per unit surface area $\frac{Q_0}{A} = \varepsilon_0 E$, where E is the electric field strength and ε_0 is the absolute permittivity of free space. This equation may be written as $\frac{Q_0}{A} = \varepsilon_0 \frac{V}{d}$ because $E = \frac{V}{d}$, where V is the p.d. between the plates and d is the spacing between the plates. Rearranging this equation gives the 'empty' capacitor's capacitance $C_0 = \frac{Q_0}{V} = \frac{A\varepsilon_0}{d}$, where A is the surface area of each plate. If the space between the plates had been filled completely with a dielectric, the capacitance would have been increased from C_0 to $C = \varepsilon_r C_0$.

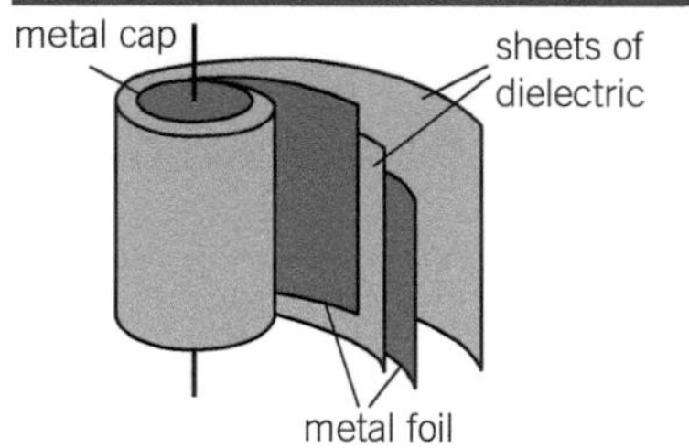

▲ **Figure 2** *Capacitor design*

dielectric substance, and Q_0 = charge stored at the same p.d. when the space is completely empty.

For a fixed p.d. V between the plates, $Q = CV$ and $Q_0 = C_0V$, where C is the capacitance of the parallel-plate capacitor with the dielectric completely between the plates and C_0 is the capacitance when the space is completely empty. Therefore, $\frac{Q}{Q_0} = \frac{C}{C_0}$, so the relative permittivity may be defined by the equation

$$\varepsilon_r = \frac{C}{C_0}$$

The relative permittivity ε_r of a substance is also called its **dielectric constant**. Typical values for ε_r are 2.3 for polythene, 2.5 for waxed paper, 7 for mica, and 81 for water! The large value of ε_r for water is the reason why ionic crystals such as sodium chloride (common salt) dissolve in water. See Topic 18.4.

Capacitor design

For a parallel-plate capacitor with dielectric filling the space between the plates, its capacitance $C = \frac{A\varepsilon_0\varepsilon_r}{d}$, where A is the surface area of each plate and d is the spacing between the plates.

This equation shows that a large capacitance can be achieved by

- making the area A as large as possible
- making the plate spacing d as small as possible
- filling the space between the plates with a dielectric which has a relative permittivity as large as possible.

Most capacitors consist of two strips of aluminium foil separated by a layer of dielectric, all rolled up as shown in Figure 2. This arrangement makes the capacitance as large as possible because the area A is as large as possible and the spacing d is as small as possible.

Application

Measuring the relative permittivity of a dielectric substance

In this investigation, a capacitor is formed by placing a dielectric sheet of uniform thickness between two parallel metal plates insulated from each other by small pieces of suitable insulation. A potential divider is used to apply a constant p.d. between the two plates. The capacitor is repeatedly charged and discharged through a microammeter using a reed switch operating at a constant frequency. Figure 3 shows the circuit.

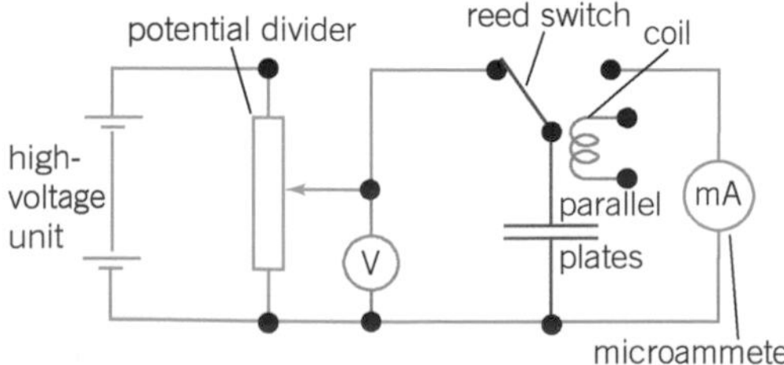

▲ **Figure 3** *Measuring relative permittivity*

The microammeter current reading is proportional to the charge gained by the capacitor each time the switch charges it. The current is measured with the dielectric in place and with the dielectric replaced by small insulating spacers of the same thickness as the dielectric sheet. The relative permittivity of the dielectric $= \frac{I}{I_0}$, where I is the current with the dielectric present and I_0 is the current without the dielectric present.

Note that a capacitance meter may be used to measure the capacitance of the arrangement directly with and without the dielectric between the plates. The ratio $\frac{C}{C_0}$ gives the relative permittivity.

Note: A suitable alternating pd applied to the reed switch coil is necessary to make the switch move to and fro repeatedly between the left hand and right hand switch terminals.

In an alternating electric field, the polar molecules rotate to align with the field as the field increases, then return to random directions as the field decreases before rotating again to align in the next half-cycle with the field in the opposite direction. Non-polar molecules are polarised in the field direction by the field so their polarity alternates as the electrons and ions in each molecule oscillate along the field direction.

Application

Capacitor applications

1 Any electronic timing circuit or time-delay circuit makes use of **capacitor discharge** through a fixed resistor.

Figure 4 shows an alarm circuit. In this circuit, the alarm rings if the input voltage to the circuit drops below a particular value after the switch is reset. The time delay between resetting the switch and the alarm ringing can be increased by increasing the resistance R or the capacitance C. This type of change to the circuit would make the discharge of C through R slower.

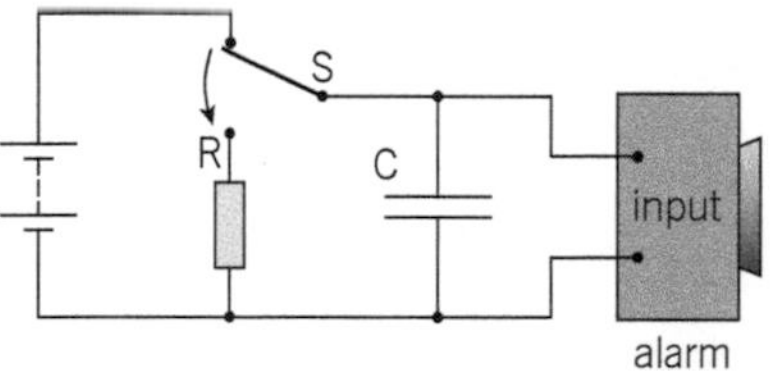

▲ **Figure 4** *A time-delayed alarm circuit*

This would then increase the time for the capacitor voltage to decrease enough to make the alarm ring.

2 Capacitor smoothing is used in applications where sudden voltage variations or glitches can have undesirable effects. For example, mains appliances being switched on or off in a building could affect computers connected to the mains supply in the building. Inside a computer there is a large capacitor that will supply current if the mains supply is interrupted. In this way, the computer circuits will continue to function normally.

Q: Explain how the time delay would be affected if the resistance R and the capacitance C in Figure 4 were both doubled.

Answer: The time delay would be greater because the rate of flow of charge would be less and also the capacitor holds more charge per volt.

Summary questions

1 A parallel-plate capacitor consists of two insulated metal plates separated by an air gap. A battery in series with a switch is connected to the plates. The capacitor is charged by closing the switch to charge the capacitor to a constant p.d. A sheet of dielectric is then inserted between the plates.

a When the sheet of dielectric is inserted, state the change that takes place in:

i the capacitance C of the capacitor

ii the charge Q stored by the capacitor.

b State and explain the change that takes place in the energy stored by the capacitor.

2 In the test in **Q1**, the switch remains open and the sheet of dielectric is then removed. State and explain how the energy stored by the capacitor changes when the dielectric is removed.

3 An air-filled parallel-plate capacitor has a capacitance of 1.4 pF.

a The space between the plates is completely filled with a sheet of dielectric that has a relative permittivity of 7.0. Calculate the capacitance of the capacitor with the dielectric present.

b Calculate the energy stored by the capacitor in **a** when the p.d. across it is 15.0 V.

4 An *electrolytic* capacitor contains a very thin layer of dielectric formed when the capacitor is first charged. The insulating property of the dielectric in a certain 100 mF electrolytic capacitor breaks down if the electric field strength across it exceeds 700 MV m^{-1}. The maximum p.d. that can be applied to the capacitor is 100 V.

a Calculate the thickness of the dielectric layer.

b The effective surface area of each capacitor plate is 1.6 m^2. Estimate the energy stored per unit volume in the capacitor at 100 V.

19.4 Charging and discharging a capacitor through a fixed resistor

Learning objectives:

- → Describe and interpret the shape of the Q–t charging curves and the shape of the Q–t discharging curves.
- → Explain which circuit components you would change to make the charge/discharge slower.
- → Define the time constant of a capacitor–resistor circuit.

Specification reference: 3.9.1

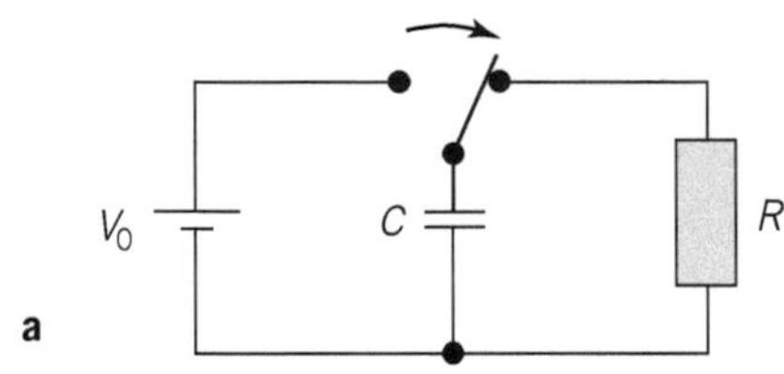

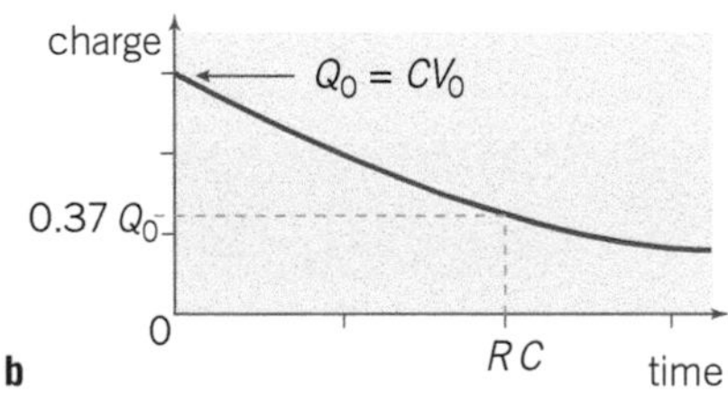

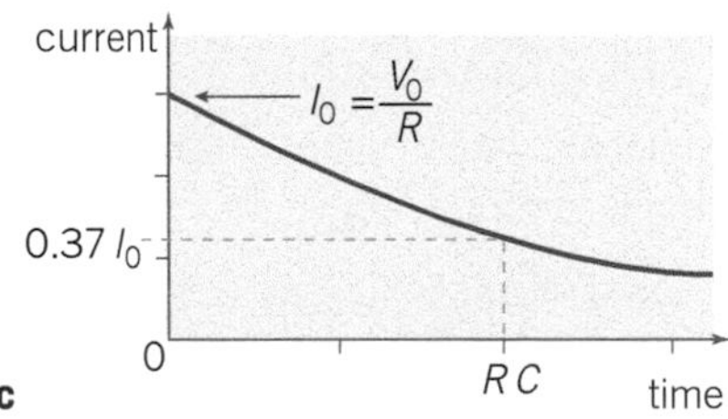

▲ **Figure 1** *Capacitor discharging*

Study tip

An exponential decrease graph starts at an intercept with the vertical axis and is asymptotic with the time axis (i.e., the curve approaches but never cuts the time axis).

Capacitor discharge through a fixed resistor

When a capacitor discharges through a fixed resistor, the discharge current decreases gradually to zero. Figure 1a shows a circuit in which a capacitor is discharged through a resistor when the switch is changed over. The reason why the current decreases gradually is that the p.d. across the capacitor decreases as it loses charge. Because the resistor is connected directly to the capacitor, the resistor current $\left(= \frac{\text{p.d.}}{\text{resistance}}\right)$ decreases as the p.d. decreases.

The situation is a bit like water emptying through a pipe at the bottom of a container. When the container is full, the flow rate out of the pipe is high because the water pressure at the pipe is high. As the container empties, the water level falls, so the water pressure at the pipe falls, and the flow rate decreases.

The graphs in Figure 1 show how the current and the charge decrease with time. Both curves have the same shape because both the current and the charge (and p.d.) decrease **exponentially**. This means that any of these quantities decreases by the same factor in equal intervals of time. For example, for initial charge Q_0, if the charge is $0.9Q_0$ after a particular time t_1, the charge will be:

- $0.9 \times 0.9Q_0$ after time $2t_1$
- $0.9 \times 0.9 \times 0.9Q_0$ after time $3t_1$ ...
- 0.9^nQ_0 after time nt_1.

To understand why the decrease is exponential, consider one small step in the discharge process of a capacitor C through a resistor R when the charge decreases from Q to $Q - \Delta Q$ in the time interval Δt.

At this stage, the current $I = \frac{\text{p.d. across the plates, } V}{\text{resistance, } R} = \frac{Q}{CR}$, because $V = \frac{Q}{C}$. The current is proportional to the charge, which is proportional to the p.d. So the curves all have the same shape. Assuming that Δt is small enough, the decrease of charge $\Delta Q = -I\Delta t$ (with a minus sign because Q decreases).

Therefore, $\Delta Q = \frac{-Q}{CR}\Delta t$, which gives

$$\frac{\Delta \boldsymbol{Q}}{\boldsymbol{Q}} = -\frac{\Delta \boldsymbol{t}}{\boldsymbol{CR}}$$

The equation tells you that the fractional drop of charge $\frac{\Delta Q}{Q}$ is the same in any short interval of time Δt during the discharge process. For example, suppose $\Delta t = 10\,\text{s}$ and $CR = 100$. Therefore, $\frac{\Delta Q}{Q} = -0.1 \left(= -\frac{\Delta t}{CR}\right)$. So the charge decreases to 0.9 of its initial value at the start of the 10 s interval. So if the initial charge is Q_0, then the charge remaining on the plates will be:

- $0.9Q_0$ after 10 s
- $0.9 \times 0.9Q_0$ after a further 10 s
- $0.9 \times 0.9 \times 0.9Q_0$ after a further 10 s, and so on.

In theory, the charge on the plates never becomes zero.

Exponential changes occur whenever the rate of change of a quantity is proportional to the quantity itself.

Rearranging the equation $\frac{\Delta Q}{Q} = -\frac{\Delta t}{CR}$ gives

$$\frac{\Delta Q}{\Delta t} = -\frac{Q}{CR}$$

For very short intervals of time (i.e., $\Delta t \rightarrow 0$), $\frac{\Delta Q}{\Delta t}$ represents the rate of change of charge and is written $\frac{dQ}{dt}$.

Therefore, $\frac{dQ}{dt} = -\frac{Q}{CR}$.

The graphical solution to this equation is shown in Figure 1b. The mathematical solution is

$$Q = Q_0 e^{\frac{-t}{RC}}$$

where Q_0 is the initial charge, and e is the exponential function (sometimes written 'exp').

The quantity RC is called the **time constant** for the circuit. At time $t = RC$ after the start of the discharge, the charge falls to 0.37 ($= e^{-1}$) of its initial value.

Time constant = RC

where R is the circuit resistance, and C is the capacitance.

The unit of RC is the second. This is because one ohm $= \frac{\text{one volt}}{\text{one ampere}}$, and one farad $= \frac{\text{one coulomb}}{\text{one volt}}$, so

$$\text{unit of } RC = \frac{\text{volts}}{\text{amperes}} \times \frac{\text{coulombs}}{\text{volts}} = \frac{\text{coulombs}}{\text{amperes}} = \text{seconds}$$

Worked example

A 2200 μF capacitor is charged to a p.d. of 9.0 V then discharged through a 100 kΩ resistor using a circuit as shown in Figure 1.

a Calculate:
 i the initial charge stored by the capacitor
 ii the time constant of the circuit.

b Calculate the p.d. after a time:
 i equal to the time constant
 ii 300 s.

Solution

a i $Q_0 = CV_0 = 2200 \times 10^{-6} \times 9.0 = 2.0 \times 10^{-2}$ C.
 ii Time constant $= RC = 100 \times 10^3 \times 2200 \times 10^{-6} = 220$ s.

b i When $t = RC$, $V = V_0 e^{-1} = 0.37 \times 9.0 = 3.3$ V.
 ii When $t = 300$ s, $\frac{t}{RC} = \frac{300}{220} = 1.36$
 Therefore, $V = V_0 e^{\frac{-t}{RC}} = 9.0 e^{-1.36} = 2.3$ V.

Study tip

You do *not* need to explain why $Q = Q_0 e^{\frac{-t}{CR}}$ is the solution of the equation

$$\frac{dQ}{dt} = -\frac{Q}{CR}$$

Notes

1 The current, the p.d., and the charge are all proportional to each other. All three quantities decrease exponentially during capacitor discharge in accordance with the equation $x = x_0 e^{\frac{-t}{CR}}$, where x represents either the current, or the charge, or the p.d.

2 The inverse function of e^x is $\ln x$, where ln is the natural logarithm. To calculate t, given x, x_0, R, and C, using the inverse function of e^x gives $\ln x = \ln x_0 - \left(\frac{t}{RC}\right)$.

3 The exponential function $e^z = 1 + z + \frac{z^2}{2 \times 1} + \frac{z^3}{3 \times 2 \times 1} + \ldots$, and so on.

It can be shown mathematically that the rate of change of this function in relation to z is the same function. This is why the function appears whenever the rate of change of a quantity is proportional to the quantity itself.

Note that $z = 1$ gives

$e = 1 + 1 + \frac{1}{2} + \frac{1}{6} + \frac{1}{24}$, etc. $= 2.718$.

You can check this on your calculator by keying in 'e^x' then pressing 1 to give 2.718. Keying in −1 instead of 1 gives 0.37 for e^{-1}.

Extension

The significance of the time constant

The time constant RC is the time taken, in seconds, for the capacitor to discharge to 37% of its initial charge. Given values of R and C, the time constant can be quickly calculated and used as an approximate measure of how quickly the capacitor discharges. However, as $5RC$ is the time taken to discharge by over 99%, $5RC$ is a better rule of thumb estimate of the time taken for the capacitor effectively to discharge.

Q: Demonstrate that $t = 5RC$ gives a value that is less than 1% of the initial value.

Answer: $0.37^5 Q_0 = 0.00690 Q_0$ $(< 0.01 Q_0)$.

Synoptic link

Wherever a quantity decreases at a rate that is proportional to the quantity, the decrease is exponential. You will meet exponential decrease in radioactive decay in Topic 8.5, where you will study what is meant by the half-life $T_{\frac{1}{2}}$ of a radioactive isotope. This is the time taken for the number of atoms of a radioactive isotope to decrease to 50% of its initial value. It can be shown that the half-life of a capacitor discharge (i.e., the time taken for the charge (or voltage) to decrease to 50% of the initial value) is equal to 0.69(3)RC. In the next topic, you will study the theory of radioactive decay in more detail to gain a deeper insight into exponential processes and mathematical modelling.

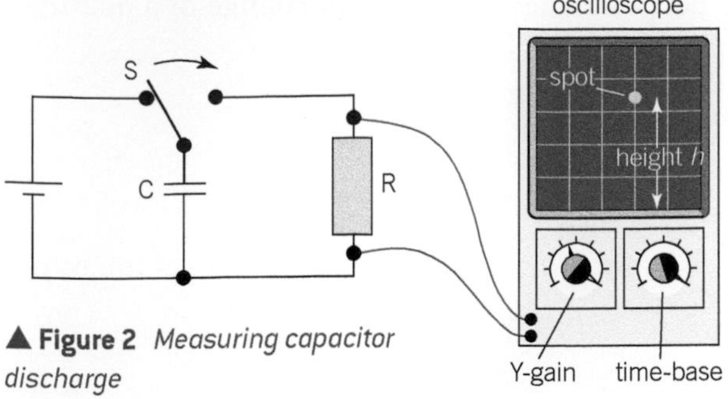

▲ **Figure 2** *Measuring capacitor discharge*

Investigating capacitor discharge

Figure 2 shows how to measure the p.d. across a capacitor as it discharges through a fixed resistor. An oscilloscope is used because it has a very high resistance so that the discharge current from the capacitor passes only through the fixed resistor. The oscilloscope is used to measure the capacitor p.d. at regular intervals. You could use a data logger or a digital voltmeter instead of the oscilloscope.

You can then use the measurements to plot a graph of voltage against time. The time taken for the voltage to decrease to 37% $\left(= \frac{1}{e}\right)$ of the initial value can be measured from the graph and compared with the calculated value of RC.

Charging a capacitor through a fixed resistor

When a capacitor is charged by connecting it to a source of constant p.d., the charging current decreases as the capacitor charge and p.d. increase. When the capacitor is fully charged, its p.d. is equal to the source p.d., and the current is zero because no more charge flows in the circuit. The graphs in Figure 3 show how the capacitor charge and current change with time.

- The capacitor charge builds up until the capacitor p.d. V is equal to the source p.d. V_0, as shown in Figure 3b. The charge Q_0 on the capacitor is then equal to CV_0. The charge curve is an inverted exponential decrease curve, which flattens out at $Q_0 = CV_0$.
- The time constant for the circuit, RC, is the time taken for the charge to reach 63% of the final charge (i.e., 37% more charge needed to be fully charged). A graph of the capacitor p.d. V against time has exactly the same shape as the charge curve because $V = \frac{Q}{C}$.
- The capacitor current I decreases exponentially to zero from its initial value I_0, as shown in Figure 3c. The current is always equal to the rate of change of charge. Therefore, the current is given by the gradient of the charge–time graph, and so it decreases exponentially.

1 At any instant during the charging process, the source p.d. V_0 = the resistor p.d. + the capacitor p.d.

Therefore, $V_0 = IR + \frac{Q}{C}$ at any instant.

2 The initial current $I_0 = \frac{V_0}{R}$, assuming that the capacitor is initially uncharged.

3 At time t after charging starts, $I = I_0\, e^{\frac{-t}{RC}}$.

4 If you combine the three equations above to eliminate I and I_0, you get $V_0 = V_0\, e^{\frac{-t}{RC}} + \frac{Q}{C}$.

So, $\frac{Q}{C} = V_0 - V_0\, e^{\frac{-t}{RC}} = V_0\left(1 - e^{\frac{-t}{RC}}\right)$.

Because $V = \frac{Q}{C}$, then $V = V_0\left(1 - e^{\frac{-t}{RC}}\right)$ and $Q = CV_0\left(1 - e^{\frac{-t}{RC}}\right)$.

Figure 3b shows how Q (and V) vary with time. For example,

- at time $t = 0$, $e^{\frac{-t}{RC}} = 1$, so $Q = 0$ and $V = 0$
- as time $t \to \infty$, $e^{\frac{-t}{RC}} \to 0$ so $V \to V_0$ and $Q \to Q_0$ $(= CV_0)$.

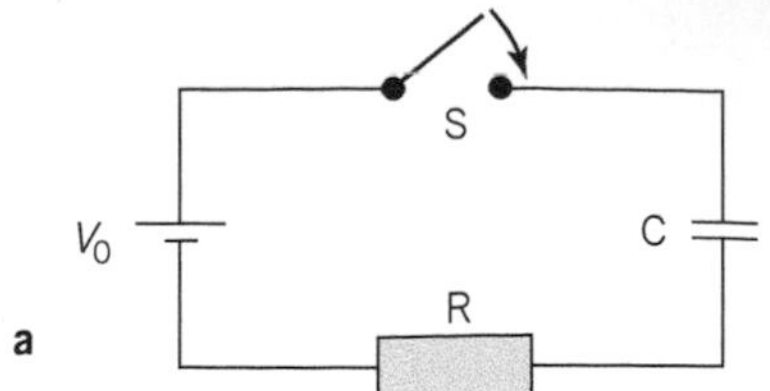

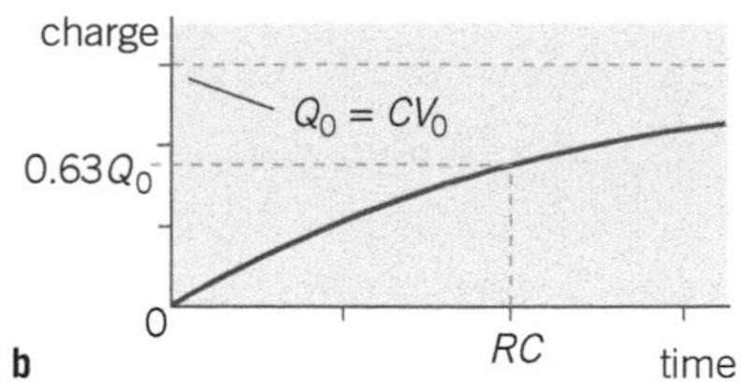

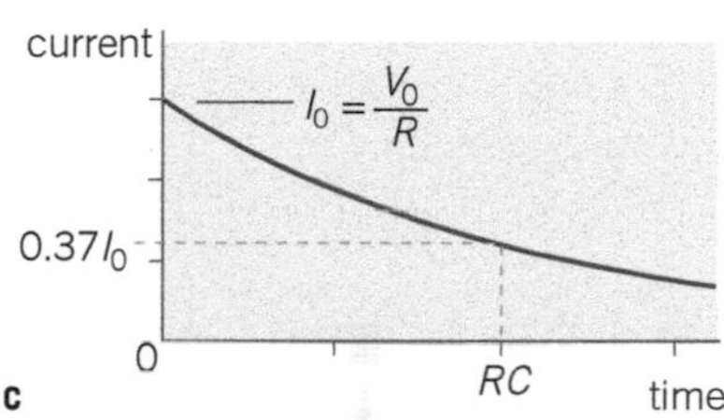

▲ **Figure 3** *Capacitor charging*

Summary questions

1 A 50 µF capacitor is charged by connecting it to a 6.0 V battery then discharged through a 100 kΩ resistor.

Calculate:

a i the charge stored in the capacitor immediately after it has been charged

ii the time constant of the circuit.

b i Estimate how long the capacitor would take to discharge to about 2 V.

ii Estimate the resistance of the resistor that you would use in place of the 100 kΩ resistor if the discharge is to be 99% completed within about 5 s.

2 A 68 µF capacitor is charged to a p.d. of 9.0 V then discharged through a 20 kΩ resistor.

a Calculate:

i the charge stored by the capacitor at a p.d. of 9.0 V

ii the initial discharge current.

b Calculate the p.d. and the discharge current 5.0 s after the discharge started.

3 A 2.2 µF capacitor is charged to a p.d. of 6.0 V and then discharged through a 100 kΩ resistor. Calculate:

a the charge and energy stored in this capacitor at 6.0 V

b the p.d. across the capacitor 0.5 s after the discharge started,

c the energy stored at this time.

4 An uncharged 4.7 µF capacitor is charged to a p.d. of 12.0 V through a 200 Ω resistor, then discharged through a 220 kΩ resistor. Calculate:

a i the initial charging current

ii the energy stored in the capacitor at 12.0 V

b the time taken for the p.d. to fall from 12.0 V to 3.0 V

c the energy lost by the capacitor in this time.

Study tip

Remember the symbol for capacitance (C) is the same as for the unit of charge, the coulomb (C).

Synoptic link

You can find out how to model rates of flow of charge in spreadsheets on a computer in Topic 27.6, Graphical and computational modelling.

19.5 The theory of radioactive decay

Learning objectives:

- → Discuss whether a radioactive source can decay completely.
- → Define exponential decrease.
- → Explain why a radioactive decay is a random process.

Specification reference: 3.9.2

The random nature of radioactive decay

An unstable nucleus becomes stable by emitting an α or a β particle or a γ photon or by electron capture. This is an unpredictable event. Every nucleus of a radioactive isotope has an equal probability of undergoing radioactive decay in any given time interval. Therefore, for a large number of nuclei of a radioactive isotope, the number of nuclei that disintegrate in a certain time interval depends only on the total number of nuclei present.

Consider a sample of a radioactive isotope X that initially contains N_0 nuclei of the isotope. Let N represent the number of nuclei of X remaining at time t after the start. Suppose in time Δt the number of nuclei that disintegrate is ΔN.

Because radioactive disintegration is a random process, ΔN is proportional to:

1. N, the number of nuclei of X remaining at time t
2. the duration of the time interval Δt.

Therefore $\Delta N = -\lambda N \Delta t$, where λ is a constant referred to as the **decay constant**. The minus sign is necessary because ΔN is a decrease.

So the rate of disintegration $\frac{\Delta N}{\Delta t} = -\lambda N$.

For a given radioactive isotope, its **activity**, A, is the rate of disintegration $\frac{\Delta N}{\Delta t}$.

Therefore, the activity A of N atoms of a radioactive isotope is given by

$$A = \lambda N$$

The solution of the equation $\frac{\Delta N}{\Delta t} = -\lambda N$ is $\mathbf{N = N_0 e^{-\lambda t}}$, where e^x is the exponential function. See below.

Figure 1 shows that a graph of N against t gives a decay curve. The number of nuclei N decreases exponentially with time. In other words,

- in one half-life, the remaining number of nuclei $N_1 = 0.5\,N_0$
- in two half-lives, the remaining number of nuclei $N_2 = 0.25\,N_0$
- in n half-lives, the remaining number of nuclei $N = 0.5^n\,N_0$.

The graph of the number of nuclei N against time t as represented by the equation $N = N_0e^{-\lambda t}$ is shown in Figure 1. It is a curve with exactly the same shape as Figure 1 in Topic 8.5.

The mass, m, of a sample of a radioactive isotope decreases from initial mass m_0 in accordance with the equation $m = m_0e^{-\lambda t}$ because the mass m is proportional to the number of nuclei, N, of the isotope in the sample.

The activity A of a sample of N nuclei of an isotope decays in accordance with the equation

$$A = A_0e^{-\lambda t}$$

This is because the activity A = the number of disintegrations per second = λN. So $A = \lambda N_0e^{-\lambda t} = A_0e^{-\lambda t}$ where $A_0 = \lambda N_0$.

Study tip

The unit of decay constant is s^{-1}. A large value of λ means fast decay and short half-life.

Synoptic link

To learn how to model the rate of decay in a computer spreadsheet, see Topic 27.6, Graphical and computational modelling.

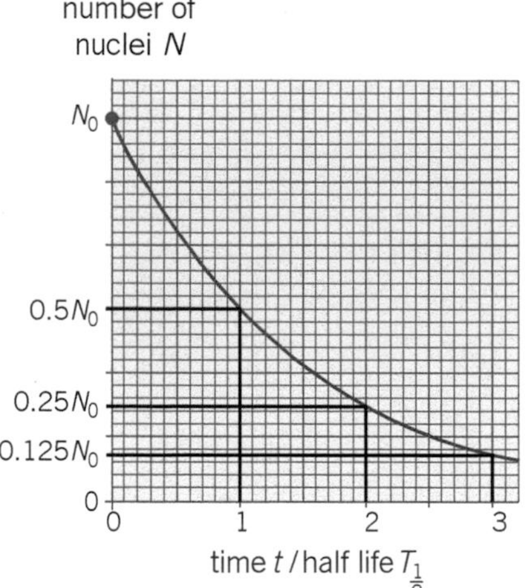

▲ **Figure 1** $N = N_0e^{-\lambda t}$

Synoptic link

The exponential decrease formula is also used in the theory of capacitor discharge, Topic 19.4.

The count rate C due to a sample of a radioactive isotope at a fixed distance from a Geiger–Müller tube is proportional to the activity of the source. Therefore, the count rate decreases with time in accordance with the equation $C = C_0 e^{-\lambda t}$, where C_0 is the count rate at time $t = 0$.

The above equations for the number of nuclei N, the activity A, and the count rate C are all of the same general form, namely $x = x_0 e^{-\lambda t}$, where x represents N or A or C and x_0 represents the initial value.

Worked example

A sample of a radioactive material initially contains 1.2×10^{20} atoms of the isotope. The decay constant for the isotope is $3.6 \times 10^{-3}\,s^{-1}$. Calculate:

a the number of atoms of the isotope remaining after 1000 s
b the activity of the sample after 1000 s.

Solution

a $N_0 = 1.2 \times 10^{20}$, $\lambda = 3.6 \times 10^{-3}\,s^{-1}$, $t = 1000\,s$,
$\lambda t = 3.6 \times 10^{-3} \times 1000 = 3.6$
$\therefore N = N_0 e^{-\lambda t} = 1.2 \times 10^{20} e^{-3.6} = 1.2 \times 10^{20} \times 2.7 \times 10^{-2}$
$= 3.2 \times 10^{18}$

b Activity $A = \lambda N = 3.6 \times 10^{-3} \times 3.2 \times 10^{18} = 1.2 \times 10^{16}\,Bq$

The decay constant

The decay constant λ is the probability of an individual nucleus decaying per second. If there are 10 000 nuclei present and 300 decay in 20 s, the decay constant is $0.0015\,s^{-1}$ $\left(= \dfrac{\left(\frac{300}{10\,000}\right)}{20}\,s^{-1}\right)$.

In general, if the change of the number of nuclei ΔN in time Δt is given by $\Delta N = -\lambda N \Delta t$, then the probability of decay $= \dfrac{\Delta N}{N} = \lambda \Delta t$ (the minus sign is not needed here as reference is made to decay).

So the probability per unit time $= \dfrac{\Delta N}{N} \div \Delta t = \lambda$.

As explained in Topic 8.5, the half-life, $T_{\frac{1}{2}}$ of a radioactive isotope is the time taken for half the initial number of nuclei to decay. The longer the half-life, the smaller the decay constant because the probability of decay per second is smaller.

The half-life $T_{\frac{1}{2}}$ is related to the decay constant λ according to the equation

$$T_{\frac{1}{2}} = \frac{\ln 2}{\lambda}$$

As $\ln 2 = 0.693$, this equation may be written as $T_{\frac{1}{2}} = \dfrac{0.693}{\lambda}$.

Notes

You do *not* need to know the following points. But they might help you improve your understanding of this topic.

1 The exponential function appears in any situation where the rate of change of a quantity is in proportion to the quantity itself. This is because the rate of change of each term in the function sequence is equal to the previous term in the sequence.

2 The exponential function, $e^x = 1 + x + \frac{x^2}{2!} + \frac{x^3}{3!} + \ldots$
Differentiating e^x with respect to x gives e^x (i.e., $\frac{d(e^x)}{dx} = e^x$) because differentiating each term in the expression for e^x gives the previous term.
The exponential function is indicated on a calculator as 'e^x' or 'inv ln'. See Topic 27.4.

3 The natural logarithm function, $\ln x$, is the inverse exponential function. In other words, if $y = e^x$, then $\ln y = x$.
Therefore, $N = N_0 e^{-\lambda t}$ may be written $\ln N = \ln N_0 - \lambda t$.
The graph of $\ln N$ against t is therefore a straight line with:
- a gradient $= -\lambda$, and
- a y-intercept $= \ln N_0$

See Topic 27.3 for more about log-linear graphs

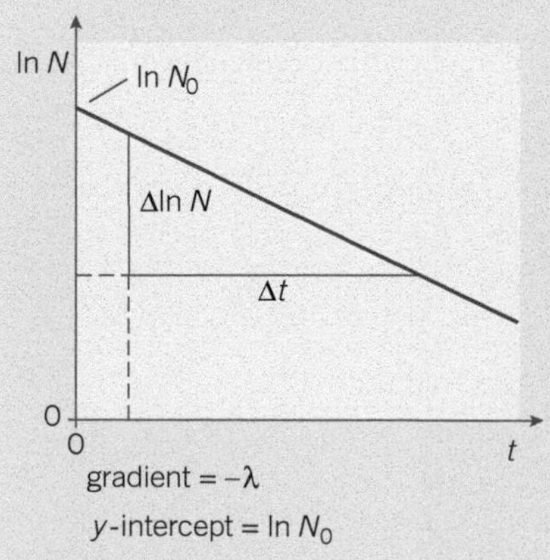

▲ **Figure 2** *Graph of ln N against t*

Hint

To calculate N at time t, given values of N_0 and $T_{\frac{1}{2}}$,

- **either** calculate λ using $\lambda = \frac{\ln 2}{T_{\frac{1}{2}}}$ then use the equation $N = N_0 e^{-\lambda t}$
- **or** calculate the number of half-lives, n, using $n = \frac{t}{T_{\frac{1}{2}}}$ then use $N = 0.5^n N_0$.

Synoptic link

Radioactive waste from a nuclear reactor contains a range of unstable **isotopes** with different half-lives. The waste products must be stored for many years until their activity is no more than background. See Topic 24.5, The thermal nuclear reactor.

Proof of $T_{\frac{1}{2}} = \frac{\ln 2}{\lambda}$

You do *not* need to know this proof. But it might help you improve your understanding of this topic. And you *do* need to know how to rearrange an exponential equation using natural logs.

Let the number of nuclei $N = N_0$ at time $t = 0$.

Therefore at time $t = T_{\frac{1}{2}}$, $N = 0.5\,N_0$.

Inserting $t = T_{\frac{1}{2}}$, $N = 0.5N_0$ into $N = N_0 e^{-\lambda t}$ gives $0.5\,N_0 = N_0 e^{-\lambda T_{\frac{1}{2}}}$.

Cancelling N_0 therefore gives $0.5 = e^{-\lambda T_{\frac{1}{2}}}$.

Taking the natural logarithm (ln) of each side gives $\ln 0.5 = -\lambda T_{\frac{1}{2}}$.

Because $\ln 0.5 = -\ln 2$, then $\ln 2 = \lambda T_{\frac{1}{2}}$.

Rearranging this equation gives $T_{\frac{1}{2}} = \frac{\ln 2}{\lambda}$.

Summary questions

$N_A = 6.02 \times 10^{23}\ \text{mol}^{-1}$, $1\ \text{MeV} = 1.6 \times 10^{-13}\ \text{J}$

1 ${}^{131}_{53}\text{I}$ is a radioactive isotope of iodine which has a half-life of 8.0 days. A fresh sample of this isotope contains 4.2×10^{16} atoms of this isotope. Calculate:

- **a** the decay constant of this isotope
- **b** the number of atoms of this isotope remaining after 24 h.

2 A radioactive isotope has a half-life of 35 years. A fresh sample of this isotope has an activity of 25 kBq. Calculate:

- **a** the decay constant in s^{-1}
- **b** the activity of the sample after 10 years.

3 The isotope ${}^{226}_{88}\text{Ra}$ has a half-life of 1620 years. An initial mass of 1.0 kg of this isotope, has an activity of 3.6×10^{13} Bq. Calculate:

- **a** the mass of this isotope remaining after 1000 years
- **b** how many atoms of the isotope will remain after 1000 years.

4 A fresh sample of a radioactive isotope has an initial activity of 40 kBq. After 48 h, its activity has decreased to 32 kBq. Calculate:

- **a** the decay constant of this isotope
- **b** its half-life.

Practice questions: Chapter 19

1 In an experiment to measure the capacitance C of a capacitor, the circuit in **Figure 1** is used to charge the capacitor then discharge it through a resistor of known resistance R.

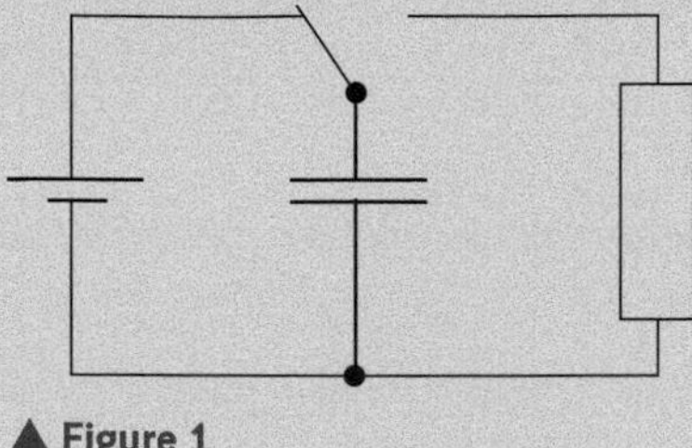

▲ Figure 1

(a) The capacitor p.d. V at time t after the discharge commences is given by $V = V_0 e^{\frac{-t}{CR}}$. Show that this equation can be rearranged into an equation of the form $\ln V = a - bt$, where a and b are constants, and determine expressions for a and b. *(4 marks)*

(b) As the capacitor discharges, its p.d. is measured every 30 seconds using a digital voltmeter. The measurements were taken three times, as shown in Table 1.

▼ Table 1

t / s	0	30	60	90	120	150	180	210	240	270	300
	4.50	3.82	3.26	2.78	2.33	2.00	1.70	1.43	1.23	1.04	0.89
V / V	4.51	3.81	3.25	2.77	2.35	2.10	1.72	1.43	1.25	1.02	0.90
	4.50	3.83	3.25	2.76	2.34	1.98	1.69	1.42	1.22	1.04	0.87
Mean V / V	4.503	3.820	3.253	2.760	2.340	2.027	1.703				
ln V	1.505	1.340	1.180	1.017	0.850	0.707	0.532				

(i) Copy Table 1 and complete the missing entries.

(ii) Use the measurements to plot a graph of ln V on the y-axis against t on the x-axis.

(iii) Use your graph to determine the time constant of the discharge circuit.

(iv) The resistance R of the resistor was 68 kΩ. Determine the capacitance C of the capacitor. *(10 marks)*

(c) (i) Discuss the reliability of the measurements.

(ii) Estimate the accuracy of your value of capacitance, given the resistor value is accurate to within 1%. *(4 marks)*

2 **Figure 2** shows a 2.0 μF capacitor connected to 150 V supply.

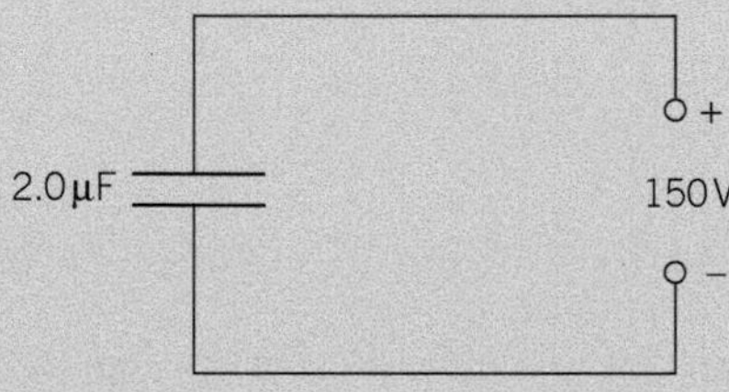

▲ Figure 2

(a) Calculate the charge on the capacitor. *(2 marks)*

(b) (i) Suggest a graph that could be drawn in order to calculate the energy stored in the capacitor by finding the area under the graph.

(ii) Calculate the energy stored by the capacitor when it has a p.d. of 150 V across it. *(3 marks)*

(c) The charged capacitor is removed from the power supply and discharged by connecting a 220 kΩ resistor across it.

(i) Calculate the maximum discharge current.

(ii) Show that the current will have fallen to 10% of its maximum value in a time of approximately 1 s. *(5 marks)*

AQA, 2002

3 A capacitor of capacitance 330 μF is charged to a potential difference of 9.0 V. It is then discharged through a resistor of resistance 470 kΩ.

(a) Calculate:

(i) the energy stored by the capacitor when it is fully charged *(2 marks)*

(ii) the time constant of the discharging circuit *(1 mark)*

(iii) the p.d. across the capacitor 60 s after the discharge has begun. *(3 marks)*

(b) The capacitor is charged using a 9.0 V battery with negligible internal resistance in series with a 1.0 kΩ resistor. Calculate:

(i) the time constant for this circuit, and sketch graphs to show how the capacitor p.d. and the current change with time during charging *(5 marks)*

(ii) the energy supplied by the battery and the energy supplied to the capacitor during the charging process, and explain the difference. *(3 marks)*

AQA, 2004

4 A student used a voltage sensor connected to a data logger to plot the discharge curve for a 4.7 μF capacitor. **Figure 3** shows the graph she obtained.

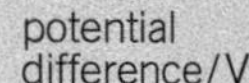

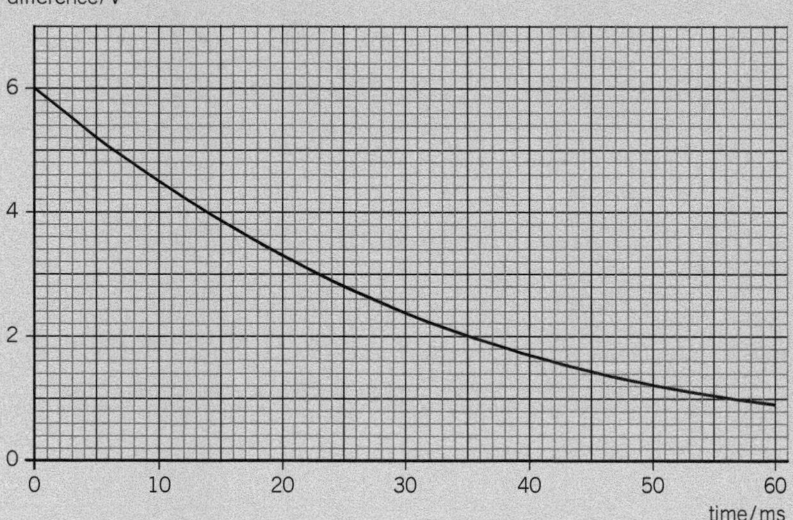

▲ **Figure 3**

Use data from the graph to calculate:

(a) the initial charge stored *(2 marks)*

(b) the energy stored when the capacitor had been discharging for 35 ms *(3 marks)*

(c) the time constant for the circuit *(3 marks)*

(d) the resistance of the circuit through which the capacitor was discharging. *(2 marks)*

AQA, 2002

5 A radioactive isotope of cobalt has a half-life of 5.3 years and it decays to form a stable isotope. A sample of the isotope has an activity of 850 kBq.

(a) Calculate the number of atoms of the cobalt isotope in the sample.

(b) Show that the activity of the sample after a full year would be about 750 kBq.

6 **(a)** (i) Alpha and beta emissions are known as *ionising radiations*. State and explain why such radiations can be described as *ionising*.

(ii) Explain why beta particles have a greater range in air than alpha particles.

(4 marks)

(b) **Figure 4** shows the variation with time of the number of radon (^{220}Ra) atoms in a radioactive sample.

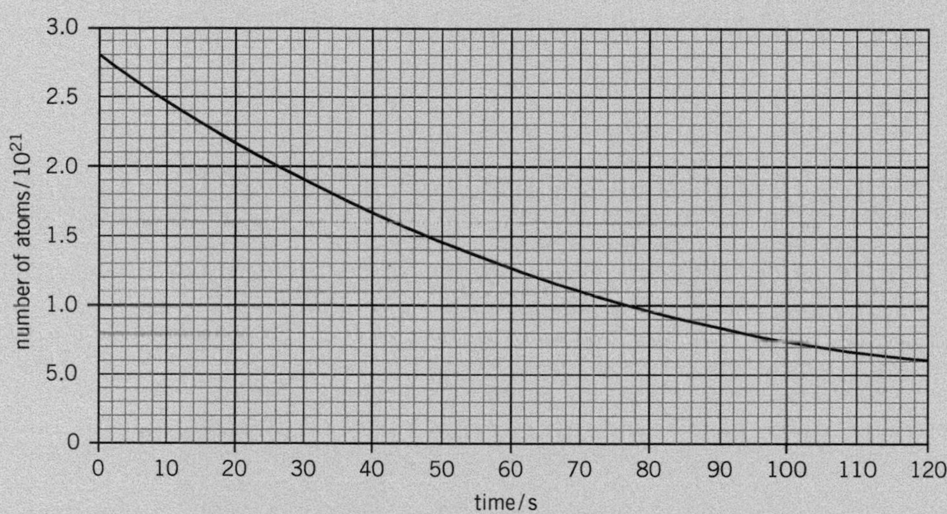

▲ **Figure 4**

(i) Use the graph to show that the half-life of the decay is approximately 53 s. Show your reasoning clearly.

(ii) The decay constant of $^{220}_{88}$Ra is $1.3 \times 10^{-2}\,s^{-1}$. Use data from the graph to find the activity of the sample at a time $t = 72$ s. *(6 marks)*

(c) (i) State *two* origins of background radiation.

(ii) Suggest why it should be unnecessary to allow for background radiation when measuring the activity of the sample described in part (b)(ii). *(3 marks)*

AQA, 2005

7 A torch bulb produces a flash of light when a 270 μF capacitor is discharged across it.

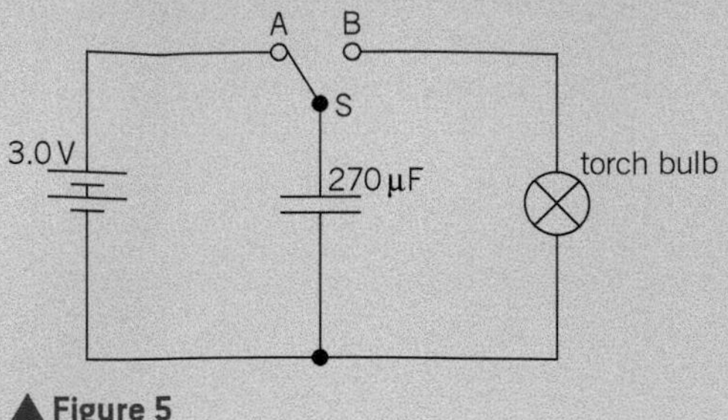

▲ Figure 5

(a) The capacitor is charged to a p.d. of 3.0 V from the battery, as shown in **Figure 5**. Calculate:
  (i) the energy stored in the capacitor
  (ii) the work done by the battery. *(3 marks)*

(b) The capacitor is discharged by moving switch S in the diagram from A to B. The discharge circuit has a total resistance of 1.5 Ω.
  (i) Show that almost all of the energy stored in the capacitor is released when the capacitor p.d. has decreased from 3.0 V to 0.3 V.
  (ii) Emission of light from the torch bulb ceases when the p.d. falls below 2.0 V. Calculate the duration of the light flash.
  (iii) Assuming that the torch bulb produces photons of average wavelength 500 nm, estimate the number of photons released during the light flash. *(8 marks)*

AQA, 2006

8 A student uses a system shown in **Figure 6** to measure the contact time of a metal ball when it bounces on a metal block.

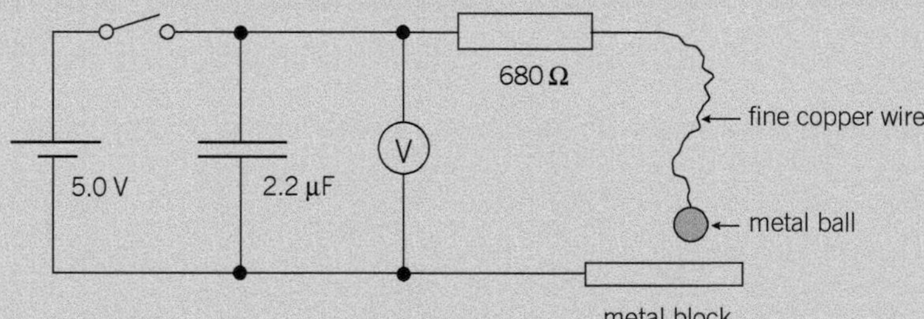

▲ Figure 6

The student charges the capacitor by closing the switch, records the voltmeter reading, and then opens the switch. The student then releases the ball and measures the p.d. after the ball has rebounded from the metal block.
In one test the student records an initial p.d. of 5.0 V and a final p.d. of 2.2 V.

(a) Calculate the time for which the ball is in contact with the block. *(3 marks)*

(b) (i) Calculate the energy lost by the capacitor during the discharge.
  (ii) State where this energy is dissipated and the form it will take. *(4 marks)*

AQA, 2002

Answers to the Practice Questions and Section Questions are available at
www.oxfordsecondary.com/oxfordaqaexams-alevel-physics

20 Magnetic fields
20.1 Current-carrying conductors in a magnetic field

Learning objectives:

- → Measure the strength of a magnetic field.
- → State the factors that the magnitude of the force on a current-carrying wire depends on.
- → Determine the direction of the force on a current-carrying wire in a magnetic field.

Specification reference: 3.10.1

Magnetic field patterns

Magnetism is a topic with a long scientific history stretching back thousands of years when lodestone was used by explorers as a navigational aid. Scientific research over the past 50 years or so has led to many applications such as particle accelerators, powerful microwave transmitters, magnetic discs and tape, superconducting magnets, and magnetic resonance scanners. Magnetism is a valuable scientific tool used by archaeologists, astronomers, and geologists. In short, magnetism has always been a fascinating topic and continues to be so.

A magnetic field is a force field surrounding a magnet or current-carrying wire which acts on any other magnet or current-carrying wire placed in the field. The magnetic field of a bar magnet is strongest at its ends which are referred to as **north-seeking** and **south-seeking** *poles* according to which direction, north or south, each end points when the magnet is free to align itself with the horizontal component of the Earth's magnetic field. A **line of force** (or magnetic field line) of a magnetic field is a line along which a north pole would move in the field.

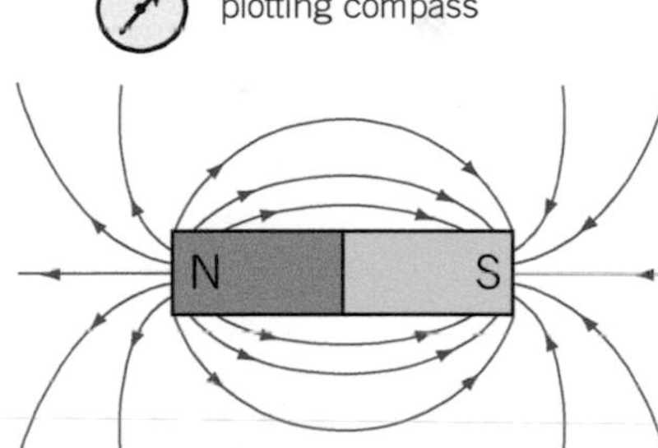

▲ **Figure 1** *The magnetic field near a bar magnet*

Extension

The Earth's magnetism

The Earth's magnetic field is not unlike the field of a giant bar magnet inside the Earth. It is caused by circulation currents in the molten iron in the Earth's core. The Earth's magnetic poles are known to drift gradually and a magnetic compass points to the Earth's magnetic north pole. At the present time, the Earth's north magnetic pole is in Northern Canada. As a result of studying the magnetism of rocks, scientists discovered that the continents are drifting across the Earth's surface on giant *tectonic plates* and that the Earth's magnetic field can reverse quite suddenly!

Q: Describe the direction of the Earth's magnetic field at the magnetic north pole.

Answer: it is vertically downwards.

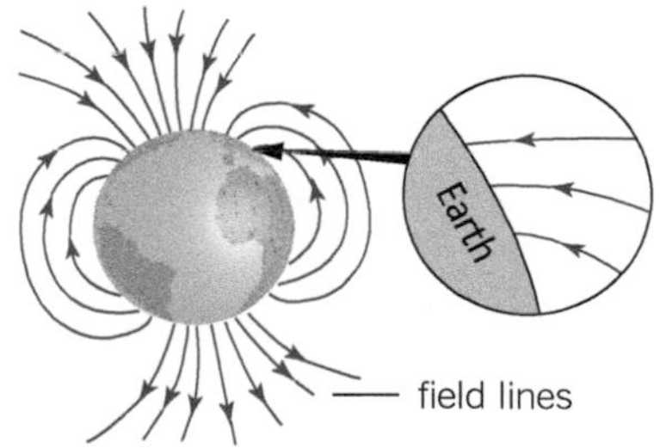

▲ **Figure 2** *The Earth's magnetic field*

The force on a current-carrying wire in a magnetic field

A current-carrying wire placed at a non-zero angle to the lines of force of an external magnetic field experiences a force due to the field. This effect is known as the **motor effect**. The force is perpendicular to the wire and to the lines of force.

The motor effect can be tested using the simple arrangement shown in Figure 3. The wire is placed between opposite poles of a U-shaped magnet so it is at right angles to the lines of force of the magnetic field.

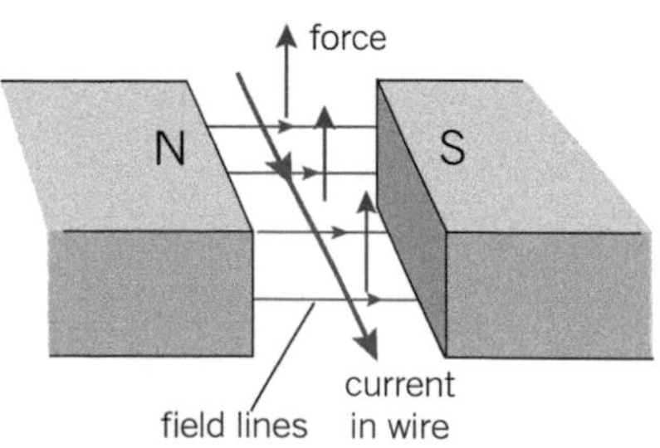

▲ **Figure 3** *The motor effect*

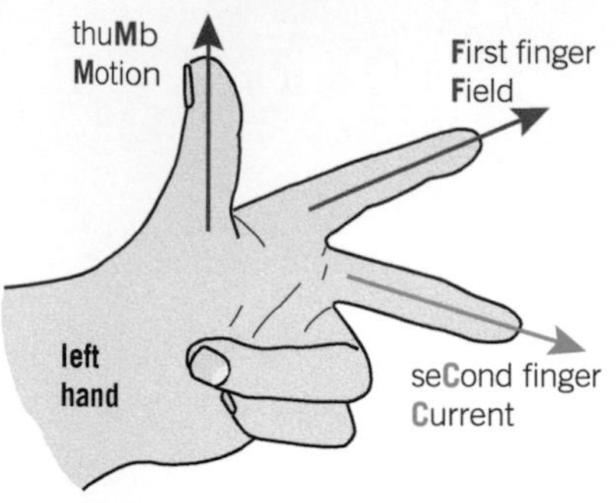

▲ **Figure 4** *Fleming's left-hand rule*

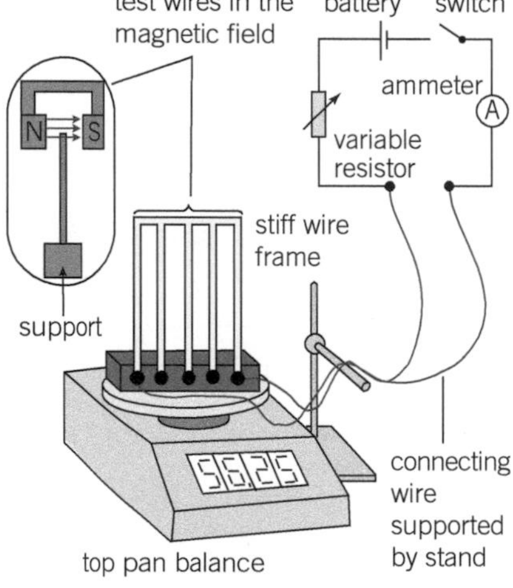

▲ **Figure 5** *Measuring the force on a current-carrying wire in a magnetic field*

Study tip

$F = BIl$ applies only when B and I are at right angles.

When a current flows, the section of the wire in the magnetic field experiences a force that pushes it out of the field. The magnitude of the force depends on the current, the strength of the magnetic field, the length of the wire, and the angle between the lines of force of the field and the current direction.

The force is:

- greatest when the wire is at right angles to the magnetic field
- zero when the wire is parallel to the magnetic field.

The direction of the force can be related to the direction of the field and to the direction of the current using **Fleming's left-hand rule** shown in Figure 4. If the current is reversed or if the magnetic field is reversed, the direction of the force is reversed.

The magnitude of the force on a current-carrying wire in a magnetic field can be investigated using the arrangement shown in Figure 5. The stiff wire frame is connected in series with a switch, an ammeter, a variable resistor, and a battery. When the switch is closed, the magnet exerts a force on the wire which can be measured from the change of the top pan balance reading.

The tests above show that the force F on the wire is proportional to:

1 the current I

2 the length l of the wire.

The **magnetic flux density** B of the magnetic field, sometimes referred to as the strength of the magnetic field, is defined as the force per unit length per unit current on a current-carrying conductor at right angles to the magnetic field lines.

Therefore, for a wire of length l carrying a current I in a uniform magnetic field B **at 90° to the field lines**, the force F on the wire is given by

$$F = BIl$$

1 The unit of B is the tesla (T), equal to $1\,\text{N}\,\text{m}^{-1}\,\text{A}^{-1}$.
2 The direction of the force is given by Fleming's left-hand rule. See Figure 4.

Worked example

A straight horizontal wire XY of length 5.0 m is in a uniform horizontal magnetic field of magnetic flux density 120 mT. The wire is at an angle of 90° to the field lines which are due north in direction. The wire conducts a current of 14 A from east to west. Calculate the magnitude of the force on the wire and state its direction.

Solution

$B = 120\,\text{mT} = 0.12\,\text{T}$

$F = BIl = 0.12 \times 14 \times 5.0 = 8.4\,\text{N}$

The force on the wire is vertically downwards.

Note: for a straight wire at angle θ to the magnetic field lines, the force on the wire is due to the component of the magnetic field perpendicular to the wire, $B\sin\theta$. Questions will be limited to situations where the field is perpendicular to the current, ie $\theta = 90°$ so $F = BIl \sin 90° = BIl$ as $\sin 90° = 1$.

The couple on a coil in a magnetic field

Consider a rectangular current-carrying coil in a uniform horizontal magnetic field, as shown in Figure 6. The coil has n turns of wire and can rotate about a vertical axis.

- The long sides of the coil are vertical. Each wire down each long side experiences a force BIl where l is the length of each long side. Each long side therefore experiences a horizontal force $F = (BIl)n$ in opposite directions at right angles to the field lines.
- The pair of forces acting on the long sides form a couple as the forces are not directed along the same line. The torque of the couple = Fd, where d is the perpendicular distance between the line of action of the forces on each side. See Topic 1.4. If the plane of the coil is at angle α to the field lines, then $d = w\cos\alpha$ where w is the width of the coil.
- Therefore, the torque = $Fw\cos\alpha = BIlnw\cos\alpha = BIAn\cos\alpha$, where the coil area $A = lw$. If $\alpha = 0$ (i.e., the coil is parallel to the field), the torque = $BIAn$ because $\cos 0 = 1$.
- If $\alpha = 90°$ (i.e., the coil is perpendicular to the field), the torque = 0 because $\cos 90° = 0$.

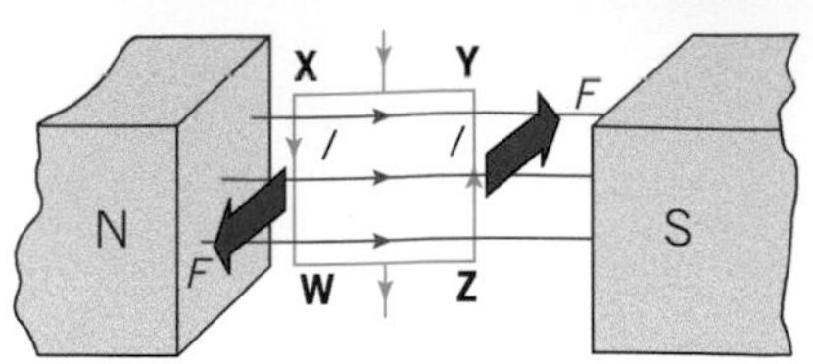

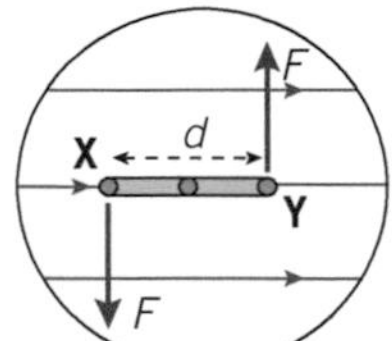

a *Top view of coil parallel to field*

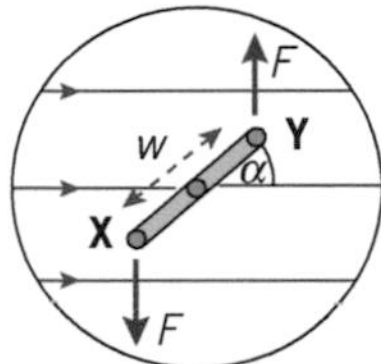

b *Top view of coil at angle α to field*

▲ **Figure 6** *Couple on a coil*

Application

The electric motor

The **simple electric motor** consists of a coil of insulated wire which spins between the poles of a U-shaped magnet.

When a direct current passes round the coil:

- the wires at opposite edges of the coil are acted on by forces in opposite directions
- the force on each edge makes the coil spin about its axis.

Current is supplied to the coil via a *split-ring commutator*. The direction of the current round the coil is reversed by the split-ring commutator each time the coil rotates through half a turn. This ensures that the current along an edge changes direction when it moves from one pole face to the other. As shown in Figure 7, the result is that the force on each edge continues to turn the coil in the same direction.

In a practical electric motor, several evenly spaced 'armature' coils are wound on an iron core. Each coil is connected to its own section of the commutator. The result is that each coil in sequence experiences a torque when it is connected to the voltage supply so the armature is repeatedly pushed round. Because the iron core makes the field radial, each coil is in the plane of the field (i.e., $\alpha = 0$) for most of the time. As a result, the torque is steady and the motor runs more smoothly. In addition, the iron core makes the field much stronger so the torque of the motor is much greater.

By using an electromagnet connected to the same voltage supply as the coils, an electric motor can operate with alternating current or with direct current. This is because the magnetic field reverses each time the armature current reverses when an ac supply is used. So the turning effect on the armature is unchanged in direction.

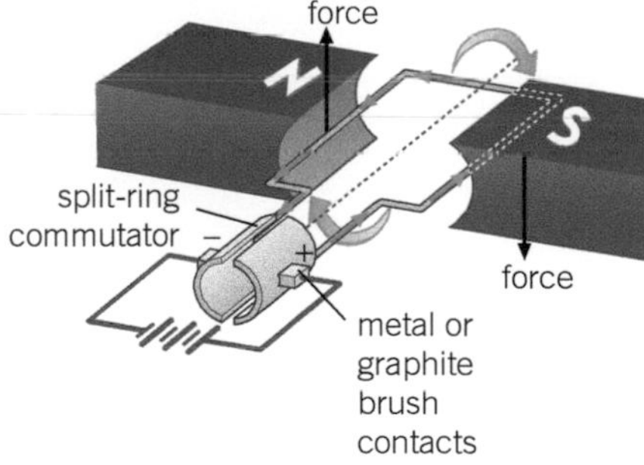

▲ **Figure 7** *In an electric motor*

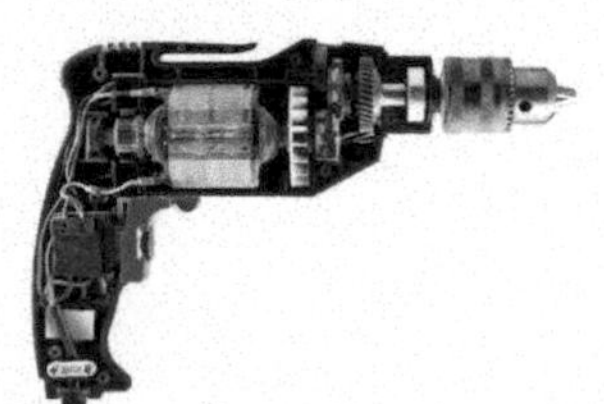

▲ **Figure 8** *A practical electric motor*

Summary questions

1 Table 1 relates the force on a current-carrying wire which is at right angles to the lines of force of a magnetic field and the current. Copy and complete Table 1 by working out the missing data in each column.

▼ **Table 1**

	(a)	(b)	(c)	(d)
B/T	0.20 T vertically down	0.20 T vertically down	?	0.1 T horizontal due ?
I/A	3.0 A horizontal due north	?	3.0 A horizontal due north	2.0 A vertically up
l/m	0.040 m	0.040 m	0.040 m	0.040 m
F/N	?	0.036 N horizontal due south	0.024 N horizontal due west	? horizontal due east

2 **a** A straight vertical wire of length 0.10 m carries a downward current of 4.0 A in a uniform horizontal magnetic field of flux density 55 mT that acts due north. Determine the magnitude and direction of the force on the wire.

b A straight horizontal wire of length 50 mm carrying a constant current is in a uniform magnetic field of flux density 140 mT which acts vertically downwards. The wire experiences a force of 28 mN in a direction which is due north. Determine the magnitude and the direction of the current in the wire.

3 A rectangular coil of width 60 mm and of length 80 mm has 50 turns. The coil is placed horizontally in a uniform horizontal magnetic field of flux density 85 mT with its shorter side parallel to the field lines. A current of 8.0 A is passed through the coil. Sketch the arrangement and determine the force on each side of the coil.

4 The Earth's magnetic field at a certain position on the Earth's surface has a horizontal component of 18 μT due north and a downwards vertical component of 55 μT. Calculate:

a the magnitude of the Earth's magnetic field at this position

b the magnitude and direction of the force on a vertical wire of length 0.80 m carrying a current of 4.5 A downwards.

20.2 Moving charges in a magnetic field

Learning objectives:

- → Describe what happens to charged particles in a magnetic field.
- → Explain why a force acts on a wire in a magnetic field when a current flows along the wire.
- → State the equation used to find the force on a moving charge.

Specification reference: 3.10.2

Electron beams

Figure 1 shows a vacuum tube designed to show the effect of a magnetic field on an electron beam. The production of the electron beam is explained in Topic 20.3. The path of the beam can be seen where it passes over the fluorescent screen in the tube. The beam is deflected downwards when a magnetic field is directed into the plane of the screen. Each electron in the beam experiences a force due to the magnetic field. The beam follows a circular path because the direction of the force on each electron is perpendicular to the direction of motion of the electron (and to the field direction). The direction of the force on an electron in the beam can also be worked out using Fleming's left-hand rule, provided we remember the convention that the current direction is opposite to the direction in which the electrons move.

The reason why a current-carrying wire in a magnetic field experiences a force is that the electrons moving along the wire are pushed to one side by the force of the field. If the electrons in Figure 1 had been confined to a wire, the whole wire would have been pushed downwards.

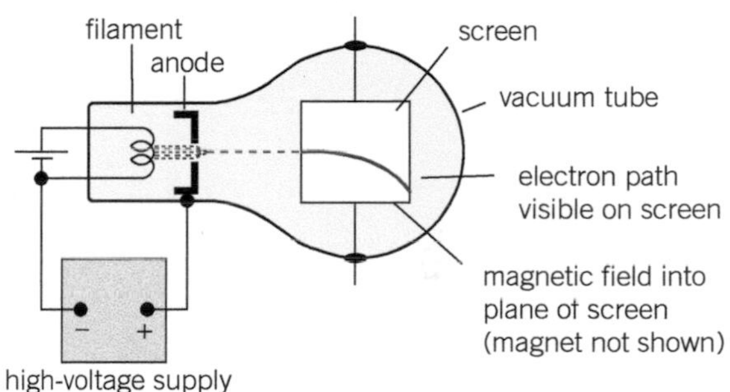

▲ **Figure 1** *An electron deflection tube*

Extension

A high-energy collision

▲ **Figure 2** *Charged particles in a magnetic field*

Magnetic fields are used in particle physics detectors to separate different charged particles out and, as will be explained in Topic 20.3, to measure their momentum from the curvature of the tracks they create. All charged particles moving across the lines of a magnetic field are acted on by a force due to the field. Positively charged particles such as protons are pushed in the opposite direction to negatively charged particles such as electrons. Figure 2 shows charged particles curving across a magnetic field. The positively charged particles curve in the opposite direction to the negatively charged particles. The particles were created by a collision between a fast-moving incoming particle and the nucleus of an atom.

Q: What type of event created the two oppositely curved tracks near the top of the picture?

Answer: A pair production event.

Synoptic link

The positron was the first antimatter particle to be discovered when a β^+ particle track in a magnetic field was found that curved in the opposite direction to the β^- tracks. See Topic 7.4, Particles and antiparticles.

Force on a moving charge in a magnetic field

A beam of charged particles crossing a vacuum tube is an electric current across the tube. Suppose each charged particle has a charge Q and moves at speed v. In a time interval t, each particle travels a distance vt. Its passage is equivalent to current $I = \frac{Q}{t}$ along a wire of length $l = vt$.

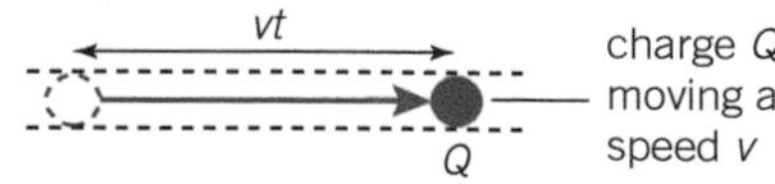

▲ **Figure 3** *Force on a moving charge*

Study tip

Stationary charges in a magnetic field experience no magnetic force. Also when applying Fleming's left-hand rule to charged particles, the current direction for negative particles is in the opposite direction to the direction of motion of the particles.

If the particles pass through a uniform magnetic field in a direction at right angles to the field lines, each particle experiences a force F due to the field. If the particles were confined to a wire, the force would be given by $F = BIl$. For moving charges, the same equation applies where $I = \frac{Q}{t}$ and $l = vt$.

Therefore, for a charged particle moving across a uniform magnetic field in a direction at right angles to the field, $F = BIl = B\left(\frac{Q}{t}\right)(vt) = BQv$.

For a particle of charge Q moving through a uniform magnetic field at speed v in a perpendicular direction to the field, the force on the particle is given by

$$F = BQv$$

If the direction of motion of a charged particle in a magnetic field is at angle θ to the lines of the field, then the component of B perpendicular to the direction of motion of the charged particle, $B\sin\theta$, is used to give $F = BQv\sin\theta$.

- If the velocity of the charged particle is perpendicular to the direction of the magnetic field, $\theta = 90°$, so the equation becomes $F = BQv$ because $\sin 90° = 1$.
- If the velocity of the charged particle is parallel to the direction of the magnetic field, $\theta = 0$, so $F = 0$ because $\sin 0 = 0$.

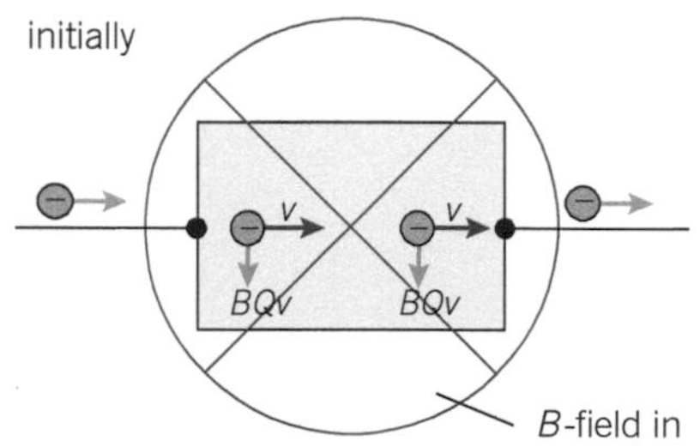

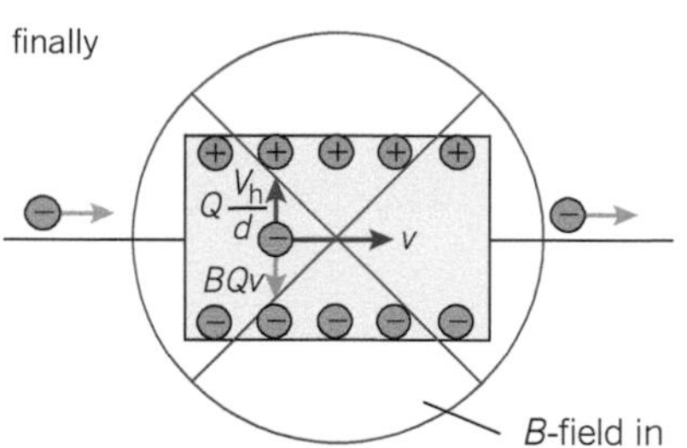

▲ **Figure 5** *The Hall voltage*

Notes

1. You do not need to know the equation $F = BQv\sin\theta$ in this course.
2. You do not need to learn about the Hall probe in this course. It is included here to show how the equations for the electric and magnetic force on a charged particle are used.

Application

The Hall probe

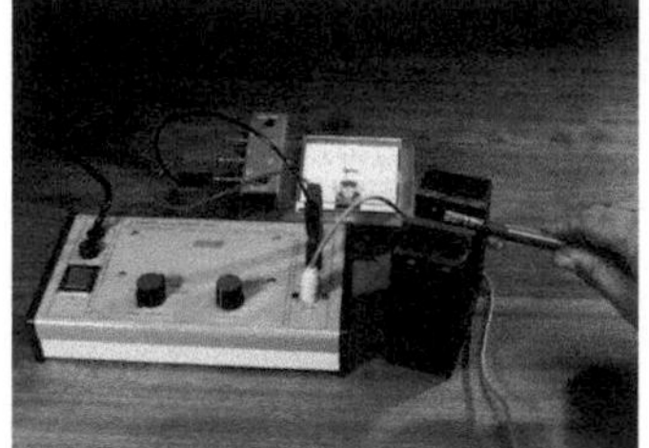

▲ **Figure 4** *Using a Hall probe*

Hall probes are used to measure magnetic flux density (Figure 4). A Hall probe contains a slice of semiconducting material. Figure 5 shows the slice in a magnetic field with the field lines perpendicular to the flat side of the slice. A constant current passes through the slice as shown. The charge carriers (which are electrons in an n-type semiconductor) are deflected by the field. As a result, a potential difference is created between the top and bottom edges of the slice. This effect is known as the Hall effect after its discoverer.

The p.d., referred to as the Hall voltage, is proportional to the magnetic flux density, provided the current is constant. This is because each charge carrier passing through the slice is subjected to a magnetic force $F_{mag} = BQv$, where v is the speed of the charge carrier. Once the Hall voltage has been created, the magnetic deflection of a charge carrier entering the slice is opposed by the force on it due to the electric field created by the Hall voltage. The electric field force $F_{elec} = \frac{QV_h}{d}$, where V_h represents the Hall voltage and d is the distance between the top and bottom sides of the slice. See Topic 18.2. Therefore, $\frac{QV_h}{d} = BQv$ gives $V_h = Bvd$. For constant current, v is constant so V_h is proportional to B.

Summary questions

$e = 1.6 \times 10^{-19}$ C

1 a In Figure 1, how would the force on the electrons in the magnetic field differ if:

i the magnetic field was reversed in direction

ii the magnetic field was reduced in strength

iii the speed of the electrons was increased.

b Calculate the force on an electron that enters a uniform magnetic field of flux density 150 mT at a velocity of 8.0×10^{6} m s^{-1} at an angle of:

i 90° **ii** 0° to the field.

2 Electrons in a vertical wire move upwards at a speed of 2.5×10^{-3} m s^{-1} into a uniform horizontal magnetic field of magnetic flux density 95 mT. The field is directed along a line from south to north as shown in Figure 6. Calculate the force on each electron and determine its direction.

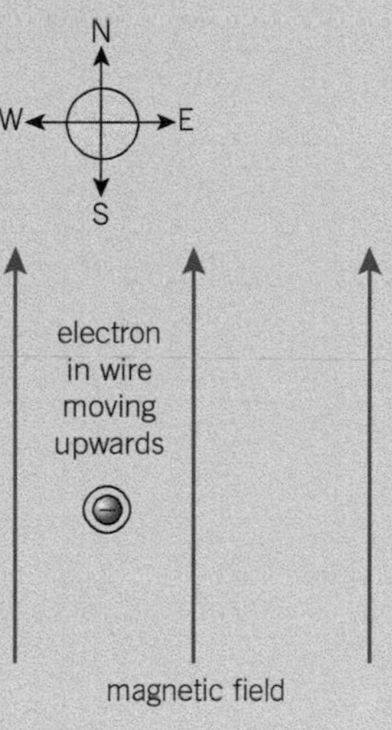

▲ **Figure 6**

3 A beam of protons and π^{+} mesons moving at the same speed is directed into a uniform magnetic field in the same direction as the field.

a Explain why the beam is not deflected by the field.

b If the particles had been directed into the field in a direction at right angles to the field lines at the same speed, state and explain what effect this would have had on the beam.

4 In a Hall probe, electrons passing through the semiconductor experience a force due to a magnetic field.

a Explain why a potential difference is created across the semiconductor as a result of the application of the magnetic field.

b When the magnetic flux density is 90 mT, each electron moving through the slice experiences a force of 6.4×10^{-20} N due to the magnetic field. Calculate:

i the mean speed of the electrons passing through the slice

ii the magnetic force on each electron if the magnetic flux density is increased to 120 mT.

20.3 Charged particles in circular orbits

Learning objectives:

- → Describe what happens to the direction of the magnetic force when electrons are deflected by a magnetic field.
- → Explain why the moving charges move in a path that is circular.
- → State the factors that affect the radius of the circular path.

Specification reference: 3.10.2

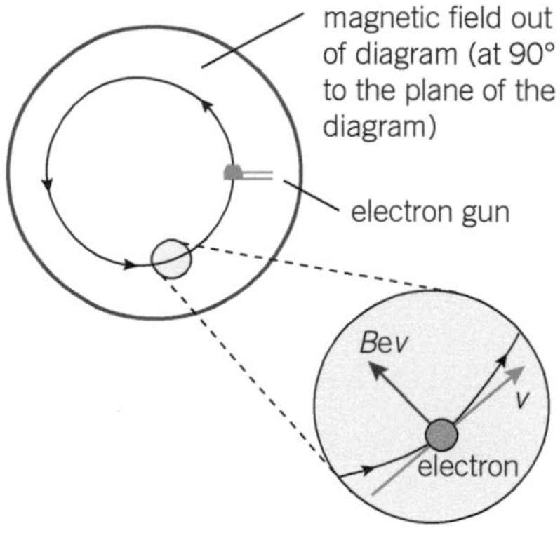

▲ **Figure 1** *A circular orbit in a magnetic field*

Synoptic link

We can apply the centripetal acceleration formula $a = \frac{v^2}{r}$ because the particle is in uniform circular motion. See Topic 15.2, Centripetal acceleration.

Study tip

Check whether a question refers to radius or diameter. Remember that equipment must be evacuated to prevent loss of speed.

Magnetic fields are used to control beams of charged particles in many devices, from television tubes to high-energy accelerators. The force of the magnetic field on a moving charged particle is at right angles to the direction of motion of the particle.

- No work is done by the magnetic field on the particle as the force always acts at right angles to the velocity of the particle. Its direction of motion is changed by the force but not its speed. The kinetic energy of the particle is unchanged by the magnetic field.
- In accordance with Fleming's left-hand rule, the magnetic force is always perpendicular to the velocity at any point along the path. The particle therefore moves on a circular path with the force always acting towards the centre of curvature of the circular path. See Topic 15.2.
- The force causes a centripetal acceleration because it is perpendicular to the velocity. Figure 1 shows the deflection of a beam of electrons in a uniform magnetic field. The path is a complete circle because the magnetic field is uniform and the particle remains in the field. In Figure 1, $Q = e$ (the charge on the electron) so $F = BeV$.

The radius, r, of the circular orbit in Figure 1 depends on the speed v of the particles and the magnetic flux density B.

At any point on the orbit, the particle is acted on by a magnetic force $F = BQv$ and it experiences a centripetal acceleration $a = \frac{v^2}{r}$ towards the centre of the circle.

Applying Newton's second law in the form $F = ma$ gives

$$BQv = \frac{mv^2}{r}$$

Rearranging this equation gives

$$r = \frac{mv}{BQ}$$

This equation shows that r decreases (so the path is more curved):

1 if B is increased or if v is decreased,
2 if particles with a larger specific charge, $\frac{Q}{m}$, are used.

Application

Thermionic devices

In Figure 1, the beam of electrons is produced by an *electron gun*. This consists of an electrically heated filament wire near a positively charged metal anode which attracts electrons emitted by the hot filament wire. This emission process is called thermionic emission. The electrons pass through a small hole in the anode to form the beam. The greater the potential difference between the anode and the filament wire, the higher the speed of the electrons when they reach the anode, so the faster they are in the beam. The oscilloscope, the cathode ray television tube, and the magnetron valve used in microwave cookers and radar systems all rely on thermionic emission.

Extension

The mass spectrometer

The mass spectrometer is used to analyse the type of atoms present in a sample. The atoms of the sample are ionised and directed in a narrow beam at the same velocity into a uniform magnetic field. Each ion is deflected in a semi-circle by the magnetic field onto a detector, as shown in Figure 2. The radius of curvature of the path of each ion depends on the specific charge $\frac{Q}{m}$ of the ion in accordance with the equation $r = \frac{mv}{BQ}$. Each type of ion is deflected by a different amount onto the detector. The detector is linked to a computer which is programmed to show the relative abundance of each type of ion in the sample.

The ions in the beam enter the magnetic field at the same velocity because they pass through a velocity selector, as shown in Figure 2. The velocity selector consists of a magnet and a pair of parallel plates at spacing d and voltage V_P due to a high-voltage supply. The magnet and the plates are aligned so each ion passing through the velocity selector is acted on by an electric field force, $F_{elec} = \frac{QV_P}{d}$, in the opposite direction to a magnetic field force $F_{mag} = B_SQv$ where B_S is the magnetic flux density of the magnet in the velocity selector. Ions moving at a certain velocity such that $B_SQv = \frac{QV_P}{d}$ experience equal and opposite forces so they pass through undeflected. All other ions are deflected and do not pass through the collimator slit. So although the beam emerging from the collimator consists of different types of ions, they all have the same speed $v = \frac{V_P}{B_Sd}$.

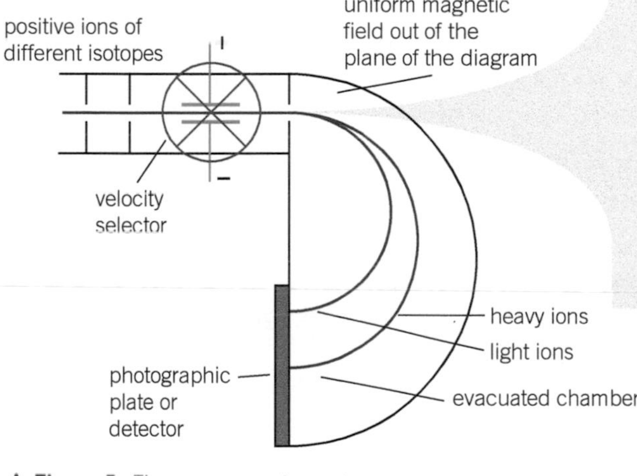

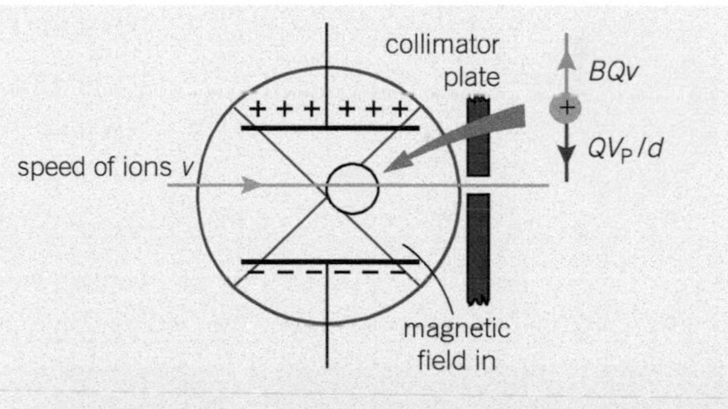

▲ **Figure 2** *The mass spectrometer*

Synoptic link

The magnetic field force is cancelled out by the electric field force. See Topic 18.2, Electric field strength, for the force on a charged particle in a uniform electric field.

Q: State whether a proton or a singly charged helium ion is deflected more easily in a mass spectrometer.

Answer: A proton.

Application

The cyclotron

The cyclotron is used in hospitals to produce high-energy beams for radiation therapy. It consists of two hollow D-shaped electrodes (referred to as 'dees') in a vacuum chamber. With a uniform magnetic field applied perpendicular to the plane of the dees, a high-frequency alternating voltage is applied between the dees.

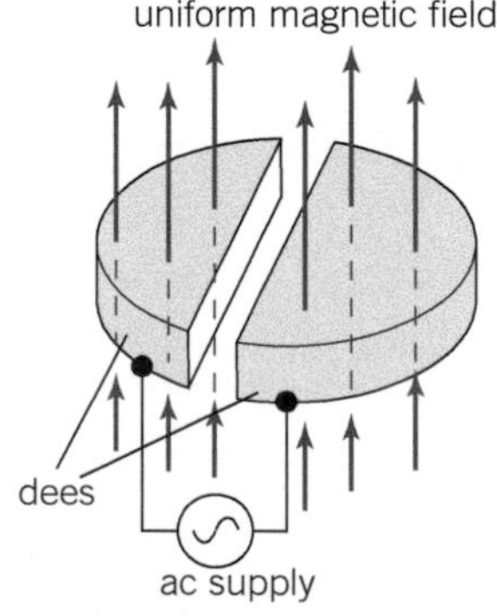

▲ **Figure 3** *The cyclotron*

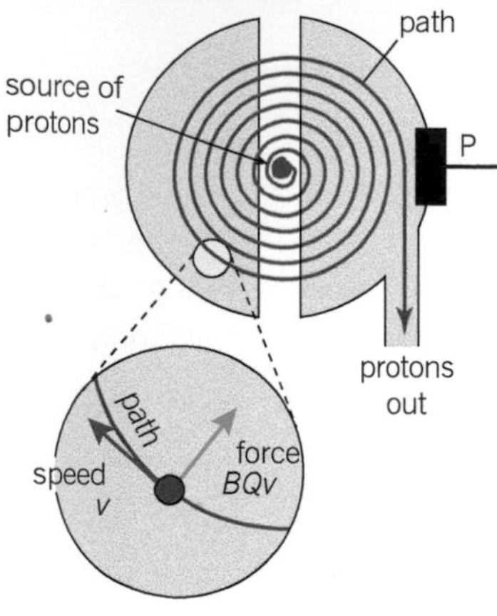

▲ **Figure 4** *Inside the cyclotron*

Charged particles are directed into the one of the dees near the centre of the cyclotron. The charged particles are forced on a circular path by the magnetic field, causing them to emerge from the dee they were directed into. As they cross into the other dee, the alternating voltage reverses so they are accelerated into the other dee where they are once again forced on a circular path by the magnetic field. On emerging from this dee, the voltage reverses again and accelerates the particles into the first dee where the process is repeated. This occurs because the time taken by a particle to move round its semi-circular path in each dee does not depend on the particle speed (provided the speed stays much less than the speed of light c). This is because $r = \frac{mv}{BQ}$ so the time taken to complete the semi-circle $= \frac{\pi r}{v} = \frac{m\pi}{BQ}$, which is independent of the particle's speed.

1 Each time a particle crosses from one dee to the other it gains speed and its radius of orbit increases, and a suitable voltage is applied at deflecting plate P. The particles emerge from the cyclotron when the radius of orbit is equal to the dee radius R and a suitable deflecting voltage is applied to the deflecting plate P.
Using the equation $r = \frac{mv}{BQ}$, it follows that the speed v of the particles on exit from the cyclotron is given by $v = \frac{BQR}{m}$.

2 The time T for one full cycle of the alternating voltage must be equal to the time taken by a particle to complete one full circle.
Hence $T = \frac{2m\pi}{BQ}$. Therefore the frequency f of the alternating voltage must be set at a value given by the equation $f = \frac{1}{T} = \frac{BQ}{2\pi m}$.

Summary questions

$e = 1.6 \times 10^{-19}$ C, $\frac{e}{m}$ for the electron $= 1.76 \times 10^{11}$ C kg^{-1}

1 A beam of electrons at a speed of 3.2×10^7 m s^{-1} is directed into a uniform magnetic field of flux density 8.5 mT in a direction perpendicular to the field lines. The electrons move on a circular orbit in the field.

a **i** Explain why the electrons move on a circular orbit.

ii Calculate the radius of the orbit.

b The flux density is adjusted until the radius of orbit is 65 mm. Calculate the flux density for this new radius.

2 A narrow beam of electrons is directed at a speed of 2.9×10^7 m s^{-1} into a uniform magnetic field.

a The beam follows a circular path of radius 35 mm in the magnetic field. Calculate the flux density of the magnetic field.

b The speed of the electrons in the beam is halved by reducing the anode voltage. Calculate the new radius of curvature of the beam in the field.

3 The first cyclotron, used to accelerate protons, was 0.28 m in diameter and was in a magnetic field of flux density 1.1 T.

a Show that protons emerge from this cyclotron at a maximum speed of 1.5×10^7 m s^{-1}.

b Calculate the maximum kinetic energy, in MeV, of a proton from this accelerator.

mass of a proton $= 1.67 \times 10^{-27}$ kg
1 MeV $= 1.6 \times 10^{-13}$ J

4 In a mass spectrometer, a beam of ions at a speed of 7.6×10^4 m s^{-1} is directed into a uniform magnetic field of flux density 680 mT.

a An ion is deflected in a semi-circular path of diameter 28 mm onto the detector. Calculate the specific charge of the ion.

b A different type of ion is deflected onto the same detector when the magnetic flux density is changed to 400 mT. Calculate the specific charge of this ion.

Practice questions: Chapter 20

1 **(a)** The equation $F = BIl$, where the symbols have their usual meanings, gives the magnetic force that acts on a conductor in a magnetic field.
Give the unit of each of the quantities in the equation: F, B, I, l.
State the condition under which the equation applies. *(2 marks)*

(b) **Figure 1** shows a horizontal copper bar of 25 mm × 25 mm square cross section and length l carrying a current of 65 A.

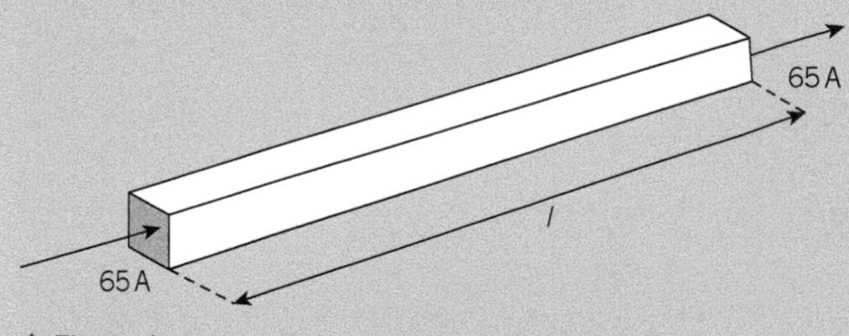

▲ **Figure 1**

(i) Calculate the minimum value of the flux density of the magnetic field in which it should be placed if its weight is to be supported by the magnetic force that acts on it.

density of copper = $8.9 \times 10^3\,\mathrm{kg\,m^{-3}}$

(ii) Copy the diagram and draw an arrow to show the direction in which the magnetic field should be applied if your calculation in part (i) is to be valid. Label this arrow M. *(5 marks)*

AQA, 2003

2 A 'bus bar' is a metal bar which can be used to conduct a large electric current. In a test, two bus bars, X and Y, of length 0.83 m are clamped at either end parallel to each other, as shown in **Figure 2**.

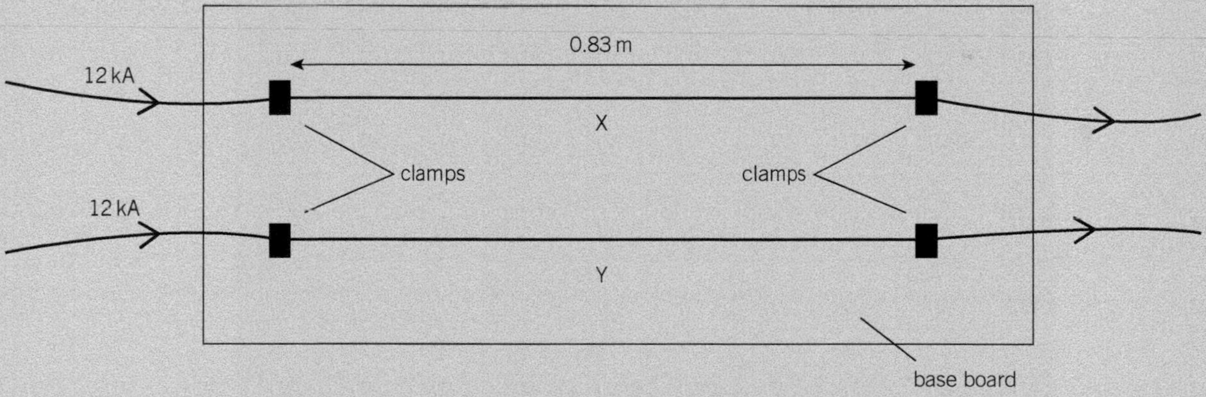

▲ **Figure 2**

(a) When a constant current of 12 kA is carried by each bus bar, they exert a force of 180 N on each other. This force is due to the magnetic field created by the current carried by each bus bar.

(i) Calculate the magnetic flux density due to the current in one bus bar at the position of the other bus bar.

(ii) The magnetic flux density at any given distance from a straight conductor is proportional to the current through the conductor. Calculate the force on each bus bar if X carries a current of 6 kA and Y carries a current of 12 kA in the same direction. *(6 marks)*

(b) When the same alternating current is passed through the two bus bars, both vibrate strongly.

(i) Explain why the bars vibrate.

(ii) State *one* way the amplitude of the vibrations could be reduced without reducing the current. *(4 marks)*

AQA, 2007

3 **(a)**

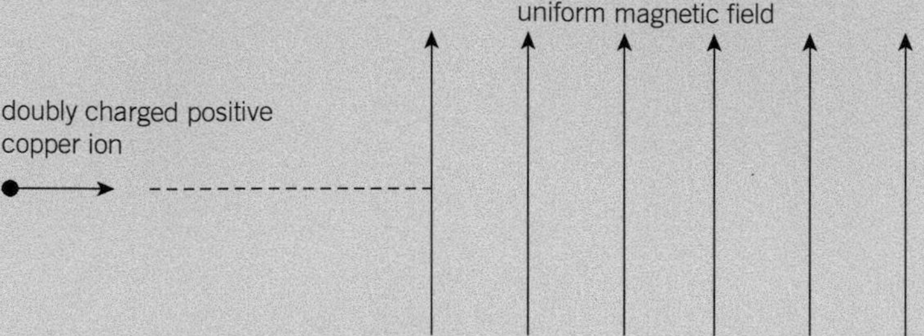

▲ **Figure 3**

Figure 3 shows a doubly charged positive ion of the copper isotope $^{63}_{29}Cu$ that is projected into a vertical magnetic field of flux density 0.28 T, with the field directed upwards. The ion enters the field at a speed of $7.8 \times 10^5\,m\,s^{-1}$.

(i) State the initial direction of the magnetic force that acts on the ion.

(ii) Describe the subsequent path of the ion as fully as you can. Your answer should include both a qualitative description and a calculation.

mass of $^{63}_{29}Cu$ ion = $1.05 \times 10^{-25}\,kg$ *(5 marks)*

(b) State the effect on the path in part (a) if the following changes are made separately.

(i) The strength of the magnetic field is doubled.

(ii) A singly charged positive Cu ion replaces the original one. *(3 marks)*

AQA, 2004

4 **Figure 4** shows a diagram of a mass spectrometer.

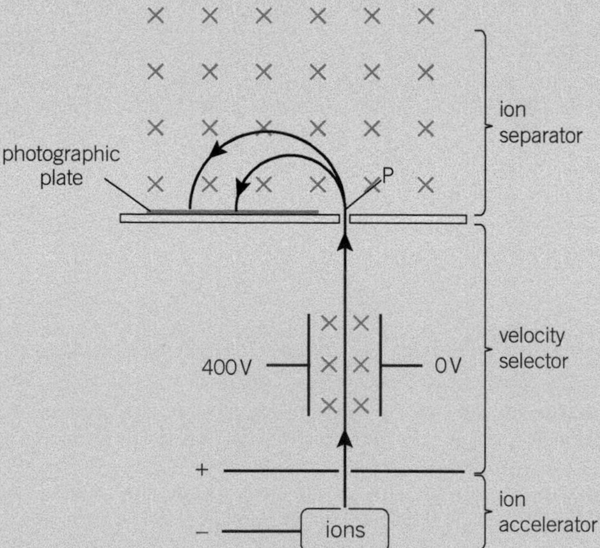

▲ **Figure 4**

(a) The magnetic field strength in the velocity selector is 0.14 T and the electric field strength is 20 000 V m^{-1}.
(i) Define the unit for magnetic flux density, the tesla.
(ii) Show that the velocity selected is independent of the charge on an ion.
(iii) Show that the velocity selected is about 140 km s^{-1}. *(5 marks)*

(b) A sample of nickel is analysed in the spectrometer. The two most abundant isotopes of nickel are $^{58}_{28}\mathrm{Ni}$ and $^{60}_{28}\mathrm{Ni}$. Each ion carries a single charge of $+1.6 \times 10^{-19}$ C.
The $^{58}_{28}\mathrm{Ni}$ ion strikes the photographic plate 0.28 m from the point P at which the ion beam enters the ion separator.
Calculate:
(i) the magnetic flux density of the field in the ion separator
(ii) the separation of the positions where the two isotopes hit the photographic plate. *(5 marks)*
AQA, 2003

5 The protons in an accelerator are directed at a solid target, causing antiprotons and negative pions, as well as other particles and antiparticles, to emerge at high speed from the target. Negative pions are particles that each have less mass than an antiproton and the same charge as an antiproton. A uniform magnetic field is used to separate the negative particles from the uncharged and positive particles, as shown in **Figure 5**.

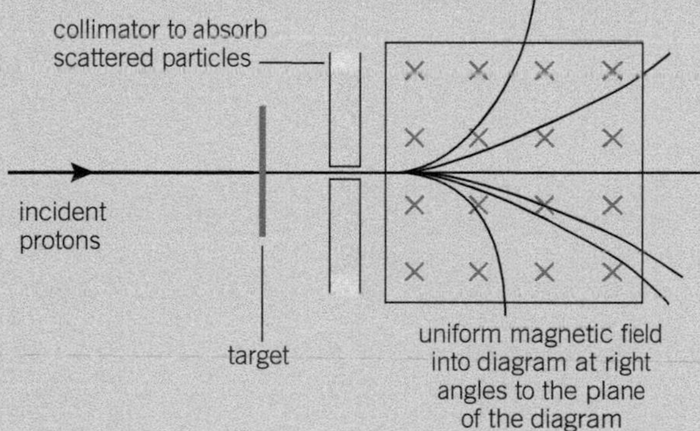

▲ **Figure 5**

(a) Show that the speed, v, of a charged particle moving in a circular path of radius r in a uniform magnetic field B is given by

$$v = \frac{BQr}{m}$$

where m is the mass of the particle and Q is its charge. *(1 mark)*

(b) An antiproton and a negative pion follow the same path in the magnetic field. Explain why they have the same momentum but different speeds. *(3 marks)*

(c) State, in terms of quarks and antiquarks, the composition of each of the following: antiproton, negative pion. *(3 marks)*
AQA, 2004

6 **Figure 6** shows the arrangement of an apparatus for determining the masses of ions. In an evacuated chamber, positive ions from an ion source pass through the slit at P with the same velocity v. After passing P, the ions enter a region over which a uniform magnetic field is applied. The ions travel in a semi-circular path of diameter d and are detected at points such as R.

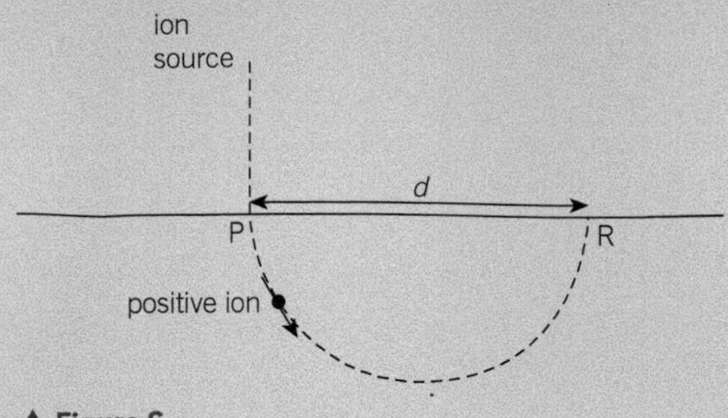

▲ **Figure 6**

(a) (i) State the direction of the applied magnetic field.

(ii) Explain why the ions travel in a semi-circular path whilst in the magnetic field.

(iii) By considering the force that acts on an ion of mass m and charge Q, having velocity v, show that the diameter d of the path of the ions is given by

$$d = \frac{2mv}{BQ}$$

where B is the flux density of the magnetic field. *(7 marks)*

(b) In an experiment using singly ionised magnesium ions travelling at a velocity of $7.5 \times 10^4\,\mathrm{m\,s^{-1}}$, d was 110 mm when B was 0.34 T. Use this result to calculate the charge to mass ratio of these ions. *(2 marks)*

(c) (i) Some ions of the same element, whilst travelling at the same velocity as each other at P, may arrive at a point that is close to, but slightly different from, R. Explain why this might happen.

(ii) Other ions of the same element, also travelling at the same velocity at P as all of the others, may travel in a path whose diameter is half that of the others. Explain why this might happen. *(3 marks)*

AQA, 2007

7 **Figure 7** shows the path of protons in a proton synchrotron.

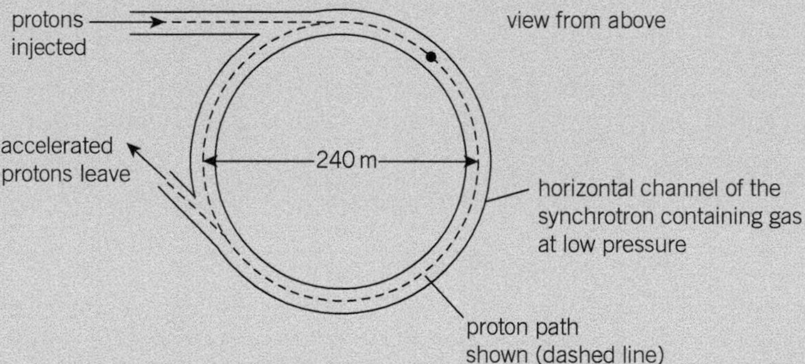

▲ **Figure 7**

The protons are injected at a speed of $1.2 \times 10^5\,\mathrm{m\,s^{-1}}$ and a magnetic field is applied to make them move in a circular path.

(a) Calculate the magnetic flux density of the field required for protons to move in the circular path when their speed is $1.2 \times 10^5\,\mathrm{m\,s^{-1}}$.

(b) Explain how the magnetic flux density required to maintain the circular path has to change as the kinetic energy of the protons increases. *(5 marks)*

AQA, 2007

21 Electromagnetic induction

21.1 Generating electricity

Learning objectives:

- → Describe what must happen to a conductor (or to the magnetic field in which it is placed) for electricity to be generated.
- → State the factors that would cause the induced e.m.f. to be greater.
- → Discuss whether an induced e.m.f. always causes a current to flow.

Specification reference: 3.10.3 and 3.10.4

Investigating electromagnetic induction

To generate electricity, all you need is a magnet and some wire, preferably connected to a sensitive meter, as shown in Figure 1. When the magnet is moved near the wire, a small current passes through the meter. This happens because an electromotive force (e.m.f.) is **induced** in the wire. This effect, known as **electromagnetic induction**, occurs whenever a wire cuts across the lines of a magnetic field. If the wire is part of a complete circuit, the **induced e.m.f.** forces electrons round the circuit. The induced e.m.f. can be increased by:

- moving the wire faster
- using a stronger magnet
- making the wire into a coil, as in Figure 3, and pushing the magnet in or out of the coil.

No e.m.f. is induced in the wire if the wire is parallel to the magnetic field lines as it moves through the field. The wire must cut across the lines of the magnetic field for an e.m.f. to be induced in the wire.

Other methods of generating an induced e.m.f. include:

1. **Using an electric motor in reverse,** as in Figure 2. The falling weight makes the motor coil turn between the poles of the magnet in the motor. The e.m.f. induced in the coil forces a current round the circuit and so causes the lamp to light. The faster the coil turns, the brighter the lamp is.
2. **Using a cycle dynamo,** as in Figure 3. When the magnet in the dynamo spins, an e.m.f. is induced in the coil. If the coil is connected to a lamp, the lamp lights because the e.m.f. forces a current round the circuit.

In both examples above, an e.m.f. is induced because there is relative motion between coil and the magnet. In the electric motor in reverse, the coil spins and the magnet is fixed. In the dynamo, the magnet spins and the coil is fixed.

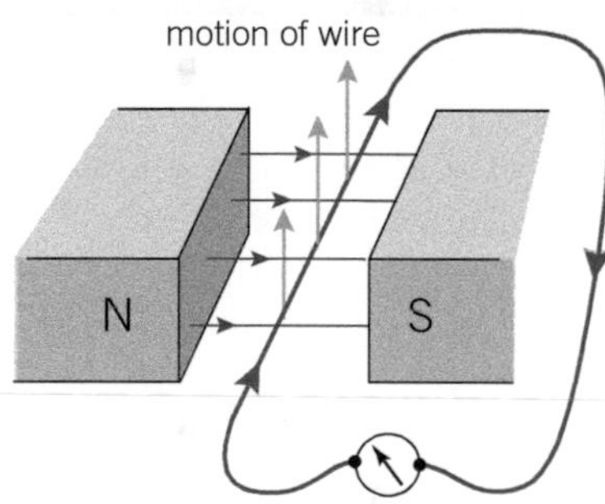

▲ **Figure 1** *Generating an electric current*

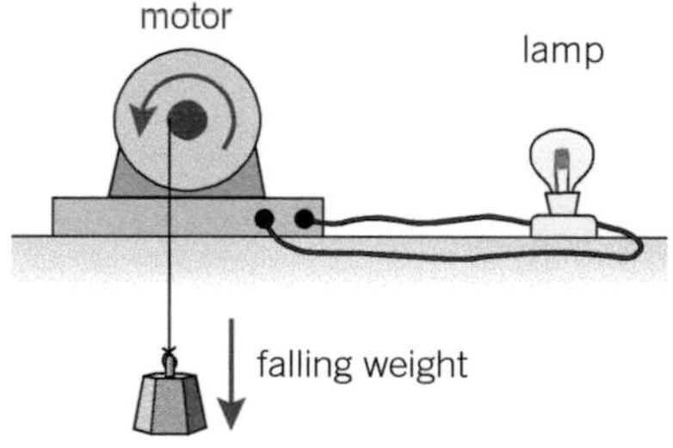

▲ **Figure 2** *A motor as a generator*

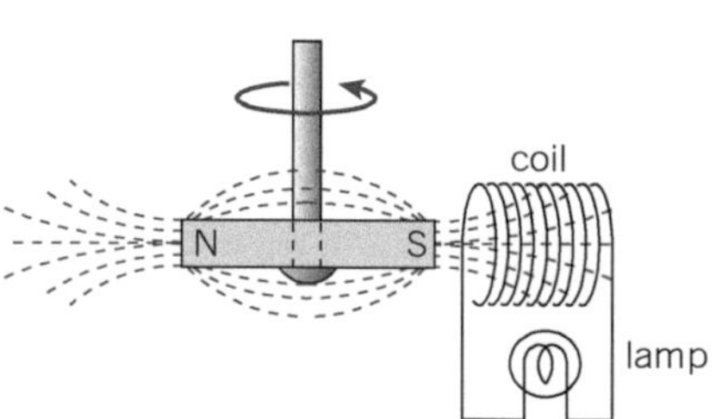

▲ **Figure 3** *A simple dynamo*

Energy changes

When a magnet is moved relative to a conductor (e.g., a wire or a coil), an e.m.f. is induced in the conductor. If the conductor is part of a complete circuit which has no other sources of e.m.f., a current passes round the circuit just as if the circuit included a battery. However, unlike the e.m.f. of a battery which is constant, the induced e.m.f. becomes zero when the relative motion between the magnet and the wires ceases.

An electric current transfers energy from the source of the e.m.f. in a circuit to the other components in the circuit. For example, when a dynamo is used to light a lamp, energy is transferred from the dynamo to the lamp. The current through the dynamo coil causes a reaction force on the coil due to the magnet. Work must therefore be done to keep the magnet spinning. The energy transferred from the coil to the lamp is equal to the work done on the coil to keep it spinning, assuming no energy is wasted as sound or due to friction or internal resistance.

The rate of transfer of energy from the source of e.m.f. to the other components of the circuit is equal to the product of the induced e.m.f. and the current. This is because:

- the induced e.m.f. is the energy transferred from the source per unit charge that passes through the source
- the current is the charge flow per second.

So the induced e.m.f. × the current = energy transferred per unit charge from the source × the charge flow per second = energy transferred per second from the source.

▲ **Figure 4** *Michael Faraday holding a bar magnet – when he demonstrated his discoveries at the Royal Institution, he was asked 'What use is electricity?' He replied, 'What use is a new baby?' No one can tell what can grow from a new discovery.*

The discovery of electromagnetic induction

Electromagnetic induction was discovered by Michael Faraday, who made his discovery public in 1831 at the Royal Institution, London. Faraday knew that a current passing along a wire produces a magnetic field near the wire and he wanted to know if a magnet could be used to produce a current. Using a magnetic compass near a loop of wire as a detector of current, he showed that the compass deflected whenever the magnet was moved in or out of the wire. He used the term 'electromotive force' (e.m.f.) to describe the voltage induced in a wire.

Understanding electromagnetic induction

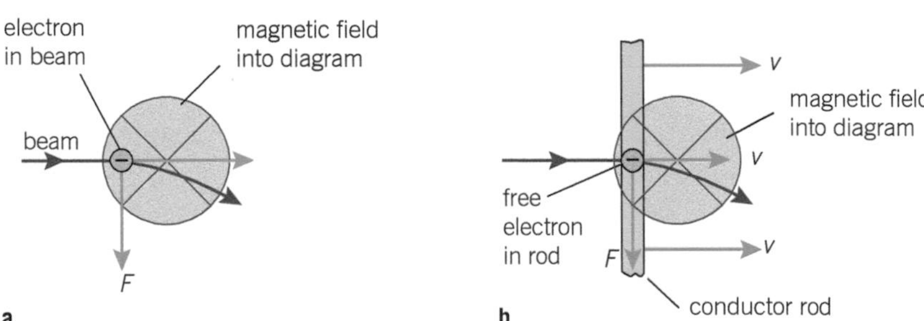

▲ **Figure 5** *Deflection of electrons in a magnetic field*

When a beam of electrons is directed across a magnetic field, each electron experiences a force at right angles to its direction of motion and to the field direction. A metal rod is a tube containing lots of free electrons. If the rod is moved across a magnetic field, as shown in Figure 5, the magnetic field forces the free electrons in the rod to one end away from the other end. So one end of the rod becomes negative and the other end positive. In this way, an e.m.f. is induced in the rod. The same effect happens if the magnetic field is moved and the rod is stationary. As long as there is relative motion between the rod and the magnetic field, an e.m.f. is induced in the rod. If the relative motion ceases, the induced e.m.f. becomes zero because the magnetic field no longer exerts a force on the electrons in the rod. Note that when the rod is part of a complete circuit, the electrons are forced round the circuit. In other words, the induced e.m.f. drives a current round the circuit.

The dynamo rule

In Figure 5, the magnetic field is into the plane of the diagram and the motion of the conductor relative to the rod is rightwards. The electrons in the rod are forced downwards. The direction of the induced current can also be worked out using **Fleming's right-hand rule**, also referred to as the **dynamo rule**, as shown in Figure 6. The direction of the induced current is, in accordance with the current convention, opposite to the direction of the flow of electrons in the conductor.

Study tip

Think carefully about which of Fleming's rules applies to generators – the riGht-hand rule is for Generators.

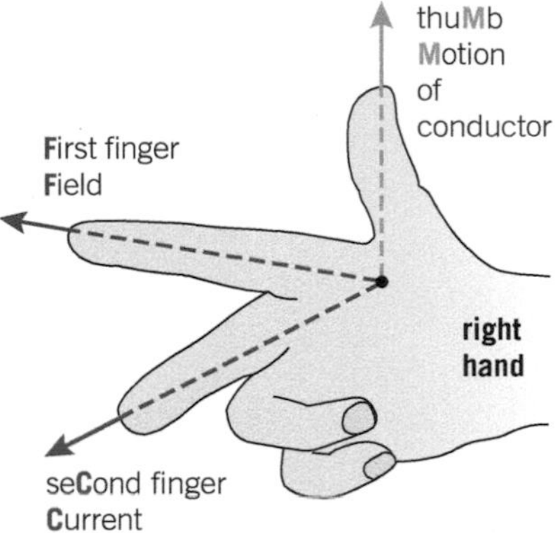

▲ **Figure 6** *Fleming's right-hand rule*

Summary questions

1 A coil of wire is connected to a sensitive meter.

- **a** Explain why the meter shows a brief reading when a magnet is pushed into the coil.
- **b** State two ways in which the meter reading could be made larger.

2 An electric motor consists of a coil of wire between the poles of a magnet. The motor is connected to a lamp. A thread wrapped round the motor spindle is used to support a weight, as shown in Figure 2 earlier in this topic.

- **a** Explain why the lamp lights when the weight descends.
- **b** What difference would have been made if the magnet had been much stronger?
- **c** Explain why a lamp connected to a dynamo lights when the dynamo turns.
- **d** Why is the dynamo easier to turn when the lamp is disconnected?

3 A horizontal rod aligned along a line from east to west is dropped through a horizontal magnetic field which is directed from south to north.

- **a** **i** What is the direction of the velocity of the rod?
  - **ii** Determine which end of the rod is positive. Explain your answer.
- **b** Explain why no e.m.f. is induced in the rod if it is aligned from north to south then dropped in the field.

21.2 The laws of electromagnetic induction

Learning objectives:

- → Define the magnetic flux and the magnetic flux linkage.
- → Relate the induced e.m.f. in a coil to the magnetic flux linkage through it.
- → State Lenz's law and the conservation law that explains it.

Specification reference: 3.10.3 and 3.10.4

Coils, currents, and fields

A magnetic field is produced in and around a coil when it is connected to a battery and a current is passed through it. A magnetic compass near the coil is deflected when current passes through it. For a long coil or solenoid, the pattern of the magnetic field lines is like the pattern for a bar magnet – except the magnetic field lines near a bar magnet loop round from the north pole to the south pole of the magnet. Figure 1 shows the magnetic field pattern of a current-carrying solenoid. The field lines pass through the solenoid and loop round outside the solenoid from one end (the north pole) to the other end (the south pole). If each end in turn is viewed from outside the solenoid:

- current passes a**N**ticlockwise (or cou**N**terclockwise) round the '**N**orth pole' end
- current passes clockwise round the 'south pole' end.

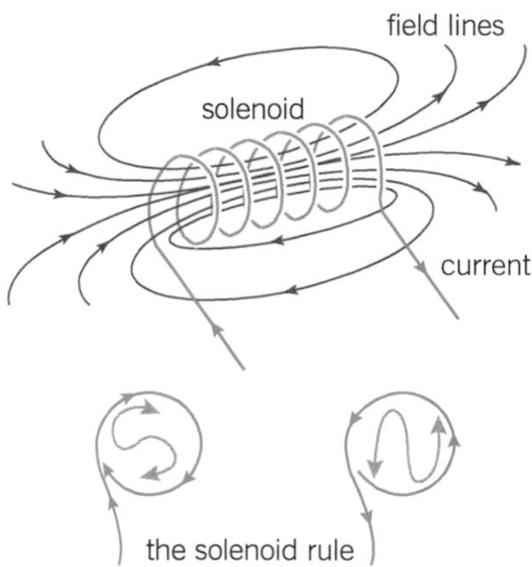

▲ **Figure 1** *The magnetic field near a solenoid*

Lenz's law

When a bar magnet is pushed into a coil connected to a meter, the meter deflects. If the bar magnet is pulled out of the coil, the meter deflects in the opposite direction. What determines the direction of the induced current? Consider the north pole of a bar magnet approaching end X of a coil, as shown in Figure 2.

The induced current passing round the circuit creates a magnetic field due to the coil. The coil field must act against the incoming north pole (N-pole), otherwise it would pull the N-pole in faster, making the induced current bigger, pulling the N-pole in even faster still, etc. Clearly, conservation of energy forbids this creation of kinetic and electrical energy from nowhere. So the induced current creates a magnetic field in the coil which opposes the incoming N-pole. The induced polarity of end X must therefore be a N-pole so as to repel the incoming N-pole. Therefore, the current must go round end X of the coil in an anticlockwise direction, as shown.

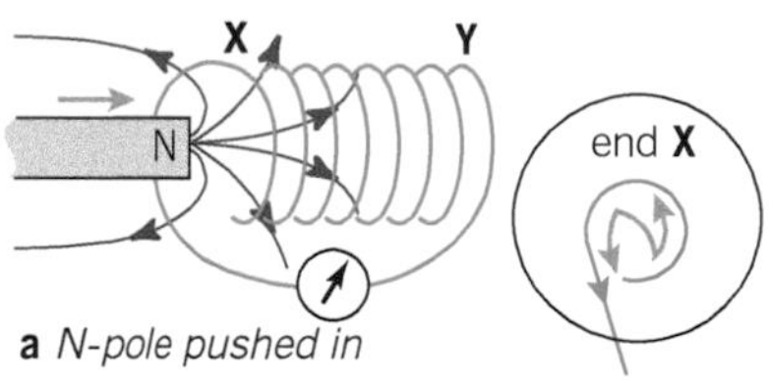

a *N-pole pushed in*

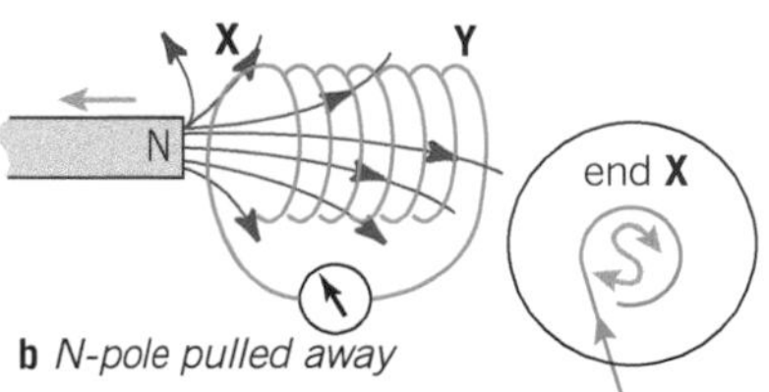

b *N-pole pulled away*

▲ **Figure 2** *Lenz's law*

If the magnet is removed from inside the coil, the induced current passes round end X of the coil in a clockwise direction. This corresponds to an induced S-pole at end X which therefore opposes the magnet moving away.

Lenz's law states that the direction of the induced current is always such as to oppose the change that causes the current.

The explanation of Lenz's law is that energy is never created or destroyed. The induced current could never be in a direction to help the change that causes it because that would mean producing electrical energy from nowhere, which is forbidden!

Faraday's law of electromagnetic induction

Consider a conductor of length l which is part of a complete circuit cutting through the lines of a magnetic field of flux density B.

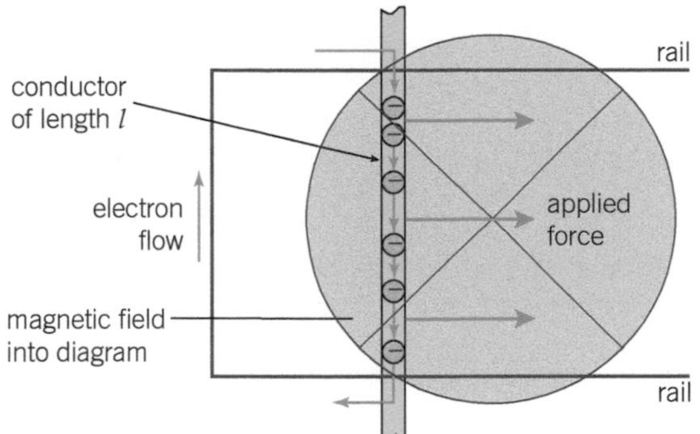

▲ **Figure 4** *Induced e.m.f. in a conductor*

An e.m.f. ε is induced in the conductor and an induced current I passes round the circuit.

The conductor experiences a force $F = BIl$ due to carrying a current in a magnetic field. The force opposes the motion of the conductor, and so an equal and opposite force must be applied to the conductor to keep it moving in the field. If the conductor moves a distance Δs in time Δt,

- the work done W by the applied force is given by $W = F\Delta s = BIl\Delta s$
- the charge transfer along the conductor in this time is $Q = I\Delta t$.

Therefore, the induced e.m.f. is $\varepsilon = \frac{W}{Q} = \frac{BIl\Delta s}{I\Delta t} = \frac{Bl\Delta s}{\Delta t}$

As $l\Delta s$ is the area A 'swept out' by the conductor in time Δt,

the induced e.m.f. is $\varepsilon = \frac{BA}{\Delta t}$

Note

Think carefully about the energy changes here. The energy transferred by the applied force (i.e., the work done) is transferred in the circuit by the electric current. The circuit needs to be complete otherwise no induced current flows, no electrical energy is transferred, and the applied force would make the rod move faster and faster.

The product of the **magnetic flux density**, B, and the area, A, swept out ($= BA$) is called the **magnetic flux**. The concept of magnetic flux is very useful for calculating induced e.m.fs. The example of the conductor cutting across the field lines shows that the induced e.m.f. is equal to the magnetic flux swept out by the conductor each second. Michael Faraday was the first person to show how induced e.m.fs could be calculated from magnetic flux changes.

- Magnetic flux $\Phi = BA$.
- **Magnetic flux linkage** through a coil of N turns $= N\Phi = NBA$ where B is the magnetic flux density perpendicular to area A.
- The unit of magnetic flux is the **weber** (Wb), equal to $1\,\mathrm{T\,m^2}$.

Note that flux density B (in teslas) is the flux per unit area passing at right angles (i.e., normally) through the area.

Therefore 1 tesla = 1 weber per square metre.

Application

Regenerative braking

A battery-powered or hybrid electric vehicle contains an alternator that can be used as an electric motor or as a generator. When the alternator is used as an electric motor, it is driven by the batteries. When the brakes are applied, the alternator is used to generate electricity which is used to recharge the battery. Some of the kinetic energy is transferred to electrical energy in the battery. The induced current through the alternator coil creates a magnetic field that acts against the magnetic field of the alternator. So the alternator experiences a braking force which helps to slow the vehicle down.

The fuel consumption of a hybrid vehicle (i.e., a vehicle with a petrol engine and an electric motor) is significantly less than that of a petrol-only vehicle. This is because some of the hybrid vehicle's kinetic energy is converted to chemical energy in its battery when the vehicle brakes. The battery supplies this energy to the electric motor when it takes over from the petrol engine at low speeds.

▲ **Figure 3** *An electric car*

Note

1 The unit of flux change per second, the weber per second, is the same as the volt. Therefore, the weber is equal to 1 volt second.

2 Whenever the flux linkage through a circuit changes, an e.m.f. is induced in the circuit. The flux can be due to a permanent magnet or due to a current-carrying wire.

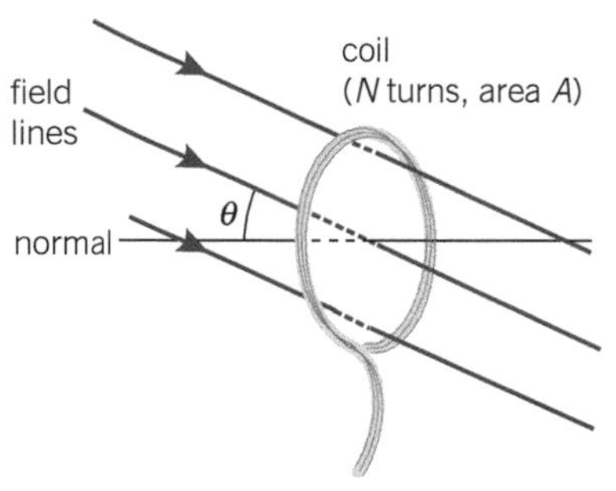

▲ **Figure 5** *Flux linkage*

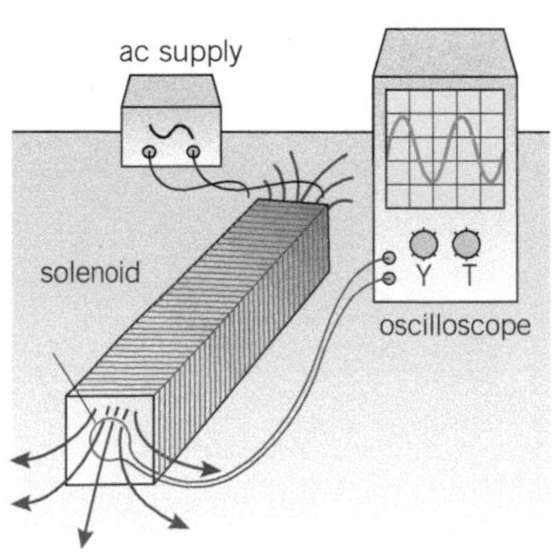

▲ **Figure 6** *A changing magnetic field*

More about flux linkage

1 When the magnetic field is along the normal (i.e., perpendicular) to the coil face, the flux linkage = $N\Phi = BAN$.
2 When the coil is turned through 180°, the flux linkage = $-BAN$.
3 When the magnetic field is parallel to the coil area, the flux linkage = 0 as no field lines pass through the coil area.

In general, when the magnetic field is at angle θ to the normal at the coil face, the flux linkage through the coil is $N\Phi = BAN\cos\theta$.

Faraday's law of electromagnetic induction states that the induced e.m.f. in a circuit is equal to the rate of change of flux linkage through the circuit.

$$\textbf{Induced e.m.f. } \varepsilon = -N\frac{\Delta\phi}{\Delta t}$$

where $N\dfrac{\Delta\Phi}{\Delta t}$ is the change of flux linkage per second.

The minus sign represents the fact that the induced e.m.f. acts in such a direction as to oppose the change that causes it (as per Lenz's law).

Examples

1 A moving conductor in a magnetic field

An e.m.f. is induced in the conductor provided the conductor cuts across the lines of the magnetic field. The direction of motion of the conductor in Figure 4 is at right angles to the field lines.

As explained earlier, the magnitude of the induced e.m.f. $\varepsilon = \dfrac{Bl\Delta s}{\Delta t}$, where l is the length of the conductor and Δs is the distance it moves in time Δt. Note that the change of flux in this time $\Delta\Phi = Bl\Delta s$ so the change of flux per second, $\dfrac{\Delta\Phi}{\Delta t}$, is equal to the magnitude of the induced e.m.f.

Because the speed of the conductor, $v = \dfrac{\Delta s}{\Delta t}$, the induced e.m.f.

$$\varepsilon = \frac{Bl\Delta s}{\Delta t} = Blv$$

$$\textbf{Induced e.m.f. } \varepsilon = Blv$$

2 A fixed coil in a changing magnetic field

Figure 6 shows a small coil on the axis of a current-carrying solenoid. The magnetic field of the solenoid passes through the small coil. If the current in the solenoid changes, an e.m.f. is induced in the small coil. This is because the magnetic field through the coil changes so the flux linkage through it changes, causing an induced e.m.f.

The flux linkage through the coil, $N\Phi = BAN$, where A is the coil area and N is the number of turns of the coil. Suppose the magnetic flux density changes from B to $B + \Delta B$ in time Δt so the flux linkage changes by an amount $N\Delta\Phi$ (= $AN\Delta B$).

The magnitude of the induced e.m.f. $= \dfrac{N\Delta\Phi}{\Delta t} = \dfrac{AN\Delta B}{\Delta t}$

Because B is proportional to the current I in the solenoid, the magnitude of the induced e.m.f. is therefore proportional to the rate of change of current in the solenoid.

3 A rectangular coil moving into a uniform magnetic field

Consider a rectangular coil of N turns, length l, and width w moving into a uniform magnetic field of flux density B at constant speed v. Figure 8 shows a similar situation. Suppose the coil enters the field at time $t = 0$.

- The time taken by the coil to enter the field completely
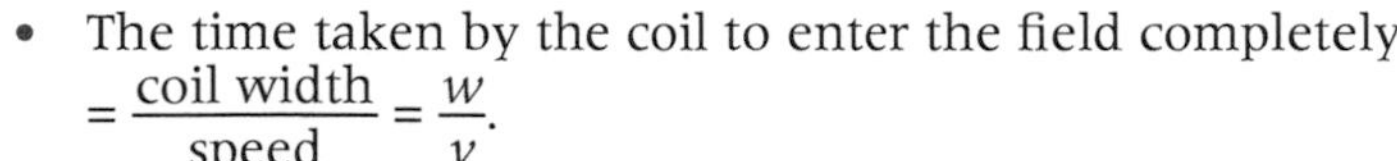
$= \frac{\text{coil width}}{\text{speed}} = \frac{w}{v}$.
During this time, the flux linkage $N\Phi$ increases steadily from 0 to $BNlw$. Therefore, the change of flux linkage per second is
$$\frac{N\Delta\Phi}{\Delta t} = \frac{BNlw}{w/v} = BNlv$$
- When the coil is completely in the field, the flux linkage through it (= $BNlw$) does not change so the induced e.m.f. is zero. In other words, the e.m.f. induced in the leading side is cancelled by the e.m.f. induced in the trailing side once the trailing side has entered the field. Figure 7 shows how the flux linkage and the induced e.m.f. change with time.

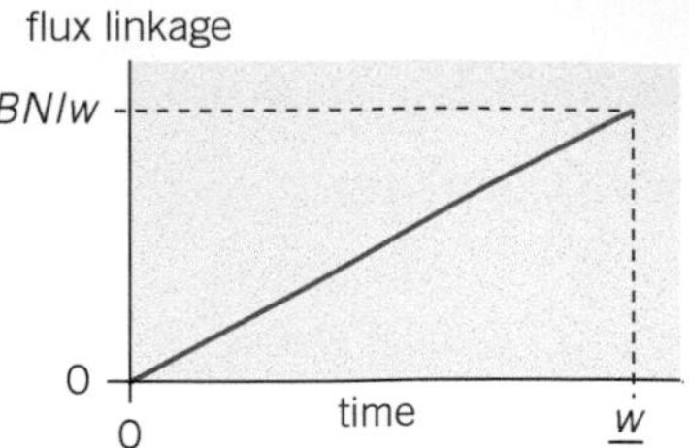

a *Flux linkage against time*

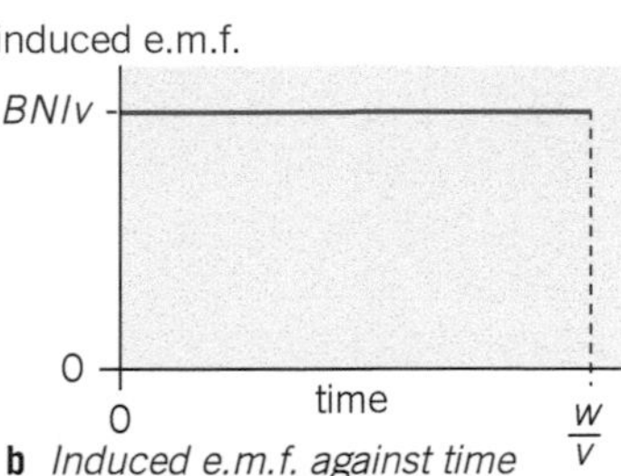

b *Induced e.m.f. against time*

▲ **Figure 7** *Flux changes*

Summary questions

1 A uniform magnetic field of flux density 72 mT is confined to a region of width 60 mm, as shown in Figure 8. A rectangular coil of length 50 mm and width 20 mm has 15 turns. The coil is moved into the magnetic field at a speed of 10 mm s^{-1} with its longer edge parallel to the edge of the magnetic field.

a Calculate:

i the flux linkage through the coil when it is completely in the field

ii the time taken for the flux linkage to increase from zero to its maximum value

iii the induced e.m.f. in the coil as it enters the field.

b **i** Sketch a graph to show how the flux linkage through the coil changes with time from the instant the coil enters the field to when it leaves the field completely.

ii Sketch a graph to show how the induced e.m.f. in the coil varies with time.

2 A rectangular coil of length 40 mm and width 25 mm has 20 turns. The coil is in a uniform magnetic field of flux density 68 mT.

a Calculate the flux linkage through the coil when the coil is at right angles to the field lines.

b The coil is removed from the field in 60 ms. Calculate the mean value of the induced e.m.f.

3 A circular coil of diameter 24 mm has 40 turns. The coil is placed in a uniform magnetic field of flux density 85 mT with its plane perpendicular to the field lines.

a Calculate:

i the area of the coil in m^2

ii the flux linkage through the coil.

b The coil was reversed in a time of 95 ms. Calculate:

i the change of flux linkage through the coil

ii the magnitude of the induced e.m.f.

4 A small circular coil of diameter 15 mm and 25 turns is placed in a fixed position on the axis of a solenoid, as shown in Figure 6. The magnetic flux density of the solenoid at this position varies with current according to the equation $B = kI$, where $k = 1.2 \times 10^{-3}\,\text{T A}^{-1}$.

a Calculate the flux linkage through the coil when the current in the solenoid is 1.5 A.

b The current in the solenoid was reduced from 1.5 A to zero in 0.20 s. Calculate the magnitude of the induced e.m.f. in the small coil.

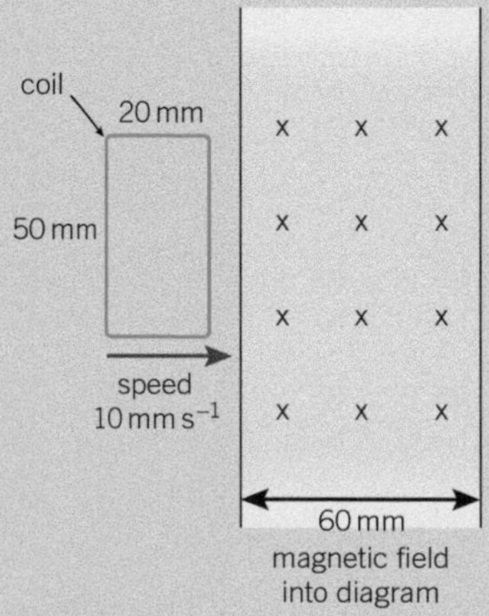

▲ **Figure 8**

21.3 The alternating current generator

Learning objectives:

- → State the two features of the output voltage waveform that change if the coil is turned faster.
- → Explain why the output alternates.
- → Explain why it is preferable for practical generators to have fixed coils and a rotating (electro)magnet.

Specification reference: 3.10.4

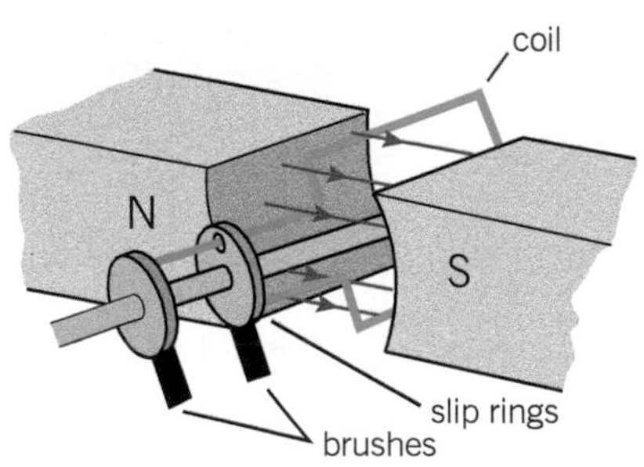

▲ **Figure 1** *The ac generator*

a

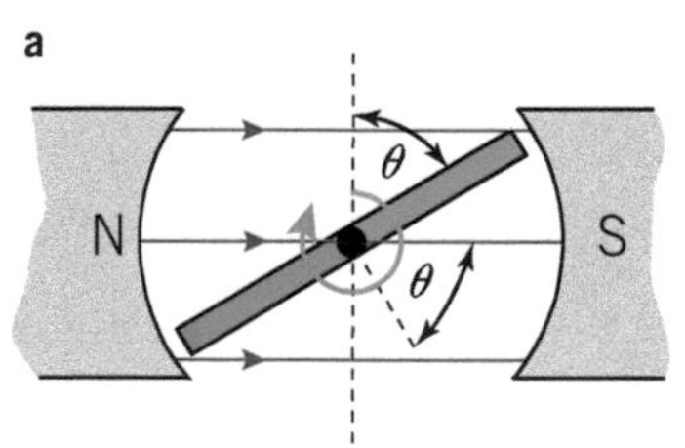

b

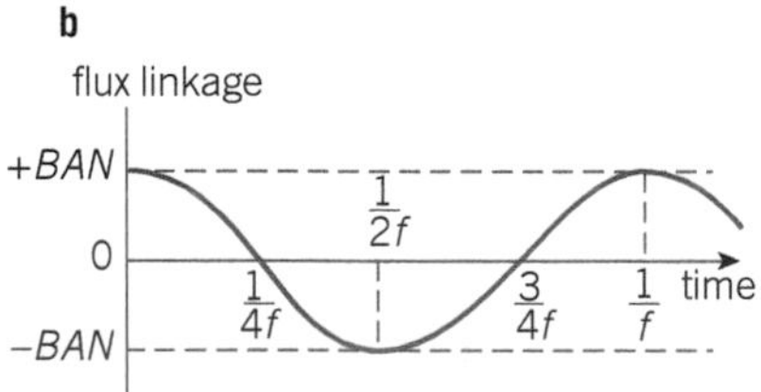

▲ **Figure 2** *Flux linkage in a spinning coil*

The alternating current (ac) generator

The simple ac generator consists of a rectangular coil that spins in a uniform magnetic field, as shown in Figure 1. When the coil spins at a steady rate, the flux linkage changes continuously. At an instant when the normal to the plane of the coil is at angle θ to the field lines, the flux linkage through the coil is $N\Phi = BAN\cos\theta$, where B is the magnetic flux density, A is the coil area, and N is the number of turns on the coil.

For a coil spinning at a steady frequency f, $\theta = 2\pi ft$ at time t after $\theta = 0$. So the flux linkage $N\Phi$ ($= BAN\cos 2\pi ft$) changes with time as shown in Figure 2.

- The gradient of the graph is the change of flux linkage per second, $\frac{N\Delta\Phi}{\Delta t}$, so it represents the induced e.m.f. It can be shown mathematically that the induced e.m.f. alternates according to the equation

$$\varepsilon = \varepsilon_0 \sin 2\pi ft$$

  where f is the frequency of rotation of the coil and ε_0 is the peak e.m.f. The above equation may be written as $\varepsilon = \varepsilon_0 \sin \omega t$ as the angular frequency of the coil is $\omega = 2\pi f$.

- The induced e.m.f. is zero when the sides of the coil move parallel to the field lines. At this position, the rate of change of flux is zero and the sides of the coil do not cut the field lines.
- The induced e.m.f. is a maximum when the sides of the coil cut the field lines at right angles. At this position, the e.m.f. induced in each wire of each side is Blv, where v is the speed of each wire and l is its length. So for N turns and two sides, the induced e.m.f. at this position is $\varepsilon_0 = 2NBlv$. This shows the peak e.m.f. can be increased by increasing the speed (i.e., the frequency of rotation) or by using a stronger magnet, a bigger coil, or a coil with more turns.

Note that $v = \omega\frac{d}{2}$, where d is the width of the coil. So $\varepsilon_0 = 2NBlv = BAN\omega$ where the coil area $A = ld$. So $\boldsymbol{\varepsilon = BAN\omega \sin \omega t}$.

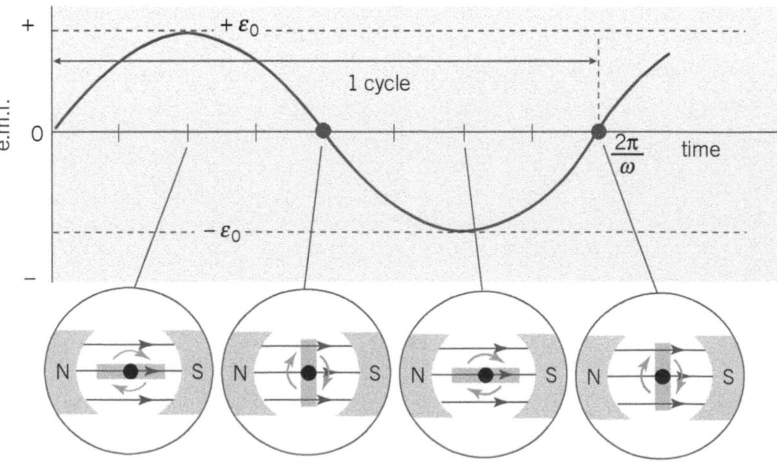

▲ **Figure 3** *E.m.f. against time for an ac generator*

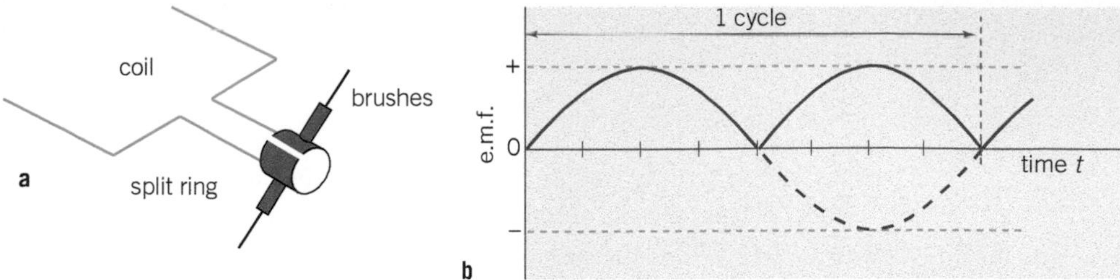

▲ **Figure 4** *The dc generator* **a** *The split-ring commutator* **b** *E.m.f. against time*

Back e.m.f. in an electric motor

An e.m.f. is induced in the spinning coil of an electric motor because the flux linkage through the coil changes. The induced e.m.f. ε is referred to as a **back e.m.f.** because it acts against the p.d. V applied to the motor in accordance with Lenz's law. At any instant, $V - \varepsilon = IR$, where I is the current through the motor coil and R is the circuit resistance.

Because the induced e.m.f. is proportional to the speed of rotation of the motor, the current changes as the motor speed changes.

- At low speed, the current is high because the induced e.m.f. is small.
- At high speed, the current is low because the induced e.m.f. is high.

Note:
Multiplying the equation $V - \varepsilon = IR$ by I throughout gives $IV - I\varepsilon = I^2R$

Rearranging this equation gives:

electrical power supplied by the source IV	=	electrical power transferred to mechanical power $I\varepsilon$	+	electrical power wasted due to circuit resistance I^2R

Note

A dc generator can be made by replacing the two slip rings of the ac generator with a split ring, as shown in Figure 4. The e.m.f. does not reverse its polarity because the connections between the split ring and the brushes reverse every half-cycle.

Study tip

For a rotating coil, the *rate of change* of flux is greatest when the flux through it is zero. The *rate of change* of flux is zero when the flux through it is greatest.

Extension

The efficiency of an electric motor

When the motor spins without driving a load, it spins at high speed so the current is very small because the back e.m.f. is large. Its speed is limited by friction in the bearings and by air resistance. So it uses very little power.

When the motor is used to drive a load, its speed is much less than when off-load so the back e.m.f. is smaller and the current is larger. The power it uses from the voltage source that is not transferred as mechanical power to the load is wasted due to the resistance heating effect of the electrical current.

$$\text{The efficiency of an electric motor} = \frac{\text{mechanical power output}}{\text{electrical power supplied}} \ (\times 100\%)$$

Q: An electric car is powered by an electric motor connected to a battery. Describe how the back e.m.f. changes when the car speeds up.

Answer: It increases in proportion to the speed.

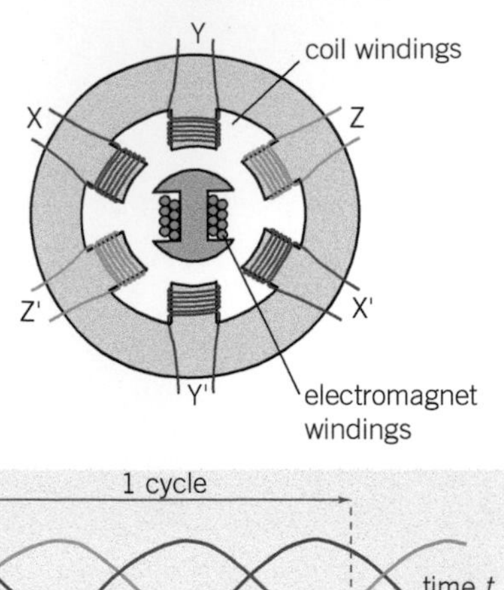

▲ **Figure 5** *Three-phase alternator*

Application

Power station alternators

A power station alternator has three sets of coils at 120° to one another (Figure 5). Each set of coils produces an alternating e.m.f. 120° out of phase with each of the other two e.m.fs. The coils are called the 'stators' because they are stationary, so they don't need slip-ring connectors and an electromagnet called the 'rotor' spins between them.

The electromagnet is supplied with current from a dc generator. So the turning of the rotor induces an alternating e.m.f. in each set of stator coils. The three phases are distributed via transformers and power lines to factories and local sub-stations. A local sub-station supplies mains electricity to local premises, a third on each of the three phases. This is why your home can sometimes suffer a blackout when other homes nearby still have electricity. This happens when a fault in the local sub-station cuts out one phase but not the others.

Summary questions

1 **a** An ac generator produces an alternating e.m.f. with a peak value of 8.0 V and a frequency of 20 Hz. Sketch a graph to show how the e.m.f. varies with time.

b The frequency of rotation of the ac generator in part **a** is increased to 30 Hz. On the same axes, sketch a graph to show how the e.m.f. varies with time at 30 Hz.

2 A rectangular coil of N turns and area A spins at a constant frequency f in a uniform magnetic field of flux density B. Complete the table to show how the flux linkage and induced e.m.f. relate to the orientation of the coil during one cycle of rotation.

Time	Orientation of coil	Flux linkage	Induced e.m.f.
0	parallel to field	?	$+\varepsilon_0$
$\frac{0.25}{f}$	perpendicular to field	$+BAN$	?
$\frac{0.50}{f}$	parallel to field	?	?
$\frac{0.75}{f}$	perpendicular to field	?	?

3 The coil of an ac generator has 80 turns, a length of 65 mm, and a width of 38 mm. It spins in a uniform magnetic field of flux density 130 mT at a constant frequency of 50 Hz.

a Calculate the maximum flux linkage through the coil.

b **i** Show that each side of the coil moves at a speed of 6.0 m s^{-1}.

ii Show that the peak voltage is 8.1 V.

4 An electric motor is to be used to move a variable load. The motor is connected in series with a battery and an ammeter.

a Explain why the motor current is very small when the load is zero.

b Explain why the motor current increases when the load is increased.

21.4 Alternating current and power

Learning objectives:

→ Define an alternating current.

→ Explain what is meant by the rms value of an alternating current.

→ Calculate the power supplied by an alternating current.

Specification reference: 3.10.5

Alternating current measurements

An alternating current (ac) is a current that repeatedly reverses its direction. In one cycle of an alternating current, the charge carriers move one way in the circuit, then reverse direction, then re-reverse direction.

The frequency f of an alternating current is the number of cycles it passes through each second. The unit of frequency is the hertz (Hz), equal to one cycle per second. Mains electricity has a frequency of 50 Hz. In a mains circuit, each cycle therefore takes 0.020 s $(= \frac{1}{50}\text{s})$. Note that the time for one full cycle, the time period T is given by $T = \frac{1}{f}$.

The peak value of an alternating current (or p.d.) is the maximum current (or p.d.) in either direction. The peak current in a circuit depends on the peak p.d. of the alternating current source and on the components in the circuit. For example, the peak p.d. in a mains circuit is 325 V. If this p.d. is applied to a 100 Ω heating element, the peak current through the heating element would be 3.25 A $\left(= \frac{325\,\text{V}}{100\,\Omega}\right)$.

Note that the peak-to-peak value is the difference between the peak value one way and the peak value in the opposite direction (i.e., twice the peak value).

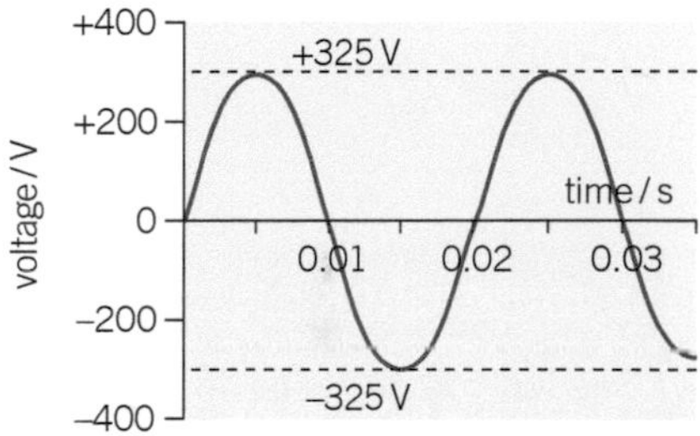

▲ **Figure 1** *The variation of mains p.d. with time*

Figure 1 shows how the mains p.d. varies with time. This type of variation is described as sinusoidal because its shape is that of a sine wave. Using transformers, we can change the peak p.d. of an alternating current but not its frequency.

Observing alternating current

1 Use an oscilloscope to display the waveform (i.e., variation with time) of the alternating p.d. from a signal generator.

- Increasing the output p.d. from the signal generator makes the oscilloscope trace taller. This shows that the peak value of the alternating p.d. has been made larger.
- Increasing the frequency of the signal generator increases the number of cycles on the screen. This is because the number of cycles per second of the alternating p.d. has increased.

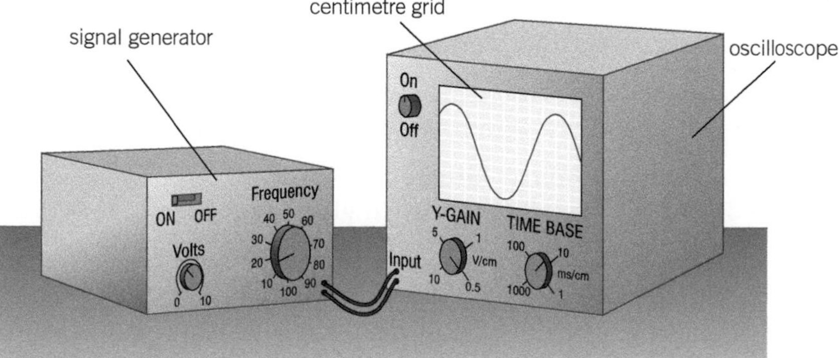

▲ **Figure 2** *Using an oscilloscope*

2 Connect the signal generator to a filament lamp and make the frequency low enough so you can see the brightness of the lamp vary.
- At very low frequency, you can see the lamp light up and then fade out repeatedly. The lamp is at its brightest each time the current is at its peak value. This happens twice each cycle corresponding to the current at its peak value in each direction.
- If the frequency is raised gradually, the lamp flickers faster and faster until the variation of brightness is too fast to notice. The mains frequency is too high to cause flickering.

The heating effect of an alternating current

Imagine an electric heater supplied with alternating current at a very low frequency. The heater would heat up, then cool down, then heat up, and so on. The heater would repeatedly heat up then go cold.

Recall that the heating effect of an electric current varies according to the square of the current (see Topic 10.2, More about resistance). This is because the electrical power *P* supplied to the heater for a current *I* is given by

$$P = IV = I^2R$$

where *R* is the resistance of the heater element.

Synoptic link

See Topic 11.10 Using an oscilloscope

Figure 3b shows how the power ($= I^2R$) varies with time.

- At peak current I_0, maximum power is supplied equal to $I_0{}^2R$.
- At zero current, zero power is supplied.

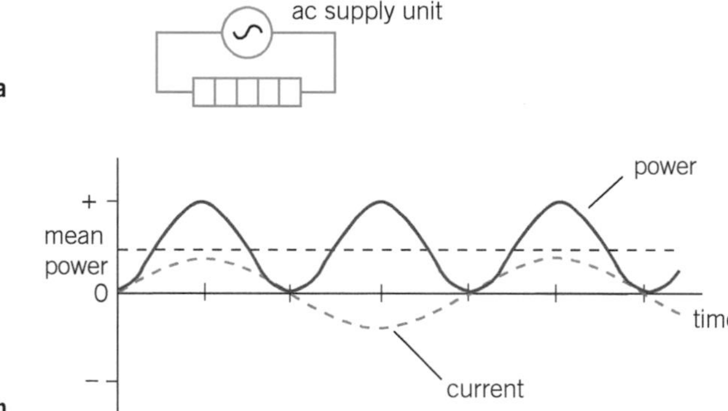

▲ **Figure 3** *Variation of power with time for an alternating current*

For a sinusoidal current, the *mean power* over a full cycle is half the peak power. You can see this from the symmetrical shape of the power curve in Figure 3b about the mean power. The mean power is therefore $\frac{1}{2}I_0{}^2R$.

The direct current that would give the same power as the mean power is called the **root mean square value** of the alternating current, I_{rms}.

The root mean square value of an alternating current is the value of direct current that would give the same heating effect as the alternating current in the same resistor.

Therefore,

$$(I_{rms})^2R = \frac{1}{2}I_0{}^2R$$

Cancelling *R* from this equation and rearranging the equation gives

$$(I_{rms})^2 = \frac{1}{2}I_0{}^2$$

Therefore,

$$I_{rms} = \frac{1}{\sqrt{2}} I_0$$

Also, the root mean square value of an alternating p.d. is given by

$$V_{rms} = \frac{1}{\sqrt{2}} V_0$$

The root mean square value of an alternating current or p.d. $= \frac{1}{\sqrt{2}} \times$ the peak value.

Note that for any resistor of known resistance in an alternating circuit, if you know the rms p.d. or current for the resistor, you can calculate the mean power supplied to it using the rms value.

$$P = (I_{rms})^2 R = \frac{(V_{rms})^2}{R} = I_{rms} V_{rms}$$

For example, if an alternating current of rms value 4 A is passed through a 5 Ω resistor, the mean power supplied to the resistor is $4^2 \times 5 = 80$ W.

Summary questions

1 An alternating current has a frequency of 200 Hz and a peak value of 0.10 A. Calculate:

- a the time for one cycle
- b the rms current.

2 An alternating current of peak value 3.0 A is passed through a 4.0 Ω resistor. Calculate:

- a i the rms current ii the rms p.d. across the resistor
- b i the peak power ii the mean power supplied to the resistor.

3 A mains electric heater gives a mean output power of 1.0 kW when the rms p.d. across it is 230 V. Calculate:

- a i the rms current ii the peak current
  iii the peak power.
- b The heater plug is fitted with a 5 A fuse. Explain why the fuse does not blow even though the peak current through it is more than 5 A.

4 A 12 V 48 W lamp lights at its normal brightness when it is connected to a source of alternating p.d.

- a Calculate:
  - i the rms current through it
  - ii the peak current through it
  - iii the peak p.d. across it.
- b The cable connecting the lamp to the source has a resistance of 0.5 Ω.
  - i Show that the rms p.d. across the cable is 2.0 V when the lamp is on.
  - ii Calculate the peak p.d. of the source when the lamp is on.

More about eddy currents

Eddy currents cause energy losses in the metal parts of electric motors and generators as well as in transformers. Applications that make use of eddy currents include

- eddy current brakes in which a metal plate or disc is stopped as a result of eddy currents opposing its motion when a magnetic field is applied to it,
- induction heating furnaces which use the heating effect of eddy currents to heat conducting substances directly.

Study tip

Remember rms $= \frac{\text{peak}}{\sqrt{2}}$

21.5 Transformers

Learning objectives:

- → Explain the purpose of transformers.
- → Describe the energy changes that take place in a transformer.
- → Discuss how the efficiency of transformers is improved by better design.

Specification reference: 3.10.6

Application

Inside a phone charger

A phone charger contains a step-down transformer that steps the mains p.d. down to a much lower p.d. (still alternating) then a diode circuit is used to convert the alternating p.d. to a direct p.d. Figure 2 shows how this is done.

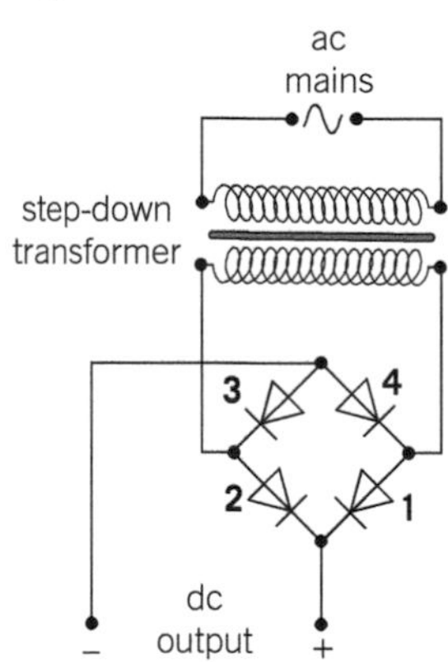

▲ **Figure 2** *Inside a phone charger*

Q Identify the diodes in Figure 2 that conduct when the transformer output p.d. is positive at the right of the diode circuit and negative at the left.

Answer: Diodes 1 and 3.

The transformer rule

A transformer changes an alternating p.d. to a different peak value. Any transformer consists of two coils: the primary coil and the secondary coil. The two coils have the same iron core. When the primary coil is connected to a source of alternating p.d., an alternating magnetic field is produced in the core. The field passes through the secondary coil. So an alternating e.m.f. is induced in the secondary coil by the changing magnetic field. The symbol for the transformer is shown in Figure 1.

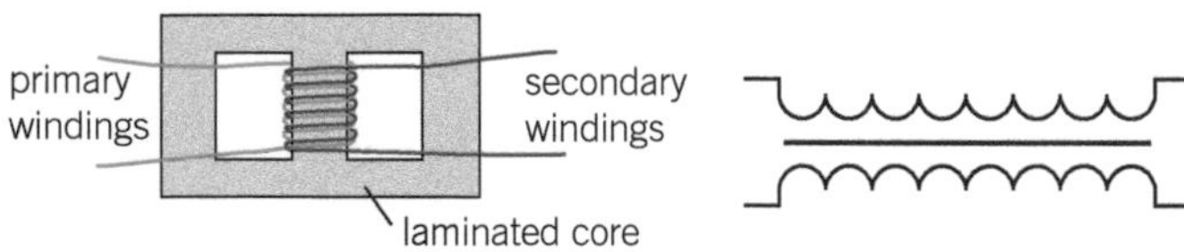

a *Practical arrangement* **b** *Transformer symbol*

▲ **Figure 1** *The transformer*

A transformer is designed so that all the magnetic flux produced by the primary coil passes through the secondary coil.

Let Φ = the flux in the core passing through each turn at an instant when an alternating p.d. V_p is applied to the primary coil.

- The flux linkage in the secondary coil = $N_s\Phi$, where N_s is the number of turns on the secondary coil. From Faraday's law, the induced e.m.f. in the secondary coil $V_s = N_s\frac{\Delta\Phi}{\Delta t}$.
- The flux linkage in the primary coil = $N_p\Phi$, where N_p is the number of turns on the primary coil. From Faraday's law, the induced e.m.f. in the primary coil = $N_p\frac{\Delta\Phi}{\Delta t}$.

The induced e.m.f. in the primary coil opposes the p.d. applied to the primary coil, V_p. Assuming the resistance of the primary coil is negligible, so all the applied p.d. acts against the induced e.m.f. in the primary coil, the applied p.d. is $V_p = N_p\frac{\Delta\Phi}{\Delta t}$.

Dividing the equation for V_s by the equation for V_p gives

$$\frac{V_s}{V_p} = N_s\frac{\Delta\Phi}{\Delta t}\Big/N_p\frac{\Delta\Phi}{\Delta t}$$

Cancelling $\frac{\Delta\Phi}{\Delta t}$ from this equation gives the **transformer rule**:

$$\frac{V_s}{V_p} = \frac{N_s}{N_p}$$

- A **step-up transformer** has more turns on the secondary coil than on the primary coil. So the secondary voltage is stepped up compared with the primary voltage (i.e., $N_s > N_p$ so $V_s > V_p$).
- A **step-down transformer** has fewer turns on the secondary coil than on the primary coil. So the secondary voltage is stepped down compared with the primary voltage (i.e., $N_s < N_p$ so $V_s < V_p$).

Transformer efficiency

Transformers are almost 100% efficient because they are designed with:

1 Low-resistance windings to reduce power wasted due to the heating effect of the current.
2 A laminated core which consists of layers of iron separated by layers of insulator. Induced currents in the core itself, referred to as **eddy currents**, are reduced in this way so the magnetic flux is as high as possible. Also, the heating effect of the induced currents in the core is reduced.
3 A core of **soft iron** which is easily magnetised and demagnetised. This reduces power wasted through repeated magnetisation and demagnetisation of the core.

$$\textbf{The efficiency of a transformer} = \frac{\textbf{power delivered by the secondary coil}}{\textbf{power supplied to the primary coil}}$$

$$= \frac{I_s V_s}{I_p V_p} \times 100\%$$

When a device (e.g., a lamp) is connected to the secondary coil, because the efficiency of a transformer is almost equal to 100%,

the electrical power supplied to the primary coil = the electrical power supplied by the secondary coil

Therefore, the current ratio is $\frac{I_s}{I_p} = \frac{V_p}{V_s} = \frac{N_p}{N_s}$

- In a step-up transformer, the voltage is stepped up and the current is stepped down.
- In a step-down transformer, the voltage is stepped down and the current is stepped up.

Worked example

A transformer is used to step down 230 V mains to 12 V. When a 12 V 48 W lamp is connected to the transformer's secondary coil, the lamp lights normally.

a The transformer has 1150 turns on its primary coil. Calculate the number of turns on its secondary coil.
b When the lamp is on, the primary current in the transformer is 0.22 A. Calculate:
 i the current in the secondary coil
 ii the efficiency of the transformer.

Solution

a Rearranging $\frac{V_s}{V_p} = \frac{N_s}{N_p}$ gives $N_s = N_p \frac{V_s}{V_p} = \frac{1150 \times 12}{230} = 60$ turns

b i The power supplied to the lamp $= I_s V_s = 48$ W

therefore $I_s = \frac{48}{V_s} = \frac{48}{12} = 4.0$ A

ii Efficiency $= \frac{I_s V_s}{I_p V_p} = \frac{48}{230 \times 0.22} = 0.95$ (= 95%)

Note

The changing magnetic flux in the core induces a **back e.m.f.** in the primary coil as well as an e.m.f. in the secondary coil. The back e.m.f. acts against the primary voltage, making the primary current very small when the secondary current is 'off'.

When the secondary current is on, the magnetic field it creates is in the opposite direction to the magnetic field of the primary current. In this situation, the back e.m.f. in the primary coil is reduced so the primary current is larger than when the secondary current is off.

Study tip

Transformers only work if the magnetic flux through them is changing; they won't work with steady dc.

A *step-down* transformer steps down the voltage but *increases* the current.

The grid system

Electricity from power stations in the United Kingdom is fed into The National Grid System, which supplies electricity to most parts of the country. The National Grid is a network of transformers and cables, underground and on pylons, which covers all regions of the UK. Each power station generates alternating current at a precise frequency of 50 Hz at about 25 kV.

Step-up transformers at the power station increase the alternating voltage to 400 kV or more for long-distance transmission via the grid system. Step-down transformers operate in stages, as shown in Figure 3. Factories are supplied with all three phases at either 33 kV or 11 kV. Homes are supplied via a local transformer sub-station with single-phase ac at 230 V.

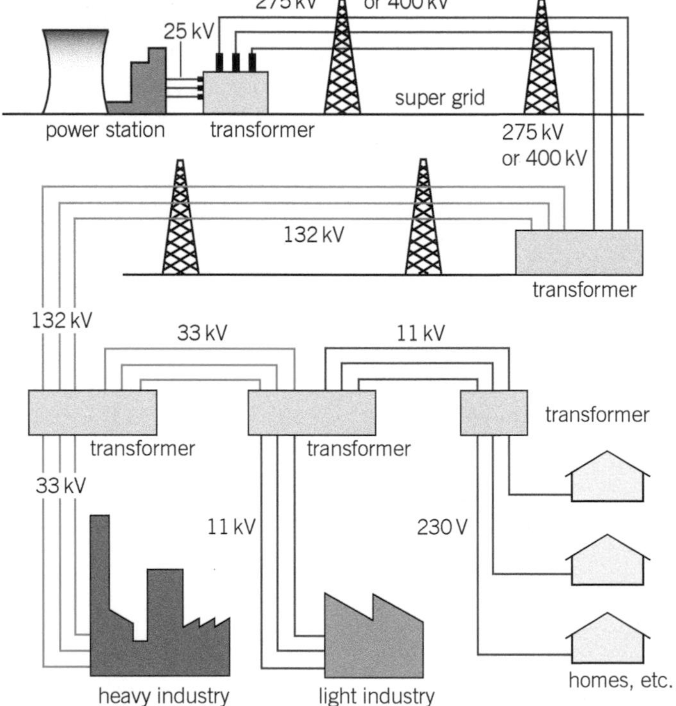

▲ **Figure 3** *The grid system*

Transmission of electrical power over long distances is much more efficient at high voltage than at low voltage. The reason is that the current needed to deliver a certain amount of power is reduced if the voltage is increased. So power wasted due to the heating effect of the current through the cables is reduced. To deliver power P at voltage V, (where V is the voltage at the end of the cable furthest from the power station), the current required is $I = \frac{P}{V}$. If the resistance of the cables is R, the power wasted through heating the cables is $I^2R = \frac{P^2R}{V^2}$. Therefore, the higher the voltage, the smaller the ratio of the wasted power to the power transmitted.

For example, for transmission of 1 MW of power through cables of resistance 500 Ω at 25 kV, the current necessary would be 40 A $\left(= \frac{1\,\text{MW}}{25\,\text{kV}}\right)$ so the power wasted would be 0.8 MW ($= I^2R = 40^2 \times 500$ W). Prove for yourself that at 400 kV the power wasted would be about 3 kW. Check your answer using the equation derived above.

Summary questions

1. **a** Explain why an alternating e.m.f. is induced in the secondary coil of a transformer when the primary coil is connected to an alternating voltage supply.

   b In terms of electrical power, explain why the current through the primary coil of a transformer increases when a device is connected to the secondary coil.

2. **a** Explain why a transformer is designed so that as much as possible of the magnetic flux produced by the primary coil of a transformer passes through the secondary coil.

   b Explain why a transformer works using alternating current but not using direct current.

3. A transformer has a primary coil with 120 turns and a secondary coil with 2400 turns.

   a Calculate the primary voltage needed for a secondary voltage of 230 V.

   b A 230 V 60 W lamp is connected to its secondary coil. Calculate the current through:

   i the secondary coil

   ii the primary coil.

   State any assumptions made in this calculation.

4. **a** Explain why transmission of electrical power over a long distance is more efficient at high voltage than at low voltage.

   b A power cable of resistance 200 Ω is to be used to deliver 2.0 MW of electrical power at 120 kV from a power station to an industrial estate. Calculate:

   i the current through the cable

   ii the power wasted in the cable.

Practice questions: Chapter 21

1 A coil is connected to a centre zero ammeter, as shown in **Figure 1**. A student drops a magnet so that it falls vertically and completely through the coil.

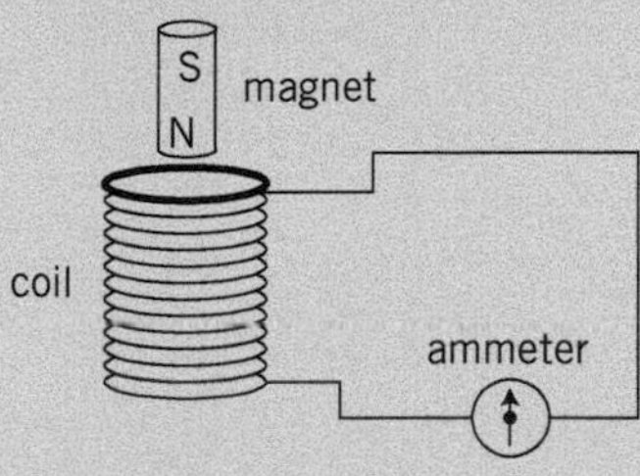

▲ **Figure 1**

(a) Describe what the student would observe on the ammeter as the magnet falls through the coil. *(2 marks)*

(b) If the coil were not present, the magnet would accelerate downwards at the acceleration due to gravity. State and explain how its acceleration in the student's experiment would be affected, if at all,

(i) as it entered the coil

(ii) as it left the coil. *(4 marks)*

(c) Suppose the student forgot to connect the ammeter to the coil, therefore leaving the circuit incomplete, before carrying out the experiment. Describe and explain what difference this would make to your conclusions in part (b). *(3 marks)*

AQA, 2004

2 Faraday's law of electromagnetic induction predicts that the induced e.m.f., ε, in a coil is given by

$$\frac{\Delta(N\Phi)}{t}$$

(a) (i) What quantity does the symbol Φ represent?

(ii) State the SI unit for Φ. *(2 marks)*

(b) In **Figure 2** the magnet forms the bob of a simple pendulum. The magnet oscillates with a small amplitude along the axis of a 240 turn coil that has a cross-sectional area of $2.5 \times 10^{-4}\,\text{m}^2$.

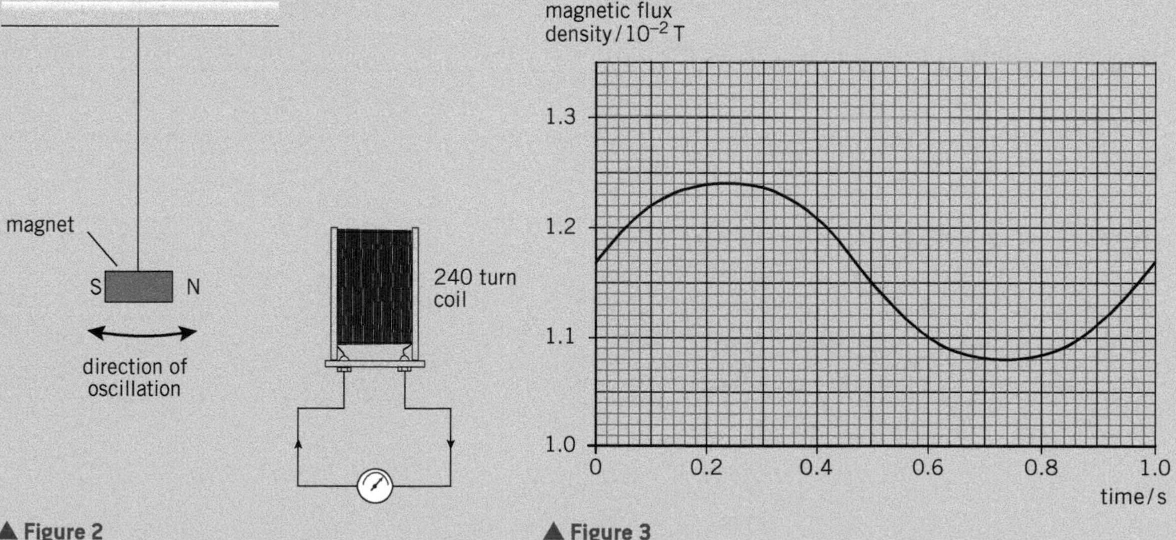

▲ **Figure 2**

▲ **Figure 3**

Figure 3 shows how the magnetic flux density, *B*, through the coil varies with time, *t*, for one complete oscillation of the magnet. The magnetic flux density through the coil can be assumed to be uniform.

(i) Calculate the maximum e.m.f. induced in the coil.

(ii) Sketch a graph to show how the induced e.m.f. in the coil varies during the same time interval.

(iii) Explain how the pendulum may be modified to double the frequency of oscillation of the magnet.

(iv) The frequency of oscillation of the magnet is increased without changing the amplitude.
Explain why this increases the maximum induced e.m.f.

(v) State *two* other ways of increasing the maximum induced e.m.f. (*11 marks*)

AQA, 2003

3 **Figure 4** shows a system used by an engineer to determine the rate of revolution of a rotating axle.

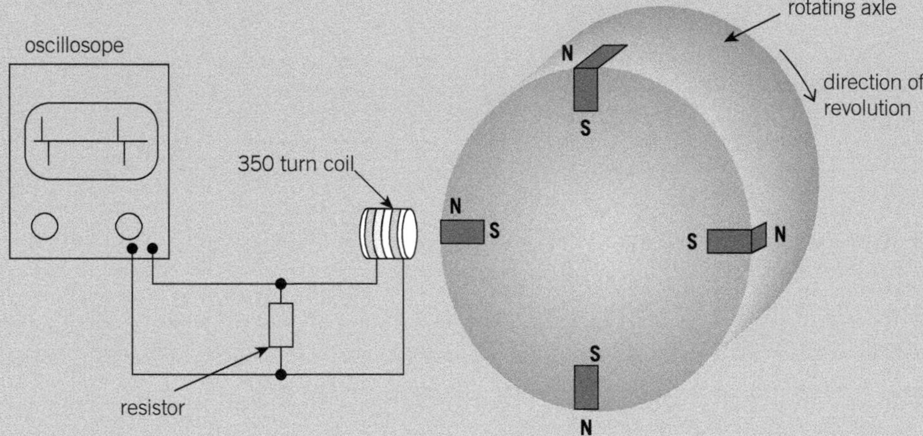

▲ **Figure 4**

Four small bar magnets are embedded in the axle as shown. The N-pole of each magnet is towards the outside of the axle. A voltage is produced between the terminals of a coil placed close to the rotating axle. The voltage produced is monitored using an oscilloscope. The waveform produced is shown in **Figure 5**.

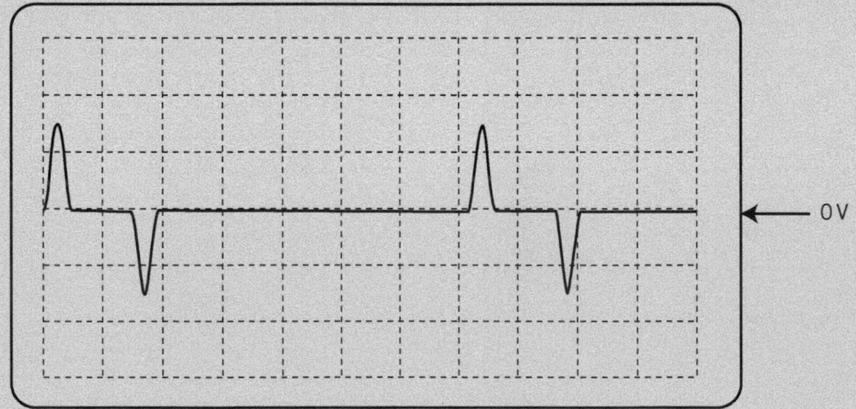

Oscilloscope grid marked in cm
The Y amplifier setting = 5 mV cm^{-1}
The time-base setting = 10 ms cm^{-1}

▲ **Figure 5**

(a) Determine the number of revolutions made by the axle in one minute. *(3 marks)*

(b) (i) Use Faraday's law to explain how the voltage pulses are produced.

(ii) The coil has 350 turns. Determine the maximum rate of change of flux through the coil. *(6 marks)*

(c) Use Lenz's law to explain the production of positive and negative voltage pulses. *(3 marks)*

(d) Draw on a copy of **Figure 5** the waveform that shows the changes you would expect to see when the rate of revolution of the axle increases. *(3 marks)*

AQA, 2007

4 **(a)** Copy and complete the diagram in **Figure 6** to show a current balance, which may be used to measure the magnetic flux density between the poles of the ceramic magnets. Clearly label the directions of the current and the magnetic force acting on the conductor in the field. *(3 marks)*

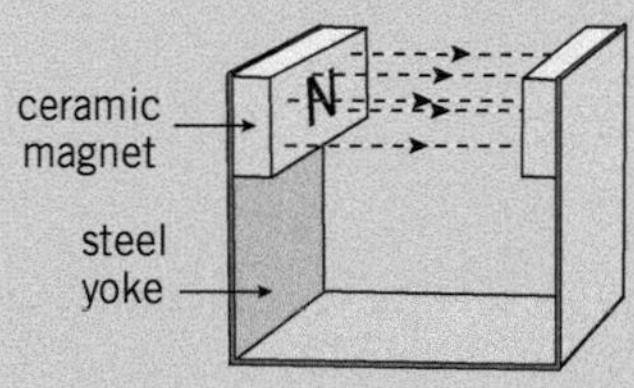

▲ **Figure 6**

(b) (i) The armature of a simple motor consists of a square coil of 20 turns and carries a current of 0.55 A just before it starts to move. The lengths of the sides of the coil are 0.15 m and they are positioned perpendicular to a magnetic field of flux density 40 mT. Calculate the force on each side of the coil.

(ii) Explain why the current falls below 0.55 A once the coil of the motor is rotating.

(iii) The resistance of the coil is 0.50 Ω. When the coil is rotating at a constant rate the minimum current in the coil is found to be 0.14 A. Calculate the maximum rate at which the flux is cut by the coil. *(8 marks)*

AQA, 2003

5 **(a)** Explain what is meant by the term *magnetic flux linkage*. State its unit. *(2 marks)*

(b) Explain, in terms of electromagnetic induction, how a transformer may be used to step down voltage. *(4 marks)*

(c) A minidisc player is provided with a mains adapter. The adapter uses a transformer with a turns ratio of 15 : 1 to step down the mains voltage from 230 V.

(i) Calculate the output voltage of the transformer.

(ii) State *two* reasons why the transformer may be less than 100% efficient. *(4 marks)*

AQA, 2004

6 **(a)** A transformer, operating from 230 V, supplies a 12 V garden lighting system consisting of eight lamps. Each lamp is rated at 30 W and they are connected in parallel.

(i) The primary coil of the transformer has 3000 turns. Calculate the number of turns on the secondary coil.

(ii) Show that the total resistance of the lamps when they are working at normal brightness is 0.60 Ω.

(iii) Calculate the power input to the transformer, assuming that the transformer is perfectly efficient. *(8 marks)*

(b) **Figure 7** shows a brass pendulum bob swinging through the magnetic field above a strong magnet. Its oscillations are observed to be quite heavily damped.

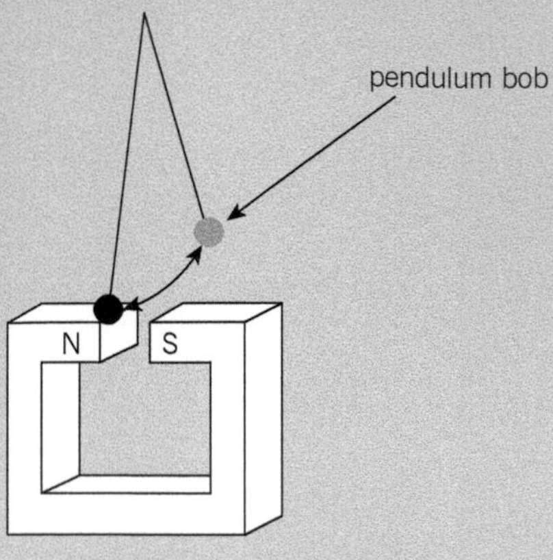

▲ Figure 7

Explain, using the principles of electromagnetic induction, why this pendulum is heavily damped. *(4 marks)*

AQA, 2006

7 (a) Electrical power is transmitted through cables of total resistance 1.8 Ω, operated with alternating current at an rms voltage of 11 kV. The power supplied to the input of the cables is 960 kW. Calculate:

(i) the peak value of the current in the cables

(ii) the percentage of the input power that is available at the output end of the cables. *(7 marks)*

(b) When public electricity supplies were first introduced, many of the power stations generated direct current. This meant that premises that were a large distance from the power station could not be supplied economically with electricity.

Discuss this with reference to the physical principles involved, stating why ac is now preferred for distributing electrical power. *(5 marks)*

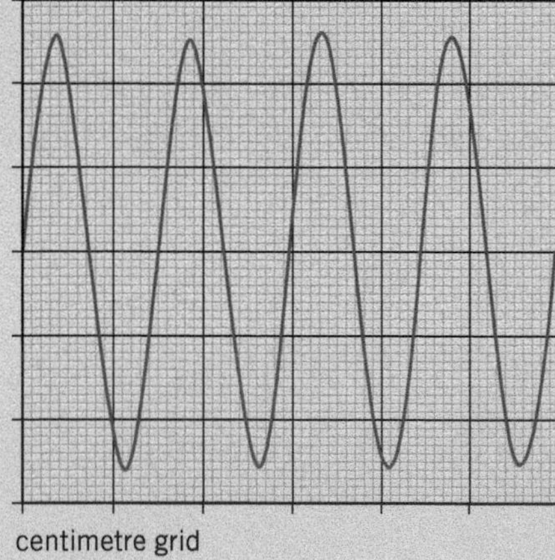

▲ Figure 8

8 An alternating current source is connected to a 470 Ω resistor to form a complete circuit. **Figure 8** shows the waveform of the alternating p.d. across the resistor.

(a) The time base setting of the oscilloscope was $20\,\text{ms}\,\text{cm}^{-1}$. Calculate the frequency of the alternating p.d. *(1 mark)*

(b) The Y-gain of the oscilloscope is $0.10\,\text{V}\,\text{cm}^{-1}$.

(i) Calculate the peak value and the rms value of the alternating p.d. *(2 marks)*

(ii) Calculate the mean power dissipated in the resistor. *(2 marks)*

Answers to the Practice Questions and Section Questions are available at www.oxfordsecondary.com/oxfordaqaexams-alevel-physics

22 Thermal physics
22.1 Internal energy and temperature

When you are outdoors in winter, you need to wrap up well, otherwise energy is transferred by heating from your body to your surroundings. Your body loses energy, and your surroundings gain energy. In summer, if you are in a very hot room, you will heat up because of energy transferred to you from the room.

Energy transfer between two objects takes place if:

- one object exerts a force on the other object and makes it move. In other words, one object does work on the other object.
- one object is hotter than the other object, so energy transfer by heating takes place by means of conduction, convection, or radiation. In other words, energy is transferred by heating because of a temperature difference between two objects.

Internal energy

The brake pads of a moving vehicle become hot if the brakes are applied for a long enough time. The work done by the frictional force between the brake pads and the wheel heats the brake pads, which gain energy from the kinetic energy of the vehicle. The temperature of the brake pads increases as a result, and the internal energy of each brake pad increases.

As explained below, the internal energy of an object is the energy of its molecules due to their individual movements and positions. The internal energy of an object due to its temperature is sometimes called **thermal energy**. However, some of the internal energy of an object might be due to other causes. For example, an iron bar that is magnetised has more internal energy than if it is unmagnetised, because of the magnetic interaction between the iron bar's atoms.

The internal energy of an object is increased because of:

- energy transfer by heating the object, or
- work done on the object, for example, work done by electricity.

If the internal energy of an object stays constant, then either:

- there is no energy transfer by heating and no work is done, or
- energy transfer by heating and work done balance each other out.

For example, the internal energy of a lamp filament increases when the lamp is switched on because work is done by the electricity supply pushing electrons through the filament. Because of this, the filament becomes hot. When it reaches its operating temperature, energy is transferred to the surroundings by heating, and the filament radiates light. Work done by the electricity supply pushing electrons through the filament is balanced by the energy transfer and light radiated from the filament.

Learning objectives:

- → Define internal energy.
- → State the lowest temperature that is possible.
- → Demonstrate the first law of thermodynamics in action.

Specification reference: 3.11.1

▲ **Figure 1** *Energy transfer by heating in winter*

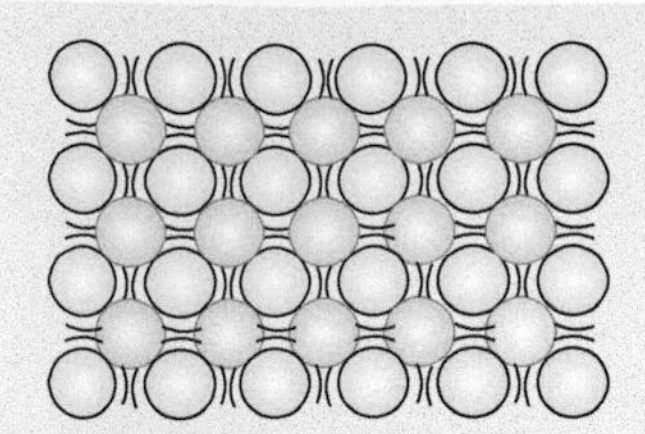

a A solid is made up of particles arranged in a 3-dimensional structure. There are strong forces of attraction between the particles. Although the particles can vibrate, they cannot move out of their positions in the structure.

When a solid is heated, the particles gain energy and vibrate more and more vigorously. Eventually they may break away from the solid structure and become free to move around. When this happens, the solid has turned into liquid – it has melted.

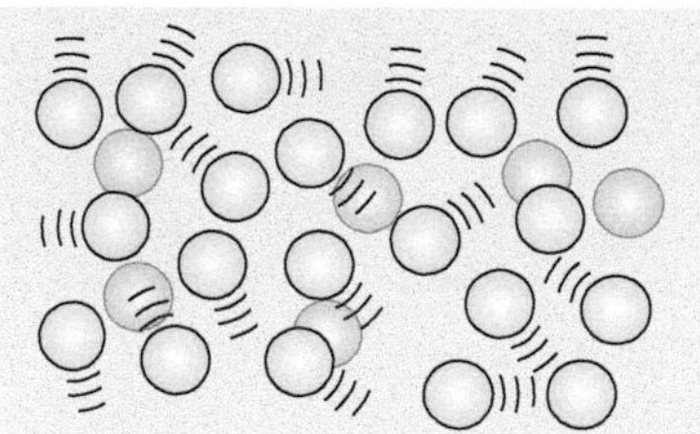

b In a liquid the particles are free to move around. A liquid therefore flows easily and has no fixed shape. There are still forces of attraction between the particles.

When a liquid is heated, some of the particles gain enough energy to break away from the other particles. The particles which escape from the body of the liquid become a gas.

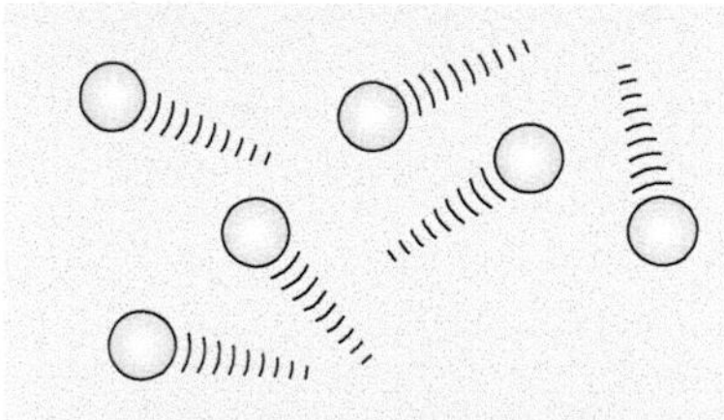

c In a gas, the particles are far apart. There are almost no forces of attraction between them. The particles move about at high speed. Because the particles are so far apart, a gas occupies a very much larger volume than the same mass of liquid.

The molecules collide with the container. These collisions are responsible for the pressure which a gas exerts on its container.

▲ **Figure 2** *Particles in a solid, a liquid, and a gas*

The first law of thermodynamics

In general, when work is done on or by an object and/or energy is transferred by heating,

the change of internal energy of the object =
the total energy transfer due to work done and heating

This general statement is called the *first law of thermodynamics* and may be written as.

Change of internal energy $\Delta U = Q + W$

where Q is the energy input to the system by heating and W is the work done ON the system. When it is applied to an object, the directions of the energy transfers (i.e., *to* or *from* the object) are very important and determine whether the overall internal energy of the object increases or decreases.

About molecules

A molecule is the smallest particle of a pure substance that is characteristic of the substance. For example, a water molecule consists of two hydrogen atoms joined to an oxygen atom.

An atom is the smallest particle of an element that is characteristic of the element. For example, a hydrogen atom consists of a proton and an electron.

- In a solid, the atoms and molecules are held to each other by forces due to the electrical charges of the protons and electrons in the atoms. The molecules in a solid vibrate randomly about fixed positions. The higher the temperature of the solid, the more the molecules vibrate. The energy supplied to raise the temperature of a solid increases the kinetic energy of the molecules. If the temperature is raised enough, the solid melts. This happens because its molecules vibrate so much that they break free from each other and the substance loses its shape. The energy supplied to melt a solid raises the potential energy of the molecules because they break free from each other.
- In a liquid, the molecules move about at random in contact with each other. The forces between the molecules are not strong enough to hold the molecules in fixed positions. The higher the temperature of a liquid, the faster its molecules move. The energy supplied to a liquid to raise its temperature increases the kinetic energy of the liquid molecules. Heating the liquid further causes it to vaporise. The molecules have sufficient kinetic energy to break free and move away from each other.
- In a gas or vapour, the molecules also move about randomly but much further apart on average than in a liquid. Heating a gas or a vapour makes the molecules speed up and so gain kinetic energy.

The internal energy of an object is the sum of the random distribution of the kinetic and potential energies of its molecules.

Increasing the internal energy of a substance increases the kinetic and/or potential energy associated with the random motion and positions of its molecules.

Temperature and temperature scales

The temperature of an object is a measure of the degree of hotness of the object. The hotter an object is, the more internal energy it has. Place your hand in cold water, and your hand's internal energy decreases because of energy transferred out by heating the water. Place your hand in warm water, and its internal energy increases because of energy transferred into it by the water heating your hand. If the water is at the same temperature as your hand, no overall energy transfer by heating takes place. Your hand is then in **thermal equilibrium** with the water. For any two objects that are at the same temperature, no overall energy transfer by heating will take place.

A temperature scale is defined in terms of *fixed points*, which are standard degrees of hotness that can be accurately reproduced.

- The **Celsius scale** of temperature, in units of °C, is defined in terms of:
  1. ice point, 0 °C, which is the temperature of pure melting ice
  2. steam point, 100 °C, which is the temperature of steam at standard atmospheric pressure.
- The **absolute scale** of temperature, in units of kelvin (K), is defined in terms of:
  1. **absolute zero**, 0 K, which is the lowest possible temperature
  2. the triple point of water, 273.16 K, which is the temperature at which ice, water, and water vapour co-exist in thermodynamic equilibrium.

Because ice point on the absolute scale is 273.15 K and steam point is 100 K higher, then

temperature in °C = absolute temperature in kelvin − 273.15

Study tip

To change from °C to kelvin, simply add on 273(.15).

The kelvin scale depends on a fundamental feature of nature, that is, the lowest temperature that is possible in nature. In contrast, the Celsius scale depends on the properties of a substance that was chosen for convenience instead of for any fundamental reason. That substance happens to be water.

Note: The symbol T is used for absolute temperature

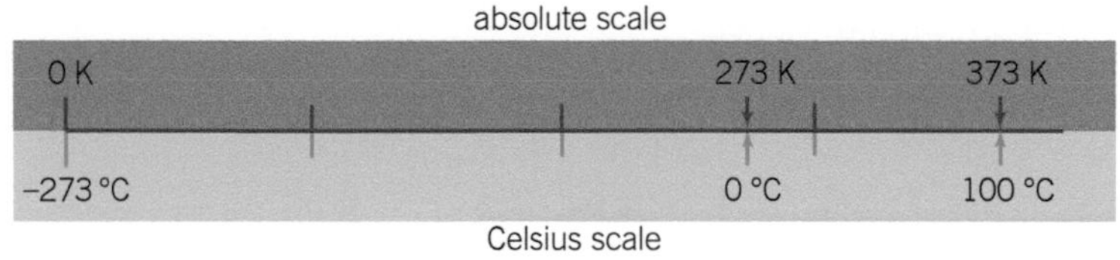

▲ **Figure 3** *Temperature scales*

About absolute zero

The absolute scale of temperature, also called the kelvin scale, is based on absolute zero, which is the lowest possible temperature. No object can have a temperature below absolute zero. *An object at absolute zero has minimum internal energy*, regardless of the substances the object consists of.

The pressure of a fixed mass of an ideal gas in a sealed container of fixed volume decreases as the gas temperature is reduced (see Topic 23.1). If the pressure measured at ice point and at steam point is plotted on a graph, as shown in Figure 4, the line between the two points always cuts the temperature axis at −273 °C, regardless of which gas is used or how much gas is used.

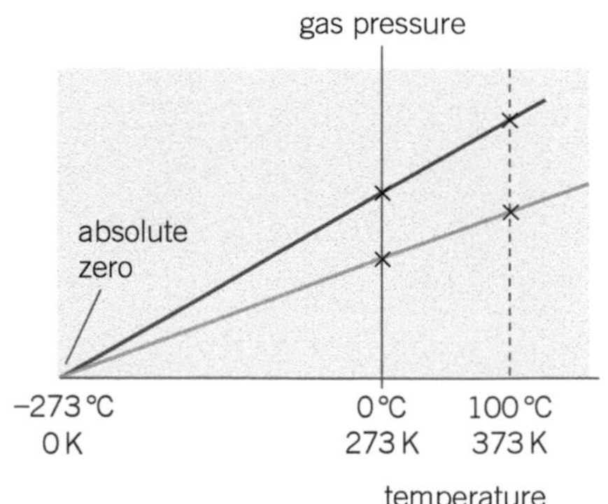

▲ **Figure 4** *Absolute zero*

Extension

A thermometer test

Use a travelling microscope to measure the interval between adjacent graduations on the scale of an accurate liquid-in-glass thermometer. You may be surprised to find that the interval distance is not the same near the middle of the scale as it is near the ends of the scale. This is because the expansion of the liquid is not directly proportional to the change of temperature.

All thermometers are calibrated in terms of the temperature measured by a gas thermometer. This is a thermometer consisting of a dry gas in a sealed container. The pressure of the gas is proportional to the absolute temperature of the gas. By measuring the gas pressure, p_{Tr}, at the triple point of water (273.16 K by definition) the temperature at which water, water vapour, and ice co-exist in thermal equilibrium and at unknown temperature T, the unknown temperature in kelvin can be calculated using

$$\frac{T}{273.16} = \frac{p}{p_{Tr}}$$

where p is the gas pressure at the unknown temperature T.

Extension

The coldest places in the world

You don't need to travel to the South Pole to find the coldest places in the world. Simply go to the nearest university physics department that has a low-temperature research laboratory. Substances have very strange properties at very low temperatures. For example, metals cooled to a few degrees within absolute zero become superconductors, which means that they have zero electrical resistance. Superfluids have been discovered that can empty themselves out of containers! Temperatures within a few microkelvin of absolute zero have been reached in these laboratories.

Q: Explain why you can't use a liquid-in-glass thermometer to measure very low temperatures.

Answer: The liquid in the thermometer would freeze.

▲ **Figure 5** *A low-temperature research laboratory*

Summary questions

1 **a** Explain why an electric motor becomes warm when it is used.

b A battery is connected to an electric motor which is used to raise a weight at a steady speed. When in operation, the electric motor is at a constant temperature, which is above the temperature of its surroundings. Describe the energy transfers that take place.

2 **a** Define internal energy.

b Describe a situation in which the internal energy of an object is constant even though work is done on the object.

3 **a** State one difference between the motion of the molecules in a solid and the molecules in a liquid.

b Describe how the motion of the molecules in a solid changes when the solid is heated.

4 **a** State each of the following temperatures to the nearest degree on the absolute scale:

i the temperature of pure melting ice

ii 20 °C

iii −196 °C.

b Gas thermometers are used to calibrate all thermometers. The pressure of a constant-volume gas thermometer is 100 kPa at a temperature of 273 K.

i Calculate the temperature, in kelvin, of the gas when its pressure is 120 kPa.

ii Calculate the pressure of the gas at 100 °C.

22.2 Specific heat capacity

Heating and cooling

Sunbathers on the hot sandy beaches of the Mediterranean Sea dive into the sea to cool off. Sand heats up much more readily than water does. Even when the sand is almost too hot to walk barefoot across, the sea water is refreshingly cool. The temperature rise of an object when it is heated depends on:

- the mass of the object
- the amount of energy supplied to it
- the substance or substances from which the object is made.

The specific heat capacity, c, of a substance is the energy needed to raise the temperature of unit mass of the substance by 1 K without change of state. The unit of c is $\mathrm{J\,kg^{-1}\,K^{-1}}$.

Specific heat capacities of some common substances are shown in Table 1.

To raise the temperature of mass m of a substance from temperature θ_1 to temperature θ_2,

the energy needed $\Delta Q = mc(\theta_2 - \theta_1)$

For example, to calculate the energy that must be supplied to raise the temperature of 5.0 kg of water from 20 °C to 100 °C, use the above equation and Table 1 for the specific heat capacity of water, and you will get $\Delta Q = 5.0 \times 4200 \times 80 = 1.7 \times 10^6\,\mathrm{J}$.

Learning objectives:

- → Explain what is meant by heating up and by cooling down.
- → State which materials heat up and cool down the fastest.
- → Define and measure specific heat capacity.

Specification reference: 3.11.1

▼ **Table 1** *Some specific heat capacities*

Substance	Specific heat capacity / $\mathrm{J\,kg^{-1}\,K^{-1}}$
Aluminium	900
Concrete	850
Copper	390
Iron	490
Lead	130
Oil	2100
Water	4200

The inversion tube experiment

In this experiment, the gravitational potential energy of an object falling in a tube is converted into internal energy when it hits the bottom of a tube. Figure 1 shows the idea. The object is a collection of tiny lead spheres.

The tube is inverted each time the spheres hit the bottom of the tube. The temperature of the lead shot is measured initially and after a particular number of inversions.

Let m represent the mass of the lead shot.

For a tube of length L, the loss of gravitational potential energy for each inversion = mgL.

Therefore, for n inversions, the loss of gravitational potential energy = $mgLn$.

The gain of internal energy of the lead shot = $mc\Delta\theta$, where c is the specific heat capacity of lead and $\Delta\theta$ is the temperature rise of the lead shot.

Assuming that all the gravitational potential energy lost is transferred to internal energy of the lead shot, $mc\Delta\theta = mgLn$.

Therefore, $c = \dfrac{gLn}{\Delta\theta}$

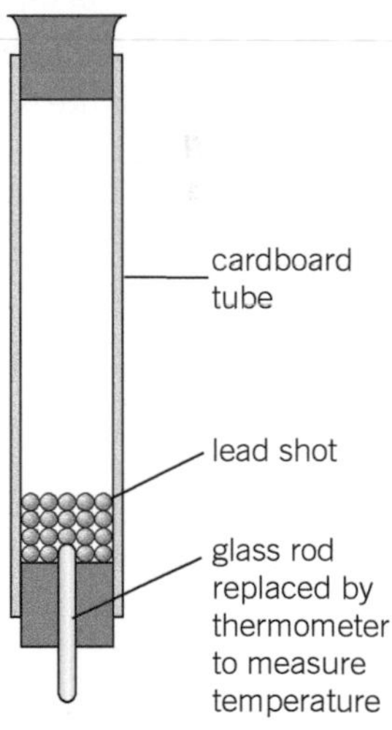

▲ **Figure 1** *The inversion tube experiment*

So, the experiment can be used to measure the specific heat capacity of lead with no other measurements than the length of the tube, the temperature rise of the lead, and the number of inversions.

Specific heat capacity measurements using electrical methods

Measurement of the specific heat capacity of a metal

A block of metal of known mass m in an insulated container is used. A 12 V electrical heater is inserted into a hole drilled in the metal and used to heat the metal by supplying a measured amount of electrical energy. A thermometer inserted into a second hole drilled in the metal is used to measure the temperature rise $\Delta\theta$ (= its final temperature − its initial temperature). A small amount of water or oil in the thermometer hole will improve the thermal contact between the thermometer and the metal.

The electrical energy supplied

= heater current I × heater p.d. V × heating time t

So, assuming no heat loss to the surroundings, $mc\Delta\theta = IVt$

Therefore, $$c = \frac{IVt}{m\Delta\theta}$$

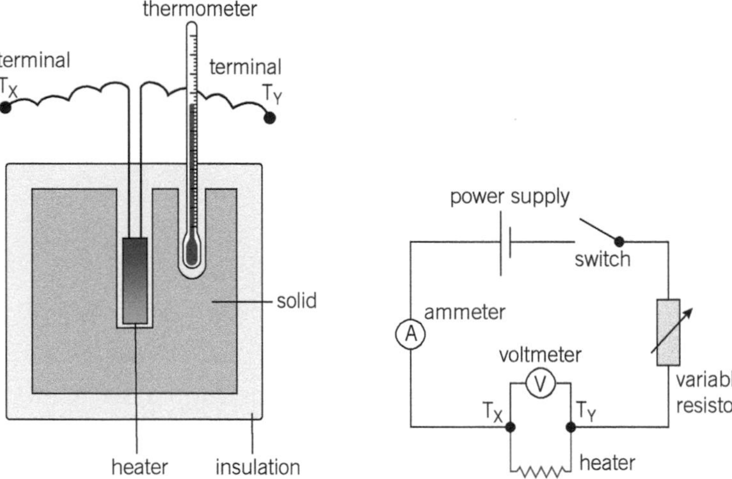

▲ **Figure 2** *Measuring c*

Measurement of the specific heat capacity of a liquid

A known mass of liquid is used in an insulated calorimeter of known mass and known specific heat capacity. A 12 V electrical heater is placed in the liquid and used to heat it directly. A thermometer inserted into the liquid is used to measure the temperature rise, $\Delta\theta$.

- The electrical energy supplied = current I × voltage V × heating time t.
- The energy needed to heat the liquid = mass of liquid (m_{liq}) × specific heat capacity of liquid (c_{liq}) × temperature rise ($\Delta\theta$).

Hint

1. Notice that the unit of mass × the unit of c × the unit of temperature change gives the joule. In other words, kg × J kg^{-1} K^{-1} × K = J.
2. The **heat capacity, *C***, of an object is the energy supplied to raise the temperature of the object by 1 K. Therefore, for an object of mass m made of a single substance of specific heat capacity c, its heat capacity $C = mc$. For example, the heat capacity of 5.0 kg of water is 21 000 J K^{-1} = 5.0 kg × 4200 J kg^{-1} K^{-1}, as the specific heat capacity of water is 4200 J kg^{-1} K^{-1}.

Study tip

A temperature change is the same in °C as it is in K. If you are given the initial and final temperatures in °C, just calculate the temperature difference in °C.

- The energy needed to heat the calorimeter = mass of calorimeter (m_{cal}) × specific heat capacity of calorimeter (c_{cal}) × temperature rise ($\Delta\theta$).

  Assuming no heat loss to the surroundings,

  $IVt = m_{liq}c_{liq}\Delta\theta + m_{cal}c_{cal}\Delta\theta$

So, c_{liq} can be calculated from this equation because all of the other quantities are known.

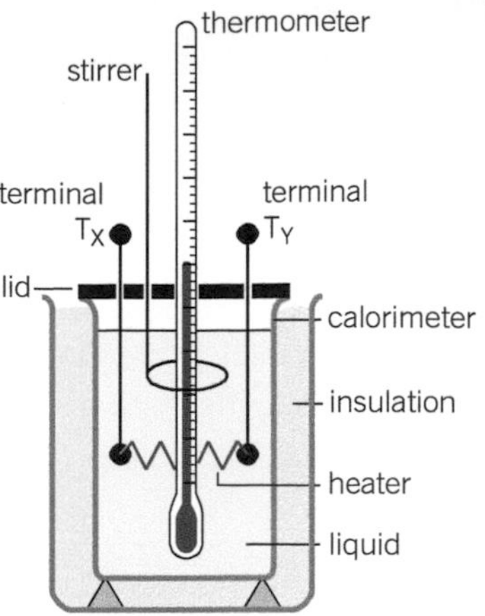

▲ **Figure 3** *Measurement of the specific heat capacity of a liquid*

Continuous flow heating

In an electric shower, water passes steadily through copper coils heated by an electrical heater. The water is hotter at the outlet than at the inlet. This is an example of continuous flow heating. For mass m of liquid passing through the heater in time t at a steady flow rate, and assuming no energy transfer by heating to the surroundings:

the electrical energy supplied per second $IV = mc\frac{\Delta\theta}{t}$

where $\Delta\theta$ is the temperature rise of the water and c is its specific heat capacity.

Note that when the outflowing water has attained a steady temperature, the temperature of the copper coils does not change, so no $mc\Delta\theta$ term is needed for the copper coils in the above equation.

For a solar heating panel, the energy gained per second by heating the liquid that flows through the panel is equal to $mc\frac{\Delta\theta}{t}$.

Study tip

Before you measure a liquid's temperature in a heating experiment, give the liquid a stir.

Study tip

If the volume flow rate is given, you need to know the density of the fluid to calculate the rate of flow of mass ($\frac{m}{t}$). See Topic 6.1, Density.

Summary questions

Use the data in Table 1 at the beginning of this topic for the following calculations.

1 Calculate:

  a the energy needed to heat an aluminium pan of mass 0.30 kg from 15 °C to 100 °C

  b the energy needed to heat 1.50 kg of water from 15 °C to 100 °C.

2 a Calculate the time taken to heat the water and pan in **Q1** from 15 °C to 100 °C using a 2.0 kW electric hot plate, assuming that no energy is transferred to the surroundings by heating.

  b Calculate the energy needed to raise the temperature of 80 kg of water in an insulated copper tank of mass 20 kg from 20 °C to 50 °C.

3 In an inversion tube experiment, 0.50 kg of lead shot at an initial temperature of 18 °C was inverted 50 times in a tube of length 1.30 m. The final temperature of the lead shot was 23 °C. Calculate:

  a the total gravitational potential energy released by the lead

  b the specific heat capacity of lead. Assume that $g = 9.81\ \mathrm{m\,s^{-2}}$.

4 An electric shower is capable of heating water from 10 °C to 40 °C when the flow rate is $0.025\ \mathrm{kg\,s^{-1}}$. Calculate the minimum power of the heater.

22.3 Change of state

Learning objectives:

→ Define latent heat.

→ Measure latent heat.

→ Explain why the temperature of a substance stays steady when it is changing state.

Specification reference: 3.11.1

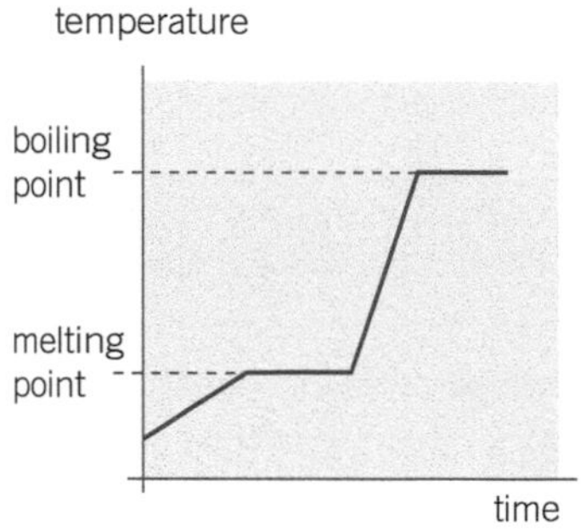

▲ **Figure 1** *Melting and boiling*

When a solid is heated and heated, its temperature increases until it melts. If it is a pure substance, it melts at a well-defined temperature, called its **melting point**. Once all the solid has melted, continued heating causes the temperature of the liquid to increase until the liquid boils. This occurs at a certain temperature, called the **boiling point**. The substance turns to a vapour as it boils away.

The three physical states of a substance, solid, liquid, and vapour, have different physical properties. For example:

- The density of a gas is much less than the density of the same substance in the liquid or the solid state. This is because the molecules of a liquid and of a solid are packed together in contact with each other. In contrast, the molecules of a gas are on average separated from each other by relatively large distances.
- Liquids and gases can flow, but solids can't. This is because the atoms in a solid are locked together by strong force bonds, which the atoms are unable to break free from. In a liquid or a gas, the molecules are not locked together. This is because they have too much kinetic energy, and the force bonds are not strong enough to keep the molecules fixed to each other.

Latent heat

When a solid or a liquid is heated so that its temperature increases, its molecules gain kinetic energy. In a solid, the atoms vibrate more about their mean positions. In a liquid, the molecules move about faster, still keeping in contact with each other, but free to move about.

1 **When a solid is heated at its melting point**, its atoms vibrate so much that they break free from each other. The solid therefore becomes a liquid due to energy being supplied at the melting point. The energy needed to melt a solid at its melting point is called **latent heat of fusion**.

   Latent heat is released when a liquid solidifies. This happens because the liquid molecules slow down as the liquid cools until the temperature decreases to the melting point. At the melting point, the molecules move slowly enough for the force bonds to lock the molecules together. Some of the latent heat released keeps the temperature at the melting point until all the liquid has solidified.

   - *Latent* means *hidden*. Latent heat supplied to melt a solid may be thought of as hidden because no temperature change takes place even though the solid is being heated.
   - Fusion is a word used for the melting of a solid because the solid *fuses* into a liquid as it melts.

2 **When a liquid is heated at its boiling point**, the molecules gain enough kinetic energy to overcome the bonds that hold them close together. The molecules therefore break away from each other to form bubbles of vapour in the liquid. The energy needed to vaporise a liquid is called **latent heat of vaporisation**.

Latent heat is released when a vapour condenses. This happens because the vapour molecules slow down as the vapour is cooled. The molecules move slowly enough for the force bonds to pull the molecules together to form a liquid.

Some solids vaporise directly when heated. This process is called **sublimation**.

In general, much more energy is needed to vaporise a substance than to melt it. For example, 2.25 MJ is needed to vaporise 1 kg of water at its boiling point. In comparison, 0.336 MJ is needed to melt 1 kg of ice at its melting point. The energy needed to change the state of unit mass (i.e., 1 kg) of a substance at its melting point (or its boiling point) is called its specific latent heat of fusion (or vaporisation).

The **specific latent heat of fusion**, l_f, of a substance is the energy needed to change the state of unit mass of the substance from solid to liquid without change of temperature.

The **specific latent heat of vaporisation** of a substance is the energy needed to change the state of unit mass of the substance from liquid to vapour without change of temperature.

So, the energy Q needed to change the state of mass m of a substance from solid to liquid (or liquid to vapour) without change of temperature is given by

$$Q = ml$$

where l is the specific latent heat of fusion (or the specific latent heat of vaporisation). The unit of specific latent heat is $\mathrm{J\,kg^{-1}}$.

Hint

Where a substance changes its state and changes its temperature, to calculate the energy transferred:

- use $Q = ml$ to calculate the energy transferred when its state changes
- use $Q = mc(T_2 - T_1)$ for the appropriate state when its temperature changes.

Synoptic link

If the volume of a substance is given, you need to know its density to find its mass. See Topic 6.1, Density.

Worked example

Calculate the energy needed to melt 5.0 kg of ice at 0 °C and heat the melted ice to 50 °C.

specific latent heat of fusion of ice = $3.36 \times 10^5\,\mathrm{J\,kg^{-1}}$

specific heat capacity of water = $4200\,\mathrm{J\,kg^{-1}\,K^{-1}}$

Solution

To melt 5.0 kg of ice, energy needed $Q_1 = ml = 5.0 \times 3.36 \times 10^5 = 1.68 \times 10^6\,\mathrm{J}$

To heat 5.0 kg of melted ice (i.e., water) from 0 °C to 50 °C, energy needed $Q_2 = mc(T_2 - T_1) = 5.0 \times 4200 \times (50 - 0) = 1.05 \times 10^6\,\mathrm{J}$

Therefore, the total energy needed $= Q_1 + Q_1 = 2.73 \times 10^6\,\mathrm{J}$

Temperature–time graphs

If a pure solid is heated to its melting point and beyond, its temperature–time graph will be as shown in Figure 2.

Assuming that no energy loss occurs during heating, and assuming that energy is transferred to the substance at a constant rate P (i.e., power supplied), then

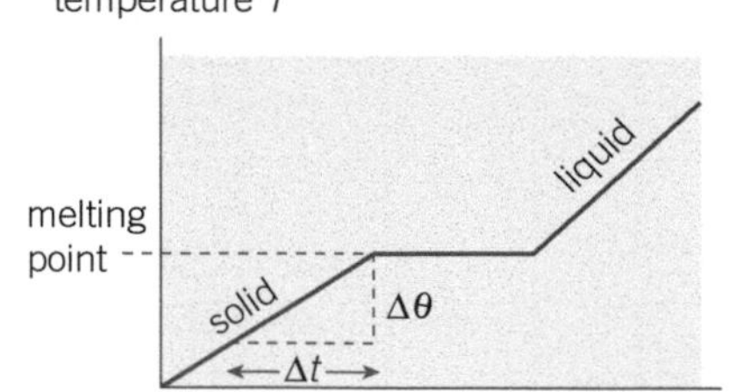

▲ **Figure 2** *Graph of temperature against time for a solid being heated*

Hint

At the melting point, P = energy supplied per second = $\frac{ml}{t}$, where l is the specific latent heat of fusion of the substance and t is the time taken to melt mass m of the substance. Therefore, the time taken to melt the substance is $t = \frac{ml}{P}$.

Study tip

For a pure substance, change of state is at constant temperature.

- Before the solid melts, $P = mc_S\left(\frac{\Delta\theta}{\Delta t}\right)_S$, where $\left(\frac{\Delta\theta}{\Delta t}\right)_S$ is the rise of temperature per second and c_S is the specific heat capacity of the solid.
  So the rise of temperature per second of the solid is $\left(\frac{\Delta\theta}{\Delta t}\right)_S = \frac{P}{mc_S}$.
- After the solid melts, $P = mc_L\left(\frac{\Delta\theta}{\Delta t}\right)_L$, where $\left(\frac{\Delta\theta}{\Delta t}\right)_L$ is the rise of temperature per second and c_L is the specific heat capacity of the liquid.
  So the rise of temperature per second of the liquid is $\left(\frac{\Delta\theta}{\Delta t}\right)_L = \frac{P}{mc_L}$.

Therefore, if the solid has a larger specific heat capacity than the liquid, the rate of temperature rise of the solid is less than that of the liquid. In other words, the liquid heats up faster than the solid.

Summary questions

1. a Explain why energy is needed to melt a solid.
   b Explain why the internal energy of the water in a beaker must be reduced to freeze the water.
2. Calculate the mass of water boiled away in a 3 kW electric kettle in 2 min.
   The specific latent heat of vaporisation of water is 2.25 MJ kg^{-1}.
3. A plastic beaker containing 0.080 kg of water at 15 °C was placed in a refrigerator and cooled to 0 °C in 1200 s.
   a Calculate how much energy each second was removed from the water in this process. The specific heat capacity of water = 4200 J kg^{-1} K^{-1}.
   b Calculate how long the refrigerator would take to freeze the water in part **a**. The specific latent heat of fusion of ice = 3.36×10^5 J kg^{-1}.

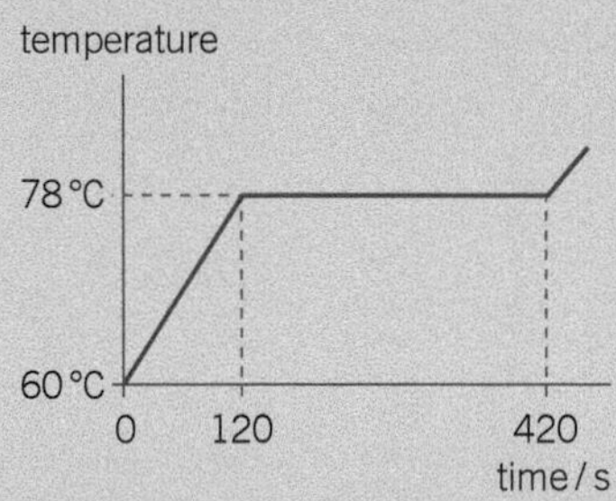

▲ **Figure 3**

4. The temperature–time graph shown in Figure 3 is obtained by heating 0.12 kg of a substance in an insulated container. The specific heat capacity of the substance in the solid state is 1200 J kg^{-1} K^{-1}.
   Calculate:
   a the energy per second supplied to the substance in the solid state if its temperature increases from 60 °C to its melting point at 78 °C in 120 s
   b the energy needed to melt the solid if it takes 300 s to melt with energy supplied at the same rate as in part **a**.

22.4 Energy transfer by heating

Energy transfer due to a temperature difference takes place by three main methods which are are thermal conduction, convection, and radiation. Evaporation from the surface of a liquid also causes transfer of energy from the liquid. This effect causes a liquid to cool because the faster molecules leave it. Conduction takes place in solids, liquids, and gases. Convection takes place in liquids and gases only. For example, energy from a hot radiator reaches other parts of a room due to thermal conduction through the radiator panel, which heats the air and causes convection currents and also radiates infrared radiation throughout the room.

Learning objectives:

- → Explain what is meant by thermal conduction, convection, and radiation and give examples of heat transfer by each method.
- → Use the equation $Q/t = kA\,\Delta\theta/L$ to solve simple problems on thermal conduction.
- → Use U-values to compare energy losses.
- → Use the equation $Q/t = U\Delta\theta$ where $U = kA/L$.

Specification reference: 3.11.2

Convection

When a fluid is heated it becomes less dense and it rises because it is less dense than the cold fluid. In a closed space, the fluid circulates as the hot fluid rises and cold fluid is drawn into the source of heating where it heats up and rises. This process is known as **convection**. Some examples are listed below.

1 A hot air balloon rises because a gas burner heats the air in the balloon and makes it less dense than the surrounding air. Without repeated bursts of heating, the hot gas in the balloon would cool and the balloon would sink as the air in it becomes more dense.

2 Ocean currents such as the Gulf Stream are convection currents of warm water that flow at or near the ocean surface from hot to colder regions.

3 Hot gases from the flames of a gas fire rise, drawing air into the gas fire. The products of combustion include carbon monoxide, which is lethal if allowed to build up. For this reason, gas fires must be well ventilated so that the products of combustion escape into the atmosphere and fresh air is drawn into the room where the heater is.

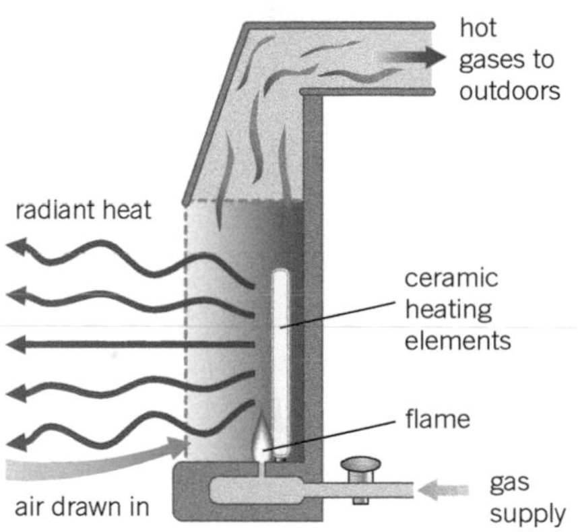

▲ **Figure 1** *A gas heater*

Thermal conduction

Some materials conduct heat much more effectively than others. Figure 2 shows how rods of different materials, which are otherwise identical, can be compared in terms of thermal conductivity. Each rod protrudes from a tank with its protruding length coated evenly with a thin wax layer. After the tank is filled with hot water, each rod conducts energy along its length, causing its wax layer to melt progressively along its length. The length of the wax layer remaining after a certain time indicates how effectively each rod conducts energy. The thermal conductivities of the rods can thus be compared.

Metals are better conductors of energy than non-metals because of the presence of conduction electrons in metals. When a metal is heated, the electrons gain energy and move about faster, transferring energy to atoms and electrons elsewhere.

Non-metals do not contain conduction electrons and therefore do not conduct as well as metals. Thermal conduction in a non-metal takes place as a result of vibrations of the atoms spreading throughout the non-metal. Heating a non-metal makes the atoms at the point of heating vibrate more, and these vibrations cause atoms to vibrate in other parts of the non-metal. This process does occur in a metal but much less than energy transfer due to conduction electrons.

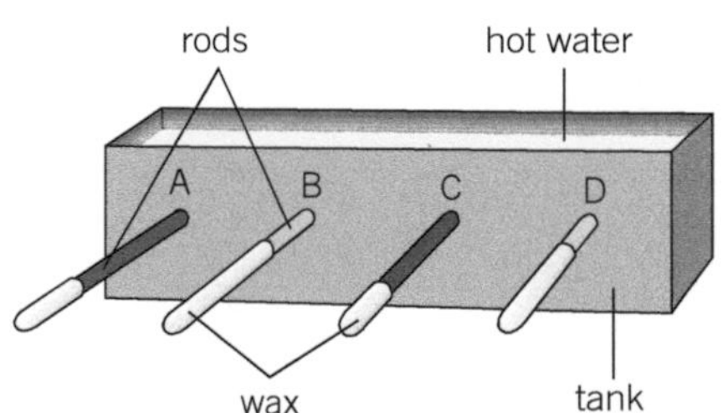

▲ **Figure 2** *Comparing thermal conductors*

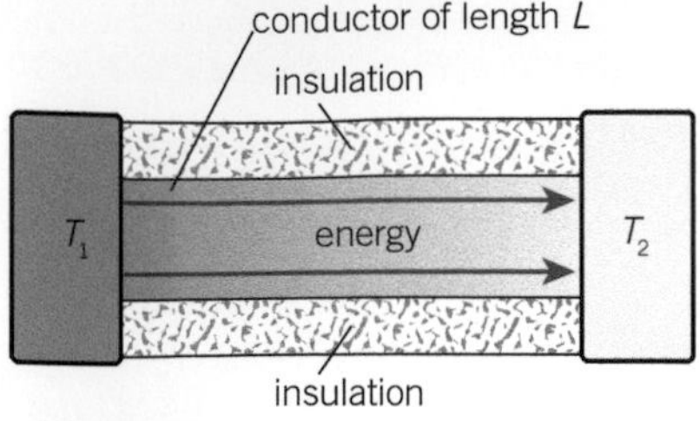

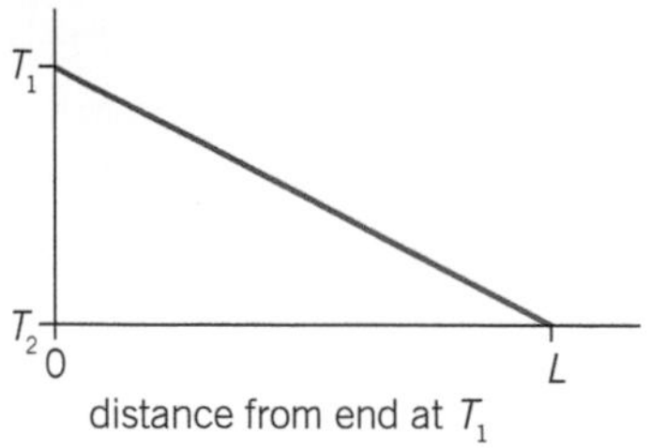

▲ **Figure 3** *Graph of temperature against position along an insulated thermal conductor*

Consider the energy flow along a conductor of uniform cross-sectional area A and of length L which has one end at fixed temperature θ_1 and the other end at a lower fixed temperature θ_2. Provided the conductor is in an insulated jacket so there is no energy loss from its sides, the temperature falls uniformly along its length from θ_1 to θ_2, as shown in Figure 3. This shows that the temperature gradient along the bar, the temperature drop per unit length, is constant.

The energy flow per second conducted along the bar, $\frac{Q}{t}$:

- is proportional to the area of cross section, A
- is proportional (and in the opposite direction) to the temperature gradient $\frac{\Delta\theta}{L}$, where $\Delta\theta = \theta_1 - \theta_2$ (i.e., the temperature difference between the ends of the bar)
- depends on the thermal conductivity of the material.

Therefore the energy flow per second, $\frac{Q}{t}$, conducted along the bar is given by the equation

$$\frac{Q}{t} = -\frac{kA(\theta_1 - \theta_2)}{L} = \frac{kA\Delta\theta}{L}$$

where k is the **coefficient of thermal conductivity** of the material. The unit of k is $\mathrm{W\,m^{-1}\,K^{-1}}$. Table 1 gives values of k for different materials.

Measurement of the thermal conductivity of copper

One end of an insulated copper bar is heated using steam, as shown in Figure 4 on the next page. The energy conducted along the bar is removed by water flow through the copper cooling coils wrapped round the cold end of the bar. Steady energy flow along the bar is attained when the temperature readings of the thermometers becomes steady.

1. The temperature gradient is measured using the thermometers at temperatures θ_1 and θ_2 at measured distance L apart.
2. The diameter d of the bar is measured and used to calculate the area of cross section using $A = \frac{\pi d^2}{4}$.
3. The rate of flow of energy along the bar $\frac{Q}{t}$ is determined by measuring the water flow rate through the cooling coils and the outflow and inflow temperatures θ_3 and θ_4.

If m is the mass of water flowing through the coils in time t, the heat flow $\frac{Q}{t}$ = the energy removed per second by the water = $\frac{mc(\theta_3 - \theta_4)}{t}$, where c is the specific heat capacity of water.

The thermal conductivity k can then be calculated using the equation

$$\frac{Q}{t} = -\frac{kA(\theta_1 - \theta_2)}{L}$$

▼ **Table 1** *Thermal conductivity values*

Material	Thermal conductivity $k/\mathrm{W\,m^{-1}\,K^{-1}}$
Aluminium	210
Brick	0.40
Cardboard	0.036
Copper	390
Glass	0.72
Felt	0.21
Plaster	0.29
Steel	450

Worked example

$k = 0.40\ \mathrm{W^{-1}\,m^{-1}\,K^{-1}}$ for brick

Calculate the energy transfer per second through a brick wall of thickness 90 mm, height 2.50 m, and length 3.60 m for a temperature difference of 20 K.

Solution

$k = 0.40\ \mathrm{W\,m^{-1}\,K^{-1}}$

$$\frac{Q}{t} = \frac{kA(\theta_1 - \theta_2)}{L} = \frac{0.40 \times (2.50 \times 3.60) \times 20}{90 \times 10^{-3}} = 800\,\mathrm{W}$$

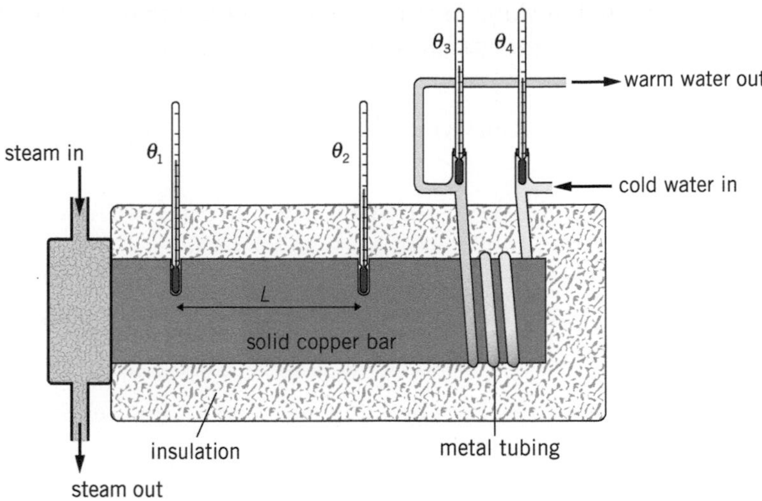

▲ **Figure 4** *Measuring the thermal conductivity of copper*

Radiation

Every object emits electromagnetic radiation due to its temperature. This radiation is known as **thermal radiation** and consists mostly of infrared radiation, although it would include visible radiation if the temperature is sufficiently high.

Thermal radiation is absorbed most effectively by matt black surfaces and least effectively by shiny silvered surfaces. **A black body** is a body that absorbs all the radiation incident on its surface. For example, a small hole in the surface of a hollow object would act as a black body as any radiation entering the hole would be completely absorbed by the surface of the cavity. The Sun and the stars may be considered as black bodies since any radiation incident on them would be absorbed.

A surface that is a good absorber of thermal radiation is also a good emitter. The radiation from a black body is referred to as **black body radiation**. The energy per second radiated from a surface depends on:

- the area, A, of the surface – the greater the area, the more energy radiated each second
- the surface temperature – the hotter an object's surface is, the more energy it radiates per second
- the nature of the surface – the surface of a black body radiates more energy per second than any other surface of the same area and at the same temperature.

U-values

The rate of energy loss from a building in winter can be reduced using different forms of insulation such as insulating the roof and the floors, fitting double-glazed windows, or, if the walls are double-brick walls, filling the gap between the walls with cavity wall insulation.

Different forms of insulation can be compared if the **U-value** of each form of insulation is known. This is the rate of energy transfer per square metre through the insulating material for a temperature difference across it of 1 K (or 1 °C).

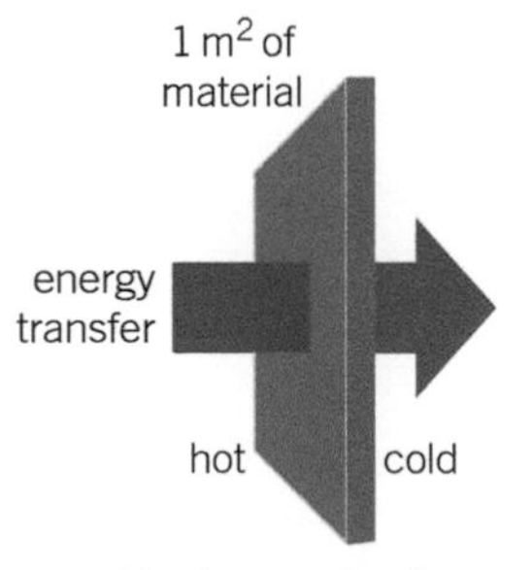

▲ **Figure 5** *U-values*

U-values are expressed in watts per square metre per kelvin (or per °C). For example, a typical single-glazed window has a U-value of $4.3\,W\,m^{-2}\,K^{-1}$.

For a given area A of insulation which has a temperature difference $\Delta\theta$ across its two surfaces,

the rate of energy transfer through the material = $UA\,\Delta\theta$

where U is the U-value of the insulation.

Note that for a single material of thermal conductivity k and thickness L, comparison on the two equations above gives $U = \frac{k}{L}$.

Table 2 gives typical U-values for a building with and without each form of insulation.

▼ **Table 2** *Typical U-values*

U-value / $W\,m^{-2}\,K^{-1}$	Without	With
Cavity wall insulation	1.6	0.6
Roof insulation	1.9	0.6
Double-glazed window	4.3	3.2

The greenhouse effect

Many scientists now believe that the increased burning of fossil fuels is causing global warming due to the increasing amounts of carbon dioxide in the Earth's atmosphere. The effect is called the greenhouse effect because the Earth's atmosphere acts like a greenhouse made of glass. In a greenhouse:

- short-wavelength infrared radiation (and light) from the Sun can pass through the glass to warm the objects inside the greenhouse
- infrared radiation from these warm objects is trapped inside by the glass because the objects emit longer wavelength infrared radiation that can't pass through the glass.

So the greenhouse stays warm. But the Earth may become too warm to avoid disastrous consequences such as rising sea levels and more extreme weather events unless the use of fossil fuels is reduced. Alternative technology such as solar panels, biofuel, and wind farms could reduce fossil fuel dependency in hot coastal regions.

Worked example

A double-glazed window of area 3.8 m^2 is used to replace a single-glazed window of the same area. Use the information in Table 2 to calculate the percentage reduction in the energy loss from the building when the temperature outside the building is 10 K less than the temperature inside.

Solution

For the single-glazed window, the rate of energy transfer $= UA\Delta\theta = 4.3 \times 3.8 \times 10 = 163\text{ W}$

For the double-glazed window, the rate of energy transfer $= UA\Delta\theta = 3.2 \times 3.8 \times 10 = 122\text{ W}$

Therefore the % reduction =

$$\frac{\text{reduction in the rate of energy transfer}}{\text{single-glazed rate of energy transfer}} \times 100\%$$

$$= \frac{163 - 122}{163} \times 100\% = 25\%$$

Summary questions

1 Figure 6 shows a cross section of a vacuum flask designed to keep a liquid hot.

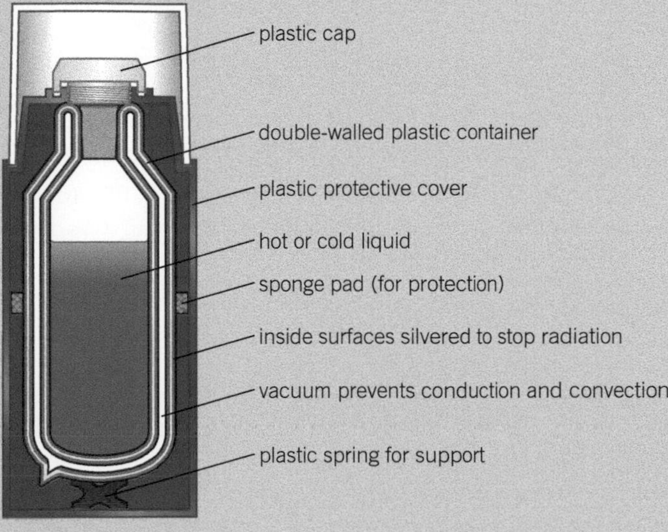

▲ **Figure 6**

- **a** Why is the glass vessel in the flask silvered?
- **b** Why is there a vacuum between the inner and outer surfaces of the vessel?
- **c** Why is a lid essential?
- **d** What would be a suitable material to surround the vessel? Give a reason for your choice.

2 Calculate the heat flow per second per square metre through:

- **a** a single-brick wall of thickness 150 mm when the temperature on one side is 25 °C and the temperature on the other side is 10 °C
- **b** a double-brick wall with a 50 mm air gap between.

thermal conductivity of air = 0.024 W m^{-1} K^{-1}

3 An aluminium (Al) pan of diameter 110 mm was used to boil water on a gas cooker. In two minutes, 0.14 kg of water was boiled away.

- **a** Calculate the energy per second needed to boil water away at this rate.
- **b** The thickness of the base of the pan was 3.2 mm. Estimate the temperature difference between the underside and the top of the base of the pan.

specific latent heat of steam = 2.3 MJ kg^{-1}
thermal conductivity of Al = 210 W m^{-1} K^{-1}

4. The external wall of a building has a U-value of 1.2 W m^{-2} K^{-1}

- **a** Explain what is meant by the U-value of a wall.
- **b** The total area of the external walls of the building is 120 m^2 and the windows have a total area of 60 m^2. The U-value of the windows is 4.5 W m^{-2} K^{-1}. Calculate the rate of energy transfer from the building when the external temperature is 15 K lower than the internal temperature.

specific heat capacity of water = $4200\,J\,kg^{-1}\,K^{-1}$
specific heat capacity of ice = $2100\,J\,kg^{-1}\,K^{-1}$
specific latent heat of fusion of ice = $3.4 \times 10^5\,J\,kg^{-1}$
specific latent heat of vaporisation of water = $2.3 \times 10^6\,J\,kg^{-1}$

1 A tray containing 0.20 kg of water at 20 °C is placed in a freezer.

(a) The temperature of the water drops to 0 °C in 10 min.
Calculate:
(i) the energy lost by the water as it cools to 0 °C
(ii) the average rate at which the water is losing energy, in $J\,s^{-1}$. *(3 marks)*

(b) (i) Estimate the time taken for the water at 0 °C to turn completely into ice.
(ii) State any assumptions you make. *(3 marks)*

AQA, 2003

2 (a) Calculate the energy released when 1.5 kg of water at 18 °C cools to 0 °C and then freezes to form ice, also at 0 °C. *(4 marks)*

(b) Explain why it is more effective to cool cans of drinks by placing them in a bucket full of melting ice rather than in a bucket of water at an initial temperature of 0 °C. *(2 marks)*

AQA, 2006

3 An electrical heater is used to heat a 1.0 kg block of metal, which is well lagged. The table shows how the temperature of the block increased with time.

Temp. / °C	20.1	23.0	26.9	30.0	33.1	36.9
Time / s	0	60	120	180	240	300

(a) Plot a graph of temperature against time. *(3 marks)*

(b) Determine the gradient of the graph. *(2 marks)*

(c) The heater provides thermal energy at the rate of 48 W. Use your value for the gradient of the graph to determine a value for the specific heat capacity of the metal in the block. *(2 marks)*

(d) The heater in part (c) is placed in some crushed ice that has been placed in a funnel, as shown in **Figure 1**.

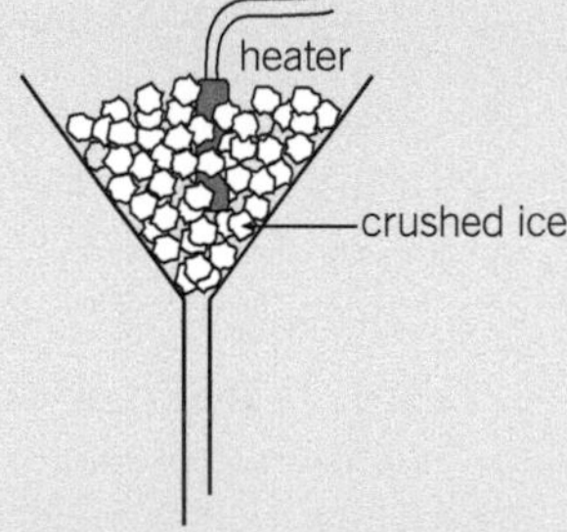

▲ **Figure 1**

The heater is switched on for 200 s and 32 g of ice are found to have melted during this time.
Use this information to calculate a value for the specific latent heat of fusion for water, stating *one* assumption made. *(3 marks)*

AQA, 2002

4 In an experiment to measure the temperature of the flame of a Bunsen burner, a lump of copper of mass 0.12 kg is heated in the flame for several minutes. The copper is then transferred quickly to a beaker, of negligible heat capacity, containing 0.45 kg of water, and the temperature rise of the water is measured.

specific heat capacity of copper = $390\,\mathrm{J\,kg^{-1}\,K^{-1}}$

(a) If the temperature of the water rises from 15 °C to 35 °C, calculate the thermal energy gained by the water. *(2 marks)*

(b) (i) State the thermal energy lost by the copper, assuming no heat is lost during its transfer.

(ii) Calculate the fall in temperature of the copper.

(iii) Calculate the temperature reached by the copper while in the flame. *(4 marks)*

AQA, 2006

5 A bicycle and its rider have a total mass of 95 kg. The bicycle is travelling along a horizontal road at a constant speed of $8.0\,\mathrm{m\,s^{-1}}$.

(a) Calculate the kinetic energy of the bicycle and rider. *(2 marks)*

(b) The brakes are applied until the bicycle and rider come to rest. During braking, 60% of the kinetic energy of the bicycle and rider is converted to thermal energy in the brake blocks. The brake blocks have a total mass of 0.12 kg and the material from which they are made has a specific heat capacity of $1200\,\mathrm{J\,kg^{-1}\,K^{-1}}$.

(i) Calculate the maximum rise in temperature of the brake blocks.

(ii) State an assumption you have made in part (b)(i). *(4 marks)*

AQA, 2004

6 A female runner of mass 60 kg generates thermal energy at a rate of 800 W.

(a) Assuming that she loses no energy to the surroundings and that the average specific heat capacity of her body is $3900\,\mathrm{J\,kg^{-1}\,K^{-1}}$, calculate:

(i) the thermal energy generated in one minute

(ii) the temperature rise of her body in one minute. *(3 marks)*

(b) In practice it is desirable for a runner to maintain a constant temperature. This may be achieved partly by the evaporation of sweat. The runner in part (a) loses energy at a rate of 500 W by this process.

Calculate the mass of sweat evaporated in one minute. *(3 marks)*

(c) Explain why, when she stops running, her temperature is likely to fall. *(2 marks)*

AQA, 2005

7 In a geothermal power station, water is pumped through pipes into an underground region of hot rocks. The thermal energy of the rocks heats the water and turns it to steam at high pressure. The steam then drives a turbine at the surface to produce electricity.

(a) Water at 21 °C is pumped into the hot rocks and steam at 100 °C is produced at a rate of $190\,\mathrm{kg\,s^{-1}}$.

(i) Show that the energy per second transferred from the hot rocks to the power station in this process is at least 500 MW.

(ii) The hot rocks are estimated to have a volume of $4.0 \times 10^6\,\mathrm{m^3}$. Estimate the fall of temperature of these rocks in one day if thermal energy is removed from them at the rate calculated in part (i) without any thermal energy gain from deeper underground.

specific heat capacity of the rocks = $850\,\mathrm{J\,kg^{-1}\,K^{-1}}$

density of the rocks = $3200\,\mathrm{kg\,m^{-3}}$ *(7 marks)*

AQA, 2006

23 Gases

23.1 The experimental gas laws

Learning objectives:

- → State the experimental gas laws.
- → Calculate the increase of the pressure of a gas when it is heated or compressed.
- → State what is meant by an isothermal change.
- → Calculate the work done in an isobaric process.

Specification reference: 3.11.3

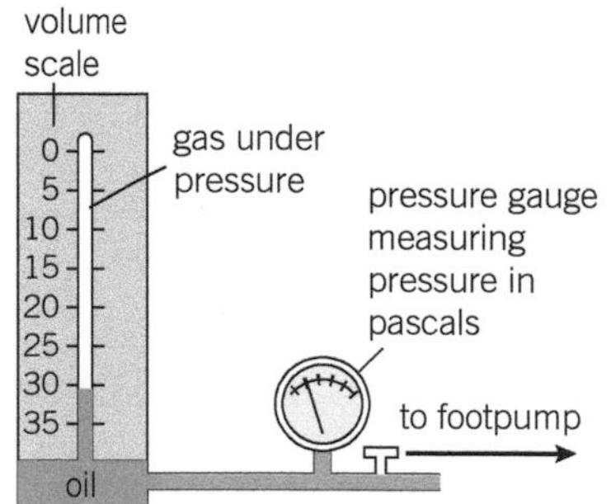

▲ **Figure 1** *Testing Boyle's law*

Hint

1. A graph of pressure against volume is a curve that tends towards each axis, as shown in Figure 2.
2. An **ideal gas** is a gas that obeys Boyle's law.

Synoptic link

A gas at very high pressure does *not* obey Boyle's law. The molecules are so close to each other that the molecules' own volume becomes significant. See Topic 23.3, The kinetic theory of gases.

When you use a cycle pump to inflate a tyre, you raise the air pressure in the tyre because the pump pushes air through a valve into the tyre. The valve lets the air in but does not allow it out. The tyre is a buffer between the wheel frame and the ground. If the tyre pressure is too low, the wheel frame will rub on the ground when you are cycling.

The **pressure** of a gas is the force per unit area that the gas exerts normally (i.e., at right angles) on a surface. Pressure is measured in pascals (Pa), where $1\,\text{Pa} = 1\,\text{N}\,\text{m}^{-2}$. The pressure of a gas depends on its temperature, the volume of the gas container, and the mass of gas in the container.

Boyle's law

The apparatus shown in Figure 1 can be used to investigate how the pressure of a fixed mass of gas depends on its volume when the temperature stays the same. Measurements using this apparatus show that the gas pressure × its volume is constant for a fixed mass of gas at constant temperature. This is called Boyle's law. Any change at constant temperature is called an *isothermal* change.

Boyle's law states that for a fixed mass of gas at constant temperature,

$$pV = \text{constant}$$

where p = gas pressure and V = gas volume.

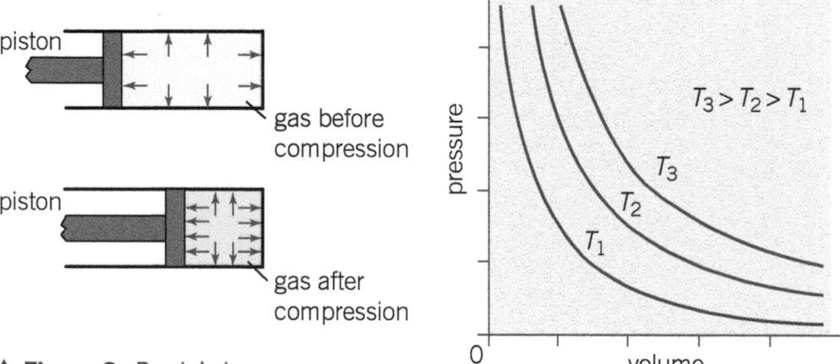

▲ **Figure 2** *Boyle's law*

The measurements plotted as a graph of pressure against $\frac{1}{\text{volume}}$ give a straight line through the origin. This is because Boyle's law can be written as $p = \text{constant} \times \frac{1}{V}$, which represents the equation $y = mx$ for a straight-line graph through the origin, if p is plotted on the y-axis and $\frac{1}{V}$ is plotted on the x-axis.

Charles's law

Using a glass tube open at one end containing dry air trapped by a suitable liquid, you can find out how the volume of a fixed mass of gas at constant pressure varies with temperature. Plotting the measurements of the volume of the gas at 0 °C and 100 °C on a graph shows you the idea behind absolute zero, which you studied in Topic 22.1. No matter how much gas is used, provided the gas is an ideal gas, its volume will be zero at absolute zero, which is −273.15 °C.

Figure 3 shows how the volume of a fixed mass of gas at constant pressure varies with absolute temperature T in kelvin. The graph is a straight line through the origin. The relationship, called **Charles's law**, between the gas volume V and the temperature T in kelvin can therefore be written as

$$\frac{V}{T} = \text{constant}$$

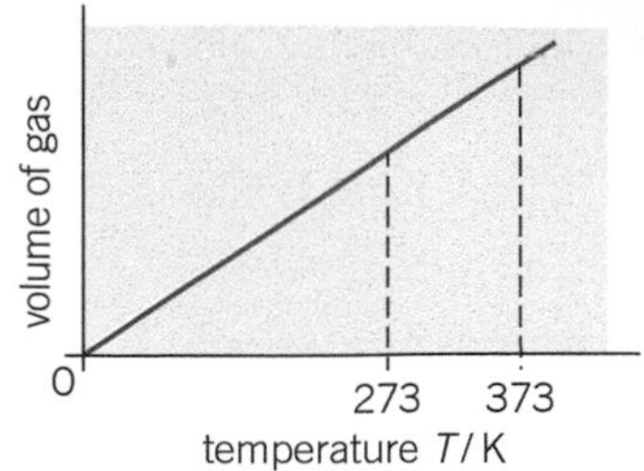

▲ **Figure 3** *Charles's law*

Any change at constant pressure is called an *isobaric* change. Suppose a gas in a piston is heated and it expands at constant pressure, The force F of the pressure p on the piston does work as the piston moves. If the piston moves a distance Δx, the work done $= F\,\Delta x = pA\,\Delta x$ since $F = pA$. Since the increase of volume of the gas $\Delta V = A\Delta x$, then the work done by the gas is given by the equation

$$\text{Work done} = p\Delta V$$

The pressure law

Figure 4 shows how the pressure of a fixed mass of gas at constant volume can be measured at different temperatures. If the measurements are plotted on a graph of pressure against temperature in kelvin, they give a straight line through the origin (as you saw in Topic 22.1). The relationship between pressure p and temperature T, in kelvin, can therefore be written as

$$\frac{p}{T} = \text{constant}$$

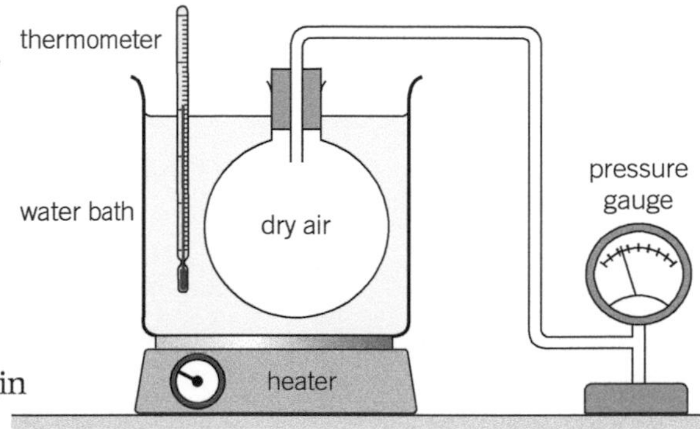

▲ **Figure 4** *The pressure law*

Extension

Deep sea diving

The extra pressure on underwater swimmers caused by a few metres of water above them is enough to give them breathing difficulties. They have to use special breathing apparatus to reach much greater depths. Gases in the lungs are compressed by the high pressure on their body, and these gases pass into their blood system. If a diver comes up too fast, dissolved nitrogen is released into their blood system, causing a life-threatening painful condition known as 'the bends'. Divers have to learn how to work out factors such as safe rates of ascent or descent.

Q: The pressure of the atmosphere at the Earth's surface is about 100 kPa, which is about the same as the pressure due to 10 m of water. In 2012, the Australian-built Deep Sea Challenger sea craft successfully descended to a depth of about 11 km. Estimate the extra pressure on it in kPa at this depth.

Answer: 110 MPa.

Summary questions

1. A hand pump of volume $2.0 \times 10^{-4}\,\text{m}^3$ is used to force air through a valve into a container of volume $8.0 \times 10^{-4}\,\text{m}^3$ which contains air at an initial pressure of 101 kPa. Calculate the pressure of the air in the container after one stroke of the pump, assuming the temperature is unchanged.
2. A sealed can of fixed volume contains air at a pressure of 101 kPa at 100 °C. The can is then cooled to a temperature of 20 °C. Calculate the pressure of the air in the can.
3. The volume of a fixed mass of gas at 15 °C is $0.085\,\text{m}^3$. The gas is then heated to 55 °C without change of pressure. Calculate the new volume of this gas.
4. A hand pump is used to raise the pressure of the air in a flask of volume $1.20 \times 10^{-4}\,\text{m}^3$, without and then with powder in the flask.
   - Without the powder in the flask, the pressure increases from 110 kPa to 135 kPa.
   - With 0.038 kg of powder in the flask, the pressure increases from 110 kPa to 141 kPa.

   a Demonstrate that the volume of air in the hand pump initially is $2.7 \times 10^{-5}\,\text{m}^3$.

   b Calculate the volume and the density of the powder.

23.2 The ideal gas law

Learning objectives:

- → Define an ideal gas.
- → Discuss whether the experimental gas laws can be combined, and, if so, how.
- → Distinguish between molar mass and molecular mass.

Specification reference: 3.11.3 and 3.11.4

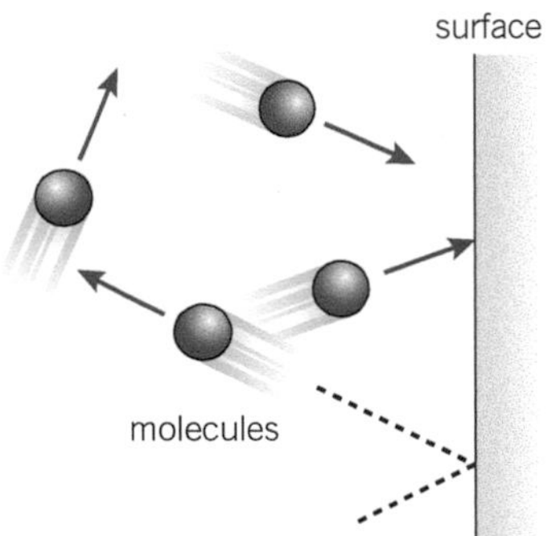

▲ **Figure 1** *Molecules in motion*

Synoptic link

Randomness occurs in radioactive decay as well as in Brownian motion of molecules. You can't predict when a random event will take place. See Topic 8.5, Radioactive decay.

Molecules in a gas

The molecules of a gas move at random with different speeds. When a molecule collides with another molecule or with a solid surface, it bounces off without losing speed. The pressure of a gas on a surface is due to the gas molecules hitting the surface. Each impact causes a tiny force on the surface. Because there are a very large number of impacts each second, the overall result is that the gas exerts a measurable pressure on the surface.

Molecules are too small to see individually. You can see the effect of individual molecules in a gas if you observe smoke particles with a microscope. If a beam of light is directed through the smoke, you will see the smoke particles as tiny specks of light wriggling about unpredictably. This type of motion is called **Brownian motion** after Robert Brown, who first observed it in 1827 with pollen grains in water. The motion of each particle is because it is bombarded unevenly and randomly by individual molecules. The particle therefore experiences forces due to these impacts, which change its magnitude and direction at random. So Brownian motion showed the existence of molecules and atoms.

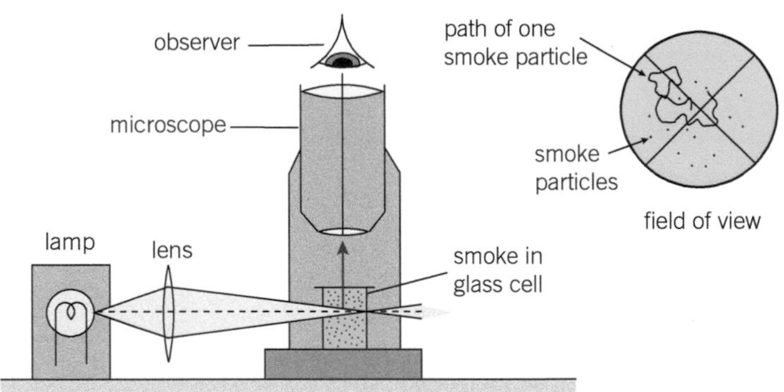

▲ **Figure 2** *Brownian motion*

The Avogadro constant

The density of oxygen gas is 16 times that of hydrogen gas at the same temperature. Therefore, the mass of a certain volume of oxygen is 16 times that of the mass of the same volume of hydrogen at the same temperature. When such measurements were first made in the 19th century, Amadeo Avogadro put forward the hypothesis that equal volumes of gases at the same temperature and pressure contain equal numbers of molecules.

How many molecules are in a particular amount of gas? Avogadro thought of the idea of counting atoms and molecules in terms of the number of atoms in 1 gram of hydrogen. Now we use 12 grams of the carbon isotope ${}^{12}_{6}C$ as the standard amount because hydrogen gas contains a small proportion of the isotope of hydrogen ${}^{2}_{1}H$, which cannot easily be removed.

- The **Avogadro constant**, N_A, is defined as the number of atoms in exactly 12 g of the carbon isotope ${}^{12}_{6}C$. The value of N_A (to four significant figures) is 6.023×10^{23}. So the mass of an atom of ${}^{12}_{6}C$ is 1.993×10^{-23} g $\left(= \frac{12\,g}{6.023 \times 10^{23}}\right)$.
- One atomic mass unit (u) is $\frac{1}{12}$ the mass of a ${}^{12}_{6}C$ atom. The mass of a carbon atom is 1.993×10^{-26} kg, so $1\,u = 1.661 \times 10^{-27}$ kg.

Molar mass

One **mole** of a substance consisting of identical particles is defined as the quantity of substance that contains N_A particles. The number of moles in a given quantity of a substance is its **molarity**. The unit of molarity is the **mol**.

The **molar mass** of a substance is the mass of 1 mol of the substance. The unit of molar mass is $kg\,mol^{-1}$. For example, the molar mass of oxygen gas is $0.032\,kg\,mol^{-1}$. So 0.032 kg of oxygen gas contains N_A oxygen molecules.

Therefore:

1 the number of moles in mass M_S of a substance $= \frac{M_S}{M}$, where M is the molar mass of the substance

2 the number of molecules in mass M_S of a substance $= \frac{N_A M_S}{M}$.

For example, because the molar mass of carbon dioxide is 0.044 kg (= 44 g), then n moles of carbon dioxide have a mass of $44n$ g and contain nN_A molecules.

The ideal gas equation

As you learnt in Topic 23.1, an **ideal gas** is a gas that obeys Boyle's law. The three experimental gas laws can be combined to give the equation

$$\frac{pV}{T} = \textbf{constant}, \text{ for a fixed mass of ideal gas}$$

where p is the pressure, V is the volume, and T is the absolute temperature. This equation takes in all situations where the pressure, volume, and temperature of a fixed mass of gas changes.

Equal volumes of ideal gases at the same temperature and pressure contain equal numbers of moles. Further measurements show that one mole of any ideal gas at 273 K and a pressure of 101 kPa has a volume of $0.0224\,m^3$. Therefore, for 1 mol of any ideal gas, the value of $\frac{pV}{T}$ for 1 mol is equal to $8.31\,J\,mol^{-1}\,K^{-1}$

$$\left(= \frac{pV}{T} = \frac{101 \times 10^3\,Pa \times 0.0224 m^3}{273 K}\right).$$

This value is called the **molar gas constant** R. A graph of pV against temperature T for n moles is a straight line through absolute zero and has a gradient equal to nR.

Hint

1 The masses in atomic mass units (to 1 u) of various atoms are: hydrogen H = 1 u, carbon C = 12 u, nitrogen N = 14 u, oxygen O = 16 u, and copper Cu = 64 u.

2 The masses in atomic mass units (to 1 u) of various molecules are: water H_2O = 18 u, carbon monoxide CO = 28 u, carbon dioxide CO_2 = 44 u, and oxygen O_2 = 32 u.

Study tip

Be clear about molar mass and molecular mass.

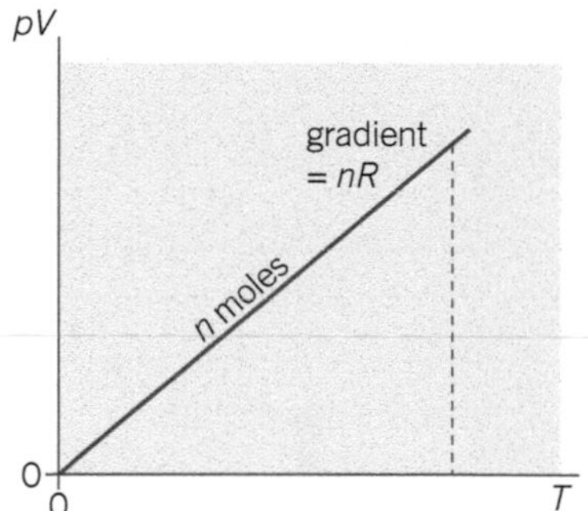

▲ **Figure 3** *A graph of pV against T for an ideal gas*

Hint

The unit of R is the joule per mole per kelvin ($J\,mol^{-1}\,K^{-1}$), which is the same as the unit of

$$\frac{\text{pressure} \times \text{volume}}{\text{absolute temp.} \times \text{no. of moles}}$$

This is because the unit of pressure (the pascal = $1\,N\,m^{-2}$) × the unit of volume (m^3) is the joule (= 1 N m).

Summary questions

$N_A = 6.02 \times 10^{23}\,\text{mol}^{-1}$
$R = 8.31\,\text{J mol}^{-1}\,\text{K}^{-1}$

1 A gas cylinder has a volume of $0.024\,\text{m}^3$ and is fitted with a valve designed to release the gas if the pressure of the gas reaches 125 kPa. Calculate:

- **a** the maximum number of moles of gas that can be contained by this cylinder at 50 °C
- **b** the pressure in the cylinder of this amount of gas at 10 °C.

2 In an electrolysis experiment, $2.2 \times 10^{-5}\,\text{m}^3$ of a gas is collected at a pressure of 103 kPa and a temperature of 20 °C. Calculate:

- **a** the number of moles of gas present
- **b** the volume of this gas at 0 °C and 101 kPa.

3 **a** Sketch a graph to show how the pressure of 2 mol of gas varies with temperature when the gas is heated from 20 °C to 100 °C in a sealed container of volume $0.050\,\text{m}^3$.

- **b** The molar mass of the gas in part **a** is $0.032\,\text{kg mol}^{-1}$. Calculate the density of the gas.

4 The molar mass of air is $0.029\,\text{kg mol}^{-1}$.

- **a** Calculate the density of air at 20 °C and a pressure of 101 kPa.
- **b** Calculate the number of molecules in $0.001\,\text{m}^3$ of air at 20 °C and a pressure of 101 kPa.

So, the combined gas law can be written as

$$pV_m = RT$$

where V_m = volume of 1 mol of ideal gas at pressure p and temperature T.

Therefore, for n moles of ideal gas,

$$\boldsymbol{pV = nRT}$$

where V = volume of the gas at pressure p and temperature T is in kelvin.

This equation is called the **ideal gas equation.**

Using the ideal gas equation

- The *mass* M_S of a substance is equal to its molar mass $M \times$ the number of moles n. Because $n = \frac{pV}{RT}$ for an ideal gas, then $M_S = M \times \left(\frac{pV}{RT}\right)$ gives the mass of ideal gas in volume V at pressure p and absolute temperature T.
- The *density of an ideal gas* of molar mass M is $\rho = \frac{\text{mass } M_S}{\text{volume } V}$

$$= \frac{nM}{V} = \frac{pM}{RT}$$

Therefore, for an ideal gas at constant pressure, its density ρ is inversely proportional to its temperature T.

- In the equation $pV = nRT$, substituting the number of moles $n = \frac{N}{N_A}$ gives

$$\boldsymbol{pV = NkT}$$

where the **Boltzmann constant** k is $\frac{R}{N_A}$, and N is the number of molecules.

Prove for yourself that $k = 1.38 \times 10^{-23}\,\text{J K}^{-1}$. We will meet k again in Topic 23.3 when we discuss how much kinetic energy a gas molecule has.

Worked example

$R = 8.31\,\text{J mol}^{-1}\,\text{K}^{-1}$

Calculate the number of moles and the mass of air in a balloon when the air pressure in the balloon is 170 kPa, the volume of the balloon is $8.4 \times 10^{-4}\,\text{m}^3$, and the temperature of the air in the balloon is 17 °C.

molar mass of air = $0.029\,\text{kg mol}^{-1}$

Solution

$T = 273 + 17 = 290\,\text{K}$

Using $pV = nRT$ gives $n = \frac{pV}{RT} = \frac{170 \times 10^3 \times 8.4 \times 10^{-4}}{8.31 \times 290}$

$= 5.9 \times 10^{-2}$ mol

Mass of air = number of moles × molar mass = $5.9 \times 10^{-2} \times 0.029$

$= 1.7 \times 10^{-3}$ kg

23.3 The kinetic theory of gases

Learning objectives:

- → Explain the increase of pressure of a gas when it is compressed or heated.
- → Describe the behaviour of a gas.
- → Discuss what the mean kinetic energy of a gas molecule depends on.

Specification reference: 3.11.4

The gas laws are experimental laws, or empirical in nature. This means they were devised by experiments and observations. They can be explained by assuming that a gas consists of point molecules moving about at random, continually colliding with the container walls. Each impact causes a force on the container. The force of many impacts is the cause of the pressure of the gas on the container walls.

Boyle's law can be explained as follows. The pressure of a gas at constant temperature is increased by reducing its volume because the gas molecules travel less distance between impacts at the walls due to the reduced volume. Therefore, there are more impacts per second, and so the pressure is greater.

The **pressure law** can be explained as follows. The pressure of a gas at constant volume is increased by raising its temperature. The average speed of the molecules is increased by raising the gas temperature. Therefore, the impacts of the molecules on the container walls are harder and more frequent. So the pressure is raised as a result.

Molecular speeds

The molecules in an ideal gas have a continuous spread of speeds, as shown in Figure 1. The speed of an individual molecule changes when it collides with another gas molecule. But the distribution stays the same, as long as the temperature does not change.

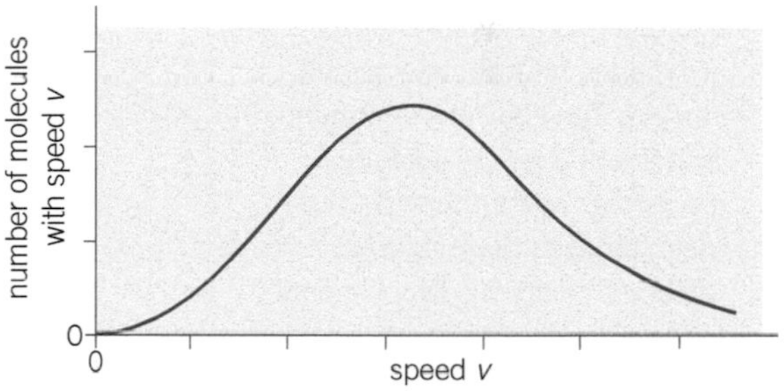

▲ **Figure 1** *Distribution of molecular speeds*

The **root mean square speed** of the molecules,

$$c_{\text{rms}} = \left[\frac{{c_1}^2 + {c_2}^2 + \cdots + {c_N}^2}{N}\right]^{1/2}$$

where $c_1, c_2, c_3, \ldots, c_N$ represent the speeds of the individual molecules, and N is the number of molecules in the gas.

If the temperature of a gas is raised, its molecules move faster, on average. The root mean square speed of the molecules increases. The distribution curve becomes flatter and broader because the greater the temperature, the more molecules there are moving at higher speeds (see Figure 2).

Hint

The root mean square speed of the molecules of a gas is *not* the same as the mean speed. The mean speed is the sum of the speeds divided by the number of molecules.

The kinetic theory equation

Kinetic theory was devised by mathematics and theories, instead of by observations and experiments like the gas laws. For an ideal gas consisting of N identical molecules, each of mass m, in a container of volume V, the pressure p of the gas is given by the equation

$$pV = \frac{1}{3}Nm(c_{\text{rms}})^2$$

where c_{rms} is the root mean square speed of the gas molecules.

To derive the **kinetic theory of gases equation**, you need to apply the laws of mechanics and statistics to the molecular model of a gas. Some assumptions must be made about the molecules in a gas:

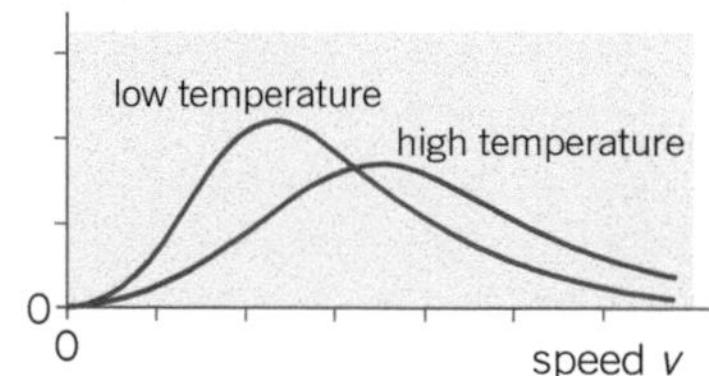

▲ **Figure 2** *The effect of temperature on the distribution of speeds*

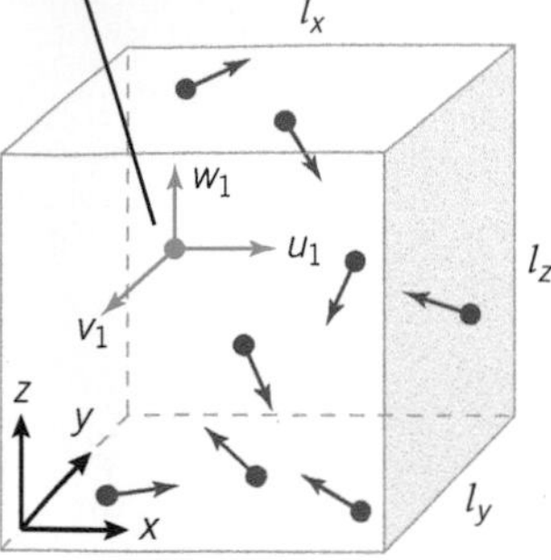

▲ **Figure 3** *Molecules in a box*

Extension

Making assumptions

In deriving the kinetic theory equation, some assumptions are made about the molecules of a gas. The ideal gas equation is then obtained by applying the further assumption that the mean kinetic energy of a gas molecule is directly proportional to the absolute temperature of the gas. Because the ideal gas law is an experimental law, the assumptions made in deriving it from the molecular model of a gas must be valid for any gas that obeys the ideal gas law. Under conditions where a gas does not obey the ideal gas equation (e.g., very high pressure), one or more of the assumptions is no longer valid. For example, a gas at very high pressure does not obey the ideal gas law. Its molecules are so close to each other that the volume of the molecules is significant and they do not behave as point molecules.

Q: When the volume of the gas decreases, there is less space for the molecules to move about in. Explain how this would affect the gas pressure.

Answer: The time between impacts would be less so the pressure would be greater.

1 The molecules are point molecules. The volume of each molecule is negligible compared with the volume of the gas.
2 They do not attract each other. If they did, the effect would be to reduce the force of their impacts on the container surface.
3 They move about in continual random motion.
4 The collisions they undergo with each other and with the container surface are elastic collisions (i.e., there is no overall loss of kinetic energy in a collision).
5 Each collision with the container surface is of much shorter duration than the time between impacts.

Part 1

Consider one molecule of mass m in a rectangular box of dimensions l_x, l_y, and l_z as shown in Figure 3. Let u_1, v_1, and w_1 represent its velocity components in the x-, y-, and z-directions, respectively.

Note that the speed, c_1, of the molecule is given by the following rule for adding perpendicular components (in this case the three velocity components u_1, v_1, and w_1):

$$c_1^2 = u_1^2 + v_1^2 + w_1^2$$

You will need to use this rule in Part 2.

- Each impact of the molecule with the shaded face in Figure 3 reverses the x-component of velocity as the impacts are elastic. So the impact changes the x-component of its momentum from $+mu_1$ to $-mu_1$. Therefore, the change of its momentum due to the impact = final momentum − initial momentum = $(-mu_1) - (mu_1) = -2mu_1$.
- As the impacts are short in duration, the time, t, between successive impacts on this face is given by the equation

$$t = \frac{\text{the total distance to the opposite face and back}}{x\text{-component of velocity}} = \frac{2l_x}{u_1}$$

Using Newton's second law therefore gives

$$\text{the force on the molecule} = \frac{\text{change of momentum}}{\text{time taken}} = \frac{-2mu_1}{\frac{2l_x}{u_1}} = \frac{-mu_1^2}{l_x}$$

Because the force F_1 of the impact on the surface is equal and opposite to the force on the molecule in accordance with Newton's third law, then

$$F_1 = \frac{+mu_1^2}{l_x}$$

- Because pressure = $\frac{\text{force}}{\text{area}}$, and the molecules do not attract each other the pressure p_1 of the molecule on the surface is given by the equation

$$p_1 = \frac{\text{force}}{\text{area of the shaded face } (l_y l_z)} = \frac{mu_1^2}{l_x l_y l_z} = \frac{mu_1^2}{V}$$

where the volume of the gas V = the volume of the box = $l_x l_y l_z$ as the molecules themselves are point molecules.

Part 2

For N molecules in the box moving at different velocities, the total pressure p is the sum of the individual pressures $p_1, p_2, p_3, \ldots, p_N$, where each subscript refers to each molecule.

$$\text{So, } p = \frac{mu_1^2}{V} + \frac{mu_2^2}{V} + \frac{mu_3^2}{V} + \dots + \frac{mu_N^2}{V}$$

$$= \frac{m}{V}(u_1^2 + u_2^2 + u_3^2 + \dots + u_N^2) = \frac{Nm\bar{u}^2}{V}$$

$$\text{where } \bar{u}^2 = \frac{u_1^2 + u_2^2 + u_3^2 + \dots + u_N^2}{N}$$

Because the motion of the molecules is random, there is no preferred direction of motion. The equation above could equally well have been derived in terms of the y-components of velocity $v_1, v_2, v_3, \dots, v_N$, or the z-components of velocity $w_1, w_2, w_3, \dots, w_N$.

That is,

$$p = \frac{Nm\bar{v}^2}{V} \qquad \text{where } \bar{v}^2 = \frac{v_1^2 + v_2^2 + v_3^2 + \dots + v_N^2}{N}$$

$$p = \frac{Nm\bar{w}^2}{V} \qquad \text{where } \bar{w}^2 = \frac{w_1^2 + w_2^2 + w_3^2 + \dots + w_N^2}{N}$$

$$\text{Therefore, } p = \frac{Nm}{3V}(\bar{u}^2 + \bar{v}^2 + \bar{w}^2)$$

The note below shows that, because the motion of the molecules is random, the *root mean square speed* of the gas molecules is given by the equation $(c_{rms})^2 = \bar{u}^2 + \bar{v}^2 + \bar{w}^2$. So

$$\mathbf{p = \frac{Nm}{3V}(c_{rms})^2 \qquad \text{or} \qquad pV = \frac{1}{3}Nm(c_{rms})^2}$$

Note

As explained in Part 1, the speed c of the nth molecule is related to its velocity components according to equations of the form:

$$c_n^2 = u_n^2 + v_n^2 + w_n^2$$

The root mean square speed of the molecules, c_{rms}, is defined by

$$(c_{rms})^2 = \frac{c_1^2 + c_2^2 + c_3^2 + \dots + c_N^2}{N}$$

$$\text{So, } (c_{rms})^2 = \frac{u_1^2 + v_1^2 + w_1^2 + u_2^2 + v_2^2 + w_2^2 + u_3^2 + v_3^2 + w_3^2 + \dots + u_N^2 + v_N^2 + w_N^2}{N}$$

$$= \bar{u}^2 + \bar{v}^2 + \bar{w}^2$$

Molecules and kinetic energy

For an ideal gas, its internal energy is due only to the kinetic energy of the molecules of the gas.

The mean **kinetic energy of the molecules of an ideal gas**

$$= \frac{\text{total kinetic energy of all the molecules}}{\text{total number of molecules}}$$

$$= \frac{\frac{1}{2}mc_1^2 + \frac{1}{2}mc_2^2 + \frac{1}{2}mc_3^2 + \dots + \frac{1}{2}mc_N^2}{N}$$

$$= \frac{\frac{1}{2}m(c_1^2 + c_2^2 + c_3^2 + \dots + c_N^2)}{N} = \frac{1}{2}m(c_{rms})^2$$

The higher the temperature of a gas, the greater the mean kinetic energy of a molecule of the gas.

Classical physics

Scientists regularly check and use each other's work. Theories are devised to understand experimental observations. Experiments then test the predictions from the theories.

The kinetic theory equation is the result of theories mainly made in the mid 19th century. Scottish physicist James Clerk Maxwell derived an equation for the distribution curve of the speeds of the molecules of a gas shown in Figure 2. Avogadro theorised that equal volumes of gas at the same temperature and pressure contain equal numbers of molecules, but couldn't tell *how many* molecules there were. In 1865, Austrian physicist Josef Loschmidt estimated the number of molecules in 1 cm^3 of gas at 0 °C and at atmospheric pressure – now known as Avogadro's number. It was redefined twice before today's definition, including by Einstein, when he developed a mathematical model of Brownian motion.

Maxwell established what we now call *classical physics*, which explains all the properties of matter and radiation by using Newton's laws, the laws of thermodynamics, and Maxwell's equations. By the late 19th century, many physicists thought that the laws of physics had all been discovered. But further discoveries were made that couldn't be explained by classical physics. Max Planck hit on a revolutionary new theory that energy is quantised, but didn't like the idea very much. A few years later, Einstein used the idea to explain another failure of classical theory, the photoelectric effect, which you learnt in Topic 13.1. Even Planck was convinced. Quantum theory took off!

Worked example

$k = 1.38 \times 10^{-23}\,\text{J K}^{-1}$, $N_A = 6.02 \times 10^{23}\,\text{mol}^{-1}$

Calculate the root mean square speed of oxygen molecules at 0 °C. The molar mass of oxygen = $0.032\,\text{kg mol}^{-1}$

Solution

$T = 273\,\text{K}$

The mass of an oxygen molecule, $m = \dfrac{0.032}{6.02 \times 10^{23}}$

$= 5.3 \times 10^{-26}\,\text{kg}$

Rearranging $\frac{1}{2}m(c_{rms})^2 = \frac{3}{2}kT$

gives $(c_{rms})^2 = \dfrac{3kT}{m}$

$= \dfrac{3 \times 1.38 \times 10^{-23} \times 273}{5.3 \times 10^{-26}}$

$= 2.13 \times 10^5\,\text{m}^2\,\text{s}^{-2}$

Therefore, the root mean square speed $c_{rms} = \sqrt{2.13 \times 10^5}$

$= 460\,\text{m s}^{-1}$ (2 s.f.)

For an ideal gas, by assuming that the mean kinetic energy of a molecule is $\frac{1}{2}m(c_{rms})^2 = \frac{3}{2}kT$, where $k = \frac{R}{N_A}$, then $3kT = m(c_{rms})^2$.

Substituting $3kT$ for $m(c_{rms})^2$ in the kinetic theory equation $pV = \frac{1}{3}Nm(c_{rms})^2$ therefore gives $pV = \frac{1}{3}N \times 3kT = NkT$

Because $Nk = \frac{NR}{N_A} = nR$, you then get the ideal gas equation $pV = nRT$.

You have derived the ideal gas equation (which is an experimental law) from the kinetic theory equation by assuming that the mean kinetic energy of an ideal gas molecule is $\frac{3}{2}kT$.

So, you can say that for an ideal gas at absolute temperature T,

the **mean kinetic energy of a molecule of an ideal gas** $= \frac{3}{2}kT$,

where $k = \frac{R}{N_A}$. Recall that the constant k is called the Boltzmann constant. Its value $\left(= \frac{R}{N_A}\right)$ is $1.38 \times 10^{-23}\,\text{J K}^{-1}$.

Notes:

Using the above equation for an ideal gas,

- the total kinetic energy of one mole $= N_A \times \frac{3}{2}kT = \frac{3}{2}RT$ (as $k = \frac{N_A}{R}$)
- the total kinetic energy of n moles of an ideal gas $= n \times \frac{3}{2}RT = \frac{3}{2}nRT$.

The total kinetic energy of n moles of an ideal gas $= \frac{3}{2}nRT$

Therefore, for n moles of an ideal gas at temperature T (in kelvin),

$$\textbf{internal energy} = \frac{3}{2}nRT$$

Summary questions

$N_A = 6.02 \times 10^{23}\,\text{mol}^{-1}$, $R = 8.31\,\text{J mol}^{-1}\,\text{K}^{-1}$, $k = 1.38 \times 10^{-23}\,\text{J K}^{-1}$

1 **a** Explain in molecular terms why the pressure of a gas in a sealed container increases when its temperature is raised.

b The molar mass of oxygen is $0.032\,\text{kg mol}^{-1}$. A cylinder of volume $0.025\,\text{m}^3$ contains oxygen gas at a pressure of 120 kPa and a temperature of 373 K. Calculate:

i the number of moles of oxygen in the cylinder

ii the total kinetic energy of all the gas molecules in the container.

2 For a hydrogen molecule (molar mass = $0.002\,\text{kg mol}^{-1}$) at 0 °C, calculate:

a its mean kinetic energy

b its root mean square speed.

3 Air consists mostly of nitrogen and oxygen in proportions 1 : 4 by mass.

a Explain why the mean kinetic energy of a nitrogen molecule in air is the same as that of an oxygen molecule in the same sample of air.

b Demonstrate that the root mean square speed of a nitrogen molecule in air is 1.07× that of an oxygen molecule in the same sample of air.

molar mass of nitrogen = $0.028\,\text{kg mol}^{-1}$
molar mass of oxygen = $0.032\,\text{kg mol}^{-1}$

4 An ideal gas of molar mass $0.028\,\text{kg mol}^{-1}$ is in a container of volume $0.037\,\text{m}^3$ at a pressure of 100 kPa and a temperature of 300 K. Calculate:

a the number of moles

b the mass of gas present

c the root mean square speed of the molecules.

Practice questions: Chapter 23

1 The escape velocity v_{esc} of an object from a planet or moon is the minimum velocity the object must have to escape from the planet. For a planet or moon of radius R, it can be shown that $v_{esc} = (2gR)^{1/2}$, where g is the gravitational field strength at the surface of the planet or moon.

(a) The radius of the Earth's moon is 1740 km, and its surface gravitational field strength is $1.62\,\mathrm{N\,kg^{-1}}$. Calculate the escape velocity from the Earth's moon. *(2 marks)*

(b) The average temperature of the lunar surface during the lunar day is about 400 K.

(i) Calculate the mean kinetic energy of a molecule of an ideal gas at 400 K.

(ii) Demonstrate that the root mean square speed of a molecule of oxygen gas at this temperature is $560\,\mathrm{m\,s^{-1}}$.

(iii) Explain why gas molecules released on the lunar surface escape into space. *(8 marks)*

(c) Astronomers have discovered the existence of water vapour in a giant gas planet orbiting a star 64 light years from Earth. The astronomers observed the spectrum of infrared light from the star and discovered absorption lines due to water vapour which are present only when the planet passes across the face of the star.

(i) Why did astronomers conclude that the absorption lines were due to the planet rather than the star?

(ii) Give *one* reason why it might not have been possible to detect such absorption lines if the planet's surface had been at the same temperature but the planet had been much smaller in diameter and in mass. *(6 marks)*

2 **(a)** (i) Sketch a graph of pressure against volume for a fixed mass of ideal gas at constant temperature. Label this graph O.

On the *same axes* sketch *two* additional curves A and B, if the following changes are made:

(ii) the same mass of gas at a lower constant temperature (label this A)

(iii) a greater mass of gas at the original constant temperature (label this B). *(3 marks)*

(b) A cylinder of volume $0.20\,\mathrm{m^3}$ contains an ideal gas at a pressure of 130 kPa and a temperature of 290 K. Calculate:

(i) the amount of gas, in moles, in the cylinder

(ii) the total kinetic energy of a molecule of gas in the cylinder

(iii) the total kinetic energy of the molecules in the cylinder. *(5 marks)*

AQA, 2005

3 **(a)** State the equation of state for an ideal gas. *(1 mark)*

(b) A fixed mass of an ideal gas is heated whilst its volume is kept constant. Sketch a graph to show how the pressure, p, of the gas varies with the absolute temperature, T, of the gas. *(2 marks)*

(c) Explain in terms of molecular motion why the pressure of the gas in part (b) varies with the absolute temperature. *(4 marks)*

(d) Calculate the average kinetic energy of the gas molecules at a temperature of 300 K. *(2 marks)*

AQA, 2004

4 **(a)** The molecular theory model of an ideal gas leads to the derivation of the equation

$$pV = \frac{1}{3}Nm(c_{rms})^2$$

Explain what each symbol in the equation represents. *(4 marks)*

(b) One assumption used in the derivation of the equation stated in part (a) is that molecules are in a state of random motion.
(i) Explain what is meant by random motion.
(ii) State *two* more assumptions used in this derivation. *(4 marks)*

(c) Describe how the motion of gas molecules can be used to explain the pressure exerted by a gas on the walls of its container. *(4 marks)*

AQA, 2002

5 The number of molecules in one cubic metre of air decreases as altitude increases. The table shows how the pressure and temperature of air compare at sea level and at an altitude of 10 000 m.

Altitude	Pressure / Pa	Temperature / K
Sea level	1.0×10^5	300
10 000 m	2.2×10^4	270

(a) Calculate the number of moles of air in a cubic metre of air at:
(i) sea level
(ii) 10 000 m. *(3 marks)*

(b) In air, 23% of the molecules are oxygen molecules. Calculate the number of extra oxygen molecules there are per cubic metre at sea level compared with a cubic metre of air at an altitude of 10 000 m. *(2 marks)*

AQA, 2006

6 **(a)** **Figure 1** shows a helium atom of mass 6.8×10^{-27} kg about to strike the wall of a container. It rebounds with the same speed.

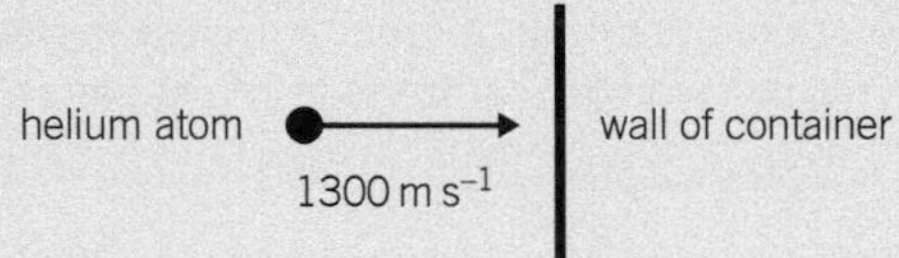

▲ **Figure 1**

(i) Calculate the momentum change of the helium atom.
(ii) Calculate the number of collisions per second on each cm^2 of the container wall that will produce a pressure of 1.5×10^5 Pa. *(5 marks)*

(b) The molar mass of gaseous nitrogen is 0.028 kg mol^{-1}. The average kinetic energy for nitrogen molecules in a sample is 8.6×10^{-21} J.
(i) Calculate the temperature of the sample.
(ii) Calculate the mean square speed of the nitrogen molecules. *(5 marks)*

AQA, 2006 and 2007

7 The graph in **Figure 2** shows the best fit line for the results of an experiment in which the volume of a fixed mass of gas was measured over a temperature range from 20 °C to 100 °C. The pressure of the gas remained constant throughout the experiment.

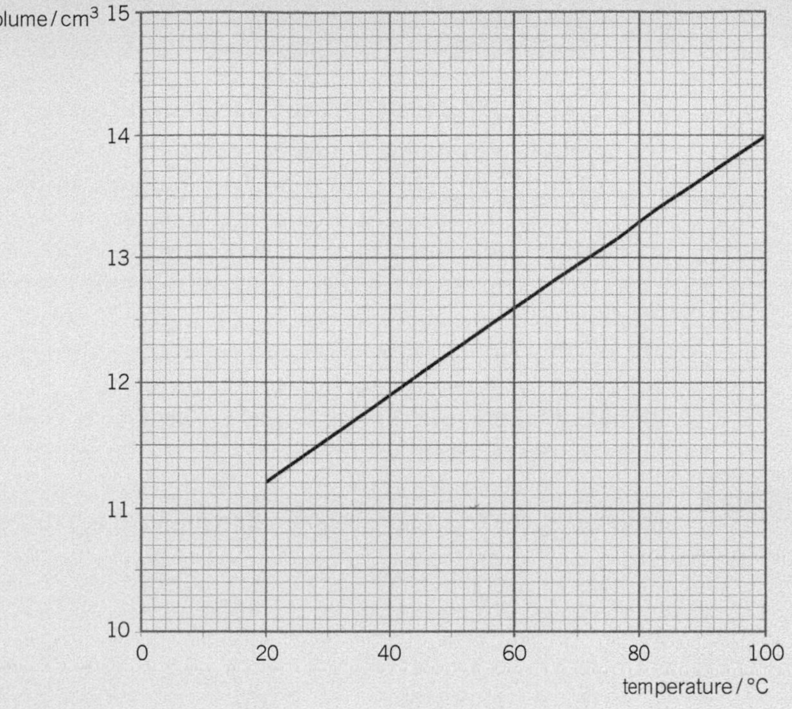

▲ **Figure 2**

(a) Use the graph in **Figure 2** to calculate a value for the absolute zero of temperature in °C.
Demonstrate clearly your method of working. *(4 marks)*

(b) Use data from the graph to calculate the mass of gas used in the experiment.
You may assume that the gas behaved like an ideal gas throughout the experiment.
gas pressure throughout the experiment = 1.0×10^5 Pa
molar mass of the gas used = $0.044\,\text{kg}\,\text{mol}^{-1}$ *(5 marks)*

(c) Use the kinetic theory of gases to explain why the pressure of an ideal gas decreases
(i) when it is expanded at constant temperature
(ii) when its temperature is lowered at constant volume. *(5 marks)*

AQA, 2005 and 2006

8 **(a)** A cylinder of fixed volume contains 15 mol of an ideal gas at a pressure of 500 kPa and a temperature of 290 K.
(i) Demonstrate that the volume of the cylinder is $7.2 \times 10^{-2}\,\text{m}^3$.
(ii) Calculate the average kinetic energy of a gas molecule in the cylinder. *(4 marks)*

(b) A quantity of gas is removed from the cylinder, and the pressure of the remaining gas falls to 420 kPa. If the temperature of the gas is unchanged, calculate the amount, in mol, of gas remaining in the cylinder. *(2 marks)*

(c) Explain in terms of the kinetic theory why the pressure of the gas in the cylinder falls when gas is removed from the cylinder. *(4 marks)*

AQA, 2003

24 Nuclear energy

24.1 Nuclear radius

Learning objectives:

- → Discuss whether more massive nuclei are wider.
- → Describe how the radius of a nucleus depends on its mass number A.
- → Describe how dense the nucleus is.

Specification reference: 3.12.1

Notes

You do *not* need to know points 2 and 3 below. But they might help you to improve your understanding of this topic.

1 The wavelength of high-energy electrons is calculated using the equation $\lambda = \frac{hc}{E}$, where E is the energy of the electrons. The speed of a high-energy electron is very close to c, the speed of light in a vacuum. Therefore, $\lambda = \frac{h}{mv} = \frac{h}{mc} = \frac{hc}{E}$, as $E = mc^2$. In practice, electrons need to be accelerated through p.ds greater than about 100 million volts to be diffracted significantly by the nucleus.

2 The angle θ_{min} depends on the radius R of the nucleus in accordance with the equation $R \sin \theta_{min} = 0.61\lambda$, where λ is the de Broglie wavelength of the electrons. The equation is derived by applying wave theory to plane waves passing at normal incidence through a circular gap.

3 Prove for yourself that values of $E = 420$ MeV and $\theta_{min} = 44°$ obtained for oxygen nuclei give a de Broglie wavelength of 3.0 fm and a nuclear radius for the oxygen nucleus of 2.6 fm. See p 497 for values of h and c. Note that 1 MeV = 1.6×10^{-13} J.

High-energy electron diffraction

In Topic 8.1, we estimated the diameter of the nucleus to be about a femtometre (1 fm = 10^{-15} m) using two methods. In this topic we will look at a much more accurate method to measure the diameter of different nuclides using high-energy electrons. When a beam of high-energy electrons is directed at a thin solid sample of an element, the incident electrons are diffracted by the nuclei of the atoms in the foil. The beam is produced by accelerating electrons through a potential difference of the order of a hundred million volts. The electrons are diffracted by the nuclei because the **de Broglie wavelength** of such high-energy electrons is of the order of 10^{-15} m which is about the same as the diameter of the nucleus. A detector is used to measure the number of electrons per second diffracted through different angles.

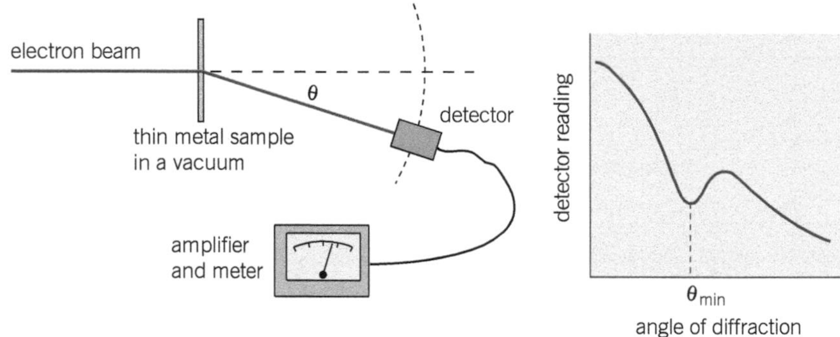

a *Outline of experiment* **b** *Typical results*

▲ **Figure 1** *High-energy electron diffraction*

The measurements show that, as the angle θ of the detector to the 'zero order' beam is increased, the number of electrons per second (i.e., the intensity of the beam) diffracted into the detector decreases then increases slightly then decreases again.

- Scattering of the beam electrons by the nuclei occurs due to their charge. This is the same as α scattering by nuclei except the electrons are attracted not repelled by the nuclei. This effect causes the intensity to decrease as angle θ increases.
- Diffraction of the beam electrons by each nucleus causes intensity maxima and minima to be superimposed on the effect above. This happens provided the de Broglie wavelength of the electrons in the beam is no greater than the dimensions of the nucleus. These superimposed intensity variations are, on a much smaller scale, similar to the concentric bright and dark fringes seen when a parallel beam of monochromatic light is directed at a circular gap or obstacle. The angle of the first minimum from the centre, θ_{min}, is measured and used to calculate the diameter of the nucleus, provided the wavelength of the incident electrons is known. See the notes in the margin.

Dependence of nuclear radius on nucleon number

Using samples of different elements, the radius R of different nuclides can be measured. By plotting a suitable graph, as explained below, it can be shown that R depends on mass number A according to $R = r_0 A^{1/3}$, where the constant $r_0 = 1.05\,\text{fm}$.

- A graph of $\ln R$ against $\ln A$ gives a straight line with a gradient of $\frac{1}{3}$ and a y-intercept equal to $\ln r_0$. This is because $\ln R = \ln A^{1/3} + \ln r_0 = \frac{1}{3}\ln A + \ln r_0$. See Topic 27.3. Plotting this graph as shown in Figure 2a therefore confirms the power of A in the equation is $\frac{1}{3}$ and it also gives a value for r_0 as the y-intercept is equal to $\ln r_0$.
- A graph of R against $A^{1/3}$ gives a straight line through the origin with a gradient equal to r_0 as shown in Figure 2b. Plotting this graph gives an accurate value of r_0.
- A graph of R^3 against A gives a straight line through the origin with a gradient equal to r_0^3. Plotting this graph would also give an accurate value of r_0.

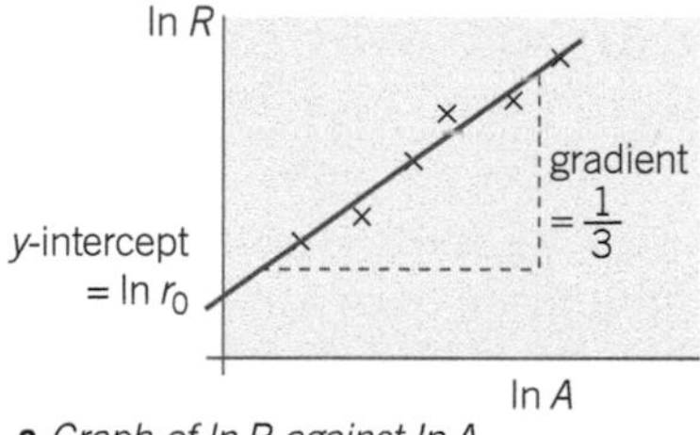

a *Graph of ln R against ln A*

b *Graph of R against $A^{\frac{1}{3}}$*

▲ **Figure 2** *Nuclear radius graphs*

Nuclear density

Assuming the nucleus is spherical,

its volume $V = \frac{4}{3}\pi R^3 = \frac{4}{3}\pi(r_0 A^{1/3})^3 = \frac{4}{3}\pi r_0^3 A$

This means that the nuclear volume V is proportional to the mass of the nucleus. In other words, the density of the nucleus is constant, independent of its radius, and is the same throughout a nucleus. From this, we can conclude that nucleons are separated by the same distance regardless of the size of the nucleus and are therefore evenly separated inside the nucleus.

We can calculate the density of a nucleus using the volume formula above ($V = \frac{4}{3}\pi r_0^3 A$) and the knowledge that its mass $m = A\text{u}$ where $1\,\text{u} = 1$ atomic mass unit $= 1.661 \times 10^{-27}\,\text{kg}$.

So the density of a nucleus $= \dfrac{A\,\text{u}}{\frac{4}{3}\pi r_0^3 A} = \dfrac{1\,\text{u}}{\frac{4}{3}\pi r_0^3} = \dfrac{1.661 \times 10^{-27}}{\frac{4}{3}\pi(1.05 \times 10^{-15})^3}$
$= 3.4 \times 10^{17}\,\text{kg}\,\text{m}^{-3}$

- A cubic millimetre of nuclear matter would have a mass of about 340 million kilograms, about the same as the total body mass of about 4 million adults.
- A neutron star is almost as dense as the nucleus of an atom. For example, a neutron star of diameter 25 km and a mass of about $4 \times 10^{30}\,\text{kg}$ (about 2 solar masses) has a density of $6 \times 10^{16}\,\text{kg}\,\text{m}^{-3}$.

Synoptic link

Log–log graphs are used to find the power n in a relationship of the form $y = kx^n$. If a graph of $\log y$ against $\log x$ is not straight, the relationship is not of this form. See Topic 27.3, Logarithms.

Summary questions

$r_0 = 1.05\,\text{fm}$

1 a Explain why a beam of high-energy electrons directed at a target is diffracted by the nuclei in the target.

b Sketch a graph to show how the intensity of the electrons varies with the angle of diffraction.

c The radius of a nucleus can be determined from high-energy electron diffraction experiments. Explain why it is important that the electrons in the beam have the same kinetic energy.

2 The volume of a $^{28}_{14}\text{Si}$ nucleus is $1.4 \times 10^{-43}\,\text{m}^3$. Use this value to calculate:

a the radius of the nucleus

b the radius of a $^{120}_{50}\text{Sn}$ nucleus.

3 a State the relationship between the radius R of a nucleus and its mass number A.

b Calculate the radius and the volume of a $^{238}_{92}\text{U}$ nucleus.

4 a Compare, without calculations, the scattering by nuclei of α particles and high-energy electrons.

b Calculate the radius and density of a $^{14}_{7}\text{N}$ nucleus.

24.2 Energy and mass

Learning objectives:

- → Explain $E = mc^2$.
- → Describe what happens to the mass of an object when it gains or loses energy.
- → Calculate the energy released in a nuclear reaction.

Specification reference: 3.12.2

In 1905, Einstein published his theory of special relativity. He showed that moving clocks run slower than stationary clocks, fast-moving objects appear shorter than when stationary, the mass of a moving object changes with its speed, and no material object can travel as fast as light. He also showed that the mass of an object increases (or decreases) when it gains (or loses) energy, E, in accordance with the equation

$$E = mc^2$$

where m is the change of its mass and c is the speed of light in free space, which is $3.0 \times 10^8\,\mathrm{m\,s^{-1}}$.

For example:

- A sealed torch that radiates 10 W of light for 10 h (= 36 000 s) would lose 0.36 MJ of energy (= 10 W × 36 000 s). Its mass would therefore decrease by 4.0×10^{-12} kg (= 0.36 MJ/c^2), an insignificant amount compared with the mass of the torch.
- A car of 1000 kg mass that speeds up from a standstill to $30\,\mathrm{m\,s^{-1}}$ would gain 450 kJ of kinetic energy so its mass when moving at $30\,\mathrm{m\,s^{-1}}$ would be 5.0×10^{-12} kg (= 450 kJ/c^2) more than when it is at rest.
- An unstable nucleus that releases a 5 MeV γ photon would lose 8.0×10^{-13} J of energy. Its mass would therefore decrease by 8.9×10^{-30} kg (= 8.0×10^{-13} J/c^2) which is not an insignificant amount compared with the mass of a nucleus.

The equation applies to all energy changes of any object. These three examples show that such changes are important in nuclear reactions but are not usually significant otherwise. A century after Einstein published his theory, the reason why the mass of an object changes when energy is transferred to or from it is still not clearly understood. However, as explained in the AS Level/Year 1 course, we know that for every type of particle there is a corresponding antiparticle with the same mass and opposite charge (if charged), and we know that:

- when a particle and its corresponding antiparticle meet, they destroy each other in a process known as **annihilation** and two gamma (γ) photons are produced, each of energy mc^2, where m is the mass of the particle or antiparticle
- a single γ photon of energy in excess of $2mc^2$ can produce a particle and an antiparticle, each of mass m, in a process known as **pair production**.

Energy changes in reactions

Reactions on a nuclear or sub-nuclear scale do involve significant changes of mass. For example, in radioactive decay, if we know the exact rest mass of each particle involved, we can calculate the energy released Q from the difference Δm in the total mass before and after the reaction. In general, for a spontaneous reaction in which no energy is supplied,

the energy released $Q = \Delta mc^2$

In any change where energy is released, such as radioactive decay, the total mass after the change is always less than the total mass before the change. This is because, in the change, some of the mass is converted to energy which is released.

1 In **alpha decay**, the nucleus recoils when the α particle is emitted so the energy released is shared between the α particle and the nucleus. Applying conservation of momentum to the recoil, you should be able to show that the energy released is shared between the α particle and the nucleus in inverse proportion to their masses.
2 In **beta decay**, the energy released is shared in variable proportions between the β particle, the nucleus, and the neutrino or antineutrino released in the decay. When the β particle has maximum kinetic energy, the neutrino or antineutrino has negligible kinetic energy in comparison. The maximum kinetic energy of the β particle is very slightly less than the energy released in the decay because of recoil of the nucleus.
3 In **electron capture**, the nucleus emits a neutrino which carries away the energy released in the decay. The atom also emits an X-ray photon when the inner-shell vacancy due to the electron capture is filled.

Notes

1 To calculate the energy corresponding to a mass difference of 1 atomic mass unit ($1\,u = 1.661 \times 10^{-27}$ kg), using $E = mc^2$ gives
$E = 1.661 \times 10^{-27}\,\text{kg} \times (3.0 \times 10^{8}\,\text{m s}^{-1})^2$
$= 1.49 \times 10^{-10}$ J
$= 931.5$ MeV.

2 When calculating Q in beta decay, assume the mass of the neutrino is negligible.

3 If the mass of each atom is given instead of the mass of its nucleus, calculate the mass of each nucleus by subtracting the mass of the electrons ($= Zm_e$) in the atom from the mass of each atom.

Worked example

The polonium isotope $^{210}_{84}\text{Po}$ emits α particles and decays to form the stable isotope of lead, $^{206}_{82}\text{Pb}$. Write down an equation to represent this process and calculate the energy released when a $^{210}_{84}\text{Po}$ nucleus emits an α particle.

mass of $^{210}_{84}\text{Po}$ nucleus = 209.936 67 u

mass of $^{206}_{82}\text{Pb}$ nucleus = 205.929 36 u

mass of α particle = 4.001 50 u

1 u is equivalent to 931.5 MeV

Solution

$^{206}_{82}\text{Pb} \rightarrow {}^{4}_{2}\alpha + {}^{206}_{82}\text{Pb}$ (+ energy released, Q)

mass difference = total initial mass – total final mass
= 209.936 67 – (205.929 36 + 4.001 50)
= 5.81×10^{-3} u

energy released Q = mass difference in u × 931.5 = 5.41 MeV

Study tip

If you are given the masses of the nuclei and particles involved in atomic mass units (u), calculate the difference between the total initial mass and the total final mass in u then multiply by 931.5 to give the energy released in MeV.

More about the strong nuclear force

As explained in your AS/A Level Year 1 course, the fact that most nuclei are stable tells us there must be an attractive force, the strong nuclear force, between any two protons or neutrons in the nucleus.

- The strength of the strong nuclear force can be estimated by working out the force of repulsion between two protons at a separation of 1 fm ($= 10^{-15}$ m), the approximate size of the nucleus. The strong nuclear force must be about the same magnitude as this force of repulsion. Prove for yourself, using Coulomb's law of force, that the force of

Synoptic link

Coulomb's law of force was covered in Topic 18.4, Coulomb's law.

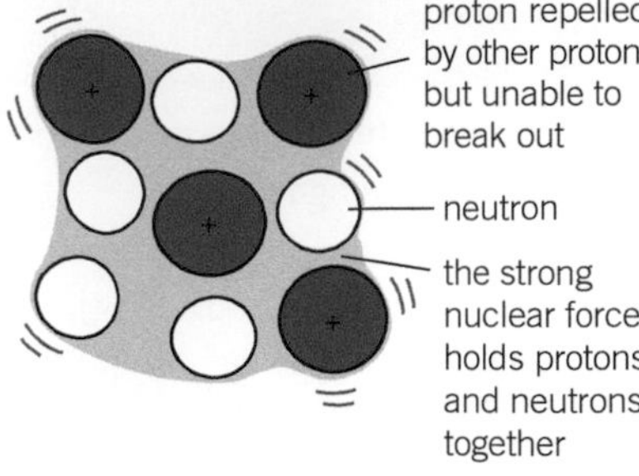

▲ **Figure 1** *The strong nuclear force*

repulsion between two protons at a separation of 10^{-15} m is of the order of 200 N. So the strong nuclear force is at least 200 N.

- The range of the strong nuclear force is no more than about 3 to 4×10^{-15} m. The diameter of a nucleus can be measured from high-energy electron scattering experiments (see Topic 24.1). The results show that nucleons are evenly spaced at about 10^{-15} m in the nucleus and therefore the strong nuclear force acts only between nearest neighbour nucleons.
- The energy needed to pull a nucleon out of the nucleus is of the order of millions of electron volts (MeV). This can be deduced because the strong nuclear force is at least about 200 N and it acts over a distance of about 2 to 3×10^{-15} m. The work done by the strong nuclear force over this distance is therefore about 7×10^{-13} J (= 200 N × 3.5×10^{-15} m) which is about 4 MeV, as 1 MeV = 1.6×10^{-13} J.
- The strong nuclear force between two nucleons must become repulsive at separations of about 0.5 fm or less, otherwise nucleons would pull each other closer and closer together and the nucleus would be much smaller than it is.

Summary questions

rest mass of an electron = 9.11×10^{-31} kg

1 u = 931.5 MeV

$g = 9.81$ m s^{-2}

1 Calculate the mass increase of:

a a 10 kg object when it is raised through a height of 2.0 m

b an electron when it is accelerated from rest through a p.d. of

i 5000 V **ii** 5 MV.

2 The bismuth isotope $^{212}_{83}$Bi emits α particles and decays to form the stable isotope of thallium $^{208}_{81}$Tl.

a Write down an equation to represent this process and calculate the energy released.

mass of $^{212}_{83}$Bi nucleus = 211.945 62 u

mass of $^{208}_{81}$Tl nucleus = 207.937 46 u

mass of α particle = 4.001 50 u

b Explain without calculation why the thallium nucleus in the above decay gains a small proportion of the energy released.

3 The strontium isotope $^{90}_{38}$Sr emits β^- particles and decays to form the stable isotope of yttrium $^{90}_{39}$Y.

a Write down an equation to represent this process and calculate the energy released.

mass of $^{90}_{38}$Sr nucleus = 89.886 40 u

mass of $^{90}_{39}$Y nucleus = 89.885 25 u

mass of β^- particle = 0.000 55 u

b Explain without calculation why the kinetic energy of the β^- particle released when the strontium nucleus decays varies from zero up to a maximum.

4 The copper isotope $^{64}_{29}$Cu decays through electron capture to form the stable isotope of nickel $^{64}_{28}$Ni.

a Write down an equation to represent this process and calculate the energy released.

mass of $^{64}_{29}$Cu nucleus = 63.913 81 u

mass of $^{64}_{28}$Ni nucleus = 63.912 56 u

mass of electron = 0.000 55 u

b State the name of the particle that takes away the energy released by the nucleus in electron capture.

24.3 Binding energy

Learning objectives:

- → Define binding energy.
- → State which nuclei are the most stable.
- → Explain why energy is released when a $^{235}_{92}U$ nucleus undergoes fission.

Specification reference: 3.12.2

Suppose all the nucleons in a nucleus were separated from one another, removing each one from the nucleus in turn. Work must be done to overcome the strong nuclear force and separate each nucleon from the others. The potential energy of each nucleon is therefore increased when it is removed from the nucleus.

The binding energy of a nucleus is the work that must be done to separate a nucleus into its constituent neutrons and protons.

When a nucleus forms from separate neutrons and protons, energy is released as the strong nuclear force does work pulling the nucleons together. The energy released is equal to the binding energy of the nucleus. Because energy is released when a nucleus forms from separate neutrons and protons, the mass of a nucleus is less than the mass of the separated nucleons.

The mass defect Δm of a nucleus is defined as the difference between the mass of the separated nucleons and the mass of the nucleus.

- Calculation of the mass defect of a nucleus of known mass: a nucleus of an isotope $^{A}_{Z}X$ is composed of Z protons and $(A - Z)$ neutrons. Therefore, for a nucleus $^{A}_{Z}X$ of mass M_{NUC},

  its mass defect $\Delta m = Zm_p + (A - Z)m_n - M_{NUC}$

  where m_p and m_n represent the masses of the proton and the neutron respectively.

- **Calculation of the binding energy of a nucleus:** the mass defect Δm is due to energy released when the nucleus formed from separate neutrons and protons. The energy released in this process is equal to the binding energy of the nucleus. Therefore,

  the binding energy of a nucleus = Δmc^2

Synoptic link

Remember 1 MeV = 1.6×10^{-13} J (and 1 eV = 1.6×10^{-19} J).
See Topic 7.4, Particles and antiparticles.

Worked example

The mass of a nucleus of the bismuth isotope $^{212}_{83}Bi$ is 211.800 12 u. Calculate the binding energy of this nucleus in MeV.

mass of a proton, m_p = 1.007 28 u
mass of a neutron, m_n = 1.008 67 u

1 u is equivalent to 931.5 MeV

Solution

Mass defect $\Delta m = 83m + (212 - 83)m_n - M_{NUC} = 1.922\,55$ u

Therefore, binding energy = 1.922 55 u × 931.5 MeV u^{-1} = 1790 MeV

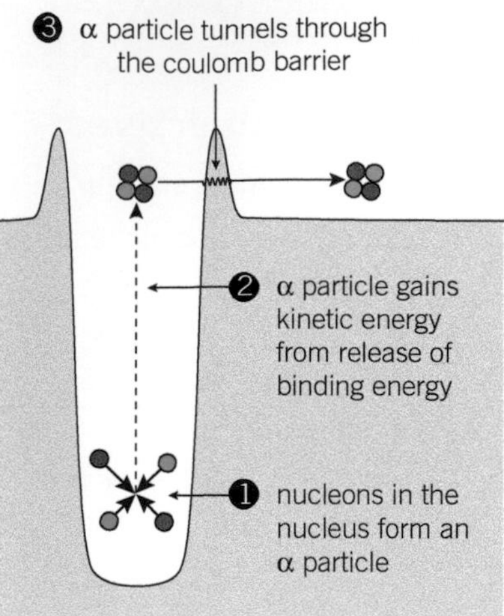

▲ **Figure 1** *Quantum tunnelling from the nucleus*

Extension

α particle tunnelling

If two protons and two neutrons inside a sufficiently large nucleus bind together as a 'cluster', they may be emitted from the nucleus as an α particle. This is because the binding energy of an α particle is very large at about 7 MeV per nucleon compared with other neutron and proton clusters that may form. The α particle therefore gains sufficient kinetic energy (equal to the binding energy of the cluster) to give it a small probability of 'quantum tunnelling' from the nucleus.

Figure 1 shows how the potential energy of an α particle varies with its distance from outside the nucleus to inside. The outer side of 'coulomb' barrier is due to the electrostatic force on the α particle. The 'well' is due to the strong nuclear force. The gain of kinetic energy of the α particle when it forms in the nucleus is sufficient for it to reach the coulomb barrier but not for it to surmount the barrier directly. However, the wave nature of the α particle gives it a small probability of tunnelling through the barrier.

Q: Explain why the probability of a decay increases the higher the kinetic energy of the α particle.

Answer: The barrier is thinner higher up so the α particle is more likely to pass through it higher up.

Notes

1. The mass of an atom of an isotope ${}^{A}_{Z}X$ is measured using a mass spectrometer. The mass of a nucleus can then be calculated by subtracting the mass of Z electrons from the atomic mass.
2. The atomic mass unit, $1\,u = 1.661 \times 10^{-27}$ kg, is defined as $\frac{1}{12}$ the mass of an atom of the carbon isotope ${}^{12}_{6}C$.
3. The energy corresponding to a mass of 1 u is $1.661 \times 10^{-27} \times (3.00 \times 10^{8})^{2}\,J = 931.5$ MeV. If you are given the mass of the nucleus in kilograms, you can convert this to atomic mass units and use the method above to calculate the mass defect. The values of the mass of the proton and the neutron are given in atomic mass units in the Useful data for A Level Physics section on page 497.

Nuclear stability

The binding energy of each nuclide is different. The **binding energy per nucleon** of a nucleus is the average work done per nucleon to remove all the **nucleons** (protons and neutrons) from a nucleus; it is therefore a measure of the stability of a nucleus. For example, the binding energy per nucleon of the ${}^{212}_{83}Bi$ nucleus is 8.4 MeV per nucleon (= 1790 MeV/212 nucleons).

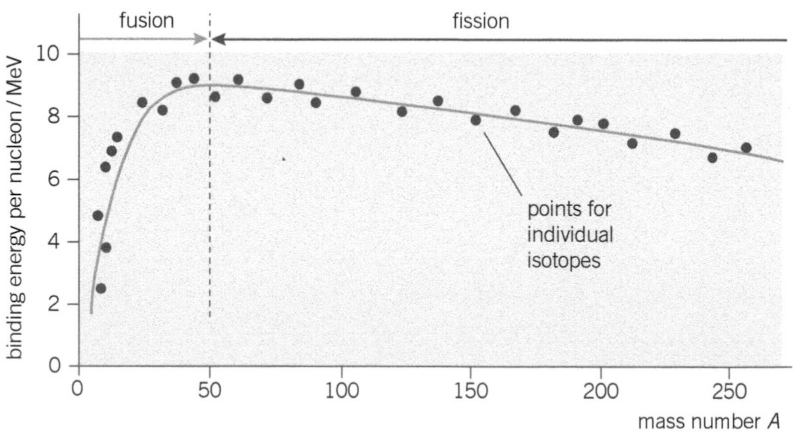

▲ **Figure 2** *Binding energy per nucleon for all known nuclides*

If the binding energies per nucleon of two different nuclides are compared, the nucleus with more binding energy per nucleon is the more stable of the two nuclei. Figure 2 shows a graph of the binding energy per nucleon against mass number A for all the known nuclides. This graph is a curve which has a maximum value of 8.7 MeV per nucleon between $A = 50$ and $A = 60$. Nuclei with mass numbers in this range are the most stable nuclei. As explained below, energy is released in:

- **nuclear fission**, the process in which a large unstable nucleus splits into two fragments which are more stable than the original nucleus. The binding energy per nucleon increases in this process, as shown in Figure 2.
- **nuclear fusion**, the process of making small nuclei fuse together to form a larger nucleus. The product nucleus has more binding energy per nucleon than the smaller nuclei. So the binding energy per nucleon also increases in this process, provided the nucleon number of the product nucleus is no greater than about 50.

Note:

The change of binding energy per nucleon is about 0.5 MeV in a fission reaction and can be more than ten times as much in a fusion reaction.

Summary questions

mass of a proton, $m_p = 1.007\,28\,\text{u}$

mass of a neutron, $m_n = 1.008\,67\,\text{u}$

1 1 u is equivalent to 931.5 MeV.

a Explain what is meant by the binding energy of a nucleus.

b Sketch a curve to show how the binding energy per nucleon of a nucleus varies with its mass number *A*, showing the approximate scale on each axis.

2 Calculate the binding energy per nucleon, in MeV per nucleon, of:

a a $^{12}_{6}\text{C}$ nucleus (mass = 12 u by definition)

b a $^{56}_{26}\text{Fe}$ nucleus (mass = 55.920 67 u).

3 a Calculate the binding energy per nucleon, in MeV per nucleon, of:

i an α particle **ii** a $^{3}_{2}\text{He}$ nucleus.

mass of α particle = 4.00150 u

mass of $^{3}_{2}\text{He}$ nucleus = 3.014 93 u

b Use the results of your calculations in part **a** to explain why an α particle rather than a $^{3}_{2}\text{He}$ nucleus is emitted by a large unstable nucleus.

4 a Copy and complete the equation below to show the fusion reaction that occurs when two protons fuse together to form a $^{2}_{1}\text{H}$ nucleus:

$$\text{p} + \text{p} \rightarrow {}^{2}_{1}\text{H} + \beta^{+}$$

b Calculate the binding energy per nucleon, in MeV, of the $^{2}_{1}\text{H}$ nucleus.

mass of $^{2}_{1}\text{H}$ nucleus = 2.013 55 u

24.4 Fission and fusion

Learning objectives:

- Describe how much energy is released in a fission or a fusion reaction.
- Explain why small nuclei can't be split.
- Explain why large nuclei can't be fused.

Specification reference: 3.12.3 and 3.12.4

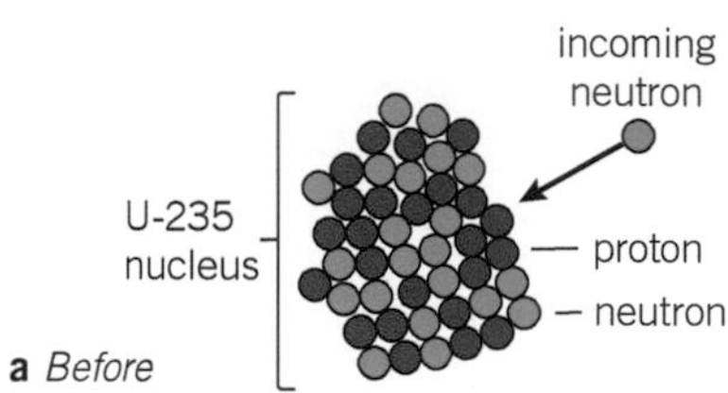

a *Before*

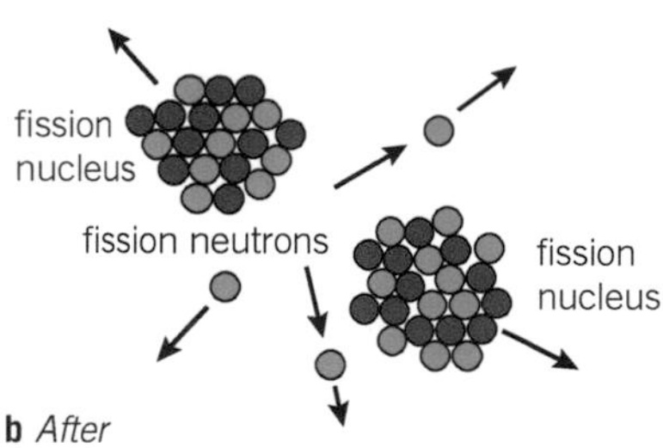

b *After*

▲ **Figure 1** *Induced fission*

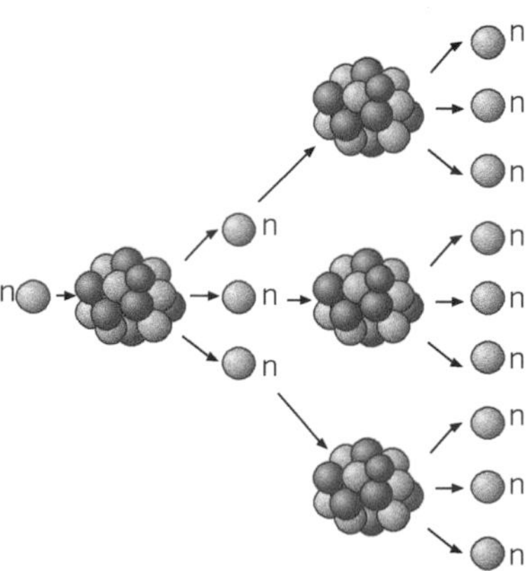

▲ **Figure 2** *A chain reaction in a nuclear reactor*

Induced fission

Fission of a nucleus occurs when a nucleus splits into two approximately equal fragments. This happens when the uranium isotope $^{235}_{92}U$ is bombarded with neutrons, a discovery made by Hahn and Strassmann in 1938. The process is known as **induced fission**. The plutonium isotope, $^{239}_{94}Pu$, is another isotope that is fissionable. This isotope is an artificial isotope formed by bombarding nuclei of the uranium isotope $^{235}_{92}U$ with neutrons.

Hahn and Strassman knew that bombarding different elements with neutrons produces radioactive isotopes. Uranium is the heaviest of all the naturally occurring elements, and scientists thought that neutron bombardment could turn uranium nuclei into even heavier nuclei. Hahn and Strassmann undertook the difficult work of analysing chemically the products of uranium after neutron bombardment to try to discover any new elements heavier than uranium. Instead, they discovered that many lighter elements such as barium were present after bombardment, even though the uranium was pure before. The conclusion could only be that uranium nuclei were split into two approximately equal fragment nuclei as a result of neutron bombardment.

Further investigations showed that each fission event releases energy and two or three neutrons.

- **Fission neutrons**, the neutrons released in a fission event, are each capable of causing a further fission event as a result of a collision with another $^{235}_{92}U$ nucleus. A **chain reaction** is therefore possible in which fission neutrons produce further fission events, which release fission neutrons, and cause further fission events, and so on. If each fission event releases two neutrons on average, after n 'generations' of fission events, the number of fission neutrons would be 2^n. Prove for yourself that fission of 6×10^{23} $^{235}_{92}U$ nuclei (i.e., 235 g of the isotope) would happen in 79 generations. Each fission event releases about 200 MeV of energy. Because each event takes no more than a fraction of a second, a huge amount of energy is released in a very short time. Using the above figures, complete fission of 235 g of $^{235}_{92}U$ would release about 10^{13} J ($= 6 \times 10^{23} \times$ 200 MeV). This is about a million times more than the energy released as a result of burning a similar mass of fossil fuel.
- **Energy is released** when a fission event occurs because the fragments repel each other (as they are both positively charged) with sufficient force to overcome the strong nuclear force trying to hold them together. The fragment nuclei and the fission neutrons therefore gain kinetic energy. The two fragment nuclei are smaller and therefore more tightly bound than the original $^{235}_{92}U$ nucleus. In other words, they have more binding energy so they are more stable than the original nucleus. The energy released is equal to the change of binding energy. The binding energy of each nucleon increases from about 7.5 MeV to about 8.5 MeV as a result of the fission event. As there are about 240 nucleons in the original nucleus, the energy released in a fission event is of the order of 200 MeV (= 240 × about 1 MeV).

- Many fission products are possible when a fission event occurs. For example, the following equation shows a fission event in which a $^{235}_{92}U$ nucleus is split into a barium $^{144}_{56}Ba$ nucleus and a krypton $^{90}_{36}Kr$ nucleus and two neutrons are released:

$$^{235}_{92}U + ^{1}_{0}n \rightarrow ^{144}_{56}Ba + ^{90}_{36}Kr + 2\,^{1}_{0}n + \text{energy released, } Q$$

- The energy released, Q, can be calculated using $E = mc^2$ in the form $Q = \Delta mc^2$, where Δm is the difference between the total mass before and after the event.
- In the above equation, the mass difference is

$$\Delta m = M_{U\text{-}238} - M_{Ba\text{-}144} - M_{Kr\text{-}90} - m_n$$

where M represents the appropriate nuclear mass and m_n is the mass of the neutron.

Study tip

Energy released in nuclear fission = change of binding energy. Fission of a given nuclide doesn't have a unique outcome.

Nuclear fusion

Fusion takes place when two nuclei combine to form a bigger nucleus. The binding energy curve (see Topic 24.3, Figure 2) shows that if two light nuclei are combined the individual nucleons become more tightly bound together. The binding energy per nucleon of the product nucleus is greater than of the initial nuclei. In other words, the nucleons become even more trapped in the nucleus when fusion occurs. As a result, energy is released equal to the increase of binding energy.

Nuclear fusion can only take place if the two nuclei that are to be combined collide at high speed. This is necessary to overcome the electrostatic repulsion between the two nuclei so they can become close enough to interact through the strong nuclear force. Some examples of nuclear fusion reactions are given below.

1 The fusion of two protons produces a nucleus of deuterium (the hydrogen isotope $^{2}_{1}H$), a β^+ particle, a neutrino, and 0.4 MeV of energy:

$$^{1}_{1}p + ^{1}_{1}p \rightarrow ^{2}_{1}H + ^{0}_{+1}\beta + \nu$$

2 The fusion of a proton and a deuterium nucleus $^{2}_{1}H$ produces a nucleus of the helium isotope $^{3}_{2}He$ and 5.5 MeV of energy:

$$^{2}_{1}H + ^{1}_{1}p \rightarrow ^{3}_{2}He$$

3 The fusion of two nuclei of helium isotope, $^{3}_{2}He$, produces a nucleus of the helium isotope $^{4}_{2}He$, two protons, and 12.9 MeV of energy:

$$^{3}_{2}He + ^{3}_{2}He \rightarrow ^{4}_{2}He + 2\,^{1}_{1}p$$

In each case, the energy released in the reaction may be calculated using $E = mc^2$ in the form $Q = \Delta mc^2$, where Δm is the difference between the total mass before and after the event.

Solar energy is produced as a result of fusion reactions inside the Sun. The temperature at the centre of the Sun is thought to be 10^8 K or more. At such temperatures, atoms are stripped of their electrons. Matter in this state is referred to as *plasma*. The nuclei of the plasma move at very high speeds because of the enormous temperature. When two nuclei collide, they fuse together because they overcome the electrostatic repulsion due to their charge and approach each other closely enough to interact through the strong nuclear force. Protons (i.e., hydrogen nuclei) inside the Sun's core fuse together in stages

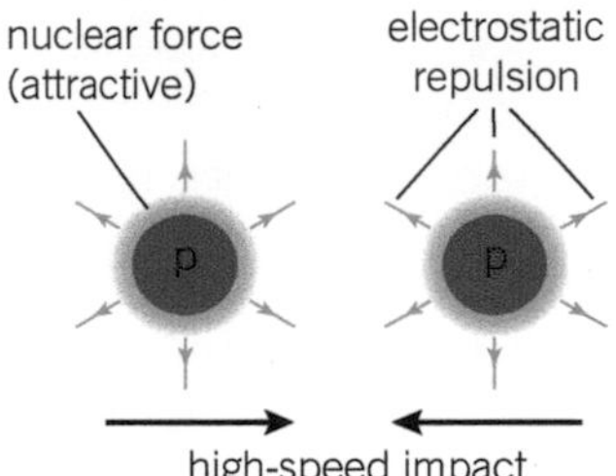

▲ **Figure 3** *Fusion of two protons*

▲ **Figure 4** *Fusion reactions inside the Sun*

▲ **Figure 5** *The JET fusion reactor. The plasma is contained in a doughnut-shaped steel container and is heated by passing a very large current through it. A magnetic field is used to confine the plasma so it does not touch the sides of its steel container, otherwise it would lose its energy.*

Note

Plasma temperature and fusion

For two nuclei to fuse together, their kinetic energy must be greater than their electrostatic potential energy at a separation of about 3 fm when they are close enough to fuse. For two protons at distance r apart, their electrostatic potential energy $= \frac{e^2}{4\pi\varepsilon_0 r}$. Hence for $r = 3$ fm, prove for yourself that the kinetic energy of *each* proton must exceed about 4×10^{-14} J.

For an ideal gas at temperature T, the mean kinetic energy of its particles $= \frac{3}{2}kT$ where k is the Boltzmann constant. Applying this equation to a plasma in which the mean kinetic energy of the particles is approximately 10^{-14} J gives a temperature T of the order of 10^9 K for fusion to occur.

(corresponding to equations 1, 2, and 3 above) to form helium ${}^4_2$He nuclei. For each helium nucleus formed, 25 MeV of energy is released. This corresponds to 6 MeV per proton, considerably more than the energy released per nucleon in a fission event.

Fusion power

Fusion reactors are still at the prototype stage even though scientific teams in several countries have been working on fusion research for more than 50 years. Prototype fusion reactors such as JET, the Joint European Torus, in the United Kingdom have produced large amounts of power but only for short periods of time. JET produces less power than it uses but the less powerful International Thermonuclear Experimental Reactor (ITER) due to start up in 2016 is designed to produce several times more power than it uses.

Energy is released in JET by fusing nuclei of deuterium ${}^2_1$H and tritium ${}^3_1$H to produce nuclei of the helium isotope ${}^4_2$He and neutrons:

$${}^2_1\text{H} + {}^3_1\text{H} \rightarrow {}^4_2\text{He} + {}^1_0\text{n} + 17.6\,\text{MeV}$$

The neutrons are absorbed by a 'blanket' of lithium surrounding the reactor vessel. The reaction between the neutrons and the lithium nuclei, as shown below, produces tritium which is then used in the main reaction. Deuterium occurs naturally in water as it forms 0.01% of naturally occurring hydrogen.

$${}^6_3\text{Li} + {}^1_0\text{n} \rightarrow {}^4_2\text{He} + {}^3_1\text{H} + 4.8\,\text{MeV}$$

Summary questions

1 u is equivalent to 931.5 MeV.

1 a Explain why the protons in a nucleus do not leave the nucleus even though they repel each other.

b Explain why the mass of a nucleus is less than the mass of the separated protons and neutrons from which the nucleus is composed.

2 a What is meant by nuclear fission?

b i The incomplete equation below represents a reaction that takes place when a neutron collides with a nucleus of the uranium isotope ${}^{235}_{92}$U. Determine the values of a and b in this equation.

$${}^{235}_{92}\text{U} + {}^1_0\text{n} \rightarrow {}^{136}_{a}\text{Xe} + {}^{b}_{36}\text{Kr} + 2\,{}^1_0\text{n} + \text{energy released, } Q$$

ii Calculate the energy, in MeV, released in this fission reaction.

masses: ${}^{235}_{92}$U nucleus 234.993 u
${}^{136}_{a}$Xe nucleus 135.877 u
${}^{b}_{36}$Kr nucleus 97.886 u, neutron 1.008 67 u

3 a What is meant by nuclear fusion?

b Hydrogen nuclei fuse together to form helium nuclei in the Sun. Two stages in this process are represented by the following equations:

$${}^1_1\text{p} + {}^1_1\text{p} \rightarrow {}^2_1\text{H} + {}^{0}_{+1}\beta$$
$${}^2_1\text{H} + {}^1_1\text{p} \rightarrow {}^3_2\text{He}$$

i Describe the reactions that these equations represent.

ii Calculate the energy released in each reaction.

masses: β particle 0.000 55 u, proton 1.007 28 u, ${}^2_1$H nucleus 2.013 55 u, ${}^3_2$He nucleus 3.014 93 u

4 a Explain why light nuclei do not fuse when they collide unless they are moving at a sufficiently high speed.

b Calculate the energy released in the following fusion reaction:

$${}^3_2\text{He} + {}^3_2\text{He} \rightarrow {}^4_2\text{He} + 2\,{}^1_1\text{p}$$

masses: proton 1.007 28 u
${}^3_2$He nucleus 3.014 93 u
α particle 4.001 50 u

c Show that about 25 MeV of energy is released when a ${}^4_2$He nucleus is formed from 4 protons.

24.5 The thermal nuclear reactor

Inside a nuclear reactor

A thermal nuclear reactor in a nuclear power station contains fuel rods spaced evenly in a steel vessel known as the **reactor core**, as shown in Figure 1. The reactor core also contains **control rods** and a **coolant** (water at high pressure in the pressurised water reactor (PWR) shown in Figure 1) as well as the fuel rods, and is connected by means of steel pipes to a **heat exchanger**. A pump is used to force the coolant through the reactor core (where it is heated) and through the heat exchanger where it is used to raise steam to drive the turbines that turn the electricity generators in the power station.

Learning objectives:

- → Explain how a nuclear reactor works.
- → Describe a thermal nuclear reactor.
- → Explain how a nuclear reactor is controlled.

Specification reference: 3.12.3 and 3.12.4

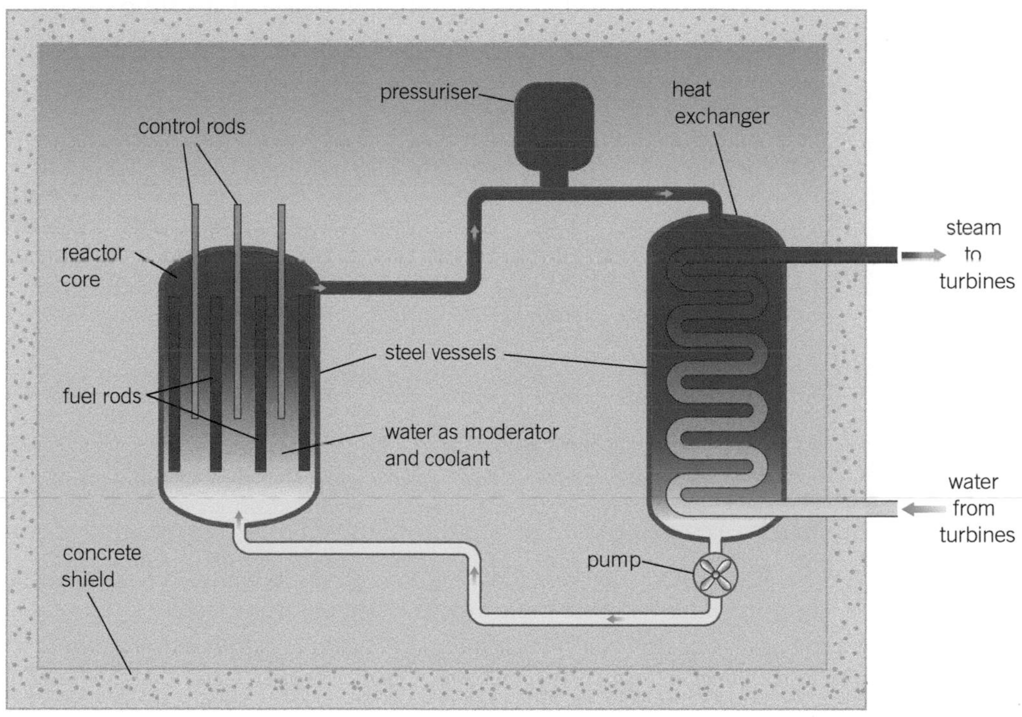

▲ **Figure 1** *Inside a nuclear reactor*

- The fuel rods contain enriched uranium which consists mostly of U-238 (the non-fissionable uranium isotope $^{238}_{92}U$) and about 2–3% U-235 (the uranium isotope $^{235}_{92}U$, which is fissionable). In comparison, natural uranium contains 99% U-238.
- The function of the control rods is to absorb neutrons. The depth of the control rods in the core is automatically adjusted to keep the number of neutrons in the core constant so that exactly one fission neutron per fission event on average goes on to produce further fission. This condition keeps the rate of release of fission energy constant. If the control rods are pushed in further, they absorb more neutrons so that the number of fission events per second and the rate of release of fission energy is reduced.
- The fission neutrons need to be slowed down significantly to cause further fission of U-235 nuclei otherwise they would be travelling too fast to cause further fission. For this reason, the fuel rods need to be surrounded by a **moderator** so the neutrons are slowed down by

Synoptic link

Neutron loss depends on the surface area. Neutron production depends on the mass of material. Below the critical mass, loss/production is too high. The same idea explains why a small object cools faster than a large object. Energy loss depends on the surface area, so temperature loss depends on surface area/mass. See Topic 22.2, Specific heat capacity.

Extension

Moderators at work

The atoms of a moderator in a nuclear reactor gain kinetic energy from fission neutrons colliding with them. The transfer of kinetic energy in such a collision is most effective if the mass of the moderator atom is as close as possible to the mass of the neutron. For this reason, and taking account of practical considerations such as chemical stability, graphite (which consists of carbon-12 atoms) and water are commonly used as moderators.

Consider an elastic head-on collision between a fission neutron and a carbon-12 nucleus. Let the neutron's speed before and after the collision be u and v, respectively, and let V be the recoil velocity of the nucleus. Conservation of momentum gives $u = 12V + v$.

Conservation of kinetic energy gives $u^2 = 12V^2 + v^2$.

Prove for yourself that combining these equations to find v in terms of u gives $v = \frac{11u}{13}$. Hence the neutron's kinetic energy after the collision is about 0.72 × its initial kinetic energy (because $\frac{v^2}{u^2} = 0.72$).

Q: Estimate how many such collisions would reduce the neutron's kinetic energy from 1 MeV to 1 eV.

Answer: 42.

repeated collisions with the moderator atoms. The reactor is described as a **thermal nuclear reactor** because the fission neutrons are slowed down to kinetic energies comparable to the kinetic energies of the moderator molecules. In the PWR, the water in the reactor core acts as the moderator as well as acting as a coolant.

- For a chain reaction to occur, the mass of the fissile material (e.g., U-235) must be greater than a minimum mass, referred to as the **critical mass**. This is because some fission neutrons escape from the fissile material without causing fission and some are absorbed by other nuclei without fission. If the mass of fissile material is less than the critical mass needed, too many of the fission neutrons escape because the surface area to mass ratio of the material is too high.
- Different types of thermal reactors are in operation throughout the world. Table 1 shows some of the features of the PWR reactor, which operates in many countries including the UK, and the Advanced Gas-cooled Reactor (AGR), which operates only in the UK.

▼ **Table 1** *Comparison of thermal reactors*

	AGR	PWR
Fuel	uranium oxide in stainless steel cans	uranium oxide in zirconium alloy cans
Moderator	graphite	water
Coolant	CO_2 gas	water
Coolant temperature / K	900	600
Typical power output / MW	1300	700

Safety features

A nuclear reactor needs to have a range of safety features to protect its workforce, the wider community, and the environment.

1. The reactor core is a thick steel vessel designed to withstand the high pressure and temperature in the core. The thick steel vessel absorbs β radiation and some of the γ radiation and neutrons from the core.
2. The core is in a building with very thick concrete walls which absorb the neutrons and γ radiation that escape from the reactor vessel.
3. Every reactor has an emergency shut-down system designed to insert the control rods fully into the core to stop fission completely.
4. The sealed fuel rods are inserted and removed from the reactor by means of remote handling devices. The rods are much more radioactive after removal than before. This is because the fuel cans
   - a before use contain U-235 and U-238 which emit only α radiation and this is absorbed by the fuel cans
   - b after use emit β and γ radiation due to the many neutron-rich fission products that form.

In addition, the spent fuel rods contain the plutonium isotope ${}^{239}_{94}Pu$ as a result of the absorption of neutrons by U-238 nuclei. This plutonium isotope is a very active α emitter and if inhaled causes lung cancer.

Extension

A nuclear future

To combat climate change, the UK government has legislated to reduce carbon emissions to 50% of 1990 levels by 2027 by introducing measures such as more non-carbon generating capacity. The new nuclear power stations will be built by the private sector. The negotiations to build the first such power station at Hinkley Point in Somerset by 2025 took five years to complete. At present, the UK's nuclear reactors provide about 10 GW of electricity, which is about 20% of total UK electricity production. Most of the present reactors will have been retired by 2028. The new nuclear power stations will provide about 16 GW by 2030, and so will make a significant contribution to the 2027 carbon emission target. However, other measures, such as more wind farms, are also likely to be needed.

Nuclear power continues to cause concern to many people. The safety features described in this topic are extremely important. Accidents at any nuclear power stations, although rare, alarm people across the world. Two such events are described below.

Chernobyl, 1986

The Chernobyl disaster in Ukraine in 1986 released radioactive materials into the atmosphere and led to the permanent evacuation of all the people in the surrounding area. The disaster was caused when the operators were testing reactor no. 4 to find out if the coolant pumps would keep operating in the event of a loss of power until the emergency diesel generators took over. When the reactor was powered down by pushing the control rods further into the core, the power fell much more than expected, so the movement of the control rods was reversed. This caused an unexpected surge in the rate of fission events, which produced a massive explosion in the reactor core. The fuel rods melted, the reactor cap was blown off, and radioactive fission products were thrown up into the atmosphere. A subsequent inquiry concluded that the main causes of the explosion were design faults in the reactor (e.g., graphite in the cap ignited in the explosion), which are not features of AGR or PWR reactors, and human error (too many control rods being moved at once).

▲ **Figure 2** *Fukushima*

Fukushima, 2011

On 11 March 2011, a tsunami created by a powerful earthquake in Japan killed over 20 000 people and caused a meltdown at the Fukushima nuclear power station. As a result of lessons learned from Chernobyl, over 100 000 people living near the Fukushima power station were evacuated. Public areas and buildings in nearby towns had to be decontaminated and food controls imposed. When the earthquake struck, planned safety procedures were put into effect. All 11 reactors at the site were shut down without stopping their cooling pumps, which prevent the reactors from overheating. However, an hour later, the tsunami hit the site and flooded the generators, causing the reactor pumps to stop. The result was that three older reactors overheated, and hydrogen gas (created by chemical reactions) and radioactive material leaked from broken fuel rods. In addition, the water supply to cooling ponds containing spent fuel rods was damaged, causing the rods in the ponds to overheat until alternative water supplies were provided. Contaminated water leaked into the sea, and traces of plutonium have been found in soil nearby. Such leakage continues to cause problems, and many evacuees are still not allowed to return home.

The new nuclear power stations in the UK are intended to help secure our future electricity supply and to reduce carbon emissions. Hundreds of nuclear reactors are operating satisfactorily in many countries. The risk of another disaster like Chernobyl and Fukushima is small, but the consequences of such a disaster could be severe.

Q: Discuss why Chernobyl is still a dangerous area today.

▲ **Figure 3** *Spent fuel rods in a cooling pond*

Summary questions

1 a What is meant by induced fission?

 b i What is the function of the control rods in a nuclear reactor?

 ii Name a suitable material from which control rods are made.

 iii Describe how the control rods are used to maintain a nuclear reactor so its power output is constant.

2 State the function of the following parts of a thermal nuclear reactor and give an example of a material used for each part:

 a the moderator

 b the coolant.

3 Explain why the mass of fissile fuel in a nuclear reactor must exceed a critical value in order for fission to be sustained in the reactor.

4 a Explain why the spent fuel rods from a nuclear reactor are more radioactive after removal from the reactor than they were before they were used in the reactor.

 b Explain why radioactive waste must be stored in secure and safe conditions.

Radioactive waste

Radioactive waste is categorised as high-, intermediate-, or low-level waste according to its activity. Most high-level radioactive waste is from nuclear power stations or from specialist users in universities and industry, or from hospitals that use radioactive isotopes for diagnosis or therapy.

Disposal of any form of radioactive waste must be in accordance with legal regulations and by approved disposal companies to ensure that the radioactive waste is stored safely in secure containers until its activity is insignificant. Disposal by dilution, for example, diluting radioactive water from nuclear power station cooling systems with large quantities of water and then dispersing it into the sea, is no longer acceptable and has been banned.

- **High-level radioactive waste**, such as spent fuel rods from a nuclear power station, contains many different radioactive isotopes, including fission fragments as well as unused uranium-235 and uranium-238 and plutonium-239. The spent fuel rods must be removed by remote control and stored underwater in cooling ponds for up to a year because they continue to release energy due to radioactive decay. In Britain, the rods are then transferred in large steel casks to the THORP reprocessing plant at Sellafield in Cumbria where the unused uranium and plutonium is then removed and stored in sealed containers for further possible use. The rest of the material (i.e., the fission products and the fuel cans) is radioactive waste and is stored in sealed containers in deep trenches at Sellafield. Such waste must be stored safely for centuries as it contains long-lived radioactive isotopes which must be prevented from contaminating food and water supplies.

  In other countries, high-level radioactive waste is stored in the same way or in underground caverns which are geologically stable. In some countries, the waste is vitrified by mixing it with molten glass and then stored as glass blocks in underground caverns.

  The long-term safe storage of high-level radioactive waste remains a major issue in Britain because no one wants such storage in their own locality, nor do people want radioactive waste to be carried through their own locality to storage facilities elsewhere.

- **Intermediate-level waste**, such as radioactive materials with low activity and containers of radioactive materials, are sealed in drums that are encased in concrete and stored in specially constructed buildings with walls of reinforced concrete.
- **Low-level waste**, such as laboratory equipment and protective clothing, is sealed in metal drums and buried in large trenches.

Practice questions: Chapter 24

1 The radioactive isotope of sodium $^{22}_{11}Na$ has a half-life of 2.6 years.
A particular sample of this isotope has an initial activity of 5.5×10^5 Bq.

(a) Explain what is meant by the *random nature* of radioactive decay. *(2 marks)*

(b) Sketch a graph of the activity of the sample of sodium over a period of 6 years. *(2 marks)*

(c) Calculate:

(i) the decay constant, in s^{-1}, of $^{22}_{11}Na$
1 year = 3.15×10^7 s

(ii) the number of atoms of $^{22}_{11}Na$ in the sample initially

(iii) the time taken, in s, for the activity of the sample to fall from 1.0×10^5 Bq to 0.75×10^5 Bq. *(6 marks)*

AQA, 2003

2 (a) Sodium-21 $\left(^{21}_{11}Na\right)$ decays to neon-21 $\left(^{21}_{10}Ne\right)$. A nucleus of neon-21 is stable.

(i) State the names of the particles emitted when a sodium-21 nucleus decays.

(ii) How many neutrons are there in a nucleus of neon-21? *(3 marks)*

(b) **Figure 1** shows how the activity A of a freshly prepared sample of sodium-21 varies as it decays. **Figure 2** shows how N, the number of sodium-21 nuclei, varies with time t during the same time interval.

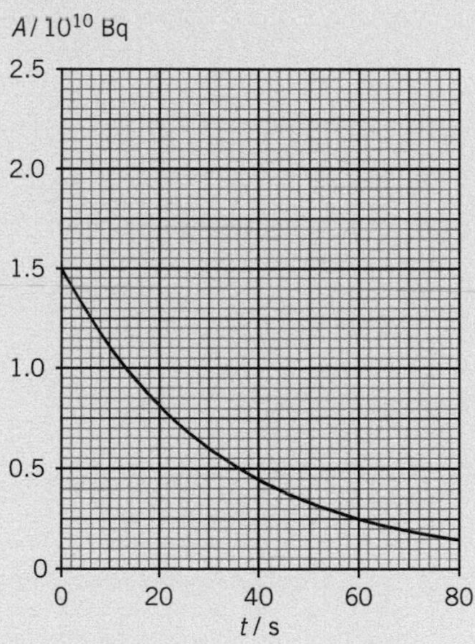

▲ **Figure 1**

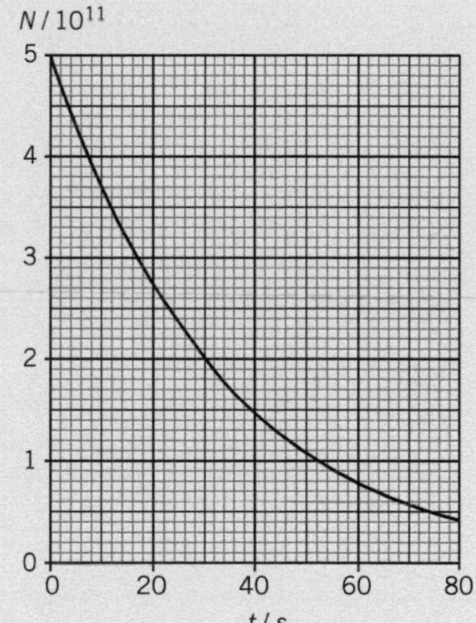

▲ **Figure 2**

(i) Use the graphs to find the number of active sodium nuclei and the corresponding activity one half-life after $t = 0$. Then find the probability of decay of a sodium-21 nucleus.

(ii) The total energy produced when a sodium-21 nucleus decays is 5.7×10^{-13} J. Calculate the number of radioactive atoms in a sample that is producing 2.6 mJ of energy each second. *(6 marks)*

AQA, 2003

3 (a) Calculate the radius of the $^{238}_{92}U$ nucleus.
$r_0 = 1.3 \times 10^{-15}$ m *(2 marks)*

(b) At a distance of 30 mm from a point source of γ rays the corrected count rate is C.
Calculate the distance from the source at which the corrected count rate is $0.10C$, assuming that there is no absorption. *(2 marks)*

(c) The activity of a source of β particles falls to 85% of its initial value in 52 s. Calculate the decay constant of the source. *(3 marks)*

AQA, 2006

4 The high-energy electron diffraction apparatus represented in **Figure 3** can be used to determine nuclear radii. The intensity of the electron beam received by the detector is measured at various diffraction angles, θ.

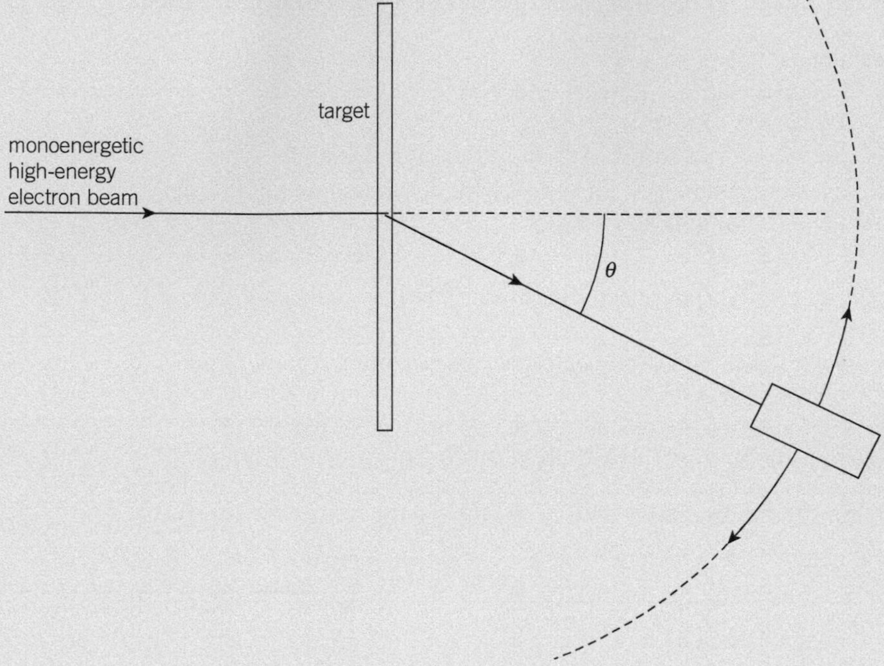

▲ **Figure 3**

(a) Sketch a graph to show how, in such an electron diffraction experiment, the electron intensity varies with the angle of diffraction, θ. *(2 marks)*

(b) (i) Use the data in the table to plot a straight-line graph that confirms the relationship

$$R = r_0 A^{1/3}$$

Element	Radius of nucleus, $R / 10^{-15}$ m	Nucleon number, A
lead	6.66	208
tin	5.49	120
iron	4.35	56
silicon	3.43	28
carbon	2.66	12

(ii) Estimate the value of r_0 from the graph. *(5 marks)*

(c) Discuss the merits of using high-energy electrons to determine nuclear radii rather than using α particles. *(3 marks)*

AQA, 2005

5 The neutron–proton model of the nucleus was first put forward by Rutherford to explain the general composition of the nucleus. The existence of the neutron was not proved experimentally until some years later.

(a) Give *two* reasons why Rutherford's neutron–proton model was considered more than an untested hypothesis when it was first put forward. (*3 marks*)

(b) The α particles from any α-emitting isotope have the same initial kinetic energy and a well-defined range in air at atmospheric pressure. The table below shows the range *R* in air and the initial kinetic energy *E* of α particles from several α-emitting isotopes.

R/mm	39	48	53	57	66	78
E/MeV	5.3	6.0	6.5	6.8	7.4	8.3

Plot a suitable graph to find out if the relationship between *R* and *E* is of the form $R = kE^n$, where *k* and *n* are constants, and determine a value for *n*. Explain your choice of graph. (*10 marks*)

6 **Figure 4** shows the general relationship between the nuclear binding energy per nucleon (*B*) and nucleon number (*A*).

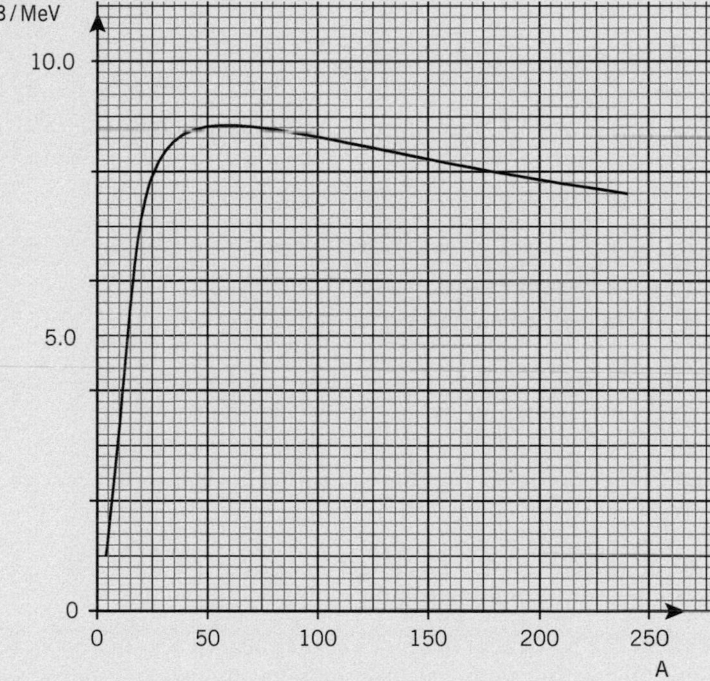

▲ **Figure 4**

(a) (i) Copy **Figure 4** and mark with the letter S to show the nucleon number and the nuclear binding energy per nucleon for the nuclide with the most stable nuclear structure.

(ii) Write down the nucleon number and the nuclear binding energy per nucleon for this nuclide.

(iii) Calculate the total binding energy of this nuclide. (*3 marks*)

(b) A fusion reaction in which two protons combine to form a deuterium nucleus is summarised by the equation:

$$^{1}_{1}\mathrm{H} + {}^{1}_{1}\mathrm{H} \rightarrow {}^{2}_{1}\mathrm{H} + {}^{0}_{1}\mathrm{e} + \nu + 1.44\ \mathrm{MeV}$$

(i) What do the symbols $^{0}_{1}\mathrm{e}$ and ν represent?

(ii) By considering charge, baryon number and lepton number for each side of the equation, show that this reaction satisfies the conservation laws for these quantities.

(iii) Subsequently two γ-ray photons are released, each with an energy of 0.51 MeV.
Calculate the wavelength of these photons. *(18 marks)*

(c) With reference to **Figure 4** explain why the fission of a heavy nucleus is likely to release more energy than when a pair of light nuclei undergo nuclear fusion. You may wish to sketch the general shape of **Figure 4** in order to aid your explanation. *(5 marks)*

AQA, 2004

7 (a) In the context of an atomic nucleus,
(i) state what is meant by *binding energy*, and explain how it arises
(ii) state what is meant by *mass difference*
(iii) state the relationship between binding energy and mass difference. *(4 marks)*

(b) Calculate the average binding energy per nucleon, in MeV nucleon^{-1}, of the zinc nucleus $^{64}_{30}Zn$.

mass of $^{64}_{30}Zn$ atom = 63.929 15 u *(5 marks)*

(c) Why would you expect the zinc nucleus to be very stable? *(1 mark)*

AQA, 2004

8 (a) With reference to the process of nuclear fusion, explain why energy is released when two small nuclei join together, and why it is difficult to make two nuclei come together. *(3 marks)*

(b) A fusion reaction takes place when two deuterium nuclei join, as represented by

$$^{2}_{1}H + ^{2}_{1}H \rightarrow ^{3}_{2}He + ^{0}_{1}n$$

mass of ^{2}H nucleus = 2.013 55 u
mass of ^{3}He nucleus = 3.014 93 u

Calculate:
(i) the mass difference produced when two deuterium nuclei undergo fusion
(ii) the energy released, in J, when this reaction takes place. *(3 marks)*

AQA, 2003

9 (a) In the context of the processes that occur in a nuclear power reactor, explain what is meant by:
(i) thermal neutrons
(ii) induced fission
(iii) a self-sustaining chain reaction. *(5 marks)*

(b) (i) Describe the process of moderation that takes place in an operational reactor.
(ii) How is the fission rate controlled in a power reactor? *(7 marks)*

10 (a) When a fuel rod has been in use in a nuclear reactor for several years, it produces less output power and presents a greater hazard than it did when first installed. Explain why this is so. *(3 marks)*

(b) Describe how the spent fuel rods are handled and processed after they have been removed from a nuclear reactor. Indicate how the active wastes are dealt with in order to reduce the hazards they could present to future generations. *(5 marks)*

Answers to the Practice Questions and Section Questions are available at www.oxfordsecondary.com/oxfordaqaexams-alevel-physics

25 Rotational dynamics
25.1 Angular acceleration

Angular speed and angular acceleration

There are many examples of rotating objects around us. Examples include the Earth spinning on its axis, the wheels of a moving car, a flywheel in motion, and the rotation of the armature coil of an electric motor.

When a rigid body rotates about a fixed axis,

- the angle it turns through is referred to as its **angular displacement**, $\Delta\theta$.

The unit of angular displacement is the radian (abbreviated as rad) where 2π radians = 360°.

For angular displacement $\Delta\theta$, the number of turns made, $n = \frac{\Delta\theta}{2\pi}$. For example, if $\Delta\theta = 13.4$ rad, $n = 2.13$ turns (to 3 significant figures) $\left(= \frac{13.4}{2\pi}\right)$.

- Its **angular speed, ω, is the change of the angle it turns through per second.**

The unit of angular speed is the radian per second (i.e., rad s^{-1}).

Notes:

1. **Angular velocity** is angular speed in a certain direction of rotation, either clockwise or anticlockwise.
2. Angular speed and angular velocity are sometimes expressed in *revolutions per minute* or *rpm*. To convert to rad s^{-1}, multiply by $\frac{2\pi}{60}$ as there are 2π radians per revolution and 60 seconds per minute.

For a rotating object that turns through angular displacement $\Delta\theta$ in a time interval Δt

$$\textbf{its average angular speed during this time} = \frac{\Delta\theta}{\Delta t}$$

1. If an object is rotating at constant angular speed ω, its angular displacement $\Delta\theta$ in one rotation is 2π radians.
   Therefore
   $$\omega, = \frac{2\pi}{T}, \text{ where } T \text{ is its period of rotation.}$$
2. If the angular speed or direction of rotation of a rotating body changes, its **angular acceleration** α is the change of its angular velocity per second.
   The unit of angular acceleration is the radian per second per second (i.e., rad s^{-2}).

   For constant angular acceleration, if the change of angular speed is $\Delta\omega$ in a time interval Δt

$$\textbf{its angular acceleration } \alpha = \frac{\Delta\omega}{\Delta t}$$

When a flywheel rotates at constant angular speed, its angular acceleration is zero because its angular speed does not change. If ω is its angular speed, its period of rotation $T = \frac{2\pi}{\omega}$. If the flywheel has a radius R, a point on the flywheel rim moves on a circular path at a speed $v = \omega R$.

Learning objectives:

→ Define angular acceleration.

→ Calculate the angular acceleration of a rotating object when it speeds up or slows down.

→ Calculate the number of turns a rotating object makes in a certain time when it accelerates uniformly.

Specification reference: 3.13.1

Hint

Angular acceleration is not the same as centripetal acceleration which is the rate of change of velocity on an object moving round a circle at constant speed.

Synoptic link

See Topic 15.1 for more about angular speed.

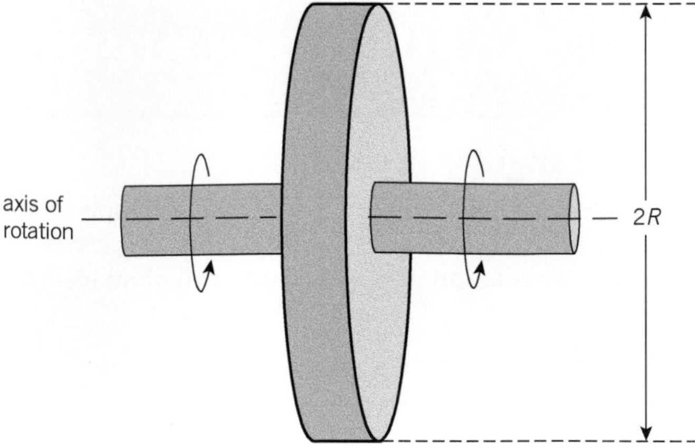

▲ **Figure 1** *A flywheel in motion*

When a flywheel speeds up, every point of the flywheel moves at increasing speed. If the flywheel speeds up from initial angular speed ω_1 to angular speed ω_2 in time t, then:

- its angular acceleration $\alpha = \frac{(\omega_2 - \omega_1)}{t}$
- the speed of any point on its rim increases from speed $u = \omega_1 R$ to speed $v = \omega_2 R$ in time t, where R is the radius of the flywheel
- the speeding-up process accelerates such a point along a circular path of radius R. Therefore the acceleration, a, of the point along its path may be calculated from

$$a = \frac{(v - u)}{t} = \frac{\omega_2 R - \omega_1 R}{t} = \frac{(\omega_2 - \omega_1)R}{t} = \alpha R$$

In general, for any rotating body, for any point (i.e., small *element*) in the body at radial distance r from the axis of rotation,

- its speed $v = \omega r$
- its acceleration $a = \alpha r$
- every part of a rotating object experiences the same angular acceleration. However, the acceleration, a, at a point in the body is tangential to its circular path (i.e., its linear acceleration) and is proportional to r (because $a = \alpha r$).

Note:

Every point of a rotating object experiences a centripetal acceleration equal to $\omega^2 r$ which acts directly towards the centre of rotation of that point.

Worked example

A flywheel is speeded up from 5.0 to 11.0 revolutions per minute in 100 s. The radius of the flywheel is 0.080 m. Calculate the angular acceleration of the flywheel and the acceleration of a point on the rim along its circular path.

Solution

Initial angular speed $\omega_1 = 5.0 \times 2\pi / 60 = 0.52\,\text{rad s}^{-1}$

Angular speed after 100 s, $\omega_2 = 11.0 \times 2\pi / 60 = 1.15\,\text{rad s}^{-1}$

Angular acceleration, $\alpha = \frac{(\omega_2 - \omega_1)}{t} = \frac{1.15 - 0.52}{100} =$ $6.3 \times 10^{-3}\,\text{rad s}^{-2}$

Acceleration tangential to the rim, $a = \alpha R = 6.3 \times 10^{-3} \times 0.080 =$ $5.0 \times 10^{-4}\,\text{m s}^{-2}$

Equations for constant angular acceleration

These may be derived in much the same way as the equations for straight-line motion with constant acceleration.

1 From the definition of angular acceleration α above, we rearrange the equation to obtain

$$\omega = \omega_0 + \alpha t$$

where ω_0 = initial angular speed and ω = angular speed at time t.

2 The average value of the angular speed is obtained by averaging the initial and final values, giving an average value of $\frac{1}{2}(\omega + \omega_0)$.

The angle which the object turns through, the angular displacement θ, is equal to the average angular speed × the time taken, which gives

$$\theta = \frac{1}{2}(\omega + \omega_0)t$$

3 The two equations above can be combined to eliminate ω to give

$$\theta = \omega_0 t + \frac{1}{2}\alpha t^2$$

4 Alternatively, the first two equations may be combined to eliminate t, giving

$$\omega^2 = \omega_0^2 + 2\alpha\theta$$

Note:

The four equations above are directly comparable with the four linear dynamics equations. The task of using the angular equations is made much easier by 'translating' between linear and angular terms, as shown in Table 1 on the next page.

Synoptic link

See Topic 2.3 for the linear dynamics equations with constant acceleration.

▼ **Table 1**

Linear equations		Angular equations
Displacement		angular displacement
s	⟷	θ
Speed or velocity		angular speed
u	⟷	ω_0
v	⟷	ω
Acceleration		angular acceleration
a	⟷	α
For example		
$v = u + at$	⟷	$\omega = \omega_0 + \alpha t$
$s = \frac{1}{2}(u + v)t$	⟷	$\theta = \frac{1}{2}(\omega_0 + \omega)t$
$s = ut + \frac{1}{2}at^2$	⟷	$\theta = \omega_0 t + \frac{1}{2}\alpha t^2$
$v^2 = u^2 + 2as$	⟷	$\omega^2 = \omega_0^2 + 2\alpha\theta$

Angular motion graphs

Angular motion can be represented graphically in a similar way to linear motion, as shown in Table 2 below.

▼ **Table 2**

Angular motion graph	Equivalent linear motion graph
1 Angular displacement θ against time t	**1 Displacement s against time t**
Gradient = angular velocity ω	gradient = velocity v
2 Angular velocity ω against time t	**2 Velocity v against time t**
Gradient = angular acceleration α	gradient = (linear) acceleration a
Area under the graph = angular displacement θ	area under the graph = displacement s

The **angular displacement graph** in Figure 2 on the next page shows how the angular displacement of a rotating object changes with time. In this example, the gradient of the line increases from zero, then becomes constant, then decreases to zero. If the graph showed how the displacement of an object varied with time, we would be able to tell from its gradient that the velocity increased from zero, reached a constant value, then decreased to zero again. The gradient of the line in Figure 2 tells us that the object's angular velocity increases from zero to a maximum value then decreases to zero. The angular velocity at any point on the line is given by the gradient of the line at that point.

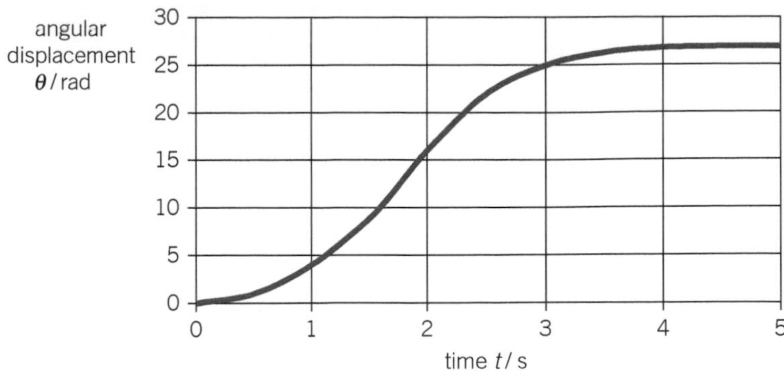

▲ **Figure 2** *Graph of angular displacement against time*

The **angular velocity graph** in Figure 3 shows how the angular velocity of a different rotating object changed with time. In this example, the angular velocity increases, then peaks, and then decreases to zero.

- The gradient of the line gives the angular acceleration of the object. The gradient is constant at first, then it decreases, and then it becomes negative. So the gradient tells us that the object has a constant angular acceleration at first. Then its angular acceleration decreases to zero, where its angular velocity is a maximum. Then its angular acceleration becomes negative so its angular velocity decreases until it stops rotating.
- The area under the line gives the angular displacement of the object. The area under the last part of the line where the angular velocity decreases is greater than the area under the first part of the line where the angular velocity increases. This tells us that the angular displacement in the last part is greater than during the first part. In other words, the object makes more turns when slowing down than it does when speeding up.

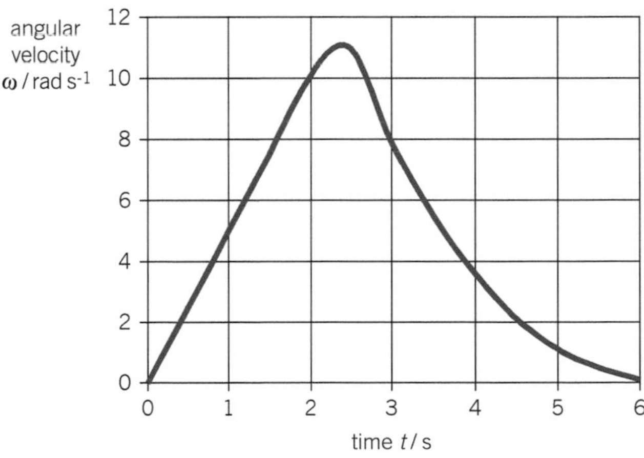

▲ **Figure 3** *Graph of angular velocity against time*

Summary questions

1 A flywheel accelerates uniformly from rest to 12 rad s^{-1} in 60 s. Calculate:
  - **a** its angular acceleration
  - **b** **i** the angle it turned through in this time
    - **ii** the number of turns it made.

2 A child's spinning top spinning at a frequency of 12 Hz decelerates uniformly to rest in 50 s. Calculate:
  - **a** its initial angular speed
  - **b** its angular deceleration
  - **c** the number of turns it makes when it decelerates.

3 A spin drier tub accelerates uniformly from rest to an angular speed of 1100 revolutions per second in 50 seconds. Calculate:
  - **a** the angle that the tub turns through in this time
  - **b** the number of turns it makes as it accelerates.

4 **a** Figure 1 shows how the angular displacement of an object changes with time. Determine the maximum angular velocity of the object.
  - **b** Figure 2 shows how the angular velocity of an object changes with time.
    - **i** Determine the maximum angular acceleration of the object.
    - **ii** Estimate the total angular displacement of the object and determine how many turns the object made.

25.2 Moment of inertia

Learning objectives:

- → Define torque.
- → Explain moment of inertia.
- → Describe how angular acceleration of a rotating object depends on its moment of inertia.

Specification reference: 3.13.1

Torque

To make a flywheel rotate, a turning force must be applied to it. The turning effect depends not only on the force, but also on where it is applied. The torque of a turning force is the moment of the force about the axis. Therefore,

Torque = force × perpendicular distance from the axis to the line of action of the force.

The unit of torque is the newton metre (N m).

The above definition gives the equation

$$\textbf{torque } T = Fr$$

where the force F that causes the torque acts at a perpendicular distance r from the axis to the line of action of the force.

The **inertia** of an object is its resistance to change of its motion. If a large torque is required to start a flywheel turning, the flywheel must have considerable inertia. In other words, its resistance to change of its motion is large.

Synoptic link

See Topic 1.3 for the moment of a force.

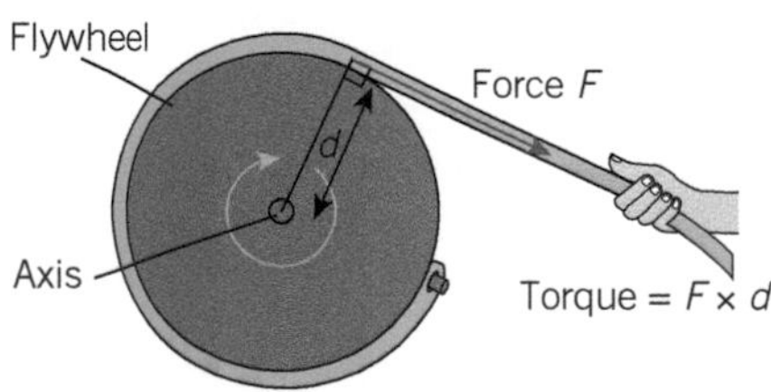

▲ **Figure 1** *Applying a torque*

Every object has the property of inertia because every object has mass. However, the inertia of a rotating body depends on the distribution of mass about the axis of rotation as well as the amount of mass. For example, the moment of inertia of a rod about an axis perpendicular to the rod is very different if the axis is through one end of the rod than if it is through the centre of the rod.

Consider a flat rigid body which can be rotated about an axis perpendicular to its plane, as shown in Figure 2. Suppose it is initially at rest and a torque is applied to make it rotate. Assuming there is no friction on its bearing, the applied torque will increase its angular speed. When the torque is removed, the angular speed stops increasing, so it turns at constant frequency once the torque is removed.

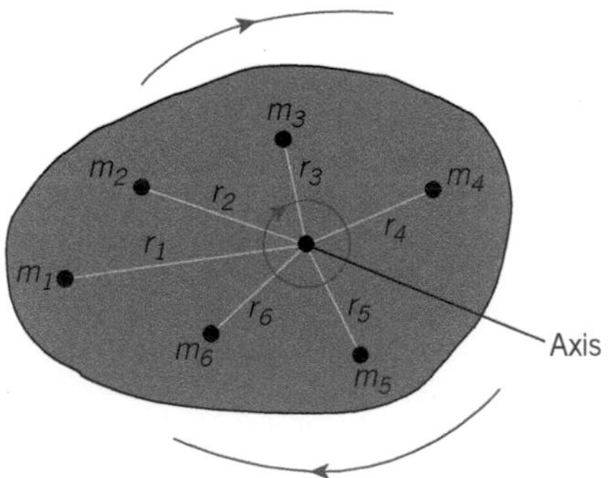

▲ **Figure 2** *A rigid body considered as a network of point masses*

The body in Figure 2 may be thought of as a network of point masses, m_1, m_2, m_3, etc., at distances r_1, r_2, r_3, etc. from the axis. Each point turns on a circular path about the axis.

At angular speed ω, the speed of each point is given by $v = \omega r$ so

- the speed of m_1 is ωr_1
- the speed of m_2 is ωr_2, etc.

When the body speeds up, every point in it accelerates. If the angular acceleration of the body is α, then the acceleration of each point mass is given by $a = \alpha r$. So

- the acceleration of m_1 is αr_1
- the acceleration of m_2 is αr_2, etc.

Using $F = ma$, the force needed to accelerate each point mass is therefore given by

- $F_1 = m_1\alpha r_1$ for m_1
- $F_2 = m_2\alpha r_2$ for m_2, etc.

The moment needed for each point mass to be given angular acceleration α is given by force $F \times$ distance r because moment = force × perpendicular distance from the point mass to the axis. Therefore

- the moment for $m_1 = (m_1\alpha r_1)r_1$
- the moment for $m_2 = (m_2\alpha r_2)r_2$, etc.

The total moment (torque T) needed to give the body angular acceleration α = the sum of the individual moments needed for all N point masses. So

$$\text{torque } T = (m_1 r_1^2)\alpha + (m_2 r_2^2)\alpha + (m_3 r_3^2)\alpha + \ldots\ldots\ldots + (m_N r_N^2)\alpha$$

$$= [(m_1 r_1^2) + (m_2 r_2^2) + (m_3 r_3^2) + \ldots\ldots\ldots + (m_N r_N^2)]\alpha$$

$$= I\alpha$$

where $I = (m_1 r_1^2) + (m_2 r_2^2) + (m_3 r_3^2) + \ldots\ldots\ldots + (m_N r_N^2)$ is the **moment of inertia** of the body about the axis of rotation.

The summation $(m_1 r_1^2) + (m_2 r_2^2) + (m_3 r_3^2) + \ldots + (m_N r_N^2)$ is written in short form as Σmr^2 (pronounced 'sigma *m r* squared).

The moment of inertia *I* of a body about a given axis is defined as $\Sigma m_i r_i^2$ for all the points in the body, where m_i is the mass of each point and r_i is its perpendicular distance from the axis.

The unit of inertia, I, is $kg\,m^2$.

In general, T is the resultant torque. For example, if a torque T_1 is applied to a flywheel which is also acted on by a frictional torque T_2, the resultant torque is $T_1 - T_2$.

Therefore when a body undergoes angular acceleration α, the resultant torque T acting on it is given by

$$T = I\alpha$$

Note

The derivation of $I = \Sigma mr^2$ is not required in this specification.

Worked example

A flywheel of moment of inertia $0.45\,kg\,m^2$ is accelerated uniformly from rest to an angular speed of $6.7\,rad\,s^{-1}$ in 4.8 s. Calculate the resultant torque acting on the flywheel during this time.

Solution

Angular acceleration

$\alpha = \dfrac{(\omega_2 - \omega_1)}{t} = \dfrac{6.7 - 0}{4.8} =$ $1.4\,rad\,s^{-2}$

Resultant torque

$T = I\alpha = 0.45 \times 1.4 = 0.63\,N\,m$

Moment of inertia and angular acceleration

When a resultant torque is applied to a body, the angular acceleration, α, of the body is given by $\alpha = \frac{T}{I}$. Therefore the angular acceleration depends not only on the torque T, but also on the moment of inertia I of the body about the given axis. This moment of inertia is determined by the distribution of mass about the axis.

Two bodies of equal mass whose mass is distributed in different ways will have different values of I. For example, compare a hoop and a disc of the same mass, as in Figure 3.

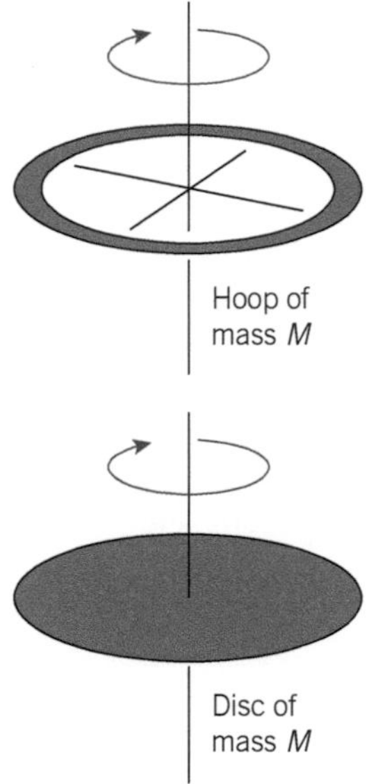

▲ **Figure 3** *Distribution of mass*

About the axis shown,

- The moment of inertia of the hoop is simply MR^2 where M is its mass and R is its radius. This is because all the mass of the hoop is at the same distance: (= R) from the axis. The moment of inertia I about the axis shown, Σmr^2, is therefore just MR^2 for the hoop.
- The moment of inertia of the disc about the same axis is less than MR^2 because the mass of the disc is distributed between the centre and the rim. Calculations show that the value of I for the disc about the axis shown is $\frac{1}{2}MR^2$.

In general,

the further the mass is distributed from the axis, the greater the moment of inertia about that axis.

The moment of inertia about a given axis of an object with a simple geometrical shape can be calculated using an appropriate mathematical formula for that shape and axis. In general terms, such formulae include geometrical factors as well as mass. For example,

1 the hoop shown in Figure 3 has a moment of inertia given by MR^2

2 the disc shown in Figure 3 has a moment of inertia given by $\frac{1}{2}MR^2$

3 a uniform beam of length L and mass M has a moment of inertia about an axis perpendicular to its length given by:

- $\frac{ML^2}{12}$ if the axis is through its centre
- $\frac{ML^2}{3}$ if the axis is through its end.

Note:

The formulae above and for other bodies with simple shapes are derived using the mathematical technique of **integration** to obtain Σmr^2 for all points in the body. Apart from a circular hoop, you do not need to know the formula for the moment of inertia of any object. In the exam, any necessary formulae will be provided.

Worked example

A solid circular disc of mass 4.0 kg and radius 0.15 m is rotating at an angular speed of 22 rad s^{-1} about an axis as shown in Figure 4 when a braking torque is applied to it which brings it to rest in 5.8 s.

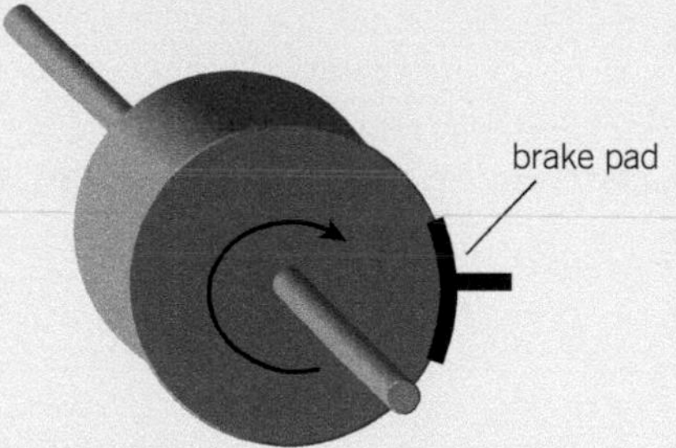

▲ **Figure 4**

Calculate:

a its angular deceleration when the braking torque is applied

b the moment of inertia of the disc about the axis shown

c the resultant torque that causes it to decelerate.

Moment of inertia of disc about the axis shown = $\frac{1}{2}MR^2$.

Solution

a Angular acceleration $\alpha = \frac{(\omega_2 - \omega_1)}{t} = \frac{22 - 0}{5.8} = 3.8\,\text{rad}\,\text{s}^{-2}$

b $I = \frac{1}{2}MR^2 = 0.5 \times 4.0 \times 0.15^2 = 0.045\,\text{kg}\,\text{m}^2$

c Resultant torque $T = I\,\alpha = 0.045 \times 3.8 = 0.17\,\text{N}\,\text{m}$

Summary questions

1 The rotating part of an electric fan has a moment of inertia 0.68 kg m^2. The rotating part is accelerated uniformly from rest to an angular speed of 3.7 rad s^{-1} in 9.2 s. Calculate the resultant torque acting on the fan during this time.

2 A solid circular disc of mass 7.4 kg and radius 0.090 m is mounted on an axis as in Figure 1. A force of 7.0 N is applied tangentially to the disc at its rim, as shown in Figure 1, to accelerate the disc from rest.

a Show that:

i the moment of inertia of the disc about this axis is 0.030 kg m^2

ii the torque applied to the disc is 0.63 N m.

b The force of 7.0 N is applied for 15.0 s. Calculate:

i the angular acceleration of the disc at the end of this time

ii the number of turns made by the disc in this time.

moment of inertia of disc about the axis shown $= \frac{1}{2}MR^2$

3 Figure 5 shows cross sections of two discs X and Y which have the same mass and radius.

State and explain which disc has the greater moment of inertia about the axis shown.

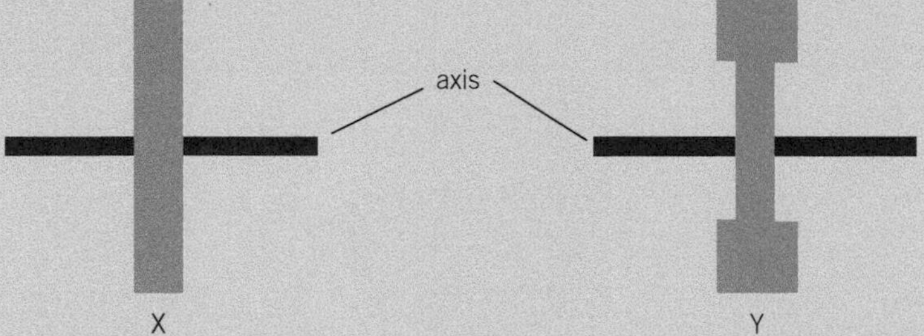

▲ **Figure 5**

4 A flywheel is accelerated by a constant torque for 18 s from rest. During this time it makes 36 turns. It then slows down to a standstill 92 s after the torque is removed, making 87 turns during this time.

a i Show that in the time it accelerates, its angular acceleration is 1.40 rad s^{-2}.

ii Show that in the time it slows down its angular deceleration is 0.13 rad s^{-2}.

b The torque applied to it when it accelerates is 26 N m.

i Show that the frictional torque that slows it down is 2.2 N m.

ii Calculate the moment of inertia of the flywheel.

25.3 Rotational kinetic energy

Kinetic energy

To make a body which is initially at rest rotate about a fixed axis, it is necessary to apply a torque to the body. The torque does work on the body and, as long as the applied torque exceeds the frictional torque, the work done increases the kinetic energy of the body and the faster the body rotates.

The kinetic energy, E_K, of a body rotating at angular speed ω is given by

$$E_K = \frac{1}{2} I \omega^2$$

where I is its moment of inertia about the axis of rotation.

To prove this equation, consider the body as a network of point masses m_1, m_2, m_3, m_4, etc. When the body rotates at angular speed ω, the speed of each point mass is given by $v = \omega r$ (see Figure 2 in Topic 25.2). So

speed of $m_1 = \omega r_1$, where r_1 is the distance of m_1 from the axis

speed of $m_2 = \omega r_2$, where r_2 is the distance of m_2 from the axis, etc.

....

speed of $m_N = \omega r_N$, where r_N is the distance of m_N from the axis.

Since the kinetic energy of each point mass is given by $\frac{1}{2}mv^2$, then

kinetic energy of $m_1 = \frac{1}{2}m_1 v_1^{\,2} = \frac{1}{2}m_1 \omega^2 r_1^{\,2}$,

kinetic energy of $m_2 = \frac{1}{2}m_2 v_2^{\,2} = \frac{1}{2}m_2 \omega^2 r_2^{\,2}$, etc.

....

kinetic energy of $m_N = \frac{1}{2}m_N v_N^{\,2} = \frac{1}{2}m_N \omega^2 r_N^{\,2}$.

Therefore the total kinetic energy $= \frac{1}{2}m_1\omega^2 r_1^{\,2} + \frac{1}{2}m_2\omega^2 r_2^{\,2} + \ldots + \frac{1}{2}m_N\omega^2 r_N^{\,2}$

$= \frac{1}{2}[(m_1 r_1^{\,2}) + (m_2 r_2^{\,2}) + \ldots\ldots\ldots + (m_N r_N^{\,2})]\omega^2$

$= \frac{1}{2}I\omega^2$ because $I = [(m_1 r_1^{\,2}) + (m_2 r_2^{\,2}) + \ldots\ldots\ldots + (m_N r_N^{\,2})]$

The equation $E_K = \frac{1}{2}I\omega^2$ enables us to calculate how much energy a rotating object stores due to its rotational motion. In addition, it shows that the kinetic energy of a rotating object is proportional to

- its moment of inertia about the axis of rotation
- the square of its angular speed.

A flywheel is used in many machines (or engines) to keep the moving parts moving when the load on the machine increases and it has to do more work. For example, when a metal press is used to make a shaped object from a sheet of thin metal, the press is able to do the necessary work because the flywheel keeps it moving. Flywheels are also used to smooth out the variations of the speed of a motor or an engine

Learning objectives:

- → State what the kinetic energy of a rotating object depends on.
- → Calculate the work done by a torque when it makes a rotating object turn.
- → Describe how to measure the moment of inertia of a flywheel.

Specification reference 3.13.1

when the load varies. For example, when a motorist changes gear, the load on the engine varies during the process. Without a flywheel in the system, the load variation would cause variations in the engine's angular speed – in other words, a jerky ride!

Flywheels are also fitted in some vehicles to store kinetic energy when the vehicle brakes are applied and it slows down. Instead of energy being transferred by heating to the surroundings, some of the vehicle's kinetic energy is transferred to an 'on-board' flywheel – to be returned to vehicle when the accelerator pedal is pressed.

Note:
You do not need to know the proof of $E_K = \frac{1}{2}I\omega^2$.

Synoptic link

See Topics 5.1, Work and energy, 5.2, Kinetic energy and potential energy, and 5.3, Power.

Work done

The **work done *W* by a constant torque *T*** when the body is turned through angle θ is given by

$$W = T\theta$$

This can be seen by considering the torque as due to a force F acting at a perpendicular distance d from the axis of rotation. The force acts through a distance $s = \theta d$ when it turns the body through angle θ. Therefore, the work done by the force $W = Fs = Fd\theta = T\theta$ as $T = Fd$.

Assuming there is no frictional torque, the applied torque $T = I\alpha$.

Therefore, the work done W by the applied torque is given by $W = T\theta = (I\alpha)\theta$.

Because the dynamics equation for angular motion $\omega^2 = \omega_0^{\ 2} + 2\alpha\theta$ gives $\alpha\theta = \frac{1}{2}\omega^2 - \frac{1}{2}\omega_0^{\ 2}$, then $W = (I\alpha)\theta = I(\frac{1}{2}\omega^2 - \frac{1}{2}\omega_0^{\ 2}) = \frac{1}{2}I\omega^2 - \frac{1}{2}I\omega_0^{\ 2} =$ the gain of kinetic energy.

Therefore, in the absence of friction, the work done by the torque is equal to the gain of rotational kinetic energy of the body.

Study tip

The number of turns for an angular displacement θ in radians $= \dfrac{\theta}{2\pi}$.

Power

For a body rotating at constant angular speed ω, the power P delivered by a torque T acting on the body is given by

$$P = T\omega$$

We can see how this equation arises by considering a constant torque T acting on a body for a time t.

If the body turns through an angle θ in this time, the work done W by the torque is given by $W = T\theta$.

Since the power P delivered by the torque is the rate of work done by the torque, then $P = \dfrac{W}{t}$.

Therefore $P = \dfrac{W}{t} = \dfrac{T\theta}{t} = T\omega$, where the angular speed $\omega = \dfrac{\theta}{t}$.

If a rotating body is acted on by an applied torque T_1 which is equally opposed by a frictional torque T_F, the resultant torque is zero so its angular speed ω is constant.

- The power P delivered by the applied torque $T = \frac{\text{work done } W}{\text{time taken } t} = \frac{T\theta}{t} = T\omega$.
- The work done per second by the frictional torque $T_F = \frac{T_F\theta}{t} = T_F\omega$.

In this situation, the rate of transfer of energy due to the applied force is equal to the rate of transfer of energy to the surroundings by the frictional force. So the rotating body does not gain any kinetic energy.

Worked example

A flywheel is rotating at an angular speed of $120\,\text{rad}\,\text{s}^{-1}$ on a fixed axle which is mounted on frictionless bearings. The moment of inertia of the flywheel and the axle about the axis of rotation is $0.068\,\text{kg}\,\text{m}^2$.

a Calculate the rotational kinetic energy of the flywheel when it rotates at $120\,\text{rad}\,\text{s}^{-1}$.

b When a braking torque of $1.4\,\text{N}\,\text{m}$ is applied to its rim, the flywheel is brought to rest.

Calculate the number of turns the flywheel makes as it decelerates to a standstill.

Solution

a $E_K = \frac{1}{2}I\omega^2 = 0.5 \times 0.068 \times 120^2 = 490\,\text{J}$

b The work done by the braking torque = loss of kinetic energy, ΔE_K

Therefore $T\theta = \Delta E_K$ so $1.4\theta = 490$

Therefore $\theta = \frac{490}{1.4} = 350\,\text{rad} = \frac{350}{2\pi} = 56$ turns

Experiment to measure the moment of inertia of a flywheel

You do not need to know the details of the experiment described below, but it reinforces and brings together many of the ideas you have covered so far. In addition, it provides an opportunity to do some practical work.

An object of known mass M hanging from a string is used to accelerate the flywheel from rest, as shown in Figure 1.

The following measurements need to be made:

1 the distance fallen, h, by the object from release to when the string unwraps itself from the axle of the flywheel
2 the diameter d of the axle
3 the time taken t for the string to unwrap.

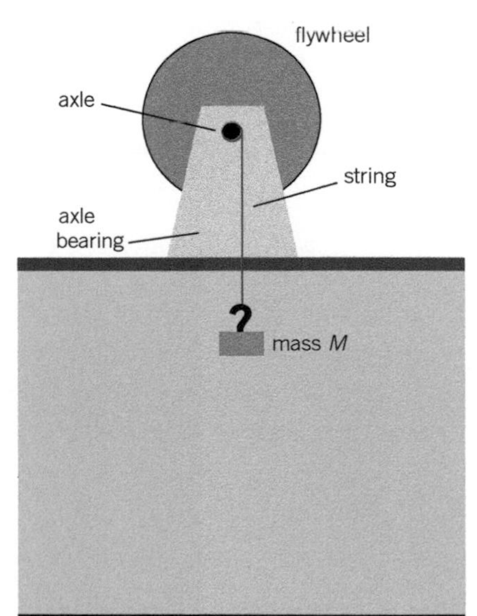

▲ **Figure 1** *Measuring the moment of inertia of a flywheel*

The measurements and the known mass M are used in the following calculations:

- the number of turns made by the flywheel as the string unwraps, $N = \frac{h}{\pi d}$
- the maximum angular speed of the flywheel, $\omega_{max} = 2 \times$ the average angular speed $= \frac{2 \times (2\pi N)}{t}$
- the speed of the object at the instant the string unwraps, $v = \frac{\omega d}{2}$
- the kinetic energy gained by the flywheel, $\Delta E_{KF} = \frac{1}{2}I\omega^2$, to be calculated in terms of I
- the kinetic energy gained by the object of mass M, $\Delta E_{KO} = \frac{1}{2}Mv^2$ where $v = \frac{\omega d}{2}$
- the potential energy lost by the object of mass M, $\Delta E_P = Mgh$.

The moment of inertia of the flywheel can be calculated from the equation below, assuming friction on the flywheel is negligible.

The total gain of kinetic energy, $\Delta E_{KF} + \Delta E_{KO}$ = the loss of potential energy, ΔE_P

$$\frac{1}{2}I\omega^2 + \frac{1}{2}Mv^2 = Mgh$$

Summary questions

1 A flywheel rotates at an angular speed of 20 rad s^{-1} on a fixed axle which is mounted on frictionless bearings. The moment of inertia of the flywheel and the axle about the axis of rotation is 0.048 kg m^2.

Calculate:

a the rotational kinetic energy of the flywheel when it rotates at 20 rad s^{-1}

b **i** the torque needed to accelerate the flywheel from rest to an angular speed of 20 rad s^{-1} in 5.0 s

ii the angle which the flywheel turns through in this time while it is being accelerated.

2 A 0.65 kg object hanging from a string is used to accelerate a flywheel on frictionless bearings from rest, as shown in Figure 1. The object falls through a vertical distance of 1.9 m in 4.6 s, which is the time the string takes to unwrap from the axle. The axle has a diameter of 8.5 mm. Calculate:

a the potential energy lost by the object in descending 1.9 m

b the kinetic energy of the object 14 s after it is released from rest

c **i** the kinetic energy gained by the flywheel

ii the moment of inertia of the flywheel.

3 A ball released at the top of a slope rolls down the slope and continues on a flat horizontal surface until it stops. Discuss the energy changes of the ball from the moment it is released to when it stops.

4 A flywheel fitted to a vehicle gains kinetic energy when the vehicle slows down and stops. The kinetic energy of the flywheel is used to make the vehicle start moving again.

a The flywheel is a uniform steel disc of diameter 0.31 m and thickness 0.08 m. Calculate:

i the mass of the disc

ii the moment of inertia of the flywheel.

density of steel = 7800 kg m^{-3}

moment of inertia of a uniform flywheel = $\frac{1}{2}MR^2$.

b **i** Calculate the kinetic energy of the flywheel when it is rotating at 3000 revolutions per minute.

ii The kinetic energy of the flywheel can be converted to kinetic energy of motion of the vehicle in 30 s. Estimate the average power transferred from the flywheel.

25.4 Angular momentum

Spin at work

▲ **Figure 1** *A spinning ice skater*

An ice skater spinning rapidly is a dramatic sight. The skater turns slowly at first, then quite suddenly goes into a rapid spin. This sudden change is brought about by the skater pulling both arms (and possibly a leg!) towards the axis of rotation. In this way the moment of inertia of the skater about the axis is reduced. As a result, the skater spins faster. To slow down, the skater only needs to stretch her arms and maybe a leg. In this way, the moment of inertia is increased. So the skater slows down.

To understand such effects, consider a rotating body with no resultant torque on it. Provided its moment of inertia stays the same, then its angular speed ω does not change. This can be seen by rewriting the equation $T = I\alpha$ where α is the angular acceleration. If the resultant torque T is zero, then the angular acceleration α is zero so the angular speed is constant.

In more general terms, the equation $T = I\,\alpha$ may be written as

$$T = \frac{\mathrm{d}}{\mathrm{d}t}(I\omega)$$

where $\frac{\mathrm{d}}{\mathrm{d}t}$ is the mathematical way of writing 'change per unit time', which you met in Topic 2.1 as $\frac{\Delta}{\Delta t}$.

The quantity $I\omega$ is the **angular momentum** of the rotating body. So the above equation tells us that resultant torque T is equal to the rate of change of angular momentum of the rotating object.

Angular momentum of a rotating object = $I\omega$

where I is the moment of inertia of the body about the axis of rotation and ω is its angular speed. The unit of angular momentum is $\mathrm{kg\,m^2\,rad\,s^{-1}}$ or N m s, as explained in Note 2 on the next page.

When the resultant torque is zero, then $\frac{\mathrm{d}}{\mathrm{d}t}(I\omega) = 0$ which means its angular momentum $I\omega$ is constant.

Learning objectives:

- → State what angular momentum is and why it is important.
- → State what is meant by the conservation of angular momentum.
- → Explain what angular impulse is.
- → Explain how the equations for angular momentum and linear momentum compare with each other.

Specification reference: 3.13.1

In the ice skater example, the moment of inertia of the ice skater suddenly decreases when she pulls the arms in. Since the angular momentum is constant, the sudden decrease in the moment of inertia causes the angular speed to increase. In specific terms, if the moment of inertia changes from I_1 to I_2 the angular speed changes from ω_1 to ω_2 such that

$$I_1\omega_1 = I_2\omega_2$$

Notes:

For a rotating object whose moment of inertia I does **not** change:

1 $T = \frac{d}{dt}(I\omega) = I\frac{d\omega}{dt} = I\alpha$ because the angular acceleration $\alpha = \frac{d\omega}{dt}$.

2 If the object undergoes uniform angular acceleration from rest to reach angular speed ω in time t, the resultant torque acting on it $T = \frac{I\omega}{t}$. Therefore its angular momentum $I\omega = Tt$. This equation shows that the unit of angular momentum can also be given as the unit of torque × the unit of time (i.e., N m s).

Note

More about units

The SI unit of angular momentum is kg m^2 rad s^{-1}. In terms of SI base units, the base unit combination for angular momentum is kg m^2 s^{-1} since the radian is defined as a ratio $(= \frac{s}{r}$ where s = the length of a circular arc of radius of curvature r and r = the radius of the circle$)$. Prove for yourself that the base unit combination of torque × time is also kg m^2 s^{-1}.

Extension

Pulsars

Pulsars are rapidly spinning stars. They were first discovered by astronomers in 1967. Regular pulses of radio energy were detected from these stars, which some astronomers called LGM stars because it seemed as if 'little green men' were trying to contact us! That hypothesis was soon abandoned when it was shown that pulsars are in fact rapidly rotating neutron stars which emit radio energy in a beam at an angle to the axis. Each time the beam sweeps round to point towards Earth, radio energy is directed towards us, rather like a light beam from a lighthouse.

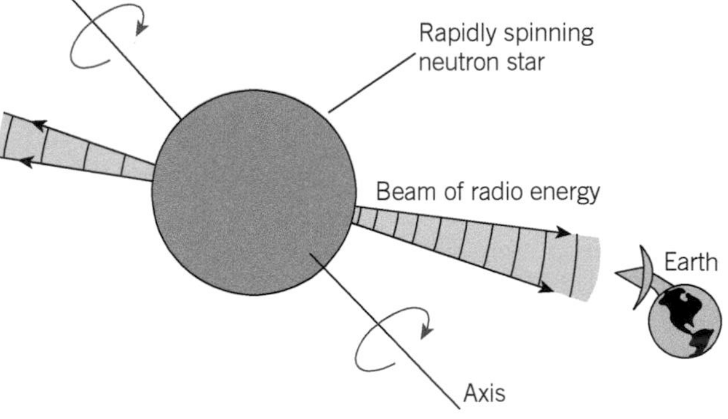

▲ **Figure 2** *A pulsar*

Neutron stars are the remnants of large stars. When a massive star runs out of fuel, a huge explosion takes place. The remnants of the explosion are pulled together by their gravitational attraction, creating a mass equivalent to the Sun shrinking to 15 km or so in diameter. The moment of inertia is therefore made much smaller so the angular speed increases. The pulse frequency from a pulsar is of the order of 1 to 10 Hz, so the rate of rotation is of that order, much much greater than the rate of the Sun, which is about once every 25 days.

Angular impulse

When the resultant torque T on a rotating object acts for a short time interval Δt, the angular velocity of the object changes rapidly. This can happen for example when a bat or a racquet hits a moving ball, changing the angular momentum or spin of the ball as well as its direction of motion.

The **angular impulse** of a short-duration torque is defined as $\boldsymbol{T\Delta t}$. If $\Delta\omega$ is the change of angular momentum due to this torque T, then

angular impulse $T\Delta t$ = change of angular momentum $I\Delta\omega$

where I is the moment of inertia of the object about the axis of rotation.

Note:

Compare the equation above with the equivalent linear equation $F\Delta t = m\Delta v$, where $F\Delta t$ is the impulse due to a force F acting on an object for a time interval Δt. As explained in Topic 4.1, the change of momentum of the object, $m\Delta v$, is equal to the impulse $F\Delta t$.

Conservation of angular momentum

In both of the above examples, the ice skater and the pulsar, the angular momentum after the change is equal to the angular momentum before the change because the resultant torque in each case is zero. In other words, the angular momentum is conserved.

Where a system is made up of more than one spinning body, then when two of the bodies interact (e.g., collide), one might lose angular momentum to the other. If the resultant torque on the system is zero, the total amount of angular momentum must stay the same.

Example

▲ **Figure 3** *Spacelab taking a satellite on board*

1 **If a spinning satellite is taken on board a space repair laboratory, the whole laboratory is set spinning.** The angular momentum of the satellite is transferred to the laboratory when the satellite is taken on board and stopped. Unless rocket motors are used to prevent it from turning, then the whole laboratory would spin. In this example,

- the total angular momentum before the satellite is taken on board = $I_1\omega_1$, where I_1 is the moment of inertia of the satellite and ω_1 is its initial angular speed
- the total angular momentum after the satellite has been taken on board and stopped = $(I_1 + I_2)\omega_2$ where ω_2 is the final angular speed and $(I_1 + I_2)$ is the total moment of inertia about the axis of rotation of the satellite and I_2 is the moment of inertia of the space lab.

According to the conservation of momentum, $(I_1 + I_2)\omega_2 = I_1\omega_1$.

Rearranging this equation gives $I_2\omega_2 = I_1\omega_1 - I_1\omega_2$.

This rearranged equation shows that the angular momentum gained by the space lab is equal to the angular momentum lost by the satellite.

2 **If a small object dropped onto a freely rotating turntable,** as shown in Figure 4, so that it sticks to the turntable, the object gains angular momentum and the turntable loses angular momentum.

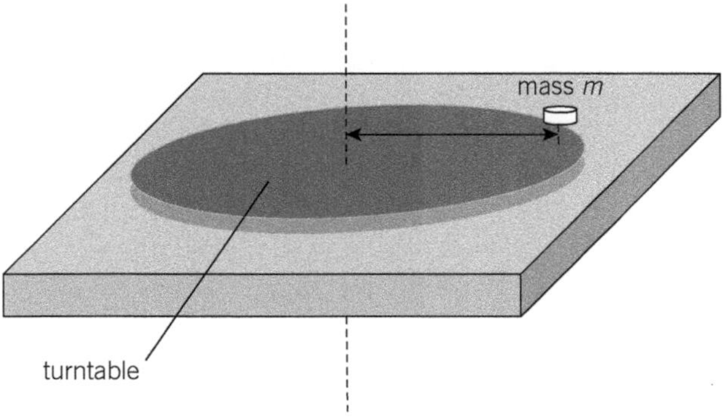

▲ **Figure 4** *Measuring the moment of inertia of a freely rotating turntable*

- The total angular momentum of the turntable before the object is dropped onto it = $I_1\omega_1$, where I_1 is the moment of inertia of the turntable and ω_1 is its initial angular speed.
- The total angular momentum after the object has been dropped onto the turntable = $(I_1 + I_2)\omega_2$, where ω_2 is the final angular speed and $(I_1 + I_2)$ is the total moment of inertia about the axis of rotation of the turntable.
- If the mass of the object is m and its perpendicular distance from the axis of rotation of the turntable is r, then the moment of inertia I_2 of the object about the axis of rotation = mr^2.

According to the conservation of momentum, $(I_1 + mr^2)\omega_2 = I_1\omega_1$.

Rearranging this equation gives $I_1 = \dfrac{mr^2 \times \omega_2}{\omega_1 - \omega_2}$.

Therefore I_1 can be found by measuring m, r, ω_1, and ω_2.

Note:

The angular momentum of a point mass is defined as its momentum × its distance from the axis of rotation. For a point mass m rotating at angular speed ω at distance r from the axis, its momentum is $m\omega r$ (because its speed $v = \omega r$) so its angular momentum is $m\omega r^2$.

For a network of point masses $m_1, m_2, \ldots, m_N$ which make up a rigid body, the total angular momentum

$$= (m_1 r_1^2\omega) + (m_2 r_2^2\omega) + \ldots + (m_N r_N^2\omega)$$
$$= [(m_1 r_1^2) + (m_2 r_2^2) + \ldots + (m_N r_N^2)]\omega = I\omega$$

Comparison of linear and rotational motion

When analysing a rotational dynamics situation, it is sometimes useful to compare the situation with an equivalent linear situation. For example, the linear equivalent of a torque T used to change the angular speed of flywheel is a force F used to change the speed of an object moving along a straight line. If the change takes place in time t

- the change of momentum of the object = Ft
- the change of angular momentum of the flywheel = Tt.

Table 1 summarises the comparison between linear and rotational motion.

Synoptic link

See Topic 4.1, Momentum and impulse.

▼ **Table 1** *Comparison between linear and rotational motion*

Linear motion	Rotational motion
Displacement s	angular displacement θ
Speed and velocity v	angular speed ω
Acceleration a	angular acceleration α
Mass m	moment of inertia I
Momentum mv	angular momentum $I\omega$
Force F $F = ma$ $F = \frac{d}{dt}(mv)$	torque $T = Fd$ $T = I\omega$ $T = \frac{d}{dt}(I\omega)$
Impulse $F\Delta t = m\Delta v$	angular impulse $T\Delta t = I\Delta\omega$
Kinetic energy $\frac{1}{2}mv^2$	kinetic energy $\frac{1}{2}I\omega^2$
Work done = Fs	work done = $T\theta$
Power = Fv	power = $T\omega$

Summary questions

1 A vehicle wheel has a moment of inertia of 5.0×10^{-2} kg m^2 and a radius of 0.30 m.

a Calculate the angular momentum of the wheel when the vehicle is travelling at a speed of 27 m s^{-1}.

b When the brakes are applied, the vehicle speed decreases from 30 m s^{-1} to zero in 9.0 s. Calculate the resultant torque on the wheel during this time.

2 A metal disc X on the end of an axle rotates freely at 240 revolutions per minute. The moment of inertia of the disc and the axle is 0.044 kg m^2.

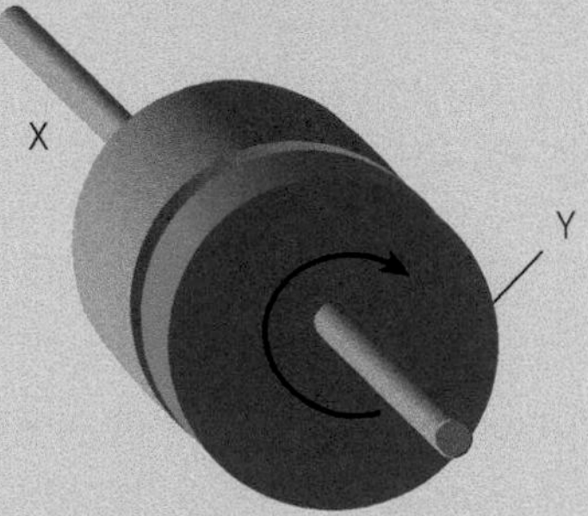

▲ **Figure 5**

a Calculate the angular momentum of the disc and the axle.

b After a second disc Y that is initially stationary is engaged by X, both discs rotate at 160 revolutions per second. Calculate the moment of inertia of Y.

c Show that the total loss of kinetic energy is 4.6 J.

3 A pulsar is a collapsed star that rotates very rapidly. Explain why a slowly rotating star that collapses rotates much faster as a result of the collapse.

4 A frictionless turntable is set rotating at a steady angular speed of 20 rad s^{-1}. A small 0.2 kg mass is dropped onto the disc from rest just above it, at a distance of 0.24 m from the centre of the disc. As a result, the angular speed of the turntable decreases to 18 rad s^{-1}. Calculate the moment of inertia of the turntable about its axis of rotation.

Practice questions: Chapter 25

1 A vehicle with wheels of diameter 0.45 m decelerates uniformly from a speed of $24\,\text{m s}^{-1}$ to a standstill in a distance of 60 m. Calculate:

(a) the angular speed of each wheel when the vehicle is moving at $24\,\text{m s}^{-1}$ *(2 marks)*

(b) (i) the time taken by the vehicle to decelerate to a standstill from a speed of $24\,\text{m s}^{-1}$ *(2 marks)*

(ii) the number of turns each wheel makes during the time the vehicle slows down *(2 marks)*

(c) the angular speed of the vehicle one second before it stops. *(2 marks)*

2 The rotating part of an electric fan has a moment of inertia of $0.92\,\text{kg m}^2$. The rotating part is accelerated uniformly from rest to an angular speed of $12.6\,\text{rad s}^{-1}$ in 14.3 s.

(a) Calculate the resultant torque acting on the fan during this time. *(3 marks)*

(b) Calculate the kinetic energy of the fan when it is rotating at $12.6\,\text{rad s}^{-1}$. *(2 marks)*

3 (a) The armature of an electric motor rotates at a constant angular speed of 1100 revolutions per minute when the current in the armature coils is constant and the torque on the coils is 0.48 N m. The moment of inertia of the armature about its axis of rotation is $0.015\,\text{kg m}^2$.

(i) Calculate the kinetic energy of the armature at this angular speed. *(2 marks)*

(ii) Calculate the power supplied to the armature to maintain its angular speed. *(2 marks)*

(b) When the current is switched off, the armature decelerates and stops after 7.5 s. Calculate the average torque acting on the armature during this time. *(3 marks)*

4 The two wheels of a bicycle each have a mass of 0.40 kg and a diameter of 0.66 m.

(a) Estimate the moment of inertia of each wheel about its axle. State any assumptions you make in your estimate. *(4 marks)*

(b) A cyclist of mass 55 kg on the bicycle travels at a speed of $18\,\text{m s}^{-1}$.

(i) Estimate the kinetic energy of the two wheels at this speed. *(2 marks)*

(ii) Estimate the ratio of the kinetic energy of the two wheels to the kinetic energy of the cyclist at this speed. *(1 mark)*

5 The four wheels of a vehicle each have a moment of inertia of $7.20 \times 10^{-2}\,\text{kg m}^2$ and a radius of 0.28 m.

▲ **Figure 1**

(a) Calculate the total angular momentum of the wheels when the vehicle is travelling at a speed of $27\,\text{m s}^{-1}$. *(3 marks)*

(b) The vehicle has a total mass of 1200 kg. When the brakes are applied, the vehicle speed decreases from $27\,\text{m s}^{-1}$ to zero in a distance of 130 m. Calculate:

(i) the time taken by the vehicle to stop *(2 marks)*

(ii) the braking force on each wheel *(1 mark)*

(iii) the torque on each wheel due to the braking force. *(2 marks)*

(c) (i) Estimate the rate of change of angular momentum of each wheel when the vehicle stops. *(3 marks)*

(ii) Explain why the torque due to the braking force on each wheel is much greater than the rate of change of angular momentum of each wheel. *(2 marks)*

6 A cylindrical satellite with its solar panels extended out spins at a constant angular velocity of 2.2 rad s^{-1}, as shown in **Figure 2**. The moment of inertia of the satellite with its panels extended is 210 kg m^2.

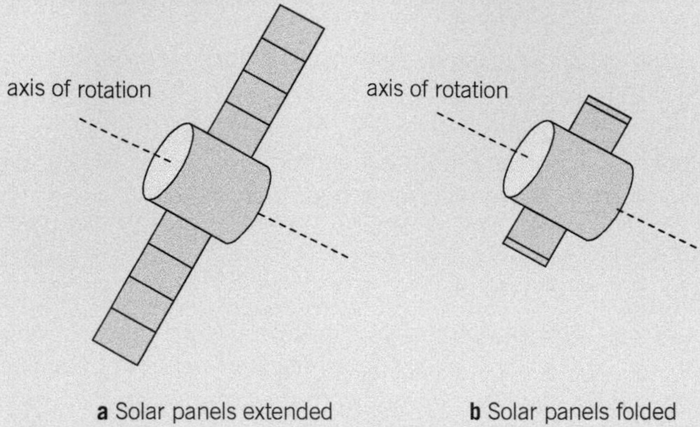

▲ **Figure 2**

(a) For this angular speed, calculate:
(i) the kinetic energy of rotation of the satellite *(2 marks)*
(ii) its angular momentum. *(2 marks)*

(b) When the solar panels are folded in, the angular speed of rotation of the satellite increases to 3.8 rad s^{-1}.
(i) Explain why the angular speed increases. *(2 marks)*
(ii) Calculate the change in the moment of inertia of the solar panels about the satellite's axis of rotation when the panels are folded in. *(4 marks)*

(c) (i) Calculate the kinetic energy of rotation of the satellite when the panels are folded in. *(2 marks)*
(ii) Explain in terms of force why the kinetic energy of rotation changes. *(2 marks)*

26 Renewable energy
26.1 Wind power

Renewable energy sources will contribute increasingly to the world's energy supplies in the future. Most of the energy we use at present is obtained from fossil fuels. But scientists think that the use of fossil fuels is causing increased climate change, due to the increasing amount of carbon dioxide in the atmosphere. Increased climate change could have disastrous consequences (e.g., rising sea levels, changing weather patterns, etc.). To cut carbon emissions, many countries are now planning to build new nuclear power stations and to develop more renewable energy resources.

Learning objectives:

→ Describe the factors that determine the power available from a wind turbine.

→ Calculate the maximum power available from the wind passing through a wind turbine.

→ Explain why not all the kinetic energy from the wind can be used.

→ Explain what is meant by wind shadows.

→ Describe the environmental effects of using wind turbines.

Specification reference: 3.13.2

Wind power

A wind turbine is an electricity generator on a tall tower, driven by large blades pushed round by the force of the wind. A typical modern wind turbine on a suitable site can generate about 2 MW of electrical power. Let's consider how this estimate is obtained.

- The kinetic energy of a mass m of wind moving at speed $v = \frac{1}{2}mv^2$.
- Therefore, the kinetic energy per unit volume of air from the wind at speed $v = \frac{1}{2}\rho v^2$, where ρ is the density (i.e., mass per unit volume) of air.
- Suppose the blades of a wind turbine sweep out an area A when they rotate. For wind at speed v, a cylinder of air of area A and length v passes every second through the area swept out by the blades. So the volume of air passing per second = vA.

Therefore, the kinetic energy per second of the wind passing through a wind turbine $= \frac{1}{2}\rho v^2 vA = \frac{1}{2}\rho v^3 A$.

For a wind turbine with blades of length 20 m, $A = \pi \times (20)^2 = 1300\,\text{m}^2$.

The density of air, $\rho = 1.2\,\text{kg}\,\text{m}^{-3}$.

The power of the wind at $v = 15\,\text{m}\,\text{s}^{-1}$

$$= \frac{1}{2} \times 1.2 \times 15^3 \times 1300$$

$$= 2.6 \times 10^6\,\text{W} = 2.6\,\text{MW}$$

▲ **Figure 1** *A wind farm*

The calculation shows that the maximum power output of a large wind turbine at a windy site could not be more than about 2 MW. To generate the same power as a 5000 MW power station, about 2500 wind turbines would need to be constructed and connected to the electricity network.

Wind turbine efficiency

The efficiency of a wind turbine is necessarily less than 100% because if a wind turbine removed all the kinetic energy from the wind passing through it, the air would stop moving and would be unable to force the turbine blades round. Air must leave the wind turbine so it must have some kinetic energy when it leaves the turbine. In other words, not all the kinetic energy of the wind passing through a wind turbine can be removed. So the electrical power a wind turbine can produce is always less than the input power to the turbine from the wind. The efficiency of a wind turbine is the ratio of the electrical power it produces to the input power of the wind (×100%). As the electrical power is always less than the power of the wind, the efficiency of an electrical turbine is always less than 100%.

Wind shadow

The wind that passes through a wind turbine is slowed down because the wind turbine has removed some of its kinetic energy. Another turbine in the path of the wind behind the first turbine is said to be in the **wind shadow** of the first turbine and would extract less energy from the wind because the wind reaching it has less kinetic energy.

Where wind turbines are near each other, they are located at positions where the prevailing wind (the usual direction of the wind at that location) does not cause other turbines to be affected by wind shadow. If too many wind turbines are located in a given area, the power available from them will be less than if they had been spread out further.

The benefits and drawbacks of wind turbines

The main benefit from the use of wind turbines is that they generate electricity without burning fossil fuels . However, they are unreliable as they do not produce electricity when there is little or no wind. Also, they create a whining noise which can upset people nearby, and some people consider them unsightly.

Summary questions

Assume the density of air = 1.3 kg m^{-3}

1 Calculate the maximum power available from a wind turbine with blades of length 15 m when the wind speed is:

   a 10 m s^{-1}

   b 20 m s^{-1}.

2 a Explain why the electrical power generated by a wind turbine can never be as much as the kinetic energy per second available from the wind passing through it.

   b The wind turbine in **Q1** generates 0.95 MW of electrical power when the wind speed is 20 m s^{-1}. Estimate the percentage efficiency of the wind turbine at this speed.

3 a Explain what is meant by the wind shadow of a wind turbine.

   b The London Array wind farm located in the Thames Estuary consists of 175 wind turbines in an area of about 100 km^2. Its maximum generating capacity is about 600 MW.

      i Assuming the average power supplied to a home is 1.2 kW, estimate how many homes could be supplied from the London Array when it is operating at maximum capacity.

      ii Some scientists reckon that the wind shadow effect of the wind turbines in a wind farm limits the power available from a wind farm to about 7 kW per square kilometre. Discuss whether or not more wind turbines should be constructed in the area covered by the London Array in order to generate more power.

4 State the main benefits and drawbacks of the use of wind turbines.

26.2 Solar power

Intensity of energy from the Sun

Solar radiation from the Sun spreads out in all directions without absorption as it travels through space. Because the planets are so small compared with their distances from the Sun, the energy per second they absorb from solar radiation is insignificant compared with the total energy per second of the solar radiation emitted by the Sun.

The **intensity I of a beam of radiation is defined as the energy per second per unit area incident normally on (i.e., perpendicular to) a surface.** The unit of intensity is the watt per square metre ($\mathrm{W\,m^{-2}}$).

Because the radiation from the Sun spreads out, the intensity of solar radiation decreases with increasing distance from the Sun. Therefore, for a solar panel of area A facing the Sun directly,

the energy per second received by the panel = intensity I × area A

The Earth is at a distance of 1.5×10^{11} m from the Sun. At this distance, measurements show that a solar panel of area $1\,\mathrm{m^2}$ directly facing the Sun receives 1400 J of energy from solar radiation each second, assuming no absorption by the Earth's atmosphere.

Imagine the solar panel as part of the surface of a sphere with the Sun at the sphere's centre, as shown in Figure 1.

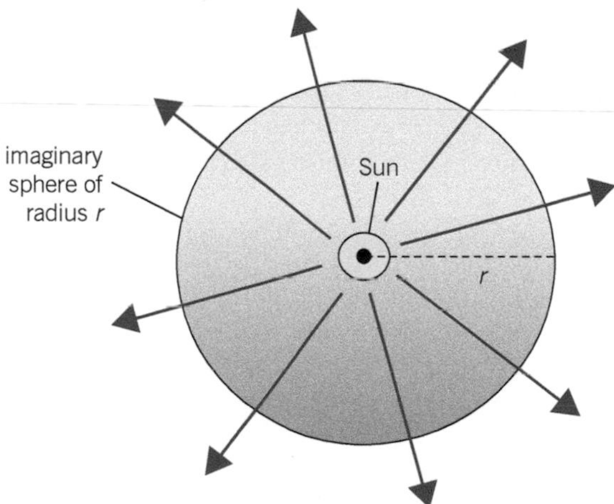

▲ **Figure 1** *Radiation from the Sun*

The area A of a sphere can be calculated from its radius r using the equation $A = 4\pi r^2$. For $r = 1.5 \times 10^{11}$, $A = 4\pi \times (1.5 \times 10^{11})^2 = 2.8(3) \times 10^{23}\,\mathrm{m^2}$.

- The total power P emitted by the Sun = the total energy per second received by the sphere of radius r = intensity at distance r from the Sun × the area of the sphere = $1400\,\mathrm{W\,m^{-2}} \times 2.8(3) \times 10^{23}\,\mathrm{m^2} = 4.0 \times 10^{26}\,\mathrm{W}$.

Learning objectives:

- → Describe the factors that affect the power available from the Sun.
- → Calculate the maximum power available from the Sun at different distances from the Sun.
- → Describe the characteristics of a solar cell.
- → Explain how solar cells may be connected together in a solar panel.

Specification reference: 3.13.3

Notes

A solar heating panel absorbs solar radiation and becomes hot enough to heat water passing through it. Figure 3a shows a solar heating panel on the roof of a house. As explained in Topic 22.2, if mass m of water passes through a solar heating panel in time t,

the energy per second gained by the water = $\left(\frac{mc\Delta\theta}{t}\right)$

where c is the specific heat capacity of water and $\Delta\theta$ is the temperature increase of the water.

A solar cell panel is an array of solar cells that produces electricity directly when it absorbs solar radiation. Figure 3b shows a solar cell panel. As explained in Topic 9.2, if a solar panel is connected an appliance,

power supplied to appliance = current in the appliance × p.d. across the appliance

a *Solar heating panel*

b *Solar cell panel*

▲ **Figure 3** *Solar panels*

In general, at distance r from the Sun,

$$\textbf{the intensity } I \textbf{ of solar radiation} = \frac{P}{4\pi r^2}$$

where P = the total power P emitted by the Sun

and $4\pi r^2$ = the area of a sphere of radius r.

The equation above is another example of the inverse square law for radiation which you met in Topic 8.3.

Worked example

Mercury orbits the Sun at a distance of 5.7×10^{10} m from the Sun. The Sun radiates energy at a rate of 4.0×10^{26} W. Calculate the intensity of solar radiation at this distance from the Sun.

Solution

$$\text{Intensity } I = \frac{P}{4\pi r^2} = \frac{4.0 \times 10^{26}\text{ W}}{4\pi \times (5.7 \times 10^{10})^2} = 9800\text{ W m}^{-2}$$

Solar cells

A solar cell or photovoltaic cell produces electricity directly when light is incident on it. The cell contains silicon which releases electrons when it absorbs photons. The released electrons create a potential difference between the terminals of a solar cell when light is directed at the solar cell.

When a solar cell is in a circuit, the current in the circuit transfers energy from the solar cell to the components in the circuit. At constant light intensity, the current is constant and depends on the resistance of the external circuit. Figure 2 shows the relationship between the current and the potential difference across the cell terminals for different light intensities.

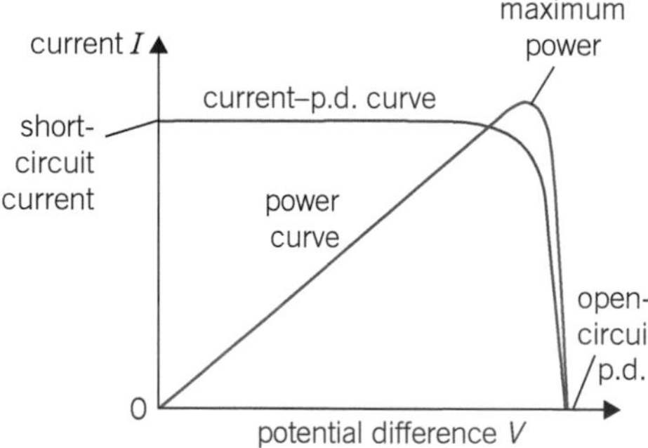

▲ **Figure 2** *Graph of current against p.d. for a solar cell (or a solar panel)*

For constant light intensity, the p.d. across the cell can be changed by altering the resistance of the external circuit.

1 On *short circuit*, the external resistance is zero and so the current is equal to the cell e.m.f. ÷ the internal resistance of the cell.

2 On *open circuit*, the external resistance is infinite so the current is zero and the p.d. is equal to the open-circuit p.d. of the solar cell.

3 If the external resistance is increased from zero to infinity, the current remains constant (because the light intensity is constant) until the potential difference approaches the open-circuit p.d. Fewer and fewer electrons are released from the silicon and as a result the current decreases and falls to zero when the p.d. is equal to the open-circuit p.d.

4 If the intensity is changed to a different constant intensity, the short-circuit current changes in proportion to the change of intensity and the open-circuit p.d. decreases slightly if the change of intensity is an increase.

The power supplied to the external circuit by a solar cell is calculated by multiplying the current by the p.d. across the cell for that current.

- The blue line in Figure 2 shows how the power varies with the cell p.d. The power supplied is greatest at the peak of the power curve.
- The peak power is at a p.d. which is between 10% and 30% below the open-circuit p.d. depending on the light intensity, and at a current which is about 10% below the short-circuit current. The lower the light intensity, the bigger the difference between the peak power p.d. and the open-circuit p.d.

Solar panels

Solar cells are connected in series and in parallel according to the current and p.d. requirements of the device which the solar panel is connected to. For example, Figure 4 consists of an array of eight solar cells connected in two rows in parallel with each row consisting of four cells in series in each row. If the open-circuit p.d. of each cell is 0.45 V and the internal resistance of each cell is 1.5 Ω,

- the open-circuit p.d. across each row = 4 × 0.45 V = 1.80 V
- the short-circuit current for each row = $\dfrac{1.80\,\text{V}}{(4 \times 1.5\,\Omega)} = 0.30\,\text{A}$

Therefore for the two rows in parallel,

- the open-circuit p.d. = 1.80 V
- the short-circuit current = 2 × 0.30 A = 0.60 A.

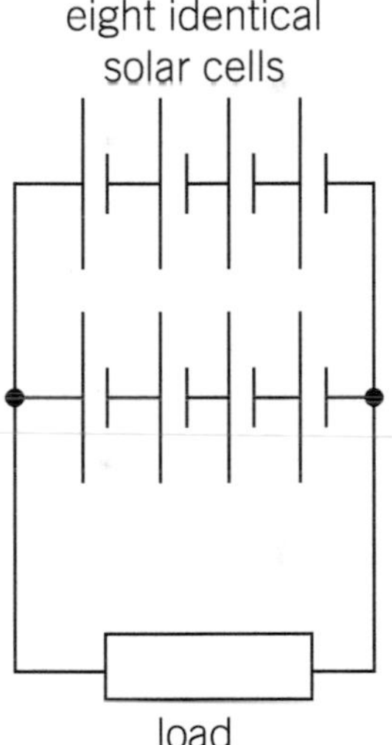

▲ **Figure 4** *A solar cell array (Note: the internal resistance of each cell is not shown)*

Note that the panel cannot produce the open-circuit p.d. and the short-circuit current at the same time. So the power value obtained from the product of these two quantities is not achievable. Since the current for peak power is about 90% of the short-circuit current and the p.d. from peak power is between 70 and 90% of the open-circuit p.d., the peak power is between approximately 65% (≈ 90% of 60%) and 80% (≈ 90% of 90%) of the power value given by the open-circuit p.d. × the short-circuit current.

When a solar panel is connected to an appliance, the power needed by the appliance can be supplied by designing a suitable array of cells that supplies a current and a p.d. to the appliance at values as close as possible to the peak power current and p.d. values.

Summary questions

1 A solar cell illuminated by light of constant intensity has a short-circuit current of 0.6 A and an open-circuit p.d. of 0. 45 V.

a In relation to solar cells, explain what is meant by:

i a short-circuit current **ii** an open-circuit p.d.

b Describe an array of solar cells similar to the one described on the previous page that would give an open-circuit p.d. of 9.0 V and a short-circuit current of 3.0 A.

2 **a** Draw a graph to show how the current from the cell varies with the p.d. across the cell when the cell is connected to a variable resistor and the p.d. is varied from zero to its open-circuit value.

b On the same axes, draw the graph you would expect if the light intensity incident on the cell had been reduced by 25%.

3 Use your graph from **Q2a** to explain why the peak power from a solar panel is less than the power value calculated by multiplying the open-circuit p.d. by the short-circuit current.

4 A solar panel which has an area of 0.40 m^2 is connected to an appliance when light of constant intensity incident on the panel is 200 W m^{-2}. The current in the panel is 1.7 A and the p.d. across its terminals is 12 V. Calculate the percentage efficiency of the solar panel.

26.3 Hydroelectric power and pumped storage

Hydroelectric power

Hydroelectric power stations make use of the gravitational potential energy released by water when it runs downhill. The water gains kinetic energy and forces the turbines in the power station to rotate and make the generators rotate and generate electricity.

A hydroelectric power station generates electrical energy from the potential energy of the water stored in the reservoir.

- Water from the rainfall in the hills above a hydroelectric power station runs into a reservoir in the hills above the power station.
- The power station generates electricity when water from the reservoir is allowed to flow downhill through pipes into the power station and then through the turbines in the power station.
- The turbines are forced to rotate by the flow of water through them so they make the electricity generators rotate and produce electricity.

When a mass m of water flows steadily through a hydroelectric power station in time t, the gravitational potential energy released per second $= \frac{mgh}{t}$ where h is the height of the reservoir above the turbines in the power station. Therefore

the maximum power available from the flow of water $= \frac{mgh}{t}$

Note that for a mass m of water, its volume $V = \frac{m}{\rho}$ where ρ is the density of water. Therefore for a volume flow rate equal to $\frac{V}{t}$, the gravitational potential energy released per second is equal to $\left(\frac{V}{t}\right)\rho gh$. Therefore at this flow rate,

the maximum power available from the water = mon ρgh

Worked example

$g = 9.8\,\text{m}\,\text{s}^{-2}$

The reservoir of a hydroelectric power station is 650 m above the power station. Calculate the maximum power available from the water from the reservoir when it flows at a rate of $25\,\text{m}^3\,\text{s}^{-1}$ through the turbines.

density of water $= 1000\,\text{kg}\,\text{m}^{-3}$

Solution

Maximum power available = mon $gh = (25\,\text{m}^3\,\text{s}^{-1}) \times 9.8\,\text{m}\,\text{s}^{-2} \times 650\,\text{m}$

$= 160\,\text{kW}$

Efficiency limits

The turbines in a hydroelectric power station cannot be 100% efficient as energy transfer devices because the water passing through them

Learning objectives:

- → Describe the energy transfers that take place when a hydroelectric power station generates electricity.
- → Calculate the change of gravitational potential energy per second.
- → Explain why the electrical energy generated is always less than the change of gravitational potential energy.
- → Describe what is meant by pumped storage.

Specification reference: 3.13.4

▲ **Figure 1** *A hydroelectric power station*

must have some kinetic energy after it passes through the turbine. If the water passing through a turbine were stopped completely by the turbine, the flow of water would stop so the turbine would stop rotating. So not all of the energy gravitational potential energy released by the water flow can be transferred as electrical energy. As explained in Topic 26.2, the same is true for a wind turbine.

Further reasons why the electrical power generated by a hydroelectric power station is less than the available power from the flow of water include:

- Friction between the moving parts in the turbine and the generator heats the moving parts and the water flowing through the turbine and therefore dissipates energy to the surroundings.
- Electrical resistance of the wires in the generator and the cables connected to the generator heats the wires and the cables due to the heating effect of the electric current in them when the generator produces electricity. Energy is therefore dissipated from the wires and cables to the surroundings.

Base-load power stations and back-up power stations

Electricity power stations supply electricity to electricity users via an electricity grid which is a network of cables and transformers that the power stations and users are connected to. The demand for electrical power from the grid system varies as the seasons change from summer to winter and back, and also changes according to the weather and the time of day. Some types of electricity power stations can be started or stopped faster than other types to meet changes of demand.

The **base-load demand** for electricity is the minimum long-term demand for electrical power.

The **back-up demand** for electricity is the extra demand for electricity needed to meet short-term increases in demand.

Hydroelectric power stations and gas-fired power stations can be started and stopped from producing electricity much faster than nuclear power stations and other fossil-fuel power stations. Therefore:

- hydroelectric power stations and gas-fired power stations can be used as back-up power stations to meet back-up demands
- nuclear power stations and coal-fired and oil-fired power stations are used to provide the base-load demand.

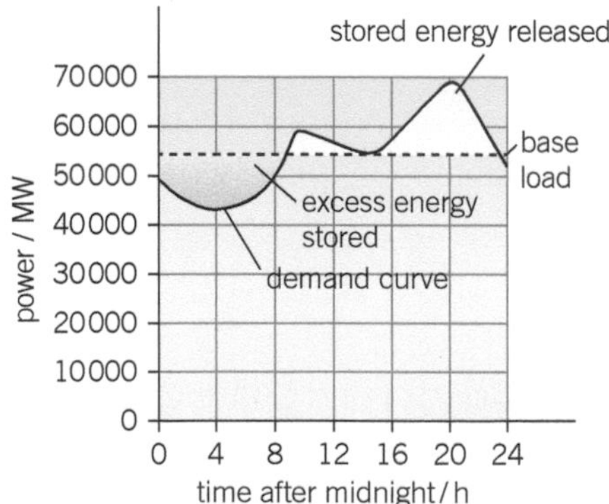

▲ **Figure 2** *Variable demand*

Pumped storage

Pumped storage schemes are hydroelectric power stations in which reversible turbines are used either to pump water uphill or to generate electricity like normal hydroelectric power stations.

At times when the demand for electricity from the grid system is low, surplus electricity can be supplied to a pumped storage hydroelectric power station from other power stations to pump water from a low-level reservoir to a high-level reservoir.

When the demand for electricity increases, a pumped storage power station can be used like a normal hydroelectric power station to generate electricity. In this way, energy that would otherwise be wasted at times of low demand is stored to be used when it is needed.

Summary questions

$g = 9.8\,\mathrm{m\,s^{-2}}$

1 A hydroelectric power station produces 250 MW of electrical power when water flows from a reservoir 1200 m above the power station at a rate of $55\,\mathrm{m^3\,s^{-1}}$ per second. Calculate the efficiency of the hydroelectric power station.

2 The power station in **Q1** is a pumped storage station.
   a Explain what is meant by pumped storage.
   b The hydroelectric power station operates at an efficiency of 60% when it pumps water to the uphill reservoir. Estimate the electric energy that can be returned to the grid as a result of 100 MW h of electrical energy supplied to it to pump water uphill.

3 State and explain why the efficiency of a hydroelectric power station must be less than 100% even if energy losses due to friction and electrical resistance are negligible.

4 a Explain the difference between a base-load power station and a back-up power station.
   b State two types of power stations that can be used as:
      i base-load power stations
      ii back-up power stations.

Tidal power

Tidal power stations are built in coastal areas where the sea level rise and fall due to tides is large and where the sea can be trapped over a large area behind a tidal barrier. The barrier is designed with doors that open to allow seawater through as the tide comes in and then close at high tide to trap the seawater behind the barrier. As the tide goes out, the trapped water is allowed to flow out to sea through turbines in the barrier that generate electricity. The method used to estimate the available power from a hydroelectric power station can be applied here, bearing in mind that the height drop is from the height of the centre of mass of the trapped seawater above the turbines (not the depth of the seawater).

For example, a tidal power station covering an area of $100\,\mathrm{km^2}$ could trap a depth of 6 m of seawater twice per day. This would mean releasing a volume of $6 \times 10^8\,\mathrm{m^3}$ of sea water over a few hours. The mass of such a volume is about 6×10^{11} kg, and if its centre of mass drops through an average height of about 2 m, the potential energy released would be about 6×10^{12} J. Prove for yourself this would give an energy transfer rate of more than 500 MW if released over about 6 hours.

▲ **Figure 3** *A tidal power station*

Practice questions: Chapter 26

1 A solar cell panel of area $1\,m^2$ can produce 200 W of electrical power on a sunny day.

(a) Calculate the area of panels that would be needed to produce 2000 MW of electrical power on a sunny day. *(1 mark)*

(b) The panels contain identical cells that each have an open-circuit p.d. of 0.45 V and a short-circuit current of 0.6 A in the conditions above. Assuming the efficiency of the cells is 25%, estimate the number of cells needed for a panel of area $1\,m^2$. *(2 marks)*

2 The maximum power that can be obtained from a wind turbine is proportional to the cube of the wind speed. When the wind speed is $10\,m\,s^{-1}$, the power output of a certain wind turbine is 1.2 MW. Calculate the power output of this wind turbine when the wind speed is $15\,m\,s^{-1}$. *(2 marks)*

3 A hydroelectric power station produces electrical power at an overall efficiency of 25%. The power station is driven by water that has descended from an upland reservoir 650 m above the power station. Calculate the volume of water passing through the power station per second when it produces 200 MW of electrical power.
density of water = $1000\,kg\,m^{-3}$ *(4 marks)*

4 At a tidal power station, water is trapped over an area of $200\,km^2$ when the tide is 3.0 m above the power station turbines. The trapped water is released gradually over a period of 6 hours.
Calculate:

(a) the mass of trapped water *(2 marks)*

(b) the average loss of potential energy per second of this trapped water when it is released over a period of 6 hours.
density of seawater = $1050\,kg\,m^{-3}$ *(3 marks)*

5 A small hydroelectric power station uses water which falls through a height of 6.3 m.

(a) Calculate the maximum power available from the water passing through this power station when water flows through it at a rate of 5.5×10^4 kg per hour. *(3 marks)*

(b) State and explain *two* factors that affect the efficiency of a hydroelectric power station. *(3 marks)*

6 (a) A wind turbine which has blades of radius 11 m is subjected to a steady stream of air at a wind speed of $12\,m\,s^{-1}$. Assume the density of air is $1.3\,kg\,m^{-3}$.

(i) Calculate the mass per second of air passing through the area swept out by the blades *(3 marks)*

(ii) The power that can be obtained by the wind turbine from the wind at this speed is about 50% of the kinetic energy per second of the incident wind. Calculate the output power from the wind turbine at this wind speed. *(3 marks)*

(b) Explain what is meant by the *wind shadow* of a wind turbine and its relevance to the concentration of wind turbines in a wind farm. *(3 marks)*

7 **Figure 1** shows the graph of current against p.d. of a solar cell when it is illuminated by light of constant intensity.

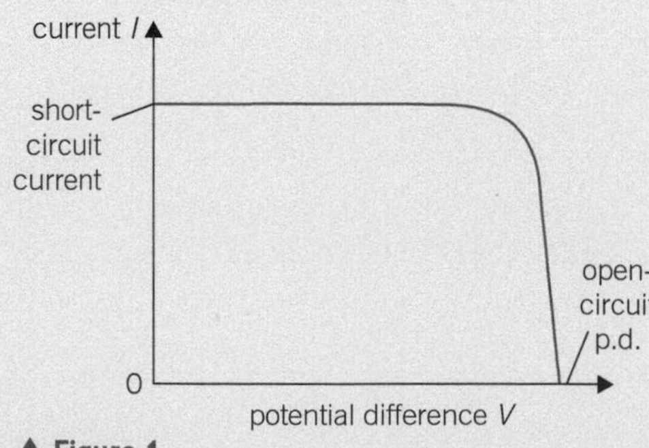

▲ **Figure 1**

(a) In terms of electrons, explain why a potential difference is generated by the cell when light is directed at it. *(3 marks)*

(b) A solar cell is connected to an external resistor and is illuminated by light of constant intensity.

(i) On a copy of Figure 1, show how the power supplied by the cell varies as the resistance of the external resistor is increased from zero. *(3 marks)*

(ii) State and explain how the maximum power the cell can supply would change if the light intensity were doubled. *(3 marks)*

(c) The intensity of solar radiation at the Earth is about $1.4\,kW\,m^{-2}$. A solar panel of area $5.4\,m^2$ on a space vehicle above the Earth produces 1.3 kW of electrical power when it faces the Sun directly.

(i) Calculate the efficiency of the solar panel in this situation. *(2 marks)*

(ii) Estimate the power output of the solar panel if the satellite had been facing the Sun directly in orbit round Mars which is 1.5 times further from the Sun than the Earth. *(2 marks)*

8 A hydroelectric power station generates 450 MW of electrical power at an efficiency of 80%. The power station is driven by water that has descended from an upland reservoir 390 m above the power station.

(a) Calculate the volume of water per second passing through the turbines of the power station when its generators produce 450 MW of electrical power.
density of water = $1000\,kg\,m^{-3}$ *(6 marks)*

(b) The turbines at the power station are also designed to pump water uphill to the reservoir using electrical power from electrical power generators elsewhere in the grid system.

(i) The pumping efficiency of the turbines is 75%. Estimate the percentage of the electrical power supplied to the power station that is returned to the grid system when it generates electricity. *(2 marks)*

(ii) State and explain two reasons why pumped storage hydroelectric stations are used even though their efficiency is significantly less than the efficiency of ordinary hydroelectric stations. *(4 marks)*

9 A wind turbine, as shown in **Figure 2**, has blades of length 22 m. When the wind speed is $15\,m\,s^{-1}$ its output power is 1.5 MW.

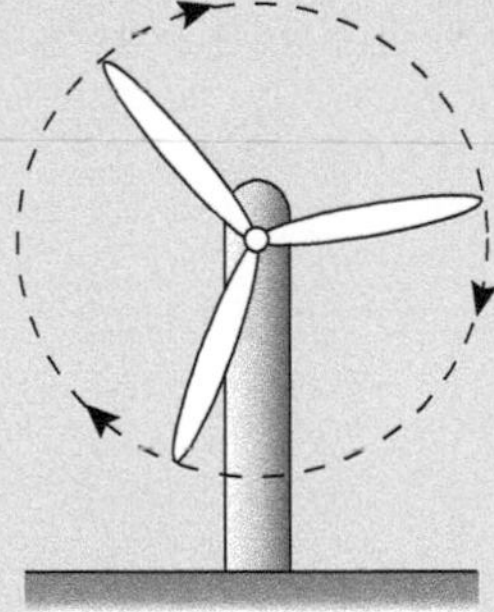

▲ **Figure 2**

(a) The volume of air passing through the blades each second can be calculated by considering a cylinder of radius equal to the length of the blade. Show that $2.3 \times 10^4\,m^3$ of air passes through the blades each second.

(b) Calculate the mass of air that passes through the blades each second.
density of air = $1.2\,kg\,m^{-3}$

(c) Calculate the kinetic energy of the air reaching the blades each second.

(d) Assuming that the power output of the turbine is proportional to the kinetic energy of the air reaching the blades each second, discuss the effect on the power output if the wind speed decreased by half. *(7 marks)*

AQA, 2005

Further practice questions: 2

Multiple choice questions

1 The London Eye is a vertical wheel that turns on a fixed horizontal axis. The wheel carries its passengers in a vertical circle of radius R at constant speed v. When a passenger of mass m is at the top of the ride, which one of the following alternatives **A–D** gives the support force on the passenger?

A $mg - \frac{mv^2}{R}$ **B** mg **C** $\frac{mv^2}{R}$ **D** $mg + \frac{mv^2}{R}$

2 An object attached to a spring oscillates vertically with an amplitude A. When its displacement is $\frac{1}{2}A$, which one of the following alternatives **A–D** gives the ratio

$$\frac{\text{its kinetic energy}}{\text{its potential energy}}$$

A 1 **B** 2 **C** 3 **D** 4

3 A simple pendulum of length L oscillates freely with an amplitude A. Which one of the following alternatives **A–D** gives a correct expression for the maximum speed of the pendulum bob?

A $A\sqrt{lg}$ **B** $A\sqrt{\frac{l}{g}}$ **C** $\frac{A}{\sqrt{lg}}$ **D** $A\sqrt{\frac{g}{l}}$

4 A spherical planet P has a radius R and uniform density ρ. The gravitational potential at the surface of P is V_s.
Another spherical planet Q has a radius $2R$ and uniform density 0.5ρ. State which *one* of the following alternatives **A–D** gives the gravitational potential at the surface of Q.

A $\frac{1}{2}V_s$ **B** V_s **C** $2V_s$ **D** $4V_s$

5 A satellite of mass m is in a circular orbit at height $2R$ above the Earth's surface, where R is the radius of the Earth.
State which *one* of the following alternatives **A–D** gives the ratio

$$\frac{\text{weight of the satellite in its orbit}}{\text{weight of the satellite at the surface of the Earth}}$$

A $\frac{1}{9}$ **B** $\frac{1}{4}$ **C** $\frac{1}{3}$ **D** $\frac{1}{2}$

6 Two small spheres carry equal and opposite charges $+Q$ and $-Q$. The force between the spheres is F when the centres of the spheres are at distance d apart.

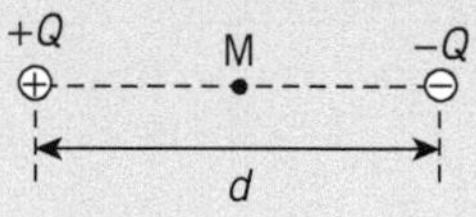

+Q M −Q
d

▲ **Figure 1**

State which *one* of the following alternatives **A–D** gives the magnitude of the electric field strength at the midpoint M of their centres.

A $\frac{F}{Q}$ **B** $\frac{2F}{Q}$ **C** $\frac{4F}{Q}$ **D** $\frac{8F}{Q}$

7 A data logger is used to measure the potential difference across the terminals of a capacitor as it discharges through a 5.0 MΩ resistor. The measurements show that the potential difference across the capacitor decreases from 5.00 V to 2.75 V in 300 s.
State which of the following alternatives **A–D** gives the capacitance of the capacitor.

A 100 μF **B** 135 μF **C** 165 μF **D** 300 μF

8 **Figure 2** shows a square coil ABCD that has been placed in a uniform horizontal magnetic field such that the magnetic field lines are parallel to the plane of the coil and the sides AD and BC of the coil are vertical. The coil is then connected in series with a battery, a resistor, and an open switch.

State which of the following alternatives **A–D** describes correctly the effect on the coil when the switch is closed.

A Each side experiences a force.
B Only sides AB and CD experience a force.
C Only the sides AD and BC experience a force.
D Opposite sides experience forces in opposite directions.

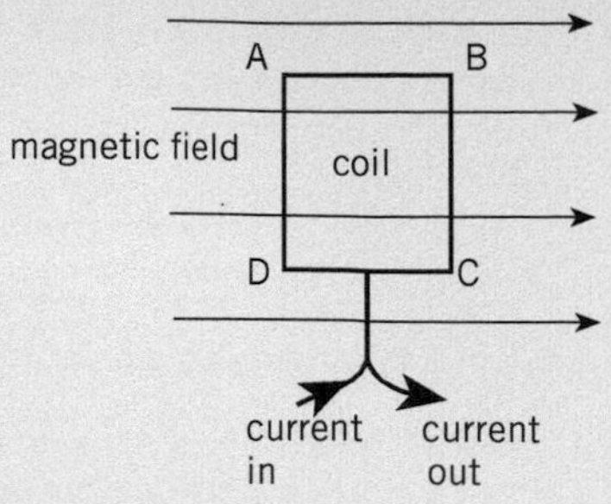

▲ **Figure 2**

9 An ac generator turns at a steady frequency of 25 Hz in a uniform 80 mT magnetic field. This produces a sinusoidal p.d. with a peak output of 4.0 V. If the frequency is reduced to 20 Hz and the magnetic flux density is increased to 100 mT, state the peak output p.d.

A 3.2 V **B** 4.0 V **C** 5.0 V **D** 6.3 V

10 A radioactive nucleus $^{A}_{Z}X$ emits alpha, beta, and gamma particles in the following sequence:

alpha beta gamma

State which of the following alternatives **A–D** represents the nucleus at the end of this sequence.

A $^{A-4}_{Z-2}X$ **B** $^{A-4}_{Z-1}X$ **C** $^{A-2}_{Z-2}X$ **D** $^{A-2}_{Z-1}X$

11 Two radioactive isotopes X and Y have half-lives of 8.0 years and 16.0 years, respectively, and form stable products. Samples of X and Y are initially pure and contain equal numbers of nuclei.
State which of the following alternatives **A–D** gives the ratio $\frac{\text{the activity of X}}{\text{the activity of Y}}$ after 16 years.

A $\frac{1}{4}$ **B** $\frac{1}{2}$ **C** 1 **D** 2

12 A gas cylinder contains *n* moles of an ideal gas at a pressure of 120 kPa and a temperature of 300 K. The temperature is then increased to 400 K and without changing the temperature of the gas in the cylinder, some gas is then released into the atmosphere until the pressure returns to 120 kPa. Which one of the following alternatives **A–D** gives the number of moles of gas released into the atmosphere?

A $\frac{n}{4}$ **B** $\frac{n}{2}$ **C** $\frac{2n}{3}$ **D** $\frac{3n}{4}$

13 Any unit in physics which is not an SI base unit can be expressed as a combination of SI base units. Which one of the following alternatives **A–D** gives the base units of specific latent heat?

A $kg^{-1}\,m^2\,s^{-1}$ **B** $kg^{-1}\,m^2\,s^{-2}$ **C** $m^2\,s^{-1}$ **D** $m^2\,s^{-2}$

14 A flywheel of moment of inertia *I* rotates with constant angular momentum *L*. Which one of the following alternatives **A–D** gives its rotational kinetic energy?

A $\frac{L}{2I}$ **B** $\frac{LI}{2}$ **C** $\frac{LI^2}{2}$ **D** $\frac{L^2}{2I}$

15 When a $^{2}_{1}H$ nucleus fuses with a $^{3}_{1}H$ nucleus, a $^{4}_{2}He$ is nucleus is formed and a neutron is ejected.
masses: neutron 1.00867 u, $^{2}_{1}H$ 2.01355 u, $^{3}_{1}H$ 3.01550 u, $^{4}_{2}He$ 4.00150 u
1 u = 931.5 MeV
State which of the following alternatives **A–D** gives the binding energy released in this event.

A 6.3 MeV **B** 17.6 MeV **C** 21.1 MeV **D** 960 MeV

In each of Questions **16** to **20**, select from the list, **A–D**, to state which relationship correctly describes the connection between y and x.

A y is proportional to x^2.

B y is proportional to $\sqrt{x}$.

C y is proportional to $\frac{1}{x}$.

D y is proportional to $\frac{1}{x^2}$.

16 y = the speed of an object falling freely from rest, x = distance fallen by the object

17 y = maximum kinetic energy of a freely oscillating object, x = amplitude of the oscillations

18 y = energy stored by a parallel-plate capacitor connected to a battery, x = perpendicular distance between the plates

19 y = intensity of gamma radiation from a point source, x = distance from the source

20 y = electric potential energy of two protons x = distance, between the protons

Longer questions

1 **Figure 3** shows a way to measure the mass of a lorry. The vehicle and its contents are driven onto a platform mounted on a spring. The platform is then made to oscillate vertically, and the mass is found from a measurement of the natural frequency of oscillation.

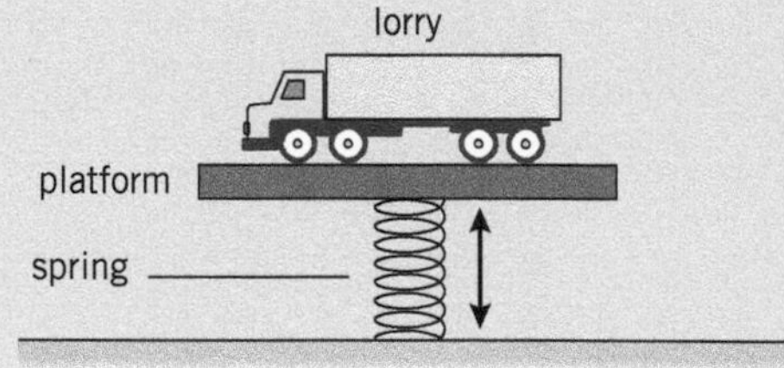

▲ **Figure 3**

(a) (i) State whether the period of oscillation increases, decreases, or remains unchanged when the amplitude of oscillation of the platform is reduced.

(ii) The spring constant k of the supporting spring is increased to four times its original value.
State the value of the ratio $\frac{\text{new oscillation period}}{\text{old oscillation period}}$

(iii) The time period of oscillation is T when a lorry is on the platform. The spring constant of the spring is k. Show that the total mass M of lorry and platform is given by $M = \frac{kT^2}{4\pi^2}$.

(iv) A lorry and its contents have a total mass of 5300 kg. The spring constant of the supporting spring k is $1.9 \times 10^5\,\text{N}\,\text{m}^{-1}$. The frequency of oscillation of the platform with the lorry resting on it is 0.91 Hz.

Calculate the mass of the platform. *(7 marks)*

(b) The driver is required to turn off the vehicle engine whilst the measurement is taking place. The driver of the lorry in part (a)(iv) fails to do this and slowly increases the frequency of vibration of his vehicle from 0.5 Hz to about 4 Hz whilst the measurement is in progress and the platform is free to move. Describe and explain how the amplitude and frequency of the platform vary as this frequency increase occurs. You should use a sketch graph to support your answer. *(4 marks)*

AQA, 2003

2 A dish on a communications satellite is used to transmit a beam of microwaves of wavelength λ. The beam spreads with an angular width $\frac{\lambda}{d}$, in radians, where d is the diameter of the dish.

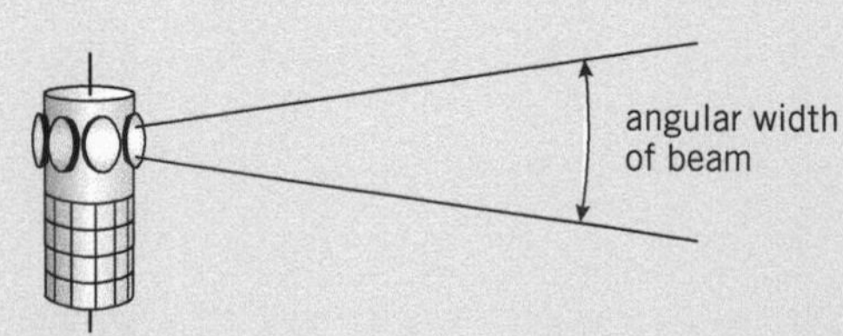

▲ **Figure 4**

(a) (i) Calculate the angular width, in degrees, of a beam of frequency 1200 MHz transmitted using a dish of diameter 1.8 m.

(ii) Show that the beam has a width of 2100 km at a distance of 15 000 km from the satellite. *(4 marks)*

(b) (i) Show that the speed, v, of a satellite in a circular orbit at height h above the Earth is given by

$$v = \sqrt{\frac{GM}{R+h}}$$

where R is the radius of the Earth and M is the mass of the Earth.

(ii) Calculate the speed and the time period of a satellite at a height of 15 000 km in a circular orbit about the Earth.

(iii) The satellite passes directly over a stationary receiver at the North Pole. Show that the beam moves at a speed of $1.3\,\text{km}\,\text{s}^{-1}$ across the Earth's surface and that the receiver can remain in contact with the satellite for no more than 27 minutes each orbit. *(9 marks)*

AQA, 2005

3 **Figure 5** shows a motor lifting a small mass. The energy required comes from a charged capacitor.

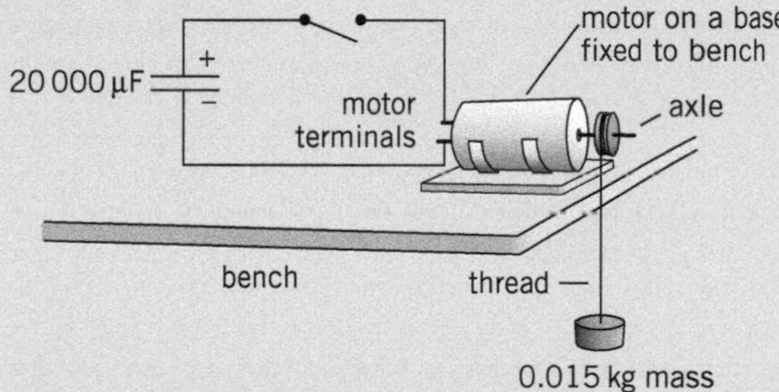

▲ **Figure 5**

The capacitor was charged to a potential difference of 4.5 V and then discharged through the motor.

(a) (i) The motor only operates when the voltage at its terminals is at least 2.5 V. Calculate the energy delivered to the motor when the potential difference across the capacitor falls from 4.5 V to 2.5 V.

(ii) The motor lifted the mass through a distance of 0.35 m. Calculate the efficiency of the transfer of energy from the capacitor to gravitational potential energy of the mass. Give your answer as a percentage.

(iii) Give *two* reasons why the transfer is inefficient. *(7 marks)*

(b) The motor operated for 1.3 s as the capacitor discharged from 4.5 V to 2.5 V. Calculate:

(i) the average useful power developed in lifting the mass

(ii) the effective resistance of the motor, assuming that it remained constant. *(5 marks)*

AQA, 2004

4 **(a)** (i) Explain why, despite the electrostatic repulsion between protons, the nuclei of most atoms of low nucleon number are stable.

(ii) Suggest why stable nuclei of higher nucleon number have greater numbers of neutrons than protons.

(iii) All nuclei have approximately the same density. State and explain what this suggests about the nature of the strong nuclear force. *(6 marks)*

(b) (i) Compare the electrostatic repulsion and the gravitational attraction between a pair of protons. The centres of the protons are separated by 1.2×10^{-15} m.

(ii) Comment on the relative roles of gravitational attraction and electrostatic repulsion in nuclear structure. *(5 marks)*

AQA, 2006

5 **(a)** **Figure 6** shows an arrangement used to investigate the energy stored by a capacitor.

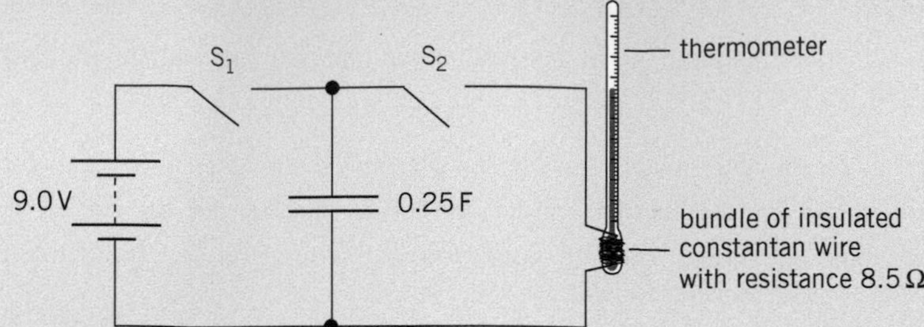

▲ **Figure 6**

The bundle of constantan wire has a resistance of 8.5 Ω. The capacitor is initially charged to a potential difference of 9.0 V by closing S_1.

(i) Calculate the charge stored by the 0.25 F capacitor.

(ii) Calculate the energy stored by the capacitor.

(iii) Switch S_1 is now opened and S_2 is closed so that the capacitor discharges through the constantan wire.
Calculate the time taken for the potential difference across the capacitor to fall to 0.10 V. *(7 marks)*

AQA, 2006

(b) The volume of constantan wire in the bundle in **Figure 6** is $2.2 \times 10^{-7} m^3$.
density of constantan = $8900 kg m^{-3}$
specific heat capacity of constantan = $420 J kg^{-1} K^{-1}$

(i) Assume that all the energy stored by the capacitor is used to raise the temperature of the wire. Use your answer to part (a)(ii) to calculate the expected temperature rise when the capacitor is discharged through the constantan wire.

(ii) Give *two* reasons why, in practice, the final temperature will be lower than that calculated in part (b)(i). *(5 marks)*

AQA, 2006

6 **(a)** A 3.0 kW electric kettle heats 2.4 kg of water from 16 °C to 100 °C in 320 s.

(i) Calculate the electrical energy supplied to the kettle.

(ii) Calculate the heat energy supplied to the water.
specific heat capacity of water = $4200 J kg^{-1} K^{-1}$

(iii) Give *one* reason why not all the electrical energy supplied to the kettle is transferred to the water. *(4 marks)*

(b) The potential difference supplied to the kettle in part (a) is 230 V.

(i) Calculate the resistance of the heating element of the kettle.

(ii) The heating element consists of an insulated conductor of length 0.25 m and diameter 0.65 mm. Calculate the resistivity of the conductor. *(5 marks)*

AQA, 2004

7 **(a)** (i) State *three* assumptions concerning the motion of the molecules in an ideal gas.

(ii) For an ideal gas at a temperature of 300 K, show that the mean kinetic energy of a molecule is 6.2×10^{-21} J. *(5 marks)*

(b) (i) When no current passes along a metal wire, conduction electrons move about in the wire like molecules in an ideal gas.
Calculate the speed of an electron which has 6.2×10^{-21} J of kinetic energy.

(ii) Describe the motion of conduction electrons in a wire when a p.d. is applied across the ends of the wire. *(6 marks)*

AQA, 2007

8 **(a)** (i) At the surface of a spherical planet of radius R, show that the gravitational potential, V_s, is related to the gravitational field of strength, g_s, by

$$V_s = -g_s R.$$

(ii) The gravitational field strength of the Moon at its surface is $1.6\,\mathrm{N\,kg^{-1}}$. Show that the gravitational potential energy of an oxygen molecule at the surface is $-1.4 \times 10^{-19}\,\mathrm{J}$.
radius of the Moon = 1700 km   molar mass of oxygen = $0.032\,\mathrm{kg\,mol^{-1}}$
(5 marks)

(b) Oxygen gas at 400 K is released on the surface of the Moon.
(i) Calculate the mean kinetic energy of an oxygen gas molecule at this temperature.
(ii) The maximum temperature of the surface of the Moon is about 400 K. Use the data from part (a)(ii) and the results of your calculations to explain why some of the oxygen gas released at the Moon's surface would escape into space.
(4 marks)
AQA, 2005

Data analysis questions

In an impact investigation by two students, a mass hanger A suspended on a thread was displaced from its equilibrium position by a certain distance and released so it collided with a ball B suspended on a thread. A horizontal metre ruler fixed in a clamp (not shown) was used to measure the horizontal displacements x_A and x_B of each thread from its equilibrium position at the level of the metre ruler, as shown in **Figure 7**. The vertical distance, d, of the ruler below the upper end of the threads was measured.

▲ **Figure 7**

The measurement of x_B was repeated without changing x_A and d for different additional masses added to the mass hanger.

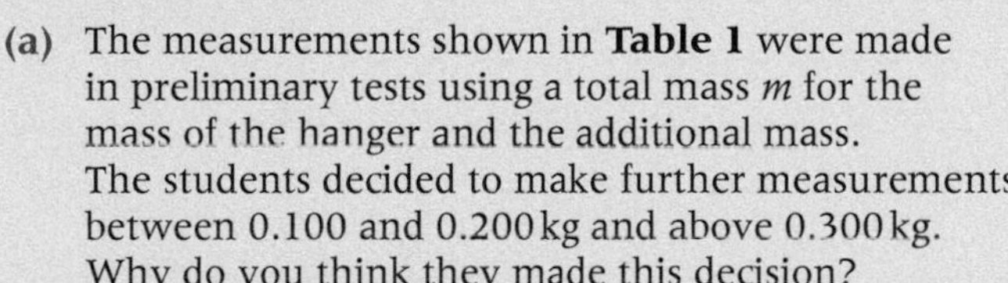

(a) The measurements shown in **Table 1** were made in preliminary tests using a total mass m for the mass of the hanger and the additional mass. The students decided to make further measurements between 0.100 and 0.200 kg and above 0.300 kg. Why do you think they made this decision?
(4 marks)

▼ **Table 1**

m / kg	x_B / mm	
0.100	60	62
0.200	77	75
0.300	80	78

(b) **Table 2** shows all their measurements.

▼ **Table 2**

m / kg	x_B / mm					$\langle x_B \rangle$/mm	θ/°
0.100	60	62	58	58	60	59.6	8.47
0.120	67	68	68	65	62	66.0	9.37
0.150	68	73	70	68	70	69.8	9.90
0.200	77	75	78	71	76	75.4	10.67
0.300	80	78	79	80	80	79.4	11.22
0.600	86	85	88	85	88		

$d = 400\,\text{mm}$, $x_A = 60\,\text{mm}$
The maximum angular displacement θ of the thread from equilibrium can be calculated using the equation $\tan\theta = \langle x_B \rangle / d$, where $\langle x_B \rangle$ is the mean value of x_B. This has been done in **Table 2** for all the rows except the last one.

Copy and complete **Table 2** by calculating $\langle x_B \rangle$ and θ for $m = 0.600\,\text{kg}$. *(2 marks)*

(c) (i) By considering the energy changes of B after the impact, show that its velocity V immediately after the impact is given by $V = \sqrt{(2gh)}$, where h is B's maximum height gain from its equilibrium position. *(2 marks)*

(ii) The height gain h can be calculated using the trigonometry formula $h = l(1 - \cos\theta)$, where l is the distance along the thread from the point of suspension of the ball to its centre. This distance was measured to be 575 mm. For each mass m, **Table 3** shows the results of these calculations except for the last row. Copy the table and complete this last row. The last two columns are for part (d) of this question.

▼ **Table 3**

m / kg	θ / °	h / mm	V / m s^{-1}		
0.100	8.47	6.27	0.351		
0.120	9.37	7.67	0.388		
0.150	9.90	8.56	0.410		
0.200	10.67	9.95	0.442		
0.300	11.22	11.00	0.465		
0.600					

(2 marks)

(d) The students found a theoretical analysis of the impact which gave the following equation relating V and m:

$$\frac{1}{V} = \frac{kM}{m} + k$$

where M is the mass of the ball and k is a constant.

(i) Plot a suitable graph to see if this relationship is correct. Show the results of any further calculations you carry out in the last two columns of your own **Table 3**.

(ii) Using your graph or otherwise, determine values for k and M. *(9 marks)*

(e) (i) What conclusions do you draw from the graph?

(ii) Use your results to evaluate your conclusions. *(3 marks)*

Extension question

The theoretical analysis is based on the diagram below in which an object (A) of mass m moving at velocity u collides with a stationary object (B) of mass M. After the impact, the two objects move apart at velocities v and V in the same direction as A's initial direction.

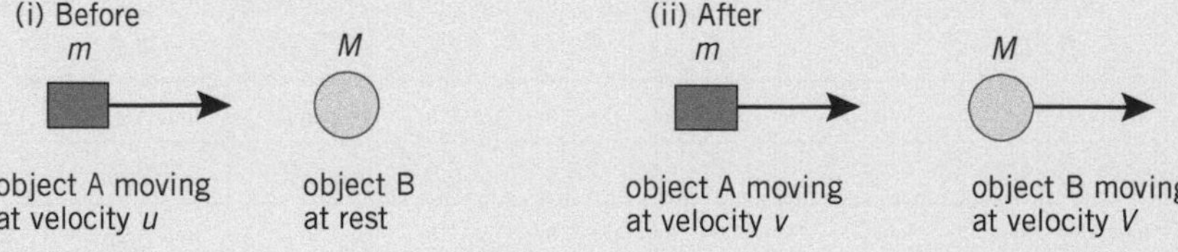

▲ **Figure 8**

The theoretical analysis assumed that the velocity of B relative to A after the collision $(V - v) = eu$, where e is a constant that depends on the two objects.

(f) Combine the equation above with the equation representing conservation of momentum to derive the theoretical equation given in part (d), where $k = \dfrac{1}{(1 + e)u}$ *(5 marks)*

Answers to the Practice Questions and Section Questions are available at www.oxfordsecondary.com/oxfordaqaexams-alevel-physics

27 Mathematical skills in Year 2 Physics

27.1 Trigonometry

In this chapter, you will consider only what you need to know at A Level Physics Year 2. If you need to check any of the mathematical skills that you need for your Year 1 course, you can use Chapter 14.

Angles and arcs

The radian

The **radian** (rad) is a unit used to express or measure angles. It is defined such that 2π radians = 360°.

When using a calculator to work out sines, cosines, tangents, or the corresponding inverse functions, always check that the calculator is in the correct 'angle' mode. This is usually indicated on the display by 'deg' for degrees or 'rad' for radians. Your calculations will be incorrect if you work in one mode when you should be working in the other mode. For example, check for yourself that sin 30° = 0.5, whereas sin(30 rad) = –0.988. You also need to know how to change from one mode to the other. Read your calculator manual or ask your teacher if you can't do this.

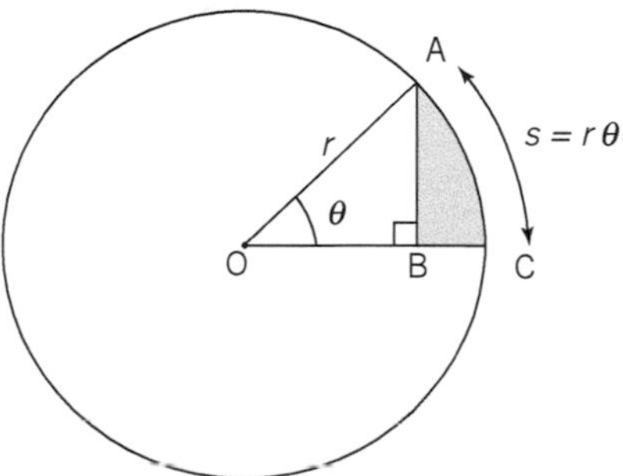

▲ **Figure 1** *Arcs and segments*

Arcs and segments

Consider an arc of length *s* on the circumference of a circle of radius *r*, as shown in Figure 1. The angle *θ*, in degrees, subtended by the arc to the centre of the circle is given by the equation

$$\theta\,/\,\text{degrees} = \frac{s}{2\pi r} \times 360$$

Because 2π radians = 360°, applying this conversion factor to the above equation for *θ* gives

$$\theta\,/\,\text{radians} = \frac{s}{r}$$

Rearranging this equation gives

$$\textbf{arc length } s = r\theta$$

where *θ* is the angle subtended in radians.

Note

Note that for $s = r$, $\theta = 1$ rad $\left(= \frac{360}{2\pi} = 57.3°\right)$

The small angle approximation

For angle *θ* less than about 10°,

sin θ ≈ tan θ ≈ θ in radians, and

cos θ ≈ 1

To explain these approximations, consider Figure 1 again. If angle *θ* is sufficiently small, then the segment OAC will be almost the same as triangle OAB, as shown in Figure 2.

- AB ≈ arc length *s* so $\sin\theta = \frac{AB}{OA} \approx \frac{s}{r} = \theta$ in radians, and therefore $\sin\theta \approx \theta$ in radians.

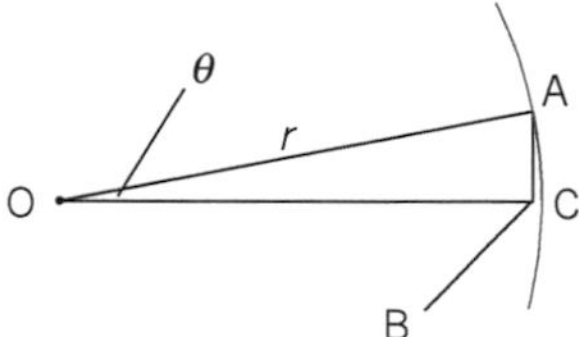

▲ **Figure 2** *The small angle approximation*

- OB ≈ radius r, so $\tan\theta = \frac{AB}{OB} \approx \frac{s}{r} = \theta$ in radians, and therefore, $\tan\theta \approx \theta$ in radians.
- Also, $\cos\theta = \frac{OB}{OA} \approx \frac{r}{r} = 1$, and therefore, $\cos\theta \approx 1$

Use a calculator to prove for yourself that for $\sin 10° = 0.1736$, $\tan 10° = 0.1763$ and $10° = 0.1745$ rad. Also, $\cos 10° = 0.9848$. So the small angle approximation is almost 99% accurate up to 10°.

The small angle approximation is used to show that the time period of a simple pendulum of length L is given by the formula $T = 2\pi\sqrt{\frac{L}{g}}$, provided the maximum angular displacement of the pendulum from equilibrium is less than about 10°. See Topic 11.1 for more about this equation.

Sine and cosine curves

Figure 3 shows how $\sin\theta$ and $\cos\theta$ change as θ increases. Notice that $\sin\theta \approx \theta$ and $\cos\theta \approx 1$ up to about 10°. The general shape of a cosine wave is the same that as that of a sine wave so we refer to them both as *sinusoidal* waveforms. In addition, notice that:

- the sine wave starts at zero and rises to a maximum of +1 from $\theta = 0$ to $\theta = \frac{\pi}{2}$ rad (= 90°), whereas the cosine wave starts at +1 and falls to zero from $\theta = 0$ to $\theta = \frac{\pi}{2}$ rad (= 90°)
- the gradient of the sine wave is zero where the cosine wave is zero (i.e., where it crosses the horizontal axis) and the gradient of the cosine wave is zero where the sine wave is zero.

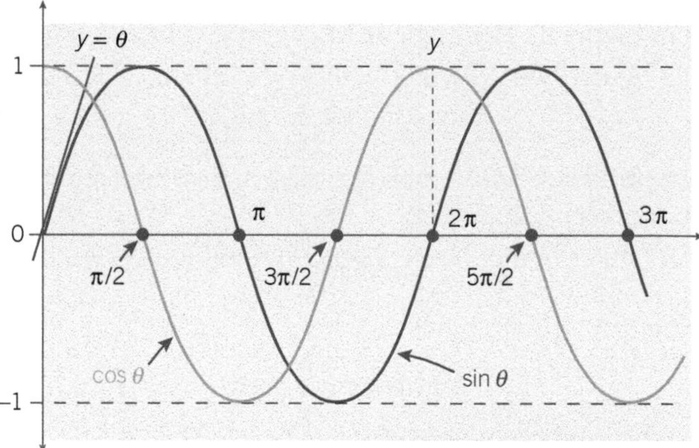

▲ **Figure 3** *Sine and cosine curves*

Equations that describe sine waves and cosine waves are often used to calculate, for example, the displacement of an oscillating particle at a certain time or of a wave at a particular position along the wave.

For example, consider the displacement x of a particle on a spring oscillating vertically in simple harmonic motion, as shown in Figure 4a. Let f represent its frequency and A its amplitude. Its displacement varies sinusoidally between a maximum value $+A$ and a minimum value $-A$. Also, suppose $x = +A$ at time $t = 0$. In other words, the object is held above the equilibrium position at displacement $x = +A$ and released at time $t = 0$.

- Its displacement–time curve will therefore be a cosine wave, as shown in Figure 4b where its time period $T = \frac{1}{f}$

- At time t later, the particle will have gone through ft cycles of oscillation, corresponding to $\theta = 2\pi ft$ radians. Hence its displacement at time t is given by $\boldsymbol{x = A \cos 2\pi ft}$

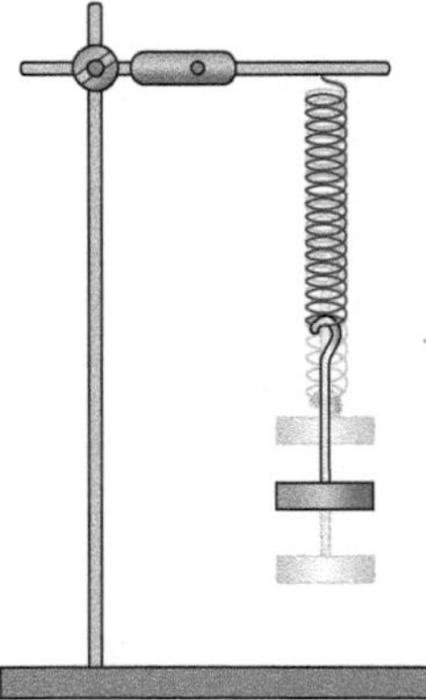

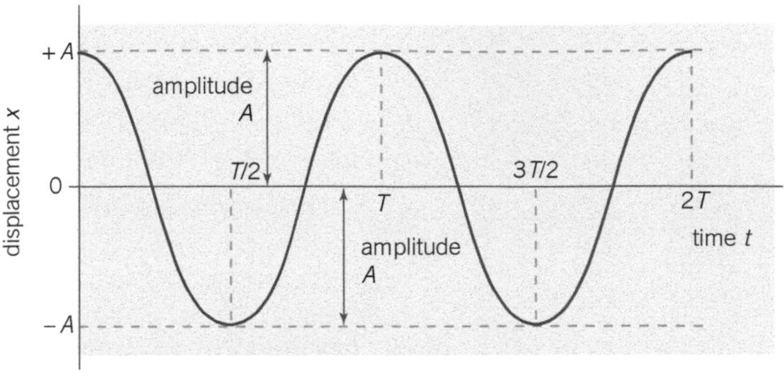

▲ **Figure 4** **a** *An object oscillating on a spring* **b** *Displacement–time curve for* $x = A \cos 2\pi ft$

Summary questions

1 **a** Convert the following angles from degrees into radians and express your answer to one further significant figure than in each question:

i 30°
ii 50°
iii 120°
iv 230°
v 300°.

b Convert the following angles from radians into degrees and express your answer to one further significant figure than in each question:

i 0.10 rad
ii 0.50 rad
iii 1.20 rad
iv 2.50 rad
v 6.00 rad.

2 **a** Measure the diameter of a 1p coin to the nearest millimetre. Calculate the angle subtended at your eye, in degrees, by a 1p coin held at a distance of 50 cm from your eye.

b **i** Estimate the angular width of the Moon, in degrees, at your eye by holding a millimetre scale at 50 cm from your eye and measuring the distance on the scale covered by the lunar disc.

ii The diameter of the Moon is 3500 km. The average distance to the Moon from the Earth is 380 000 km. Calculate the angular width of the Moon as seen from the Earth and compare the calculated value with your estimate in **bi**.

3 **a** Use the small angle approximation to calculate $\sin \theta$ for

i $\theta = 2.0°$
ii $\theta = 8.0°$.

b Show that the small angle approximation for $\sin \theta$ is more than 99% accurate for $\theta = 10°$.

4 Use your calculator to find

i $\sin \theta$
ii $\cos \theta$ for the following values of θ:

a 0.1 rad
b 10°
c 45°
d 0.25π rad.

27.2 Algebra

Linear simultaneous equations

Two linear equations with two variable quantities, x and y, in each can be solved to find the values of x and y. Such a pair of equations are referred to as **simultaneous equations** because they have the same solution. They are described as **linear** because they contain terms in x and y and do not contain any higher order terms such as x^2 or y^2.

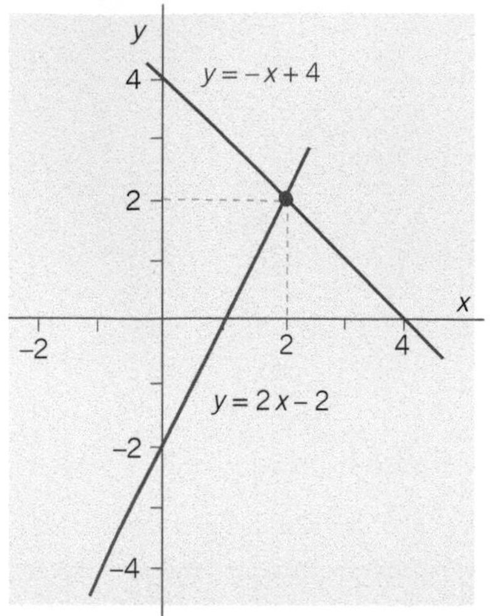

▲ **Figure 1** *A graphical solution*

The general equation for a straight-line graph is $y = mx + c$, as explained in Topic 14.4. Two straight lines on a graph can be represented by two such equations. Provided the two lines are not parallel to one another, they cross each other at a single point. The coordinates of this point are the values of x and y that fit both equations. In other words, these coordinates are the solution of a pair of simultaneous equations representing the two straight lines.

The graph approach to finding the solution of a pair of simultaneous equations is shown in Figure 1 and is described in Topic 14.4. However, plotting graphs takes time and is not as accurate as a systematic algebraic method. This method can best be explained by considering an example, as follows:

$$2x - y = 2 \qquad \textbf{(Equation 1)}$$

$$x + y = 4 \qquad \textbf{(Equation 2)}$$

Make the coefficient of x the same in both equations by multiplying one or both equations by a suitable number. In the above example, this is most easily achieved by multiplying Equation 2 throughout by 2 to give $2x + 2y = 8$.

The two equations to be solved are now

$$2x - y = 2 \qquad \textbf{(Equation 1)}$$

$$2x + 2y = 8 \qquad \textbf{(modified Equation 2)}$$

Subtracting modified Equation 2 from Equation 1 gives

$$(2x - y) - (2x + 2y) = 2 - 8$$

$$\text{therefore, } -y - 2y = -6$$

$$-3y = -6$$

$$y = \frac{-6}{-3} = 2$$

Substituting this value into Equation 1 or Equation 2 enables the value of x to be determined. Using Equation 2 for this purpose gives $x + 2 = 4$, hence $x = 4 - 2 = 2$.

The solution of the two equations is therefore $\boldsymbol{x = 2, y = 2}$.

Hint

You can check this using the equations (e.g., $\boldsymbol{x = 2, y = 2}$ is also a solution for $\boldsymbol{x + y = 4}$).

Linear simultaneous equations with two unknown quantities can arise in several parts of the A Level Physics course, for example,

- $v = u + at$ in kinematics (see Topic 2.3)
- $V = \varepsilon - Ir$ in electricity (see Topic 10.3)
- $E_{kmax} = hf - \phi$ (see Topic 13.2).

The quadratic equation

Any quadratic equation can be written in the form $ax^2 + bx + c = 0$, where a, b, and c are constants. The general solution of the quadratic equation $ax^2 + bx + c = 0$ is

$$x = \frac{-b \pm \sqrt{(b^2 - 4ac)}}{2a}$$

Note that every quadratic equation has two solutions, one given by the + sign before the square root sign in the above expression, and the other given by the – sign. For example, consider the solution of the equation $2x^2 + 5x - 3 = 0$.

As $a = 2$, $b = 5$, and $c = -3$, the solution is

$$x = \frac{-5 \pm \sqrt{(5^2 - (4 \times 2 \times -3))}}{2 \times 2} = \frac{-5 \pm \sqrt{49}}{4} = \frac{-5 \pm 7}{4} = +\,0.5 \text{ or } -3$$

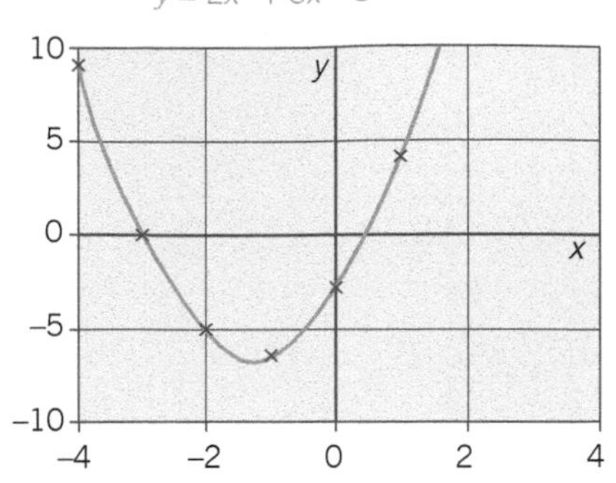

▲ **Figure 2** $y = 2x^2 + 5x - 3$

A graph of $y = 2x^2 + 5x - 3$ is shown in Figure 2. Note that the two solutions above are the values of the x-intercepts, which is where $y = 0$.

Quadratic equations occur in A Level Physics where a formula contains the square of a variable. The equation $s = ut + \frac{1}{2}at^2$ for displacement at constant acceleration is a direct example. Other examples can arise indirectly. For example, suppose the p.d. across a certain type of component varies with current I according to the equation $V = kI^2$. In a circuit with a battery of negligible internal resistance and a resistor of resistance R, the battery p.d., $V_0 = IR + kI^2$. Given values of R, k, and V_0, the current could be calculated using the solution for the quadratic equation with $a = k$, $b = R$, and $c = -V_0$.

Summary questions

1 Solve each of the following pairs of simultaneous equations.

  a $3x + y = 6$, $2y = 5x + 1$

  b $3a - 2b = 8$, $a + b = 2$

  c $5p + 2q = 18$, $q = 2p$

2 Use the data and the given equation to write down a pair of simultaneous equations and so determine the unknown quantities in each case:

  a For $v = u + at$, when $t = 3.0$ s, $v = 8.0\ \text{m s}^{-1}$ and when $t = 6.0$ s, $v = 2.0\ \text{m s}^{-1}$. Determine the values of u and a.

  b For $\varepsilon = IR + Ir$, when $R = 5.0\ \Omega$, $I = 1.5$ A and when $R = 9.0\ \Omega$, $I = 0.9$ A. Determine the values of ε and r.

3 Solve each of the following quadratic equations:

  a $2x^2 + 5x - 3 = 0$

  b $x^2 - 7x + 8 = 0$

  c $3x^2 + 2x - 5 = 0$.

4 Use the data and the given equation to write down a quadratic equation and so determine the unknown quantity in each case:

  a $s = ut + \frac{1}{2}at^2$, where $s = 20$ m, $u = 4\ \text{m s}^{-1}$ and $a = 6\ \text{m s}^{-2}$. Find t.

  b $P = V^2 \frac{R}{(R + r)^2}$, where $P = 16$ W, $V = 12$ V, $r = 2.0\ \Omega$. Find R.

27.3 Logarithms

Logarithms and powers

Any number can be expressed as any other number raised to a particular power. You can use the y^x key on a calculator to show, for example, that $8 = 2^3$ and $9 = 2^{3.17}$. In these examples, 2 is referred to as the base number and is raised to a different power in each case to generate 8 or 9. The power is defined as the **logarithm** of the number generated.

In general, for a number $\boldsymbol{n = \mathrm{b}^p}$, where b is the base number, then $\boldsymbol{p = \log_\mathrm{b} n}$ where $\log_\mathrm{b}$ means a logarithm using b as the base number.

Note:
$\log_\mathrm{b}(\mathrm{b}^p) = p$ as $\mathrm{b}^p = n$ and $\log_\mathrm{b} n = p$.

Applying the general definition above gives the following rules to remember when working with logs:

1 **For any two numbers *m* and *n*,**

$\boldsymbol{\log_\mathrm{b}(nm) = \log_\mathrm{b} n + \log_\mathrm{b} m}$

Let $p = \log_\mathrm{b} n$ and let $q = \log_\mathrm{b} m$ so $n = \mathrm{b}^p$ and $m = \mathrm{b}^q$.

Therefore, $nm = \mathrm{b}^p\mathrm{b}^q = \mathrm{b}^{p+q}$ so $\log_\mathrm{b}(nm) = p + q = \log_\mathrm{b} m + \log_\mathrm{b} n$

2 **For any two numbers *m* and *n*,**

$\boldsymbol{\log_\mathrm{b}\left(\frac{n}{m}\right) = \log_\mathrm{b} n - \log_\mathrm{b} m}$

Let $p = \log_\mathrm{b} n$ and let $q = \log_\mathrm{b} m$ so $n = \mathrm{b}^p$ and $m = \mathrm{b}^q$.

Therefore, $\frac{1}{m} = \frac{1}{b^q} = \mathrm{b}^{-q}$

so $\frac{n}{m} = \mathrm{b}^p\mathrm{b}^{-q} = \mathrm{b}^{p-q}$ so $\log_\mathrm{b}\left(\frac{n}{m}\right) = p - q = \log_\mathrm{b} n - \log_\mathrm{b} m$

3 **For any number *m* raised to a power *p*,**

$\boldsymbol{\log_\mathrm{b}(m^p) = p \log_\mathrm{b} m}$

This is because $m^p = m$ multiplied by itself p times.

$$\text{Therefore } \log_\mathrm{b} m^p = \underbrace{\{\log_\mathrm{b} m + \log_\mathrm{b} m + \ldots + \log_\mathrm{b} m\}}_{\longleftarrow\ p \text{ terms}\ \longrightarrow} = p \log_\mathrm{b} m$$

Base 10 logarithms and natural logarithms are used extensively in physics and are explained below.

Base 10 logarithms

Base 10 logs are written as $\log_{10}$ or lg (or sometimes incorrectly as log).

For example,

- $100 = 10^2$ so $\log_{10} 100 = 2$
- $50 = 10^{1.699}$ so $\log_{10} 50 = 1.699$
- $10 = 10^1$ so $\log_{10} 10 = 1$
- $5 = 10^{0.699}$ so $\log_{10} 5 = 0.699$.

The base 10 examples illustrate the product rule for logs

(i.e., $\log_b nm = \log_b n + \log_b m$)

since $\log_{10} 50 = \log_{10} 5 + \log_{10} 10 = 0.699 + 1 = 1.699$.

Uses of base 10 logs

In graphs, where a **logarithmic scale** is necessary to show the full range of a variable that covers a very wide range, as shown in Figure 2. Notice in Figure 2 that the frequency increases by ×10 in equal intervals along the horizontal axis.

In data analysis, where a relationship between two variables is of the form $y = kx^n$ and k and n are unknown constants. As explained on the previous page, for an equation of the form $y = kx^n$, then

$$\log_{10} y = \log_{10} k + \log_{10} x^n = \log_{10} k + n \log_{10} x$$

The graph of $\log_{10} y$ (on the vertical axis) against $\log_{10} x$ is therefore a straight line of gradient n with an intercept equal to $\log_{10} k$. Therefore n and k can be determined.

In certain formulae, where a ×10 scale is used. For example, the gain of an amplifier in decibels (dB) is a ×10 scale defined by the formula

$$\text{voltage gain / dB} = 10\log_{10}\left(\frac{V_{out}}{V_{in}}\right)$$

where V_{out} and V_{in} are the output and input voltages, respectively. If $V_{out} = 50V_{in}$, the gain of the amplifier is 17 dB (= $10\log_{10} 50$).

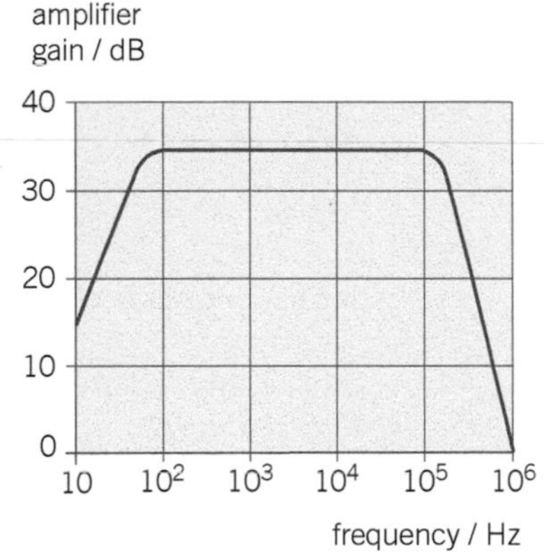

▲ **Figure 2** *Logarithmic scales*

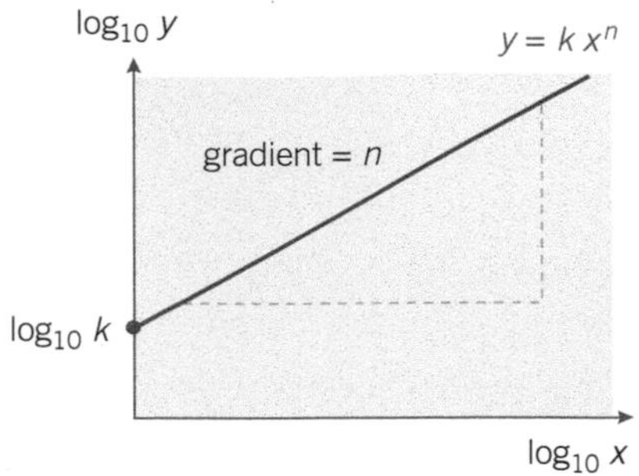

▲ **Figure 1** *Using logs to test $y = kx^n$*

Hint

Note that natural logs could be used in Figure 1 instead of base 10 logs – the gradient would still be n but the y-intercept would be ln k.

Hint

The Richter scale for earthquakes is another example. A Richter scale 8 earthquake is ten times as powerful as a Richter scale 7 earthquake.

Natural logarithms

Natural logs are written as $\log_e$ or ln, where e is the exponential number used as the base of natural logarithms and is equal to 2.718. For example,

- $2.718 = e^1$ so $\ln 2.718 = 1$
- $7.389 = e^2$ so $\ln 7.389 = 2$
- $20.009 = e^3$ so $\ln 20.009 = 3$
- In general, for any number n, if p is such that $n = e^p$, then $\ln n = p$.

Uses of natural logarithms

Natural logs are used in the equations for radioactive decay (Topic 19.5) and capacitor discharge (Topic 19.4) or any other process where the rate

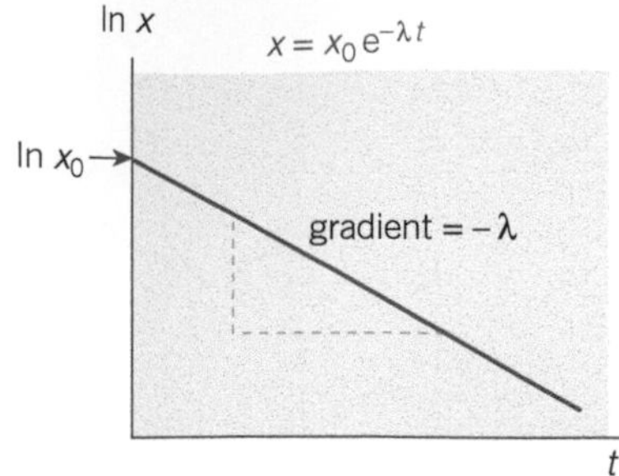

▲ **Figure 3** *A log-linear graph used to test $x = x_0e^{-\lambda t}$*

of change of a quantity is proportional to the quantity itself. For example, the rate of decrease of p.d. across a capacitor discharging through a resistor is proportional to the p.d. across the capacitor. This type of change is described as an exponential decrease because the quantity decreases by the same factor in equal intervals of time.

Applying the general rule (that if p is such that $n = e^p$, then $p = \ln n$) to the equation $x = x_0e^{-\lambda t}$ gives $\ln x = \ln x_0 - \lambda t$.

Therefore, a graph of $\ln x$ (on the vertical axis) against t (on the horizontal axis) is a straight line with a gradient equal to $-\lambda$ and a y-intercept equal to $\ln x_0$.

Comparing the equation for capacitor discharge $Q = Q_0e^{\frac{-t}{RC}}$ with the radioactive decay equation $N = N_0e^{-\lambda t}$,

- for capacitor discharge, $\ln Q = \ln Q_0 - \frac{t}{RC}$ so a graph of $\ln Q$ (on the vertical axis) against t is a straight line which has a gradient $\frac{-1}{RC}$ and $\ln Q_0$ as its y-intercept
- for radioactive decay, $\ln N = \ln N_0 - \lambda t$ so a graph of $\ln N$ (on the vertical axis) against t is a straight line which has a gradient $-\lambda$ and $\ln N_0$ as its y-intercept.

Summary questions

1 **a** Use your calculator to work out
  i $\log_{10} 3$
  ii $\log_{10} 15$.
 b Use your answers in part **a** to work out
  i $\log_{10} 45$
  ii $\log_{10} 5$.

2 The gain of an amplifier, in decibels, is given by the formula $10 \log_{10}\left(\frac{V_{out}}{V_{in}}\right)$.
 a Calculate the gain, in decibels (dB), for
  i $V_{out} = 12V_{in}$
  ii $V_{out} = 5V_{in}$.
 b Show that the gain, in decibels, of an amplifier for which $V_{out} = 60V_{in}$ is equal to the sum of the gain in part **a i** and the gain in part **a ii** above.

3 Write down the gradient and the y-intercept of a line on a graph representing the equation $\log_{10} y = n \log_{10} x + \log_{10} k$ for
 a $y = 3x^5$
 b $y = \frac{1}{2}x^3$
 c $y = x^2$.

4 **a** Use your calculator to work out
  i ln 3
  ii ln 15.
 b Use your answers in part **a** to work out
  i ln 45
  ii ln 5.

27.4 Exponential decrease

Rates of change

Consider a variable quantity y that changes with respect to a second quantity x as shown in Figure 1. The gradient of the curve at any point is the rate of change of y with respect to x at that point. This can be worked out from the graph by drawing a tangent to the curve at that point and measuring the gradient of the tangent. The rate of change of y with respect to x at point P is equal to the gradient of the tangent to the curve at P, which is $\frac{\Delta y}{\Delta x}$.

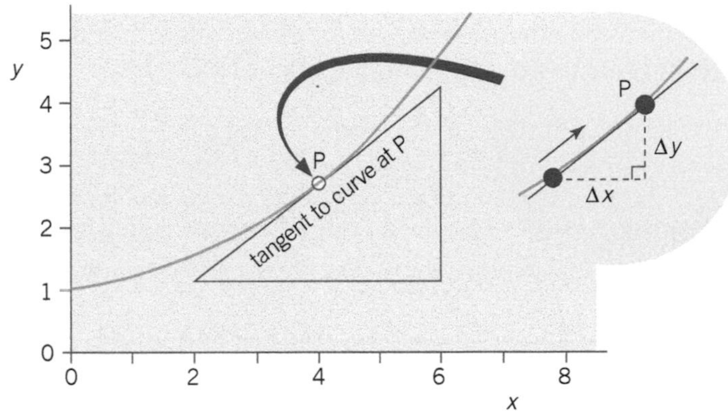

▲ **Figure 1** *Tangents and curves*

The rate of change of y with respect to x can be worked out algebraically if the equation relating y and x is known. This process is known as **differentiation.** For example,

- For $y = x^2$, then increasing x to $x + \Delta x$ increases y to $y + \Delta y$ where $y + \Delta y = (x + \Delta x)^2$

  Multiplying out $(x + \Delta x)^2$ gives $y + \Delta y = x^2 + 2x\Delta x + \Delta x^2$

  Subtracting $y = x^2$ from this equation gives $\Delta y = 2x\Delta x + \Delta x^2$

  Dividing by Δx therefore gives $\frac{\Delta y}{\Delta x} = \frac{2x\Delta x + \Delta x^2}{\Delta x} = 2x + \Delta x$

  Therefore, as $\Delta x \longrightarrow 0$, $\frac{\Delta y}{\Delta x} \longrightarrow 2x$, which is therefore the formula for the gradient at x.

  This is written $\frac{dy}{dx} = 2x$, where $\frac{dy}{dx}$ is the mathematical expression for the rate of change of y with respect to x.

- For the general expression $y = x^n$, it can be shown that $\frac{dy}{dx} = nx^{n-1}$

For example, if $y = 3x^5$, then $\frac{dy}{dx} = 15x^4$.

Hint

Note that differentiation of functions is not required in the AS or the A2 specification. The information on differentiation is provided to help you develop your understanding of exponential change in the next section. But from your studies of simple harmonic motion (Chapter 16), you should know that the gradient of $\sin x$ is $\cos x$ and the gradient of $\cos x$ is $-\sin x$. In other words, differentiating $\sin x$ gives $\cos x$ and differentiating $\cos x$ gives $-\sin x$.

Exponential change

Exponential change happens when the change of a quantity is proportional to the quantity itself. Such a change can be an increase (i.e., exponential growth) or a decrease (i.e., exponential decay).

In both cases, the quantity changes by a fixed proportion in equal intervals of time. The A Level Year 2 specification requires knowledge and understanding of the equations and graphs for exponential decrease in radioactive decay and capacitor discharging and for exponential growth in capacitor charging. The notes below concentrate on the mathematical basis of exponential decrease rather than exponential increase. This is because the equations for exponential increase can be developed from the exponential decrease equations. For example, the exponential increase equations for capacitor charging in Topic 19.4 are developed from the preceding exponential decrease equations for capacitor discharge.

In your studies of capacitor discharge (Topic 19.4) and of radioactive decay (Topic 19.5), you will have met and used the equation $\frac{dx}{dt} = -\lambda x$ and the solution of this equation, $x = x_0 e^{-\lambda t}$.

Let's consider why the equation $\frac{dx}{dt} = -\lambda x$ represents an exponential decrease, which is a change where the variable quantity x decreases with time at a rate in proportion to the quantity.

If x decreases by Δx in time Δt, the rate of change is $\frac{\Delta x}{\Delta t}$. This is written as $\frac{dx}{dt}$ in the limit $\Delta t \longrightarrow 0$.

For an exponential decrease, the rate of change is negative and is proportional to x, therefore $\frac{dx}{dt} = -\lambda x$, where λ is referred to as the decay constant.

Now consider why the solution of this equation is $x = x_0 e^{-\lambda t}$, where x_0 is a constant.

Look at the function:

$$x = x_0 \left[1 + t + \frac{t^2}{2 \times 1} + \frac{t^3}{3 \times 2 \times 1} + \frac{t^4}{4 \times 3 \times 2 \times 1} + \text{similar higher order terms}\right] \ldots$$

Applying the rules of **differentiation** to it gives:

$$\frac{dx}{dt} = x_0 \left[0 + 1 + t + \frac{t^2}{2 \times 1} + \frac{t^3}{3 \times 2 \times 1} + \text{similar higher order terms}\right] \ldots$$

which is the same as x.

So $\frac{dx}{dt} = x$ if x is the above function.

It can be shown that the function in brackets above may be written as n^t, where n is a specific number which is referred to as the exponential number e.

Therefore, $e^t = 1 + t + \frac{t^2}{2 \times 1} + \frac{t^3}{3 \times 2 \times 1} + \frac{t^4}{4 \times 3 \times 2 \times 1}$ + similar higher order terms

To show that the solution of the equation $\frac{dx}{dt} = -\lambda x$ is $x = x_0 e^{-\lambda t}$, divide both sides of the equation by $-\lambda$ to give $\frac{dx}{-\lambda dt} = x$.

Substituting z for $-\lambda t$ therefore gives $\frac{dx}{dz} = x$, which has the solution $x = x_0 e^z = x_0 e^{-\lambda t}$.

The half-life $T_{½}$ of an exponential decrease is the time taken for x to decrease from x_0 to $\frac{1}{2}x_0$.

Note

The value of e, the exponential number, can be worked out by substituting $t = 1$ in the above expression for e^t, giving $e = 1 + 1 + \frac{1}{2} + \frac{1}{6} + \text{etc.} = 2.718$ to four significant figures.

Substituting $x = \frac{1}{2}x_0$ and $t = T_{½}$ into $x = x_0 e^{-\lambda t}$ gives $\frac{x_0}{2} = x_0 e^{-\lambda T_{½}}$

Applying logs to both sides gives $\ln x_0 - \ln 2 = \ln x_0 - \lambda T_{½}$ which simplifies to $\lambda T_{½} = \ln 2$. Therefore,

$$T_{½} = \frac{\ln 2}{\lambda} = \frac{0.693}{\lambda}$$

The time constant τ of an exponential decrease is the time taken for x to decrease from x_0 to $\frac{x_0}{e}$ (= 0.368 x_0 as $\frac{1}{e} = 0.368$).

Substituting $x = \frac{x_0}{e}$ and $t = \tau$ into $x = x_0 e^{-\lambda t}$ gives $\frac{x_0}{e} = x_0 e^{-\lambda \tau}$.

Applying natural logs to both sides gives $\ln x_0 - \ln e = \ln x_0 - \lambda\tau$, which simplifies to $\tau = \frac{1}{\lambda}$ as $\ln e = 1$.

For capacitor discharge, $\lambda = \frac{1}{CR}$, therefore $\tau = \frac{1}{\lambda} = CR$ (see Topic 19.4).

Note

Note that knowledge of the equation for the exponential function $e^t = 1 + t + \frac{t^2}{2 \times 1}$ + etc., is not required in this specification. The information is provided to help you develop your understanding of the exponential number e and the exponential function e^{-x}.

Testing exponential decrease

As explained in Topic 27.3, $\ln(e^{-\lambda t}) = -\lambda t$.

Therefore, $\ln x = \ln(x_0 e^{-\lambda t}) = \ln x_0 + \ln(e^{-\lambda t}) = \ln x_0 - \lambda t$.

Suppose two physical variables x and t are thought to relate to each other through an equation of the form $x = x_0 e^{-\lambda t}$. If so, a graph of $\ln x$ on the y–axis against t on the x-axis would be a straight line in accordance with the equation $\ln x = \ln x_0 - \lambda t$, where the gradient is $-\lambda$ and the y-intercept is $\ln x_0$.

Summary questions

1 **a** For each exponential decrease equation, write down the initial value at $t = 0$ and the decay constant:

- **i** $x = 2e^{-3t}$
- **ii** $x = 12e^{-t/5}$
- **iii** $x = 4e^{-0.02t}$.

b For each exponential decrease equation above, work out the half-life.

2 A radioactive isotope has a half-life of 720 s and it decays to form a stable product. A sample of the isotope is prepared with an initial activity of 12.0 kBq. Calculate the activity of the sample after:

- **a** 1 min
- **b** 5 min
- **c** 1 h.

3 A capacitor of capacitance 22 μF discharges from a p.d. of 12.0 V through a 100 kΩ resistor.

a Calculate:

- **i** the time constant of the discharge circuit
- **ii** the half-life of the exponential decrease.

b Calculate the capacitor p.d.

- **i** 2.0 s and
- **ii** 5.0 s after the discharge started.

4 A certain exponential decrease process is represented by the equation

$x = 1000e^{-5t}$

a **i** Calculate the half-life of the process.

ii Calculate x when $t = 0.5$ s.

b Show that the above equation can be rearranged as an equation of the form $\ln x = a + bt$ and determine the values of a and b.

27.5 Areas and integration

From the Year 1 course, you should recall that the area under a line on a graph can give useful information if the product of the y-variable and the x-variable represents another physical variable. For example, the tension against extension graph for a spring is a straight line through the origin and the area under the line represents the work done to stretch the spring. See Topic 6.2.

Table 1 gives some further examples where the area under a graph has physical significance

▼ **Table 1** *Areas in graphs*

Examples (y-variable first)	Area between the line and the x-axis	Equation	Units
Power against time	energy transferred	energy transferred = power × time	$1\,W = 1\,J\,s^{-1}$
Potential difference against charge (or stored)	electrical energy transferred	electrical energy transferred = p.d. × charge	$1\,V = 1\,J\,C^{-1}$
Force against time	change of momentum (or impulse)	change of momentum = force × time	$1\,kg\,m\,s^{-1} = 1\,N\,s$
Force against distance	work done	work done = force × distance	$1\,J = 1\,N\,m$

To find the area under the line, we can either:

- count the squares of the grid under the line and multiply the number of squares by the amount of the physical variable that one square of area represents, or
- use the mathematical process known as **integration** as outlined below. The notes below are intended to give a deeper understanding of how areas under curves can be calculated. They are not part of the Oxford AQA International specification.

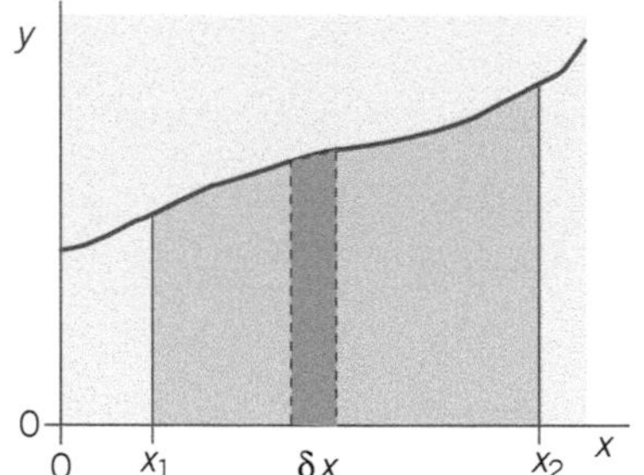

▲ **Figure 1** *Integration*

Consider Figure 1, which shows a y-variable that changes as the x-variable changes. A small increase of the x-variable, δx (δ for small) gives little or no change of the y-variable. The area under that section of the curve, $\delta A = y\,\delta x$ as it is a strip of width δx and height y. Note that rearranging $\delta A = y\,\delta x$ gives $y = \dfrac{\delta A}{\delta x}$.

Therefore the total area under the line from x_1 to x_2 in Figure 1 is equal to the area of all the strips, each of width δx, from x_1 to x_2. The process of adding the individual strip areas together to give the total area is called **integration.**

In mathematical terms,

Total area $A = \int_{x_1}^{x_2} \delta A = \int_{x_1}^{x_2} y\,\delta x$, where $\int$ is the mathematical symbol for integration.

Note

Note that integration of functions is not required in the specification. The information on integration is provided to help you develop your understanding of how areas under graphs can be found precisely if the equation of the line is known.

As $y = \frac{\delta A}{\delta x}$, then differentiating A in terms of x gives y. If we know the formula for y in terms of x, we can find the formula for A in terms of x by using the differentiation formula in Topic 27.4 in reverse.

For example, if $y = 2x$, then using the process of reverse differentiation gives $A = x^2$.

Force-field curves, such as the inverse-square law of force between two point charges, give areas that represent potential energy. We can use the ideas outlined above to obtain an exact formula for the potential energy of two point charges at a certain distance apart.

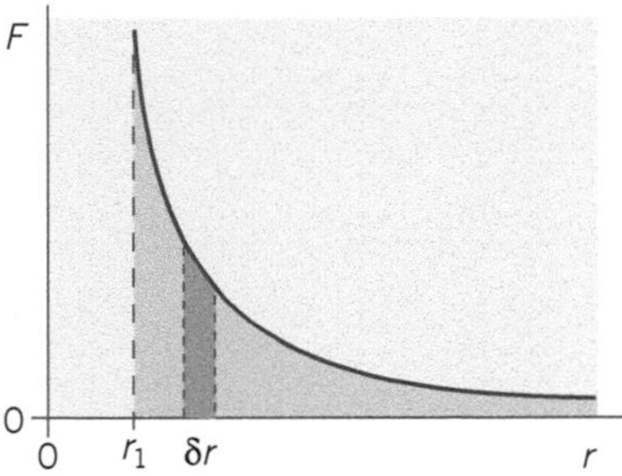

▲ **Figure 2** *The inverse-square law of force*

Consider the two point charges q_1 and q_2 at distance apart r. The force F between the charges is given by Coulomb's law

$$F = \frac{q_1 q_2}{4\pi\varepsilon_0} \times \frac{1}{r^2}$$

If the charges move such that their distance apart changes by δr, the work they do in this movement = $F\delta r$. This is represented on Figure 2 by the narrow strip of width δr.

Since the work they do reduces their potential energy, their change of potential energy $\delta E_P = -F\delta r$.

When the distance apart decreases to r_1 from infinite separation, the potential energy changes from zero at infinite separation to E_P at distance apart r_1. Since E_P is represented by the total area under the line from r = infinity to $r = r_1$

$$E_P = \int_{\text{infinity}}^{r_1} \delta E_P = \int_{\text{infinity}}^{r_1} -F\delta r$$

Because the force is given by the inverse-square law above,

$$E_P = \int_{\text{infinity}}^{r_1} \frac{-k}{r^2}\delta r$$

where $k = \frac{q_1 q_2}{4\pi\varepsilon_0}$.

Therefore, $E_P = \frac{k}{r}$ because differentiating $\frac{1}{r}$ gives $\frac{-1}{r^2}$.

If you need to, look again at Topic 27.4.

The potential energy of the charges at distance r_1 apart is $E_P = \frac{q_1 q_2}{4\pi\varepsilon_0 r_1}$ because the potential energy at infinity is zero.

Note

The inverse-square law also applies to the gravitational force $F = \frac{GMm}{r^2}$ between two point masses M and m. The constant k is written as $-GMm$.

(The minus sign represents the attractive nature of the force.)

Therefore, for a small mass m at distance r from the centre of a spherical planet of mass M at or beyond its surface, the gravitational potential energy $E_P = -\frac{GMm}{r}$.

Summary questions

1 For a velocity–time graph, what physical variable is represented by:

a the gradient? **b** the area under the line?

2 What physical variable is represented by:

a the area under a graph of acceleration against time?

b the area under a graph of current against time?

c What physical variable is represented by the area under a graph of pressure against volume for a gas?

d State the unit of

i pressure

ii volume

iii pressure × volume.

3 For the electric field near a point charge, what physical variable is represented by

a the area under the graph of electric field strength E against distance r in Figure 3a?

b the gradient of the graph of electric potential V against distance r in Figure 3b?

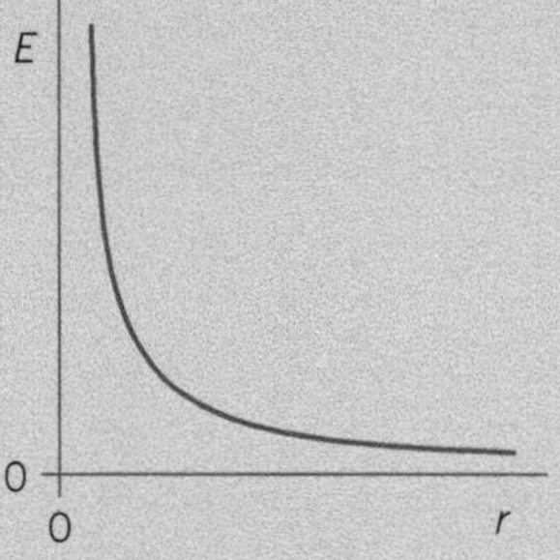

▲ **Figure 3a**

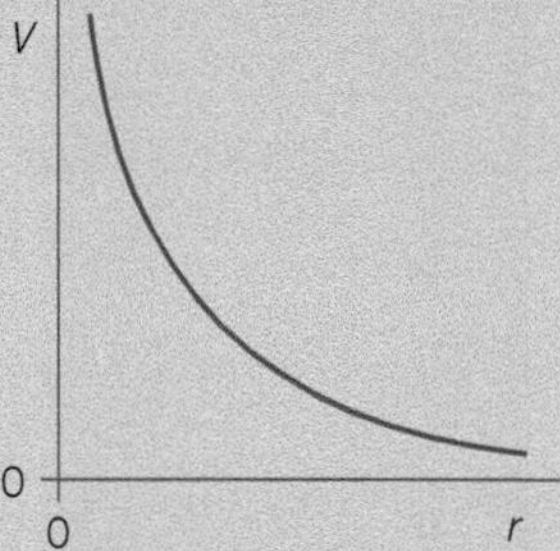

▲ **Figure 3b**

4 a For the gravitational field strength near a spherical object, what physical variable is represented by

i the area under the graph of gravitational field strength g against distance r in Figure 4a?

ii the gradient of the graph of gravitational potential V_{grav} against distance r in Figure 4b?

b Which of the graphs shown in Figures 3 and 4 are inverse-square curves?

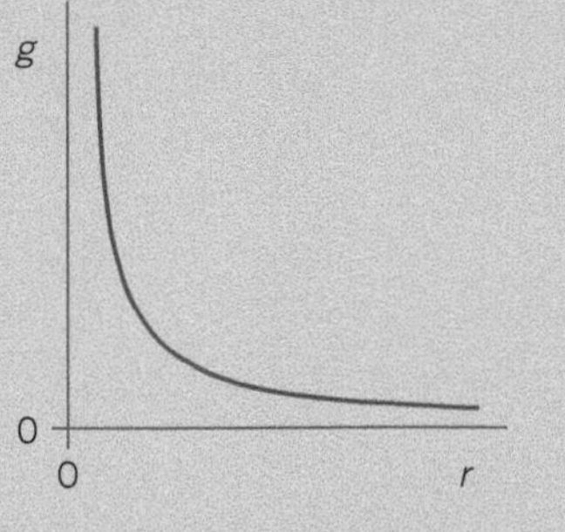

▲ **Figure 4a**

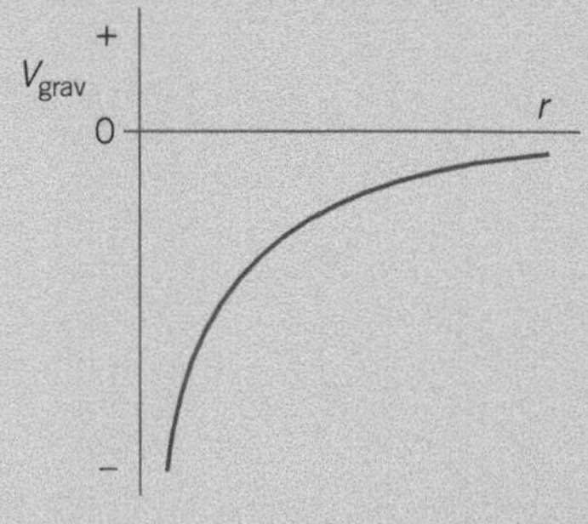

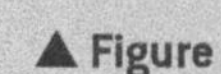

+
V_{grav}
0
r
−

▲ **Figure 4b**

27.6 Graphical and computational modelling

Graphical methods

Useful physical quantities in equations can be worked out directly and indirectly by measuring rates of change from graph gradients. For example, the acceleration of an object can be determined from its displacement–time graph by

- measuring the gradient at two different points on the graph to give the velocity at each point
- dividing the change of velocity between the two points by the time between them.

Figure 1 is a graph of height (h) against time (t) for a projectile. The graph can be used to find g, the acceleration of free fall, by measuring the gradient of the curve at two different times. The gradient at A is zero so, in this example, only the gradient at a second point B needs to be found.

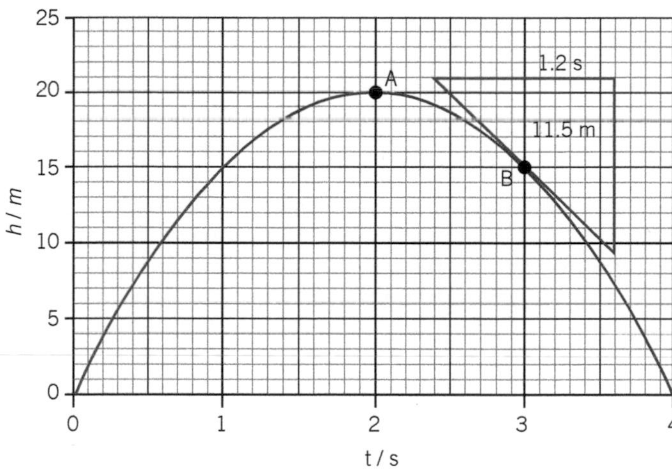

▲ **Figure 1** *Projectile motion*

- The velocity at A = 0.
- The velocity 1.0 s later at B = gradient of the *blue* triangle $= -\dfrac{11.5\,\text{m}}{1.2\,\text{s}}$ $= -9.6\,\text{m}\,\text{s}^{-1}$ to two significant figures.

Therefore, the vertical acceleration $= g = -\dfrac{9.6\,\text{m}\,\text{s}^{-1}}{1.0\,\text{s}} = -9.6\,\text{m}\,\text{s}^{-2}$.

Modelling a physical system

Physicists often try to understand the equations for a physical system by modelling the system. For example, consider the dice model of radioactive decay, which you studied in Topic 8.5. In this model, a large number of dice (1000 to start with) are considered, each one representing a radioactive nucleus. Each time the dice are thrown, all the dice that show '1' uppermost are removed before the remaining dice are thrown. Each throw should result in a sixth of the dice being removed (because each of the six numbers on a dice are equally probable as the uppermost number). Table 1 shows how the number of dice decreases after each of nine throws.

▼ **Table 1**

	Number of dice at the start of each throw, N	Number of dice removed	Number of dice remaining
1st throw	1000	167	833
2nd throw	833	139	694
3rd throw	694	116	578
4th throw	578	96	482
5th throw	482	80	402
6th throw	402	67	335
7th throw	335	56	279
8th throw	279	47	232
9th throw	232	39	193

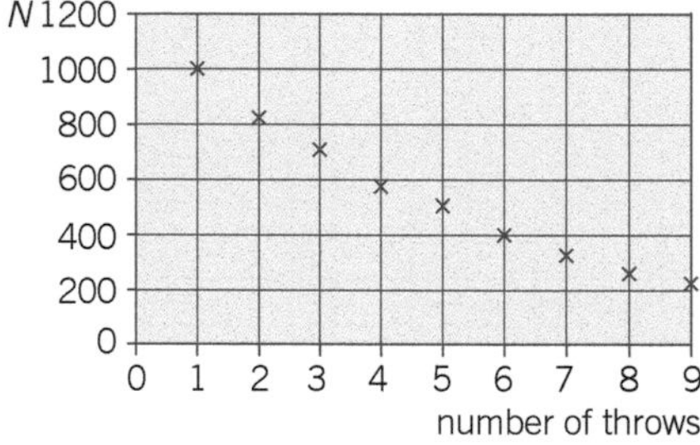

▲ **Figure 2** *Number of dice, N, against number of throws*

Figure 2 shows how the number of dice, N, decreases with the number of throws. A curve drawn through the points would be similar to a radioactive decay curve.

- After about four throws, the number of dice remaining is about half the initial number.
- After about seven to eight throws, the number of dice remaining is about a quarter of the initial number.

The curve is like a radioactive decay curve, so the dice model seems to work – but it isn't good enough to tell us with confidence.

To improve the dice model, you could use a *spreadsheet* to carry out the calculations and to display the results. The use of a spreadsheet also has the advantage that we can change the half-life and use much bigger numbers.

Before setting up the spreadsheet, you need to recall that the rate of change of the number of nuclei N of a radioactive isotope decreases with time t according to the equation

$$\frac{\Delta N}{\Delta t} = -\lambda N$$

where λ is the decay constant, which is the probability per second of a nucleus decaying. See Topic 19.5.

Rearranging this equation gives the change of the number of nuclei of the isotope:

$$\Delta N = -\lambda N \Delta t$$

Comparing this with the dice model, you can see that the product $\lambda \Delta t$ tells us the fraction of the nuclei that decay in time Δt. So $\lambda \Delta t$ is like the fraction $\frac{1}{6}$ in the dice model because this tells us that $\frac{1}{6}$ of the dice are removed after each throw.

By choosing suitable values for λ, Δt, and the initial number of nuclei N_0, you can use a spreadsheet to calculate the number of nuclei remaining after any number of time intervals.

You can set up your own radioactive decay spreadsheet by following the instructions below.

Step 1 Choose values of λ, N_0, and Δt. To compare the results with the dice experiment, choose λ and Δt such that $\lambda\Delta t = \frac{1}{6}$ (= 0.1667 to four significant figures).

- Insert λ = in cell A1, Δt = in cell C1, and N_0 = in cell E1.
- Insert a value of 16.67 (s^{-1}) for λ in cell B1, a value of 0.01 s for Δt in cell D1, and a value of 1000 for N_0 in cell F1.
- Insert column headings for time t, the initial number of nuclei N_i, ΔN, and the final number of nuclei N into cells A3 to D3.

Note

Don't include the unit symbol when you enter a value into a cell.

Step 2 Calculate ΔN and N after the first time interval.

- Insert 0 into cell A4, +F1 into cell B4 and the expression +(B4)*(D1)*(B1) for ΔN into cell C4.
- Insert the expression +B4 – C4 into cell D4 to calculate the final number of nuclei N.

The values of t, N_i, ΔN, and N after the first interval should now be in cells A4 and D4.

Step 3 Repeat the calculation for the second time interval 0.01 s to 0.02 s.

- Insert +A4+D1 into cell A5 to calculate t.
- Insert +D4 into cell B5, and copy cells C4 and D4 into cells C5 and D5.

The values of t, N_i, ΔN, and N after the second interval should now be in cells C5 and D5.

Step 4 Copy cells A5 to D5 into the rows below down to row 49.

- The values of t, N_i, ΔN, and N for successive intervals should now appear in cells C6 to C49 and D6 to D49.

To display the results as a graph, select the block of cells from A4 to B24 at the same time, and select a graph option from the *Insert* menu across the top of the spreadsheet. A graph showing N against t should then appear on the screen, as shown in Figure 3 on the next page.

Study tip

You *don't* need to know how to design your own spreadsheets in your A Level Physics course.

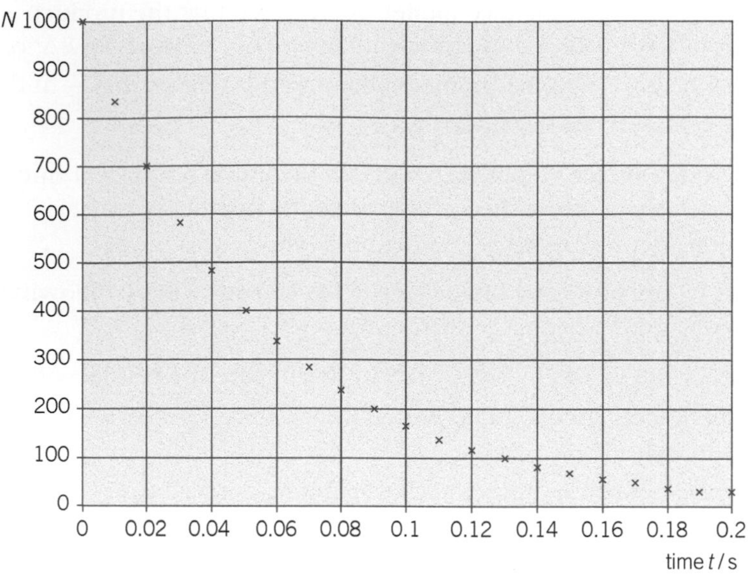

▲ **Figure 3** *A radioactive decay spreadsheet graph*

About spreadsheets

A spreadsheet consists of a matrix of cells labelled alphabetically across the top and numerically down the left-hand side. Data entered in a cell can be processed mathematically and displayed in another cell.

The basic rules for designing a spreadsheet are:

- A number entered into a cell can be processed mathematically and displayed in another cell. For example, if a number is entered into cell A1, the square of the number in cell A1 can be displayed in cell B1 by entering `+A1*A1` (or `+A1^2`) into cell B1.
- The contents of a cell can be copied into other cells. For example, suppose the content of cell B1 is copied into cells B2, B3, B4, and B5. Cells B2 to B5 will then display the square of the numbers entered into cells A2 to A5. See Figure 4.
- Data from a specific cell (e.g., A5) can be copied into other cells by prefixing the cell address with the dollar symbol $. For example, if `+((A1) – ($A$1))` is entered into cell C1 and copied into cells C2 to C5, cell C1 will display 0, and cells C2 to C5 will display the number in cells A2 to A5 minus the number in cell A1, as shown in Figure 4.

	A	B	C
1	0.5	0.25	0
2	1.5	2.25	1
3	2.5	6.25	2
4	3.5	12.25	3
5	4.5	20.25	4

▲ **Figure 4** *A simple spreadsheet*

Summary questions

1. Determine a value of g from Figure 1 by using the gradient given at B and measuring the gradient at 0.6 s.
2. **a** Estimate from Figure 3 the half-life $T_{½}$ of the isotope.

   b Describe and explain what difference it makes to the graph in Figure 3 if the decay constant is halved.

Useful data for A Level Physics

Data

Fundamental constants and values

Quantity	Symbol	Value	Units
Speed of light in vacuo	c	3.00×10^{8}	$\mathrm{m\,s^{-1}}$
Permeability of free space	μ_0	$4\pi \times 10^{-7}$	$\mathrm{H\,m^{-1}}$
Permittivity of free space	ε_0	8.85×10^{-12}	$\mathrm{F\,m^{-1}}$
Magnitude of charge of electron	e	1.60×10^{-19}	C
Planck constant	h	6.63×10^{-34}	J s
Gravitational constant	G	6.67×10^{-11}	$\mathrm{N\,m^{2}\,kg^{-2}}$
Avogadro constant	N_A	6.02×10^{23}	$\mathrm{mol^{-1}}$
Molar gas constant	R	8.31	$\mathrm{J\,K^{-1}\,mol^{-1}}$
Boltzmann constant	k	1.38×10^{-23}	$\mathrm{J\,K^{-1}}$
Stefan constant	σ	5.67×10^{-8}	$\mathrm{W\,m^{-2}\,K^{-4}}$
Wien constant	α	2.90×10^{-3}	m K
Electron rest mass (equivalent to 5.5×10^{-4} u)	m_e	9.11×10^{-31}	kg
Electron charge/mass ratio	$\frac{e}{m_e}$	1.76×10^{11}	$\mathrm{C\,kg^{-1}}$
Proton rest mass (equivalent to 1.007 28 u)	m_p	$1.67(3) \times 10^{-27}$	kg
Proton charge/mass ratio	$\frac{e}{m_p}$	9.58×10^{7}	$\mathrm{C\,kg^{-1}}$
Neutron rest mass (equivalent to 1.008 67 u)	m_n	$1.67(5) \times 10^{-27}$	kg
Gravitational field strength	g	9.81	$\mathrm{N\,kg^{-1}}$
Acceleration due to gravity	g	9.81	$\mathrm{m\,s^{-2}}$
Atomic mass unit (1 u is equivalent to 931.5 MeV)	u	1.661×10^{-27}	kg

Astronomical data

Body	Mass / kg	Mean radius / m
Sun	1.99×10^{30}	6.96×10^{8}
Earth	5.98×10^{24}	6.37×10^{6}

Geometrical equations

arc length = $r\theta$

circumference of circle = $2\pi r$

area of circle = πr^2

surface area of cylinder = $2\pi rh$

volume of cylinder = $\pi r^2 h$

area of sphere = $4\pi r^2$

volume of sphere = $\frac{4}{3}\pi r^3$

Mechanics and materials

moments: moment $= Fd$

velocity and acceleration: $v = \frac{\Delta s}{\Delta t}$ $\quad a = \frac{\Delta v}{\Delta t}$

equations of motion: $v = u + at$

$v^2 = u^2 + 2as$

$s = \frac{(u+v)}{2}t$

$s = ut + \frac{at^2}{2}$

force: $F = ma$

$F = \frac{\Delta(mv)}{\Delta t}$

impulse: $F\,\Delta t = \Delta(mv)$

work, energy and power $W = F s \cos\theta$

$E_K = \frac{1}{2}mv^2 \quad \Delta E_p = mg\Delta h$

$P = \frac{\Delta W}{\Delta t}, P = Fv$

$\text{efficiency} = \frac{\text{useful output power}}{\text{input power}}$

density $\rho = \frac{m}{V}$

Hooke's law $F = k\Delta L$

Young modulus $= \frac{\text{tensile stress}}{\text{tensile strain}}$

tensile stress $= \frac{F}{A}$

tensile strain $= \frac{\Delta L}{L}$

energy stored $E = \frac{1}{2}F\Delta L$

Particles, radiation and radioactivity

the inverse square law for γ radiation $I = \frac{I_0}{r^2}$

Electricity

current and pd $I = \frac{\Delta Q}{\Delta t} \quad V = \frac{W}{Q} \quad R = \frac{V}{I}$

resistivity $\rho = \frac{RA}{L}$

resistors in series $R_T = R_1 + R_2 + R_3 +$

resistors in parallel $\frac{1}{R_T} = \frac{1}{R_1} + \frac{1}{R_2} + \frac{1}{R_3} +$

energy transferred $E = IVt$

power $P = VI = I^2R = \frac{V^2}{R}$

emf $\varepsilon = \frac{E}{Q} \quad \varepsilon = I(R + r)$

Oscillations and waves

for a mass-spring system $T = 2\pi\sqrt{\frac{m}{k}}$

for a simple pendulum $T = 2\pi\sqrt{\frac{l}{g}}$

wave speed $c = f\lambda$ period $T = \frac{1}{f}$

first harmonic $f = \frac{1}{2l}\sqrt{\frac{T}{\mu}}$

fringe spacing $w = \frac{\lambda D}{s}$ diffraction grating $d\sin\theta = n\lambda$

refractive index of a substance $s, n = \frac{c}{c_s}$

for two different substances of refractive indices n_1 and n_2,

law of refraction $n_1 \sin\theta_1 = n_2 \sin\theta_2$

critical angle $\sin\theta_c = \frac{n_2}{n_1}$ for $n_1 > n_2$

photon energy $E = hf = \frac{hc}{\lambda}$

photoelectricity $hf = \phi + E_{k(max)}$

energy levels $hf = E_1 - E_2$

de Broglie Wavelength $\lambda = \frac{h}{p} = \frac{h}{mv}$

Circular motion and periodic motion

magnitude of angular speed $\omega = \frac{v}{r}$

$\omega = 2\pi f$

centripetal acceleration $a = \frac{v^2}{r} = \omega^2 r$

centripetal force $F = \frac{mv^2}{r} = m\omega^2 r$

acceleration $a = -\omega^2 x$

displacement $x = A\cos(\omega t)$

speed $v = \pm\omega\sqrt{A^2 - x^2}$

maximum speed $v_{max} = \omega A$

maximum acceleration $a_{max} = \omega^2 A$

for a mass-spring system $T = 2\pi\sqrt{\frac{m}{k}}$

for a simple pendulum $T = 2\pi\sqrt{\frac{l}{g}}$

Gravitational fields and satellites

force between point masses $F = \frac{Gm_1m_2}{r^2}$

gravitational field strength $g = \frac{F}{m}$

magnitude of gravitational field strength in a radial field $g = \frac{GM}{r^2}$

work done $\Delta W = m\Delta V$

gravitational potential $V = -\frac{GM}{r}$

$g = -\frac{\Delta V}{\Delta r}$

Exponential change

time constant RC

time to halve $\ln 2\,RC$

capacitor charging $Q = Q_0\left(1 - e^{-t/RC}\right)$

capacitor discharging $Q = Q_0 e^{-t/RC}$

radioactive decay $\frac{\Delta N}{\Delta t} = -\lambda N, N = N_0 e^{-\lambda t}$

activity $A = \lambda N$

half-life $T_{½} = \frac{\ln 2}{\lambda}$

Electric fields and capacitance

force between point charges in a vacuum $F = \frac{1}{4\pi\varepsilon_0}\frac{Q_1Q_2}{r^2}$

force on a charge $E = \frac{F}{Q}$

field strength for a uniform field $E = \frac{V}{d}$

field strength for a radial field $E = \frac{Q}{4\pi\varepsilon_0 r^2}$

electric potential $\Delta W = Q\Delta V$

$V = \frac{1}{4\pi\varepsilon_0}\frac{Q}{r}$

$E = \frac{\Delta V}{\Delta r}$

capacitance $C = \frac{Q}{V}$

$C = \frac{A\varepsilon_0\varepsilon_r}{d}$

capacitor energy stored $E = \frac{1}{2}QV = \frac{1}{2}CV2 = \frac{1}{2}\frac{Q^2}{C}$

Magnetic fields

force on a current $F = BIl$

force on a moving charge $F = BQv$

magnetic flux $\Phi = BA$

magnetic flux linkage $N\Phi = BAN\cos\theta$

magnitude of induced emf $\varepsilon = N\frac{\Delta\Phi}{\Delta t}$

$N\Phi = BAN\cos\theta$

emf induced in a rotating coil $\varepsilon = BAN\omega\sin\omega t$

alternating current $I_{rms} = \frac{I_0}{\sqrt{2}}$ $V_{rms} = \frac{V_0}{\sqrt{2}}$

transformer equations $\frac{N_s}{N_p} = \frac{V_s}{V_p}$

efficiency $= \frac{I_sV_s}{I_pV_p}$

Thermal physics

energy to change temperature $Q = mc\Delta\theta$

energy to change state $Q = ml$

gas law $pV = nRT$

$pV = NkT$

Kinetic theory model $pV = \frac{1}{3}Nm\left(c_{rms}\right)^2$

kinetic energy of gas molecule $\frac{1}{2}m\left(c_{rms}\right)^2 = \frac{3}{2}kT = \frac{3RT}{2N_A}$

thermodynamics $Q = \Delta U + W$

$W = p\Delta V$

Nuclear physics

nuclear radius $R = R_0 A^{⅓}$

energy-mass equation $E = mc^2$

Energy sources

moment of inertia $I = mr^2$

$I = \Sigma mr^2$

angular kinetic energy $E_k = \frac{1}{2}I\omega^2$

equations of angular motion $\omega = \omega_0 + \alpha t$

$\omega^2 = \omega_0^2 + 2\alpha\theta$

$\theta = \omega_0 t + \frac{1}{2}\alpha t^2$

$\theta = \frac{\left(\omega_0 + \omega\right)}{2}t$

torque $T = I\alpha$

$T = Fr$

angular momentum angular momentum $= 1\omega$

angular impulse $T\Delta t = \Delta(I\omega)$

work done $W = T\theta$

power $P = T\omega$

maximum power available from a turbine $E = \frac{1}{2}\pi r^2 p v^3$

Solar intensity $I = \frac{P}{4\pi r^2}$

isothermal change $pV = \text{constant}$

Glossary

A glossary of practical terms is given at the end of this glossary. *Italics* is used to denote items appearing elsewhere in the Glossary

A

absolute scale: temperature scale in kelvin (K) defined in terms of *absolute zero*, 0 K, and the triple point of water, 273.16 K, which is the temperature at which ice, water, and water vapour are in thermal equilibrium.

absolute temperature, T: in kelvin = temperature in °C + 273(.15).

absolute zero: the lowest possible temperature, the temperature at which an object has minimum internal energy.

acceleration: change of velocity per unit time.

acceleration of free fall: acceleration of an object acted on only by the force of gravity.

acoustic impedance: the product of the density of a substance and the speed of sound or ultrasound in it.

activity, A: of a radioactive isotope, the number of nuclei of the isotope that disintegrate per second. The unit of activity is the becquerel (Bq), equal to 1 disintegration per second.

alpha (α) decay: change in an unstable nucleus when it emits an α particle, which is a particle consisting of two protons and two neutrons.

alpha radiation: particles that are each composed of two protons and two neutrons. An alpha (α) particle is emitted by a heavy unstable nucleus which is then less unstable as a result. Alpha radiation is easily absorbed by paper, has a range in air of no more than a few centimetres, and is more ionising than beta (β) or gamma (γ) radiation.

amplitude: the maximum displacement from equilibrium of an oscillating object. For a transverse wave, it is the distance from the middle to the peak of the wave.

angular acceleration, a: the rate of change of angular velocity of a rotating object.

angular displacement: the angle an object in circular motion turns through. If its time period is T and its frequency is f, its angular displacement in time t, in radians = $2\pi ft = 2\pi t/T$.

angular frequency, ω: for an object oscillating at frequency f in simple harmonic motion, its angular frequency = $2\pi f$.

angular impulse: the product of torque and time for which the torque acts.

angular momentum, $I\omega$: the product of the moment of inertia and the angular velocity of a rotating object.

angular speed, ω: the rate of change of angular displacement of an object in circular (or orbital or spinning) motion.

annihilation: when a particle and its antiparticle meet, they annihilate or destroy each other and two gamma photons are created.

antimatter: *antiparticles* that each have the same rest mass and, if charged, have equal and opposite charge to the corresponding particle. See *annihilation* and *pair production*.

antineutrino: the antiparticle of the *neutrino.*

antinode: fixed point in a stationary wave pattern where the amplitude is a maximum.

antiparticle: there is an antiparticle for every type of particle. A particle and its corresponding antiparticle have equal rest mass and, if charged, equal and opposite charge.

atomic mass unit, u: correctly referred to as the unified atomic mass constant; one-twelfth of the mass of an atom of the carbon isotope ${}^{12}_{6}C$, equal to 1.661×10^{-27} kg.

atomic number, Z: of an atom of an element is the number of protons in the nucleus of the atom. It is also the order number of the element in the Periodic Table.

Avogadro constant, N_A: the number of atoms in 12 g of the carbon isotope ${}^{12}_{6}C$. N_A is used to define the mole. Its value is 6.02×10^{23} mol^{-1}.

B

back e.m.f.: e.m.f. induced in the spinning coil of an electric motor or in any coil in which the current is changing (e.g., the primary coil of a transformer). A back e.m.f. acts against the change of applied p.d.

background radiation: radiation due to naturally occurring radioactive substances in the environment (e.g., in the ground or in building materials or elsewhere in the environment). Background radiation is also caused by cosmic radiation.

back-up demand: the extra demand for electrical power needed to meet short-term increases in demand.

base load demand: the minimum long-term demand for electrical power.

base units: the units that define the SI system (e.g., the metre, the kilogram, the second, the ampere).

beta (β) decay: change in a nucleus when a neutron changes into a proton, and a β^- particle and an antineutrino are emitted if the nucleus is neutron rich, or a proton changes to a neutron, and a β^+ particle and a neutrino are emitted if the nucleus is proton rich.

beta-minus (β^-) radiation: electrons (β^-) emitted by unstable neutron-rich nuclei (i.e., nuclei with a neutron/proton ratio greater than for stable nuclei). β^- radiation is stopped by about 5 mm of aluminium, has a range in air of up to a metre, and is less ionising than alpha (α) radiation and more ionising than gamma (γ) radiation.

beta-plus (β^+) radiation: *positrons* (β^+) emitted by unstable proton-rich nuclei (i.e., nuclei with a neutron/proton ratio smaller than for stable nuclei). Positrons emitted in solids or liquids travel no further than about 2 mm before they are annihilated.

binding energy of a nucleus: the work that must be done to separate a nucleus into its constituent neutrons and protons. Binding energy = mass defect × c^2. Binding energy in MeV = mass defect in u × 931.5.

binding energy per nucleon: the average work done per nucleon to separate a nucleus into its constituent parts. The binding energy per nucleon of a nucleus = the binding energy of a nucleus ÷ mass number *A*. The binding energy per nucleon is greatest for iron nuclei of mass number about 56. The binding energy curve is a graph of binding energy per nucleon against mass number *A*.

boiling point: the temperature at which a pure liquid at atmospheric pressure boils.

Boltzmann constant, *k*: the molar gas constant divided by the Avogadro number $\left(\text{i.e., } \frac{R}{N_A}\right)$. See *kinetic energy of the molecules of an ideal gas*.

Boyle's law: for a fixed mass of gas at constant temperature, its pressure × its volume is constant. A gas that obeys Boyle's law is said to be an *ideal gas*.

braking distance: the distance travelled by a vehicle in the time taken to stop it.

breaking stress: see *ultimate tensile stress*.

brittle: snaps without stretching or bending when subject to *stress*.

Brownian motion: the random and unpredictable motion of a particle such as a smoke particle caused by molecules of the surrounding substance colliding at random with the particle. Its discovery provided evidence for the existence of atoms.

C

capacitance, *C*: the charge stored per unit p.d. of a capacitor. The unit of capacitance is the farad (F), equal to 1 coulomb per volt. For a capacitor of capacitance *C* at p.d. *V*, the charge stored $Q = CV$.

capacitor discharge: through a fixed resistor of resistance *R*, time constant = *RC*, exponential decrease equation for current or charge or p.d. is given by $x = x_0 e^{\frac{-t}{RC}}$.

capacitor energy: energy stored by the capacitor $E = \frac{1}{2}QV = \frac{1}{2}CV^2 = \frac{1}{2}\frac{Q^2}{C}$.

Celsius scale: temperature, in degrees Celsius or °C, is defined as absolute temperature in kelvin −273.15. This definition means that the temperature of pure melting ice (ice point) is 0 °C, and the temperature of steam at standard atmospheric pressure (steam point) is 100 °C.

centre of mass: the centre of mass of a body is the point through which a single force on the body has no turning effect.

centripetal acceleration: 1. For an object moving at speed *v* (or angular speed *ω*) in uniform circular motion, its centripetal acceleration $a = \frac{v^2}{r} = \omega^2 r$ towards the centre of the circle. 2. For a satellite in a circular orbit, its centripetal acceleration $\frac{v^2}{r} = g$.

centripetal force: the resultant force on an object that moves along a circular path. For an object of mass *m* moving at speed *v* along a circular path of radius *r*, the centripetal force = $\frac{mv^2}{r}$ towards the centre of the circle.

chain reaction: a series of reactions in which each reaction causes a further reaction. In a nuclear reactor, each *fission* event is due to a neutron colliding with a $^{235}_{92}U$ nucleus, which splits and releases two or three further neutrons that can go on to produce further fission. A steady chain reaction occurs when one fission neutron on average from each fission event produces a further fission event.

charge carriers: charged particles that move through a substance when a p.d. is applied across it.

Charles's law: for a fixed mass of an ideal gas at constant pressure, its volume is directly proportional to its *absolute temperature*.

circuit rule for current (Kirchhoff's 1st Law): 1. The current passing through two or more components in series is the same through each component. 2. At a junction, the total current in = the total current out.

circuit rules for p.d. (Kirchhoff's 2nd Law): 1. For two or more components in series, the total p.d. across all the components is equal to the sum of the p.ds across each component. 2. The sum of the e.m.fs round a complete loop in a circuit = the sum of the p.ds round the loop.

coherent: two sources of waves are coherent if they emit waves with a constant phase difference.

collisions: see *elastic collision*.

conservation of angular momentum: for a system of interacting objects, the total angular momentum of the objects is constant, provided no external torque acts.

conservation of momentum: for a system of interacting objects, the total momentum of the objects remains constant provided no external resultant force acts on the system.

contrast medium: a substance used in X-ray imaging that is a good absorber of X-rays.

control rods: rods made of a neutron-absorbing substance such as cadmium or boron that are moved in or out of the core of a nuclear reactor to control the rate of *fission* events in the reactor.

coolant: a fluid that is used to prevent a machine or device from becoming dangerously hot. The coolant of a nuclear reactor is pumped through the core of the reactor to transfer thermal energy from the core to a heat exchanger.

Coulomb's law of force: for two point charges Q_1 and Q_2 at distance apart r, the force F between the two charges is given by the equation $F = \frac{Q_1 Q_2}{4\pi\varepsilon_0 r^2}$, where ε_0 is the *permittivity of free space*.

count rate: the number of counts per unit time detected by a Geiger–Müller tube. Count rates should always be corrected by measuring and subtracting the background count rate (i.e., the count rate with no radioactive source present).

couple: pair of equal and opposite forces acting on a body but not along the same line.

critical angle: the angle of incidence of a light ray must exceed the critical angle for *total internal reflection* to occur.

critical mass: the minimum mass of the fissile *isotope* (e.g., the uranium isotope $^{235}_{92}U$) in a nuclear reactor necessary to produce a *chain reaction*. If the mass of the fissile isotope in the reactor is less than the critical mass, a chain reaction does not occur because too many *fission* neutrons escape from the reactor or are absorbed without fission.

critical temperature: of a superconducting material, temperature at and below which the resistivity of the material is zero.

cycle: interval for a vibrating particle (or a wave) from a certain displacement and velocity to the next time the particle (or wave) that has the same displacement and velocity.

D

damped (oscillations): oscillations that reduce in *amplitude* due to the presence of resistive forces such as friction and drag. 1. For a lightly damped system, the amplitude of oscillations decreases gradually. 2. For a heavily damped system displaced from equilibrium then released, the system slowly returns to equilibrium without oscillating. 3. For a critically damped system, the system returns to equilibrium in the least possible time without oscillating.

de Broglie hypothesis: every matter particle has a dual wave-particle nature characterised by its *de Broglie wavelength*, λ, which is equal to the Planck constant, h, divided by its momentum.

de Broglie wavelength: a particle of matter has a wave-like nature, which means that it can behave as a wave. For example, electrons directed at a thin crystal are diffracted by the crystal. The de Broglie wavelength, λ, of a matter particle depends on its momentum, p, in accordance with de Broglie's equation $\lambda = \frac{h}{p} = \frac{h}{mv}$, where h is the Planck constant.

decay constant, λ: the probability of an individual nucleus decaying per second.

decay curve: an exponential decrease curve showing how the mass or *activity* of a radioactive *isotope* decreases with time.

de-excitation: process in which an atom loses energy by photon emission, as a result of an electron inside an atom moving from an outer shell to an inner shell or in which an excited nucleus emits a gamma photon.

density of a substance: mass per unit volume of the substance.

dielectric: material that increases the capacity of a parallel-plate capacitor to store charge when placed between the plates of the capacitor. Polythene and waxed paper are examples of dielectrics.

dielectric constant: see *relative permittivity*.

differentiation: mathematical process of finding the gradient of a line from its equation.

diffraction: the spreading of waves when they pass through a gap or round an obstacle. X-ray diffraction is used to determine the structure of crystals, metals, and long molecules. Electron diffraction is used to probe the structure of materials. High-energy electron scattering is used to determine the diameter of the nucleus.

diffraction grating: a plate with many closely ruled parallel slits on it.

dispersion: splitting of a beam of white light by a glass prism into colours.

displacement: distance in a given direction.

dissipative forces: forces that transfer energy which is wasted.

dose equivalent: a comparative measure of the effect of each type of *ionising radiation*, defined as the energy that would need to be absorbed per unit mass of matter from 250 kV X-radiation to have the same effect as a certain 'dose' of the ionising radiation. The unit of dose equivalent is the sievert (Sv).

drag force: the force of fluid resistance on an object moving through the fluid.

ductile: stretches easily without breaking.

dynamo rule: see *Fleming's right-hand rule*.

E

eddy currents: induced currents in the metal parts of ac machines.

efficiency: the ratio of useful energy transferred (or the useful work done) by a machine or device to the energy supplied to it.

effort: the force applied to a machine to make it move.

elastic collision: an elastic collision is one in which the total kinetic energy after the collision is equal to the total kinetic energy before the collision.

elastic limit: point beyond which a wire is permanently stretched.

elasticity: property of a solid that enables it to regain its shape after it has been deformed or distorted.

electric field strength, E: at a point in an electric field, is the force per unit charge on a small positively charged object at that point in the field.

electric potential, V: at a point in an electric field, is the work done per unit charge on a small positively charged object to move it from infinity to that point in the field.

electrical conductor: an object that can conduct electricity.

electrically insulating materials: an electrical insulator is a material that cannot conduct electricity. A thermal insulator is a material that is a poor conductor of energy.

electrolysis: process of electrical conduction in a solution or molten compound due to ions moving to the oppositely charged electrode.

electrolyte: a solution or molten compound that conducts electricity.

electromagnetic induction: the generation of an e.m.f. when the *magnetic flux linkage* through a coil changes or a conductor cuts across magnetic field lines.

electromagnetic radiation: see *electromagnetic wave*.

electromagnetic wave: an electric and magnetic wavepacket or *photon* that can travel through free space.

electromotive force (e.m.f.): the amount of electrical energy per unit charge produced inside a source of electrical energy.

electron: a *particle* of rest mass 9.11×10^{-31} kg and electric charge -1.60×10^{-19} C (to 3 significant figures).

electron capture: a proton-rich *nucleus* captures an inner-shell electron to cause a *proton* in the nucleus to change into a *neutron*. An electron *neutrino* is emitted by the nucleus. An X-ray *photon* is subsequently emitted by the atom when the inner shell vacancy is filled.

electron volt: amount of energy equal to 1.6×10^{-19} J defined as the work done when an electron is moved through a p.d. of 1 V.

endoscope: optical fibre device used to see inside cavities.

energy: the capacity to do work. See *work done*.

energy levels: the energy of an electron in an electron shell of an atom or the allowed energies of a nucleus.

equilibrium: state of an object when at rest or in uniform motion.

equipotential: a line or surface in a field along which the electric or gravitational potential is constant.

escape velocity: the minimum velocity an object must be given to escape from the planet when projected vertically from the surface.

excitation: process in which an atom absorbs energy without becoming ionized as a result of an electron inside an atom moving from an inner shell to an outer shell.

excited state: an atom which is not in its ground state (i.e., its lowest energy state).

explosion: when two objects fly apart, the two objects carry away equal and opposite *momentum*.

exponential change: exponential change happens when the change of a quantity is proportional to the quantity itself. For an exponential decrease of a quantity x, $\frac{dx}{dt} = -\lambda x$, where λ is referred to as the decay constant. The solution of this equation is $x = x_0 e^{-\lambda t}$ where x_0 is an initial value of x.

F

Faraday's law of electromagnetic induction: the induced e.m.f. in a circuit is equal to the rate of change of *magnetic flux linkage* through the circuit. For a changing magnetic field in a fixed coil of area A and N turns, the induced e.m.f. $= -NA\frac{\Delta B}{\Delta t}$.

field line: see *line of force or a field line*.

first harmonic: pattern of stationary waves on a string when it vibrates at its lowest possible frequency.

fission: the splitting of a $^{235}_{92}$U *nucleus* or a $^{235}_{94}$Pu nucleus into two approximately equal fragments. Induced fission is fission caused by an incoming *neutron* colliding with a $^{235}_{92}$U nucleus or a $^{235}_{94}$Pu nucleus.

fission neutrons: neutrons released when a *nucleus* undergoes *fission* and which may collide with nuclei to cause further fission.

Fleming's left-hand rule: rule that relates the directions of the force, magnetic field, and current on a current-carrying conductor in a magnetic field.

Fleming's right-hand rule: rule that relates the directions of the induced current, magnetic field, and velocity of the conductor when the conductor cuts across magnetic field lines and an e.m.f. is induced in it.

fluorescence: glow of light from a substance exposed to ultraviolet radiation; the atoms de-excite in stages and emit visible photons in the process.

force: = rate of change of *momentum*

$$= \frac{\text{change of momentum}}{\text{time taken}}$$

(= mass × acceleration for fixed mass).

forced vibrations: vibrations (oscillations) of a system subjected to an external *periodic force*.

free body force diagram: a diagram of an object showing only the forces acting on the object.

free electrons: electrons in a conductor that move about freely inside the metal because they are not attached to a particular atom.

free vibrations: vibrations (oscillations) where there is no damping and no *periodic force* acting on the system, so the *amplitude* of the oscillations is constant.

frequency: of an oscillating object is the number of cycles of oscillation per second.

friction: force opposing the motion of a surface that moves or tries to move across another surface.

fundamental mode of vibration: see *first harmonic*.

fusion (nuclear): the fusing together of light nuclei to form a heavier *nucleus*.

fusion (thermal): the fusing together of metals by melting them together.

G

gamma (γ) radiation: electromagnetic radiation emitted by an unstable nucleus when it becomes more stable. See *pair production*.

geostationary satellite: a satellite that stays above the same point on the Earth's equator as it orbits the Earth because its orbit is in the same plane as the equator, its period is exactly 24 h, and it orbits in the same direction as the Earth's direction of rotation.

gold leaf electroscope: a device used to detect electric charge.

gravitational constant G: the constant of proportionality in *Newton's law of gravitation*.

gravitational field: the region surrounding an object in which it exerts a gravitational force on any other object.

gravitational field strength g: the force per unit mass on a small mass placed in the field. 1. $g = \frac{F}{m}$, where F is the gravitational force on a small mass m. 2. At distance r from a point mass M, $g = \frac{GM}{r^2}$. 3. At or beyond the surface of a sphere of mass M, $g = \frac{GM}{r^2}$, where r is the distance to the centre. 4. At the surface of a sphere of mass M and radius R, $g_s = \frac{GM}{R^2}$.

gravitational force: an attractive force that acts equally on any two objects due to their mass.

gravitational potential, V: at a point in a gravitational field, is the work done per unit mass to move a small object from infinity to that point. At distance r from the centre of a spherical object of mass M, $V = -\frac{GM}{r}$.

gravitational potential energy: at a point in a gravitational field, is the work done to move a small object from infinity to that point. The change of gravitational potential energy of a mass m moved through height h near the Earth's surface is given by $\Delta E_p = mg\Delta h$.

grid system: the network of transformers and cables that is used to distribute electrical power from power stations to users.

ground state: the lowest energy state of an atom.

H

half-life, $T_{\frac{1}{2}}$: the time taken for the mass of a radioactive *isotope* to decrease to half the initial mass or for its *activity* to halve. This is the same as the time taken for the number of nuclei of the isotope to decrease to half the initial number.

Hall probe: a device used to measure *magnetic flux density*.

heat, Q: energy transfer due to a difference of temperature.

heat capacity: the energy needed to raise the temperature of an object by 1 K.

heat exchanger: a steel vessel containing pipes through which hot *coolant* in a sealed circuit is pumped, causing water passing through the steel vessel in separate pipes to turn to steam which is used to drive turbines.

Hooke's law: the extension of a spring is proportional to the force needed to extend it up to a limit referred to as its limit of proportionality.

I

ideal gas: a gas under conditions such that it obeys *Boyle's law*.

ideal gas equation: $pV = nRT$, where p is the gas pressure, V is the gas volume, n is the number of moles of gas, T is the absolute temperature, and R is the *molar gas constant*.

impulse: of a force acting on an object, force × time for which the force acts.

induced e.m.f.: see *electromagnetic induction*.

induced fission: see *fission*.

inertia: resistance of an object to change of its motion.

integration: mathematical process of finding the area under a curve from its mathematical equation.

intensity of radiation: at a surface, is the radiation energy per second per unit area at normal incidence to the surface. The unit of intensity is $J\,s^{-1}\,m^{-2}$ or $W\,m^{-2}$.

interference: formation of points of cancellation and reinforcement where two *coherent* waves pass through each other.

internal energy: of an object, is the sum of the random distribution of the kinetic and potential energies of its molecules.

internal resistance: resistance inside a source of electrical energy – the loss of p.d. per unit current in the source when current passes through it.

inverse-square laws: 1. Force: *Newton's law of gravitation* and *Coulomb's law of force* between electric charges are inverse-square laws because the force between two point objects (masses in the case of gravitation and charge in the case of charges) is inversely proportional to the square of the distance between the two objects. Because these two laws are inverse-square laws, the field strength due to a point mass or a point charge varies with distance according to the inverse of the square of the distance to the point object. 2. Intensity: the intensity of γ radiation from a point source varies with the inverse of the square of the distance from the source. The same rule applies to radiation from any point source that spreads out equally in all directions and is not absorbed. 3. Intensity: of solar radiation is inversely proportional to the square of the distance from the centre of the Sun.

ion: a charged atom.

ionisation: process of creating ions.

ionising radiation: radiation that produces ions in the substances it passes through. It destroys cell membranes and damages vital molecules such as DNA directly or indirectly by creating 'free radical' ions which react with vital molecules.

isotopes: of an element are atoms which have the same number of protons in each nucleus but different numbers of neutrons.

K

Kepler's third law: for any planet, the cube of its mean radius of orbit r is directly proportional to the square of its time period T. Using *Newton's law of gravitation*, it can be shown that $\frac{r^3}{T^2} = \frac{GM}{4\pi^2}$.

kinetic energy: the energy of a moving object due to its motion. For an object of mass m moving at speed v, its kinetic energy $E_K = \frac{1}{2}mv^2$, provided $v \ll c$ (the speed of light in free space).

kinetic energy of the molecules of an ideal gas: 1. Mean kinetic energy of a molecule of an *ideal gas* = $\frac{3}{2}kT$, where the *Boltzmann constant* $k = \frac{R}{N_A}$. 2. Total kinetic energy of n moles of an ideal gas = $\frac{3}{2}nRT$.

kinetic theory equation: $pV = \frac{1}{3}Nm(c_{rms})^2$.

kinetic theory of a gas: 1. Assumptions: a gas consists of identical point molecules which do not attract one another. The molecules are in continual random motion colliding elastically with each other and with the container. 2. The pressure p of N molecules of such a gas in a container of volume V is given by the equation $pV = \frac{1}{3}Nm(c_{rms})^2$, where m is the mass of each molecule and $(c_{rms})^2$ is the mean square speed of the gas molecules. 3. Assuming that the mean kinetic energy of a gas molecule $\frac{1}{2}m(c_{rms})^2 = \frac{3}{2}kT$, where $k = \frac{R}{N_A}$, it can be shown from $pV = \frac{1}{3}Nm(c_{rms})^2$ that $pV = nRT$, which is the *ideal gas equation*.

L

laser: device that produces a parallel *coherent* beam of monochromatic light.

latent heat of fusion: the energy needed to change the state of a solid to a liquid without change of temperature. See *specific latent heat of fusion*.

latent heat of vaporisation: the energy needed to change the state of a liquid to a vapour without change of temperature. See *specific latent heat of vaporisation*.

Lenz's law: when a current is induced by electromagnetic induction, the direction of the induced current is always such as to oppose the change that causes the current.

light-dependent resistor: resistor which is designed to have a resistance that changes with light intensity.

limit of proportionality: the limit beyond which, when a wire or a spring is stretched, its extension is no longer proportional to the force that stretches it.

line of force or a field line: the direction of a line of force or a field line indicates the direction of the force. An electric field line is the path followed by a free positive test charge. The gravitational field lines of a single mass point towards that mass.

linear: two quantities are said to have a linear relationship if the change of one quantity is proportional to the change of the other.

load: the force to be overcome by a machine when it shifts or raises an object.

log graphs: 1. For $y = kx^n$, $\log_{10}y = \log_{10}k + n\log_{10}x$. The graph of $\log_{10}y$ (on the vertical axis) against $\log_{10}x$ is therefore a straight line of gradient n with an intercept equal to $\log_{10}k$. 2. For $x = x_0e^{-\lambda t}$, $\ln x = \ln x_0 - \lambda t$. The graph of $\ln x$ (on the vertical axis) against t is a straight line with a gradient equal to $-\lambda$ and a y-intercept equal to $\ln x_0$.

logarithmic scale: a scale such that equal intervals correspond to a change by a constant factor or multiple (e.g., ×10).

logarithms: for a number $n = b^p$ where b is the base number, then $p = \log_b n$
$\log(nm) = \log n + \log m$
$\log(\frac{n}{m}) = \log n - \log m$
$\log(m^p) = p\log m$
natural logs: for $n = e^p$, then $\ln n = p$
base 10 logs: for $n = 10^p$, then $\log_{10}n = p$.

longitudinal waves: waves with a direction of vibration parallel to the direction of propagation of the waves.

M

magnetic flux, ϕ: $\phi = BA$ for a uniform magnetic field of flux density B that is perpendicular to an area A. The unit of magnetic flux is the weber (Wb).

magnetic flux density, B: the magnetic force per unit length per unit current on a current-carrying conductor at right angles to the field lines. The unit of magnetic flux density is the tesla (T). B is sometimes referred to as the magnetic field strength.

magnetic flux linkage, $N\phi$: through a coil of N turns, $= N\phi = NBA$, where B is the *magnetic flux density* perpendicular to area A. The unit of magnetic flux and of flux linkage is the weber (Wb), equal to $1\,\text{T m}^2$ or $1\,\text{V s}$.

magnetic force: 1. $F = BIl\sin\theta$ gives the force F on a current-carrying wire of length l in a uniform magnetic field B at angle θ to the field lines, where I is the current. The direction of the force is given by *Fleming's left-hand rule* where the field direction is the direction of the field component perpendicular to the wire. 2. $F = BQv\sin\theta$ gives the force F on a particle of charge Q moving through a uniform magnetic field B at speed v in a direction at angle θ to the field. If the velocity of the charged particle is perpendicular to the field, $F = BQv$. The direction of the force is given by *Fleming's left-hand rule*, provided the current is in the direction that positive charge would move in. 3. $BQv = \frac{mv^2}{r}$ gives the radius of the orbit of a charge moving in a direction at right angles to the lines of a magnetic field.

mass: measure of the inertia or resistance to change of motion of an object.

mass defect: of a *nucleus*, is the difference between the mass of the separated nucleons (i.e., protons and neutrons from which the nucleus is composed) and the nucleus.

mass number: see *nucleon number.*

matter waves: the wave-like behaviour of particles of matter.

mean kinetic energy: for a molecule in a gas at absolute temperature T, its mean kinetic energy = $\frac{3}{2}kT$, where k is the *Boltzmann constant* $\left(=\frac{R}{N_A}\right)$.

melting point: the temperature at which a pure substance melts.

metastable state: an excited state of the nuclei of an *isotope* that lasts long enough after α or β emission for the isotope to be separated from the parent isotope (e.g., technetium $^{99}_{43}Tc$).

modal dispersion: the lengthening of a light pulse as it travels along an optical fibre, due to rays that undergo less *total internal reflection.*

moderator: substance in a *thermal nuclear reactor* that slows the *fission* neutrons down so they can go on to produce further fission.

molar gas constant, R: see *ideal gas equation.*

molar mass: the mass of one mole of a substance.

molarity: the number of moles in a certain quantity of a substance. The unit of molarity is the mol.

mole: one mole of a substance consisting of identical particles is the quantity of substance that contains N_A particles of the substance.

moment: of a force about a point, force × perpendicular distance from the line of action of the force to the point.

moment of inertia I: the sum of $m_ir_i^2$ for all the points in a body where m_i is the mass of each point and r_i is its perpendicular distance from the axis of rotation.

momentum: mass × velocity. The unit of momentum is kg m s^{-1}.

motive force: the force that drives a vehicle.

motor effect: the force on a current-carrying conductor due to a magnetic field.

muon: a *particle* which is negatively charged and has a greater rest mass than the electron.

N

natural frequency: the frequency of free oscillations of an oscillating system.

negative temperature coefficient: the resistance of a semiconductor decreases when its temperature is increased.

neutrinos: uncharged particles with a very low rest mass compared with the electron.

neutron: an uncharged particle that has a rest mass of 1.674×10^{-27} kg. Neutrons are in every atomic nucleus except that of hydrogen ${}^{1}_{1}\text{H}$.

Newton's law of gravitation: the gravitational force F between two point masses m_1 and m_2 at distance r apart is given by $F = \frac{Gm_1m_2}{r^2}$.

Newton's laws of motion:
First law: an object continues at rest or in uniform motion unless it is acted on by a resultant force.
Second law: the rate of change of momentum of an object is proportional to the resultant force on it.
Third law: when two objects interact, they exert equal and opposite forces on one another.
Newton's second law may be written as $F = \frac{\Delta p}{\Delta t}$ where p is the momentum ($= mv$) of the object and F is the force in newtons. For constant mass, $\Delta p = m\Delta v$ so $F = \frac{m\Delta v}{\Delta t} = ma$.

node: fixed point in a stationary wave pattern where the *amplitude* is zero.

nuclear fission: see *fission*.

nuclear fusion: see *fusion (nuclear)*.

nucleon: a *neutron* or a *proton* in the *nucleus*.

nucleon number, A: the number of neutrons and protons in a nucleus, also referred to as *mass number*.

nucleus: the relatively small part of an atom where all the atom's positive charge and most of its mass is concentrated.

nuclide: of an isotope, ${}^{A}_{Z}\text{X}$, a *nucleus* composed of Z *protons* and $(A - Z)$ *neutrons*, where Z is the proton number (and also the atomic number of element X) and A is the mass number (or nucleon number, i.e., the number of protons and neutrons in a nucleus).

O

Ohm's law: The p.d. across a metallic conductor is proportional to the current, as long as the physical conditions do not change.

optical fibre: a thin flexible transparent fibre used to carry light pulses from one end to the other.

P

pair production: when a gamma *photon* changes into a particle and an *antiparticle*.

pascal: unit of pressure or *stress* equal to $1\,\text{N}\,\text{m}^{-2}$.

path difference: the difference in distances from two *coherent* sources to an interference fringe.

period of a wave: time for one complete cycle of a wave to pass a point.

periodic force: a *force* that varies regularly in magnitude with a definite time period.

permittivity of free space, ε_0: the charge per unit area in coulombs per square metre on oppositely charged parallel plates in a vacuum when the *electric field strength* between the plates is 1 volt per metre. See *Coulomb's law of force*.

phase difference: in radians, for two objects oscillating with the same time period, T_p, the phase difference $= \frac{2\pi\Delta t}{T_p}$, where Δt is the time between successive instants when the two objects are at maximum displacement in the same direction.

photoelectric effect: emission of electrons from a metal surface when the surface is illuminated by light of frequency greater than a minimum value known as the *threshold frequency*.

photon: electromagnetic radiation consists of photons. Each photon is a wavepacket of electromagnetic radiation. The energy of a photon $E = hf$, where f is the frequency of the radiation and h is the Planck constant.

piezo-electric effect: property of certain solids whereby a p.d. applied between opposite faces causes a change of distance between the two faces. Also, a p.d. is created when opposite faces are squeezed.

plane-polarised waves: *transverse waves* that vibrate in one plane only.

plastic deformation: deformation of a solid beyond its *elastic limit*.

polarised: the positive charge and the negative charge of a polarised molecule are displaced in opposite directions.

positive temperature coefficient: the resistance of a metal increases when its temperature is increased.

positron: a particle of antimatter that is the *antiparticle* of the electron.

potential difference (p.d.): work done or energy transfer per unit charge between two points when charge moves from one point to the other.

potential divider: two or more resistors in series connected to a source of p.d.

potential energy: the energy of an object due to its position.

potential gradient: at a point in a field, is the change of potential per unit change of distance along the field line at that point. The potential gradient = – the field strength at any point.

power: rate of transfer of energy = $\frac{\text{energy transferred}}{\text{time taken}}$

pressure: the force per unit area that a gas or a liquid or a solid at rest exerts normally on (i.e., at right angles to) a surface. Pressure is measured in pascals (Pa), where $1\,\text{Pa} = 1\,\text{N}\,\text{m}^{-2}$.

pressure law: for a fixed mass of an *ideal gas* at constant volume, its pressure is directly proportional to its *absolute temperature*.

principle of conservation of energy: in any change, the total amount of energy after the change is always equal to the total amount of energy before the change.

principle of conservation of momentum: when two or more bodies interact, the total *momentum* is unchanged, provided no external forces act on the bodies.

principle of moments: for an object in equilibrium, the sum of the clockwise moments about any point = the sum of the anticlockwise moments about that point.

progressive waves: waves which travel through a substance or through space if electromagnetic.

projectile: a projected object in motion acted on only by the force of gravity.

proton: a particle that has equal and opposite charge to the *electron* and has a rest mass of $1.673 \times 10^{-27}\,\text{kg}$ which is about 1836 times that of the electron. Protons are in every atomic nucleus. The nucleus of hydrogen, ${}^{1}_{1}\text{H}$, is a single proton. The proton is the only stable baryon.

proton number: see *atomic number*.

R

radial field: a field in which the field lines are straight and converge or diverge as if from a single point.

radian: 1 radian = $\frac{360}{2\pi}$ degrees, so 2π radians = 360°.

reactor core: the fuel rods and the control rods together with the moderator substance are in a steel vessel through which the *coolant* (which is also the moderator in a pressurised water reactor) is pumped.

refraction: change of direction of a wave when it crosses a boundary where its speed changes.

refractive index: $\frac{\text{speed of light in free space}}{\text{speed of light in the substance}}$

relative permittivity: ratio of the charge stored by a parallel-plate capacitor with *dielectric* filling the space between its plates to the charge stored without the dielectric for the same p.d.

renewable energy: energy from a source that is continually renewed. Examples include hydroelectricity, tidal power, geothermal power, solar power, wave power, and wind power.

resistance: $\frac{\text{p.d.}}{\text{current}}$

resistivity: resistance per unit length × area of cross section.

resonance: the amplitude of vibration of an oscillating system subjected to a periodic force is largest when the periodic force has the same frequency as the resonant frequency of the system. For a lightly damped system, the frequency of the periodic force = natural frequency of the oscillating system. At resonance, the system vibrates such that its velocity is in phase with the periodic force.

resonant frequency: the frequency of an oscillating system in *resonance*.

rest energy: energy due to rest mass m_0, equal to m_0c^2, where c is the speed of light in free space.

root mean square speed, c_{rms}: square root of the mean value of the square of the molecular speeds of the molecules of a gas.

$$c_{\text{rms}} = \left(\frac{c_1^2 + c_2^2 + \ldots + c_N^2}{N}\right)^{\frac{1}{2}}$$

where $c_1, c_2, c_3, \ldots, c_N$ represent the speeds of the individual molecules and N is the number of molecules in the gas.

Rutherford's α-particle scattering experiment: demonstrated that every atom contains a positively charged nucleus which is much smaller than the atom and where all the positive charge and most of the mass of the atom is located.

S

satellite: a small object in orbit round a larger object.

satellite motion: for a *satellite* moving at speed v in a circular orbit of radius r round a planet, its *centripetal acceleration*, $\frac{v^2}{r} = g$. Substituting $v = \frac{2\pi r}{T}$, where T is its time period, and $g = \frac{GM}{r^2}$, where M is the mass of the planet, $T^2 = \left(\frac{4\pi^2}{GM}\right)r^3$. See *geostationary satellite*. See *Kepler's third law*.

scalar: a physical quantity with magnitude only.

semiconductor: a substance in which the number of charge carriers increases when the temperature is raised.

SI system: the scientific system of units.

simple electric motor: an electric motor with an armature consisting of a single coil of insulated wire.

simple harmonic motion: motion of an object if its acceleration is proportional to the displacement of the object from equilibrium and is always directed towards the equilibrium position. 1. The acceleration, a, of an object oscillating in simple harmonic motion is given by $a = -(2\pi f)^2 x = -\omega^2 x$, where x = displacement from equilibrium, f = frequency of oscillations, and ω = the angular frequency = $2\pi f$. 2. The solution of this equation depends on the initial conditions. If $x = 0$ and the object is moving in the + direction at time $t = 0$, then $x = A \sin(2\pi ft)$. If the object is at maximum displacement, $+A$, at time $t = 0$, then $x = A \cos(2\pi ft)$.

simple harmonic motion applications: 1. For a simple pendulum of length L, its time period $T = 2\pi \left(\frac{L}{g}\right)^{\frac{1}{2}}$. 2. For an oscillating mass m on the end of a vertical spring, its time period $T = 2\pi \left(\frac{m}{k}\right)^{\frac{1}{2}}$, where k is the spring constant.

sinusoidal curves: any curve with the same shape as a sine wave (e.g., a cosine curve).

specific charge: charge/mass value of a charged particle.

specific heat capacity, *c*: of a substance, is the energy needed to raise the temperature of 1 kg of the substance by 1 K without change of state. To raise the temperature of mass m of a substance from T_1 to T_2, the energy needed, $Q = mc(T_2 - T_1)$, where c is the specific heat capacity of the substance.

specific latent heat of fusion: of a substance, is the energy needed to change the state of unit mass of a solid to a liquid without change of temperature.

specific latent heat of vaporisation: for a substance, is the energy needed to change the state of unit mass of a liquid to a vapour without change of temperature. To change the state of mass m of a substance without change of temperature, the energy needed $Q = ml$, where l is the specific latent heat of fusion or vaporisation of the substance.

spectrometer: instrument used to measure light wavelengths very accurately.

speed: change of distance per unit time.

stationary waves: wave pattern with *nodes* and *antinodes* formed when two or more *progressive waves* of the same frequency and amplitude pass through each other.

stiffness constant: the force per unit extension needed to extend a wire or a spring.

stopping distance: thinking distance + braking distance.

strain: extension per unit length of a solid when deformed.

stress: force per unit area of cross section in a solid perpendicular to the cross section.

strong interaction: interaction between two hadrons.

strong nuclear force: force that holds the *nucleons* together. It has a range of about 2–3 fm and is attractive down to distances of about 0.5 fm. Below this distance, it is a repulsive force.

sublimation: the change of state when a solid changes to a vapour directly.

superconductor: a material that has zero electrical resistance.

superposition: the effect of two waves adding together when they meet.

T

temperature: the degree of hotness of an object. Defined in terms of fixed points (e.g., the triple point of water = 273.16 K).

terminal speed: the maximum speed reached by an object when the *drag force* on it is equal and opposite to the force causing the motion of the object.

thermal energy: the internal energy of an object due to temperature.

thermal equilibrium: when no overall energy transfer by heating occurs between two objects at the same temperature.

thermal nuclear reactor: nuclear reactor which has a moderator in the core.

thermistor: resistor which is designed to have a resistance that changes with temperature.

thinking distance: the distance travelled by a vehicle in the time it takes the driver to react.

threshold frequency: minimum frequency of light that can cause *photoelectric effect.*

time constant: the time taken for a quantity that decreases exponentially to decrease to 0.37 $\left(= \frac{1}{e}\right)$ of its initial value. For the discharge of a capacitor through a fixed resistor, the time constant = resistance × capacitance.

time period (or period): time taken for one complete cycle of oscillations.

torque *T*: the moment of a force about an axis. The force × the perpendicular distance from the axis to the line of action of the force.

total internal reflection: a light ray travelling in a substance is totally internally reflected at a boundary with a substance of lower *refractive index* if the angle of incidence is greater than a certain value known as the *critical angle.*

transducer: any device that is designed to convert energy from one form to another.

transformer: converts the amplitude of an alternating p.d. to a different value. It consists of two insulated coils, the primary coil and the secondary coil, wound round a soft iron laminated core. **step-down transformer**: a transformer in which the rms p.d. across the secondary coil is less than the rms p.d. applied to the primary coil. **step-up transformer**: a transformer in which the rms p.d. across the secondary coil is greater than the rms p.d. applied to the primary coil.

transformer efficiency: for an ideal transformer (i.e., one that is 100% efficient), the output power (= secondary voltage × secondary current) = the input power (= primary voltage × primary current). Transformer inefficiency is due to: resistance heating of the current in each coil, the heating effect of eddy currents (i.e., unwanted induced currents) in the core, and heating caused by repeated magnetisation and demagnetisation of the core.

transformer rule: the ratio of the secondary voltage to the primary voltage is equal to the ratio of the number of secondary turns to the number of primary turns.

transverse waves: waves with a direction of vibration perpendicular to the direction of propagation of the waves.

types of light spectra:

continuous spectrum – continuous range of colours corresponding to a continuous range of wavelengths.

line emission spectrum – characteristic coloured vertical lines, each corresponding to a certain wavelength.

line absorption spectrum – dark vertical lines against a continuous range of colours, each line corresponding to a certain wavelength.

U

ultrasound: sound waves at frequencies of more than about 18 kHz (the upper frequency limit of the human ear).

ultimate tensile stress: tensile stress needed to break a solid material.

uniform circular motion: motion of an object moving at constant speed along a circular path.

uniform field: a region where the field strength is the same in magnitude and direction at every point in the field. 1. The **electric field** between two oppositely charged parallel plates is uniform. The electric field strength $E = \frac{V}{d}$, where V is the p.d. between the plates and d is the perpendicular distance between the plates. 2. The **gravitational field** of the Earth is uniform over a region which is small compared to the scale of the Earth. 3. The **magnetic field** inside a solenoid carrying a constant current is uniform along and near the axis.

universal constant of gravitation: see *gravitational constant.*

useful energy: energy transferred to where it is wanted when it is wanted.

V

vector: a physical quantity with magnitude and direction.

velocity: change of displacement per unit time.

W

wave–particle duality: 1. Matter particles have a wave-like nature (e.g., electrons directed at a thin crystal are diffracted by the crystal) and particle-like behaviour (e.g., electrons in a beam deflected by a magnetic field). See *de Broglie wavelength*. 2. Photons have a particle-like nature, as shown in the photoelectric effect, as well as a wave-like nature as shown in *diffraction* experiments.

wavefronts: lines of constant phase (e.g., wave crests).

wavelength: the least distance between two adjacent vibrating particles with the same displacement and velocity at the same time (e.g., distance between two adjacent wave peaks).

weak interaction: interaction between two leptons.

weak nuclear force: force responsible for *beta decay*.

weight: the force of gravity acting on an object.

wind shadow: the region of air downwind of a turbine in which the wind is slower than the air unaffected by the wind turbine.

work done: work is energy transferred by means of a force. Work = force × distance moved in the direction of the force. The work done W by a force F when its point of application moves through displacement s at angle θ to the direction of the force is given by $W = Fs\cos\theta$.

work function of a metal: minimum amount of energy needed by an electron to escape from a metal surface.

X

X-rays: electromagnetic radiation of wavelength less than about 1 nm. X-rays are emitted from an X-ray tube as a result of fast-moving electrons from a heated filament as the cathode being stopped on impact with the metal anode. X-rays are ionising and they penetrate matter. Thick lead plates are needed to absorb a beam of X-rays.

Y

yield point: point at which the *stress* in a wire suddenly drops when the wire is subjected to increasing *strain*.

Young modulus: tensile stress/*strain* (assuming the limit of proportionality has not been exceeded). The unit of the Young modulus is the pascal (Pa), which is equal to $1\,N\,m^{-2}$.

Young's fringes: parallel bright and dark fringes observed when light from a narrow slit passes through two closely spaced slits.

Glossary of practical terms

accepted value: value of the most accurate measurement available, sometimes referred to as the true value.

accuracy: closeness of a measurement to the accepted (or true) value (if known).

dependent variable: the variable of which the value is measured for each and every change in the *independent variable*.

error bar: representation of an uncertainty on a graph.

errors: 1. **Random** errors cause readings to be spread about the accepted (or true) value, due to the results varying in an unpredictable way from one result to the next. 2. **Systematic** errors cause readings to differ from the accepted (or true) value by a consistent amout each time a measurement is made.

independent variable: physical quantities whose values are selected or controlled by the experimenter.

linearity: an instrument that gives readings that are directly proportional to the magnitude of the quantity being measured.

mean value of a set of readings: sum of the readings divided by the number of readings.

percentage uncertainty: $= \dfrac{\text{uncertainty}}{\text{mean value}} \times 100\%$

precision of a measurement: precise measurements are ones in which there is very little spread about the mean value. Precision depends only on the extent of random error and gives no indication of how close the results are to the accepted (or true) value.

precision of an instrument: the smallest non-zero reading that can be measured using the instrument, also sometimes referred to as the instrument sensitivity or resolution.

range of a set of readings: the maximum and minimum values of a set of readings.

range of an instrument: the minimum and maximum reading that can be obtained using the instrument.

random error: See *errors*.

reliability: an experiment or measurement is reliable if a consistent value is obtained each time it is repeated under identical conditions. The reliability of an experiment is increased if random and systematic *errors* have been considered and eliminated, and, where approriate, a more precise best-fit line has been obtained.

repeatable: an experiment or measurement that gives the same results when it is repeated by the original experimenter using the same method and equipment.

reproducible: an experiment or measurement that gives the same results when it is repeated by another person or by using different equipment or techniques.

sensitivity of an instrument: output response per unit input quantity.

systematic error: See *errors*.

uncertainty of a measurement: the interval within which the true value can be expected to lie, with a given level of confidence or probability.

valid measurement: measurements that give the required information by an acceptable method.

zero error: any indication that a measuring system gives a false reading when the true value of a measured quantity is zero. A zero error may result in a systematic uncertainty.

Answers to summary questions

Answers to the Practice Questions and Section Questions are available at www.oxfordsecondary.com/oxfordaqaexams-alevel-physics

1.1

1 a 3.7 N at 33° to 3.1 N b 17.1 N at 21° to 16 N
  c 1.4 N at 45° to 3 N and 1 N

2 a 14.0 N in the same direction
  b 6.0 N in the direction of the 10 N force
  c 10.8 N at 22° to the 10 N force
  d 13.5 N at 9.8° to the 10 N force

3 6.1 kN vertically up, 2.2 kN horizontal

4 a 268 N b 225 N

1.2

1 a 7.3 N b 7.3 N at 31.5° to the vertical

2 b i 2.7 N ii 4.7 N

3 a 139 N b 95 N

4 a 73° b 6.8 N

1.3

1 300 N

2 b 6.2 N

3 0.27 m

4 6.75 N

1.4

1 0.51 N at 100 mm mark, 0.69 N at the 800 mm mark

2 a 122 N at 1.0 m end and 108 N at the other end, both vertically upwards
  b 122 N at 1.0 m end and 108 N at the other end, both vertically downwards

3 620 kN, 640 kN

4 a 100 N, 50 N b 150 N

1.5

1 The centre of mass is higher if the upper shelves are filled instead of the lower shelves. If tilted, it will topple over at a smaller angle with the upper shelves full than if they were empty.

2 89 N

3 a 48°
  b Yes, they will raise the overall centre of mass so it will topple on a less steep slope.

1.6

1 a 50 N b 250 N

2 a 1800 N m b 1800 N

3 a 6.0 kN b 10.8 kN

4 1.5 N in the cord at 40° to the vertical, 1.9 N in the other cord

1.7

1 a 15 N b 3.0 N c 10.8 N

2 7 N

3 a 6.8 N b 52°

4 18.0 N

5 a 16.9 kN b 16.9 kN

6 a 6.2 N b 12.2 N

7 Move it further 50 mm away from the pivot.

8 a 6.8 N b 9.8 N

9 a 2200 N b 3100 N

10 b 950 N at X, 750 N at Y

11 a 2820 kN b 1660 kN and 1540 kN (to 3 sf)

12 a 8.0 N and 16.0 N b 38 N and 76 N

13 b 11 kN, 11 kN

14 b 28.4 N

2.1

1 a 80 km h^{-1} b 22 m s^{-1}

2 a 2.5 m s^{-1} b 3.0 m s^{-1}

3 a 2.5×10^4 km h^{-1} b 7.0×10^3 m s^{-1}

4 a 45 000 m b 28.3 m s^{-1}

5 b i 4.0 km
  ii 30 m s^{-1} then 25 m s^{-1} in the opposite direction

2.2

1 a 1.5 m s^{-2} b −2.5 m s^{-2}

2 a 0.45 m s^{-2} b 7.9 m s^{-1}

3 b 0.60 m s^{-2}, 0, −0.40 m s^{-2}

2.3

1 a 2.0 m s^{-2} b 221 m

2 a 43 s b −0.93 m s^{-2}

3 a i 0.2 m s^{-2} ii 90 m
  b i −0.75 m s^{-2} ii 8.0 s
  d 3.0 m s^{-1}

4 a 5.0 m s^{-2} b 7.5 m
  c 18 m d 6.4 m s^{-1}

2.4

1 a 4.0 m b 8.8 m s^{-1}

2 a 3.2 s b 31 m s^{-1}

3 a i 3.9 s ii 38 m s^{-1}

4 a 1.6 m s^{-2} b 3.6 m s^{-1} c 0.64 m

2.5

1 a i 83 s ii 127 s

2 a 600 s

3 a

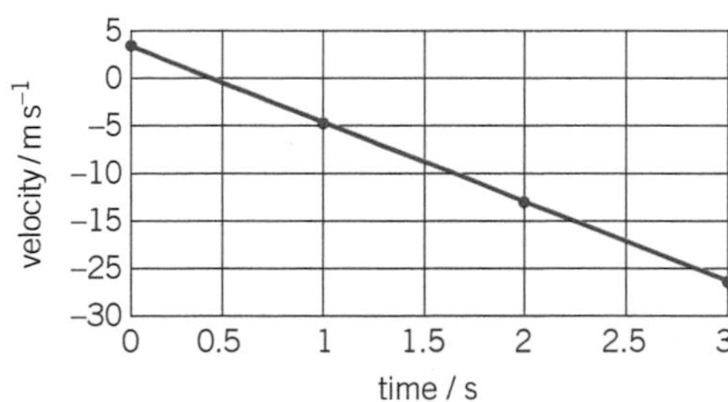

4 a i 0.61 s ii 5.9 m s^{-1} iii 0.43 s iv 4.2 m s^{-1}

2.6

1 a i 52 s ii 0.49 m s^{-2}

b i 406 m ii −1.04 m s^{-2}

2 a 15 m b −0.13 m s^{-2}

c 0.67 m s^{-1} downwards, 13.4 m from the start

3 a i 80 m ii 8.0 m s^{-1}

b i 65 s ii −0.12 m s^{-2}

4 a i 180 m s^{-1} ii 2.7 km

b 4.4 km c 290 m s^{-1}

2.7

1 a 32 m s^{-1} b 2.8 s c 39 m

2 a 3.0 s b 49 m

c 34 m s^{-1} (at 62° to the horizontal)

3 a 0.20 s b 11.7 m s^{-1}

4 a 354 m b i 1020 m ii 1020 m

c 146 m s^{-1}

2.8

1 a 470 mm b 3.0 m s^{-2} c 2.7 m s^{-1}

2 a 2.02 s b 50.5 m

3 a 3.5 m s^{-1}, 3.0 m s^{-1} b i 150 m ii 20 m

3.1

1 a 0.24 m s^{-2} b 190 N c 0.024

2 a 2.4 m s^{-2} b 12 000 N

3 a 360 N b 23 s

4 a -1.3×10^5 m s^{-2} b 260 N

3.2

1 a 5400 N b 7700 N

2 a 60 N b 270 N

3 a 11.8 kN b 11.8 kN c 12.3 kN d 12.3 kN

4 a 1.0 m s^{-2} b 12.5 N

3.3

1 a i 0.04 m s^{-1} ii 1.5 N

3 a 0.14 m s^{-2} b 520 m

3.4

1 a i 7.2 m ii 33.7 m b 4.1 m

2 a 6.75 m s^{-2} b 6750 N

4 b 76 m

3.5

1 a 2.0g b 24 kN

2 a 80 ms b 375 kN

3 a 7.5 m s^{-2} b 6750 N

4 b i 6.4g ii 4250 N

4.1

1 a i 1.2×10^{-18} kg m s^{-1} ii 0.050 kg m s^{-1}

iii 14 kg m s^{-1}

b i 6.0 kg ii 20 m s^{-1}

2 a 3.6×10^5 kg m s^{-1} b 60 s

3 a 5.4×10^6 kg m s^{-1} b 45 s

4 a 9.0×10^3 kg m s^{-1}

b i -8.4×10^3 kg m s^{-1} ii 1.0 m s^{-1}

4.2

1 a 1600 kg m s^{-1} b 3200 N

2 a 3000 kg m s^{-1} b 7.5 kN

3 a -4.2×10^{-23} kg m s^{-1} b -1.9×10^{-13} N

4 a -2.1×10^{-23} kg m s^{-1} b -9.5×10^{-14} N

4.3

1 0.72 m s^{-1}

2 0.7 m s^{-1} in the same direction

3 0.05 m s^{-1} in the direction the 1.0 kg trolley was moving in

4 −0.63 m s^{-1} in the opposite direction to its initial direction

4.4

2 a $9.0\,m\,s^{-1}$ in the same direction b 24 kJ

3 a $1.1\,m\,s^{-1}$ in the reverse direction b 20 J

4 a i $1.0\,m\,s^{-1}$

b The driver of the 250 kg car experiences a force from his/her car which slows him/her down. The other driver experiences a force from his/her car that accelerates him/her.

4.5

1 $0.35\,m\,s^{-1}$

2 a $0.25\,m\,s^{-1}$ – the mass of A and X was greater than the mass of B, so B moved away faster

3 a i $0.10\,m\,s^{-1}$ ii 15 mJ b $0.19\,m\,s^{-1}$

4 a $9.0\,m\,s^{-1}$ b i 1.1 J ii 81 J

5.1

1 a 200 J b 4.5 J

2 a 48 J b 24 J c 0

3 a 1000 J b 600 J c 400 J

4 a 2.4 N b 0.12 J

5.2

1 a 9.0 J b 9.0 J c 1.8 m

2 a 15.7 kJ b 5.8 kJ c 9.9 kJ d 20 N

3 a 590 kJ b 2.4 kJ c 470 kJ d 122 kJ e 1.6 kN

5.3

1 a 1.1 kJ b $62.5\,J\,s^{-1}$

2 500 MW

3 a 156 MJ b 140 MJ c 12 MW

4 122 m

5.4

1 a $450\,J\,s^{-1}$ b $1800\,J\,s^{-1}$

2 a 480 J b 50 J c 10%

3 a 63 W b 31.5%

4 a 600 b 3.7 MJ c 8%

6.1

1 a $8.0 \times 10^{-4}\,m^3$ b $3.1 \times 10^{3}\,kg\,m^{-3}$

2 a 6.3 kg b $2.0 \times 10^{-3}\,m^3$ c $3.1 \times 10^{3}\,kg\,m^{-3}$

3 a $9.6 \times 10^{-6}\,m^3$ b $7.5 \times 10^{-2}\,kg$

4 a i 0.29 kg ii 0.12 kg b $2.3 \times 10^{3}\,kg\,m^{-3}$

6.2

1 a 0.40 m b 12.5 N

2 a 20 N b 100 mm c $200\,N\,m^{-1}$

3 a 40 N b 200 mm

4 a $12.3\,N\,m^{-1}$ b $8.8 \times 10^{-2}\,J$ c 2.2 N

6.3

1 $1.0 \times 10^{9}\,Pa$

2 $1.3 \times 10^{11}\,Pa$

3 a $9.4 \times 10^{8}\,Pa$ b $1.1 \times 10^{-2}\,m$

6.4

1 a $3.3 \times 10^{6}\,Pa$ b $2.8 \times 10^{-4}\,m$ c 0.21 J

2 a 2.3 mm b $1.7 \times 10^{-2}\,J$

3 a 470 kN b 47 J

4 a 10.0 J b 4.2 J

7.1

1 a i 6p, 6n ii 8p, 8n iii 92p, 143n iv 11p, 13n v 29p, 34n

b i $^{235}_{92}U$ ii $^{12}_{6}C$ and $^{16}_{8}O$

2 a neutron b electron c neutron

3 a $+3.2 \times 10^{-19}\,C$ b 63 c $3.04 \times 10^{6}\,C\,kg^{-1}$

4 a $2.67 \times 10^{-26}\,kg$ b 8 neutrons and 10 electrons

7.2

1 a electrostatic force b strong nuclear force

c strong nuclear force d electrostatic force

2 a $^{225}_{88}U$ Ra, $^{4}_{2}\alpha$ b $^{65}_{29}Cu$, $^{0}_{-1}\beta$

3 a Each beta emission increases the proton number by 1 and the alpha emission decreases it by 2, so the number of protons at the end is unchanged, which means the final nuclide is a bismuth nucleus.

b i 209 ii 82p, 127n

4 a A hypothesis is an untested idea or theory.

b i It is uncharged – it hardly interacts with matter.

ii the Sun, a nuclear reactor, a beta-emitting isotope

7.3

1 b i $5.1 \times 10^{14}\,Hz$ ii $1.5 \times 10^{6}\,Hz$

3 a $7.0 \times 10^{14}\,Hz$ b $4.6 \times 10^{-19}\,J$

4 a $4.7 \times 10^{14}\,Hz$, $3.1 \times 10^{-19}\,J$ b 4.8×10^{15}

7.4

1 a 939 MeV (5 sf values for c, e, and h give 938.26 MeV)

2 The rest energy of an electron and a positron is less than 2 MeV but not for a proton–antiproton pair, so pair production can happen for an electron and positron.

3 a 0.511 MeV b i 1.180 MeV

4 a A proton in the nucleus changes into a neutron, and a positron and a neutrino are created and emitted from the nucleus.

b No photon is involved in positron emission, and no neutrino is emitted in pair production.

8.1

4 b 3×10^{-14} m

c 4.5×10^{-14} m

8.2

1 a 42%

b 58%

8.3

1 a $^{235}_{92}\text{U} \rightarrow {}^{234}_{90}\text{Th} + {}^{4}_{2}\alpha$

b $^{228}_{90}\text{Th} \rightarrow {}^{224}_{88}\text{Ra} + {}^{4}_{2}\alpha$

2 a $^{64}_{29}\text{Cu} \rightarrow {}^{64}_{30}\text{Zn} + {}^{0}_{-1}\beta\ (+\bar{\nu})$

b $^{32}_{15}\text{P} \rightarrow {}^{32}_{16}\text{S} + {}^{0}_{-1}\beta\ (+\bar{\nu})$

3 a $^{213}_{84}\text{Po}$, $^{290}_{82}\text{Pb}$, $^{209}_{83}\text{Bi}$

b i 83 p + 130 n

ii 83 p + 126 n

4 a 3.2 counts per second

b 160 mm

8.4

1 Ionisation is a process that causes an atom to become charged as a result of the uncharged atom gaining or losing one of more electrons.

If α radiation from a source outside the body is incident on the skin, the radiation is absorbed by the outer layers of the skin whereas β radiation from a source outside the body penetrates into the body where it destroys or damages body cells.

2 a Ionising radiation is hazardous because it can destroy or damage living cells because it creates ions in the cells which can destroy cell membranes causing the cells to die. Ionising radiation can also damage DNA molecules in cells causing the cells to divide and grow uncontrollably and create tumours.

b i A film badge is designed to record how much of each type of ionising radiation the wearer is exposed to when the badge is worn.

ii A film badge contains a photographic film in a light–proof wrapper. The film is mostly covered in different areas by absorbing materials such as different metals of different thicknesses. The film is blackened by ionising radiation and the extent of the blackening depends on the material over the film. By comparing the blackening of each part of the film, its exposure to different types of ionising radiation can be estimated and the total radiation dose received by the wearer can then be calculated.

3 a i The lead lining absorbs the radiation from the source that reaches the inside surface of the box. Lead lining at least 5 mm thick would absorb α and β particles and most of the γ radiation from the source. Lead is used because it absorbs radiation more effectively than most other materials for the same thickness.

ii Long-handled tongs enables the source to be at a distance from the user when the user is moving the source. α radiation from the source would therefore not reach the user and the intensity of β and γ radiation reaching the user would be much weaker than if the source was closer.

b The source at a fixed position needs to be as far away as possible from the user and should be used only for as long as necessary. It should be returned to its storage box using long-handled tongs as soon as the necessary measurements have been used. The storage box should then be returned without delay to its storage safe.

4 a X-ray machines in hospitals or radioactive isotopes used in hospitals and environmental monitoring or nuclear reactors in nuclear power stations.

b People in a building where radon gas is present would inhale air containing radon gas when they breathe in. The radon gas particles would enter the lungs and the lung tissues would be exposed to a radiation from the radon nuclei when they decay. The ionising effect of this radiation would destroy or damage cells in the lung tissue and in the blood cells pas through the blood vessels in the lung tissue.

8.5

1 a 40 s
 b 6 mg

2 a i 9.0×10^{14}
 ii $2.2(5) \times 10^{14}$
 b i 9.0×10^{14}
 ii 15.8×10^{14}
 c 1.3×10^{3} J

3 a 1.8×10^{10}
 b 1.4×10^{8}

4 a 2.5×10^{-4} kg

8.6

1 b 1320 years

2 a i 7.3×10^{11}
 ii 8.7×10^{11}
 iii 1.9×10^{-13} kg

4 b i 8.8 g
 ii 2.4 J s^{-1}

8.7

1 a A metastable state is a long-lived excited state of a radioactive isotope which decays to the ground state of the isotope by emitting a gamma photon.
 b Since the half-life of the metastable isotope is 6 hours, the number of metastable nuclei after 48 hours is 0.5^8 the initial number of nuclei which is 0.4% of the initial number of metastable nuclei. So the rate of emission of gamma photons is 0.4% of the initial rate. The half-life of the isotope in its ground state is extremely long in comparison so the rate of emission of β particles is insignificant compared with the rate of emission of gamma photons and scarcely contributes to the total rate of emission of gamma photons and beta particles.

2 a 82 p, 126 n b 208

4 a

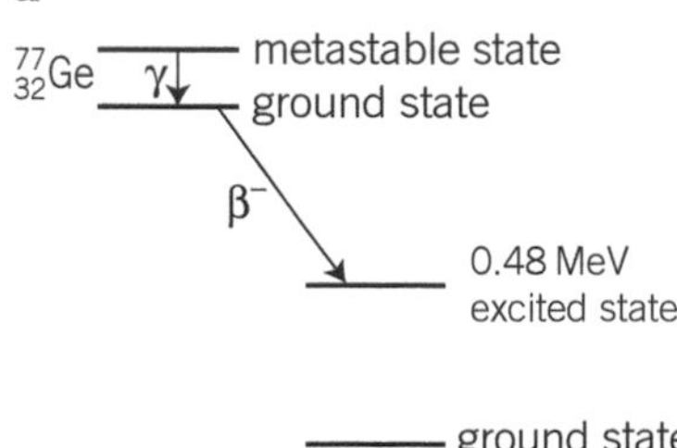

 b 0.21 MeV, 0.27 MeV

9.1

1 a i 3.5 C ii 210 C b i 3.0 A ii 0.15 A

2 a 3.8×10^{15} b 1.9×10^{22}

3 a 72 mC b 4.5×10^{17}

4 a 1600 s b 8000 s

9.2

1 a 29 kJ b 720 J

2 a 2.0 A b 22 kJ

3 i 48 kJ ii 3.5 A

4 a 12 kJ b 4.5 W c 2700 s

9.3

1 a 6.0 Ω, 10 V, 0.125 mA, 160 Ω, 2.5 mA
 b 7.5 Ω

2 31 Ω

3 0.11 mΩ

4 a 1.8×10^{-6} Ω m b 33 mm

9.4

1 a 0.25 A, 12 Ω
 b The filament would become brighter and hotter until it melts and breaks as a result.

2 a 0.03 mA b 0.38 mA

3 a

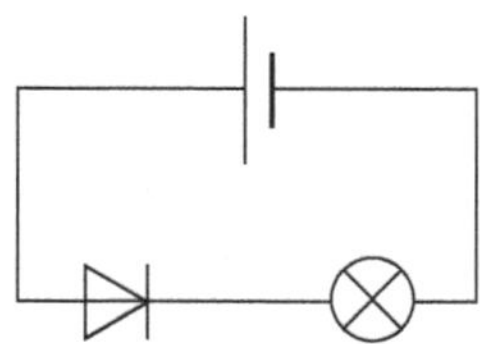

 b The diode would then be 'reverse biased' so the current in the circuit would be negligible.

4 a 30.4 Ω b 46 °C

10.1

1 a 1.0 A, 4.0 A b 5.0 A c 30 W

2 b i 2.0 V ii 0.20 A

3 b i 4.0 V ii 2.0 V iii 10 Ω

4 a 3.6 V b 30 Ω

10.2

1 a 16 Ω b 3.0 Ω c 4 Ω

2 a 2 Ω b 6 Ω c 1.0 A d 4.0 W

3 a 3.6 Ω b 0.83 A
 c 2 Ω: 0.5 W, 4 Ω: 1.0 W, 9 Ω: 1.0 W d 2.5 W

4 a 14.4 W b 2.4 Ω

10.3

1 a 6.0 Ω b 2.0 A c 3.0 V d 9.0 V

2 a 0.5 A b 1.25 V c 0.63 W d 0.13 W

3 a 2.0 Ω b 1.5 V

4 a 12 V, 2 Ω

10.4

1 a 12.0 Ω b 0.25 A

c 4 Ω: 0.25 A, 1.0 V, 24 Ω: 0.08 A, 2.0 V
12 Ω: 0.17 A, 2.0 V

2 a 20.0 Ω b 1.05 A c 1.05 A, 15.8 V

3 i 2.0 W ii 2.0 W

10.5

1 The brightness of the light bulb increases gradually from zero to maximum brightness.

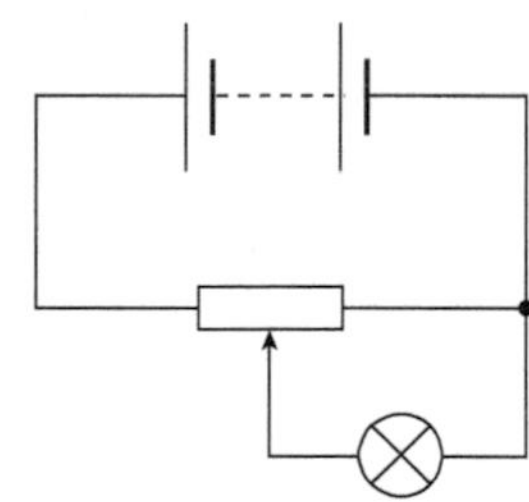

2 a i 0.5 A ii 8.0 Ω: 4.0 V, 4.0 Ω: 2.0 V

b i 3.0 V ii 4.0 V

3 a i 2.8 V iii 6.4 kΩ

11.1

1 His velocity increases uniformly in a downward direction until the rope becomes taut then decreases non-uniformly to zero as the rope stretches until the jumper is at the lowest point. The velocity then increases non-uniformly in an upward direction until the rope becomes slack then the velocity decreases uniformly to zero at the highest point.

2 a Free vibrations are vibrations where no frictional forces are present and the amplitude of vibration is constant.

b Fix a metre rule vertically so the vertical position of the free end of the rule to be tested can be measured. Depress the free end of the rule vertically by a measured distance then release the free end and measure its maximum downward displacement at regular intervals. If the rule is oscillating freely, these measurements should not change.

3 a 0.48 s b 2.1 Hz

4 a i 993 mm ii +20 mm

b 26 N m^{-1}

11.2

1 a i 1.50 s

ii 0.56 m

iii 0.029 J

b See Figure 2

2 a i 60 N m^{-1}

ii 0.54 s

b i 75 mJ

ii 75 mJ

iii 0.50 m s^{-1}

3 a i light ii heavy

4 b 82 mm d 44 mm

11.3

1 a Resonance is the effect of a periodic force acting on an oscillating system when the frequency of the periodic force is such that the system oscillates at its maximum amplitude. This occurs for light damping when the frequency of the periodic force is equal to the natural frequency of oscillation of the system.

b For a mass on a spring, when the system is in resonance at constant amplitude, the periodic force is in phase with the velocity of the mass so the periodic force acts against the damping force and causes the amplitude of the oscillations to increase until the damping force is equal and opposite to the periodic force.

2 a 27 N m^{-1}

b 1.7 Hz

3 The rotating motion of the drum causes a periodic force to act on the panel to make it vibrate. When the drum frequency is equal to the natural frequency of vibration of the panel, the panel vibrates in resonance with the drum so its amplitude of vibration becomes very large. The panel vibrations creates sound waves in the surrounding air and the sound is loudest when resonance occurs.

4 b 2.8 m s^{-1}

11.4

1 longitudinal c; transverse a, b, d

2 See Topic 11.4 Figure 2

3 See Topic 11.4 Figure 3

4 a A polarised wave is a transverse wave in which the vibrations are in the same plane

b The intensity of the light transmitted through the polaroids gradually decreases to zero when the polaroid has been turned through 90°. As the polaroid is turned further in the same direction of rotation, the intensity increases from zero to the same brightness as at the start.

11.5

1 a 0.10 m b 1.9×10^{-2} m

2 a 10 GHz b 5.0×10^{14} Hz

3 1.0 V, 1.0 kHz

4 a i amplitude = 8 mm, wavelength = 47 mm

ii 180° iii 270°

b +8 mm

11.6

1

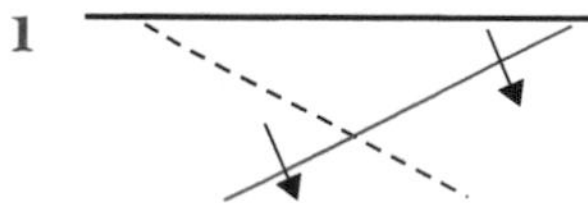

2

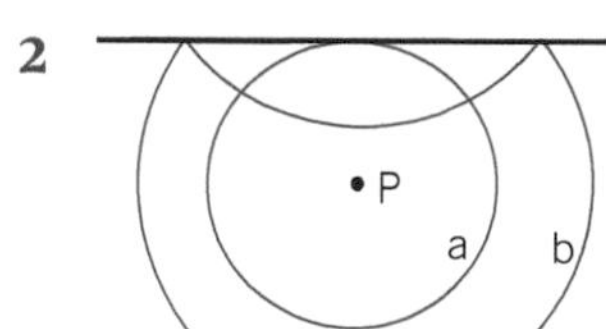

3

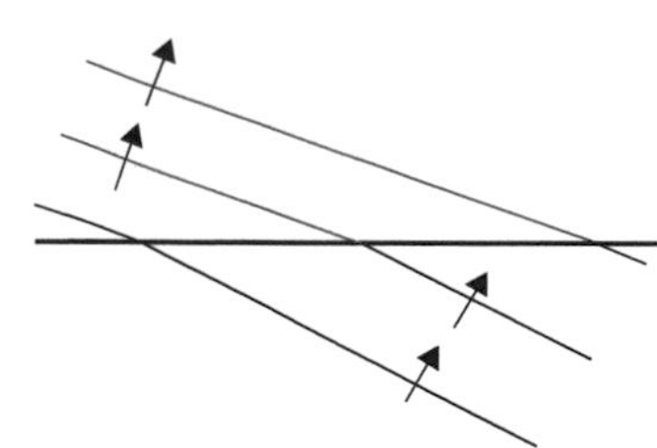

4 a less diffraction b more diffraction

c more diffraction d no change

11.7

1 a

b

2 a The distance along a wavefront between adjacent points of cancellation and reinforcement decreases

b The distance along a wavefront between adjacent points of cancellation and reinforcement increases because the wavelength increases

3 a The signal strength increases to a maximum when the detector is nearest to the middle of the gap then it decreases as it is moved further up. It decreases to a minimum then increases slightly as it is moved further through successive points of minimum intensity.

b If the slit is made narrower without moving the detector, the signal strength increases as the slit narrows

4 a The detector moved to a point of cancellation where the waves from each slit cancel each other out because they are out of phase by 180°.

b The detector moved to a point of reinforcment where the waves from each slit reinforce each other because they are in phase.

11.8

1 b 8.0 m

2 a 2.0 m

b i 180° ii 225° iii 0

4 b 30 mm

11.9

1 a 1.6 m b 410 m s^{-1}

2 a 0.4 m b 0.53 m

3 a 2.4×10^{-4} kg m^{-1}

b 0.20 mm

4 Nylon has a lower density than steel therefore the mass per unit length μ of a nylon wire is less than that of a steel wire of the same diameter. Since the speed of the waves on a wire is inversely proportional to $\frac{1}{\mu^{0.5}}$, the speed of the waves on the nylon wire is greater than on a steel wire of the same diameter. For a nylon wire and a steel wire of the same length, the wavelength of identical stationary wave patterns on each wire is the same. As the frequency of vibration is equal to the $\frac{\text{speed}}{\text{wavelength}}$, the frequency of vibration is therefore greater for the nylon wire than for the steel wire.

11.10

1 a i $5.0\,V\,cm^{-1}$ ii 2.4 cm b i 1.63 V ii 1.15 V

2 a 22 ms b 45.5 Hz

3 a 12.5 V, 8.8 V b 10 ms, 100 Hz

4 A straight horizontal line 2.0 cm above the centre line would be seen.

The waveform seen on the screen has a peak height above the centre of 2.83 cm, and 1.5 cycles would appear on the screen.

11.11

1 a i 0.14 mm ii 0.62 mm

b At this frequency, their wavelength in the body is significant compared with the transducer and with body structures so diffraction is significant. As a result, they spread out and weaken too much so they do not form distinct notes.

2 a i See Figure 1, Topic 11.1.

ii The back block prevents ultrasound waves created at the two surfaces of the disc from cancelling each other out. The pad also damps the vibrations of the disc rapidly after each pulse is emitted.

b i A is due to reflection at the cornea. B is due to reflection at the front surface of the eye lens. C is due to reflection at the back surface of the eye lens. The furthest pulse (at the far right of the screen) is due to reflection at the retina.

ii about 13 mm

3 a i 0.999 ii 1.73×10^{-3}

b Gel and skin have similar acoustic impedances so a gel–skin interface hardly reflects any ultrasound waves from the probe. If the gel was not used, trapped air between the probe and the body would reflect most of the ultrasound waves because air and skin have very different acoustic impedances.

c i $1.64 \times 10^{6}\,kg\,m^{-2}\,s^{-1}$ ii 1.6×10^{-5}

4 a A B scan gives a 2-dimensional image whereas an A scan only gives information about the distances from the probe to reflecting boundaries in one direction. A B scan requires a multi-transducer probe whereas an A scan uses a probe with a single transducer.

b Ultrasound waves are non-ionising unlike X-rays so they would not harm the baby whereas X-rays might.

12.1

2 550 nm

3 0.9 mm

4 0.75 m

12.2

3 1.1 mm

12.3

2 a 6 mm

12.4

1 a 10.9°, 22.2° b 5

2 a 2 b 0.58 (= 35′)

3 a 1090 b 69.9°

4 a $599\,mm^{-1}$ b 3

12.5

1 a i 14.9° ii 28.9° iii 40.6°

b i 58.7° ii See diagram below

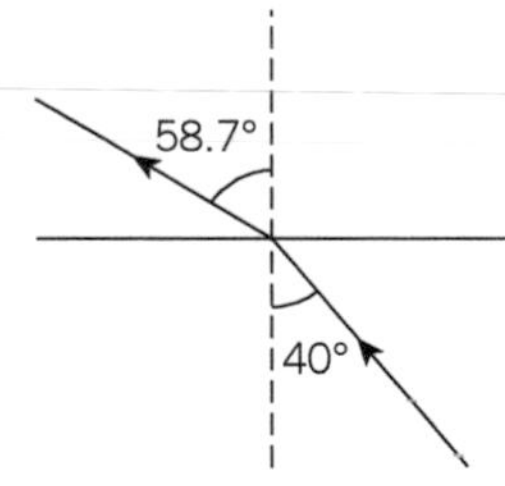

2 a i 19.5° ii 35.2° b i 59.4° ii 74.6°

3 a 1.53 b 34.5°

4 b ii 72.6°

12.6

1 a i 0.040 m ii 0.030 m b 18.5°

2 a i $1.97 \times 10^{8}\,m\,s^{-1}$ ii $2.26 \times 10^{8}\,m\,s^{-1}$

3 a 25.4° b 35°

4 a i red 49.5° ii blue 50.8°

12.7

1 b i 41° ii 49°

2 b i 34° ii 34°

3 a 65° b i 30° ii 45°

13.1

2 a i 6.7×10^{14} Hz, 4.4×10^{-19} J

ii 2.0×10^{14} Hz, 1.3×10^{-19} J

3 a 1.7×10^{14} Hz b 2.7×10^{-19} J

4 a 3.1×10^{-19} J b 1.6×10^{-19} J c 2.5×10^{14} Hz

13.2

2 a 1.6×10^{12}

3 a 3.4×10^{-19} J b 1.5×10^{15} c 2.5×10^{12}

4 a i 4.0×10^{14} Hz ii 2.7×10^{-19} J

b 2.7×10^{-19} J

13.3

1 Similarity – energy is absorbed by the atom. Difference – an electron stays in the atom when excitation occurs but leaves the atom when ionisation occurs.

2 a 1.66×10^{-18} J b 1.6 eV

3 a less b more, increases, decreases

4 a An electron inside the atom moves to an outer shell.

b The electron has insufficient kinetic energy to excite or ionise the atom and so cannot be absorbed.

13.4

1 a 9.0 eV

b There are 3 energy levels below the 5.7 eV level. There are three possible transitions to the ground state, two to the first excited state, and one to the second excited state. The energy changes for these transitions are all different. Therefore there are six possible photon energies.

2 a The energy of an electron in the atom increases in excitation and decreases in de-excitation. Excitation can occur through photon absorption or electron collision. De-excitation only occurs through photon emission.

b i There are two excited levels at 1.8 eV and 4.6 eV above the ground state.

ii photon energies 1.8 eV, 2.8 eV, and 4.6 eV.

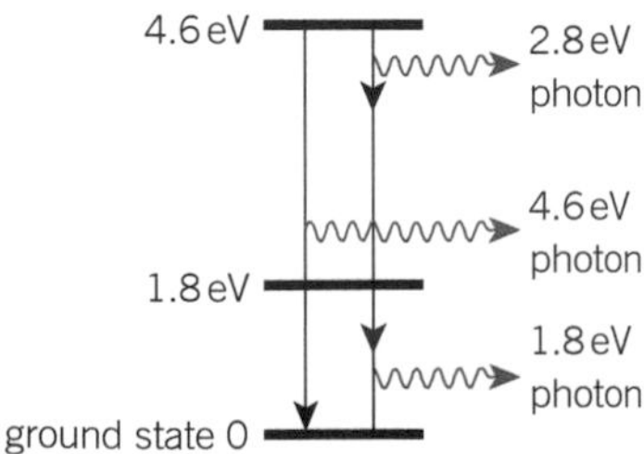

3 a The intermediate level could be at 0.6 eV or at 3.2 eV. The diagram shows one of these alternatives.

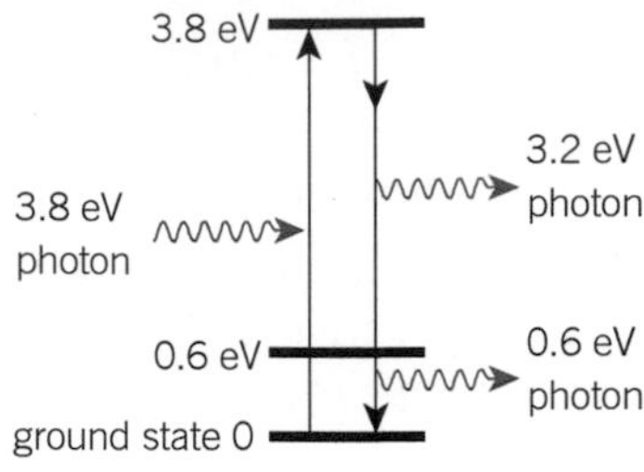

b An electron in the atom absorbs the 3.8 eV photon and moves to an outer shell. It moves to an inner shell and emits a 3.2 eV photon. An electron moves from this shell to the ground state, emitting a 0.6 eV photon.

4 When the electricity supply is switched off, excitation by collision of the gas atoms in the tube ceases because the supply of electrons to the tube is cut off. Therefore the mercury atoms no longer emit ultraviolet radiation and so the coating atoms can no longer emit light.

13.5

1 A line spectrum consists of discrete coloured lines, whereas a continuous spectrum has a continuous spread of colours. A line spectrum is due to photons of certain energies only, whereas a continuous spectrum is due to a continuous spread of photon energies.

2 a 1.28×10^{-19} J b 1.55×10^{-6} m

3 a 3.21×10^{-19} J b 2.0 eV

4 The energy levels of the atoms of an element are unique to those atoms. The photons emitted by them have energies equal to the energy differences between the levels. Therefore the photon energies are characteristic of the atoms and can be used to identify the element.

13.6

1 a i 0.15 W

b i The intensity increases because more electrons strike the anode each second so more X-ray photons are produced.

ii The beam becomes more penetrating because the maximum energy of the X-ray photons in the beam is increased. This happens because the electrons strike the anode with more kinetic energy when the tube voltage is increased.

2 a X-ray photons are emitted in a continuous range of wavelengths upwards from a minimum wavelength that depends on the anode voltage. The intensity of this continuous range of wavelengths increases from zero at the minimum wavelength to a maximum at a longer wavelength then gradually to zero as the wavelength increases. At certain wavelengths, sharp intensity peaks are produced at wavelengths that do not change as the voltage is increased.

b Excitation of atoms in the anode occurs when electrons from the filament collide with them. As a result, electrons in the atoms move temporarily from the innermost shells of the atom to higher energy levels. When these electrons return to their original levels, they emit X-ray photons at energies which are characteristic of the anode atoms.

3 a i Sharpness is to do with how clearly the edges of structures in an X-ray image can be seen.

ii Contrast is to do with the relative darkening of an X-ray film in different areas of the film. Poor contrast occurs if the lightest areas are not much lighter than the darkest areas.

b A scattering grid placed between the patient and the film prevents X-rays scattered in the body of the patient from reaching the film. Without the grid, such X-rays would darken the shadow areas of the film, which would reduce the contrast and sharpness of the image.

4 A contrast medium is necessary for an X-ray image of a soft body organ such as the stomach. If the organ is filled with the contrast medium, it absorbs X-rays more effectively. This improves the contrast of the image so the edges and structures of the image can be seen more clearly.

13.7

3 a 3.6×10^{-11} m b 2.0×10^{-14} m

4 a 1.3×10^{-27} kg m s^{-1}, 1.5×10^{3} m s^{-1}

b 1.3×10^{-27} kg m s^{-1}, 0.78 m s^{-1}

14.1

1 a i 0.500 m ii 320 cm iii 95.6 m

b i 450 g ii 1.997 kg iii 5.4×10^{7} g

c i 2.0×10^{-3} m^2 ii 5.5×10^{-5} m^2

iii 5.0×10^{-6} m^2

2 a i 1.50×10^{11} m ii 3.15×10^{7} s

iii 6.3×10^{-7} m iv 2.57×10^{-8} kg

v 1.50×10^{5} mm vi 1.245×10^{-6} m

b i 35 km ii 650 nm iii 3.4×10^{3} kg

iv 870 MW (= 0.87 GW)

3 a i 20 m s^{-1} ii 20 m s^{-1} iii 1.5×10^{8} m s^{-1}

iv 3.0×10^{4} m s^{-1}

b i $6.0 \times 10^{3}\,\Omega$ ii 5.0 Ω iii $1.7 \times 10^{6}\,\Omega$

iv $4.9 \times 10^{8}\,\Omega$ v 3.0 Ω

4 a i 301 ii 2.8×10^{9} iii 1.9×10^{-23}

iv 1.2×10^{-3} v 2.0×10^{4} vi 7.9×10^{-2}

b i 1.6×10^{-3} ii 5.8×10^{-6}

iii 1.7 iv 3.1×10^{-2}

14.2

1 a 1.57 m b i 1.57 m ii 1.05 m iii 0.26 m

2 a i 68° ii 41° iii 22° iv 61°

b i 17 cm ii 16 m (15.6 m to 3 sf) iii 4.8 mm

iv 101 cm

3 a 49 mm b i 35 km ii 31°

4 a i 3.9 N, 4.6 N ii 3.4 N, 9.4 N iii 4.8 N, 5.7 N

b 4.0 N, 30° to 3.5 N

14.3

1 a 0.2 b 0.1, 0.25 c < 0.1

2 b i 2 ii −1 iii 4.25 iv $\frac{1}{3}$

3 b i 0.25 ii ±2.8 iii ±0.5 iv 4

4 a 3.1×10^{-7} m^3

b 6.2×10^{-3} m

c 2.5 s

d 17 m s^{-1}

14.4

1 a i 3 ii −3 iii 1

b i −4 ii 8 iii 2

c i −1 ii 5 iii 5

d i −1.5 ii 3 iii 2

2 a i $y = 2x - 8$ ii −8

b i 3 m s^{-2} ii 5 m s^{-1}

3 a (1, 4) b $y = 4x$

4 a $x = 2, y = 0$ b $x = 3, y = 5$ c $x = 2, y = 0$

14.5

3 b i use gradient $= \frac{A}{\rho L}$ ii use gradient $= \frac{VA}{\rho}$

4 b ii $u = y$-intercept, $\frac{1}{2}a$ = gradient

14.6

1 b i gradient ii area under line

2 b resistance = $\frac{1}{\text{gradient}}$

3 b The power is constant and is represented by the gradient of the line.

4 b i acceleration

ii distance fallen

c velocity

15.1

1 a 1.75×10^{-3} rad b 0.105 rad

c 6.28 rad

2 a 20 ms

b i 0.31 rad

ii 310 rad

3 a $465\,\text{m}\,\text{s}^{-1}$

b i 0.0042°

ii 7.3×10^{-5} rad

4 a $7.0\,\text{km}\,\text{s}^{-1}$

b i 0.050°

ii 8.7×10^{-4} rad

15.2

1 a $0.23\,\text{m}\,\text{s}^{-1}$

b i $7.9 \times 10^{-4}\,\text{m}\,\text{s}^{-2}$

ii 5.1×10^{-2} N

2 a $0.53\,\text{m}\,\text{s}^{-1}$, $0.66\,\text{m}\,\text{s}^{-2}$

b 9.9×10^{-2} N

3 a i $3.0 \times 10^{4}\,\text{m}\,\text{s}^{-1}$

ii $6.0 \times 10^{-3}\,\text{m}\,\text{s}^{-2}$

b i $7.9 \times 10^{3}\,\text{m}\,\text{s}^{-1}$

ii 5.1×10^{3} s

4 a $8.4\,\text{m}\,\text{s}^{-1}$

b $88\,\text{m}\,\text{s}^{-2}$

c 175 N

15.3

1 a $6.7\,\text{m}\,\text{s}^{-2}$

b 3.8 kN

2 a $4.1\,\text{m}\,\text{s}^{-2}$

b 3.0 kN

3 a $40\,\text{m}\,\text{s}^{-1}$

15.4

1 a $30\,\text{m}\,\text{s}^{-1}$

b i $11.3\,\text{m}\,\text{s}^{-2}$

ii 690 N

2 a $25\,\text{m}\,\text{s}^{-1}$

b $20\,\text{m}\,\text{s}^{-2}$

c 2000 N

3 a $13\,\text{m}\,\text{s}^{-1}$

b $13\,\text{m}\,\text{s}^{-2}$

c 240 N

4 −0.04 N

16.1

1 a +25 mm, changing direction from up to down

b 0, moving down

c −25 mm, changing direction from down to up

d 0, moving up

2 a 0.5 Hz

b i $-0.25\,\text{m}\,\text{s}^{-2}$

ii 0

iii $0.25\,\text{m}\,\text{s}^{-2}$

3 a 0.5 Hz

b $-0.32\,\text{m}\,\text{s}^{-2}$

4 a −32 mm $0.32\,\text{m}\,\text{s}^{-2}$

b 0, 0

16.2

1 a 0.33 Hz

b $0.25\,\text{m}\,\text{s}^{-2}$

2 a i 12 mm

ii 0.63 s

b 6.5 mm

3 a 2.1 Hz

b 0.057 m

4 a 3.7 Hz

b i −8.2 mm towards maximum negative displacement

ii −0.7 mm towards maximum positive displacement

16.3

1 a i 0.33 s

ii 3.1 Hz

b acceleration: i 0 ii $-3.7\,\mathrm{m\,s^{-2}}$
iii $-7.5\,\mathrm{m\,s^{-2}}$
speed: i $0.39\,\mathrm{m\,s^{-1}}$
ii $0.34\,\mathrm{m\,s^{-1}}$ iii 0

2 a i 3.0 Hz
ii 0.33 s
b $f_2 < f_1 \therefore m_2 > m_1$

3 a i 70 mm
ii $21\,\mathrm{N\,m^{-1}}$
b ii 0.53 s

4 a i 1.25 N
ii $2.5\,\mathrm{m\,s^{-2}}$
b ii 1.1 Hz, +47 mm

5 a i 2.0 s
ii 1.0 s
b 5.0 s

17.1

1 a $1.3 \times 10^{-6}\,\mathrm{N}$
b 5.4 mm

2 a 780 N
b $6.0 \times 10^{24}\,\mathrm{kg}$

3 a 54 N
b 0.24 N

4 a i 16.6 N
ii 0.2 N
b 16.4 N towards the centre of the Earth

17.2

1 a i 33 N
ii 160 N
b i $16\,\mathrm{N\,kg^{-1}}$ ii $4.0\,\mathrm{N\,kg^{-1}}$

17.3

1 a 235 J

2 a $2.0\,\mathrm{MJ\,kg^{-1}}$
b i $-61\,\mathrm{MJ\,kg^{-1}}$
ii $2.2 \times 10^{9}\,\mathrm{J}$

3 a i −250 J
ii −200 J
iii −200 J
b i 50 J
ii 0

4 b $5\,\mathrm{N\,kg^{-1}}$
c 25 MJ

17.4

1 a $7.35 \times 10^{22}\,\mathrm{kg}$

2 a i $272\,\mathrm{N\,kg^{-1}}$
ii $5.9 \times 10^{-3}\,\mathrm{N\,kg^{-1}}$

3 a $0.028\,\mathrm{N\,kg^{-1}}$
c $7.1 \times 10^{6}\,\mathrm{J}$

4 $2.8\,\mathrm{MJ\,kg^{-1}}$, 1410 MJ

17.5

2 a $3.4 \times 10^{6}\,\mathrm{m}$
b $3.0\,\mathrm{m\,s^{-2}}$
c $5.2 \times 10^{23}\,\mathrm{kg}$

3 b i $9.5\,\mathrm{N\,kg^{-1}}$
ii $7.9\,\mathrm{km\,s^{-1}}$
iii 5200 s

4 b ii 7100 s

18.1

2 b i 75 nA
ii 1.9×10^{11}

18.2

1 a $1.4 \times 10^{-3}\,\mathrm{N}$
b $4.0 \times 10^{4}\,\mathrm{V\,m^{-1}}$

2 a i negative
ii $1.3 \times 10^{-7}\,\mathrm{C}$
b i $7.3 \times 10^{-3}\,\mathrm{N}$
ii towards the metal surface

3 a i $9.0 \times 10^{4}\,\mathrm{V\,m^{-1}}$
ii $7.2 \times 10^{-14}\,\mathrm{N}$
b 80 mm

4 a 2.9 kV
b i $5.6 \times 10^{-15}\,\mathrm{N}$

18.3

1 a i $-8.0 \times 10^{-18}\,\mathrm{J}$
ii $+7.2 \times 10^{-17}\,\mathrm{J}$
b $+8.0 \times 10^{-17}\,\mathrm{J}$

2 a $-1.8 \times 10^{-3}\,\mathrm{J}$ b $+1.2 \times 10^{-3}\,\mathrm{J}$

3 a i $250\,\mathrm{V\,m^{-1}}$
ii $8.0 \times 10^{-17}\,\mathrm{N}$ (towards the negative plate)
b $-8.0 \times 10^{-19}\,\mathrm{J}$

4 b i $3000\,\mathrm{V\,m^{-1}}$

18.4

1 a 3.7×10^{-11} N
  b 2.6×10^{-10} N
2 a i 69 mm
    ii 3.6×10^{-6} N
  b 2.5×10^{-5} N repulsion
3 a 6.1 nC, negative
  b 2.2×10^{-2} N
4 a 2.7 nC, attract
  b 6.2×10^{-2} m, repel

18.5

1 a 5.3×10^{6} V m^{-1} b 10 mm
2 a i 3.7×10^{8} V m^{-1}
    ii 5.6×10^{-3} N towards Q_1
3 a i 4.5×10^{8} V m^{-1} towards Q_2
    ii 2.6×10^{8} V m^{-1} away from Q_1
  b ii 11 mm from Q_1, 9 mm from Q_2
4 a -9.0×10^{6} V
  b ii 2.0×10^{9} V m^{-1} directly towards Q_2

19.1

1 a 5.0 μF
  b 2.2 V
  c 9.9 mC
  d 1.4 μF
2 a 264 μC
  b 106 s
3 a 27.5 μC
  b 5.5 μF
4 a 0.91 μC
  b 0.22 μF
  c 700 μC
  d 7.4 V

19.2

1 a 30 μC, 45 μJ
  b 60 μC, 180 μJ
2 a 0.45 C, 2.0 J
  b 10 W
3 a i 6.6 μC, 9.9 μJ
    ii 6.6 μC, 19.8 μJ
4 a 56 μC, 338 μJ
  b i 2.2 μF; 18 μC, 4.7 μF; 38 μC
    ii 8.2 V
  c 2.2 μF; 73 μJ, 4.7 μF; 157 μJ

19.3

1 a i C increases
    ii Q increases
  b The energy stored $= \frac{1}{2}QV$, so the energy stored increases because Q increases and V is unchanged.
2 The capacitor is isolated from the battery, so the charge stored does not change. The capacitance decreases when the dielectric is removed. The energy stored $= \frac{1}{2}\frac{Q^2}{C}$, so it increases because C decreases and Q is unchanged. The increase of energy is due to the work done to overcome the electrostatic attraction between the charge on each dielectric surface and the opposite type of charge on the adjacent plate.
3 a 9.8 pF
  b 1100 pJ
4 a 0.14 μm
  b 2.2 MJ m^{-3}

19.4

1 a i 300 μC
    ii 5.0 s
  b i 5 s approx.
    ii 20 kΩ
2 a i 0.61 mC
    ii 0.45 mA
  b 0.23 V, 11 μA
3 a 13 μC, 40 μJ
  b 0.62 V, 0.42 μJ
4 a i 60 mA ii 0.34 mJ
  b 1.4 s c 0.32 mJ

19.5

1 a 1.0×10^{-6} s^{-1}
  b 3.9×10^{16}
2 a 6.3×10^{-10} s^{-1}
  b 20.5 kBq
3 a 0.65 kg
  b 1.7×10^{24}

4 a $1.3 \times 10^{-6}\,s^{-1}$
  b 149 hours

20.1

1 a 2.4×10^{-2} N; west
  b 4.5 A; east to west
  c 0.20 T; vertically down
  d 8.0×10^{-3} N; due south

2 a 22 mN due east
  b 4.0 A west to east

3 Short sides, zero. Long sides, 2.72 N vertically up on one side and vertically down on the other side

4 a 58 μT
  b 6.5×10^{-5} N due east

20.2

1 b i 1.9×10^{-13} N
    ii 0

2 3.8×10^{-23} N horizontal due east

4 b i $4.4\,m\,s^{-1}$
    ii 8.5×10^{-20} N

20.3

1 a ii 21 mm
  b 2.8 mT

2 a 4.7 mT
  b 17.5 mm

3 b 1.2 MeV

4 a $8.0 \times 10^{6}\,C\,kg^{-1}$
  b $1.4 \times 10^{7}\,C\,kg^{-1}$

21.1

1 a An e.m.f. is induced in the coil as the wire is pushed in. Therefore a current is induced in the circuit during this time so the meter deflects.
  b Move the magnet faster or use a coil with more turns or use a stronger magnet

2 a When the weight falls, the coil turns so an e.m.f. is generated in the coil which causes an induced current in the circuit so the lamp lights.
  b The descent would be slower because the magnetic forces acting on the coil due to the current would be larger and they would oppose the motion of the coil more.
  c The dynamo generates an e.m.f. when it turns and the e.m.f. causes a current in the lamp circuit.
  d When the dynamo generates a current, work must be done to keep it turning against the force of the magnetic field due to the induced current which opposes the motion of the rotating magnet.

3 a i Vertically downwards
    ii The end pointing east because the conduction electrons in the rod experience a force towards the opposite end.
  b The field is parallel to the rod so no e.m.f. can be induced in the rod. The electrons experience a force to the east which is across the rod's width not towards one end.

21.2

1 a i 1.1 mWb
    ii 2.0 s
    iii 0.54 mV

2 a 1.4 mWb
  b 23 mV

3 a i $4.5 \times 10^{-4}\,m^2$
    ii 1.5(4) mWb
  b i 3.1 mWb
    ii 33 mV

4 a 8.0 μWb
  b 40 μV

21.3

2 flux linkage 0, 0, $-BAN$, induced e.m.f. 0, $-\mathcal{E}_0$, 0

3 a 26 mWb

21.4

1 a 5 ms
  b 0.0707 A

2 a i 2.12 A
    ii 8.49 V
  b i 36 W
    ii 18 W

3 a i 4.35 A
    ii 6.15 A
    iii 2.0 kW

b The fuse melts only if the rms current exceeds 5 A. The rms current in this case is less than 5 A, so the fuse does not melt.

4 a i 4.0 A

ii 5.66 A

iii 17.0 V

b i $V_{rms} = I_{rms} R = 4.0 \times 0.5 = 2.0$ V

ii 19.8 V

21.5

1 a 11.5 V

b i 0.26 A

ii 5.2 A

2 b i 17 A (16.7 A to 3 sf)

ii 56 kW

22.1

4 a i 273 K

ii 293 K

iii 77 K

b i 328 K

ii 137 kPa

22.2

1 a 23 kJ

b 535 kJ

2 a 280 s

b 10.3 MJ

3 a 320 J

b 130 J kg^{-1} K^{-1}

4 3.2 kW (3.15 kW to 3 sf)

22.3

2 0.16 kg

3 a 4.2 J s^{-1} b 6400 s

4 a 22 J s^{-1} b 6.5 kJ

22.4

1 a The vessel is silvered to reduce energy loss by radiation from its outer surface to outer container.

b Energy transfer across the gap due to conduction and convection is eliminated because there is no material in the gap.

c The lid prevents energy transfer due to evaporation from the surface of the liquid in the flask.

d Felt or polystyrene could be used, as such material is a good thermal insulator and so reduces energy loss due to thermal conduction from the flask to the outer container.

2 a 40 J s^{-1} b 5.3 J s^{-1}

3 a 2.7 J s^{-1} b 4.3 K

4 a 6.2 kJ s^{-1}

23.1

1 126 kPa

2 79 kPa

3 0.097 m^3

4 b 2.3×10^{-5} m^3, 1600 kg m^{-3}

23.2

1 a 1.1 moles

b 109 kPa

2 a 9.3×10^{-4} moles

b 2.1×10^{-5} m^3

3 b 1.3 kg m^{-3}

4 a 1.2 kg m^{-3}

b 2.5×10^{22}

23.3

1 b i 0.97 mol

ii 4.5 kJ

2 c 5.7×10^{-21} J

d 1.8×10^3 m s^{-1}

4 a 1.48 mol

b 4.2×10^{-2} kg

c 5.2×10^2 m s^{-1}

24.1

2 a 3.2 fm

b 5.2 fm

3 b 6.5 fm, 1.2×10^{-42} m^3

4 b 2.5 fm, 3.4×10^{17} kg m^{-3}

24.2

1 a 2.18×10^{-15} kg

b i 8.89×10^{-33} kg

ii 8.89×10^{-30} kg

2 6.2 MeV

3 0.56 MeV

4 1.68 MeV

24.3

1 a 7.4 MeV
  b 8.8 MeV
2 a i 7.1 MeV
    ii 2.6 MeV
3 a $p + p \rightarrow {}^2_1H + {}^{0}_{+1}\beta + \nu$
  b 1.12 MeV

24.4

2 b i $a = 56$, $b = 98$
    ii 206 MeV
3 b ii 0.43 MeV, 5.5 MeV
4 b 12.9 MeV

24.5

1 a Induced fission is the splitting of a large nucleus into two approximately equal smaller nuclei as a result of a neutron colliding with the large nucleus which absorbs the neutron and splits.
  b The control rods absorb neutrons and they are used to keep the number of free neutrons in the reactor constant so that the rate of fission events and the rate of release of energy is constant.
2 a The neutrons released by fission events need to be slowed down so they can induce further fission events to maintain a chain reaction of events. The moderator is used to reduce the speed of the neutrons released by fission events to a speed at which they can induce further fission events.
  b The moderator becomes very hot as a result of the energy transferred to it by fission neutrons when they collide with the atoms of the moderator. The coolant is used to transfer energy from the moderator to a heat exchanger which is used to raise steam to drive the turbines and generators in the power station.
3 The rate of fission events and therefore the rate of production of fission neutrons in a reactor is proportional to the mass of the fissile isotope in the reactor. The rate of loss of fission neutrons from the fissile material is proportional to the surface area of the fissile material. If the mass of fissile material is too small, its surface area to mass ratio is too large to sustain the rate of fission events as the rate of loss of neutrons is greater than the rate of production. The critical mass of fissile material is the mass at which the rate of production is equal to the rate of loss of fission neutrons.
4 a See Topic 24.5
  b See Topic 24.5

25.1

1 a $0.20\,\text{rad s}^{-2}$
  b i 360 rad ii 57
2 a $75\,\text{rad s}^{-1}$
  b $1.5\,\text{rad s}^{-2}$
  c 29 rad, 4.6 turns
3 a 1.4×10^5 rad b 22 000
4 a $17\,\text{rad s}^{-1}$
  b i $5.0\,\text{rad s}^{-2}$ ii 31 rad, 4.9 turns

25.2

1 0.27 N m
2 b i $21\,\text{rad s}^{-2}$ ii 376 turns
3 Although the two discs have the same mass and radius, Y has the greater moment of inertia because a greater proportion of its mass is nearer its rim than X.
4 b ii $17\,\text{kg m}^2$

25.3

1 a 9.6 J b i 19 N m ii 50 rad
2 a 12.1 J b 0.2 J
  c i 11.9 J ii $6.3 \times 10^{-4}\,\text{kg m}^2$
3 As the ball rolls down the slope, it loses potential energy and gains kinetic energy due to its increasing speed (translational kinetic energy) and it gains rotational kinetic energy. When it rolls along the flat surface, it gradually loses its rotational and translational kinetic energy due to the frictional forces acting on it.
4 a i 47 kg ii $0.57\,\text{kg m}^2$
  b i 28 kJ ii 2.0 kW

25.4

1 a 4.5 N m s b 0.50 N m
2 a 1.1 N m s b $0.022\,\text{kg m}^2$
3 When the star collapses, its moment of inertia about its axis of rotation decreases because more of its mass is near its axis of rotation. Its angular momentum is conserved because there is no resultant torque acting on it when it collapses. Since angular momentum equals moment of

inertia × its angular velocity, its angular velocity therefore increases because its moment of inertia decreases.

4 $2.6 \times 10^{-2}\,kg\,m^2$

26.1

1 a 460 kW b 1.8 MW

2 a The wind is slowed down but not stopped by a wind turbine. The wind leaving a turbine has some kinetic energy so not all its initial kinetic energy has been removed. Stopping the wind would remove all its kinetic energy but the wind turbine would not work if it stopped the air moving through it completely.

b 52%

3 a The wind shadow of a wind turbine is the region the wind passes through after passing through the wind turbine, where the wind speed is less than the wind speed before the wind turbine.

b i 500 000 homes

ii The maximum power from the London array at $7\,kW\,m^{-2}$ would be 700 MW. To increase the amount of power from 600 MW to 700 MW, the number of turbines would need to be increased by one-sixth of the present number of turbines, which is 29 more turbines. However, the present turbines would be affected by the wind shadow of the extra turbines, so the extra power generated would be less than 100 MW.

4 The main environmental benefit is the reduction in the use of fossil fuel as a result of generating electrical power using wind turbines instead. The drawbacks include their unreliability due to variable wind speeds and the noise from the turbines when they are in operation.

26.2

1 a i A short-circuit current is the current when the light intensity is constant and the cell terminals are short circuited (i.e., by connecting together by a conductor).

ii An open-circuit p.d. is the p.d. across the cell terminals when it is not connected to a circuit.

b The array would need to have 5 rows of cells in parallel, each row with 20 cells in series.

2 a See Figure 2 in Topic 26.2.

b The flat section should be 25% lower and slightly shorter.

3 As the p.d. increases, the current remains constant until the p.d. approaches the open-circuit p.d. As the power is the product of the current and the p.d. the power is proportional to the p.d, so the power increases linearly from zero until the the current starts to decrease. As the p.d. increases further towards the open-circuit p.d. the current decreases gradually and the power increases less rapidly until the current falls sharply to zero and the power falls sharply to zero. The peak power is at the point just before where the current starts to fall sharply, and this is where the product of the current and the p.d. is a maximum. The p.d. at this point is less than the open-circuit p.d. and the current is less than the short-circuit current, so the peak power is less than the power value given by the product of the open-circuit p.d. and the short-circuit current.

4 26%

26.3

1 39%

2 a A pumped storage scheme is a hydroelectric power station in which the turbines can be used to generate electricity from the downhill flow of water through them or to pump water uphill from a reservoir near the power station to an uphill reservoir, thereby storing energy as gravitational potential energy.

b 23 MWh

3 The water leaving the turbines cannot be stopped so it must have some kinetic energy when it leaves the turbine. Therefore the amount of electrical energy from the power station must be less than the amount of gravitational potential energy released by the water from the reservoir. Since the efficiency is the percentage of the gravitational potential energy that is transferred from the power station as electrical energy, the efficiency can never be as great as 100%.

4 a A base-load power station cannot be started quickly enough when the demand for electricity from the grid system increases, whereas a back-up power station can be started quickly enough to meet an increasing demand.

b i A nuclear or oil-fired (or coal-fired) power station can be used as a base-load power station ii A pumped storage scheme or a gas-fired power station can be used as a back-up power station when demand is high.

27.1

1 a i 0.524 rad
ii 0.873 rad
iii 2.094 rad
iv 4.014 rad
v 5.236 rad
b i 5.73°
ii 28.7°
iii 68.8°
iv 143.2°
v 343.8°

2 a 20 mm, 2.3°
b ii 0.5°

3 a i 0.035
ii 0.140

4 a i 0.0998
ii 0.995
b i 0.1736
ii 0.9848
c i 0.7071
ii 0.7071
d i 0.7071
ii 0.7071

27.2

1 a $x = 1, y = 3$
b $a = 2.4, b = -0.4$
c $p = 2, q = 4$

2 a $u = 14\,\mathrm{m\,s^{-1}}, a = -2.0\,\mathrm{m\,s^{-2}}$
b $r = 1.0\,\Omega, \varepsilon = 9.0\,\mathrm{V}$

3 a 0.5 or −3
b 1.4 or 5.6
c $-\frac{10}{6} = -1.67$ (to 3 sf) or 1

4 a $t = -\frac{20}{6} = -3.33$ (to 3 sf) or 2 s
b $R = 1$ or $4\,\Omega$

27.3

1 a i 0.477
ii 1.176
b i 1.653
ii 0.699

2 a i 10.8 dB
ii 7.0 dB
b 17.8 dB

3 a $n = 5, k = 3$
b $n = 3, k = \frac{1}{2}$
c $n = 2, k = 1$

4 a i 1.10
ii 2.71
b i 3.81
ii 1.61

27.4

1 a i 2, 3
ii 12, 0.2
iii 4, 0.02
b i 0.23 s ii 3.5 s iii 35 s

2 a 11.3 kBq
b 9.0 kBq
c 0.38 kBq

3 a i 2.2 s
ii 1.52 s
b i 4.83 V
ii 1.24 V

4 a i 0.14 s
ii 82
b $a = 6.9, b = -5$

27.5

2 d i Pa or $\mathrm{N\,m^{-2}}$ ii $\mathrm{m^3}$ iii J

27.6

1 The gradient at 0.6 s = $14.4\,\mathrm{m\,s^{-1}}$. This gives a value of g equal to $10.0\,\mathrm{m\,s^{-2}}$.

2 a $0.038\,\mathrm{s^{-1}}$
b N decreases at a slower rate, and the half-life is doubled.

Index